GEOMETRIC FORMULAS: PERIMETER AND AREA

Figure	Name	Perimeter	Area
	Square	$P = 4s$	$A = s^2$
	Rectangle	$P = 2l + 2w$	$A = lw$
	Triangle	$P = a + b + c$	$A = \dfrac{1}{2}bh$
	Trapezoid	$P = a + b + c + d$	$A = \dfrac{1}{2}h(b + d)$
	Circle	$C = 2\pi r$	$A = \pi r^2$

GEOMETRIC FORMULAS: VOLUME

Figure	Name	Volume
	Rectangular solid	$V = lwh$
	Cylinder	$V = Bh$ where B is the area of the base
	Pyramid	$V = \frac{1}{3}Bh$ where B is the area of the base
	Cone	$V = \frac{1}{3}Bh$ where B is the area of the base (If the base is a circle, then $B = \pi r^2$)
	Sphere	$V = \dfrac{4}{3}\pi r^3$

Here are your FREE passcodes to BCA/iLrn Tutorial and InfoTrac College Edition—only available with NEW copies of Thomson•Brooks/Cole textbooks.

www.brookscole.com

www.brookscole.com is the World Wide Web site for Brooks/Cole and is your direct source to dozens of online resources.

At *www.brookscole.com* you can find out about supplements, demonstration software, and student resources. You can also send email to many of our authors and preview new publications and exciting new technologies.

www.brookscole.com
Changing the way the world learns®

We would like to dedicate this book to our families:
David and Sherah
Blaine, Blaire, and Trey
and in loving memory of Noel Cochener and Rose Jolly

This book would not have been a reality had it not been for their patience, understanding, support, and love.

Tell me, and I'll forget,
Show me, and I may not remember.
Involve me, and I'll understand.
                    Native American saying

Beginning and Intermediate Algebra with Graphing Calculators

An Integrated Approach
Second Edition

Deborah Jolly Cochener
Austin Peay State University

Bonnie MacLean Hodge
Austin Peay State University

R. David Gustafson
Rock Valley College

THOMSON
BROOKS/COLE

Australia • Canada • Mexico • Singapore •
Spain • United Kingdom • United States

THOMSON

BROOKS/COLE

Publisher: Robert W. Pirtle

Editor: Jennifer Huber

Assistant Editor: Rachael Sturgeon

Editorial Assistant: Sarah Woicicki

Technology Project Manager: Christopher Delgado

Marketing Manager: Greta Kleinert

Marketing Assistant: Jessica Bothwell

Advertising Project Manager: Bryan Vann

Project Manager, Editorial Production: Andy Marinkovich

Print/Media Buyer: Kristine Waller

Permissions Editor: Kiely Sexton

Production Service: Ellen Brownstein, Chapter Two

Text Designer: Kim Rokusek, Rokusek Design

Photo Researcher: Sue C. Howard

Copy Editor: Ellen Brownstein

Illustrator: Lori Heckelman

Cover Designer: Didona Design

Cover Photo: © Michael Rosenfeld/Getty Images

Compositor: G & S Typesetters

Printer: Quebecor World, Versailles

Printed in the United States of America

1 2 3 4 5 6 7 06 05 04

For more information about our products, contact us at:
Thomson Learning Academic Resource Center
1-800-423-0563

For permission to use material from this text, contact us by:
Phone: 1-800-730-2214   **Fax:** 1-800-730-2215
Web: http://www.thomsonrights.com

Library of Congress Control Number: 2003117038

ISBN 0-534-39071-4

Annotated Instructor's Edition: ISBN 0-534-46315-0

Brooks/Cole–Thomson Learning
10 Davis Drive
Belmont, CA 94002
USA

Asia
Thomson Learning
5 Shenton Way #01-01
UIC Building
Singapore 068808

Australia/New Zealand
Thomson Learning
102 Dodds Street
Southbank, Victoria 3006
Australia

Canada
Nelson
1120 Birchmount Road
Toronto, Ontario M1K 5G4
Canada

Europe/Middle East/Africa
Thomson Learning
High Holborn House
50/51 Bedford Row
London WC1R 4LR
United Kingdom

Latin America
Thomson Learning
Seneca, 53
Colonia Polanco
11560 Mexico D.F.
Mexico

Spain/Portugal
Paraninfo
Calle Magallanes, 25
28015 Madrid
Spain

Contents

CHAPTER 4

An Extended Look at Applications and Inequalities

228

CHAPTER 5

Factoring

293

CHAPTER 6

Rational Expressions and Equations

346

CHAPTER 10

Quadratics

599

CHAPTER 11

Functions and Their Inverses

662

CHAPTER 12

Systems of Equations and Inequalities

727

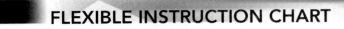

FLEXIBLE INSTRUCTION CHART

This text provides flexibility in designing elementary and intermediate algebra courses. Possible course sequences of chapters/topics are delineated in the chart below.

Standard Approach

Elementary Algebra

Chapter 1   Review of Fractions (5 sections)
Chapter 2   Exponents and Polynomials (7 sections)
Chapter 3   Solutions of Equations in One and
    Two Variables (7 sections)
Chapter 4   An Extended Look at Applications and
    Inequalities (6 sections)
Chapter 5   Factoring (5 sections)
Chapter 6   Rational Expressions and Equations
    (6 sections)
Total Sections: 36

Intermediate Algebra

Chapter 7   Making the Transition from Elementary to
    Intermediate Algebra (5 sections)
Chapter 8   Functions (5 sections)
Chapter 9   Rational Exponents and Radicals (7 sections)
Chapter 10   Quadratics (5 sections)
Chapter 11   Functions and Their Inverses (6 sections)
Chapter 12   Systems of Equations and Inequalities
    (6 sections)

Total Sections: 34

An Early Functions Approach

Elementary Algebra

Chapter 1   Review of Fractions (5 sections)
Chapter 2   Exponents and Polynomials (7 sections)
Chapter 3   Solutions of Equations in One and
    Two Variables (7 sections)
Section 8.1   Relations and Functions
Section 8.3   Slope of a Nonvertical Line
Section 8.4   Equations of Lines (3 sections)
Chapter 4   An Extended Look at Applications and
    Inequalities (6 sections)
Chapter 5   Factoring (5 sections)
Optional   Sections 6.1–6.3 could be covered in the
    course to show students uses of factoring. (3 sections)
Total Sections: 33–36

Intermediate Algebra

Chapter 7   Making the Transition from Elementary to
    Intermediate Algebra (5 sections)
Chapter 6   Rational Expressions and Equations
    (6 sections)
Section 8.2   Functions and Graphs
Section 8.5   Variation (2 sections)
Chapter 9   Rational Exponents and Radicals (7 sections)
Chapter 10   Quadratics (5 sections)
Chapter 11   Functions and Their Inverses (6 sections)
Chapter 12   Systems of Equations and Inequalities
    (6 sections)

Total Sections: 37

An Early Functions Approach Coupled with Systems of Two Equations in Two Variables

Elementary Algebra

Chapter 1   Review of Fractions (5 sections)
Chapter 2   Exponents and Polynomials (7 sections)
Chapter 3   Solutions of Equations in One and
    Two Variables (7 sections)
Section 8.1   Relations and Functions
Section 8.3   Slope of a Nonvertical Line
Section 8.4   Equations of Lines (3 sections)
Chapter 4   An Extended Look at Applications and
    Inequalities (6 sections)
Chapter 5   Factoring (5 sections)
Section 12.1   Solving Systems of Two Equations in
    Two Variables
Section 12.2   Applications of Systems of Two Equations
    in Two Variables
Section 12.6   Systems of Inequalities (3 sections)
Total Sections: 36

Intermediate Algebra

Chapter 7   Making the Transition from Elementary to
    Intermediate Algebra (5 sections)
Chapter 6   Rational Expressions and Equations
    (6 sections)
Section 8.2   Functions and Graphs
Section 8.5   Variation (2 sections)
Chapter 9   Rational Exponents and Radicals (7 sections)
Chapter 10   Quadratics (5 sections)
Chapter 11   Functions and Their Inverses (6 sections)
Section 12.3   Solving Systems of Three Equations in
    Three Variables
Section 12.4   Using Matrices to Solve Systems
    of Equations
Section 12.5   Using Determinants to Solve Systems
    of Equations (3 sections)
Total Sections: 34

Preface

To the Instructor and Student

The second edition of *Beginning and Intermediate Algebra with Graphing Calculators: An Integrated Approach* combines the topics of beginning and intermediate algebra. This type of book has several advantages:

- By combining topics, much overlap and redundancy can be eliminated. The instructor will have time to teach for mastery of the material.
- For many students, the purchase of a single book will save money.
- A combined approach in one book will enable some colleges to cut back on the number of hours needed for mathematics remediation.

However, there are three concerns inherent in a combined approach:

- The first half of the book must include enough beginning algebra to ensure that students who complete the first half of the book and then transfer to another college will have the necessary prerequisite to enroll in an intermediate algebra course.
- The beginning algebra material should not get too difficult too fast.
- Intermediate algebra students beginning in the second half of the book must get some review of basic topics so that they can compete with students who are continuing from the first course.

Unlike many other texts, this book uses an *integrated approach*, which addresses each of the previous three concerns by

- including a full course in beginning algebra in the first six chapters,
- delaying the presentation of intermediate algebra topics until after Chapter 7 and
- providing a review of basic topics in Chapter 7 for those who begin the course at the intermediate algebra level.

A chart is provided to address the variations in course content from state to state and school to school, including

- an early functions approach, with an in-depth examination of linear functions in the elementary half of the text, and
- an early functions approach that also includes systems of equations in two variables in an elementary algebra course.

The approach throughout the text is fourfold and cyclic, using

- numerical approaches, specifically through the use of tables;
- graphical approaches, done both on graph paper and with the graphing calculator, providing students with opportunities to model data and to interpret models;
- symbolic approaches, which help students link the numerical and graphical approaches;
- verbal approaches in which students are required to express mathematical concepts in their own words; and

- continuous requests for students to demonstrate proficiency in the evaluation of expressions, the simplification of expressions, the graphing of functions, and the solving of equations.

Every effort has been made to provide the student with opportunities to become active participants in the formulation of their algebraic foundation, so that they may apply the principles of mathematics to other disciplines. The calculator is presented as a tool to be used for the confirmation of analytical processes and for the discovery of numerical and graphical patterns that lead to the formulation of fundamental algorithms.

▌ CONTENT CHANGES IN THE SECOND EDITION

Organizational changes were made in the Second Edition to better reflect pedagogical trends:

Chapter 2 Negative exponents have been moved from the transition chapter (Chapter 7) to this chapter so that all of the laws of exponents are addressed in the same chapter.

Chapter 3 Absolute value equations have been moved from the transition chapter (Chapter 7) to the end of this chapter.

Chapter 4 Linear inequalities, compound inequalities, and absolute value inequalities were moved from the transition chapter to be included in this chapter.

The resulting effect of these major shifts is that Chapter 7 is now purely a transition chapter, reviewing the elementary algebra skills typically found in the first six chapters. Other changes include:

Chapter 5 Factoring the sum and difference of cubes is now introduced in Section 5.3 and a new section was added to the chapter that summarizes factoring techniques.

Chapter 6 Section 6.2 (Addition and Subtraction of Rational Expressions) and Section 6.3 (Multiplication and Division of Rational Expressions) have been reversed, resulting in the concepts of the multiplication and division of rational expressions being introduced before the concepts of the addition and subtraction of these expressions.

Chapter 7 New sections were added to reflect a more complete review of the first half of the text. No new information is introduced in this chapter.

Chapter 8 A chart is provided in this preface so that instructors have the option of introducing and exploring functions in more depth at the elementary algebra level through the study of linear functions.

Chapter 9 Section 9.3 (Arithmetic Operations with Radicals) was split into two sections to address the addition and subtraction of radicals in one section and their multiplication and division in a subsequent section.

In addition to these organizational changes, the following changes or additions are included in the Second Edition:

- A Quick Reference Guide for calculator keystroking is included on the inside cover of the text.
- A Calculator Reference Guide for the TI-83 Plus and the TI-86 has been added as an appendix.
- Answers to **all** Vocabulary/Notation and Concept questions are included in Appendix F of the student edition of the text.

- Calculator screen shots appear in the student solution manual rather than in Appendix F.
- Exercise sets have been expanded to provide ample opportunity for drill and conceptual understanding.

▚ GOALS OF THE BOOK

In addition to using a truly integrated approach, our goal has been to write a book that

- is enjoyable to read,
- is easy to understand,
- is relevant, and
- will develop the necessary skills for success in future academic courses or on the job.

The NCTM standards, the AMATYC Crossroads, and the current trends in mathematics reform have been reinforced by emphasizing graphing and problem solving.

▚ CALCULATORS

The use of a Texas Instruments TI-83 Plus or TI-86 graphing calculator is assumed throughout the text. Students who use the TI-82 or TI-85 will have no problem making the transition. We believe that students should learn calculator skills at this level of mathematics so they will be better able to use the technology in science and business classes as they progress through their individual curricula.

▚ ANCILLARIES FOR THE INSTRUCTOR

Annotated Instructor's Edition   0-534-46315-0
This special version of the complete student text contains a Resource Integration Guide and all answers printed next to their respective exercises. Teaching Tips are also provided throughout the text.

Test Bank   0-534-46319-3
The *Test Bank* includes 8 tests per chapter as well as 3 final exams. The tests are made up of a combination of multiple-choice, free-response, true/false, and fill-in-the-blank questions.

Instructor's Resource Manual   0-534-46318-5
The *Instructor's Resource Manual* provides teaching notes, transparency masters, and worked out solutions to all of the problems in the text.

Text-Specific Videotapes   0-534-46320-7
The text-specific videotape set, available at no charge to qualified adopters of the text, features 10- to 20-minute problem-solving lessons that cover each section of every chapter.

BCA/iLrn Instructor Version   0-534-28036-6
With a balance of efficiency and high-performance, simplicity and versatility, *BCA/iLrn* gives you the power to transform the teaching and learning experience. *BCA/iLrn Instructor Version* is made up of two components, *BCA/iLrn Testing* and *BCA/iLrn Tutorial*. *BCA/iLrn Testing* is a revolutionary, Internet-ready, text-

specific testing suite that allows instructors to customize exams and track student progress in an accessible, browser-based format. *BCA/iLrn* offers full algorithmic generation of problems and free-response mathematics. *BCA/iLrn Tutorial* is a text-specific, interactive tutorial software program that is delivered via the Web (at http://bca.brookscole.com) and is offered in both student and instructor versions. Like *BCA/iLrn Testing*, it is browser-based, making it an intuitive mathematical guide, even for students with little technological proficiency. *BCA/iLrn Tutorial* allows students to work with real math notation in real time, providing instant analysis and feedback. The tracking program built into the instructor version of the software enables instructors to monitor student progress carefully. The complete integration of the testing, tutorial, and course management components simplifies your routine tasks. Results flow automatically to your gradebook and you can communicate easily with individuals, sections, or entire courses.

TLE Labs— to upgrade any *BCA/iLrn Student Tutorial*
0-534-46331-2
Think of *TLE Online Lessons* as electronic labs that upgrade any *BCA/iLrn Tutorial*. *TLE Online Lessons* introduce and explore key concepts. *BCA/iLrn Tutorials* reinforce those concepts with unlimited practice. Each of the 15 *TLE* lessons, scheduled one per week in a traditional semester course, covers the key concepts of the course. With the opportunity to explore these core concepts interactively at their own pace, students are solidly prepared for the work of the traditional course. Because they are better prepared, they enjoy the class more and, consequently, perform better in the course overall. When they perform better in the course, they are better prepared to tackle the next level of mathematics that awaits them.

WebTutor ToolBox on WebCT and Blackboard
0-534-59746-7 WebCT; 0-534-59683-5 Blackboard
Preloaded with content and available free via PIN code when packaged with this text, *WebTutor ToolBox* for WebCT pairs all the content of this text's rich Book Companion Web Site with all the sophisticated course management functionality of a WebCT product. You can assign materials (including online quizzes) and have the results flow automatically to your gradebook. *ToolBox* is ready to use as soon as you log on or you can customize its preloaded content by uploading images and other resources, adding Web links, or creating your own practice materials. Students have access only to student resources on the Web site. Instructors can enter a PIN code for access to password-protected Instructor Resources.

ANCILLARIES FOR THE STUDENT

Student Solutions Manual   0-534-46317-7
The *Student Solutions* manual provides worked out solutions to the odd-numbered problems in the text.

Brooks/Cole Mathematics Web site   http://mathematics.brookscole.com
The Book Companion Web Site offers a range of learning resources that can help your students get the most from their beginning algebra class. Chapter-specific quizzes and tests allow students to assess their understanding and receive immediate feedback. Students also have access to *Web Explorations* and InfoTrac® College Edition *Exercises*, which feature additional explorations of concepts. Crossword puzzles, additional graphing-calculator exercises, and matching exercises also are available to help students with their understanding of the concepts and vocabulary presented in the text.

InfoTrac® College Edition

In addition to robust tutorial services, your students also receive anytime, anywhere access to InfoTrac® College Edition. This online library offers the full text of articles from almost 5,000 scholarly and popular publications, updated daily and going back as far as 22 years. Both adopters and their students receive unlimited access for four months.

BCA/iLrn Tutorial Student Version   0-534-20860-9

Free access to this text-specific, interactive, Web-based tutorial system is included with the *Second Edition*. So sophisticated, it's simple, *BCA/iLrn Tutorial* allows students to work with real math notation in real time, providing instant analysis and feedback. The entire textbook is available in PDF format through *BCA/iLrn Tutorial*, as are section-specific video tutorials, unlimited practice problems, and additional student resources such as a glossary, Web links, and more. And, when students get stuck on a particular problem or concept, they need only log on to *vMentor*™, where they can talk (using their own computer microphones) to tutors who will guide them skillfully through the problem using the interactive whiteboard for illustration. *BCA/iLrn Tutorial* is also offered on a dual-platform CD-ROM.

vMentor

Packaged free with every text and accessed seamlessly through *BCA/iLrn Tutorial, vMentor* provides live, on-line tutorial help that can substantially improve student performance, increase test scores, and enhance technical aptitude. Your students will have access, via the Web, to highly qualified tutors with thorough knowledge of our textbooks. When students get stuck on a particular problem or concept, they need only log on to *vMentor*, where they can talk (using their own computer microphones) to *vMentor* tutors, who will guide them skillfully through the problem using the interactive whiteboard for illustration.

Interactive Video Skillbuilder **CD-ROM**

The *Interactive Video Skillbuilder* CD-ROM contains more than eight hours of video instruction. The problems worked during each video lesson are shown next to the viewing screen so that students can try working them before watching the solution. To help students evaluate their progress, each section contains a 10-question Web quiz (the results of which can be emailed to the instructor) and each chapter contains a chapter test, with answers to each problem on each test.

Also included are *MathCue* tutorial and testing software. Keyed to the text, *MathCue* includes these features:

- *MathCue Skill Builder*   Presents problems to solve, evaluates answers, and tutors students by displaying complete solutions with step-by-step explanations. Students may view partial solutions to get started on a problem, see a continuous record of progress in a session, and back up to review missed problems.
- *MathCue Quiz*   Generates large numbers of quiz problems keyed to problem types from each section of the book. Students receive immediate feedback after entering their answers.
- *MathCue Chapter Test*   Provides large numbers of problems keyed to problem types from each chapter. Feedback is given only after the test is completed.
- *MathCue Solution Finder*   Allows students to enter their own basic problems and receive step-by-step help as if they were working with a tutor.
- Create customized *MathCue* sessions   Students work default Skill Builders, Quizzes, and Tests, or customize sessions to include desired problems from one or more sections or chapters. These customized session setups can be saved and run repeatedly. *Special note for instructors*: Instructors can create targeted, cus-

tomized *MathCue* session files to send to students as email attachments for use as make-up assignments, extra-credit assignments, specialized review, quizzes, or for other purposes.

- Print or email score reports   Score reports for any *MathCue* session can be printed or sent to instructors via *MathCue*'s secure email score system.

Acknowledgments

We are grateful to the following people, who reviewed the manuscript at various stages of development. They all had valuable suggestions that have been incorporated into the text.

Patrick Guy Adams
College of the Redwoods

Sue Christiansen
Lansing Community College

Gladys H. Crates
Chattanooga State Community College

Ruth Ann Edwards
Craven Community College

Susan Fife
Tomball College

Roger Jay
Tomball College

Susan Jones
Nashville State Technical College

James Lapp
College of the Redwoods

Charlotte Salas
Lansing Community College

Kathryn C. Wetzel
Amarillo College

Kevin Yokoyama
College of the Redwoods

We are grateful to our editor, Jennifer Huber, for her encouragement and support throughout this project. We also thank the many people at Brooks/Cole whose contributions made the Second Edition possible, especially Rachael Sturgeon, Andy Marinkovich, and Vernon Boes, and Lori Heckelman, the artist for this book. Thanks also to Ellen Brownstein, not only for her guidance and good care of our project but also for the patience, support, and understanding she showed throughout production, no matter the challenge before us. Our thanks also go to Rosemary Karr, Collin County Community College, and Ian Crewe for their excellent suggestions for changes to the Second Edition.

We are particularly grateful to David Cochener for his meticulous attention to detail in reading the manuscript and working problems, and for his support during some difficult moments.

Deborah J. Cochener
Bonnie M. Hodge
R. David Gustafson

INDEX OF APPLICATIONS

Applications that are examples are shown with **boldface** page numbers.
Applications that are exercises are shown with lightface page numbers.

Basic Arithmetic and Graphing Calculator Concepts

© Keren Su/CORBIS

InfoTrac Project

Do a subject guide search on "Siberia" and find the newspaper article, "Cold wave breathes fire in Siberia." Write a summary of the article and find the difference between the record low in Western Siberia and the lowest temperature the houses have been built to withstand.

The formula $C = \frac{5}{9}(F - 32)$ is used to convert degree measurement in Fahrenheit to degree measurement in Celsius. Using this formula, convert the following Fahrenheit temperatures to degrees Celsius: $212°$, $32°$, $98.6°$, $-10°$, $-20°$, $-30°$, $-40°$, $-50°$, and $-60°$. Did you discover anything unusual? Write a statement about your findings.

The formula to convert Celsius degrees to Fahrenheit degrees is $F = \frac{9}{5}(C + 32)$. Use this formula to convert the following temperatures in degrees Celsius to degrees Fahrenheit: $125°$, $46°$, $-20°$, and $-55°$.

Complete this project after studying Section 1.5.

PERSPECTIVE

All Dark Things

The ancient Egyptians developed two systems of writing. Each symbol in one system, hieroglyphics, was a picture of an object. Because hieroglyphic writing was usually inscribed on stone, many examples still survive. For daily purposes, Egyptians used hieratic writing. Simpler than hieroglyphics, hieratic writings were done with ink on papyrus sheets. Papyrus, made from plants, is very delicate and quickly dries and crumbles. Few Egyptian papyri survive.

Those that do survive provide important clues to the content of ancient mathematics. One, the Rhind Papyrus, was discovered in 1858 by a British archeologist, Henry Rhind. Also known as the Ahmes Papyrus after its ancient author, it begins with a description of its contents: *Directions for Obtaining the Knowledge of All Dark Things.*

The Ahmes Papyrus and another, the Moscow Papyrus, together contain 110 mathematical problems and their solutions. Many of these were probably for education, because they represented situations that the scribes, priests, and other workers in government and temple administration were expected to be able to solve.

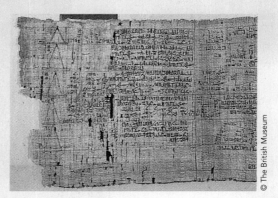

The Ahmes Papyrus

© The British Museum

Algebra is an extension of arithmetic and allows us to reason from the specific to the general. Learning algebra provides us with powerful problem-solving strategies and enhances our logical thinking skills.

1.1 REVIEW OF FRACTIONS

- SETS OF NUMBERS
- MULTIPLYING FRACTIONS
- ADDING FRACTIONS
- MIXED NUMBERS

- SIMPLIFYING FRACTIONS
- DIVIDING FRACTIONS
- SUBTRACTING FRACTIONS
- DECIMAL FRACTIONS

 ## SETS OF NUMBERS

We begin our discussion with the set of real numbers. A **set** is a well-defined collection of objects. For example, the set {1, 2, 3, 4, 5} contains the numbers 1, 2, 3, 4, and 5. The members, or elements, in a set are listed within braces { }.

Two basic sets of numbers are the **natural numbers** and the **whole numbers**.

Natural Numbers

The **natural numbers** (or the **positive integers**) are the numbers 1, 2, 3, 4, 5, ... This would be written as {1, 2, 3, 4, 5, . . . } when using set notation.

Whole Numbers

The **whole numbers** are the numbers 0, 1, 2, 3, 4, 5, ... This would be written as {0, 1, 2, 3, 4, 5, . . . } when using set notation.

The three dots in the definitions indicate that the elements in the lists continue forever.

We can use whole numbers to describe many real-life situations. For example, some cars get 30 miles per gallon of gas, some students might have paid $1,750 in tuition, and the temperature can be 0°.

Integers

The **integers** are the numbers . . . , −5, −4, −3, −2, −1, 0, 1, 2, 3, 4, 5, ... This would be written as {. . . , −5, −4, −3, −2, −1, 0, 1, 2, 3, 4, 5, . . .} when using set notation.

Numbers that show a loss or a downward direction are called **negative integers**, denoted as −1, −2, −3, and so on. For example, a debt of $1,500 can be denoted as −$1,500, and a temperature of 20° below zero can be denoted as −20°. The set of negative integers together with the set of whole numbers form the set of **integers**. Another way to define the set of integers would be as the set of natural numbers, their opposites, and 0.

Because the set of natural numbers and the set of whole numbers are included within the set of integers, we say that these sets are **subsets** of the set of integers.

Integers cannot describe every real-life situation. For example, a student might study $3\frac{1}{2}$ hours, or a television might cost $217.36. To describe these situations, we need fractions. A **fraction** is a number that can be expressed as a quotient of integers, $\frac{p}{q}$, where $q \neq 0$. It is called a **rational number**.

Rational Numbers

A **rational number** is any number that can be written as a fraction with an integer in its numerator and a nonzero integer in its denominator.

Fractions can also be used to indicate division. For example, the division of $8 \div 2$ can be written as a fraction, $\frac{8}{2}$. Fractions are also used to indicate parts of a whole. In Figure 1-1, a rectangle has been divided into 5 equal parts, and 3 of the parts are shaded. The fraction $\frac{3}{5}$ indicates how much of the figure is shaded. The denominator of the fraction shows the total number of equal parts into which the whole is divided, and the numerator shows the number of these equal parts that are being considered.

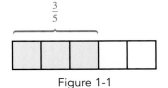

Figure 1-1

Some examples of rational numbers are

$$\frac{3}{2}, \frac{17}{12}, -\frac{43}{8}, 0.25, \quad \text{and} \quad -0.66666\ldots$$

TECHNOLOGY TIP

Your calculator will display an error message if you attempt to divide by 0.

The decimals 0.25 and $-0.66666\ldots$ are rational numbers because they can be written as fractions: 0.25 is the fraction $\frac{1}{4}$ and $-0.66666\ldots$ is the fraction $-\frac{2}{3}$.

Because every integer can be written as a fraction with a denominator of 1, every integer is also a rational number. Because every integer is a rational number, the set of integers is a subset of the rational numbers.

WARNING! Because division by 0 is undefined, expressions such as $\frac{6}{0}$ and $\frac{0}{0}$ do not represent any number.

SIMPLIFYING FRACTIONS

A **prime number** is a whole number greater than 1 whose only factors are 1 and the number itself. The set of prime numbers can be represented as {2, 3, 5, 7, 11, 13, 17, 19, . . .}. A fraction is in **lowest terms** when the numerator and the denominator have no common factors (other than 1). To reduce the fraction to lowest terms (often called *simplifying*) we can factor the numerator and the denominator into **primes** and divide out common factors. The following example illustrates the **fundamental property of fractions**.

The Fundamental Property of Fractions

If a represents a real number, and b and x represent real numbers, then

$$\frac{a \cdot x}{b \cdot x} = \frac{a}{b} \quad (b \neq 0 \text{ and } x \neq 0)$$

The fraction $\frac{6}{15}$ **is not** in lowest terms because 6 and 15 have a common factor of 3.

$$\frac{6}{15} = \frac{2 \cdot 3}{5 \cdot 3} \quad \text{Factor the numerator and denominator into a product of prime factors.}$$

$$= \frac{2 \cdot \overset{1}{\cancel{3}}}{5 \cdot \underset{1}{\cancel{3}}} \quad \text{Divide out the common factor of 3.}$$

$$= \frac{2}{5}$$

Therefore, $\frac{6}{15}$ simplifies to $\frac{2}{5}$. From Figure 1-2 we can see that $\frac{6}{15}$ and $\frac{2}{5}$ are equal fractions because they represent the same part of the rectangle.

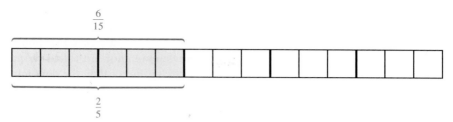

Figure 1-2

EXAMPLE 1

a. To simplify (reduce) $\frac{6}{30}$, we factor the numerator and denominator into a product of primes and divide out (cancel) the common factors of 2 and 3.

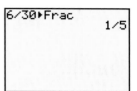

$$\frac{6}{30} = \frac{\overset{1}{\cancel{2}} \cdot \overset{1}{\cancel{3}}}{\underset{1}{\cancel{2}} \cdot \underset{1}{\cancel{3}} \cdot 5} = \frac{1}{5}$$

b. To simplify (reduce) $\frac{33}{40}$, we factor the numerator and denominator into a product of primes. Because the numerator and denominator have no common factors, the fraction $\frac{33}{40}$ is simplified.

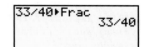

$$\frac{33}{40} = \frac{3 \cdot 11}{2 \cdot 2 \cdot 2 \cdot 5} = \frac{33}{40}$$

The number 1 is the **identity element for multiplication**. This means that *any* number or expression multiplied by 1 gives a result equal to the original number or expression. Symbolically, this may be stated as $a \cdot 1 = a$.

Basic Properties of Fractions

1. $\dfrac{a}{0}$ is undefined.

2. If $a \neq 0$, then $\dfrac{0}{a} = 0$.

3. $\dfrac{a}{1} = a$.

4. If $a \neq 0$, then $\dfrac{a}{a} = 1$.

WARNING!   Remember that a fraction is simplified only when its numerator and denominator have no common factors.

▌ MULTIPLYING FRACTIONS

Multiplying Fractions

If a, b, c, and d represent real numbers, then $\dfrac{a}{b} \cdot \dfrac{c}{d} = \dfrac{a \cdot c}{b \cdot d}$ ($b \neq 0$ and $d \neq 0$).

Simply stated, when multiplying two fractions, write the product of the numerators over the product of the denominators. Remember, you must then reduce/simplify the fraction to lowest terms. Fractions such as $\frac{8}{21}$ with a numerator that is smaller than the denominator are called **proper fractions**. Fractions such as $\frac{8}{3}$ with a numerator that is larger than the denominator are called **improper fractions**.

EXAMPLE 2

Multiply: $\frac{3}{7} \cdot \frac{13}{5}$.

SOLUTION

$$\frac{3}{7} \cdot \frac{13}{5} = \frac{3 \cdot 13}{7 \cdot 5}$$

$$= \frac{39}{35}$$

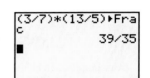

TECHNOLOGY TIP

When using the graphing calculator, it is suggested that students initially enclose all fractions within parentheses. This is necessary when multiplying and/or dividing fractions.

EXAMPLE 3

European travel   Out of 36 students in a history class, three-fourths have signed up for a trip to Europe. There are 30 places available on the chartered flight. Will there be room for one more student?

SOLUTION   We first find three-fourths of 36 by multiplying 36 by $\frac{3}{4}$.

$$\frac{3}{4} \cdot 36 = \frac{3}{4} \cdot \frac{36}{1} \qquad \text{Write 36 as } \frac{36}{1}.$$

$$= \frac{3 \cdot 36}{4 \cdot 1} \qquad \text{Multiply the numerators and multiply the denominators.}$$

$$= \frac{3 \cdot \overbrace{3 \cdot 3 \cdot 2 \cdot 2}^{36}}{\underbrace{2 \cdot 2}_{4} \cdot 1} \qquad \text{To simplify fractions, factor the numerator and denominator.}$$

$$= \frac{3 \cdot 3 \cdot 3 \cdot \overset{1}{2} \cdot \overset{1}{2}}{\underset{1}{2} \cdot \underset{1}{2} \cdot 1} \qquad \text{Divide out the common factors of 2.}$$

$$= \frac{27}{1}$$

$$= 27$$

Twenty-seven students plan to go on the trip. Because there is room for 30 passengers, there is plenty of room for one more.

▌▌▌ DIVIDING FRACTIONS

One number is called the **reciprocal** (or **multiplicative inverse**) of another if their product is 1. For example, $\frac{3}{5}$ is the reciprocal of $\frac{5}{3}$ because

$$\frac{3}{5} \cdot \frac{5}{3} = \frac{15}{15} = 1$$

Dividing Fractions

To divide two fractions, we multiply the first fraction by the reciprocal of the second fraction. In symbols, if a, b, c, and d are real numbers, then

$$\frac{a}{b} \div \frac{c}{d} = \frac{a}{b} \cdot \frac{d}{c} = \frac{a \cdot d}{b \cdot c} \qquad (b \neq 0, c \neq 0, \text{ and } d \neq 0)$$

EXAMPLE 4

a. $\dfrac{3}{5} \div \dfrac{6}{5} = \dfrac{3}{5} \cdot \dfrac{5}{6}$ Multiply $\frac{3}{5}$ by the reciprocal of $\frac{6}{5}$.

$= \dfrac{3 \cdot 5}{5 \cdot 6}$ Multiply the numerators and multiply the denominators.

$= \dfrac{3 \cdot 5}{5 \cdot 2 \cdot 3}$ Factor the denominator.

$= \dfrac{\overset{1}{\cancel{3}} \cdot \overset{1}{\cancel{5}}}{\underset{1}{\cancel{5}} \cdot 2 \cdot \underset{1}{\cancel{3}}}$ Divide out the common factors of 3 and 5.

$= \dfrac{1}{2}$

```
(3/5)/(6/5)▸Frac
                1/2
```

b. $\dfrac{15}{7} \div 10 = \dfrac{15}{7} \div \dfrac{10}{1}$ Write 10 as the improper fraction $\frac{10}{1}$.

$= \dfrac{15}{7} \cdot \dfrac{1}{10}$ Multiply $\frac{15}{7}$ by the reciprocal of $\frac{10}{1}$.

$= \dfrac{15 \cdot 1}{7 \cdot 10}$ Multiply the numerators and multiply the denominators.

$= \dfrac{3 \cdot \overset{1}{\cancel{5}} \cdot 1}{7 \cdot 2 \cdot \underset{1}{\cancel{5}}}$ Factor the numerator and denominator, and divide out the common factor of 5.

$= \dfrac{3}{14}$

```
(15/7)/10▸Frac
             3/14
```

STUDY TIP

Mathematics cannot be learned by passive observation; it requires active participation on your part. Take the time now to make a 3 × 5 note card for each boldfaced definition that has been presented thus far. Place the word(s) on the front of the card and the definition on the back. As you work on learning these definitions, tackle only a few at each study session and take advantage of all those "hidden" study times: waiting for a professor to begin class, waiting for your child to come out of school, etc. These brief study moments are excellent opportunities for painlessly learning the properties and definitions that are fundamental to a mathematics course.

▌ ADDING FRACTIONS

Adding Fractions with the Same Denominator

To add two fractions with the same denominator, we add the numerators and keep the common denominator. In symbols, if a, b, and d are real numbers, then

$$\frac{a}{d} + \frac{b}{d} = \frac{a+b}{d} \qquad (d \neq 0)$$

TECHNOLOGY TIP

When entering addition or subtraction problems involving fractions, it is not necessary to enclose each fraction in parentheses.

For example,

$$\frac{3}{7} + \frac{2}{7} = \frac{3+2}{7} \qquad \text{Add the numerators and keep the common denominator.}$$

$$= \frac{5}{7}$$

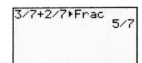

To add fractions that do not have common denominators, we must find the least common denominator. The **least common denominator** is the smallest number that can be divided evenly by each of the original denominators.

$$\frac{1}{3} + \frac{1}{5} \qquad \text{The least common denominator is 15.}$$

Thus each fraction must be rewritten with the denominator of 15.

$$\frac{1}{3} = \frac{1 \cdot 5}{3 \cdot 5} = \frac{5}{15} \quad \text{and} \quad \frac{1}{5} = \frac{1 \cdot 3}{5 \cdot 3} = \frac{3}{15}$$

Now add the two new fractions that are equivalent to the original fractions:

$$\frac{1}{3} + \frac{1}{5} = \frac{5}{15} + \frac{3}{15} = \frac{5+3}{15} = \frac{8}{15}$$

EXAMPLE 5   Add $\dfrac{3}{10}$ and $\dfrac{5}{28}$.

SOLUTION   To find the least common denominator (LCD), we find the prime factorization of both denominators and use each prime factor the greatest number of times it appears in either factorization:

$$\left.\begin{array}{l} 10 = 2 \cdot 5 \\ 28 = 2 \cdot 2 \cdot 7 \end{array}\right\} \text{LCD} = 2 \cdot 2 \cdot 5 \cdot 7 = 140$$

Because 140 is the smallest number that 10 and 28 divide exactly, we write both fractions as equivalent fractions with the least common denominator of 140.

$$\frac{3}{10} + \frac{5}{28} = \frac{3 \cdot \mathbf{14}}{10 \cdot \mathbf{14}} + \frac{5 \cdot \mathbf{5}}{28 \cdot \mathbf{5}}$$ Write each fraction as a fraction with a denominator of 140.

$$= \frac{42}{140} + \frac{25}{140}$$

$$= \frac{42 + 25}{140}$$ Add numerators and keep the denominator.

$$= \frac{67}{140}$$

```
3/10+5/28▶Frac
            67/140
```

Since 67 is a prime number that does not divide 140, $\frac{67}{140}$ is in lowest terms and cannot be simplified.

Verify with your calculator.

▊ SUBTRACTING FRACTIONS

Subtracting Fractions with the Same Denominator

To subtract two fractions with the same denominator, we subtract the numerators and keep the common denominator. In symbols, if a, b, and d are real numbers, then

$$\frac{a}{d} - \frac{b}{d} = \frac{a - b}{d} \qquad (d \neq 0)$$

For example,

$$\frac{7}{9} - \frac{2}{9} = \frac{7 - 2}{9}$$

$$= \frac{5}{9}$$

To subtract fractions with unlike denominators, we write them as equivalent fractions with a common denominator. For example, to subtract $\frac{2}{5}$ from $\frac{3}{4}$, find the LCD of 20 and proceed as follows:

$$\frac{3}{4} - \frac{2}{5} = \frac{3 \cdot \mathbf{5}}{4 \cdot \mathbf{5}} - \frac{2 \cdot \mathbf{4}}{5 \cdot \mathbf{4}}$$ Convert each fraction to an equivalent fraction with a denominator of 20.

$$= \frac{15}{20} - \frac{8}{20}$$

$$= \frac{15 - 8}{20}$$ Subtract the numerators and keep the denominator.

$$= \frac{7}{20}$$

EXAMPLE 6

Subtract 5 from $\frac{23}{3}$.

SOLUTION

$$\frac{23}{3} - 5 = \frac{23}{3} - \frac{5}{1}$$ Write 5 as the improper fraction $\frac{5}{1}$.

$$= \frac{23}{3} - \frac{5 \cdot 3}{1 \cdot 3}$$ Write $\frac{5}{1}$ as a fraction with a denominator of 3.

$$= \frac{23}{3} - \frac{15}{3}$$

$$= \frac{23 - 15}{3}$$ Subtract the numerators and keep the denominator.

$$= \frac{8}{3}$$

▚ MIXED NUMBERS

The **mixed number** $3\frac{1}{2}$ represents the sum of 3 and $\frac{1}{2}$. We can write $3\frac{1}{2}$ as an improper fraction as follows:

$$3\frac{1}{2} = \frac{3}{1} + \frac{1}{2}$$

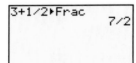

$$= \frac{6}{2} + \frac{1}{2}$$ $3 = \frac{6}{2}$.

$$= \frac{6 + 1}{2}$$ Add the numerators and keep the denominator.

$$= \frac{7}{2}$$

EXAMPLE 7

Add $2\frac{1}{4}$ and $1\frac{1}{3}$.

SOLUTION

We can change each mixed number to an improper fraction and then add the fractions.

$$2\frac{1}{4} = \frac{2}{1} + \frac{1}{4} \qquad\qquad 1\frac{1}{3} = \frac{1}{1} + \frac{1}{3}$$

$$= \frac{8}{4} + \frac{1}{4} \qquad\qquad = \frac{3}{3} + \frac{1}{3}$$

$$= \frac{9}{4} \qquad\qquad\qquad = \frac{4}{3}$$

TECHNOLOGY TIP

Before performing operations with fractions that are mixed numbers, you may prefer to convert the mixed numbers to improper fractions first.

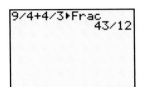

$$2\frac{1}{4} + 1\frac{1}{3} = \frac{9}{4} + \frac{4}{3}$$

$$= \frac{9 \cdot 3}{4 \cdot 3} + \frac{4 \cdot 4}{3 \cdot 4}$$

$$= \frac{27}{12} + \frac{16}{12}$$

$$= \frac{43}{12}$$

Finally, we change $\frac{43}{12}$ to a mixed number by dividing 43 by 12 and expressing the remainder as a fractional part of 12.

$$\frac{43}{12} = 3 + \frac{7}{12} = 3\frac{7}{12}$$

 WARNING!   Many students in beginning algebra classes believe a fraction is not simplified when left as an improper fraction. A fraction is considered simplified when the numerator and denominator have **no** common factors. Improper fractions are acceptable.

EXAMPLE 8

Fencing land   The three sides of a triangular piece of land measure $33\frac{1}{4}$, $57\frac{3}{4}$, and $72\frac{1}{2}$ meters as in Figure 1-3. How much fencing will be needed to enclose the area?

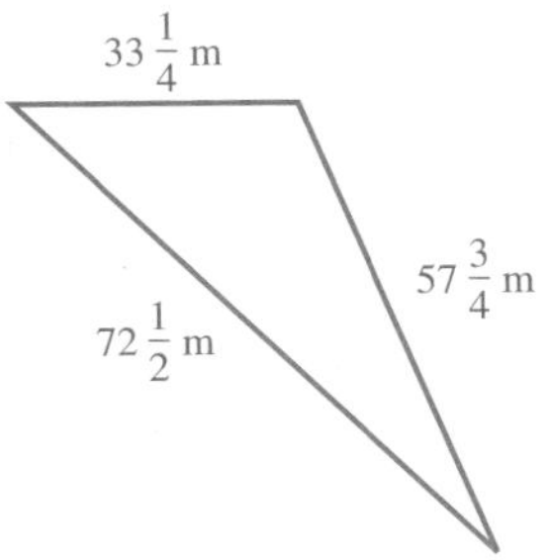

SOLUTION   We can find the sum of the lengths by adding the whole number parts and the fractional parts of the dimensions separately:

$$33\frac{1}{4} + 57\frac{3}{4} + 72\frac{1}{2} = 33 + 57 + 72 + \frac{1}{4} + \frac{3}{4} + \frac{1}{2}$$

$$= 162 + \frac{1}{4} + \frac{3}{4} + \frac{2}{4}$$

Change $\frac{1}{2}$ to $\frac{2}{4}$ to obtain an equivalent fraction with the common denominator.

$$= 162 + \frac{6}{4}$$

Add the fractions by adding the numerators and keeping the denominator.

$$= 162 + \frac{3}{2}$$

$\frac{6}{4} = \frac{2 \cdot 3}{2 \cdot 2} = \frac{\overset{1}{\cancel{2}} \cdot 3}{\underset{1}{\cancel{2}} \cdot 2} = \frac{3}{2}$

$$= 162 + 1\frac{1}{2}$$

Change $\frac{3}{2}$ to a mixed number.

$$= 163\frac{1}{2}$$

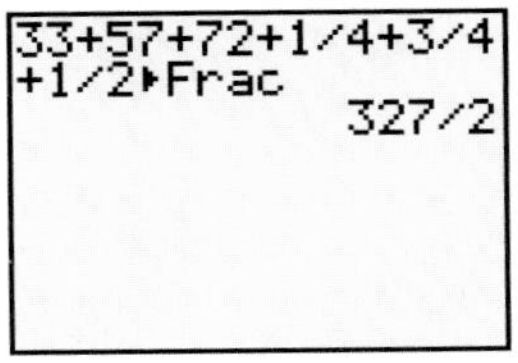

It will require $163\frac{1}{2}$ meters of fencing to enclose the triangular area.

Figure 1-3

$33\frac{1}{4}$ m

$57\frac{3}{4}$ m

$72\frac{1}{2}$ m

Due to the size of the mixed numbers, it was easier to enter the numbers as displayed instead of first converting to improper fractions.

Commutative Properties

If a and b are real numbers, then

$$a + b = b + a \qquad \text{Commutative property of addition.}$$
$$ab = ba \qquad \text{Commutative property of multiplication.}$$

In the previous example, $33\frac{1}{4} + 57\frac{3}{4} + 72\frac{1}{2}$ is rewritten as $33 + 57 + 72 + \frac{1}{4} + \frac{3}{4} + \frac{1}{2}$ using the Commutative Property of Addition.

EXAMPLE 9

When multiplication or division involves mixed numbers, it is best to convert the mixed number to an improper fraction. Recall that we multiply the whole number by the denominator of the fraction, add the numerator, and place the result over the denominator.

a. $\left(4\frac{1}{3}\right)\left(2\frac{3}{5}\right)$

b. $\left(5\frac{1}{2}\right) \div \left(1\frac{3}{5}\right)$

SOLUTION a. $\left(4\frac{1}{3}\right)\left(2\frac{3}{5}\right) = \frac{13}{3} \cdot \frac{13}{5} = \frac{13 \cdot 13}{3 \cdot 5} = \frac{169}{15}$

b. $\left(5\frac{1}{2}\right) \div \left(1\frac{3}{5}\right) = \frac{11}{2} \div \frac{8}{5} = \frac{11}{2} \cdot \frac{5}{8} = \frac{11 \cdot 5}{2 \cdot 8} = \frac{55}{16}$

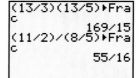

TECHNOLOGY TIP

Mixed numbers such as $(4\frac{1}{3})(2\frac{3}{5})$ can also be entered on the calculator as $(4 + 1/3)(2 + 3/5)$.

DECIMAL FRACTIONS

Rational numbers can always be converted to decimal fractions. For example, to change $\frac{1}{4}$ and $\frac{5}{22}$ to decimal fractions, we use long division:

```
   0.25              0.22727 . . .
4)1.00           22)5.00000
  8                 44
  20                60
  20                44
   0               160
                   154
                    60
                    44
                    16
```

The decimal fraction 0.25 is called a **terminating decimal**, and the decimal fraction 0.2272727 . . . (often written as $0.2\overline{27}$) is called a **repeating decimal** because it repeats a block of digits, 27. Every rational number can be converted to either a terminating or a repeating decimal.

Terminating Decimals

$\dfrac{1}{2} = 0.5$

$\dfrac{3}{4} = 0.75$

$\dfrac{5}{8} = 0.625$

Repeating Decimals

$\dfrac{1}{3} = 0.33333 \ldots$ *or* $0.\overline{3}$

$\dfrac{1}{6} = 0.16666 \ldots$ *or* $0.1\overline{6}$

$\dfrac{5}{22} = 0.2272727 \ldots$ *or* $0.2\overline{27}$

```
1/3
        .3333333333
1/6
        .1666666667
5/22
        .2727272727
```

When performing computations with fractions it is acceptable to use the *equivalent* terminating decimal, for example, $\frac{1}{2}$ or 0.5. **Do not** round repeating decimals to replace fractions when computing: $\frac{1}{3} \neq 0.3$.

Some numbers cannot be expressed as a quotient of integers or as a terminating or repeating decimal. Such numbers are called **irrational** numbers. Examples of these types of numbers would be $\sqrt{2}$ and π. We will discuss irrational numbers in more detail in Section 1.5. It is important to mention irrational numbers at this point because the set of irrational numbers combined with the set of rational numbers determine the set of **real** numbers. All of the properties defined thus far are applicable to the set of real numbers.

Real Numbers

A real number is any number that is either rational or irrational.

EXERCISE 1.1

Note: Read the information in Technology 1.1 for your particular calculator before beginning the exercises.

VOCABULARY AND NOTATION   *In Exercises 1–7, fill in the blank to make the statement true.*

1. The _____ numbers or positive integers are the set of numbers {1, 2, 3, . . .}.

2. The _____ numbers are the set of numbers {0, 1, 2, 3, . . .}.

3. A number that can be written as the quotient of two integers is called a _____ number.

4. A number greater than 1 whose only factors are 1 and itself is called a _____ number.

5. The number 1 is the _____ element for multiplication, and 0 is the _____ element for addition.

6. Another name for reciprocal is _____

7. Division by _____ is undefined.

CONCEPTS

8. Identify the rational numbers that are in lowest terms.

   a. $\dfrac{4}{5}$ **b.** $\dfrac{6}{8}$ **c.** $\dfrac{5}{2}$

9. Simplify.

a. $\dfrac{5}{0}$ **b.** $\dfrac{9}{5}$

10. Which of the following are correct?

a. $\dfrac{4}{5} \div \dfrac{2}{3} = \dfrac{5}{4} \cdot \dfrac{2}{3}$ **b.** $\dfrac{3}{7} \div 10 = \dfrac{3}{7} \cdot \dfrac{1}{10}$

c. $1\dfrac{1}{2} \div \dfrac{2}{5} = \dfrac{3}{2} \cdot \dfrac{5}{2}$ **d.** $\dfrac{4}{5} \div 4\dfrac{2}{3} = \dfrac{4}{5} \cdot 4\dfrac{3}{2}$

11. Determine the least common denominator of each pair.

a. $\dfrac{4}{5} + \dfrac{3}{2}$ **b.** $\dfrac{7}{8} + \dfrac{5}{16}$ **c.** $\dfrac{5}{9} + \dfrac{7}{15}$

12. Describe how you would find the least common denominator of two fractions.

13. Which of the following are displayed correctly on the calculator?

a. $\dfrac{4}{5} + \dfrac{3}{7}$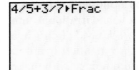

b. $\dfrac{4}{5} \cdot \dfrac{3}{7}$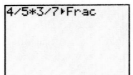

c. $\dfrac{4}{5} \div \dfrac{3}{7}$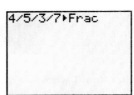

14. Explain why the commutative property is not valid for division.

15. Enter the fraction 2/3 on the calculator. Is the decimal number displayed equal to 2/3? Explain.

16. Explain how to convert a mixed number such as $4\frac{17}{19}$ to an improper fraction using only the calculator.

17. Explain how to convert an improper fraction into a mixed number.

PRACTICE *In Exercises 18–21, reduce each fraction to lowest terms by factoring numerators and denominators into primes and dividing out common factors. Use the ➤Frac option of the graphing calculator to verify your results.*

18. $\dfrac{24}{18}$ **19.** $\dfrac{35}{14}$

20. $\dfrac{72}{64}$ **21.** $\dfrac{26}{21}$

In Exercises 22–45, perform the indicated operation. Make sure your result is reduced to lowest terms. Use the ➤Frac option of the graphing calculator to verify your results.

22. $\dfrac{5}{12} \cdot \dfrac{18}{5}$ **23.** $\dfrac{5}{4} \cdot \dfrac{12}{10}$

24. $12 \cdot \dfrac{5}{6}$ **25.** $9 \cdot \dfrac{7}{12}$

26. $2\dfrac{2}{3} \cdot 1\dfrac{5}{8}$ **27.** $8\dfrac{2}{9} \cdot 7\dfrac{3}{4}$

28. $\dfrac{2}{12} \div \dfrac{8}{13}$ **29.** $\dfrac{4}{7} \div \dfrac{20}{21}$

30. $6 \div \dfrac{3}{14}$ **31.** $23 \div \dfrac{46}{5}$

32. $1\dfrac{12}{30} \div 7$ **33.** $4\dfrac{2}{8} \div 17$

34. $\dfrac{3}{5} + \dfrac{3}{5}$ **35.** $\dfrac{4}{7} - \dfrac{2}{7}$

36. $\dfrac{1}{6} + \dfrac{1}{24}$ **37.** $\dfrac{17}{25} - \dfrac{2}{5}$

38. $\dfrac{5}{14} - \dfrac{4}{21}$ **39.** $\dfrac{2}{33} + \dfrac{3}{22}$

40. $\dfrac{17}{3} + 4$ **41.** $\dfrac{13}{9} - 1$

42. $4\dfrac{3}{5} + \dfrac{3}{5}$ **43.** $2\dfrac{1}{8} + \dfrac{3}{8}$

44. $3\dfrac{3}{4} - 2\dfrac{1}{2}$ **45.** $15\dfrac{5}{6} + 11\dfrac{5}{8}$

APPLICATIONS *In Exercises 46–50, use the Student Characteristics for Fall 2002 at The University of Texas, which are given in Illustration 1. Use an arithmetic expression to model the given question/direction and determine the result.*

46. What fractional part of the Fall 2002 student body at The University of Texas was classified as American Indian and African American?

47. What fractional part of the student body was classified as Asian American and white?

48. How many of the students enrolled in the Fall of 2002 were Hispanic?

49. How many of the students enrolled in the Fall of 2002 were foreign students?

50. How many of the students enrolled in the Fall of 2002 declared no ethnicity?

The University of Texas at Austin
Office of Institutional Research

STUDENT CHARACTERISTICS
FALL 2002 ENROLLMENT OF 52,261 STUDENTS

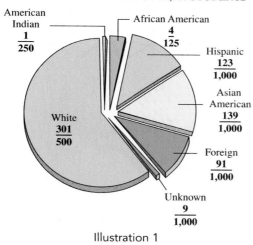

Illustration 1

In Exercises 51–56, write an arithmetic expression to model the direction/question and determine the result.

51. Data analysis   Seven hundred fifty students at the local community college were surveyed when they enrolled for classes to determine how they learn about Web sites. The survey determined that one-fifth of the students learn about Web sites from their friends.

  a. Write an arithmetic expression to determine how many students learned about Web sites from their friends. Simplify the expression.

  b. Write an arithmetic expression to determine how many students learned about Web sites from other sources. Simplify the expression.

52. Data analysis   Three hundred fifteen students graduated from Northeast High School. One-fifteenth of these students all attended the same state university.

  a. Write an arithmetic expression to determine how many students went to this university. Simplify the expression.

  b. Write an arithmetic expression to determine how many students did NOT attend this university. Simplify the expression.

  c. One-third of the class did not attend any school of higher education at all. Write an arithmetic expression to determine how many students did not pursue higher education. Simplify the expression.

53. Energy consumption   Illustration 2 indicates household energy use. What fractional part of the energy consumption is represented by the combination of home heating and water heating?

HOUSEHOLD ENERGY USE
(How energy is used in U.S. households, in percentage)

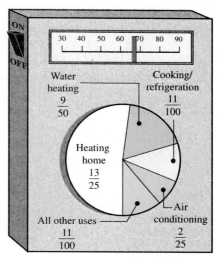

SOURCE: U.S. Energy Information Administration

Illustration 2

54. Energy consumption   Illustration 2 in Exercise 53 indicates household energy use. What fractional part of the energy consumption is represented by the combination of air conditioning and cooking/refrigeration?

55. Revenue   Based on the federal government revenues for the fiscal year 2004 in Illustration 3, what fractional part of the revenues comes from sources other than excise taxes and social insurance receipts?

56. Revenue   Based on the federal government outlays for the fiscal year 2004 as displayed in Illustration 3, what fractional part of the outlay is determined from sources other than federal operations and net interest?

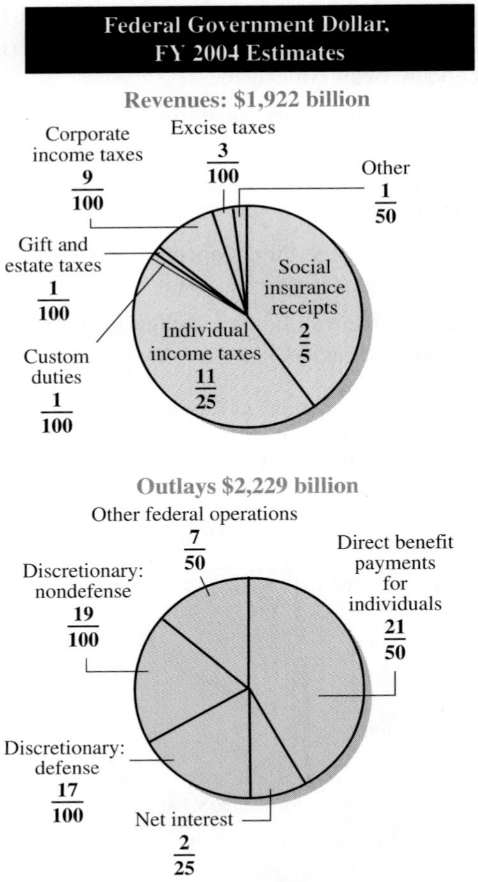

Federal Government Dollar, FY 2004 Estimates

Revenues: $1,922 billion

Corporate income taxes $\frac{9}{100}$

Excise taxes $\frac{3}{100}$

Other $\frac{1}{50}$

Gift and estate taxes $\frac{1}{100}$

Social insurance receipts $\frac{2}{5}$

Custom duties $\frac{1}{100}$

Individual income taxes $\frac{11}{25}$

Outlays $2,229 billion

Other federal operations $\frac{7}{50}$

Discretionary: nondefense $\frac{19}{100}$

Direct benefit payments for individuals $\frac{21}{50}$

Discretionary: defense $\frac{17}{100}$

Net interest $\frac{2}{25}$

SOURCE: Office of Management and Budget, *Budget of the United States of Government, FY* 2003

Illustration 3

TECHNOLOGY 1.1

Introduction to the Graphing Calculator

Technology sections are structured for independent reading. They are not designed to be part of the classroom lecture material.

Read the box headings carefully to be sure that you are following the directions for your specific calculator model. All the TI-86 information begins on page 21.

Touring the TI-83 Plus

Take a few minutes to study the TI-83 Plus graphing calculator. The keys are color coded and positioned in a way that is user friendly. Notice there are dark blue, black, and gray keys, along with a single yellow key and a single green key.

Dark blue keys: On the right and across the top of the calculator are the dark blue keys. At the top right are four directional cursor keys. These may be used to move the cursor on the screen in the direction of the arrow printed on the key. The four operation symbols (addition, subtraction, multiplication, and division) are also in dark blue. Notice the key marked **ENTER**. This will be used to activate entered commands; thus there is no key on the face of the calculator with the equal sign printed on it. Below the screen are five keys labelled [Y=], [WINDOW] [ZOOM], [TRACE], and [GRAPH]. These keys are positioned together below the screen because they are used for graphing.

Black keys: The majority of the keys on the calculator are black. Notice the [X,T,θ,n] key in the second row and second column. It will be used frequently in algebra to enter the variable **X**. The **ON** key is the black key located in the bottom-left position.

Gray keys: The 12 gray keys that are clustered at the bottom center are used to enter digits, a decimal point, or a negative sign.

Yellow key: The yellow key is the **2nd** key located in the upper-left position.

To access a symbol in yellow (printed above any of the keys) first press the yellow [2nd] key, and then the key BELOW the symbol (feature) to be accessed. For example, to turn the calculator **OFF**, notice that the word **OFF** is printed in yellow above the **ON** key. Therefore, press the **2nd** key and the **ON** key to turn the calculator off. This would be denoted as [2nd] [OFF]. The word "2nd" is the cue to look above the keypad for the command. These keystrokes are done *sequentially*, not simultaneously.

Green key: Alphabet letters (printed in green above some of the keys), or any other symbol and/or word printed in green above a key, are accessed by first pressing the green **ALPHA** key and then the key below the desired letter/symbol/word. Again, the keystrokes are sequential.

Catalog feature: Press [2nd] [CATALOG] to display an alphabetical list of available calculator operations. Use the [▲] and [▼] cursor keys to scroll through this list. Operations may be accessed by placing the pointer adjacent to the operation and pressing [ENTER]. To exit the catalog, press [CLEAR].

Courtesy Texas Instruments Incorporated

*Note: The TI-83 Plus has an **A**utomatic **P**ower **D**own (**APD**) feature that turns the calculator off when no keys have been pressed for several minutes. When this happens, press [ON] to access the last screen used.*

Let's Get Started!: TI-83 Plus

Turn the calculator on by pressing [ON]. If the display is not clear, press [2nd] [▲] to darken the screen or [2nd] [▼] to lighten the screen. Notice that when the [2nd] key is pressed, an arrow pointing up appears on the blinking cursor.

To ensure the calculator is in the desired mode, press [MODE]. All of the options on the far left should be highlighted. If not, use the [▼] to place the blinking cursor on the appropriate entry and press [ENTER]. Exit MODE by pressing [CLEAR]. This accesses the home screen, which is where expressions are entered. Press [CLEAR] until the screen is cleared except for the blinking cursor in the top left corner.

Integer Operations: TI-83 Plus

When entering integers on the calculator, you must differentiate between a subtraction sign and a negative sign. Notice that the subtraction sign appears as a blue key on the right side of the calculator, whereas the negative sign appears as a gray key next to the [ENTER] key and is labeled ($-$).

EXAMPLE **Simplify:** $-8 - 2$ **Keystrokes:** [($-$)] [8] [$-$] [2] [ENTER]

Screen display:

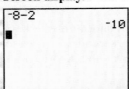

Observe the difference in the *size* and *position* of the negative sign as compared to the subtraction sign.

Absolute Value: TI-83 Plus

Absolute value is accessed by pressing [MATH] [▶] (NUM) [1:abs(] or through the [CATALOG]; press [2nd] [CATALOG] [ENTER]. The absolute value operation is displayed as **abs(**. Be sure to enter the right parenthesis.

EXAMPLE **Simplify:** $|-3 - 2|$ **Keystrokes:** [2nd] [CATALOG] [ENTER] [($-$)] [3] [$-$] [2] [)] [ENTER]

Screen display:

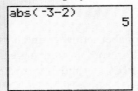

Note: This would be read as "the absolute value of the quantity negative three minus two." Keep this in mind as you enter the expression.

Square Roots: TI-83 Plus

The square root symbol is located above the [x^2] key. To access it, press [2nd] [√].

EXAMPLE **Simplify:** $\sqrt{25}$ **Keystrokes:** [2nd] [√] [2] [5] [)] [ENTER]

Screen display:

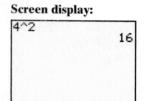

Powers of Numbers: TI-83 Plus

A number may be squared (raised to the second power) either by pressing the [x^2] key after entering the number or by using the caret [^] and then entering the desired exponent.

EXAMPLE   **Simplify: 4^2**

Keystrokes: [4] [x^2] [ENTER]   or   **Keystrokes: [4] [^] [2] [ENTER]**

Screen display: **Screen display:**

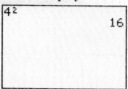

To raise to the third power (or higher), use the caret key.

EXAMPLE   **Simplify: $4(3)^5$**   **Keystrokes: [4] [(] [3] [)] [^] [5]**

Screen display:

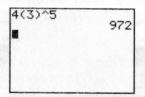

Fraction Basics: TI-83 Plus

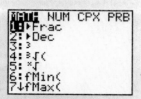

The calculator can be used to perform arithmetic operations with fractions. Pressing the [MATH] key reveals a math menu. Take a minute to use the [▼] cursor to scroll down the menu. Notice that there are ten options available. Use of the first option, **1:►Frac**, will now be illustrated.

Enter the fraction 1/2 by pressing [1] [÷] [2] [MATH]. Under the MATH menu, **1:►Frac** is highlighted. Because it is highlighted, press [ENTER] to select option 1. Press [ENTER] again to activate the "convert to a fraction" command. Notice that the calculator simply displays the fraction.

Reenter 1/2 by using the following keystrokes: [1] [÷] [2] [ENTER]. This time the MATH menu and the **►Frac** command were not accessed. The graphing calculator converts the expression to a decimal (as will any scientific calculator). However, if a fraction is desired rather than a decimal, press [MATH] [ENTER] (again, the **►Frac** option is chosen because it is highlighted). The calculator now displays **Ans ►Frac**, and the decimal is converted back to the fractional form once [ENTER] is pressed.

Note: From this point on, the **►Frac** option will be denoted by the keystrokes [MATH] [1:►Frac].

Enter the fraction 12/24 and access the ►**FRAC** option. How does the calculator display this fraction? The graphing calculator will *always* display fractions in reduced form.

Read the mixed number "$-3\frac{1}{2}$" aloud. Did you say the word "and"? This is the clue as to the way mixed numbers are entered on the calculator. Enter $-3\frac{1}{2}$ by pressing **[(−)] [3] [+] [(−)] [1] [÷] [2] [MATH] [1:►Frac] [ENTER]**. The display should correspond to the one at the left. Because the entire fraction is negative, both the integer part and the rational part must be entered as negative numbers. This can become cumbersome; therefore, we suggest that mixed numbers be entered as improper fractions.

Order of Operations: TI-83 Plus

The calculator performs the operations of multiplication and division before the operations of addition and subtraction. Moreover, it does the multiplication and division in the order in which they appear in the expressions (from left to right). It then performs the additions and subtractions in the order in which they appear from left to right. Which of the screens below is the correct display for the problem $\frac{8+16}{2}$?

a. 8+16/2■

b. (8+16)/2■

The correct screen is "b." The problem should be read as "the quantity 8 plus 16 divided by 2."

Addition and Subtraction of Fractions: TI-83 Plus

Examining the problem as displayed by the calculator, 1/2 + 1/3, we see that the order of operations clearly dictates that the two division operations be completed before the addition. Therefore, no parentheses are necessary when entering this problem on the calculator.

EXAMPLE $\quad \dfrac{1}{2}+\dfrac{1}{3}$ **Keystrokes: [1] [÷] [2] [+] [1] [÷] [3] [MATH] [1:►Frac] [ENTER]**

Screen display:

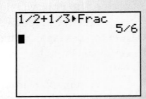

1/2+1/3►Frac
                5/6
■

Multiplication and Division of Fractions: TI-83 Plus

Parentheses are critical on this problem to ensure that the calculator addresses the division of two fractions. If the problem were entered with no parentheses, 2/3/1/5, the calculator would perform the operations of division from left to right: 2 divided by 3 = 2/3, 2/3 divided by 1 = 2/3, 2/3 divided by 5 = 2/15.

EXAMPLE $\quad \dfrac{2}{3} \div \dfrac{1}{5}$ **Keystrokes: [(] [2] [÷] [3] [)] [÷] [(] [1] [÷] [5] [)] [MATH] [1:►Frac] [ENTER]**

Screen display:

```
(2/3)/(1/5)▶Frac
              10/3
```

Touring the TI-86

Take a few minutes to study the TI-86 graphing calculator. The keys are color coded and positioned in a way that is user friendly.

Gray keys: The 12 gray keys clustered at the bottom center are used to enter digits, a decimal point, and a negative sign. At the top right are 4 directional cursor keys. These may be used to move the cursor on the screen in the direction of the arrow.

Black keys: The majority of the keys on the calculator are black. The four operation symbols (division, multiplication, subtraction, and addition) are located in the column on the far right. Notice the key marked **ENTER** at the bottom-right position. This will be used to activate commands that have been entered. The equal sign on the face of the calculator is NOT used for computation. The key marked **x-VAR** in the second row, second column will be used frequently to enter the variable **x**. The **ON** key is in the bottom-left position.

Below the screen are 5 black keys labeled **F1**, **F2**, **F3**, **F4**, and **F5**. These are menu keys and will be addressed as they are needed.

Yellow-orange key: The only key of this color is located in the top-left position and is labeled **2nd**.

Blue key: This key is labeled **ALPHA** and is located at the top left of the keypad.

Above most of the keys are words and/or symbols printed in either yellow-orange or blue. To access a symbol in yellow-orange (printed above any of the keys), first press the yellow-orange **[2nd]** key, and then the key BELOW the symbol (feature) you wish to access. For example, to turn the calculator **OFF**, notice that the word **OFF** is printed in yellow-orange above the **ON** key. Therefore, press the **2nd** key and the **ON** key to turn the calculator off. These keystrokes are done *sequentially*, not simultaneously. Thus, the previous command for turning the calculator off would appear as **[2nd] [OFF]**. The word "2nd" is the cue to look above the keypad for the command. A feature that is accessed from a menu will be written in parentheses, ( ). When a menu key is indicated, the current function of the key will follow in parentheses to correspond to the display at the bottom of the screen. For example, press **[GRAPH]** to display the graph menu. The notation **[F2](WIND)** denotes access to the WINDOW submenu. Users should be aware that the menu denoted as WIND on the TI-86 corresponds to the WINDOW menu referred to for the TI-83 Plus. To remove the menu at the bottom of the screen, press **[EXIT]**.

To access a symbol printed in blue above some of the keys, first press the blue **ALPHA** key and then the key *below* the desired letter or symbol. Again, the keystrokes are sequential.

*Note: The TI-86 has an **Automatic Power Down** (APD) feature that turns the calculator off when no keys have been pressed for several minutes. When this happens, press [ON] to access the last screen used.*

Let's Get Started!: TI-86

Turn the calculator on by pressing **[ON]**. If the display is not clear, press **[2nd]** and hold down **[▲]** to darken the screen or **[2nd]** **[▼]** to lighten it. Notice that when the **[2nd]** key is pressed, an arrow pointing up appears on the blinking cursor.

To ensure the calculator is in the desired mode, press **[2nd]** **[MODE]**. All of the options on the far left should be highlighted. If not, use the **[▼]** to place the blinking cursor on the appropriate entry and press **[ENTER]**. Exit **MODE** by pressing **[EXIT]**, **[CLEAR]** or **[2nd]** **[QUIT]**. This accesses the home screen, which is where expressions are entered.

Integer Operations: TI-86

When entering integers on the calculator, you must differentiate between a subtraction sign and a negative sign. Notice the subtraction sign is in the right column, whereas the negative sign appears as a gray key next to the **[ENTER]** key and is labeled **(−)**.

EXAMPLE   **Simplify: −8 − 2**

Keystrokes: [(−)] [8] [−] [2] [ENTER]

Screen display:

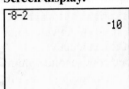

Observe the difference in *size* and *position* of the negative sign as compared to the subtraction sign.

Absolute Value: TI-86

The TI-86 does not have absolute value on its keypad. However, a key can be created using the **[CUSTOM]** key. Up to 15 frequently used functions can be customized (and accessed with only two keystrokes) provided the function desired is listed under **CATALOG**. To customize a function, press **[2nd]** **[CATLG-VARS]** **[F1](CATLG)** followed by **[F3](CUSTM)**. Place the arrow next to **abs** in the list, then press **[F1]**. The function **abs** is now listed under **PAGE▼**. The same procedure can be used to add other functions to the **CUSTOM** menu as necessary. Each time, select an open slot in the menu by choosing the menu key below the open slot. Pressing **[MORE]** accesses additional slots for customizing. When finished, press **[EXIT]** until the menus at the bottom of the screen clear and the home screen is displayed.

Pressing **[CUSTOM]** reveals the customized menu; to access a customized function, simply press the **F** key below the desired function. Pressing **[EXIT]** removes this menu.

EXAMPLE   **Simplify |−3 − 2|** (Make sure you are at the home screen; press **[EXIT]** if necessary.)

Keystrokes:
[CUSTOM] [F1](abs) [(] [(−)] [3] [−] [2] [)] [ENTER]

Note: Absolute value can also be accessed by pressing **[2nd]** **[MATH]** **[F1](NUM) [F5]** (abs).

Screen display:

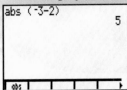

Square Roots: TI-86 The square root symbol is located above the [x^2] key. To access it, press [**2nd**] [$\sqrt{}$].

EXAMPLE   **Simplify: $\sqrt{25}$**   **Keystrokes: [2nd] [$\sqrt{}$] [2] [5] [ENTER]**

Screen display:

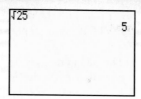

Powers of Numbers: TI-86 A number may be squared (raised to the second power) either by pressing the [x^2] key after entering the number or by using the caret [^] and then entering the desired exponent.

EXAMPLE   **Simplify: 4^2**

Keystrokes: [4] [x^2] [ENTER]   or   **Keystrokes: [4] [^] [2] [ENTER]**

Screen display: **Screen display:**

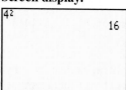

To raise to the third power (or higher), use the caret key.

EXAMPLE   **Simplify: $4(3)^5$**   **Keystrokes: [4] [(] [3] [)] [^] [5]**

Screen display:

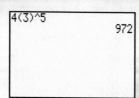

Fraction Basics: TI-86 Since the ➤**Frac** option is used frequently, it will be added to the CUSTOM menu. To do this, press [**2nd**] [**CATLG-VARS**] [**F1**](**CATLG**) [**F3**](**CUSTM**). Cursor down the list to place the arrow next to ➤**Frac**, found near the bottom of the list. Press [**F2**] or another open **F** key to create the custom key. (*Hint:* Since ➤**Frac** was found at the bottom of the list, pressing [▲] [**2nd**] [**M2**](**PAGE▲**) until the arrow is positioned next to the desired selection would be quicker than scrolling through the entire list.)

   Note: To delete a customized entry, press [**2nd**] [**CATLG-VARS**] [**F1**] (**CATLG**) [**F3**](**CUSTM**) [**F4**](**BLANK**) and then the **F** key that contains the command to be deleted.

   Press [**EXIT**] until the blinking cursor is displayed at the HOME screen.

   Enter the decimal number 0.42 by pressing [.] [4] [2] [**CUSTOM**] [**F2**](➤**Frac**) [**ENTER**]. Notice that the calculator displays the fraction reduced to lowest terms.

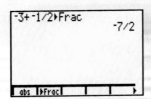

Recall that a mixed number is actually the sum of an integer and a fraction. Therefore, to enter a mixed number, simply enter the expression as an indicated sum. To enter $-3\frac{1}{2}$ and express it as a fraction, press **[(−)] [3] [+] [(−)] [1] [÷] [2] [CUSTOM] [F2](▶Frac)**. Because the entire fraction is negative, both the integer part and the rational part must be entered as negative numbers. After pressing ENTER, your display should correspond to the one at the left.

Order of Operations: TI-86

The calculator performs the operations of multiplication and division before the operations of addition and subtraction. Moreover, it does the multiplication and division in the order in which they appear in the expressions (from left to right). It then performs the additions and subtractions in the order in which they appear from left to right. Which of the screens below is the correct display for the problem $\dfrac{8+16}{2}$?

a.

b.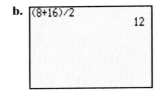

The correct screen is "b." The problem should be read as "the quantity 8 plus 16 divided by 2."

Addition and Subtraction of Fractions: TI-86

Examining the problem as displayed by the calculator, $1/2 + 1/3$, we see that the order of operations clearly dictates that the two division operations be completed before the addition. Therefore, no parentheses are necessary when entering this problem on the calculator.

EXAMPLE $\dfrac{1}{2} + \dfrac{1}{3}$

Keystrokes: [1] [÷] [2] [+] [1] [÷] [3] [CUSTOM] [F2](▶Frac) [ENTER]

Screen display:

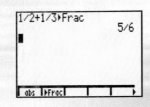

Multiplication and Division of Fractions: TI-86

Parentheses are critical on this problem to ensure that the calculator addresses the division of two fractions. If the problem were entered with no parentheses, $2/3/1/5$, the calculator would perform the operations of division from left to right: 2 divided by 3 = 2/3, 2/3 divided by 1 = 2/3, 2/3 divided by 5 = 2/15.

EXAMPLE $\dfrac{2}{3} \div \dfrac{1}{5}$

Keystrokes: [(] [2] [÷] [3] [)] [÷] [(] [1] [÷] [5] [)] [CUSTOM] [F1](▶Frac) [ENTER]

Screen display:

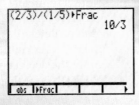

1.2 EXPONENTS AND ORDER OF OPERATIONS

- NATURAL-NUMBER EXPONENTS
- ORDER OF OPERATIONS

⚫ NATURAL-NUMBER EXPONENTS

To show how many times a number is to be used as a factor in a product, we use an **exponent**. For example, 3 is the exponent in the expression 2^3. The exponent indicates that 2 is to be used as a factor three times:

$$2^3 = \overbrace{2 \cdot 2 \cdot 2}^{\text{3 factors of 2}}$$
$$= 8$$

 WARNING!   Note that $2^3 = 8$. This is not the same as $2 \cdot 3 = 6$.

In algebra, we use letters, called **variables**, to represent real numbers. For example:

If x represents the number 4, then $x = 4$.

 WARNING!   We usually do not use a times sign, ×, to indicate multiplication. It might be mistaken for the variable x.

The exponent 5 in the expression x^5 indicates that x is to be used as a factor five times.

$$x^5 = \overbrace{x \cdot x \cdot x \cdot x \cdot x}^{\text{5 factors of } x}$$

In the expression x^5 (called an **exponential expression**), 5 is the **exponent** and x is the **base**. In expressions such as x or y, the exponent is understood to be 1:

$$x^1 = x \quad \text{and} \quad y^1 = y$$

In general, we have the following definition.

Natural-Number Exponent
If n is a natural number, then

$$x^n = \overbrace{x \cdot x \cdot x \cdot \ldots \cdot x}^{n \text{ factors of } x}$$

EXAMPLE 1

a. $4^2 = 4 \cdot 4 = 16$ Read 4^2 as "4 squared" or
as "4 to the second power."

b. $5^3 = 5 \cdot 5 \cdot 5 = 125$ Read 5^3 as "5 cubed" or as
"5 to the third power."

c. $6^4 = 6 \cdot 6 \cdot 6 \cdot 6 = 1296$ Read 6^4 as "6 to the fourth
power."

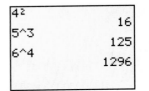

The base of an exponential expression can be a variable.

EXAMPLE 2

a. $y^6 = y \cdot y \cdot y \cdot y \cdot y \cdot y$ Read y^6 as "y to the sixth power."

b. $x^3 = x \cdot x \cdot x$ Read x^3 as "x cubed" or "x to the third power."

c. $z^2 = z \cdot z$ Read z^2 as "z squared" or as "z to the second power."

d. $a^1 = a$ Read a^1 as "a to the first power."

TECHNOLOGY TIP

You cannot perform symbolic operations on most graphing calculators. If you enter y^6 on your calculator, the calculator will determine the value of y^6 using the value that is stored in y.

▉ ORDER OF OPERATIONS

The order in which we do arithmetic is important. For example, if we simplify $2 + 3 \cdot 4$ by multiplying first, we get a result of 14. However, if we add first, we get a result of 20.

$$2 + \mathbf{3 \cdot 4} = 2 + \mathbf{12} \qquad 2 + 3 \cdot 4 = \mathbf{5} \cdot 4$$
$$= 14 \qquad\qquad = 20$$

TECHNOLOGY TIP

The graphing calculator is programmed to follow the order of operations. Regardless of the displayed grouping symbols, you must enter **only** the parentheses on your graphing calculator. The calculator **reserves** brackets [ ] for matrix operations and braces { } for list operations. This means that a problem may have nested layers of parentheses such as ((( ))).

To eliminate the possibility of getting different answers, we will agree to do multiplications before additions. The correct calculation of $2 + 3 \cdot 4$ is

$$2 + \mathbf{3 \cdot 4} = 2 + \mathbf{12}$$
$$= 14$$

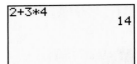

To indicate addition should be done before multiplication (to override the order of operations), we use **grouping symbols** such as parentheses ( ), brackets [ ], or braces { }. In the expression $(2 + 3)4$, the parentheses indicate that the addition should be done first:

$$(\mathbf{2 + 3})4 = \mathbf{5} \cdot 4$$
$$= 20$$

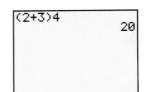

Unless grouping symbols indicate otherwise, exponential expressions are always simplified before multiplications. The expression $5 + 4 \cdot 3^2$ should be simplified in the following way:

$$5 + 4 \cdot \mathbf{3^2} = 5 + 4 \cdot \mathbf{9}$$
$$= 5 + \mathbf{36}$$
$$= 41$$

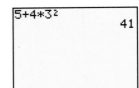

To guarantee that calculations will have a single correct result, we use the following set of **priority rules**.

TECHNOLOGY TIP

These "priority rules" are designated as the **E**quation **O**perating **S**ystems in your calculator manual. Read the page of your manual carefully to determine if your calculator gives "implied" multiplication a higher priority than the order of operations that is listed at the right.

Order of Mathematical Operations

Use the following steps to do all calculations within each pair of grouping symbols, working from the innermost pair to the outermost pair.

1. Find the values of any exponential expressions.
2. Do all multiplications and divisions as they are encountered while working from left to right.
3. Do all additions and subtractions as they are encountered while working from left to right.

When all grouping symbols have been removed, repeat the rules above to finish the calculation.

In a fraction, simplify the numerator and the denominator separately. Then perform the division indicated by the fraction bar, if possible.

WARNING!   Note that $4(2)^3 \neq (4 \cdot 2)^3$:

$$4(2)^3 = 4 \cdot 2 \cdot 2 \cdot 2 = 4(8) = 32 \quad \text{and} \quad (4 \cdot 2)^3 = 8^3 = 8 \cdot 8 \cdot 8 = 512$$

```
4(2)^3
              32
(4*2)^3
              512
```

Likewise, $4x^3 \neq (4x)^3$ because

$$4x^3 = 4 \cdot x \cdot x \cdot x \quad \text{and} \quad (4x)^3 = (4x)(4x)(4x) = 64x^3$$

EXAMPLE 3

Simplify $5^3 + 2(8 - 3 \cdot 2)$.

SOLUTION   We do the work within the parentheses first and then simplify.

```
5^3+2(8-3*2)
              129
```

$$
\begin{aligned}
5^3 + 2(8 - 3 \cdot 2) &= 5^3 + 2(8 - 6) && \text{Do the multiplication within the parentheses.} \\
&= 5^3 + 2(2) && \text{Do the subtraction within the parentheses.} \\
&= 125 + 2(2) && \text{Find the value of the exponential expression.} \\
&= 125 + 4 && \text{Do the multiplication.} \\
&= 129 && \text{Do the addition.}
\end{aligned}
$$

EXAMPLE 4

Simplify $\dfrac{3(3 + 2) + 5}{17 - 3(4)}$.

SOLUTION   We simplify the numerator and the denominator of the fraction separately and then simplify the fraction.

$$\frac{3(3 + 2) + 5}{17 - 3(4)} = \frac{3(5) + 5}{17 - 3(4)}$$

$$= \frac{15 + 5}{17 - 12}$$

$$= \frac{20}{5}$$

$$= 4$$

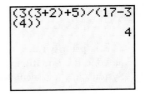

TECHNOLOGY TIP

There is no way to "draw" the long fraction bar that acts as a grouping symbol in the expression $\frac{3(3 + 2) + 5}{17 - 3(4)}$. Parentheses must be used to designate the numerator and denominator.

The series of equality statements that show $\frac{3(3 + 2) + 5}{17 - 3(4)}$ is ultimately equal to 4 demonstrates the **transitive property of equality**.

Transitive Property of Equality

If $a = b$ and $b = c$, then $a = c$.

An **arithmetic expression** contains only constants (no variables) whereas an **algebraic expression** contains variables and/or constants. Example 4 illustrates the **simplification** of an arithmetic expression using the order of operations. The word **simplify** means to remove all grouping symbols and perform the indicated operations. The example below illustrates the **evaluation** of an algebraic expression. To **evaluate** means to "find the value of." To evaluate an algebraic expression, substitute the given values for the variables into the expression. Then, simplify (evaluate) the resulting arithmetic expression.

EXAMPLE 5   If $x = 3$ and $y = 4$, evaluate

a. $3y + x^2$ and   **b.** $3(y + x^2)$.

SOLUTION   **a.** $3y + x^2 = 3(4) + 3^2$ Substitute 3 for x and 4 for y.

$= 3(4) + 9$ Evaluate the exponential expression.

$= 12 + 9$ Do the multiplication.

$= 21$ Do the addition.

TECHNOLOGY TIP

Use the STO▶ feature to store variable values and then compute the value of the expression. The colon key can be used to separate commands so that the complete problem may be keyed in before pressing ENTER. (See "Technology 1.2: Using the STOre Key," page 32.)

b. $3(y + x^2) = 3(4 + 3^2)$ Substitute 3 for x and 4 for y.

$= 3(4 + 9)$ Evaluate the exponential expression.

$= 3(13)$ Parentheses indicate that addition is performed next.

$= 39$ Do the multiplication.

EXAMPLE 6   If $x = 4$ and $y = 3$, evaluate $\dfrac{3x^2 - 2y}{2(x + y)}$.

SOLUTION

$$\dfrac{3x^2 - 2y}{2(x + y)} = \dfrac{3(4^2) - 2(3)}{2(4 + 3)}$$ Substitute 4 for x and 3 for y.

$$= \dfrac{3(16) - 2(3)}{2(7)}$$ Find the value of 4^2 in the numerator and do the addition in the denominator.

$$= \dfrac{48 - 6}{14}$$ Do the multiplications.

$$= \dfrac{42}{14}$$ Do the subtraction.

$$= 3$$ Divide.

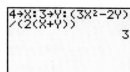

STUDY TIP

Although you are only on Section 1.2, you should begin to prepare **now** for your first major exam. When you finish tonight's homework (and as you complete each subsequent homework assignment) go back and work a few problems from each of the Exercise Sets that have been previously assigned. If you do this, you will find that much of your preparation for the test will have been accomplished by the time your instructor announces the test date.

EXAMPLE 7   **Test scores**   A student made the following scores on the four tests given during the fall semester in his elementary algebra class:

80, 72, 88, 60

What is his average going into the final?

SOLUTION   To find an average, simply add the four scores and divide by 4 (the number of scores added).

$$\dfrac{80 + 72 + 88 + 60}{4} = \dfrac{300}{4}$$ Simplify the numerator.

$$= 75$$ Divide.

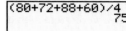

To confirm this result with your calculator, make sure you enclose the entire numerator in parentheses so that the entire numerator is simplified before the division is performed.

EXERCISE 1.2

NOTE: *Read the information in Technology 1.2 for your particular calculator before beginning the exercises.*

VOCABULARY AND NOTATION *In Exercises 1–6, fill in the blank to make a true statement.*

1. An _____ is used to show how many times a number is used as a factor.

2. _____ are used in algebra to represent real numbers.

3. In the exponential expression x^3, x is the _____ and 3 is the _____

4. The _____ property of equality states *if a = b and b = c, then a = c.*

5. An _____ expression contains only constants.

6. An _____ expression contains constants and/or variables.

CONCEPTS

7. Write each expression as a product of several factors using the definition of natural-number exponent.
   a. x^2
   b. $3z^4$
   c. $(5x)^2$
   d. $5(2x)^3$

8. The symbols $3x$ and x^3 have different meanings. Explain.

9. Students often say that x^n means, "x multiplied by itself n times." Explain why this is not correct.

10. If x were greater than 1, would raising x to higher and higher powers produce bigger numbers or smaller?

11. What would happen in Exercise 10 if x was a number between 0 and 1?

12. The following grouping symbols are used in problems involving the order of operations rules: ( ), { }, [ ]. Which of these symbols cannot be used on the calculator when performing orders of operations?

13. Explain the difference between the directions *evaluate* and the directions *simplify*.

PRACTICE *In Exercises 14–17, find the value of each expression.*

14. 4^2

15. 5^2

16. 6^3

17. 7^3

In Exercises 18–23, find the value of each expression if $x = 3$ and $y = 2$. Verify your computations using the STOre feature on the calculator. Copy the screen display to justify your work.

18. $4x^2$

19. $4y^3$

20. $(5y)^3$

21. $(2y)^4$

22. $2x^y$

23. $3y^x$

In Exercises 24–51, simplify each expression by doing the indicated operations. Verify your computations with the calculator. Copy the screen display to justify your work.

24. $3 \cdot 5 - 4$

25. $4 \cdot 6 + 5$

26. $3(5 - 4)$

27. $4(6 + 5)$

28. $3 + 5^2$

29. $4^2 - 2^2$

30. $(3 + 5)^2$

31. $(5 - 2)^3$

32. $2 + 3 \cdot 5 - 4$

33. $12 + 2 \cdot 3 + 2$

34. $64 \div (3 + 1)$

35. $16 \div (5 + 3)$

36. $(7 + 9) \div (2 \cdot 4)$

37. $(7 + 9) \div 2 \cdot 4$

38. $24 \div 4 \cdot 3 + 3$

39. $36 \div 9 \cdot 4 - 2$

40. $(100 \div 10) \cdot (10 \div 100)$

41. $100 \div [10 \cdot (10 \div 100)]$

42. $3^2 + 2(1 + 4) - 2$

43. $4 \cdot 3 + 2(5 - 2) - 2^3$

44. $5^2 - (7 - 3)^2$

45. $3^3 - (3 - 1)^3$

46. $(2 \cdot 3 - 4)^3$

47. $(3 \cdot 5 - 2 \cdot 6)^2$

48. $\dfrac{(3 + 5)^2 + 2}{2(8 - 5)}$

49. $\dfrac{25 - (2 \cdot 3 - 1)}{2 \cdot 9 - 8}$

50. $\dfrac{(5 - 3)^2 + 2}{4^2 - (8 + 2)}$

51. $\dfrac{(4^2 - 2) \div 7}{5(2 + 4) - 3^2}$

In Exercises 52–61, evaluate each expression given that
$x = 3$, $y = 2$, *and* $z = 4$. *Verify your computation by using the STOre feature on your calculator.*

52. $2x - y$ **53.** $2z + y$

54. $(3 + x)y$ **55.** $(4 + z)y$

56. $xyz + z^2 - 4x$ **57.** $zx + y^2 - 2z$

58. $3x^2 + 2y^2$ **59.** $3x^2 + (2y)^2$

60. $\dfrac{2x + y^2}{y + 2z}$ **61.** $\dfrac{2z^2 - y}{2x - y^2}$

APPLICATIONS *In Exercises 62–63, find the area of a square whose side has the given measure. The formula for the area of a square is $A = s^2$.*

62. $s = 12$

63. $s = 13$

In Exercises 64–65, find the volume of a cube whose sides have the given measure. The formula for the volume of a cube is $V = s^3$.

64. $s = 8$

65. $s = 6$

In Exercises 66–70, use a numerical expression to model the situation and simplify using the order of operations.

66. Averaging grades A student earns grades of 77, 85, 83, and 87 on four major exams. What is his average in the course?

67. College enrollment Full-time enrollments for a regional group of community colleges are 8791, 2385, 5468, 6101, 1282, and 7181 students for the fall term. What is the total full-time enrollment for the community colleges for the fall term?

68. Entertainment A family of four spends Saturday at a local amusement park. Entrance fees are $28.75, cokes are $2.50, burgers are $4.25, and chips are $1.50. If the family eats lunch consisting of drinks, burgers, and chips and then takes a break in the afternoon for everyone to have a coke, what is the total cost of the outing?

69. Entertainment An afternoon matinee birthday party for 12 middle school children costs $45 for tickets, $30 for popcorn, and $21 for drinks. What is the average cost per child?

70. Advertising An advertisement in the classified section of the newspaper reads:
   Wanted: person at least 18 years old to deliver phone books. Earn $50 plus $.15 per book delivered.
If 1000 phone books are delivered, how much is earned?

REVIEW *In Exercises 71–74, perform the indicated operation. Reduce answers and verify with your calculator.*

71. $\dfrac{7}{8} + \dfrac{7}{12}$ **72.** $\dfrac{6}{11} \div \dfrac{9}{8}$

73. $5\dfrac{3}{8} - 2\dfrac{5}{6}$

74. Enrollment If there are 21 girls in Sherah's sorority's pledge class and $\frac{5}{7}$ of them went to private school when in high school, how many of the pledges went to private school?

TECHNOLOGY 1.2

Using the STOre Key

The graphing calculator can be used to evaluate expressions. To "evaluate an expression" means to find the value of the expression for assigned values of the variable. Evaluating with a calculator can be accomplished in two ways. First, by substituting the values in by hand and then entering the resulting expression in the calculator. The second approach is to store the values under different variable names by using the STO▶ key. Remember, read the box headings carefully to be certain you are following the directions for your specific calculator model. All the TI-86 information begins at the bottom of this page.

STOre: TI-83 Plus

To store a value, simply enter the value, press **[STO▶]**, and then the desired variable. For example, to store $x = 5$, press **[5] [STO▶] [X,T,θ,n] [ENTER]**. The value 5 is now stored under the variable x. This value will remain stored in x until another value replaces it. To check and see what value is actually stored under a specific variable, enter the variable at the prompt (blinking cursor) and then press **[ENTER]**. The value is then displayed on the screen.

Note: Because x is used as a variable so often in algebra, it has its own key on the TI-83 Plus. The calculator is in function mode; therefore, when **[X,T,θ,n]** is pressed the screen displays the variable x.

EXAMPLE **Evaluate $3x^2 + 6x + 2$ when $x = -11$.**

Solution: Store -11 for x, by pressing **[(−)] [1] [1] [STO▶] [X,T,θ,n]**. The colon ":" key (located above the gray decimal key) is used to separate commands that are to be entered on the same line, so now press **[ALPHA] [:]** before entering the expression to be evaluated.

Press **[3] [X,T,θ,n] [x²] [+] [6] [X,T,θ,n] [+] [2]** to enter the expression. At this point, the calculator has been instructed to store -11 in x and then evaluate $3x^2 + 6x + 2$ for this value of x. The calculator will not perform the instructions until **[ENTER]** is pressed. The polynomial evaluates to 299. Your screen should look like the one at the left.

```
-11→X:3X²+6X+2
              299
```

To store a value under a variable other than x, the **[ALPHA]** key must be used to access the letters of the alphabet. Pressing the **[ALPHA]** key, followed by another key, allows access to the uppercase letters and symbols written in green above the keypads.

EXAMPLE **Evaluate $a^2 - 3a + 5$ when $a = -7$.**

Keystrokes: **[(−)] [7] [STO▶] [ALPHA] [A] [ALPHA] [:] [ALPHA] [A] [x²] [−] [3] [ALPHA] [A] [+] [5] [ENTER]**

Screen Display:

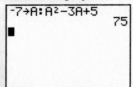

Note: Be sure to use the gray **[(−)]** key for the negative sign on a number and use the blue **[−]** key to indicate subtraction.

STOre: TI-86

The [x-VAR] key is used to display the variable x. Note that the variable x is displayed as a lowercase letter on the TI-86, instead of an uppercase x as on the TI-83 Plus screens that appear in this text.

To store a value, simply enter the value, press [STO▶], and then press the desired variable. For example, to store 5 for x, press [5] [STO▶] [x-VAR] [ENTER]. The value 5 is now stored under the variable x. This value will remain stored in x until another value replaces it. To determine the value that is actually stored under a specific variable, enter the variable at the prompt (blinking cursor) and then press [ENTER]. The value is then displayed on the screen.

Note: Pressing the STO▶ key automatically initiates the [ALPHA] key, allowing access to the remaining letters of the alphabet. Press STO▶ and observe that the blinking cursor has an **A** in it for **ALPHA**. Letters will automatically be recorded in uppercase format. To disengage the ALPHA feature, press the [ALPHA] key.

EXAMPLE

Evaluate $3x^2 + 6x + 2$ when $x = -11$.

Solution   Store -11 for x by pressing [(−)] [1] [1] [STO▶] [x-VAR]. The colon ":" key (located above the gray decimal key) is used to separate commands that are to be entered on the same line. Press [2nd] [:] before entering the expression to be evaluated. At this point, you should note that the blinking cursor has an **A** in it for ALPHA. Disengage the ALPHA feature by pressing [ALPHA] now. Press [3] [x-VAR] [x²] [+] [6] [x-VAR] [+] [2] to enter the expression that is to be evaluated. At this point, the calculator has been instructed to store -11 in x and to evaluate $3x^2 + 6x + 2$ for this value of x. The calculator will not perform the instructions until the [ENTER] key is pressed. The polynomial evaluates to 299. Your screen should look like the one at the left.

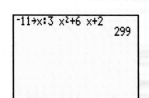

The TI-86 has both uppercase and lowercase alphabet letters that can be used as variables. To access a lowercase letter when using the STO▶ feature, press [2nd] [ALPHA] after pressing the [STO▶] key.

When not using the STO▶ feature, uppercase letters are accessed by [ALPHA] and lowercase letters are accessed by [2nd] [ALPHA].

EXAMPLE

Evaluate $a^2 - 3a + 5$ when $a = -7$.

Solution   The keystrokes would need to be [(−)] [7] [STO▶] [A] [2nd] [:] [A] (*observe that the cursor is still blinking with an uppercase A inside of it*) [ALPHA] (*the ALPHA feature is now disengaged*) [x²] [−] [3] [ALPHA] [A] [+] [5] [ENTER].

The TI-86 has the capability of recognizing combinations of letters as independent variables. For example, the expression $4AC$ means "4 times A times C." However, the TI-86 would recognize AC as representing only *one* unknown, not a product of two unknowns. Thus, the expression "4 times A times C" must be displayed on the TI-86 as 4A*C or 4A C. Variables can be separated with a space to denote the multiplication of two different quantities. By being able to combine letters to form new variables, the TI-86 has a limitless list of variables available for use. Uppercase letters are reserved for user variable names. Note that the TI-86 would recognize 4a*c as an entirely different product because the variables are lowercase.

Lowercase letters should be used with care as some are reserved for system variables. Examples of these are:

a	coefficient of regression	b	coefficient of regression
c	speed of light	e	natural log base
g	force of gravity	h	Planck's constant
k	Boltzman's constant	n	number of items in a sample
u	atomic mass unit		

The variables r, t, x, y, and θ are updated by graph coordinates based on the graphing mode.

1.3 ADDING AND SUBTRACTING REAL NUMBERS

- NUMBER LINE
- ADDING REAL NUMBERS

- ABSOLUTE VALUE OF A REAL NUMBER
- SUBTRACTING REAL NUMBERS

NUMBER LINE

TECHNOLOGY TIP

Your graphing calculator has both a subtraction key and a negative key. The negative key is displayed as $(-)$. Be sure you use this key to designate negative numbers.

The basic operations of addition, subtraction, multiplication, and division can be applied to all real numbers. As we address these basic operations in the next two sections, the rules of operations with signed numbers will focus on the set of integers. However, these results are applicable to *all* real numbers.

We use the number line shown in Figure 1-4 to represent the set of real numbers.

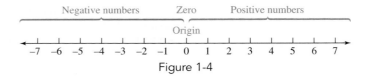

Figure 1-4

The number line continues forever to the left and to the right. Numbers to the left of 0 are negative, and we say that the line continues to negative infinity. Numbers to the right of 0 are positive, and we say that the line continues to positive infinity.

 WARNING! The number 0 is neither positive nor negative.

ABSOLUTE VALUE OF A REAL NUMBER

STUDY TIP

When taking notes, leave a 2-inch margin on the left. This space allows you to add any comments you may need later. Review your notes as soon as possible when the class is over.

On a number line, the distance between a number and 0 is called the **absolute value** of the number. For example, the distance between 5 and 0 is 5 units (see Figure 1-5). Thus, the absolute value of 5, denoted as $|5|$, is 5:

$$|5| = 5 \qquad \text{Read as "the absolute value of 5 is 5."}$$

The distance between -6 and 0 is 6. Thus,

$$|-6| = 6 \qquad \text{Read as "the absolute value of } -6 \text{ is 6."}$$

Figure 1-5

Because the absolute value of a real number represents the distance on the number line that the number is from 0, without regard to direction, the absolute value of every real number is either positive or 0. That is, if x is a real number, then

$$|x| \geq 0$$

EXAMPLE 1

Simplify   **a.** $|6|$,   **b.** $|-3|$,   **c.** $|0|$,   **d.** $|2 + 5|$.

SOLUTION   **a.** $|6| = 6$, because 6 is 6 units from 0.

b. $|-3| = 3$, because -3 is 3 units from 0.

c. $|0| = 0$, because 0 is 0 units from 0.

d. $|2 + 5| = |7| = 7$

Each of the previous problems can be verified with your graphing calculator. Locate the **ABS** key on your calculator or look for the **ABS** feature under the MATH menu. Each example is verified below:

<table>
<tr><td>a.</td><td>abs(6)</td><td></td></tr>
<tr><td></td><td></td><td>6</td></tr>
<tr><td>b.</td><td>abs(-3)</td><td></td></tr>
<tr><td></td><td></td><td>3</td></tr>
<tr><td>c.</td><td>abs(0)</td><td></td></tr>
<tr><td></td><td></td><td>0</td></tr>
</table>

<table>
<tr><td>d.</td><td>abs(2+5)</td><td></td></tr>
<tr><td></td><td></td><td>7</td></tr>
</table>

Example **d**, $|2 + 5|$, should be read as "the absolute value of the quantity 2 plus 5." For this reason, the parentheses are critical. **Absolute value symbols serve as a grouping symbol.**

▌▌▌ ADDING REAL NUMBERS

Because the positive direction on the number line is to the right, positive numbers can be represented by arrows pointing to the right. Negative numbers can be represented by arrows pointing to the left.

Using this idea of direction, the following examples can be represented with a number line interpretation.

$2 + 3 = 5$

$(-2) + (-3) = -5$

$$(-6) + 2 = -4$$

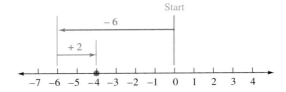

$$7 + (-4) = 3$$

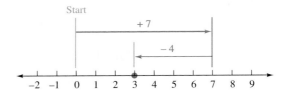

EXPLORING THE CONCEPT

Use your calculator to determine each pair of sums and respond to the questions that follow.

a. 4.5 + 7.2

$$\left(-\frac{1}{2}\right) + \left(-\frac{2}{3}\right)$$

How do the signs of each of the numbers being added relate to the sign of the sum?

b. −8.2 + 5.9

$$\frac{1}{5} + \left(-\frac{8}{7}\right)$$

Compare the absolute values of the terms. Based on the absolute value, which term is used to designate the sign of the sum?

c. 3.7 + (−1.9)

$$\left(-\frac{1}{8}\right) + \frac{7}{8}$$

Again, compare the absolute values of the terms. Based on the absolute value, which term is being used to designate the sign of the sum?

d. −6 + 6

$$\left(\frac{1}{4}\right) + \left(-\frac{1}{4}\right)$$

What happens when you add a number and its opposite?

CONCLUSIONS

a.
```
4.5+7.2
            11.7
-1/2+-2/3▶Frac
            -7/6
```

The signs of the terms are the same and the sum retains this same sign.

b.
```
-8.2+5.9
            -2.3
1/5+-8/7▶Frac
          -33/35
```

The term with the larger absolute value is used to determine the sign of the sum.

c.
```
3.7+-1.9
              1.8
-1/8+7/8▶Frac
              3/4
```

d.
```
-6+6
              0
(1/4)+(-1/4)
              0
```

The term with the larger absolute value is used to determine the sign of the sum.

Adding a number and its opposite produces a sum of 0.

Additive Inverse

Because $a + (-a) = 0$, the numbers a and $-a$ are called **negatives** or **additive inverses**.

Adding Real Numbers

1. To find the sum of two real numbers with the same sign, we add their absolute values and keep their common sign.
2. To find the sum of two real numbers with unlike signs, we subtract their absolute values (the smaller from the larger) and use the sign of the number with the larger absolute value.

EXAMPLE 2

a. $(+4) + (+6) = +(4 + 6)$
$$= 10$$

b. $(-4) + (-6) = -(4 + 6)$
$$= -10$$

c. $-\dfrac{1}{2} + \left(-\dfrac{3}{2}\right) = -\left(\dfrac{1}{2} + \dfrac{3}{2}\right)$
$$= -\dfrac{4}{2}$$
$$= -2$$

a.
b.
c.
```
4+6
              10
-4+-6
             -10
-1/2+(-3/2)▶Frac
              -2
```

EXAMPLE 3

a. $6 + (-9) = -(9 - 6)$
$$= -3$$

b. $-\dfrac{2}{3} + \left(\dfrac{1}{2}\right) = -\left(\dfrac{2}{3} - \dfrac{1}{2}\right)$
$$= -\left(\dfrac{4}{6} - \dfrac{3}{6}\right)$$
$$= -\dfrac{1}{6}$$

a.
b.
```
6+-9
              -3
-2/3+(1/2)▶Frac
             -1/6
```

EXAMPLE 4

a. $[(+3) + (-7)] + (-4) = [-4] + (-4)$ Do the work within the brackets first.

$$= -8$$

b. $3 + [(-7) + (-4)] = 3 + [-11]$ Do the work within the brackets first.

$$= -8$$

a.
```
(3+-7)+-4
              -8
```
b.
```
3+(-7+-4)
              -8
```

TECHNOLOGY TIP

1. Remember to use ONLY parentheses as grouping symbols on the calculator.
2. Textbooks will frequently place parentheses around a negative number: $5 + (-4)$. This is done for clarity to distinguish between the negative sign on the 4 and the operation symbol that precedes it. The calculator needs no such distinction since the negative sign is smaller and elevated when compared to the operation symbol. The negative sign on the calculator is also reserved for the operation of negation only.

The **associative property of addition**, as illustrated in Example 4, guarantees that three real numbers can be regrouped in an indicated addition.

Associative Property of Addition

If a, b, and c are real numbers, then $(a + b) + c = a + (b + c)$.

EXAMPLE 5

If $x = -4$, $y = 5$, and $z = -13$, evaluate **a.** $x + y$ and **b.** $2y + z$.

SOLUTION   We substitute -4 for x, 5 for y, and -13 for z and simplify.

a. $x + y = (-4) + (5)$

$$= 1$$

b. $2y + z = 2 \cdot 5 + (-13)$

$$= 10 + (-13)$$

$$= -3$$

a.
```
-4→X:5→Y: -13→Z:X
+Y
              1
```
b.
```
2Y+Z
             -3
```

TECHNOLOGY TIP

Because all the variable values for both examples were initially stored for the first example, we did not have to reenter the values for the second example.

Words and phrases such as *found*, *gain*, *credit*, *up*, *increase*, *forward*, *rises*, *in the future*, and *to the right* indicate a positive direction. Words and phrases such as *lost*, *loss*, *debit*, *down*, *decrease*, *backward*, *falls*, *in the past*, and *to the left* indicate a negative direction.

EXAMPLE 6

Account balance   The treasurer of a math club opens a checking account by depositing $350 in the bank. The bank debits the account $9 for check printing, and the treasurer writes a check for $22. Find the account balance after these transactions.

SOLUTION   The deposit can be represented by $+350$. The debit of $9 can be represented by -9, and the check written for $22 can be represented by -22. The balance in the account after these transactions is the sum of 350, -9, and -22.

$$350 + (-9) + (-22) = 341 + (-22)$$ Work from left to right.
$$= 319$$

The account balance is $319.

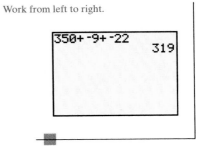

▎ SUBTRACTING REAL NUMBERS

The numbers 4 and -4 are opposites or additive inverses of one another. In fact, -4 can be read as either "negative 4" or "the opposite of 4." Thus, $-(-4)$ could be read as "the opposite of negative 4." This idea of a double negative will be important as we discuss subtracting real numbers.

Double Negative Rule

If a is a real number, then $-(-a) = a$.

EXPLORING THE CONCEPT

Perform each of the addition problems and the corresponding subtraction problems on your graphing calculator and then respond to the questions.

Addition	Subtraction
a. $(-6) + 5$	$(-6) - (-5)$
b. $6 + 5$	$6 - (-5)$
c. $(-6) + (-5)$	$(-6) - 5$
d. $6 + (-5)$	$6 - 5$

e. How is each pair of addition and subtraction problems the same? different?

f. How does the double negative rule apply in **a** and **b**?

CONCLUSIONS

a.
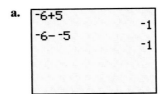

b.
```
6+5
            11
6- -5
            11
```

c.
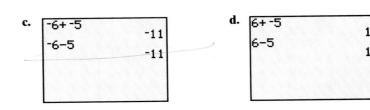

d.
```
6+ -5
             1
6-5
             1
```

e. The results are the same in each pair of problems. They are different in the following ways:

a and **b**: you are adding a positive versus subtracting a negative.
c and **d**: you are adding a negative versus subtracting a positive.

f. Subtracting a negative 5 is the same as adding a positive 5: $-(-5) = +5$

Subtracting a negative number is the same as adding the opposite number.

Subtracting Real Numbers

If a and b are two real numbers, then $a - b = a + (-b)$.

EXAMPLE 7

Simplify   **a.** $12 - 4$,   **b.** $-13 - 5$,   and   **c.** $-14 - (-6)$.

SOLUTION   **a.** $12 - 4 = 12 + (-4)$
                          $= 8$

b. $-13 - 5 = -13 + (-5)$
                 $= -18$

c. $-14 - (-6) = -14 + 6$ Use the double negative rule.
                    $= -8$

a. | 12-4
                                        8
b. | -13-5
                                       -18
c. | -14--6
                                        -8

EXAMPLE 8

If $x = -5$ and $y = -3$, evaluate   **a.** $\dfrac{y - x}{7 + x}$ and   **b.** $\dfrac{6 + x}{y - x} - \dfrac{y - 4}{7 + x}$.

SOLUTION   We can substitute -5 for x and -3 for y into each expression and simplify.

a. $\dfrac{y - x}{7 + x} = \dfrac{-3 - (-5)}{7 + (-5)}$

$= \dfrac{-3 + [-(-5)]}{2}$ To subtract -5, add the opposite of -5.

$= \dfrac{-3 + 5}{2}$ $-(-5) = +5$

$= \dfrac{2}{2}$

$= 1$

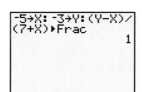

-5→X: -3→Y: (Y-X)/
(7+X)▶Frac
                    1

b. $\dfrac{6+x}{y-x} - \dfrac{y-4}{7+x} = \dfrac{6+(-5)}{-3-(-5)} - \dfrac{-3-4}{7+(-5)}$

$$= \dfrac{1}{-3+5} - \dfrac{-3+(-4)}{2} \qquad -(-5) = +5$$

$$= \dfrac{1}{2} - \dfrac{-7}{2}$$

$$= \dfrac{1-(-7)}{2}$$

$$= \dfrac{1+[-(-7)]}{2} \qquad \text{To subtract } -7, \text{ add the opposite of } -7.$$

$$= \dfrac{1+7}{2} \qquad -(-7) = +7$$

$$= \dfrac{8}{2}$$

$$= 4$$

Calculator display:
```
(6+X)/(Y-X)-(Y-4
)/(7+X)▶Frac
                4
```

EXAMPLE 9   Simplify   **a.** $3 - [4 + (-6)]$ and   **b.** $[-5 + (-3)] - [-2 - (+5)]$.

SOLUTION   **a.** $3 - [4 + (-6)] = 3 - (-2)$ Do the addition within the brackets first.

$$= 3 + [-(-2)] \qquad \text{To subtract } -2, \text{ add the opposite of } -2.$$

$$= 3 + 2 \qquad -(-2) = 2$$

$$= 5$$

Calculator display:
```
3-(4+ -6)
               5
```

b. $[-5 + (-3)] - [-2 - (+5)]$

$$= [-5 + (-3)] - [-2 + (-5)] \qquad \text{To subtract } +5, \text{ add the opposite of } 5.$$

$$= -8 - (-7) \qquad \text{Do the work within the brackets.}$$

$$= -8 + [-(-7)] \qquad \text{To subtract } -7, \text{ add the opposite of } -7.$$

$$= -8 + 7 \qquad -(-7) = 7$$

$$= -1$$

Calculator display:
```
( -5+ -3)-( -2-5)
              -1
```

EXAMPLE 10   Simplify $[-6 - (-4)] - [-7 + 10]$.

SOLUTION $[-6 - (-4)] - [-7 + 10] = [-6 + 4] - [-7 + 10]$

$$= -2 - 3$$

$$= -2 + (-3)$$

$$= -5$$

Calculator display:
```
( -6- -4)-( -7+10)
              -5
```

EXAMPLE 11

Temperature change   At noon the temperature was 7° above zero. At midnight the temperature was 4° below zero. Find the difference between these two temperatures.

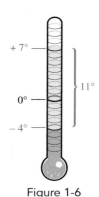

SOLUTION   A temperature of 7° above zero can be represented as +7. A temperature of 4° below zero can be represented as −4. To find the difference between these two temperatures, we set up a subtraction problem and simplify.

$$7 - (-4) = 7 + [-(-4)] \quad \text{To subtract} -4, \text{add the opposite of} -4.$$
$$= 7 + 4 \quad\quad -(-4) = 4$$
$$= 11$$

The difference between the temperatures is 11°.
Figure 1-6 shows this difference.

```
7- -4
              11
```

Figure 1-6

EXERCISE 1.3

VOCABULARY AND NOTATION   *In Exercises 1– 4, fill in the blanks to make a true statement.*

1. The distance between 0 and a number *x* on the number line is called the _____ of *x*.

2. If $a + (-a) = 0$, then *a* and −*a* are called additive _____.

3. The _____ property of addition guarantees that $(a + b) + c = a + (b + c)$.

4. The _____ rule, $-(-a) = a$, is used in the subtraction of real numbers.

CONCEPTS

5. Determine each absolute value. Verify with your calculator, copying the screen display to justify your work.
   a. $|-4|$ **b.** $-|-6|$ **c.** $-|5|$

6. Find each sum. Verify with your calculator, copying the screen display to justify your work.
   a. $4 + 11$ **b.** $(-4) + (-11)$
   c. $(-4) + 11$ **d.** $4 + (-11)$

7. Find each difference. Verify with your calculator, copying the screen display to justify your work.
   a. $8 - 4$ **b.** $-8 - 4$
   c. $8 - (-4)$ **d.** $-8 - (-4)$

8. Explain the process for adding integers.

9. Explain the process for subtracting integers.

10. Many students believe that the operation involving absolute value will always yield a positive result. Describe a situation when this will not be the case.

PRACTICE   *In Exercises 11–20, find each sum and verify with your calculator. Copy your screen display to justify your work.*

11. $|9 + (-11)|$ **12.** $|10 + (-13)|$
13. $(-5) + (-7)$ **14.** $(-6) + (-4)$
15. $(-0.4) + 0.9$ **16.** $(-1.2) + -5.3$

17. $\frac{1}{5} + \left(+\frac{1}{7}\right)$ **18.** $\frac{2}{3} + \left(-\frac{1}{4}\right)$

19. $\left(-\frac{3}{4}\right) + \left(+\frac{2}{3}\right)$ **20.** $\frac{3}{5} + \left(-\frac{2}{3}\right)$

48. $\left(\frac{7}{3} - \frac{5}{6}\right) - \left[\frac{5}{6} - \left(-\frac{7}{3}\right)\right]$

In Exercises 21–32, simplify each expression and verify with your calculator. Copy your screen display to justify your work.

21. $5 + [4 + (-2)]$

22. $-6 + [(-3) + 8]$

23. $-2 + (-4 + 5)$

24. $5 + [-4 + (-6)]$

25. $[-4 + (-3)] + [2 + (-2)]$

26. $[3 + (-1)] + [-2 + (-3)]$

27. $-4 + (-3 + 2) + (-3)$

28. $5 + [2 + (-5)] + (-2)$

29. $-|-9 + (-3)| + (-6)$

30. $-|8 + (-4)| + 7$

31. $-5.2 + |-2.5 + (-4)|$

32. $6.8 + |8.6 + (-1.1)|$

In Exercises 49–60, let $x = -4$, $y = 5$, and $z = -6$ and evaluate each expression (use fractions where applicable). Verify your computation by using the STOre feature of the calculator. Copy your screen display to justify your work.

49. $(x + y) + 3$ **50.** $(y + 5) + x$

51. $|2x + y|$ **52.** $3|x + y + z|$

53. $x - y - z$ **54.** $x - (y - z)$

55. $z - (x - y) + 10$ **56.** $3 - [x + (-3)]$

57. $\dfrac{y - x}{3 - z}$ **58.** $\dfrac{y - z}{3y + x}$

59. $\dfrac{|y + x|}{y - x}$ **60.** $\dfrac{|y - x|}{|x - y|}$

In Exercises 33–38, find each difference and verify with your calculator. Copy your screen display to justify your work.

33. $\frac{5}{3} - \frac{7}{6}$ **34.** $-\frac{5}{9} - \frac{5}{3}$

35. $-5 - \left(-\frac{3}{5}\right)$ **36.** $\frac{7}{8} - (-3)$

37. $|-6.7 - (-2.5)|$ **38.** $|17.5 - 25.3|$

In Exercises 61–64, let $x = -2.34$, $y = 3.47$, and $z = 0.72$. Use the STOre feature on your calculator to evaluate each expression and round each answer to the nearest tenth.

61. $x^3 - y + z^2$

62. $y - z^2 - x^2$

63. $x^2 - y^2 - z^2$

64. $z^3 - x^2 - y^3$

In Exercises 39–48, simplify each expression and verify with your calculator. Copy your screen display to justify your work.

39. $3 - [(-4) - 3]$ **40.** $-5 - [4 - (-2)]$

41. $(5 - 3) + (3 - 5)$

42. $(3 - 5) - [5 - (-3)]$

43. $5 - [4 + (-2) - 5]$ **44.** $3 - [-(-2) + 5]$

45. $[5 - (-34)] - [-2 + (-23)]$

46. $-5 + \{-3 - [-2 - (+4)]\}$

47. $\left(\frac{5}{2} - 3\right) - \left(\frac{3}{2} - 5\right)$

APPLICATIONS *In Exercises 65–78, use an arithmetic expression with signed numbers to model each problem. Use your calculator to perform the computation. Copy your screen display to justify your work.*

65. Temperature The temperature fell from zero to $-14°$ in one night. By 5:00 P.M. the next day, the temperature had risen 10°. What was the temperature at 5:00 P.M.?

66. History In 1897, Joseph Thompson discovered the electron. Fifty-four years later, the first fission reactor was built. Nineteen years before the reactor, James Chadwick discovered the neutron. In what year was the neutron discovered?

67. History The Greek mathematician Euclid was alive in 300 B.C. The English mathematician Sir Isaac Newton was alive in A.D. 1700. How many years apart did they live?

68. Military science   An army retreated 2300 meters. After regrouping, it moved forward 1750 meters. The next day it gained another 1875 meters. What was the army's net gain?

69. Sports   A football player gained and lost the following yardage on six consecutive plays: $+5$, $+7$, -5, $+1$, -2, -6. How many yards were gained or lost?

70. Temperature   Find the difference between a temperature of $32°$ above zero and a temperature of $27°$ above zero.

71. Temperature   Find the difference between a temperature of $3°$ below zero and a temperature of $21°$ below zero.

72. Business   The owner of a small business has a gross income of $97,345.32. However, he paid $37,675.66 in expenses, plus $7537.45 in taxes, $3723.41 in health care premiums, and $5767.99 in pension payments. Find his profit.

73. Real estate   A woman sold her house for $115,000. Her fees at closing were $78 for preparing a deed, $446 for title work, $216 for revenue stamps, and a sales commission of $7612.32. In addition, there was a deduction of $23,445.11 to pay off her old mortgage. As a part of the deal, the buyer agreed to pay half the title work. How much money did the woman receive?

74. Carpentry   John is assembling a set of metal shelves. The bolts that were packaged with the shelves are $\frac{3}{8}$ inches long. He finds that the bolts are $\frac{5}{16}$ inches too long. What length bolt is needed?

75. Farming   The Mitchells harvested $28\frac{1}{2}$ tons of wheat, sold $19\frac{3}{4}$ tons, and lost $5\frac{1}{10}$ tons to spoilage. How much wheat was left?

76. Medical   A premature baby weighs $3\frac{1}{2}$ lb at birth. From the chart below, determine the weight at the end of the week.

Day 1	2	3	4	5	6	7
loss:	loss:	gain:	gain:	gain:	gain:	gain:
$\frac{1}{8}$ lb	$\frac{1}{4}$ lb	$\frac{1}{16}$ lb	$\frac{1}{8}$ lb	$\frac{5}{16}$ lb	$\frac{1}{16}$ lb	$\frac{3}{16}$ lb

77. Distance   If the distance between two points, x and y, on the number line is defined to be $|x - y|$, find the distance between each pair of points.

x	y
$\frac{1}{2}$	$-\frac{2}{3}$
$-\frac{4}{5}$	$-\frac{1}{8}$

78. Enrollment   Illustration 1 shows the enrollment in music, art, and theatre courses at a small liberal arts college over a two-year period.

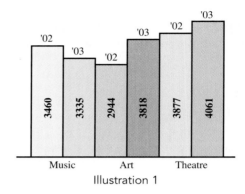

Illustration 1

a. What is the difference in enrollment from 2002 to 2003 for each department?

b. Which department had the greatest increase of students? How many students were gained?

c. What is the total change in student enrollment for all three departments combined?

79. Explain why, using a concrete everyday application, the sum of two negative numbers is always negative and the sum of two positive numbers is always positive.

80. Explain why, using a concrete everyday application, the sum of a positive number and a negative number could be either positive or negative.

81. Think of two numbers. First add the absolute values of the two numbers and write your answer. Second, add the two numbers, take the absolute value of that sum, and write that answer. Do your two answers agree? Try this several different times, using different combinations of positive and negative numbers. Can you find two numbers that produce different answers? When do you get answers that agree, and when don't you?

REVIEW   *In Exercises 82–83, simplify each expression. Verify with your calculator, copying the screen display to justify your work.*

82. $(7 - 3)^2$ **83.** $2 + 5 \cdot 3^2$

In Exercises 84–85, evaluate each expression. Use the STOre feature of the calculator to justify your work. Let $x = 3$ and $y = 2$.

84. $(4x)^y$ **85.** $6x^2 - 2y^2$

86. Grades An algebra class has 24 students. One-third of the class made an A. How many students did not make an A?

87. Averages If the five highest grades on a biology test were 90, 89, 82, 76, and 71, find the average. Write the arithmetic expression necessary to determine the average and then use your calculator for the computation. Copy your screen display to justify your work.

1.4 MULTIPLYING AND DIVIDING REAL NUMBERS

- MULTIPLYING REAL NUMBERS
- DIVIDING REAL NUMBERS

MULTIPLYING REAL NUMBERS

Because the times sign, $\times$, looks like the letter x, it is seldom used in algebra. Instead, a dot, parentheses, or no symbol at all is used to denote multiplication. Each of the following expressions indicates the **product** obtained when two real numbers, x and y, are multiplied.

$$x \cdot y \qquad (x)(y) \qquad x(y) \qquad (x)y \qquad xy$$

To develop rules for multiplying real numbers, we rely on the definition of multiplication. The expression $5 \cdot 4$ indicates that 4 is to be used as a term in a sum five times. That is

$$5(4) = 4 + 4 + 4 + 4 + 4 \qquad \text{Read } 5(4) \text{ as "5 times 4."}$$
$$= 20$$

Likewise, the expression $5(-4)$ indicates that -4 is to be used as a term in a sum five times. Thus,

$$5(-4) = (-4) + (-4) + (-4) + (-4) + (-4)$$
$$= -20 \qquad \text{Read } 5(-4) \text{ as "5 times negative 4."}$$

We realize that the order in which we multiply two whole numbers does not affect the outcome: $2 \cdot 3 = 3 \cdot 2 = 6$. This applies to all real numbers, not just whole numbers, and is called the **commutative property of multiplication**. Thus, if we are given the problem $(-5)(4)$, we can apply the commutative property to rewrite the product as $4(-5)$:

$$4(-5) = (-5) + (-5) + (-5) + (-5) = -20$$

Thus far we can conclude that

a. The product of two positive numbers is a positive number.
b. The product of a positive number and a negative number (or the product of a negative number and a positive number) is a negative number.

STUDY TIP

In order to be successful in algebra you must attend class regularly and be an active participant. Sit near the front of the class and participate in class discussions.

EXPLORING THE CONCEPT

Use your calculator to examine the pattern that is developed below to help you determine what happens when two negative numbers are multiplied together.

$$5(-4) = ?$$
$$4(-4) = ?$$
$$3(-4) = ?$$
$$2(-4) = ?$$
$$1(-4) = ?$$ Multiplicative identity: any number multiplied by 1 is that number itself.
$$0(-4) = ?$$ The product of any number and 0 is 0.
$$-1(-4) = ?$$

CONCLUSION   The pattern of numbers above is -20, -16, -12, -8, -4, 0, 4. Because $-1(-4) = 4$, the product of two negative numbers is a positive number.

Rules for Multiplying Signed Numbers

1. The product of two real numbers with like signs is the product of the absolute values.
2. The product of two real numbers with unlike signs is the negative of the product of their absolute values.
3. Any number multiplied by 0 is 0: $a \cdot 0 = 0 \cdot a = 0$.
4. Any number multiplied by 1 is that number itself: $a \cdot 1 = 1 \cdot a = a$.

EXAMPLE 1

Find each product: **a.** $4(-7)$,   **b.** $(-5)(-4)$,   **c.** $(-7)(6)$,   **d.** $8(6)$, **e.** $(-3)(5)(-4)$,   **f.** $(-4)(-2)(-3)$.

SOLUTION   **a.** $4(-7) = -(4 \cdot 7)$
$$= -28$$

b. $(-5)(-4) = +(5 \cdot 4)$
$$= +20$$

c. $(-7)(6) = -(7 \cdot 6)$
$$= -42$$

d. $8(6) = +(8 \cdot 6)$
$$= +48$$

e. $(-3)(5)(-4) = (-15)(-4)$
$$= 60$$

f. $(-4)(-2)(-3) = 8(-3)$
$$= -24$$

a.
```
4(-7)
            -28
```
b.
```
(-5)(-4)
             20
```
c.
```
(-7)(6)
            -42
```

d.
```
8(6)
             48
```
e.
```
(-3)(5)(-4)
             60
```
f.
```
(-4)(-2)(-3)
            -24
```

EXAMPLE 2   If $x = -3$, $y = 2$, and $z = 4$, evaluate   **a.** $y + xz$ and   **b.** $x(y - z)$.

SOLUTION   We substitute -3 for x, 2 for y, and 4 for z in each expression and simplify.

a. $y + xz = 2 + (-3)(4)$
$$= 2 + (-12)$$
$$= -10$$

b. $x(y - z) = -3[2 - 4]$
$$= -3[2 + (-4)]$$
$$= -3(-2)$$
$$= 6$$

TECHNOLOGY TIP

Be careful! TI-86 graphing calculators **do not** recognize the implied multiplication xz as x "times" z. If your calculator gives an error message when xz is entered, enter $x * z$ instead.

a.
```
-3→X:2→Y:4→Z:Y+X
Z
                -10
```
b.
```
X(Y-Z)
                  6
```

EXAMPLE 3   If $x = -2$ and $y = 3$, evaluate   **a.** $x^2 - y^2$ and   **b.** $-x^2$.

SOLUTION   **a.** We substitute -2 for x and 3 for y and simplify.

$$x^2 - y^2 = (-2)^2 - 3^2$$
$$= 4 - 9 \qquad \text{Simplify the exponential expressions first.}$$
$$= -5 \qquad \text{Do the subtraction.}$$

a.
```
-2→X:3→Y:X²-Y²
                 -5
```
b.
```
-X²
                 -4
```

b. We substitute -2 for x and simplify.

$$-x^2 = -(-2)^2$$
$$= -4 \qquad \text{Simplify } (-2)^2.$$

EXAMPLE 4   Find each product: **a.** $\left(-\dfrac{2}{3}\right)\left(-\dfrac{6}{5}\right)$ and   **b.** $\left(\dfrac{3}{10}\right)\left(-\dfrac{5}{9}\right)$.

SOLUTION   **a.** $\left(-\dfrac{2}{3}\right)\left(-\dfrac{6}{5}\right) = +\left(\dfrac{2}{3} \cdot \dfrac{6}{5}\right)$

$$= +\dfrac{2 \cdot 6}{3 \cdot 5}$$

$$= +\dfrac{\overset{1}{2} \cdot 2 \cdot \overset{}{\cancel{3}}}{\underset{1}{\cancel{3}} \cdot 5}$$

$$= +\dfrac{4}{5}$$

b. $\left(\dfrac{3}{10}\right)\left(-\dfrac{5}{9}\right) = -\dfrac{3}{10} \cdot \dfrac{5}{9}$

$$= -\dfrac{3 \cdot 5}{10 \cdot 9}$$

$$= -\dfrac{\overset{1}{\cancel{3}} \cdot \overset{1}{\cancel{5}}}{2 \cdot \underset{1}{\cancel{3}} \cdot 3 \cdot \underset{1}{\cancel{5}}}$$

$$= -\dfrac{1}{6}$$

a.
```
(-2/3)(-6/5)▶Fra
c
              4/5
```
b.
```
(3/10)(-5/9)▶Fra
c
             -1/6
```

EXAMPLE 5   **Temperature** If the temperature is dropping $4°$ each hour, how much warmer was it 3 hours ago?

SOLUTION A temperature drop of 4° per hour can be represented by −4° per hour. Three hours ago can be represented by −3. The temperature 3 hours ago is the product $(-3)(-4)$.

$$(-3)(-4) = +12$$

The temperature was 12° warmer 3 hours ago.

▮ DIVIDING REAL NUMBERS

In general, the rule

$$\frac{a}{b} = c \quad \text{if and only if} \quad c \cdot b = a$$

is true for the division of any real number, *a*, by any nonzero real number, *b*. For example,

$$\frac{8}{4} = 2 \quad \text{because} \quad 2 \cdot 4 = 8$$

EXPLORING THE CONCEPT
Use your calculator to perform each division problem and verify the results by computing the corresponding multiplication problem.

a. $\dfrac{10}{2}$

b. $\dfrac{-10}{-2}$

c. $\dfrac{10}{-2}$

d. $\dfrac{-10}{2}$

e. Compare the rules for signed numbers to the operation of division.

CONCLUSION

a.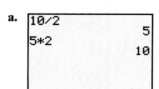

b.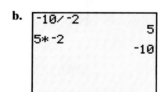

c.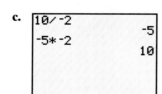

d.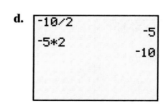

e. Parts **a** and **b** indicate that the division of two numbers with like signs produces a positive result. Parts **c** and **d** indicate that the division of two numbers with different signs produces a negative result.

Rules for Dividing Signed Numbers

1. The quotient of two real numbers with like signs is the quotient of their absolute values.
2. The quotient of two real numbers with unlike signs is the negative of the quotient of their absolute values.
3. $\dfrac{a}{0}$ is undefined; $\dfrac{0}{0}$ is undefined (indeterminate).
4. If $a \neq 0$, then $\dfrac{0}{a} = 0$.

The expression $\dfrac{8}{0}$ is undefined because there is no number x for which $0 \cdot x = 8$.

The expression $\dfrac{0}{0}$ presents a different problem, however, because $\dfrac{0}{0}$ seems to equal any number. For example, $\dfrac{0}{0} = 17$ because $0 \cdot 17 = 0$. Similarly, $\dfrac{0}{0} = 5$ because $0 \cdot 5 = 0$. Since "no answer" and "any answer" are both unacceptable, division by 0 is not defined.

EXAMPLE 6   Find each quotient: **a.** $\dfrac{36}{18}$,   **b.** $\dfrac{-44}{11}$,   **c.** $\dfrac{27}{-9}$,   **d.** $\dfrac{-64}{-8}$,   and   **e.** $\dfrac{6}{0}$.

Verify results using the graphing calculator.

SOLUTION   **a.** $\dfrac{36}{18} = +\dfrac{36}{18} = 2$ The quotient of real numbers with like signs is the quotient of their absolute values.

b. $\dfrac{-44}{11} = -\dfrac{44}{11} = -4$ The quotient of real numbers with unlike signs is the negative of the quotient of their absolute values.

c. $\dfrac{27}{-9} = -\dfrac{27}{9} = -3$ The quotient of real numbers with unlike signs is the negative of the quotient of their absolute values.

d. $\dfrac{-64}{-8} = +\dfrac{64}{8} = 8$ The quotient of real numbers with like signs is the quotient of their absolute values.

e. $\dfrac{6}{0}$ is undefined.

a.
```
36/18
            2
```
b.
```
-44/11
           -4
```
c.
```
27/-9
           -3
```

d.
```
-64/-8
            8
```

e.
```
ERR:DIVIDE BY 0
1:Quit
2:Goto
```

EXAMPLE 7   If $x = -64$, $y = 16$, and $z = -4$, evaluate   **a.** $\dfrac{yz}{-x}$,   **b.** $\dfrac{z^3 y}{x}$,   and   **c.** $\dfrac{x + y}{-z^2}$.

SOLUTION   We substitute -64 for x, 16 for y, and -4 for z in each expression and simplify.

a. $\dfrac{yz}{-x} = \dfrac{16(-4)}{-(-64)}$

$= \dfrac{-64}{+64}$

$= -1$

b. $\dfrac{z^3 y}{x} = \dfrac{(-4)^3(16)}{-64}$

$= \dfrac{(-64)(16)}{(-64)}$

$= 16$

c. $\dfrac{x+y}{-z^2} = \dfrac{-64+16}{-(-4)^2}$

$= \dfrac{-48}{-16}$

$= 3$

a.
```
-64→X: 16→Y: -4→Z:
YZ/-X
                  -1
```
b.
```
Z^3Y/X
                  16
```
c.
```
(X+Y)/-Z²
                   3
■
```

EXAMPLE 8

If $x = -50$, $y = 10$, and $z = -5$, evaluate   **a.** $\dfrac{xyz}{x - 5z}$ and   **b.** $\dfrac{3xy + 2yz}{2(x + y)}$.

Verify your results using the graphing calculator.

SOLUTION   We substitute -50 for x, 10 for y, and -5 for z in each expression and simplify.

a. $\dfrac{xyz}{x - 5z} = \dfrac{(-50)(10)(-5)}{-50 - 5(-5)}$

$= \dfrac{(-500)(-5)}{-50 + 25}$

$= \dfrac{2500}{-25}$

$= -100$

b. $\dfrac{3xy + 2yz}{2(x + y)} = \dfrac{3(-50)(10) + 2(10)(-5)}{2(-50 + 10)}$

$= \dfrac{-150(10) + (20)(-5)}{2(-40)}$

$= \dfrac{-1500 - 100}{-80}$

$= \dfrac{-1600}{-80}$

$= 20$

a.
b.
```
-50→X: 10→Y: -5→Z:
XYZ/(X-5Z)
                -100
(3XY+2YZ)/(2(X+Y
))
                  20
```

EXAMPLE 9

Stock reports   In its annual report, a publicly held corporation reports its performance on a per-share basis. When a company with 35 million shares outstanding loses \$2.3 million, what will be the per-share loss?

SOLUTION   A loss of \$2.3 million can be represented by $-2{,}300{,}000$. Because there are 35 million shares, the per-share amount lost can be represented by the quotient $\dfrac{-2{,}300{,}000}{35{,}000{,}000}$.

```
-2300000/3500000
0
        -.0657142857
```

If we use the calculator to perform the division, the result is a loss of approximately 6.6 cents per share.

Remember these facts about dividing real numbers.

Division

1. $\dfrac{a}{0}$ is undefined.

2. If $a \neq 0$, then $\dfrac{0}{a} = 0$.

3. $\dfrac{a}{1} = a$.

4. If $a \neq 0$, then $\dfrac{a}{a} = 1$.

EXERCISE 1.4

VOCABULARY AND NOTATION *Fill in the blank to make a true statement.*

1. A _____ is the result obtained when two real numbers are multiplied.

2. A _____ is the result obtained when two real numbers are divided.

3. Division by 0 is _____.

CONCEPTS

4. Explain the difference between products and factors.

5. Explain the difference between the divisors of a number and the factors of a number.

6. Find each product and verify with your calculator.
   a. $4(9)$ b. $(-4)(9)$
   c. $4(-9)$ d. $(-4)(-9)$

7. In your own words, explain the rule for multiplication of signed numbers.

8. Find each quotient and verify with your calculator.
   a. $\dfrac{10}{5}$ b. $-\dfrac{10}{5}$
   c. $\dfrac{10}{-5}$ d. $\dfrac{-10}{-5}$

9. Explain the rule for dividing signed numbers.

10. Explain how you would decide whether the product of several numbers is positive or negative.

11. If the quotient of two numbers is undefined, what would their product be?

12. If x^5 is a negative number, can you determine if x is negative? Explain.

PRACTICE *In Exercises 13–24, simplify each expression. Confirm each result with the graphing calculator, making sure to record the screen display.*

13. $\left(\dfrac{1}{2}\right)(32)$

14. $\left(-\dfrac{3}{4}\right)(12)$

15. $\left(-\dfrac{3}{4}\right)\left(-\dfrac{8}{3}\right)$

16. $\left(-\dfrac{12}{5}\right)\left(\dfrac{15}{2}\right)$

17. $(3)(-4)(-6)$ 18. $(-1)(-3)(-6)$

19. $(-2)(3)(4)$ 20. $(5)(0)(-3)$

21. $\dfrac{8 - 12}{-2}$ 22. $\dfrac{16 - 2}{2 - 9}$

23. $\dfrac{20 - 25}{7 - 12}$ 24. $\dfrac{2(15)^2 - 2}{-2^3 + 1}$

In Exercises 25–46, let $x = -1$, $y = 2$, and $z = -3$. Evaluate each expression and confirm results by using the STOre feature of the graphing calculator.

25. x^2 26. $-z^2$

27. $-xz$ 28. $y - xz$

29. $z - xy$

30. $(x + y)z$

31. $y(x - z)$

32. $xy + yz$

33. $zx - zy$

34. y^2z^2

35. z^3y

36. $y(x - y)^2$

37. $z(y - x)^2$

38. $\dfrac{yz}{x}$

39. $\dfrac{zx}{y}$

40. $\dfrac{z + x}{y}$

41. $\dfrac{xyz}{y - 1}$

42. $\dfrac{z - xy}{x + y}$

43. $\dfrac{x^2y^3}{yz}$

44. $\dfrac{2x^2 + 2y}{x + y}$

45. $\dfrac{y^2 + z^2}{y + z}$

46. $\dfrac{2x^2 - 2z^2}{x + z}$

In Exercises 47–54, evaluate each expression if $x = \frac{1}{2}$, $y = -\frac{2}{3}$, and $z = -\frac{3}{4}$. Confirm results by using the STOre feature of the graphing calculator. Express all results as fractions (when applicable).

47. $x + y$

48. $y + z$

49. $x + y + z$

50. $y + x - z$

51. $(x + y)(x - y)$

52. $(x - z)(x + z)$

53. $(x + y + z)(xyz)$

54. $xyz(x - y - z)$

55. If the product of five numbers is negative, how many of the factors could be negative?

56. If x^6 is a positive number, can you determine if x is positive?

APPLICATIONS   In Exercises 57–63, (a) write an arithmetic expression for each problem, and (b) use your graphing calculator to simplify each expression. Copy the screen display to verify results.

57. **Mowing lawns**   Rafael worked all day mowing lawns and was paid $8.00 per hour. If he had $94 at the end of an 8-hour day, how much did he have before he started working?

58. **Gambling**   In Las Vegas, Robert lost $30 per hour playing the slot machines for 15 hours. What product of signed numbers represents the change in his financial condition?

59. **Dieting**   A man lost 37.5 pounds. If he lost 2.5 pounds each week, how long had he been dieting?

60. **Saving for college**   A student has saved $15,000 to attend graduate school. If she estimates that her expenses will be $613.50 a month while in school, does she have enough to complete an 18-month master's degree program?

61. **Earnings per share**   Over a 5-year period, a corporation reported profits of $18 million, $21 million, and $33 million in 3 of the years. In the other 2 years, it reported losses of $5 million and $71 million. Find the average gain (or loss) per year.

62. Describe two situations in which negative numbers are useful.

63. **Stock market**   A stock originally went on the market for $2.40. During the next 6-month period, the stock gained $\frac{3}{4}$ point, gained 2 points, dropped $1\frac{1}{2}$ points, dropped $\frac{7}{8}$ point, rose $\frac{1}{4}$ point, and dropped $\frac{3}{8}$ point, as indicated in Illustration 1. If each point represents $1.00, what is the stock's value at the end of the 6-month period?

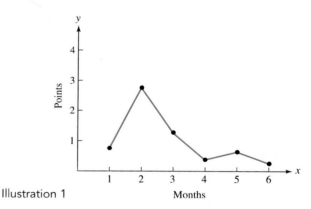

Illustration 1

REVIEW   In Exercises 64–68, simplify and verify your results with your calculator.

64. 4^4

65. $\left(-\dfrac{2}{3}\right) + \left(\dfrac{4}{5}\right)$

66. $[-4 - (6 - 1)] + [3 - 7]$

67. Evaluate for $y = -2$ and $x = -\frac{1}{2}$

  a. $x + y$ b. $x - y$

68. A pattern for a Halloween costume requires $1\frac{1}{8}$ yards of blue material, $\frac{2}{3}$ yard of red material, $\frac{1}{2}$ yard of white material, and $\frac{3}{4}$ yard of yellow material. What is the total amount of material purchased?

1.5 BASIC APPLICATIONS

- IRRATIONAL NUMBERS
- ROUNDING DECIMALS
- BUSINESS

- GEOMETRY
- COMMONLY USED FORMULAS IN MATHEMATICS

▐ IRRATIONAL NUMBERS

In the first section, we talked about studying problem situations and developing our tools for solving these problems. Many application problems can be solved using various mathematical formulas: Determining the volume of a swimming pool is often necessary to compute the amount of chemicals needed to treat the water; converting Fahrenheit temperatures to Celsius is necessary for scientific applications; computing interest on loans or savings accounts is necessary for personal finances.

We will begin by examining some basic geometric formulas for perimeter, area, and volume. To do this, we need to discuss the set of irrational numbers in more detail than was done in Section 1.1.

Irrational Numbers

An **irrational number** is a number that cannot be expressed as a fraction with an integer in its numerator and a nonzero integer in its denominator. They are non-repeating, nonterminating decimals.

Numbers such as $\sqrt{2}$ and π are examples of **irrational numbers**, because they cannot be written as fractions with an integer numerator and an integer denominator. We can find *decimal approximations* for the values of irrational numbers with a calculator.

EXPLORING THE CONCEPT

Enter each of the following on your graphing calculator. Determine which numbers are rational and which are irrational.

a. $\sqrt{2}$ **b.** π **c.** $\dfrac{17}{20}$ **d.** $\sqrt{36}$

CONCLUSIONS

a.
b.
c.
d.

```
√(2)
        1.414213562
π
        3.141592654
17/20
                .85
√(36)
                  6
```

a. irrational

b. irrational

c. rational

d. rational

```
1234/1235▶Frac
    .9991902834
■
```

You must be careful not to make generalizations about the decimal numbers displayed on your calculator. For any denominator of more than four digits, the calculator will always display a decimal approximation. For example, the fraction $\frac{1234}{1235}$ is a rational number because it is a quotient of two integers. However, if you enter the fraction $\frac{1234}{1235}$ on your calculator followed by the convert to fraction command, it will display the decimal equivalent because the denominator has four digits.

When a result is an irrational number, an approximation is given on the calculator. The symbols used for "is approximately equal to" are ≈ or ≐ . Therefore, $\sqrt{2} \approx 1.41$ or $\sqrt{2} \doteq 1.41$ when expressed to two decimal places.

 WARNING!   You must know the definitions of rational and irrational numbers and not rely on calculator displays.

Recall that the set of real numbers is comprised of the sets of rational numbers and irrational numbers. Figure 1-7 shows how the various sets of numbers are interrelated. Examples of numbers within each set are displayed in the boxes.

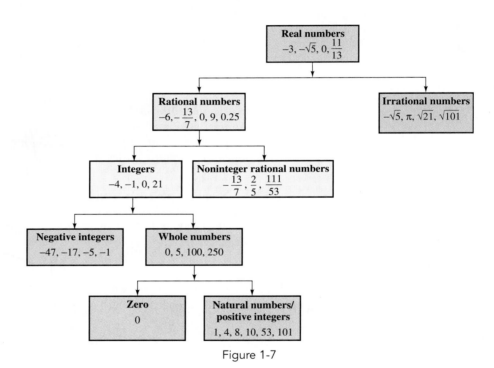

Figure 1-7

EXAMPLE 1

List the numbers in the set $\{-3, 0, \frac{1}{2}, 1.25, \sqrt{3}, 5\}$ that are

a. natural numbers   b. whole numbers

c. negative integers   d. rational numbers

e. irrational numbers   f. real numbers

SOLUTION

a. The only natural number is 5.

b. The whole numbers are 0 and 5.

c. The only negative integer is −3.

d. The rational numbers are $-3, 0, \frac{1}{2}, 1.25$, and 5. (1.25 is rational because 1.25 can be written in the form $\frac{5}{4}$.)

e. The only irrational number is $\sqrt{3}$.

f. All of the numbers are real numbers.

▌ GEOMETRY

Substituting numbers for variables is often required when finding perimeters and areas of geometric figures. The **perimeter** of a geometric figure is the distance around it, and the **area** of the figure is the amount of surface that it encloses. The perimeter of a circle is called its **circumference**. Perimeter and circumference are always expressed in linear units, the same units in which the measurements are given. Area is always expressed in square units or units².

Table 1-1 shows the formulas for the perimeter and area of several geometric figures.

TABLE 1-1

Figure	Name	Perimeter	Area
 Square figure with sides s	Square	$P = 4s$	$A = s^2$
Rectangle figure with length l, width w	Rectangle	$P = 2l + 2w$	$A = lw$
Triangle figure with sides a, c, height h, base b	Triangle	$P = a + b + c$	$A = \dfrac{1}{2}bh$
Trapezoid figure with sides a, c, d, height h, base b	Trapezoid	$P = a + b + c + d$	$A = \dfrac{1}{2}h(b + d)$
Circle figure with radius r	Circle	$C = 2\pi r$	$A = \pi r^2$

TABLE 1-2

Figure	Name	Volume
	Rectangular solid	$V = lwh$
	Cylinder	$V = Bh$ where B is the area of the base
	Pyramid	$V = \frac{1}{3}Bh$ where B is the area of the base
	Cone	$V = \frac{1}{3}Bh$ where B is the area of the base (If the base is a circle, then $B = \pi r^2$)
	Sphere	$V = \frac{4}{3}\pi r^3$

The **volume** of a three-dimensional geometric solid is the amount of space it encloses. Volume is always expressed in cubic units. Table 1-2 shows the formulas for the volume of several solids.

EXAMPLE 2

Prefab 4-foot-high fencing is $4.99 per linear foot. Find the cost of fencing the rectangular lot displayed:

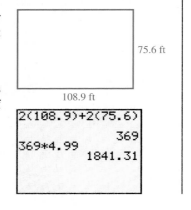

75.6 ft

108.9 ft

SOLUTION   The formula for the perimeter of a rectangle is $P = 2l + 2w$. Substitute in the known values of $l = 108.9$ and $w = 75.6$.

$$P = 2(108.9) + 2(75.6)$$
$$= 217.8 + 151.2$$
$$= 369$$

```
2(108.9)+2(75.6)
                369
369*4.99
             1841.31
```

We need 369 linear feet of fencing. Because the fencing is $4.99 per linear foot, we have

$$369(4.99) = \$1841.31$$

The cost of fencing the lot is $1841.31.

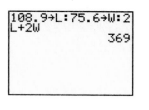

EXAMPLE 3

Winter driving   Find the number of cubic feet of road salt in a conical pile that is 18.75 feet high and covers a circular area of 28.60 feet in diameter. Use a calculator and round the answer to two decimal places.

SOLUTION   We can find the area of the circular base of the cone shown in Figure 1-8 by substituting 14.3 for the radius in the formula for the area of a circle.

$$A = \pi r^2$$
$$= \pi (\mathbf{14.3})^2$$
$$\approx 642.4242817 \qquad \text{Use a calculator.}$$

We then substitute 642.4242817 for B and 18.75 for h in the formula for the volume of a cone.

$$V = \frac{1}{3} Bh$$

$$\approx \frac{1}{3} (\mathbf{642.4242817})(\mathbf{18.75})$$

$$\approx 4{,}015.151761 \qquad \text{Use a calculator.}$$

To two decimal places, there are 4,015.15 cubic feet of salt in the pile.

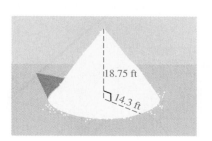

Figure 1-8

▌▌ ROUNDING DECIMALS

When decimal fractions are long, we often **round** them to a specific number of decimal places. This gives an *approximate value* of the decimal fraction. For example, the decimal fraction 25.36124 rounded to one place (or to the nearest tenth) is 25.4. Rounded to two places (or to the nearest one-hundredth), the decimal is 25.36. To round decimals, we use the following rules.

Rounding Decimals

1. Determine to how many decimal places you wish to round.
2. Look at the first digit to the right of that decimal place.
3. If that digit is 4 or less, drop it and all digits that follow. If it is 5 or greater, add 1 to the digit in the position in which you wish to round, and drop all the digits that follow.

▌▌▌ COMMONLY USED FORMULAS IN MATHEMATICS

Distance (d) can be determined as it relates to rate of speed (r) and time (t). This is given by the formula $d = rt$. For example, if we were riding in a car traveling at a speed of 50 mph for 3 hours, we would have traveled 150 miles.

Suppose we want to determine the distance traveled by a train going 72 mph at 2-hour intervals for the first 12 hours of the excursion. We could display the information in a table like the one below.

Time (hours)	2	4	6	8	10	12	t
Distance (miles)	144	288	432	576	720	864	$72t$

This table can be created analytically (paper-and-pencil computation), via the calculator by entering the arithmetic expression as in Example 2, or via the STOre key, also illustrated in Example 2. The calculator itself will also generate a table.

First, go to the table setup menu. You must designate the start value (or minimum). The table above has a start value of 2 hours. The table increment must also be designated. The Greek letter delta, Δ, is used in mathematics to represent the change in value. Thus $\Delta t = 2$ would represent time intervals of 2-hour increments. Your calculator may have ΔTbl to designate the table increment.

To generate your table, go to the y= screen and enter the expression 72x. The calculator will only recognize x as a variable when generating a table. The screens below should match your screens. Access the TABLE screen to display the table seen below. TI-86 screens will vary slightly.

TI-83 Plus: press [2nd] [TBLSET]
TI-86: press [TABLE], [F2](TBLST)

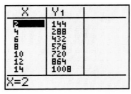

TI-83 Plus: press [Y=]
TI-86: press [GRAPH], [F1](y,(x))

TI-83 Plus: press [2nd] [TABLE]
TI-86: press [TABLE], [F1](TABLE)

It displays the same information as the table created analytically above.

EXAMPLE 4

The formula for converting Fahrenheit temperature to Celsius is given as

$$C = \frac{5}{9}(F - 32).$$

a. Use the conversion formula to determine the equivalent Celsius temperature for 23° F.

b. Use the calculator to generate a conversion table for 23° F, 33° F, 43° F, and 53° F. Approximate the equivalent Celsius temperature to the nearest tenth of a degree for each of the given Fahrenheit temperatures.

SOLUTION   **a.** $C = \dfrac{5}{9}(F - 32)$

Substitute 23 for F (the Fahrenheit temperature).

$$C = \frac{5}{9}(23 - 32)$$

$$C = \frac{5}{9}(-9)$$

$$C = -5$$

Therefore, 23° Fahrenheit is equivalent to −5° Celsius.

b. The calculator will be evaluating the expression $\frac{5}{9}(F - 32)$ for the different values of F. In order to generate a table, the expression must be entered at the y1= prompt. Enter it now.

Access the table setup and set the table to start at 23 with increments of 10. Display the table:

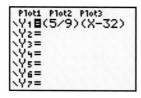

We can conclude the following approximations of temperatures:
23°F = −5° C, 33° F ≈ 0.6° C, 43° F ≈ 6.1° C, and 53° F ≈ 11.7° C.

▮ BUSINESS

The basic formula for computing simple interest is $i = prt$, where i represents interest earned, p represents the principal (the amount invested or borrowed), r is the annual rate (percent), and t is the amount of time (in years) the money is invested. A **percent** represents parts out of 100. Thus, 6% would be $\frac{6}{100}$ or 0.06; $6\frac{1}{4}$% would be $\frac{6.25}{100}$ or 0.0625. All percents should be converted to fractions or the equivalent terminating decimal (**do not round**) before performing computations.

EXAMPLE 5

Paying tuition   Juan signs a one-year note to borrow $8,500 for tuition. If the rate of interest is $6\frac{1}{4}$%, how much interest will he pay?

SOLUTION   For the privilege of using the bank's money for one year, Juan must pay $6\frac{1}{4}$% of $8,500. We calculate the interest, i, as follows:

$$i = 6\frac{1}{4}\% \text{ of } 8,500$$

$$= 0.0625 \cdot 8,500 \qquad \text{The word } of \text{ means } times.$$

$$= 531.25$$

Juan will pay $531.25 interest.

EXERCISE 1.5

VOCABULARY AND NOTATION   *In Exercises 1–6, fill in the blank to make a true statement.*

1. An _____ number is a number that cannot be expressed as a quotient of integers.

2. The _____ of a geometric figure is the distance around it.

3. The _____ of a geometric figure is the amount of surface it encloses.

4. The perimeter of a circle is called the _____.

5. The _____ of a three-dimensional geometric solid is the amount of space it encloses.

6. A _____ represents the number of parts out of 100.

CONCEPTS

7. List the numbers in the set $\left\{-3, -\dfrac{1}{2}, 0, \sqrt{2}, 3, 5.\overline{6}\right\}$ that are

   a. natural numbers

   b. whole numbers

   c. negative integers

   d. rational numbers

   e. irrational numbers

8. Determine whether perimeter, area, or volume would be used to determine the measure of the indicated quantity.

   a. water for a swimming pool

   b. wallpaper to cover two walls of a bedroom

   c. carpeting for a bedroom

   d. wallpaper border for a kitchen

   e. paint for the exterior of a house

   f. sand in an hour glass

   g. concrete edging for flower bed/garden

   h. compost for flower bed/garden

   i. weed killer for flower bed/garden

9. True or False: Two-thirds $\left(\dfrac{2}{3}\right)$ is equal to 0.67. Explain.

10. Which is more accurate when computing the area of a circle, 3.14 or the value displayed by the π key on your calculator? Explain.

11. Convert each of the following percentages to a decimal representation.

   a. 5%   **b.** $4\frac{1}{2}\%$ **c.** $8\frac{3}{4}\%$

PRACTICE   *In Exercises 12–15, find the perimeter of each figure. Perimeter is considered a linear measure, therefore your results should be expressed in the same units as the given measurement. If the calculator is used, make sure screens are copied exactly as they appear.*

12.

13.
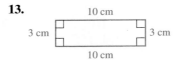

14. 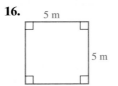   **15.**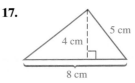

In Exercises 16–19, find the area of each figure. Area is always expressed in square units, or units². If the calculator is used, make sure screens are copied exactly as they appear.

16. 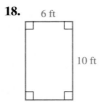   **17.**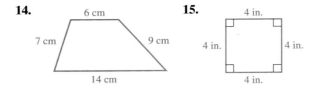

18.   **19.**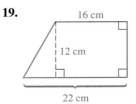

In Exercises 20 and 21, find the circumference of each circle. Use the π key on your calculator. Circumference is considered a linear measurement. If the calculator is used, make sure screens are copied exactly as they appear.

20.

14 m

21.

21 cm

In Exercises 22 and 23, find the area of each circle. Use the π key on your calculator. Make sure results are expressed in square units. If the calculator is used, make sure screens are copied exactly as they appear.

22.

42 ft

23.

7 m

In Exercises 24–29, find the volume of each figure. Use the π key on your calculator. Make sure results are expressed in cubic units. If the calculator is used, make sure screens are copied exactly as they appear. Round results to the nearest thousandth.

24.
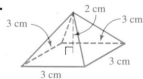
2 cm
3 cm   3 cm
3 cm
3 cm

25.
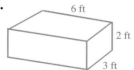
6 ft
2 ft
3 ft

26.

6 m

27.

14 in.
|←— 12 in. —→|

28.
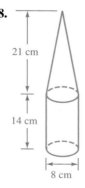
21 cm
14 cm
8 cm

29.
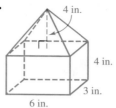
4 in.
4 in.
3 in.
6 in.

APPLICATIONS   *In Exercises 30–54, use the π key on your calculator where applicable. Make sure solutions are expressed in the appropriate units. Copy calculator screens exactly as they appear and round results to two decimal places.*

30. Circumference Find the circumference of the bicycle wheel shown in Illustration 1.

12 in.
Illustration 1

31. Area   A built-in kitchen eating area has the dimensions pictured in Illustration 2. Find the amount of floor area occupied.

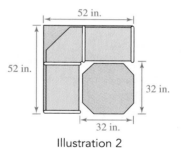
52 in.
52 in.
32 in.
32 in.
Illustration 2

32. Area   Find the area of the 25-inch television screen in Illustration 3 if the width is 17 inches and the height is 18.5 inches.

25 in.
Illustration 3

33. Volume Find the volume, in cubic inches, of the lawn cart shown in Illustration 4.

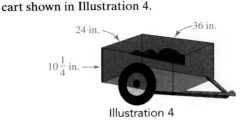

24 in.

36 in.

$10\frac{1}{4}$ in.

Illustration 4

34. Volume Find the volume of the hot water heater with the dimensions pictured in Illustration 5.

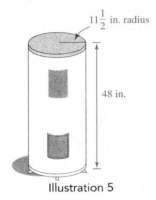

$11\frac{1}{2}$ in. radius

48 in.

Illustration 5

35. Volume Find the volume of the can of tomato sauce shown in Illustration 6.

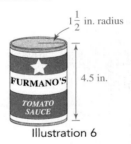

$1\frac{1}{2}$ in. radius

4.5 in.

FURMANO'S

TOMATO SAUCE

Illustration 6

36. Shipping Find the volume of the crate that was used to ship the stove shown in Illustration 7. Assume there is no extra space between the stove and the crate.

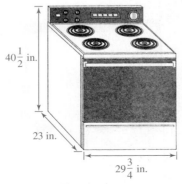

$40\frac{1}{2}$ in.

23 in.

$29\frac{3}{4}$ in.

Illustration 7

37. Construction A chicken coop is to be constructed with a tin roof and sides that are covered in poultry netting. If the coop has the dimensions indicated in Illustration 8, will one roll of the netting be sufficient? Justify your response.

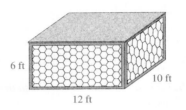

6 ft

10 ft

12 ft

POULTRY NETTING
24 in. H x 50 ft L

Illustration 8

38. Storing solvents A hazardous solvent fills a rectangular tank with dimensions of 12 by 9.5 by 7.3 inches. For disposal, it must be transferred to a cylindrical canister 7.5 inches in diameter and 18 inches high. Find how much solvent will be left over.

39. Wallpapering One roll of wallpaper covers about 33 square feet. At $27.50 per roll, how much would it cost to paper two walls 8.5 feet high and 17.3 feet long? (*Hint:* Wallpaper can only be purchased in full rolls.)

40. Sealing asphalt A rectangular parking lot is 253.5 feet long and 178.5 feet wide. A 55-gallon drum of asphalt sealer covers 4000 square feet and costs $97.50. Find the cost to seal the parking lot. Sealer can only be purchased in full drums.

41. Installing carpet What will it cost to carpet a 23-by-17.5-foot living room and a 17.5-by-14-foot dining room with carpet that costs $29.79 per square yard? One square yard is 9 square feet. (*Hint:* Carpet can be purchased only in full square yards.)

42. Temperature Use your calculator to generate a table to compute the Celsius temperature for each Fahrenheit temperature to the nearest tenth of a degree.

Fahrenheit temp. x	−20	−15	−10	−5	0
Celsius temp. $\frac{5}{9}(x-32)$					

43. Scroll through the table constructed in Exercise 42 to answer the following questions:

a. When it is 35° F, what is the Celsius temperature (to the nearest degree)?

b. When it is 10° C, what is the Fahrenheit temperature?

44. Distance The formula for finding the distance between two points a and b on a number line is $|a - b|$. For each pair of points specified below, find the distance between them.

a.

a at -14, b at -8

b.

a at $-8\frac{1}{2}$, b at $2\frac{3}{5}$

c.

a at $\frac{2}{5}$, b at $7\frac{5}{9}$

45. Business Blaire deposits her summer job earnings of $2,300 in a savings account earning $1\frac{1}{4}\%$ simple interest. How much interest will she have

a. at the end of 2 years?

b. at the end of 6 months?

46. Ecology Approximately 242,000,000 old tires are disposed of each year. If 77.7% are dumped illegally, how many old tires are dumped illegally?

47. Banking If Clarksville, Tennessee, has 52,000 people with bank accounts, according to Illustration 9, how many people do their bank transactions by phone?

OFF-SITE BANKING

Less than half of all banking transactions are done at bank branch offices. Automated teller machines (ATMs) are the most popular off-site option:

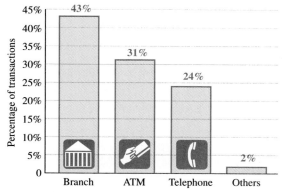

SOURCE: Bank Administration Institute

Illustration 9

48. Speed skating In tryouts for the Olympics, a speed skater had times of 44.47, 43.24, 42.77, and 42.05 seconds. Find the average time.

49. Statistics An elementary school has an enrollment of 871 students. Based on Illustration 11, how many of these children do not have snacks?

WHO CHOOSES SNACKS FOR KIDS?

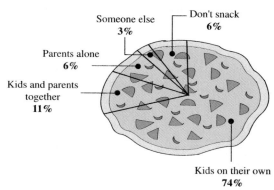

SOURCE: Statistics based on *USA Today*

Illustration 10

50. Dairy production If a Holstein cow produces 7600 gallons of milk with a $3\frac{1}{2}\%$ butterfat content, and a Guernsey cow produces 6500 gallons of milk that is 5% butterfat, which cow produces more butterfat?

51. Cost of gasoline Diego drove his car 15,675.2 miles last year, averaging 25.5 miles per gallon of gasoline. The average cost of gasoline was $1.27 per gallon. Find the fuel cost to drive the car.

52. Feeding dairy cows Each year a typical dairy cow will eat 12,000 pounds of food, which is 57% silage. To feed a herd of 30 cows, how much silage will a farmer use in a year?

53. Comparing bids Two contractors bid on a home remodeling project. The first bids $9350 for the entire job. The second contractor will work for $27.50 per hour, plus $4500 for materials. He estimates that the job will take 150 hours. Which contractor has the lower bid?

54. Choosing a furnace A high-efficiency home heating system can be installed for $4170, with an average monthly heating bill of $57.50. A regular furnace can be installed for $1730, but monthly heating bills average $107.75. After three years, which system is more expensive?

REVIEW

55. Simplify each of the following. Verify the results with your calculator.

    a. -3^2 **b.** $(-3)^2$

    c. 3^2 **d.** $4 \cdot 3^2$

56. Evaluate each of the following for $x = -\dfrac{1}{2}$ and $y = -\dfrac{2}{3}$.

Express as fractions.

    a. $-x^2$ **b.** xy **c.** xy^3

57. Area   A college student purchased 4 gallons of paint to paint her efficiency apartment. If she uses wallpaper in the bathroom, she will only use $2\frac{7}{9}$ gallons of paint. How much paint will be left over?

58. Perimeter   An irregularly shaped lot has sides of $14\frac{1}{2}$ yards, $11\frac{2}{3}$ yards, $17\frac{5}{8}$ yards, and $15\frac{1}{3}$ yards. How much chainlink fencing will be needed to enclose the lot?

CHAPTER SUMMARY

CONCEPTS

Section 1.1

All results that are fractions should be expressed in lowest terms.

A fraction is completely reduced when the numerator and the denominator have no common prime factors.

Fractions may be added or subtracted as long as they have a common denominator.

Fractions may be multiplied by finding the product of the numerator over the product of the denominator.

To divide one fraction by another, multiply the first fraction by the reciprocal of the second fraction.

Before performing arithmetic operations with mixed numbers, first convert them to improper fractions.

Fractions may be converted to their terminating decimal equivalent before performing arithmetic operations. Fractions that represent repeating decimals may not be converted to decimals before performing arithmetic operations.

Section 1.2

If n is a natural number,

$$\overbrace{}^{n \text{ factors of } x}$$

then $x^n = x \cdot x \cdot \ldots \cdot x$

REVIEW EXERCISES

Review of Fractions

1. Simplify each of the expressions below arithmetically. Confirm results with the calculator, copying screen displays exactly as they appear. Express as fractions.

   a. $\dfrac{1}{3} + \dfrac{1}{7}$

   b. $\dfrac{2}{3} - \dfrac{1}{7}$

   c. $\dfrac{31}{15} \cdot \dfrac{10}{62}$

   d. $\dfrac{18}{21} \div \dfrac{6}{7}$

   e. $3\dfrac{2}{3} + 4\dfrac{1}{2}$

   f. $6\dfrac{4}{5} - 2\dfrac{1}{3}$

   g. $3\dfrac{1}{2} \cdot 4\dfrac{1}{3}$

   h. $3\dfrac{3}{4} \div 1\dfrac{5}{8}$

2. **Flu shots** Faculty at a local university are required to get flu shots on one of 3 days. The first day, $\frac{1}{5}$ of the faculty receive shots and $\frac{2}{3}$ receive their shots on the second day.

   a. What fractional part of the faculty received flu shots on the third day?

   b. If the faculty totaled 600, how many received shots each day?

Exponents and Order of Operations

3. Simplify each of the following expressions. Confirm results with the graphing calculator, copying screen displays exactly as they appear.

   a. 3^4

   b. $\left(\dfrac{2}{3}\right)^2$

Order of Mathematical Operations

Within each pair of grouping symbols (working from the innermost pair to the outermost pair), do the following operations:

1. Find the values of any exponential expressions.
2. Do all multiplications and divisions as they are encountered while working from left to right.
3. Do all additions and subtractions as they are encountered while working from left to right.

When all of the grouping symbols have been removed, repeat the above rules to finish the calculation.

Only the parentheses keys may be used as grouping symbols on your graphing calculator. Replace all [ ] and { } with parentheses when entering expressions to be evaluated on the calculator.

In a fraction simplify the numerator and denominator separately. Then simplify the fraction, if possible.

4. Simplify each of the following expressions. Confirm results with the graphing calculator, copying screen displays exactly as they appear.

   a. $4 + (8 \div 4)$

   b. $(4 + 8) \div 4$

   c. $(5 - 2)^2 + 5^2 + 2^2$

   d. $\dfrac{4 \cdot 3 + 3^4}{31}$

   e. $\left(\dfrac{3}{2}\right)^2 - \dfrac{1}{3}$

   f. $\dfrac{4}{3} \cdot \dfrac{9}{2} + \dfrac{1}{2} \cdot 18$

5. For each of the following, write the arithmetic expression whose calculator entry is displayed below. Your expression should be written as it would be displayed in your text.

   a. `(-1/4)^3`■

   b. `(4-8)/(-12+1)`■

   c. `4-8/-12+1`

6. **Average study time**   Four students recorded time spent working on a take-home exam: 5.2, 4.5, 9.5, and 8 hours. Find the average time spent. Make sure you write an arithmetic expression to model the problem and then simplify the expression. Confirm your calculations with the graphing calculator, copying your screen display to justify the results.

Section 1.3

Adding and Subtracting Real Numbers

The absolute value of a number x, denoted by $|x|$, is the distance between x and 0 on the number line,

$$|x| \geq 0$$

To find the sum of two real numbers with the same sign, we add their absolute values and keep their common sign.

7. Simplify each of the given expressions. Confirm your calculations with the graphing calculator, copying the screen display exactly as it appears.

   a. $1 - [5 - (-3)]$

   b. $\left|\dfrac{3}{7} - \left(-\dfrac{4}{7}\right)\right|$

   c. $\dfrac{3}{7} - \left|-\dfrac{4}{7}\right|$

   d. $\dfrac{2}{3} - \left(\dfrac{1}{3} - \dfrac{2}{3}\right)$

   e. $-\dfrac{3}{14} \cdot \left(-\dfrac{7}{6}\right)$

   f. $-\dfrac{1}{2} \cdot \dfrac{4}{3}$

To add two real numbers that have unlike signs, we subtract their absolute values (the smaller from the larger) and use the sign of the number with the greatest absolute value.

If a and b are two real numbers, then

$$a - b = a + (-b)$$

On the graphing calculator, be careful to observe the difference between the operation key of subtraction and the key used for a negative sign.

8. Simplify each of the given expressions. Confirm your calculations with the graphing calculator, copying the screen display exactly as it appears.

a. $\left(\dfrac{-3 + (-3)}{3}\right)\left(-\dfrac{15}{5}\right)$ **b.** $\dfrac{-2 - (-8)}{5 + (-1)}$

c. $\dfrac{[-3 + (-4)]^2}{10 + (-3)}$ **d.** $\left(\dfrac{-10}{2}\right)^2 - (-1)^3$

Section 1.4

Multiplying and Dividing Real Numbers

The product of two real numbers with like signs is the product of their absolute values.

The product of two real numbers with unlike signs is the negative of the product of their absolute values.

The quotient of two real numbers with like signs is the quotient of their absolute values.

The quotient of two real numbers with unlike signs is the negative of the quotient of their absolute values.

Division by zero is undefined.

TI-86 graphing calculators do not recognize xy as a product of x and y, and you will have to enter the expression as $x * y$.

In Review Exercises 9–24, let $x = 2$, $y = -3$, and $z = -1$. Evaluate each expression below both analytically (paper-and-pencil computation) and using the STOre feature of the graphing calculator. Be sure to copy your screen display exactly as it appears.

9. $y + z$ **10.** $x + y$

11. $x + (y + z)$ **12.** $x - y$

13. $x - (y - z)$ **14.** $(x - y) - z$

15. xy **16.** yz

17. $x(x + z)$ **18.** xyz

19. $y^2z + x$ **20.** $yz^3 + (xy)^2$

21. $\dfrac{xy}{z}$ **22.** $\dfrac{|xy|}{3z}$

23. $\dfrac{3y^2 - x^2 + 1}{y|z|}$ **24.** $\dfrac{2y^2 - xyz}{x^2|yz|}$

Section 1.5

Basic Applications

Mathematical formulas can be used to solve many application problems.

25. Packaging   Four steel bands surround the shipping crate in Illustration 1. Find the total length of strapping needed.

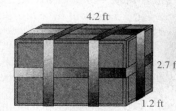

4.2 ft

2.7 ft

1.2 ft

Illustration 1

If using π in a formula, be sure to use the π key on your calculator instead of a decimal or fractional approximation.

The graphing calculator can be used for more than just computation. It can be used to generate tables to display basic information relating to a specific formula.

Percentages should always be converted to the decimal or fractional equivalent before performing the operations of multiplication or division.

26. Petroleum storage Find the volume of the cylindrical storage in Illustration 2. Round the result to the nearest thousandth.

27. Temperature Use the formula to convert Fahrenheit degrees to Celsius to determine the Celsius temperature when the temperature is 95° F.

28. Area of a Classroom Twenty-five students are in a classroom that has a dimension of 40′ by 40′ by 9′. How many square feet of floor space is there per student?

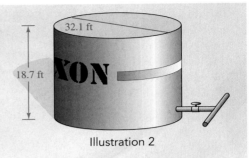

Illustration 2

CHAPTER TEST

In Problems 1–8, simplify each expression (use fractions where applicable). Verify your results with the calculator, copying the screen display exactly as it appears.

1. $3(4 - 2) - 2(2 - 6)$

2. $6 + 4 \cdot 3 - 2(-7)$

3. $-13.7 - |-13.7|$

4. $\dfrac{7}{8} \cdot \dfrac{24}{21}$

5. $\dfrac{18}{35} \div \dfrac{9}{14}$

6. $\dfrac{24}{16} + 3$

7. $\dfrac{17 - 5}{36} - \dfrac{2(13 - 5)}{12}$

8. $\dfrac{|-7 - (-6)|}{-7 - |-6|}$

In Problems 9–14, write an arithmetic expression to model the given data. Simplify the expression both analytically (paper and pencil) and by using the calculator, copying the screen display exactly as it appears.

9. Find the area of a rectangle 12.8 feet wide and 23.56 feet long. Round the result to the nearest hundredth.

10. Find the area of the figure in Illustration 1.

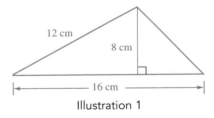

Illustration 1

11. Find the volume of the solid in Illustration 2. Round results to the nearest thousandth.

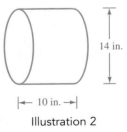

Illustration 2

12. A student has $5,000 in a certificate of deposit that earns $3\frac{1}{2}$% simple interest and $1,500 in a savings account that earns 1% simple interest. What is the total amount of interest earned?

13. If a small city has 150,000 homes (or apartment units), according to Illustration 3, how many are heated by natural gas?

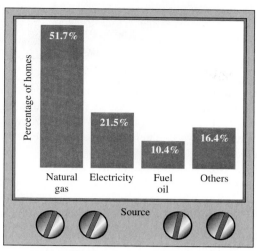

SOURCE: Energy Information Administration

Illustration 3

14. The ottoman in Illustration 4 is $14\frac{1}{2}$ inches long, $14\frac{1}{2}$ inches wide, and $18\frac{1}{2}$ inches high. What is the volume of stuffing required to fill a single ottoman?

Illustration 4

In Problems 15–20, let $x = -2$, $y = 3$, and $z = 4$. Evaluate each expression both analytically and by using the calculator. Be sure to copy screen displays exactly as they appear to justify your results.

15. $xy + z$

16. $x(y + z)$

17. $\dfrac{z + 4y}{2x}$

18. $|x^y - z|$

19. $x^3 + y^2 + z$

20. $|x| - 3|y| - 4|z|$

CAREERS & MATHEMATICS

© Walter Hodges/CORBIS

Computer Systems Analyst

Computer systems analysts help businesses and scientific research organizations develop analytical systems to process and interpret data. Using such techniques as cost accounting, sampling, and mathematical model building, they analyze information and often present results graphically by charts and diagrams. They may also prepare cost–benefit analyses to help management decide whether proposed solutions are satisfactory.

Once the system is accepted, systems analysts adapt its logical requirements to the capabilities of computer machinery. They work closely with programmers to rid the system of errors.

Systems analysts solve a wide range of problems in many different industries. Because the work is so varied and complex, analysts usually specialize in either business or scientific and engineering applications.

Qualifications

Businesses prefer a college degree in accounting, business management, or economics. Scientific organizations prefer a college degree in the physical sciences, mathematics, or engineering. Advanced degrees and degrees in computer science and information science are becoming more important in both employment areas.

Job Outlook

Through the year 2010, the demand for systems analysts is expected to grow more rapidly than the average for all other occupations. As competition increases worldwide, systems analysts will help organizations take advantage of technological advances in automation, telecommunications, and scientific research. Because personal computers are now found on almost every desktop, systems analysts will be needed to design computer networks that facilitate the sharing of data, while providing adequate security for sensitive information.

Example Application

The process of arranging records into a sequential order, called **sorting**, is a common and important task in electronic data processing. In any sorting operation, records must be compared with other records to determine which one should precede the other. One sorting technique, called a **selection sort**, requires C comparisons to sort N records into their proper order, where C and N are related by the following formula:

$$C = \frac{N(N-1)}{2}$$

How many comparisons are required to

a. sort 20 records?

b. sort 10,000 records?

Solution   **a.** Substitute 20 for N in the formula

$$C = \frac{N(N-1)}{2}$$

and calculate C:

$$C = \frac{20(20 - 1)}{2}$$

$$C = \frac{20 \cdot 19}{2}$$

$$C = 190$$

Sorting 20 records requires 190 comparisons.

b. Substitute 10,000 for N in the formula

$$C = \frac{N(N - 1)}{2}$$

and calculate C:

$$C = \frac{10,000(10,000 - 1)}{2}$$

$$C = 49,995,000$$

Sorting 10,000 records requires almost 50 million comparisons. A selection sort is not efficient for large lists.

EXERCISES

1. How many comparisons are required to sort 500 records?

2. If the computer time required to sort 20 records were 0.5 second, how long would it take to sort 10,000 records?

3. Another important data-processing task is that of finding a particular entry in a large list. In a **sequential search**, an average of C comparisons is required to find an entry in a list of N items. C and N are related by the formula

$$C = \frac{(N + 1)}{2}$$

    How many comparisons are needed (on the average) to search a list of 25 items?

4. How many comparisons are needed (on the average) to search 10,000 items?

(*Answers:* **1.** 124,750 **2.** ≈36.5 hr **3.** 13 **4.** 5,000.)

Exponents and Polynomials

© Tom Stewart/CORBIS

 InfoTrac Project

Do a subject guide search on "HIV," choose the subdivision "prevention," and find the article "Success for AIDS prevention in the U.S. but catastrophe elsewhere." Write a summary of the article.

After rounding to the nearest hundred thousand, convert to scientific notation the figure representing the number of people who have escaped HIV infection because of prevention efforts in the U.S. Convert the amount of money saved to scientific notation and calculate the cost of treating one AIDS patient. Assuming that 4 million of the 7 million infected people in Asia will develop AIDS, use scientific notation to calculate the cost of treating them.

Complete this project after studying Section 2.2.

PERSPECTIVE

A Father of Algebra

One important figure in the history of mathematics, François Vieta (1540–1603) was one of the first mathematicians to use notation close to that which we use today. Trained as a lawyer, Vieta served in the parliament of Brittany and as the personal lawyer of Henry of Navarre. If he had continued as a successful lawyer, Vieta might now be forgotten. Fortunately, he lost his job.

When political opposition forced him out of office in 1584, Vieta had time to devote himself entirely to his hobby, mathematics. He used the time to study the writings of earlier mathematicians, and

adapted and improved their ideas. Vieta was the first to use letters to represent unknown numbers, but he did not use our modern notation for exponents. To us, his notation seems awkward. For example, what we would write as

$$(x + 1)^3 = x^3 + 3x^2 + 3x + 1$$

Vieta would have written as

$$\overline{x + 1} \text{ cubus aequalis x cubus + x quad.}$$
$$3 + x \text{ in } 3 + 1$$

In this chapter, we shall develop rules for finding products, powers, and quotients of certain exponential expressions and then discuss how to add, subtract, multiply, and divide polynomials. We will examine the simplification of arithmetic expressions, the transition from arithmetic to algebraic expressions, the simplification of algebraic expressions, and the evaluation of algebraic expressions. To **simplify** an expression means to remove grouping symbols; to perform all arithmetic operations of addition, subtraction, multiplication, and division; and to combine like terms.

2.1 WHOLE-NUMBER EXPONENTS

- DEFINITION OF NATURAL-NUMBER EXPONENTS
- PROPERTIES OF NATURAL-NUMBER EXPONENTS
- ZERO EXPONENT

DEFINITION OF NATURAL-NUMBER EXPONENTS

We have used natural number exponents to indicate repeated multiplication. For example,

$$2^5 = 2 \cdot 2 \cdot 2 \cdot 2 \cdot 2 = 32 \qquad (-7)^3 = (-7)(-7)(-7) = -343$$
$$x^4 = x \cdot x \cdot x \cdot x \qquad\qquad -y^5 = -y \cdot y \cdot y \cdot y \cdot y$$

The expression t^5 is in **exponential form** whereas $t \cdot t \cdot t \cdot t \cdot t$ is written in **expanded form**.

These examples suggest a definition for x^n, where n is a natural number.

Natural-Number Exponents

If n is a natural number, then

$$x^n = \overbrace{x \cdot x \cdot x \cdot \ldots \cdot x}^{n \text{ factors of } x}$$

In the exponential expression x^n, x is called the **base**, and n is called the **exponent**. If an exponent is a natural number, it tells how many times its base is to be used as a factor in a product. An exponent of 1 indicates that its base is to be used as a factor one time, an exponent of 2 indicates that its base is to be used as a factor two times, and so on.

$$3^1 = 3 \qquad (-y)^1 = -y \qquad (-4z)^2 = (-4z)(-4z) \qquad \text{and} \qquad (t^2)^3 = t^2 \cdot t^2 \cdot t^2$$

WARNING! A negative sign is **not** part of a base unless parentheses are used.

EXAMPLE 1

Show that -2^4, read "the opposite of 2 to the fourth power," and $(-2)^4$, read "negative 2 quantity to the fourth power," are different numbers.

```
-2^4
             -16
(-2)^4
              16
```

SOLUTION   We write each expression without exponents and observe different results.

$$-2^4 = -2 \cdot 2 \cdot 2 \cdot 2 \qquad (-2)^4 = (-2)(-2)(-2)(-2)$$
$$= -16 \qquad\qquad\qquad = 16$$

Since $-16 \neq 16$, it follows that $-2^4 \neq (-2)^4$.

EXAMPLE 2

Write each expression in expanded form using the definition of natural-number exponent.

a. r^3, **b.** $(-2s)^4$, and **c.** $\left(\dfrac{1}{3}ab\right)^5$.

SOLUTION   **a.** $r^3 = r \cdot r \cdot r$

b. $(-2s)^4 = (-2s)(-2s)(-2s)(-2s)$

c. $\left(\dfrac{1}{3}ab\right)^5 = \left(\dfrac{1}{3}ab\right)\left(\dfrac{1}{3}ab\right)\left(\dfrac{1}{3}ab\right)\left(\dfrac{1}{3}ab\right)\left(\dfrac{1}{3}ab\right)$

▮ PROPERTIES OF NATURAL-NUMBER EXPONENTS

TECHNOLOGY TIP

Remember, many graphing calculators cannot perform symbolic operations. If you enter $(x^2)\wedge3$, the calculator will evaluate the expression for the value that is stored in x.

Because of the definition of natural-number exponents, we can develop a rule for multiplying exponential expressions. To multiply x^3 by x^2, for example, we note that the expression x^2 means that x is to be used as a factor two times and the expression x^3 means that x is to be used as a factor three times. Thus,

$$x^2 x^3 = \overbrace{x \cdot x}^{\text{2 factors of } x} \cdot \overbrace{x \cdot x \cdot x}^{\text{3 factors of } x}$$

$$= \underbrace{x \cdot x \cdot x \cdot x \cdot x}_{\text{5 factors of } x}$$

$$= x^5$$

In general,

$$\underbrace{}_{m \text{ factors of } x} \quad \underbrace{}_{n \text{ factors of } x}$$

$$x^m \cdot x^n = \overbrace{x \cdot x \cdot x \cdot \ldots \cdot x}^{} \cdot \overbrace{x \cdot x \cdot x \cdot \ldots \cdot x}^{}$$

$$= \underbrace{x \cdot x \cdot x \cdot x \cdot x \cdot x \cdot \ldots \cdot x \cdot x \cdot x}_{m + n \text{ factors of } x}$$

$$= x^{m+n}$$

This discussion justifies the rule for multiplying exponential expressions: *To multiply two exponential expressions with the same base, we keep the base and add the exponents.*

Product Rule for Exponents

If *m* and *n* are natural numbers, then

$$x^m x^n = x^{m+n}$$

a. $x^3 x^4 = x^{3+4}$
$= x^7$

b. $y^2 y^4 = y^{2+4}$
$= y^6$

c. $z z^3 = z^1 z^3$
$= z^{1+3}$
$= z^4$

d. $x^2 x^3 x^6 = (x^2 x^3) x^6$
$= (x^{2+3}) x^6$
$= x^5 x^6$
$= x^{5+6}$
$= x^{11}$

e. $(2y^3)(3y^2) = 2(3) y^3 y^2$
$= 6 y^{3+2}$
$= 6 y^5$

f. $(4x)(-3x^2) = 4(-3) x x^2$
$= -12 x^{1+2}$
$= -12 x^3$

WARNING!   The product rule for exponents applies only to exponential expressions with the same base. An expression such as $x^2 y^3$ cannot be simplified, because x^2 and y^3 have different bases.

To find another rule of exponents, we consider the expression $(x^3)^4$, which can be written in expanded form as $x^3 \cdot x^3 \cdot x^3 \cdot x^3$. Because each of the four factors of x^3 contains three factors of x, there are $4 \cdot 3$, or 12, factors of x. Thus, the product can be written as x^{12}.

$$(x^3)^4 = x^3 \cdot x^3 \cdot x^3 \cdot x^3$$

$$= \overbrace{x \cdot x \cdot x}^{} \cdot \overbrace{x \cdot x \cdot x}^{} \cdot \overbrace{x \cdot x \cdot x}^{} \cdot \overbrace{x \cdot x \cdot x}^{\text{12 factors of } x}$$
$$\underbrace{}_{x^3} \underbrace{}_{x^3} \underbrace{}_{x^3} \underbrace{}_{x^3}$$

$$= x^{12}$$

In general,

$$(x^m)^n = \overbrace{x^m \cdot x^m \cdot x^m \cdot \ldots \cdot x^m}^{n \text{ factors of } x^m}$$

$$= \overbrace{x \cdot x \cdot x \cdot x \cdot x \cdot x \cdot x \cdot \ldots \cdot x}^{mn \text{ factors of } x}$$

$$= x^{mn}$$

This discussion justifies a rule for raising an exponential expression to a power: *To raise an exponential expression to a power, we keep the base and multiply the exponents.*

Power to a Power Rule

If m and n are natural numbers, then

$$(x^m)^n = x^{mn}$$

EXPLORING THE CONCEPT   Use your graphing calculator to explore the relationship between each pair of numbers.

a. $(2^4)^5; (2^5)^4$ **b.** $(4^2)^4; (4^4)^2$

c. Which property justifies the results of each pair of exponential expressions?

```
(2^4)^5
        1048576
(2^5)^4
        1048576
■
```
```
(4²)^4
        65536
(4^4)²
        65536
```

Since $(2^4)^5 = 2^{4 \cdot 5}$ and $(2^5)^4 = 2^{5 \cdot 4}$, it is the commutative property of multiplication that justifies the results. Thus $(x^m)^n = (x^n)^m$.

EXAMPLE 4   **a.** $(2^3)^7 = 2^{3 \cdot 7}$
$= 2^{21}$

b. $(y^5)^2 = y^{5 \cdot 2}$
$= y^{10}$

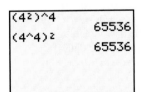

c. $(z^7)^7 = z^{7 \cdot 7}$
$= z^{49}$

d. $(u^x)^y = u^{x \cdot y}$
$= u^{xy}$

In Example 5, both the product and power rules of exponents are applied.

EXAMPLE 5

a. $(x^2x^5)^2 = (x^7)^2$
$= x^{14}$

b. $(yy^6y^2)^3 = (y^9)^3$
$= y^{27}$

c. $(z^2)^4(z^3)^3 = z^8z^9$
$= z^{17}$

d. $(x^3)^2(x^5x^2)^3 = x^6(x^7)^3$
$= x^6x^{21}$
$= x^{27}$

$(x^6)(x^{21})$

To find two more power rules for exponents, we consider the expressions $(2x)^3$ and $\left(\frac{2}{x}\right)^3$.

$$(2x)^3 = (2x)(2x)(2x)$$
$$= 2 \cdot x \cdot 2 \cdot x \cdot 2 \cdot x$$
$$= 2 \cdot 2 \cdot 2 \cdot x \cdot x \cdot x \qquad \text{Commutative property of multiplication.}$$
$$= 2^3 \cdot x^3$$
$$= 8x^3$$

$$\left(\frac{2}{x}\right)^3 = \left(\frac{2}{x}\right)\left(\frac{2}{x}\right)\left(\frac{2}{x}\right) \qquad (x \neq 0)$$
$$= \frac{2 \cdot 2 \cdot 2}{x \cdot x \cdot x}$$
$$= \frac{2^3}{x^3}$$
$$= \frac{8}{x^3}$$

These examples suggest that *to raise a product to a power, we raise each factor of the product to that power*, and *to raise a fraction to a power, we raise both the numerator and denominator to that power*.

Product to a Power and Quotient to a Power

If n is a natural number, then

$$(xy)^n = x^ny^n \qquad \text{and if } y \neq 0, \text{ then} \qquad \left(\frac{x}{y}\right)^n = \frac{x^n}{y^n}$$

EXAMPLE 6

a. $(ab)^4 = a^4b^4$

b. $(3c)^3 = 3^3c^3$
$= 27c^3$

c. $(x^2y^3)^5 = (x^2)^5(y^3)^5$
$$= x^{10}y^{15}$$

d. $(-2x^3y)^2 = (-2)^2(x^3)^2y^2$
$$= 4x^6y^2$$

e. $\left(\dfrac{4}{k}\right)^3 = \dfrac{4^3}{k^3}$
$$= \dfrac{64}{k^3}$$

f. $\left(\dfrac{3x^2}{2y^3}\right)^5 = \dfrac{3^5(x^2)^5}{2^5(y^3)^5}$
$$= \dfrac{243x^{10}}{32y^{15}}$$

To find a rule for dividing exponential expressions, we consider the fraction

$$\frac{4^5}{4^2}$$

where the exponent in the numerator is greater than the exponent in the denominator. We can simplify the fraction as follows:

$$\frac{4^5}{4^2} = \frac{4\cdot4\cdot4\cdot4\cdot4}{4\cdot4}$$

$$= \frac{\overset{1}{\cancel{4}}\cdot\overset{1}{\cancel{4}}\cdot4\cdot4\cdot4}{\underset{1}{\cancel{4}}\cdot\underset{1}{\cancel{4}}}\qquad \tfrac{4}{4}=1.$$

$$= 4^3$$

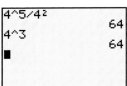

The result of 4^3 has a base of 4 and an exponent of $5 - 2$, or 3. This suggests that *to divide exponential expressions with the same base, we keep the base and subtract the exponents.*

Quotient Rule for Exponents

If m and n are natural numbers and $x \neq 0$, then

$$\frac{x^m}{x^n} = x^{m-n}$$

EXAMPLE 7

If there are no divisions by 0, then

a. $\dfrac{x^4}{x^3} = x^{4-3}$
$$= x^1$$
$$= x$$

b. $\dfrac{8y^2y^6}{4y^3} = \dfrac{8y^8}{4y^3}$
$$= \dfrac{8}{4}y^{8-3}$$
$$= 2y^5$$

c. $\dfrac{a^3a^5a^7}{a^4a} = \dfrac{a^{15}}{a^5}$
$$= a^{15-5}$$
$$= a^{10}$$

d. $\dfrac{(a^3b^4)^2}{ab^5} = \dfrac{a^6b^8}{ab^5}$
$$= a^{6-1}b^{8-5}$$
$$= a^5b^3$$

▐▐ ZERO EXPONENT

If we apply the quotient rule to the fraction $\dfrac{5^3}{5^3}$, where the exponents in the numerator and the denominator are equal, we obtain

$$\frac{5^3}{5^3} = 5^{3-3} = 5^0$$

However, because any nonzero number divided by itself is equal to 1, we have

$$\frac{5^3}{5^3} = 1$$

To make the results of 5^0 and 1 consistent, we shall define 5^0 to be equal to 1. In general, we have

Zero Exponent

If x is any nonzero real number, then

$$x^0 = 1.$$

0^0 is undefined.

EXAMPLE 8

STUDY TIP

How do you cope with getting stuck? Whether you are at the beginning or middle of an exercise, getting stuck can be a frustrating experience. However, do not let this defeat you. Answer the following questions:

1. Have you thoroughly reviewed both the text and the notes before attempting the problem?

2. Have you spent adequate time attempting the problem or did you just "throw in the towel" when you could not figure it out at first glance? This is **not** getting stuck — it is giving up before you get started!

If, after carefully analyzing a similar problem, you are still unable to complete the problem, move on to something else. Come back later and tackle the problem again. If you are still unsuccessful, seek the help of your instructor or a tutor — preferably **before** the class meets again.

Simplify each expression and verify with the graphing calculator where applicable.

a. $\left(\dfrac{1}{13}\right)^0 = 1$ **b.** $(-0.115)^0 = 1$

c. $\dfrac{4^2}{4^2} = 4^{2-2}$ **d.** $\dfrac{x^5}{x^5} = x^{5-5}$ $(x \neq 0)$

$\qquad\quad = 4^0$ $\qquad\quad = x^0$

$\qquad\quad = 1$ $\qquad\quad = 1$

e. $3x^0 = 3(\mathbf{1})$ **f.** $(3x)^0 = 1$

$\qquad = 3$

g. $\dfrac{6^n}{6^n} = 6^{n-n}$ **h.** $\dfrac{y^m}{y^m} = y^{m-m}$ $(y \neq 0)$

$\qquad\quad = 6^0$ $\qquad\quad = y^0$

$\qquad\quad = 1$ $\qquad\quad = 1$

a. ```(1/13)^0```
$\qquad\qquad\qquad$ 1
b. ```(-0.115)^0```
$\qquad\qquad\qquad$ 1
c. ```4²/4²```
$\qquad\qquad\qquad$ 1

It is clear from parts **e** and **f** that $3x^0 \neq (3x)^0$.

The rules for natural number and zero exponents are summarized as follows:

Exponents

Definitions

If n is a natural number, then

$$x^n = \overbrace{x \cdot x \cdot x \cdot \ldots \cdot x}^{n \text{ factors of } x}$$

$$x^0 = 1 \quad (x \neq 0)$$

0^0 is undefined

Properties

If m and n are natural numbers and there are no divisions by 0:

Product Rule: $x^m \cdot x^n = x^{m+n}$

Power to a Power Rule: $(x^m)^n = x^{mn}$

Product to a Power and Quotient to a Power:

$$(xy)^n = x^n y^n \qquad \left(\frac{x}{y}\right)^n = \frac{x^n}{y^n}$$

Quotient Rule: $\dfrac{x^m}{x^n} = x^{m-n}$

Recall the process of evaluating an algebraic expression. To evaluate an algebraic expression, substitute the given value(s) for the variable(s) and simplify the resulting arithmetic expression following the order of operations.

EXAMPLE 9

Evaluate each expression below and verify your results with the calculator using the STOre feature.

a. $5x^2$, $x = -3$

b. $(-3x)^4$, $x = -1$

SOLUTION   **a.** $5x^2$, $x = -3$

$= 5(-3)^2$ Substitute -3 into the expression for x.

$= 5(9)$ Simplify following the order of operations.

$= 45$

b. $(-3x)^4$, $x = -1$

$(-3 \cdot -1)^4$ Substitute -1 into the expression for x.

$= (3)^4$ Simplify following the order of operations.

$= 81$

a.
```
-3→X:5X²
              45
```
b.
```
-1→X:(-3X)^4
              81
```

EXERCISE 2.1

VOCABULARY AND NOTATION   *In Exercises 1–2, fill in the blank to make a true statement.*

1. In the exponential expression x^n, x is called the _____ and n is called the _____.

2. If m and n are natural numbers, then

a. $x^m x^n =$ _____

b. $(x^m)^n =$ _____

c. $(xy)^n =$ _____

d. $\left(\dfrac{x}{y}\right)^n =$ _____ $(y \neq 0)$

e. $\dfrac{x^m}{x^n} = $ _____, provided $x \neq 0$

f. $x^0 = $ ___ provided $x \neq 0$

CONCEPTS *In Exercises 3–10, identify the base and the exponent in each expression.*

3. 4^3

4. $(-5)^2$

5. $(-x)^2$

6. $-x^4$

7. x

8. (xy)

9. $2x^3$

10. $-3y^6$

In Exercises 11–16, write each expression in exponential form.

11. $2 \cdot 2 \cdot 2$

12. $5 \cdot 5$

13. $x \cdot x \cdot x \cdot x \cdot x$

14. $y \cdot y \cdot y \cdot y \cdot y \cdot y$

15. $(2x)(2x)(2x)$

16. $(-4y)(-4y)$

PRACTICE *In Exercises 17–20, let $x = -2$ and $y = -3$. Evaluate each expression both analytically and with the calculator, copying screen displays exactly as they appear.*

17. $3x^3$

18. $-4x^5$

19. $(-2y)^4$

20. $(3y)^5$

In Exercises 21–30, use the product rule to simplify each expression.

21. x^4x^3

22. y^5y^2

23. x^5x^5

24. yy^3

25. $y^3(y^2y^4)$

26. $(y^4y)y^6$

27. $4x^2(3x^5)$

28. $-2y(y^3)$

29. $(-y^2)(4y^3)$

30. $(-4x^3)(-5x)$

In Exercises 31–40, use the power to a power and product to a power rules to simplify each expression.

31. $(3^2)^4$

32. $(4^3)^3$

33. $(y^5)^3$

34. $(b^3)^6$

35. $(a^3)^7$

36. $(b^2)^3$

37. $(x^2x^3)^5$

38. $(y^3y^4)^4$

39. $(2x^5)^2(x^7)^3$

40. $(2y^3y)^2(y^2)^2$

In Exercises 41–56, use the product to a power and quotient to a power rules to simplify each expression.

41. $(xy)^3$

42. $(uv^2)^4$

43. $(r^3s^2)^2$

44. $(a^3b^2)^3$

45. $(4ab^2)^2$

46. $(3x^2y)^3$

47. $(-2r^3s^3t)^3$

48. $(-3x^2y^4z)^2$

49. $\left(\dfrac{a}{b}\right)^3$

50. $\left(\dfrac{r^2}{s}\right)^4$

51. $\left(\dfrac{x^2}{y^3}\right)^5$

52. $\left(\dfrac{u^4}{v^2}\right)^6$

53. $\left(\dfrac{-2a}{b}\right)^5$

54. $\left(\dfrac{2t}{3}\right)^4$

55. $\left(\dfrac{b^2}{3a}\right)^3$

56. $\left(\dfrac{a^3b}{c^4}\right)^5$

In Exercises 57–74, simplify each expression.

57. $\dfrac{x^5}{x^3}$

58. $\dfrac{a^6}{a^3}$

59. $\dfrac{a^4a^4}{a^3}$

60. $\dfrac{c^3c^4}{c^2}$

61. $\dfrac{x^3y^4}{yy^2}$

62. $\dfrac{b^4b^5}{b^2b^3}$

63. $\dfrac{c^{12}c^5}{(c^5)^2}$

64. $\dfrac{a^3a^8}{(a^2)^3}$

65. $\dfrac{15(a^2b)(a^3b^4)}{3(ab)(a^2b^2)}$

66. $\dfrac{18(x^3y^2)(xy^3)}{6(x^2y^2)(xy^2)}$

67. $\dfrac{(2x^2y^3)^3}{(xy)^2}$

68. $\dfrac{(3a^4b^2)^2}{(a^2b)^3}$

69. $\dfrac{(ab^2)^3}{(ab)^2}$

70. $\dfrac{(m^3n^4)^3}{(mn^2)^3}$

71. $\dfrac{20(r^4s^3)^4}{6(rs^3)^3}$

72. $\dfrac{35(r^3s^2)^2}{49r^2s^4}$

73. $\left(\dfrac{y^3y}{2yy^2}\right)^3$

74. $\left(\dfrac{3t^3t^4t^5}{4t^2t^6}\right)^3$

In Exercises 75–78, simplify each expression.

75. $2x^0$

76. $(2x)^0$

77. $\dfrac{x^0 - 5x^0}{2x^0}$

78. $\dfrac{4a^0 + 2a^0}{3a^0}$

79. Is the operation of raising to a power commutative? That is, is $a^b = b^a$? Explain. Use your calculator to explore this idea, letting a and b assume various positive integer values.

80. Is the operation of raising to a power associative? That is, is $(a^b)^c = a^{(bc)}$? Explain. Use your calculator to explore this idea, letting a, b, and c assume various positive integer values.

APPLICATIONS   *In Exercises 81–84, write a simplified algebraic expression involving only one exponent to represent the area of the figure. (Refer to the formulas on the inside cover as necessary.)*

81.

$3x^2$

$5x$

82.

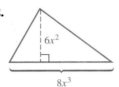

$6x^2$

$8x^3$

83.

$3x^2$

84.

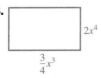

$2x^4$

$\frac{3}{4}x^3$

REVIEW

85. Mental arithmetic: Perform the following computations without using pencil and paper or calculator.
   a. $-4 + -5$ b. $-5(-6)$
   c. $-8 - (-5)$ d. $12 - 15$
   e. $\frac{1}{2}(-6)$ f. $-\frac{1}{4}(-12)$

86. For the set of numbers $\{-3, -2, -1.5, -\frac{1}{4}, 0, \sqrt{3}, \sqrt{4}, 5, 7.2\}$, identify the following groups of numbers:
   a. whole numbers
   b. irrational numbers
   c. rational numbers
   d. negative integers
   e. positive integers
   f. natural numbers

87. Convert each percentage to a decimal number.
   a. 45%   b. 3%   c. 2.5%

88. Use your calculator to convert each of the following fractions to a decimal and round the result to the nearest thousandth:
   a. $\frac{17}{19}$ b. $\frac{2}{3}$ c. $\frac{5}{7}$

2.2 NEGATIVE INTEGER EXPONENTS AND SCIENTIFIC NOTATION

- REVIEW OF BASIC EXPONENT CONCEPTS
- NEGATIVE INTEGER EXPONENTS
- SCIENTIFIC NOTATION
- USING SCIENTIFIC NOTATION TO SIMPLIFY COMPUTATIONS

▐ REVIEW OF BASIC EXPONENT CONCEPTS

In Section 2.1 we developed the rules for natural-number exponents. These rules are valid for both integer and rational number exponents. This section shall reexamine integer exponents, relying on your knowledge of positive integer (natural number) exponents and the zero exponent discussed in Section 2.1 to explore negative integer exponents.

Points to remember

1. In the expression x^n, x is the base and n is the exponent.

2. When n is a natural number then $x^n = \overbrace{x \cdot x \cdot x \cdot x \cdot \ldots \cdot x}^{n \text{ factors of } x}$.

3. In -2^4, 2 is the base that is raised to the power of 4. The negative sign indicates we want the opposite of 2^4.

$$-2^4 = -2 \cdot 2 \cdot 2 \cdot 2 = -16$$

4. In $(-2)^4$, (-2) is the base that is raised to the power of 4.

$$(-2)^4 = (-2)(-2)(-2)(-2) = +16$$

5. If $x \neq 0$, then $x^0 = 1$. 0^0 is undefined.

▉ NEGATIVE INTEGER EXPONENTS

When we first examined exponents in the previous section we considered quotients such as $\dfrac{6^5}{6^2}$.

$$\frac{6^5}{6^2} = \frac{\overset{1}{\cancel{6}} \cdot \overset{1}{\cancel{6}} \cdot 6 \cdot 6 \cdot 6}{\underset{1}{\cancel{6}} \cdot \underset{1}{\cancel{6}}} = \frac{6^3}{1} \; or \; 6^3 \qquad \text{Simplify using properties of fractions.}$$

or

$$\frac{6^5}{6^2} = 6^{5-2} = 6^3 \qquad \text{Use the quotient rule for exponents.}$$

In these problems, the exponent in the numerator was always larger than the exponent in the denominator. We will now investigate what happens when the exponent in the denominator is larger than the exponent in the numerator.

EXPLORING
THE CONCEPT

Use your calculator to simplify each pair of expressions below. Use the ►frac feature to display all results as fractions.

a. $\dfrac{6^2}{6^5}, 6^{-3}$

b. $\dfrac{4^4}{4^5}, 4^{-1}$

CONCLUSIONS

a.

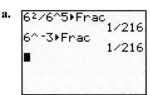

b.

If we apply the quotient rule to the fraction $\dfrac{6^2}{6^5}$, where the exponent in the numerator is less than the exponent in the denominator, we obtain

$$\frac{6^2}{6^5} = 6^{2-5} = 6^{-3}$$

However, we know that

$$\frac{6^2}{6^5} = \frac{\overset{1}{\cancel{6}} \cdot \overset{1}{\cancel{6}}}{\underset{1}{\cancel{6}} \cdot \underset{1}{\cancel{6}} \cdot 6 \cdot 6 \cdot 6} = \frac{1}{6^3}$$

To make the results of 6^{-3} and $\dfrac{1}{6^3}$ consistent, we define 6^{-3} to be equal to $\dfrac{1}{6^3}$.

In general,

Negative Exponent

If x is any nonzero number and n is a natural number, then $x^{-n} = \dfrac{1}{x^n}$ and $\dfrac{1}{x^{-n}} = x^n$.

EXAMPLE 1

Express each quantity without using negative exponents. Assume that no denominators are zero.

a. $3^{-5} = \dfrac{1}{3^5}$

$\quad = \dfrac{1}{243}$

b. $x^{-4} = \dfrac{1}{x^4}$

EXPLORING THE CONCEPT

Use your calculator to simplify each pair of expressions below. Use the ▸frac feature to display all results as fractions.

a. $\dfrac{1}{3^{-1}}, \dfrac{3^{-1}}{1}$

b. $\dfrac{2}{3^{-1}}, \dfrac{3^{-1}}{2}$

CONCLUSIONS

a.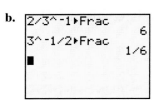

```
1/3^-1▸Frac
             3
3^-1/1▸Frac
           1/3
```

b.

```
2/3^-1▸Frac
             6
3^-1/2▸Frac
           1/6
```

EXAMPLE 2

Use the definition of a negative exponent to simplify each quantity. Results should be expressed with no negative exponents; assume that no denominators are zero.

a. $\dfrac{2}{3^{-1}} = \dfrac{2}{\dfrac{1}{3^1}}$ Definition of negative exponent.

$\quad = 2 \div \dfrac{1}{3^1}$ Express the division.

$\quad = 2 \cdot \dfrac{3^1}{1}$ Dividing is the same as multiplying by the reciprocal.

$\quad = 2 \cdot 3$

$\quad = 6$

b. $\dfrac{3^{-1}}{2} = \dfrac{\frac{1}{3}}{2}$ Definition of a negative exponent.

$= \dfrac{1}{3} \div 2$ Express the division.

$= \dfrac{1}{3} \cdot \dfrac{1}{2}$ Dividing is the same as multiplying by the reciprocal.

$= \dfrac{1}{6}$

c. $\dfrac{2}{x^{-3}} = \dfrac{2}{\frac{1}{x^3}}$ Definition of a negative exponent.

$= 2 \div \dfrac{1}{x^3}$ Express the division.

$= 2 \cdot \dfrac{x^3}{1}$ Dividing is the same as multiplying by the reciprocal.

$= 2x^3$

Because of the definition of negative and zero exponents, the product, power, and quotient rules discussed in Section 2.1 are also true for all integer exponents. We restate the properties of exponents for integer exponents.

Exponents

Definitions

If n is a natural number then

$$x^n = \overbrace{x \cdot x \cdot x \cdot \ldots \cdot x}^{n \text{ factors of } x}$$

$x^0 = 1 \quad (x \neq 0)$

0^0 is undefined

$x^{-n} = \dfrac{1}{x^n} \quad (x \neq 0)$

$\dfrac{1}{x^{-n}} = x^n \quad (x \neq 0)$

Properties

If m and n are natural numbers and there are no divisions by 0:

Product Rule: $x^m \cdot x^n = x^{m+n}$

Power to a Power Rule: $(x^m)^n = x^{mn}$

Product to a Power and Quotient to a Power:

$$(xy)^n = x^n y^n \qquad \left(\dfrac{x}{y}\right)^n = \dfrac{x^n}{y^n}$$

Quotient Rule: $\dfrac{x^m}{x^n} = x^{m-n}$

Do not leave a negative exponent in a result.

EXAMPLE 3   Use the properties of exponents to simplify each expression. Assume that no denominator is zero. All results should be expressed with positive exponents.

a. $(x^{-3})^2 = x^{-6}$ $(x^m)^n = x^{mn}$

$\qquad\quad = \dfrac{1}{x^6}$ $x^{-n} = \dfrac{1}{x^n}$

b. $\dfrac{x^3}{x^7} = x^{3-7}$ $\dfrac{x^m}{x^n} = x^{m-n}$

$\qquad\; = x^{-4}$

$\qquad\; = \dfrac{1}{x^4}$ $x^{-n} = \dfrac{1}{x^n}$

c. $\dfrac{y^{-4}y^{-3}}{y^{-20}} = \dfrac{y^{-7}}{y^{-20}}$ $x^m \cdot x^n = x^{m+n}$

$\qquad\qquad = y^{-7-(-20)}$ $\dfrac{x^m}{x^n} = x^{m-n}$

$\qquad\qquad = y^{-7+20}$

$\qquad\qquad = y^{13}$

d. $\dfrac{12a^3b^4}{4a^5b^2} = \dfrac{12}{4} \cdot \dfrac{a^3}{a^5} \cdot \dfrac{b^4}{b^2}$ Multiplication of fractions.

$\qquad\qquad = 3 \cdot a^{3-5} \cdot b^{4-2}$ Division of integers and $\dfrac{x^m}{x^n} = x^{m-n}$.

$\qquad\qquad = 3a^{-2}b^2$ Simplify.

$\qquad\qquad = \dfrac{3}{1} \cdot \dfrac{1}{a^2} \cdot \dfrac{b^2}{1}$ $x^{-n} = \dfrac{1}{x^n}$

$\qquad\qquad = \dfrac{3b^2}{a^2}$ Multiplication of fractions.

e. $\left(-\dfrac{x^3y^2}{xy^{-3}}\right)^{-2} = (-x^{3-1}y^{2-(-3)})^{-2}$ Order of operations: simplify inside grouping symbols first.

$\qquad\qquad\qquad$ $\dfrac{x^m}{x^n} = x^{m-n}$

$\qquad\qquad = (-x^2y^5)^{-2}$ Simplify.

$\qquad\qquad = \dfrac{1}{(-x^2y^5)^2}$ $x^{-n} = \dfrac{1}{x^n}$

$\qquad\qquad = \dfrac{1}{x^4y^{10}}$ $(xy)^n = x^ny^n$

The result cannot be simplified further.

SCIENTIFIC NOTATION

Scientists often deal with extremely large and extremely small numbers. For example, the distance from the earth to the sun is approximately 150,000,000 kilometers, and ultraviolet light emitted from a mercury arc has a wavelength of approximately 0.000025 centimeters. The large number of zeros in these numbers makes them difficult to read and hard to remember. To make such numbers easier to work with, scientists use a compact form of notation called **scientific notation**.

Scientific Notation

A number is written in **scientific notation** if it is written as the product of a number between 1 (including 1) and 10 and an integer power of 10.

EXPLORING
THE CONCEPT

Change your calculator mode from normal display to **SCI**entific notation display so that all results will be displayed in scientific notation. Enter each of the following on your calculator:

a. 30,000   **b.** 3×10^4

c. Compare the display to the definition of scientific notation and explain the calculator's approach to displaying numbers in scientific notation.

CONCLUSION

a.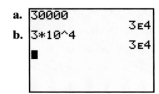

b.

c. Instead of 3×10^4 the calculator displays 3 E 4. The notation E 4 indicates "times ten to the fourth power."

TECHNOLOGY TIP

To change from NORMAL to SCIentific mode, press [MODE]. (This option is located above the [MORE] key on the TI-86 and is accessed by first pressing [2nd].) Cursor right [▶] so that the option SCI is covered by the blinking cursor. Press [ENTER] to choose this option. Return to the home screen by pressing [CLEAR]. To return to NORMAL mode, access [MODE], press [ENTER] (to choose NORMAL, the cursor is blinking on that option), then press [CLEAR] to return to the home screen.

EXAMPLE 4

Change 150,000,000 to scientific notation.

SOLUTION

To write the number 150,000,000 in scientific notation, we must write it as a product of a number between 1 and 10 and some power of 10. Note that the number 1.5 lies between 1 and 10.

To obtain the number 150,000,000, the decimal point in the number 1.5 must be moved eight places to the right. Because multiplying a number by 10 moves the decimal point one place to the right, we can accomplish this by multiplying 1.5 by 10 eight times. To see this, we count from the decimal point in 1.5 to where the decimal point should be in 150,000,000:

$$1 . \underbrace{5\ 0\ 0\ 0\ 0\ 0\ 0\ 0}_{\text{8 places to the right}} 0.$$

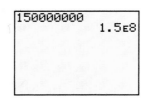

The number 150,000,000 written in scientific notation is 1.5×10^8.

EXAMPLE 5

Change 0.000025 to scientific notation.

SOLUTION   To write the number 0.000025 in scientific notation, we write it as a product of a number between 1 and 10 and some power of 10. To obtain the number 0.000025, the decimal point in the number 2.5 must be moved five places to the left. We can accomplish this by dividing 2.5 by 10^5, which is equivalent to multiplying 2.5 by $\dfrac{1}{10^5}$, or 10^{-5}. To do this, we count from the decimal point in 2.5 to where the decimal point should be in 0.000025:

$$. \underbrace{0\ 0\ 0\ 0\ 2}_{\text{5 places to the left}} . 5$$

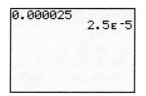

The number 0.000025 written in scientific notation is 2.5×10^{-5}.

We can change a number written in scientific notation to **standard notation**. For example, to write the number 9.3×10^7 in standard notation, we multiply 9.3×10^7.

$$9.3 \times 10^7 = 9.3 \times \mathbf{10{,}000{,}000}$$
$$= 93{,}000{,}000$$

Change your calculator back to NORMAL MODE for Example 6.

EXAMPLE 6

Write   **a.** 3.4×10^5 and   **b.** 2.1×10^{-4} in standard notation.

SOLUTION   **a.** $3.4 \times 10^5 = 3.4 \times \mathbf{100{,}000}$
$$= 340{,}000$$

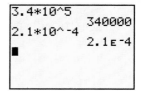

b. $2.1 \times 10^{-4} = 2.1 \times \dfrac{1}{10^4}$
$$= 2.1 \times \dfrac{1}{10{,}000}$$
$$= 0.00021$$

Notice that the calculator expresses very small numbers in scientific notation even when set to NORMAL MODE. The same is true for very large numbers.

To summarize, each of the following numbers is written in both scientific and standard notation. In each case, the exponent gives the number of places that the decimal point moves, and the sign of the exponent indicates the direction that it moves.

$$5.32 \times 10^5 = 5\ 3\ 2\ 0\ 0\ 0.$$ 5 places to the right.

$$2.37 \times 10^6 = 2\ 3\ 7\ 0\ 0\ 0\ 0.$$ 6 places to the right.

$$8.95 \times 10^{-4} = 0.\ 0\ 0\ 0\ 8\ 9\ 5$$ 4 places to the left.

$$8.375 \times 10^{-3} = 0.\ 0\ 0\ 8\ 3\ 7\ 5$$ 3 places to the left.

$$9.77 \times 10^0 = 9.77$$ No movement of the decimal point.

EXAMPLE 7

Write 432.0×10^5 in scientific notation.

SOLUTION

The number 432.0×10^5 **is not** written in scientific notation because 432.0 is not a number between 1 and 10. To write the number in scientific notation, we proceed as follows:

$$\mathbf{432.0} \times 10^5 = \mathbf{4.32} \times \mathbf{10^2} \times 10^5$$ Write 432.0 in scientific notation.

$$= 4.32 \times 10^7$$ $10^2 \times 10^5 = 10^7$.

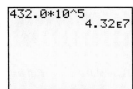

The calculator is set to scientific mode.

TECHNOLOGY TIP

Be sure your calculator is set to the desired mode.

▮ USING SCIENTIFIC NOTATION TO SIMPLIFY COMPUTATIONS

An advantage of scientific notation becomes apparent when simplifying fractions such as

$$\frac{(320)(25,000)}{0.000040}$$

that contain very large or very small numbers. Although we can simplify this fraction using ordinary arithmetic, scientific notation provides an easier way. First we write each number in scientific notation. The commutative and associative properties allow us to move factors as desired. Then we write the result in standard form, if desired.

TECHNOLOGY TIP

The calculator will automatically display very large or very small numbers in scientific notation regardless of the mode setting.

$$\frac{(320)(25,000)}{0.000040} = \frac{(3.2 \times 10^2)(2.5 \times 10^4)}{4.0 \times 10^{-5}}$$

$$= \frac{(3.2)(2.5)}{4.0} \times \frac{10^2 10^4}{10^{-5}}$$

$$= \frac{8.0}{4.0} \times 10^{2+4-(-5)}$$

$$= 2.0 \times 10^{11}$$

$$= 200,000,000,000$$

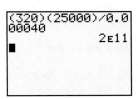

EXAMPLE 8

Speed of light   In a vacuum, light travels 1 meter in approximately 0.000000003 second. How long does it take for light to travel 500 kilometers? (*Hint:* 1 kilometer = 1,000 meters.)

SOLUTION   Because 1 kilometer is equal to 1000 meters, the length of time for light to travel 500 kilometers ($500 \cdot 1000$ meters) is given by

$$
\begin{aligned}
(0.000000003)(500)(1000) &= (3 \times 10^{-9})(5 \times 10^{2})(1 \times 10^{3}) \\
&= 3(5) \times 10^{-9+2+3} \\
&= 15 \times 10^{-4} \\
&= 1.5 \times 10^{1} \times 10^{-4} \\
&= 1.5 \times 10^{-3} \\
&= 0.0015
\end{aligned}
$$

Light travels 500 kilometers in approximately 0.0015 second.

EXERCISE 2.2

VOCABULARY   *Fill in the blank to make a true statement.*

1. A number written in _____ is the product of a number between 1 (including 1) and 10 and an integer power of 10.

CONCEPTS   *In Exercises 2–6, classify each statement as true or false.*

2. The number 23.5×10^{3} is correctly written in scientific notation.

3. The number 0.234×10^{4} is correctly written in scientific notation.

4. The number 2.5 E 4 is correctly written in standard notation as 25,000.

5. The number $-3.6\text{E}-2$ is correctly written in standard notation as -0.036.

6. All rational numbers can be written in scientific notation.

7. If $x \neq 0$ and n is a natural number, then $x^{-n} = \underline{\quad}$.

8. Evaluate each expression below. Verify your results with your calculator.

   a. 2^{-3} **b.** 3^{-1}

   c. $\dfrac{1}{2^{-3}}$ **d.** $\dfrac{1}{3^{-1}}$

9. Simplify each expression below. Results should include only positive exponents.

   a. $x^{-4} \cdot x^{7}$ **b.** $(x^{-4})^{7}$

   c. $\dfrac{x^{4}}{x^{7}}$ **d.** $\left(\dfrac{x}{y}\right)^{-7}$

   e. $(xy)^{-3}$ **f.** $(x^{2}y^{-3})^{-4}$

10. Convert each pair of expressions written in scientific notation to standard notation using your calculator.

    a. 4.2×10^{3} and   **b.** 1.513×10^{2} and
    4.2×10^{-3} 1.513×10^{-2}

PRACTICE   *In Exercises 11–16, simplify each expression below. Results should be expressed without parentheses or negative exponents and should be verified with your calculator.*

11. $2^{5} \cdot 2^{-2}$ 12. $10^{2} \cdot 10^{-4}$

13. $\dfrac{4^{-5}}{4^{-3}}$ 14. $\dfrac{2^{-3}}{2^{3}}$

15. $\dfrac{2^{3} \cdot 2^{-1}}{2^{2}}$ 16. $\dfrac{3^{-2} \cdot 3^{4}}{3^{3}}$

In Exercises 17–40, use the laws of exponents to simplify each expression. Answers should be expressed without using parentheses or negative exponents.

17. x^{-2}

18. y^{-3}

19. $(2y)^{-4}$

20. $(-3x)^{-1}$

21. $(ab^2)^{-3}$

22. $(m^2n^3)^{-2}$

23. $\dfrac{y^4y^3}{y^{-4}y^{-2}}$

24. $\dfrac{x^{12}x^{-7}}{x^3x^4}$

25. $(x^{-4}x^3)^3$

26. $(y^{-2}y)^3$

27. $(-2x^3y^{-2})^{-5}$

28. $(3u^{-2}v)^{-3}$

29. $\left(\dfrac{a^3}{a^{-4}}\right)^2$

30. $\left(\dfrac{a^4}{a^{-3}}\right)^3$

31. $\left(\dfrac{x^5}{y^{-2}}\right)^{-2}$

32. $\left(\dfrac{x^{-2}}{y^3}\right)^{-3}$

33. $\left(\dfrac{4x^2}{3x^{-5}}\right)^4$

34. $\left(\dfrac{-3x^4x^{-3}}{x^{-3}x^7}\right)^3$

35. $\left(\dfrac{12y^3z^{-2}}{3y^{-4}z^3}\right)^2$

36. $\left(\dfrac{6xy^3}{3x^{-1}y}\right)^3$

37. $\left(\dfrac{9u^2v^3}{18u^{-3}v}\right)^4$

38. $\left(\dfrac{14u^{-2}v^3}{21u^{-3}v}\right)^4$

39. $\dfrac{16(x^{-2}yz)^{-2}}{(2x^{-3}z^0)^4}$

40. $\dfrac{(2x^{-2})^{-2}}{4(x^2y)^{-1}}$

In Exercises 41–48, write each number in scientific notation and verify with your calculator.

41. 23,000

42. 4750

43. 0.062

44. 0.00073

45. 42.5×10^2

46. 0.3×10^3

47. 0.25×10^{-2}

48. 25.2×10^{-3}

In Exercises 49–56, write each number in standard notation and verify with your calculator.

49. 2.3×10^2

50. 3.75×10^4

51. 1.14×10^{-3}

52. 4.9×10^{-2}

53. 25×10^6

54. 0.07×10^3

55. 0.51×10^{-3}

56. 617×10^{-2}

In Exercises 57–59, convert each calculator display to the indicated notation.

57.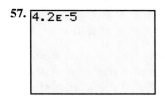

  a. scientific notation

  b. standard notation

58.

  a. scientific notation

  b. standard notation

59.

  a. scientific notation

  b. standard notation

60. "Round-off" error can occur when performing calculator computations. When this happens, the calculator will yield a value that is very close to the anticipated value and yet not the exact value. For example, you may expect the result to be 5 but the calculator may display 5.0001 or 4.9999 instead. If the calculator displayed the value $-1.0000001 \text{ E} -11$ as a result of a round-off error, what value is being represented?

In Exercises 61–66, use scientific notation to simplify each expression. Verify your work with your calculator. Express results in both standard and scientific notation.

61. $(3.4 \times 10^2)(2.1 \times 10^3)$

62. $(4.1 \times 10^{-3})(3.4 \times 10^4)$

63. $\dfrac{9.3 \times 10^2}{3.1 \times 10^{-2}}$

64. $\dfrac{7.2 \times 10^6}{1.2 \times 10^8}$

65. $\dfrac{(12,000)(3600)}{0.0003}$

66. $\dfrac{(0.0004)(0.0012)}{80,000}$

APPLICATIONS

67. Distance to Alpha Centauri   The distance from the earth to the nearest star outside our solar system is approximately 25,700,000,000,000 miles. Express this number in scientific notation.

68. Distance to Venus   The distance from Venus to the sun is approximately 6.7×10^7 miles. Express this number in standard notation.

69. Speed of sound   The speed of sound in air is 33,100 centimeters per second. Express this number in scientific notation.

70. Length of a meter   One meter is approximately 0.00622 mile. Use scientific notation to express this number.

71. Distance to Mars   The distance from Mars to the sun is approximately 1.14×10^8 miles. Express this number in standard notation.

72. Angstrom   One angstrom is 1×10^{-7} millimeter. Express this number in standard notation.

73. Distance between Mercury and the sun   The distance from Mercury to the sun is approximately 3.6×10^7 miles. Use scientific notation to express this distance in feet. (*Hint*: 5280 feet = 1 mile.)

74. Speed of sound   The speed of sound in air is approximately 3.3×10^4 centimeters per second. Use scientific notation to express this speed in kilometers per second. (*Hint*: 100 centimeters = 1 meter, and 1000 meters = 1 kilometer.)

75. Mass of a proton   The mass of one proton is approximately 1.7×10^{-24} gram. Use scientific notation to express the mass of 1 million protons.

76. Light year   One light year is approximately 5.87×10^{12} miles. Use scientific notation to express this distance in feet. (*Hint*: 5280 feet = 1 mile.)

REVIEW

77. Evaluate $xy^2 - 4(3 - y) + 2y$ for $x = 5$ and $y = -2$.

78. Use the STOre feature of the graphing calculator to evaluate $\dfrac{2x^2 - 3x + 4}{x + 5}$ for $x = -3$.

79. A computer that costs \$2,300 is on sale for 15% off if purchased online. If the customer pays a 3% shipping and handling charge, what is the final cost for the computer?

80. Find the area of the triangle:

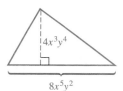

2.3   ALGEBRAIC EXPRESSIONS AND POLYNOMIALS

- TRANSLATING EXPRESSIONS
- ALGEBRAIC TERMS
- POLYNOMIALS
- $P(x)$ NOTATION

▮ TRANSLATING EXPRESSIONS

Variables and numbers can be combined with the operations of arithmetic to produce **algebraic expressions**. For example, if x and y are variables, the algebraic expression $x + y$ represents the **sum** of x and y, and the algebraic expression $x - y$ represents their **difference**.

There are many ways to read the **sum** $x + y$. Some of them are:

- the **sum of** x and y
- x **increased by** y
- x **plus** y

- y **more than** x
- y **added to** x

There are many ways to read the **difference** $x - y$. Some of them are:

- the result obtained when y is **subtracted from** x
- the result of **subtracting** y **from** x
- x **less** y
- y **less than** x
- x **decreased by** y
- x **minus** y

EXAMPLE 1   Let x represent a certain number. Write an expression that represents   **a.** the number that is 5 more than x and   **b.** the number 12 decreased by x.

SOLUTION   **a.** The number "5 more than x" is the number found by adding 5 to x. It is represented by $x + 5$.

b. The number "12 decreased by x" is the number found by subtracting x from 12. It is represented by $12 - x$.

EXAMPLE 2   **Income taxes**   Luciano worked x hours preparing his income taxes. He worked 3 hours less than that preparing his son's return. Write an expression that represents   **a.** the number of hours he spent preparing his son's return   and   **b.** the total number of hours he worked.

SOLUTION   **a.** Because Luciano worked x hours on his return and 3 hours less on his son's return, he worked $(x - 3)$ hours on his son's return.

b. Because he worked x hours on his return and $(x - 3)$ hours on his son's return, the total time he spent on taxes was $x + (x - 3)$ hours.

There are several ways to indicate the **product** xy in words. Some of them are:

- x **multiplied by** y
- the **product of** x and y
- x **times** y

EXAMPLE 3   Let x represent a certain number. Denote a number that is   **a.** twice as large as x,   **b.** 5 more than 3 times x,   and   **c.** 4 less than $\frac{1}{2}$ of x.

SOLUTION   **a.** The number "twice as large as x" is found by multiplying x by 2. It is represented by $2x$.

b. The number "5 more than 3 times x" is found by adding 5 to the product of 3 and x. It is represented by $3x + 5$.

c. The number "4 less than $\frac{1}{2}$ of x" is found by subtracting 4 from the product of $\frac{1}{2}$ and x. It is represented by $\frac{1}{2}x - 4$.

EXAMPLE 4

Stock valuation Jim owns *x* shares of Transitronic stock, valued at $29 a share, *y* shares of Positone stock, valued at $32 a share, and 300 shares of Baby Bell stock, valued at $42 a share.   **a.** How many shares of stock does he own? **b.** What is the value of his stock?

SOLUTION   **a.** Because there are *x* shares of Transitronic, *y* shares of Positone, and 300 shares of Baby Bell, his total number of shares is $x + y + 300$.

b. The value of *x* shares of Transitronic is $29x$, the value of *y* shares of Positone is $32y$, and the value of 300 shares of Baby Bell is $42(300)$. The total value of the stock is $\$(29x + 32y + 12{,}600)$.

If *x* and *y* represent two numbers and $y \neq 0$, the **quotient** obtained when *x* is divided by *y* is denoted by each of the following expressions:

$$x \div y, \qquad x/y, \qquad \text{and} \qquad \frac{x}{y}$$

EXAMPLE 5

Let *x* and *y* represent two numbers. Write an algebraic expression that represents the sum obtained when 3 times the first number is added to the quotient obtained when the second number is divided by 6.

SOLUTION   Three times the first number, *x*, is denoted as $3x$. The quotient obtained when the second number, *y*, is divided by 6 is the fraction $\frac{y}{6}$. Their sum is expressed as $3x + \frac{y}{6}$.

EXAMPLE 6

Cutting a rope A 5-foot section is cut from the end of a rope that is *l* feet long. The remaining rope is then divided into three equal parts. Find the length of each of the equal pieces.

SOLUTION   After a 5-foot section is cut from one end of *l* feet of rope, the rope that remains is $(l - 5)$ feet long. When that remaining rope is cut into 3 equal pieces, each piece will be $\frac{l-5}{3}$ feet long. See Figure 2-1.

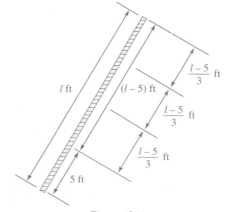

Figure 2-1

▓ ALGEBRAIC TERMS

Numbers without variables, such as 7, 21, and 23, are called **constants**. Expressions such as 37, xyz, or $32t$, which are constants, variables, or products of constants and variables, are called **algebraic terms**. Ungrouped addition and subtraction signs are term indicators.

Consider the terms in each algebraic expression below:

Expression	Numbers of Terms	Terms
$5x + 5y$	2	$5x, 5y$
$4(xy - 7)$	1	$4(xy - 7)$
$3x + 5(y - 8)$	2	$3x, 5(y - 8)$
$3 + 5y - 4x + 2$	4	$3, 5y -4x, 2$

Numbers that are part of an indicated product are called **factors** of that product. For example, the product $7x$ has two factors, 7 and x. When we speak of the coefficient in a term such as $7x$, we mean the **numerical coefficient**, which in this case is 7. For example, the numerical coefficient of the term $12xyz$ is 12. The coefficient of such terms as x, ab, or rst is understood to be 1. Thus,

$$x = 1x, \qquad ab = 1ab, \qquad \text{and} \qquad rst = 1rst$$

EXAMPLE 7

a. The expression $5x + y$ has two terms. The numerical coefficient of its first term is 5. The numerical coefficient of its second term is 1.

b. The expression $-17wxyz$ has one term, which contains the five factors -17, $w, x, y,$ and z. Its numerical coefficient is -17.

c. The expression 37 has one term, the constant 37. Its numerical coefficient is 37.

▓ POLYNOMIALS

An algebraic expression that is the sum of one or more terms containing whole-number exponents on its variables is called a **polynomial**. The expressions

$$8xy^2t \qquad 3x + 2 \qquad 4y^2 - 2y + 3 \qquad \text{and} \qquad 3a - 4b - 4c + 8d$$

are examples of polynomials.

 WARNING!   An expression such as $2x^3 - 3y^{-2}$ is not a polynomial, because the second term contains a negative exponent on a variable base.

A polynomial with exactly one term is called a **monomial**. A polynomial with exactly two terms is called a **binomial**. A polynomial with exactly three terms is called a **trinomial**. Here are some examples of each:

Monomials	Binomials	Trinomials
$5x^2y$	$3u^3 - 4u^2$	$-5t^2 + 4t + 3$
$-6x$	$18a^2b + 4ab$	$27x^3 - 6x - 2$
29	$-29z^{17} - 1$	$-32r^6 + 7y^3 - z$

The monomial $7x^6$ is called a *monomial of sixth degree* or a *monomial of degree 6* because the variable x occurs as a factor six times. The monomial $3x^3y^4$ is a monomial of the seventh degree because the variables x and y occur as factors a total of seven times. Other examples are:

$-2x^3$ is a monomial of degree 3
$47x^2y^3$ is a monomial of degree 5
$18x^4y^2z^8$ is a monomial of degree 14
8 is a monomial of degree 0 because $8 = 8x^0$

These examples illustrate the following definition.

Degree of a Monomial

If a is a nonzero constant, the **degree of the monomial** ax^n is n. The degree of a monomial with several variables is the sum of the exponents of those variables, ax^my^n has degree $m + n$.

We define the degree of a polynomial by considering the degree of each of its terms.

Degree of a Polynomial

The **degree of a polynomial** is the same as the degree of its term with the highest degree.

For example,

- $x^2 + 2x$ is a binomial of degree 2 because the degree of its first term is 2 and the degree of its other term is less than 2.
- $3x^3y^2 + 4x^4y^4 - 3x^3$ is a trinomial of degree 8 because the degree of its second term is 8 and the degree of each of its other terms is less than 8.
- $25x^4y^3z^7 - 15xy^8z^{10} - 32x^8y^8z^3 + 4$ is a polynomial of degree 19 because its second and third terms are of degree 19. Its other terms have degrees less than 19.

▮ $P(x)$ NOTATION

Polynomials that contain a single variable are often denoted by symbols such as

$P(x)$ Read as "P of x."

$Q(t)$ Read as "Q of t."

$R(z)$ Read as "R of z."

where the letter within the parentheses represents the variable of the polynomial.

WARNING!   The symbol $P(x)$ does not indicate the product of P and x. Instead, it represents a polynomial with the variable x.

The symbols $P(x)$, $Q(t)$, and $R(z)$ could represent the polynomials

$$P(x) = 3x + 4, \qquad Q(t) = 3t^2 + 4t - 5, \qquad \text{or} \qquad R(z) = -z^3 - 2z + 3$$

The symbol $P(x)$ is convenient, because it provides a way to indicate the value of a polynomial in x at different values of x. If $P(x) = 3x + 4$, for example, then $P(1)$ represents the value of the polynomial $P(x) = 3x + 4$ when $x = 1$.

$$P(x) = 3x + 4$$
$$P(1) = 3(1) + 4$$
$$= 7$$

Likewise, if $Q(t) = 3t^2 + 4t - 5$, then $Q(-2)$ represents the value of the polynomial $Q(t) = 3t^2 + 4t - 5$ when $t = -2$.

$$Q(t) = 3t^2 + 4t - 5$$
$$Q(-2) = 3(-2)^2 + 4(-2) - 5$$
$$= 3(4) - 8 - 5$$
$$= 12 - 8 - 5$$
$$= -1$$

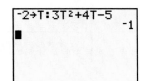

EXAMPLE 8

Consider the polynomial $P(z)$ where $P(z) = 3z^2 + 2$. Find   **a.** $P(0)$,   **b.** $P(2)$, **c.** $P(-3)$,   and   **d.** $P(s)$.

SOLUTION   **a.** $P(z) = 3z^2 + 2$
$$P(0) = 3(0)^2 + 2$$
$$= 2$$

b. $P(z) = 3z^2 + 2$
$$P(2) = 3(2)^2 + 2$$
$$= 3(4) + 2$$
$$= 12 + 2$$
$$= 14$$

c. $P(z) = 3z^2 + 2$
$$P(-3) = 3(-3)^2 + 2$$
$$= 3(9) + 2$$
$$= 27 + 2$$
$$= 29$$

d. $P(z) = 3z^2 + 2$
$$P(s) = 3s^2 + 2$$

EXAMPLE 9

Trajectory   The polynomial $h(t) = -16t^2 + 28t + 8$ gives the height (in feet) of an object t seconds after it has been thrown upward. Find the height of the object at   **a.** 1 second   and   **b.** 2 seconds.

SOLUTION   **a.** To find the height at 1 second, we find $h(1)$.

$$h(t) = -16t^2 + 28t + 8$$
$$h(1) = -16(1)^2 + 28(1) + 8$$
$$= -16 + 28 + 8$$
$$= 20$$

b. To find the height at 2 seconds, we find $h(2)$.

$$h(t) = -16t^2 + 28t + 8$$
$$h(2) = -16(2)^2 + 28(2) + 8$$
$$= -16(4) + 28(2) + 8$$
$$= -64 + 56 + 8$$
$$= 0$$

At 1 second, the object is 20 feet above the ground.

At 2 seconds, the height is 0 feet. That is when the object strikes the ground.

```
1→T: -16T²+28T+8
                20
2→T: -16T²+28T+8
                 0
```

EXAMPLE 10

Consider $P(x)$ where $P(x) = x^3 + 1$. Find **a.** $P(2t)$, **b.** $P(-3y)$, **c.** $P(s^4)$, and **d.** $P(x) + P(a)$.

SOLUTION

a.
$$P(x) = x^3 + 1$$
$$P(2t) = (2t)^3 + 1$$
$$= 8t^3 + 1$$

b.
$$P(x) = x^3 + 1$$
$$P(-3y) = (-3y)^3 + 1$$
$$= -27y^3 + 1$$

c.
$$P(x) = x^3 + 1$$
$$P(s^4) = (s^4)^3 + 1$$
$$= s^{12} + 1$$

d.
$$P(x) + P(a) = x^3 + 1 + a^3 + 1$$
$$= x^3 + a^3 + 2$$

EXERCISE 2.3

VOCABULARY AND NOTATION   *In Exercises 1–7 fill in the blank to make a true statement.*

1. Variables and numbers combined with arithmetic operations produce _____ .

2. Numbers without variables are _____ because they never change in value.

3. Expressions that are composed of factors of numbers and variables are called _____ .

4. In the term $4x$, 4 is called the _____ of the term.

5. An algebraic expression that is the sum of one or more terms containing whole-number exponents on the variables is called a _____ .

6. A polynomial consisting of one term is called a _____ , of two terms a _____ , and of three terms a _____ .

7. The _____ of a polynomial is the same as the _____ of its term with the highest _____ .

CONCEPTS

8. Identify each of the following as a sum, product, quotient, or difference.

    a. xy

    b. $x - y$

    c. $x + y$

    d. $\dfrac{x}{y}$

9. Which of the following pairs of algebraic expressions is *not* equivalent?

    a. $\dfrac{1}{2}x, \dfrac{x}{2}$

    b. $3x, x^3$

    c. $x + 2, 2 + x$

    d. $x - 2, 2 - x$

10. State the degree of each monomial specified below.

    a. $4x^5$

    b. $3x^3y^3$

    c. $3xy^2$

    d. $3^2x^3y^2$

    e. $5x^3y^2z^4$

11. Using the polynomial $3x^2 - 5x + 6$,

    a. determine if the polynomial is a monomial, binomial, or trinomial.

b. determine the degree of the polynomial.

c. determine the numerical coefficient of the second term.

12. State the number of terms in each algebraic expression and also state the numerical coefficient of the second term.

a. $-4c + 3d$

b. $3ab + bc - cd - ef$

c. $-4xyz + 7xy - z$

d. $5uvw - 4uv + 8uw$

13. If $P(x) = 3x + 5$, without performing any computation circle the line in the TABLE that allows you to determine $P(-2)$.

PRACTICE   *In Exercises 14–21, let x, y, and z represent three real numbers. Write an algebraic expression to denote each quantity.*

14. The sum of x and y.

15. The product of x and y.

16. The product of x and twice y.

17. The sum of twice x and twice y.

18. The difference obtained when x is subtracted from y.

19. The quotient obtained when y is divided by x.

20. The quotient obtained when the sum of x and y is divided by z.

21. y decreased by x.

In Exercises 22–27, write each algebraic expression as an appropriate English phrase.

22. $x + 3$

23. $y - 2$

24. $\dfrac{x}{y}$

25. xz

26. $\dfrac{5}{x + y}$

27. $\dfrac{3x}{y + z}$

In Exercises 28–35, classify each polynomial as a monomial, a binomial, or a trinomial, if possible.

28. $3x + 7$

29. $3y - 5$

30. $3y^2 + 4y + 3$

31. $3xy$

32. $3z^2$

33. $3x^4 - 2x^3 + 3x - 1$

34. $5t - 32$

35. $9x^2y^3z^4$

In Exercises 36–47, state the degree of each polynomial.

36. $3x^4$

37. $3x^5 - 4x^2$

38. $-2x^2 + 3x^3$

39. $-5x^5 + 3x^2 - 3x$

40. $3x^2y^3 + 5x^3y^5$

41. $-2x^2y^3 + 4x^3y^2z$

42. $-5r^2s^2t - 3r^3st^2$

43. $4r^2s^3t^3 - 5r^2s^8 + 3$

44. $x^{12} + 3x^2y^3z^4$

45. 17^2x

46. 38

47. -25

In Exercises 48–55, let $P(x) = x^2 - 2x + 3$. Find and simplify each of the following analytically (by hand), and verify 48–51 with your calculator.

48. $P(2)$

49. $P(0)$

50. $P(-1)$

51. $P(-2)$

52. $P(w)$

53. $P(t)$

54. $P(-y)$

55. $P(2t)$

In Exercises 56–67, let $P(x) = 5x - 2$. Find and simplify each of the following analytically (by hand), and verify 56–57 with your calculator.

56. $P\left(\dfrac{1}{5}\right)$

57. $P\left(\dfrac{1}{10}\right)$

58. $P(u^2)$

59. $P(-v^4)$

60. $P(-4z^6)$

61. $P(10x^7)$

62. $P(x + h)$

63. $P(x - h)$

64. $P(x) + P(h)$

65. $P(x) - P(h)$

66. $P(2y + z)$

67. $P(-3r + 2s)$

APPLICATIONS

68. Course load   A man enrolls in college for c hours of credit, and his sister enrolls for 4 more hours than her brother. Write an expression that represents the number of hours the sister is taking. Could the sister be taking 3 hours of courses? Why or why not?

69. Mileage   An antique Ford has 25,000 more miles on its odometer than a new car. If the new car has traveled m miles, find an expression that represents the mileage on the Ford.

70. T-bills   Write an expression that represents the value of t T-bills, each worth $9987.

71. Real estate   Write an expression that represents the value of a vacant lots if each lot is worth $35,000.

72. Cutting rope   A rope x ft long is cut into 5 equal pieces. Write an expression to represent the length of each piece.

73. Plumbing   A plumber cuts a pipe that is 12 ft long into x equal pieces. Find the length of each piece.

74. Comparing assets   A girl had d dollars, and her brother had $5 less than three times that amount.

 a. Write an expression to represent the brother's amount of money.

 b. What is the smallest dollar amount (integer value) that the girl could have?

 c. What is the smallest dollar amount the brother could have?

75. Comparing investments   Wendy has x shares of stock. Her sister has 8 fewer shares than twice Wendy's shares.

 a. Write an expression to represent the sister's stock shares.

 b. Is it possible for Wendy to own only 3 shares of stock?

 c. Why or why not?

In Exercises 76–79, assume that the height in feet of an object t seconds after it is thrown is given by $h(t) = -16t^2 + 120t + 16.$

76. Find the height from which the object was thrown: i.e. $h(0)$.

77. Find the height at 2 seconds: i.e. $h(2)$.

78. Find the height at 6 seconds: i.e. $h(6)$.

79. Find the height at 7 seconds: i.e. $h(7)$.

REVIEW

80. Complete each table.

a.

x	0	1	2	3
2^x				

b.

x	0	1	2	3
3^x				

81. Based on each table, determine the missing base value.

a.

x	0	1	2	3
x	1	9	81	729

b.

x	0	1	2	3
x	−1	−16	−256	−4,096

82. Simplify. Verify your results with your calculator.

 a. 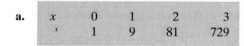 $25 + 2[18 - 4(5 - 2)]$

 b. $(6 - 4)^2 + (8 + -6)^3$

 c. 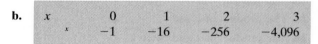 $-2(3 - 4) - 5(2^2 - 3^2)$

2.4 ADDING AND SUBTRACTING POLYNOMIALS

- DISTRIBUTIVE PROPERTY
- ADDING AND SUBTRACTING LIKE TERMS
- ADDING AND SUBTRACTING MULTIPLES OF POLYNOMIALS

▍▍▍ DISTRIBUTIVE PROPERTY

The **distributive property of multiplication over addition** (often simply referred to as the **distributive property**) allows us to distribute the operation of multiplication.

Examine the problem $2(3 + 7)$. It can be evaluated in two different ways. We can add and then multiply, or we can multiply each number within the parentheses by 2 and then add.

$$2(3 + 7) = 2(10) \qquad \text{and} \qquad 2(3 + 7) = 2 \cdot 3 + 2 \cdot 7$$
$$= 20 \qquad\qquad\qquad\qquad = 6 + 14$$
$$= 20$$

Illustration of distributive property

Either way, the result is 20. In general, we have the following property:

Distributive Property of Multiplication over Addition

If a, b, and c are real numbers, then $a(b + c) = ab + ac$. Because of the definition of subtraction, $a - b = a + (-b)$, this property can be extended: $a(b - c) = ab - ac$.

Multiplication is commutative, therefore the distributive property can be written in the form

$$(b + c)a = ba + ca \qquad\qquad (b - c)a = ba - ca$$

EXAMPLE 1

Use the distributive property to write $3(x + 2)$ without using parentheses.

SOLUTION

$$3(x + 2) = 3x + 3 \cdot 2$$
$$= 3x + 6$$

The distributive property can be extended to three or more terms. For example, if a, b, c, and d are real numbers, then

$$a(b + c + d) = ab + ac + ad$$

▍ ADDING AND SUBTRACTING LIKE TERMS

Terms with identical variable factors are called **like terms**. For example

Like terms	*Unlike terms*
$4x$ and $5x$	$4x$ and $5x^2$
$3xyz^2$ and $-2xyz^2$	$\dfrac{1}{2}ab^2c$ and $\dfrac{1}{3}a^2bc$

Because of the distributive property, we can combine like terms by adding their coefficients and using the same variables and exponents. For example,

$$2y + 5y = (2 + 5)y \qquad \text{and} \qquad -3x^2 + 7x^2 = (-3 + 7)x^2$$
$$= 7y \qquad\qquad\qquad\qquad\qquad = 4x^2$$

Likewise,

$$4x^3y^2 + 9x^3y^2 = 13x^3y^2 \qquad \text{and} \qquad 4r^2s^3t^4 + 7r^2s^3t^4 = 11r^2s^3t^4$$

Remember, the word *simplify* means to remove all grouping symbols and perform the indicated operations, which includes combining like terms.

EXAMPLE 2

Simplify.

a. $5xy^3 + 7xy^3 = (5 + 7)xy^3 = 12xy^3$

b. $-7x^2y^2 + 6x^2y^2 + 3x^2y^2 = (-7 + 6 + 3)x^2y^2$
$$= (-1 + 3)x^2y^2$$
$$= 2x^2y^2$$

c. $(2x^2)^2 + (3x)^4 = 4x^4 + 81x^4$
$$= (4 + 81)x^4$$
$$= 85x^4$$

d. $8x^2 - 3x^2 = (8 - 3)x^2 = 5x^2$

e. $6x^3y^2 - 9x^3y^2 = (6 - 9)x^3y^2 = (6 + -9)x^3y^2 = -3x^3y^2$

f. $-3r^2st^3 - 5r^2st^3 = (-3 - 5)r^2st^3 = -8r^2st^3$

Because of the distributive property, we can also remove parentheses enclosing several terms when the sign preceding the parentheses is a $-$ sign. The negative sign indicates that we want the negative, or opposite, of the polynomial. Because a polynomial is a sum of monomials, we must get the opposite of each *term* of the polynomial. It is comparable to multiplying through the polynomial by negative 1.

$$-(3x^2 + 3x - 2) = -1(3x^2 + 3x - 2)$$
$$= -1(3x^2) + (-1)(3x) + (-1)(-2)$$
$$= -3x^2 + (-3x) + 2$$
$$= -3x^2 - 3x + 2$$

EXAMPLE 3

Simplify.

a. $(3x - 4) - (5x + 7) = 3x - 4 - 5x - 7$

            Remove the second parentheses by multiplying the second polynomial by -1.

$$= 3x + (-4) + (-5x) + (-7) \quad a - b = a + -b$$
$$= 3x + (-5x) + (-4) + (-7) \quad \text{Commutative property of addition.}$$
$$= -2x + -11 \quad \text{Combine like terms.}$$
$$= -2x - 11 \quad a - b = a + -b$$

b. $(3x^2 - 4x - 6) - (2x^2 - 6x + 12)$

$$= 3x^2 - 4x - 6 - 2x^2 + 6x - 12$$

            Remove parentheses by multiplying the second polynomial by -1.

$$= 3x^2 + (-4x) + (-6) + (-2x^2) + 6x + (-12) \quad a - b = a + -b$$
$$= 3x^2 + (-2x^2) + (-4x) + 6x + (-6) + (-12) \quad \text{Commutative property of addition.}$$
$$= x^2 + 2x + (-18) \quad \text{Combine like terms.}$$
$$= x^2 + 2x - 18 \quad a - b = a + -b$$

c. $(-4rt^3 + 2r^2t^2) - (-3rt^3 + 2r^2t^2)$

$\quad = -4rt^3 + 2r^2t^2 + 3rt^3 - 2r^2t^2$ Remove parentheses by multiplying the second polynomial by -1.

$\quad = -4rt^3 + 2r^2t^2 + 3rt^3 + (-2r^2t^2)$ $a - b = a + -b$

$\quad = -4rt^3 + 3rt^3 + 2r^2t^2 + (-2r^2t^2)$ Commutative property of addition.

$\quad = -rt^3$ Combine like terms.

EXAMPLE 4

Find the difference when the polynomial $(3x^2 - 4x + 2)$ is subtracted from the polynomial $(8x^2 + 2x - 5)$.

SOLUTION $(8x^2 + 2x - 5) - (3x^2 - 4x + 2)$

$\quad = 8x^2 + 2x - 5 - 3x^2 + 4x - 2$ Remove grouping by multiplying the second polynomial by -1.

$\quad = 8x^2 + 2x + (-5) + (-3x^2) + 4x + (-2)$ $a - b = a + (-b)$

$\quad = 8x^2 + (-3x^2) + 2x + 4x + (-5) + (-2)$ Commutative property of addition.

$\quad = 5x^2 + 6x - 7$ Combine like terms.

ADDING AND SUBTRACTING MULTIPLES OF POLYNOMIALS

Because of the distributive property, we can remove parentheses enclosing several terms when a monomial precedes the parentheses. We simply multiply every term within the parentheses by that monomial. For example, to add $3(2x + 5)$ and $2(4x - 3)$, we proceed as follows:

$3(2x + 5) + 2(4x - 3) = 6x + 15 + 8x - 6$ Distributive property.

$\qquad\qquad\qquad\quad = 6x + 8x + 15 - 6$ Commutative property of addition.

$\qquad\qquad\qquad\quad = 14x + 9$

This is consistent with the order of operations. We cannot simplify what is grouped because the parentheses contain expressions that are not like terms and cannot be combined. We must multiply first.

EXAMPLE 5

Simplify.

a. $3(x^2 + 4x) + 2(x^2 - 4) = 3x^2 + 12x + 2x^2 - 8$

$\qquad\qquad\qquad\qquad\quad = 5x^2 + 12x - 8$

b. $8(y^2 - 2y + 3) - 4(2y^2 + y - 3) = 8y^2 - 16y + 24 - 8y^2 - 4y + 12$

$\qquad\qquad\qquad\qquad\quad = 8y^2 + (-16y) + 24 + (-8y^2) + (-4y) + 12$

$\qquad\qquad\qquad\qquad\quad = 8y^2 + (-8y^2) + (-16y) + (-4y) + 24 + 12$

$\qquad\qquad\qquad\qquad\quad = 0y^2 + (-20y) + 36$

$\qquad\qquad\qquad\qquad\quad = -20y + 36$

c. $-4x(xy^2 - xy + 3) - x(xy^2 - 2) + 3(x^2y^2 + 2x^2y)$

$$= -4x^2y^2 + 4x^2y - 12x - x^2y^2 + 2x + 3x^2y^2 + 6x^2y$$

$$= -4x^2y^2 + 4x^2y + (-12x) + (-x^2y^2) + 2x + 3x^2y^2 + 6x^2y$$

$$= -4x^2y^2 + (-x^2y^2) + 3x^2y^2 + 4x^2y + 6x^2y + (-12x) + 2x$$

$$= -2x^2y^2 + 10x^2y + (-10x)$$

$$= -2x^2y^2 + 10x^2y - 10x$$

EXAMPLE 6

Cindy has $2,000 to invest for one year. She will invest in two different accounts. One account earns 4% interest and the other earns $2\frac{1}{2}\%$. She decides to invest x dollars at 4% and the rest of her money, $2,000 - x$, at $2\frac{1}{2}\%$. Write a simplified algebraic expression to represent the total interest earned after one year.

SOLUTION

Recall, $i = prt$; interest earned is a product of the principal, rate, and time. The amount of interest earned at 4% is represented as .04x. The amount of interest earned at $2\frac{1}{2}\%$ is represented as $0.025(2,000 - x)$.

$.04x + 0.025(2,000 - x)$	Total amount of interest earned.
$.04x + 50 - .025x$	Distributive property.
$.015x + 50$	Combine like terms.

The expression $.015x + 50$ represents the total amount of interest earned from the two investments.

EXERCISE 2.4

VOCABULARY AND NOTATION *In Exercises 1–2, fill in the blank to make a true statement.*

1. The _____ property allows the expression $3(x + y)$ to be written as $3x + 3y$.

2. _____ terms have identical variable factors.

CONCEPTS *In Exercises 3–10 tell whether the terms are like or unlike terms. If they are like terms, find their sum.*

3. $3y, 4y$

4. $3x^2, 5x^2$

5. $3x, 3y$

6. $3x^2, 6x$

7. $3x^3, 4x^3, 6x^3$

8. $-2y^4, -6y^4, 10y^4$

9. $32x^5y^3, -21x^5y^3, -11x^5y^3$

10. $-x^2y, xy, 3xy^2$

In Exercises 11–18, simplify each expression, if possible. To simplify means to combine like terms.

11. $4y + 5y$

12. $-2x + 3x$

13. $-8t^2 - 4t^2$

14. $15x^2 + 10x^2$

15. $32u^3 - 16u^3$

16. $25xy^2 - 7xy^2$

17. $18x^5y^2 - 11x^5y^2$

18. $17x^6y - 22x^6y$

PRACTICE *In Exercises 19–24, simplify each expression, if possible.*

19. $3rst + 4rst + 7rst$

20. $-2ab + 7ab - 3ab$

21. $-4a^2bc + 5a^2bc + 7a^2bc$

22. $3x^2y - 5xy^2 - 7x^2y$

23. $3mn^3 - 5mn^3 + 3m^3n$

24. $6abc + 6bca + 7cba$

In Exercises 25–28, find the polynomial in simplified form that represents the perimeter of each figure.

25.

26.

27.

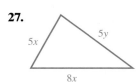

28.

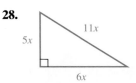

In Exercises 29–43, simplify each expression.

29. $(3x + 7) + (4x - 3)$

30. $(2y - 3) + (4y + 7)$

31. $(4a + 3) - (2a - 4)$

32. $(5b - 7) - (3b + 5)$

33. $(3x^2 - 3x - 2) + (3x^2 + 4x - 3)$

34. $(2b^2 + 3b - 5) - (2b^2 - 4b - 9)$

35. $(3a^2 - 2a + 4) - (a^2 - 3a + 7)$

36. $2(x + 3) + 3(x + 3)$

37. $5(x + y) + 7(x + y)$

38. $-8(x - y) + 11(x - y)$

39. $-4(a - b) - 5(a - b)$

40. $2(x^2 - 5x - 4) - 3(x^2 - 5x - 4) + 6(x^2 - 5x - 4)$

41. $-5(y^2 - 2y - 6) + 6(2y^2 + 2y - 5)$

42. $2a(ab^2 - b) - 3b(a + 2ab) + b(b - a + a^2b)$

43. $3y(xy + y) - 2y^2(x - 4 + y) + 2(y^3 + y^2)$

In Exercises 44–47, find the simplified polynomial that represents the perimeter of each figure

44.

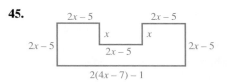

45.

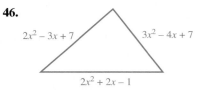

46.

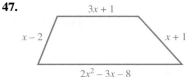

47.

48. Use the STOre feature of your calculator to determine the following:

   a. the value of the perimeter in Exercise 44 based on the polynomial expression you determined to represent the perimeter if $x = 4$.

   b. the value of the length and width of the rectangle if $x = 4$.

   c. Conclusion: Is $x = 4$ a valid value for the given rectangle? Why or why not?

49. Use the STOre feature of your calculator to determine the following:

   a. the value of the perimeter in Exercise 45 based on the polynomial expression you determined to represent the perimeter if $x = 9$.

   b. the values of the sides of the figure if $x = 9$.

   c. Conclusion: Is $x = 9$ a valid value for the given figure? Why or why not?

50. Use the STOre feature of your calculator to determine the following:

   a. the value of the perimeter of the figure in Exercise 46 based on the polynomial you determined to represent the perimeter if $x = 2$.

   b. the values of each side if $x = 2$.

   c. Conclusion: Is $x = 2$ a valid value for the given figure? Why or why not?

51. Use the STOre feature of your calculator to determine the following:

   a. the value of the perimeter in Exercise 47 based on the polynomial you determined to represent the perimeter if $x = 2$.

   b. the values of the sides of the trapezoid if $x = 2$.

   c. Conclusion: Is $x = 2$ a valid value for the given trapezoid? Why or why not?

52. Find the sum when $3y^2 - 5y + 7$ is added to the sum of $-3y^2 - 7y + 4$ and $5y^2 + 5y - 7$.

53. Find the sum when $x^2 + x - 3$ is added to the sum of $2x^2 - 3x + 4$ and $3x^2 - 2$.

54. Find the difference when $-3z^3 - 4z + 7$ is subtracted from the sum of $2z^2 + 3z - 7$ and $-4z^3 - 2z - 3$.

55. Find the difference when $t^3 - 2t^2 + 2$ is subtracted from the sum of $3t^3 + t^2$ and $-t^3 + 6t - 3$.

56. Find the difference when $32x^2 - 17x + 45$ is subtracted from the sum of $23x^2 - 12x - 7$ and $-11x^2 + 12x + 7$.

57. Find the sum when $3x^2 + 4x - 7$ is added to the sum of $-2x^2 - 7x + 1$ and $-4x^2 + 8x - 1$.

APPLICATIONS

58. Rental cars   A rental car agency charges a flat fee of $37 per day plus $0.21 per mile. Write a polynomial expression to model the cost for one day's rental if the car is driven x miles.

59. Portrait studio   A photographer charges a $75 sitting fee and then $10 per unit purchased for studio photographs. (A unit consists of one 8 × 10, or a sheet of two 5 × 7s, or a sheet of 8 wallets.) Write a polynomial expression when x represents the number of units purchased to model the cost of having pictures made.

60. River travel   A boat traveling x mph upstream against a 3 mph current travels a distance of $5(x - 3)$ miles in 5 hours ($d = rt$). When the boat returns downstream in 4 hours, the distance traveled is $4(x + 3)$ miles. Write a

polynomial expression and simplify it to model the total distance traveled.

61. Investments   George has $5,000 to invest in two simple interest-bearing accounts. He plans to invest x dollars in a certificate of deposit that earns 3% interest and the remaining money, $5,000 - x$ dollars, in a money market account that earns $1\frac{1}{2}\%$. Write a simplified polynomial expression to represent the total interest earned in the two accounts. (*Hint: i = prt*; time is 1 year in this problem.)

62. Geometry   The length of a rectangle is 4 more than the width. If the width is x and the length is $4 + x$, write a polynomial expression to model the perimeter and simplify this expression.

63. Travel   In order to reach her destination, a woman drives x hours at 45 mph to the train station and then rides the train for $x + 2$ hours traveling at a speed of 72 mph. Write a simplified polynomial expression to model the total distance traveled.

64. Distance   Two cars start at City A and drive in opposite directions for x hours. One vehicle is traveling 55 mph and the other, 65 mph. Write a polynomial expression to model the total distance between the two cars and simplify your expression.

65. Pages in a manual   A calculator manual is being assembled as indicated below. Find an expression to model the total number of pages (write the expression in simplified form).

Preface	x pages
table of contents	2 fewer pages than the preface
body of text	40 times as many pages as the preface
index	4 more pages than the preface

66. Geometry   Find the circumference of the given circle, expressing the circumference in terms of π and x.

$4x + 5$

REVIEW

67. Write the expanded form of the exponential expression 4^3.

68. Insert parentheses to make the statement true.

   a. $14 + 3 \cdot 7 - 5 = 20$

   b. $33 - 8 - 10 = 15$

   c. $18 - 5 + 6 - 9 = 10$

   d. $4 + 2 \cdot 3^4 = 486$

69. Tax The sales tax for Clarksville, TN is $9\frac{3}{4}\%$.

   a. Represent this percentage as a decimal.

   b. Find the amount of sales tax on a purchase of $299.99.

 c. Explain how you determine if it is appropriate to round the results in **b.**

70. If the polynomial $P(x) = 2x^2 + 11x + 5$ represents the area of a rectangle whose sides measure $2x + 1$ and $x + 5$, find the area if $x = 2$.

2.5   MULTIPLYING POLYNOMIALS

- MULTIPLYING BY A MONOMIAL
- MULTIPLYING POLYNOMIALS
- THE FOIL METHOD

▐ MULTIPLYING BY A MONOMIAL

In Section 2.1, we multiplied certain monomials by other monomials. To multiply $4x^2$ by $-2x^3$, for example, we use the commutative property of multiplication to group the numerical factors and the variable factors together and multiply.

$$
\begin{aligned}
4x^2(-2x^3) &= 4 \cdot x^2 \cdot -2 \cdot x^3 \\
&= 4 \cdot -2 \cdot x^2 \cdot x^3 && \text{Commutative property of multiplication} \\
&= (4 \cdot -2)(x^2 \cdot x^3) && \text{Associative property of multiplication} \\
&= -8x^5 && \text{Product Rule for exponents}
\end{aligned}
$$

EXAMPLE 1

Simplify.

a. $3x^5(2x^5) = 3 \cdot x^5 \cdot 2 \cdot x^5$

            $= 3 \cdot 2 \cdot x^5 \cdot x^5$

            $= 6x^{10}$

b. $-2a^2b^3(5ab^2) = -2 \cdot a^2 \cdot b^3 \cdot 5 \cdot a \cdot b^2$

                 $= -2 \cdot 5 \cdot a^2 \cdot a \cdot b^3 \cdot b^2$

                 $= -10a^3b^5$

c. $-4y^5z^2(2y^3z^3)(3yz) = -4 \cdot y^5 \cdot z^2 \cdot 2 \cdot y^3 \cdot z^3 \cdot 3 \cdot y \cdot z$

                       $= -4 \cdot 2 \cdot 3 \cdot y^5 \cdot y^3 \cdot y \cdot z^2 \cdot z^3 \cdot z$

                       $= -24y^9z^6$

The previous example suggests the following rule:

Multiplying Two Monomials

To multiply two monomials, first multiply the numerical factors and then multiply the variable factors using the product rule for exponents.

To find the product of a monomial and a polynomial with more than one term, we use the distributive property. To multiply $x + 4$ by $3x$, for example, we proceed as follows:

$$3x(x + 4) = 3x \cdot x + 3x \cdot 4$$
$$= 3x^2 + 12x$$

EXAMPLE 2

Simplify.

a. $2a^2(3a^2 - 4a) = 2a^2 \cdot 3a^2 - 2a^2 \cdot 4a$
$$= 6a^4 - 8a^3$$

b. $-2xz^2(2x - 3z + 2x^2z^2) = -2xz^2(2x) - (-2xz^2)(3z) + (-2xz^2)(2x^2z^2)$
$$= -4x^2z^2 + 6xz^3 - 4x^3z^4$$

The results of Example 2 suggest the following rule:

Multiplying Polynomials by Monomials

To multiply a polynomial with more than one term by a monomial, use the distributive property to remove parentheses and simplify.

We must use the Distributive Property more than once to multiply a polynomial by another polynomial. For example, to multiply $3x^2 + 3x - 5$ by $2x + 3$, we proceed as follows:

$$(2x + 3)(3x^2 + 3x - 5) = 2x(3x^2) + 2x(3x) + 2x(-5) + 3(3x^2) + 3(3x) + 3(-5)$$
$$= 6x^3 + 6x^2 + (-10x) + 9x^2 + 9x + (-15)$$
$$= 6x^3 + 6x^2 + 9x^2 + (-10x) + 9x + -15$$
$$= 6x^3 + 15x^2 - x - 15$$

MULTIPLYING POLYNOMIALS

EXAMPLE 3

Simplify.

a. $(2x - 4)(3x + 5) = 2x(3x) + 2x(5) - 4(3x) - 4(5)$
$$= 6x^2 + 10x - 12x - 20$$
$$= 6x^2 - 2x - 20$$

b. $(3x - 2y)(2x + 3y) = 3x(2x) + 3x(3y) - 2y(2x) - 2y(3y)$

$$= 6x^2 + 9xy - 4xy - 6y^2$$

$$= 6x^2 + 5xy - 6y^2$$

c. $(3y + 1)(3y^2 + 2y + 2) = 3y(3y^2) + 3y(2y) + 3y(2) + 1(3y^2) + 1(2y) + 1(2)$

$$= 9y^3 + 6y^2 + 6y + 3y^2 + 2y + 2$$

$$= 9y^3 + 6y^2 + 3y^2 + 6y + 2y + 2$$

$$= 9y^3 + 9y^2 + 8y + 2$$

The results of Example 3 suggest the following rule:

Multiplying Polynomials

To multiply one polynomial by another, multiply each term of one polynomial by each term of the other polynomial and combine like terms.

It is often convenient to organize the work vertically.

EXAMPLE 4

a. Multiply:

$$2x - 4$$
$$\underline{3x + 2}$$

$3x(2x - 4) \rightarrow$ $6x^2 - 12x$

$2(2x - 4) \rightarrow$ $\underline{ + 4x - 8}$

$$6x^2 - 8x - 8$$

b. Multiply:

$$3a^2 - 4a + 7$$
$$\underline{2a + 5}$$

$2a(3a^2 - 4a + 7) \rightarrow$ $6a^3 - 8a^2 + 14a$

$5(3a^2 - 4a + 7) \rightarrow$ $\underline{ + 15a^2 - 20a + 35}$

$$6a^3 + 7a^2 - 6a + 35$$

c. Multiply:

$$3y^2 - 5y + 4$$
$$\underline{- 4y^2 - 3}$$

$-4y^2(3y^2 - 5y + 4) \rightarrow$ $-12y^4 + 20y^3 - 16y^2$

$-3(3y^2 - 5y + 4) \rightarrow$ $\underline{ - 9y^2 + 15y - 12}$

$$-12y^4 + 20y^3 - 25y^2 + 15y - 12$$

▮ THE FOIL METHOD

The mnemonic FOIL (First Outer Inner Last) is often used to quickly multiply two *binomials*. Remember, when multiplying two polynomials each term of one polynomial must multiply each term of the other polynomial. To use the FOIL method to multiply $2a - 4$ by $3a + 5$, we

1. multiply the **F**irst terms $2a$ and $3a$ to obtain $6a^2$,
2. multiply the **O**uter terms $2a$ and 5 to obtain $10a$,
3. multiply the **I**nner terms -4 and $3a$ to obtain $-12a$, and
4. multiply the **L**ast terms -4 and 5 to obtain -20.

Then we simplify the resulting polynomial, if possible.

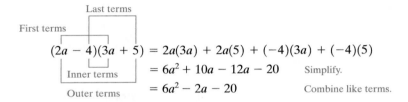

$$(2a - 4)(3a + 5) = 2a(3a) + 2a(5) + (-4)(3a) + (-4)(5)$$
$$= 6a^2 + 10a - 12a - 20 \qquad \text{Simplify.}$$
$$= 6a^2 - 2a - 20 \qquad \text{Combine like terms.}$$

Be sure to observe the distributive property at work here as the term $2a$ is distributed to $3a$ and 5, and the term -4 is also then distributed to $3a$ and 5.

EXAMPLE 5

Use the FOIL method to find each product, observing the distributive property at work within the method.

a. $(3x + 4)(2x - 3) = 3x(2x) + 3x(-3) + 4(2x) + 4(-3)$
$$= 6x^2 - 9x + 8x - 12$$
$$= 6x^2 - x - 12$$

b. $(2y - 7)(5y - 4) = 2y(5y) + 2y(-4) + (-7)(5y) + (-7)(-4)$
$$= 10y^2 - 8y - 35y + 28$$
$$= 10y^2 - 43y + 28$$

c. $(2r - 3s)(2r + t) = 2r(2r) + 2r(t) - 3s(2r) - 3s(t)$
$$= 4r^2 + 2rt - 6rs - 3st$$

EXAMPLE 6

Simplify each expression. The Associative Property of Multiplication allows us to choose whether to first multiply 3 times $(2x - 3)$ or the two binomials. We will multiply the two binomials first.

a. $3(2x - 3)(x + 1) = 3(2x^2 + 2x - 3x - 3)$ Multiply the binomials.

$\qquad\qquad\qquad\quad = 3(2x^2 - x - 3)$ Combine like terms.

$\qquad\qquad\qquad\quad = 6x^2 - 3x - 9$ Use the distributive property to remove parentheses.

b. $(x + 1)(x - 2) - 3x(x + 3) = x^2 - 2x + x - 2 - 3x^2 - 9x$ Follow the order of operations and perform multiplication before subtraction.

$\qquad\qquad\qquad\qquad\qquad = -2x^2 - 10x - 2$ Combine like terms.

The products discussed in Example 7 are called **special products**.

EXAMPLE 7

Use the distributive property (or the mnemonic FOIL) to find each special product.

Simplify.

a. $(x + y)^2 = (x + y)(x + y)$

$\qquad\qquad = x^2 + xy + xy + y^2$

$\qquad\qquad = x^2 + 2xy + y^2$

The square of the sum of two terms such as $x + y$ has three terms: the square of the first term, plus twice the product of the two terms, and the square of the second term.

Simplify.

b. $(x - y)^2 = (x - y)(x - y)$

$\qquad\qquad = x^2 - xy - xy + y^2$

$\qquad\qquad = x^2 - 2xy + y^2$

The square of the difference of two terms such as $x - y$ has three terms: the square of the first term, minus twice the product of the two terms, and the square of the second term.

Simplify.

c. $(x + y)(x - y) = x^2 - xy + xy - y^2$

$\qquad\qquad\qquad = x^2 - y^2$

The product of a sum and a difference of two terms such as $x + y$ and $x - y$ is a binomial. It is the product of the first term squared minus the product of the second term squared.

Binomials that have the same terms but different signs between them are often called **conjugate binomials**. For example, the conjugate of $x + y$ is $x - y$, and the conjugate of $ab - c$ is $ab + c$.

Because the special products discussed in Example 7 occur so often, it is wise to learn their forms.

Special Products

$$(x + y)^2 = x^2 + 2xy + y^2$$
$$(x - y)^2 = x^2 - 2xy + y^2$$
$$(x + y)(x - y) = x^2 - y^2$$

 WARNING!   Note that $(x + y)^2 \neq x^2 + y^2$ and $(x - y)^2 \neq x^2 - y^2$.

EXPLORING THE CONCEPT

A frequent student error is to simplify $(x + y)^2$ as $x^2 + y^2$ and $(x - y)^2$ as $x^2 - y^2$. Enter the following expressions on your calculator to numerically justify that $(x + y)^2 \neq x^2 + y^2$ and $(x - y)^2 \neq x^2 - y^2$.

a. $(5 + 7)^2; 5^2 + 7^2$

b. $(-3 + -8)^2; (-3)^2 + (-8)^2$

c. $(8 - 2)^2; 8^2 - 2^2$

d. $(4 - 11)^2; 4^2 - 11^2$

CONCLUSION

a.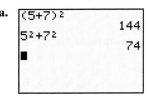

b.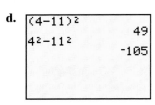

c.

```
(8-2)²
          36
8²-2²
          60
```

d.

```
(4-11)²
          49
4²-11²
        -105
```

EXERCISE 2.5

VOCABULARY AND NOTATION   *In Exercises 1–3, fill in the blank to make a true statement.*

1. Binomials that have the same terms but different signs between the terms are called _____ binomials.

2. To multiply $2a^2(3a^2 - 4a)$ we use the _____ property.

3. The _____ and _____ properties are used when multiplying two monomials.

CONCEPTS

4. Which of the following is equal to $(x + y)^2$?

    a. $x^2 + y^2$ **b.** $(x + y)(x + y)$

5. Find each product.

    a. $(2x)(3x)$ **b.** $(-4x^2)(-x)$

    c. $\left(\dfrac{1}{2}x\right)\left(\dfrac{4}{5}x\right)$

6. Find each product.

    a. $3(x + 40)$ **b.** $-3(a - 2)$

    c. $-4(t + 7)$ **d.** $6(x^2 - 3)$

PRACTICE   *In Exercises 7–14, simplify.*

7. $(3x^2)(4x^3)$ **8.** $(-2a^3)(3a^2)$

9. $(3b^2)(-2b)(4b^3)$ **10.** $(3y)(2y^2)(-y^4)$

11. $(x^2y^3)^5$ **12.** $(a^3b^2c)^4$

13. $(a^3b^2c)(abc^3)^2$ **14.** $(xyz^3)(xy^2z^2)^3$

In Exercises 15–22, simplify.

15. $3x(x - 2)$

16. $4y(y + 5)$

17. $-2x^2(3x^2 - x)$

18. $-4x^2(3x^2 - x)$

19. $2x^2(3x^2 + 4x - 7)$

20. $3y^3(2y^2 - 7y - 8)$

21. $\dfrac{1}{4}x^2(8x^5 - 4)$

22. $-\dfrac{2}{3}r^2t^2(9r - 3t)$

In Exercises 23–32, simplify.

23. $(a + 4)(a + 5)$

24. $(y - 3)(y + 5)$

25. $(3x - 2)(x + 4)$

26. $(t + 4)(2t - 3)$

27. $(2a + 4)(3a - 5)$

28. $(2b - 1)(3b + 4)$

29. $(3x - 5)(2x + 1)$

30. $(2y - 5)(3y + 7)$

31. $(2t + 3s)(3t - s)$

32. $(3a - 2b)(4a + b)$

In Exercises 33–42, simplify the following special products.

33. $(x + 4)(x + 4)$

34. $(a + 3)(a + 3)$

35. $(x + 5)^2$

36. $(y - 6)^2$

37. $(r + 4)(r - 4)$

38. $(b + 2)(b - 2)$

39. $(4x + 5y)(4x - 5y)$

40. $(6p + 5q)(6p - 5q)$

41. $(2a - 3b)^2$

42. $(3a + 2b)^2$

In Exercises 43–44, find the volume of each figure.

43.

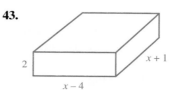

44.

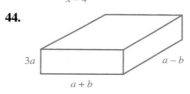

In Exercises 45–48, find the area of each figure.

45.

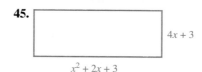

$4x + 3$

$x^2 + 2x + 3$

46.

$3x + y$

$2x^2 - 3xy + y^2$

47.

$x - 2y$

$x^2 + 2xy + 4y^2$

48.

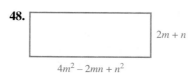

$2m + n$

$4m^2 - 2mn + n^2$

In Exercises 49–58, simplify each expression.

49. $2t(t + 2) + 3t(t - 5)$

50. $3y(y + 2) + (y + 1)(y - 1)$

51. $3xy(x + y) - 2x(xy - x)$

52. $(a + b)(a - b) - (a + b)(a + b)$

53. $(x + y)(x - y) + x(x + y)$

54. $(2x - 1)(2x + 1) + x(2x + 1)$

55. $(x + 2)^2 - (x - 2)^2$

56. $(x - 3)^2 - (x + 3)^2$

57. $(2s - 3)(s + 2) + (3s + 1)(s - 3)$

58. $(3x + 4)(2x - 2) - (2x + 1)(x - 3)$

APPLICATIONS *In Exercises 59–68, write a simplified algebraic expression to model each problem.*

59. The length of a rectangle is 5 more than the width. If the width is x, then the length is $5 + x$. Find the area of the rectangle.

60. The side of a square is $2x - 3$. Find the area.

61. The sides of a rectangular prism are $x, x + 2,$ and $x - 2$ in length. Find the volume.

62. Find the product of three consecutive integers represented by $x, x + 1,$ and $x + 2$.

63. Find the product of three consecutive even integers represented by $x, x + 2,$ and $x + 4$.

64. If two consecutive even integers are x and $x + 2$, find the difference of their squares.

65. If two consecutive odd integers are x and $x + 2$, find the sum of their squares.

66. Area Find the polynomial to represent the area of the desk calendar in Illustration 1.

$4x - 1$

Illustration 1

67. Area If the top of a desk has a width of $2x^2 - 5x$ and a depth of $3x^2 + 2x$, find the polynomial that models the amount of the desktop that is *not* occupied by the calendar in Exercise 66.

68. Volume Find the polynomial that models the volume of the space occupied by the microwave in Illustration 2.

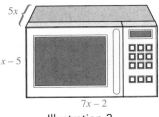

Illustration 2

REVIEW

69. Distance If a car is traveling 65 mph, then the formula $d = 65t$ can be used to determine the distance traveled at any given time. Use your calculator to complete the table of values relating time to distance that corresponds to the following table. (Replace the variable t, time, with the variable x when using the calculator.)

x	.5	1	1.5	2
$65x$				

Record the process with a step-by-step description.

70. Cutting a board The board below has been cut into two pieces. What was the length of the board (in feet) before it was cut?

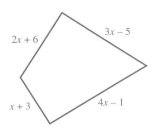

71. Distance A car traveling from Nashville, TN to Louisville, KY must travel at two different speeds: 70 mph in Tennessee and 60 mph in Kentucky. Write the expression that models the total distance traveled, using the variable t for time.

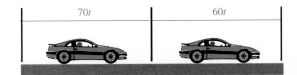

72. Perimeter Find the perimeter of the quadrilateral.

2.6 DIVIDING POLYNOMIALS BY MONOMIALS

- DIVIDING A MONOMIAL
  BY A MONOMIAL

- DIVIDING A POLYNOMIAL
  BY A MONOMIAL

▐ DIVIDING A MONOMIAL BY A MONOMIAL

Recall that to simplify a fraction, we write both its numerator and denominator as the product of several factors and then divide out all common factors. For example, to simplify $\frac{4}{6}$, we can write

$$\frac{4}{6} = \frac{2 \cdot 2}{2 \cdot 3} = \frac{\overset{1}{2} \cdot 2}{\underset{1}{2} \cdot 3} = \frac{2}{3} \qquad \tfrac{2}{2} = 1.$$

To simplify the fraction $\frac{20}{25}$, we can write

$$\frac{20}{25} = \frac{4 \cdot 5}{5 \cdot 5} = \frac{4 \cdot \overset{1}{5}}{\underset{1}{5} \cdot 5} = \frac{4}{5} \qquad \tfrac{5}{5} = 1.$$

To simplify algebraic fractions, we can either use the method just illustrated for simplifying arithmetic fractions or use the rules of exponents.

EXAMPLE 1   Simplify   **a.** $\dfrac{x^2 y}{x y^2}$ and   **b.** $\dfrac{-8a^3 b^2}{4ab^3}$.

SOLUTION   *Method for arithmetic fractions*

a.
$$\frac{x^2 y}{x y^2} = \frac{xxy}{xyy}$$
$$= \frac{\overset{1}{x}\,\overset{1}{x}\,\overset{}{y}}{\underset{1}{x}\,y\,\underset{1}{y}}$$
$$= \frac{x}{y}$$

b.
$$\frac{-8a^3 b^2}{4ab^3} = \frac{-2 \cdot 4aaabb}{4abbb}$$
$$= \frac{-2 \cdot \overset{1\,1}{4aa}\overset{1\,1}{abb}}{\underset{1\,1\,1\,1}{4abbb}}$$
$$= \frac{-2a^2}{b}$$

Using the rules of exponents

a.
$$\frac{x^2 y}{x y^2} = x^{2-1} y^{1-2}$$
$$= x^1 y^{-1} \qquad y^{-1} = \tfrac{1}{y}$$
$$= \frac{x}{y}$$

b.
$$\frac{-8a^3 b^2}{4ab^3} = \frac{-2^3 a^3 b^2}{2^2 ab^3}$$
$$= -2^{3-2} a^{3-1} b^{2-3}$$
$$= -2^1 a^2 b^{-1} \qquad b^{-1} = \tfrac{1}{b}$$
$$= \frac{-2a^2}{b}$$

▍ DIVIDING A POLYNOMIAL BY A MONOMIAL

To divide a polynomial with more than one term by a monomial, we divide each term of the polynomial by the monomial.

EXAMPLE 2   Simplify $\dfrac{9x + 6y}{3xy}$.

SOLUTION $\dfrac{9x + 6y}{3xy} = \dfrac{9x}{3xy} + \dfrac{6y}{3xy}$ Express the division of each term of the polynomial by the monomial.

$= \dfrac{3}{y} + \dfrac{2}{x}$ Simplify each fraction.

EXAMPLE 3   Simplify $\dfrac{6x^2y^2 + 4x^2y - 2xy}{2xy}$.

SOLUTION $\dfrac{6x^2y^2 + 4x^2y - 2xy}{2xy} = \dfrac{6x^2y^2}{2xy} + \dfrac{4x^2y}{2xy} - \dfrac{2xy}{2xy}$ Express the division of each term of the polynomial by the monomial.

$= 3xy + 2x - 1$ Simplify each fraction.

EXAMPLE 4   Simplify $\dfrac{12a^3b^2 - 4a^2b + a}{6a^2b^2}$.

SOLUTION $\dfrac{12a^3b^2 - 4a^2b + a}{6a^2b^2} = \dfrac{12a^3b^2}{6a^2b^2} - \dfrac{4a^2b}{6a^2b^2} + \dfrac{a}{6a^2b^2}$ Express the division of each term of the polynomial by the monomial.

$= 2a - \dfrac{2}{3b} + \dfrac{1}{6ab^2}$ Simplify each fraction.

EXAMPLE 5   Simplify $\dfrac{(x - y)^2 - (x + y)^2}{xy}$.

SOLUTION $\dfrac{(x - y)^2 - (x + y)^2}{xy} = \dfrac{x^2 - 2xy + y^2 - (x^2 + 2xy + y^2)}{xy}$ Multiply the binomials in the numerator.

$= \dfrac{x^2 - 2xy + y^2 - x^2 - 2xy - y^2}{xy}$ Remove parentheses.

$= \dfrac{-4xy}{xy}$ Combine like terms.

$= -4$ Divide out xy.

STUDY TIP

As you work through an exercise section, write down questions for your instructor as they occur to you. Keep the list on a separate piece of paper and write down the answers to *all* of your questions.

EXERCISE 2.6

CONCEPTS   *In Exercises 1–4, perform each division by simplifying each fraction.*

1. $\dfrac{8x}{2}$ **2.** $\dfrac{4x}{12}$ **3.** $\dfrac{xy}{yx}$ **4.** $\dfrac{ab}{bc}$

In Exercises 5–8, perform each division using the distributive property.

5. $\dfrac{4x + 12y}{4}$ **6.** $\dfrac{6x - 3y}{3}$

7. $\dfrac{7x + 9xy}{x}$ **8.** $\dfrac{5ab - 6bc}{b}$

PRACTICE   *In Exercises 9–34, simplify each expression following the order of operations. Write all results with positive exponents.*

9. $\dfrac{r^3 s^2}{rs^3}$ **10.** $\dfrac{y^4 z^3}{y^2 z^2}$

11. $\dfrac{8x^3 y^2}{4xy^3}$ **12.** $\dfrac{-3y^3 z}{6yz^2}$

13. $\dfrac{12u^5 v}{-4u^2 v^3}$ **14.** $\dfrac{16rst^2}{-8rst^3}$

15. $\dfrac{-16r^3 y^2}{-4r^2 y^4}$ **16.** $\dfrac{35xyz^2}{-7x^2 yz}$

17. $\dfrac{-65rs^2 t}{15r^2 s^3 t}$ **18.** $\dfrac{112u^3 z^6}{-42u^3 z^6}$

19. $\dfrac{x^2 x^3}{xy}$ **20.** $\dfrac{(xy)^2}{x^2 y^3}$

21. $\dfrac{(a^3 b^4)^3}{ab^4}$ **22.** $\dfrac{(a^2 b^3)^3}{a^6 b^6}$

23. $\dfrac{15(r^2 s^3)^2}{-5(rs^2)^3}$ **24.** $\dfrac{-5(a^2 b)^3}{10(ab)^3}$

25. $\dfrac{-32(x^3 y)^3}{128(xy^2)^3}$ **26.** $\dfrac{68(a^6 b^7)^2}{-96(abc^2)^3}$

27. $\dfrac{(5a^2 b)^3}{(2a^2 b)^3}$ **28.** $\dfrac{-(4x^3 y^3)^2}{x^2 y^4}$

29. $\dfrac{-(3x^3 y^4)^3}{-(9x^4 y^5)^2}$ **30.** $\dfrac{(2r^3 s^2 t)^2}{-(4r^2 s^2 t)^2}$

31. $\dfrac{(a^2 a^3)^4}{(a^4)^3}$ **32.** $\dfrac{(b^3 b^4)^5}{(bb^2)^2}$

33. $\dfrac{(z^3 z^4)^3}{(z^3)^2}$ **34.** $\dfrac{(t^3 t^5)}{(t^2)^3}$

In Exercises 35–48, simplify.

35. $\dfrac{6x + 9y}{3xy}$ **36.** $\dfrac{8x + 12y}{4xy}$

37. $\dfrac{5x - 10y}{25xy}$ **38.** $\dfrac{2x - 32}{16x}$

39. $\dfrac{3x^2 + 6y^3}{3x^2 y^2}$ **40.** $\dfrac{4a - 9b^2}{12ab}$

41. $\dfrac{15a^3 b^2 - 10a^2 b^3}{5a^2 b^2}$

42. $\dfrac{9a^4 b^3 - 16a^3 b^4}{12a^2 b}$

43. $\dfrac{4x - 2y + 8z}{4xy}$

44. $\dfrac{5a^2 + 10b^2 - 15ab}{5ab}$

45. $\dfrac{12x^3 y^2 - 8x^2 y - 4x}{4xy}$

46. $\dfrac{12a^2 b^2 - 8a^2 b - 4ab}{4ab}$

47. $\dfrac{-25x^2 y + 30xy^2 - 5xy}{-5xy}$

48. $\dfrac{-30a^2 b^2 - 15a^2 b - 10ab^2}{-10ab}$

In Exercises 49–58, simplify.

49. $\dfrac{5x(4x - 2y)}{2y}$

50. $\dfrac{9y^2(x^2 - 3xy)}{3x^2}$

51. $\dfrac{(-2x)^2 + (3x^2)^2}{6x^2}$

52. $\dfrac{(-3x^2 y)^3 + (3xy^2)^3}{27x^3 y^4}$

53. $\dfrac{4x^2 y^2 - 2(x^2 y^2 + xy)}{2xy}$

54. $\dfrac{-5a^3 b - 5a(ab^2 - a^2 b)}{10a^2 b^2}$

55. $\dfrac{(3x - y)(2x - 3y)}{6xy}$

56. $\dfrac{(2m - n)(3m - 2n)}{-3m^2 n^2}$

57. $\dfrac{(a + b)^2 - (a - b)^2}{2ab}$

58. $\dfrac{(x - y)^2 + (x + y)^2}{2x^2y^2}$

APPLICATIONS   *In Exercises 59–64, write an expression involving division to model each problem. Simplify your expression.*

59. Perimeter   The perimeter of a square is $8x + 12$. Find an algebraic expression to represent the length of one side.

60. Perimeter   The perimeter of an equilateral triangle is $6x^2 - 3x + 9$. Find an algebraic expression to represent the length of one side.

61. Area   Find an algebraic expression to represent the width of a rectangle whose length is $4x^2$ if the area is $8x^3 + 2x^2 - 12$.   (*Hint: If $A = lw$, then $w = \frac{A}{l}$.*)

62. Area   Find an algebraic expression to represent the length of a rectangle whose width is $3x^3$ if the area is $6x^4 - 6x^3$. (*Hint: If $A = lw$, then $w = \frac{A}{l}$.*)

63. Volume   The volume of a swimming pool is $18x^4 - 12x^2 + 8$. If the length is $2x$ and the width is $3x^2$, find an algebraic expression to represent the depth. (*Hint: $V = lwh$; $h = \frac{V}{lw}$.*)

64. Volume   The volume of a box of cereal is $6x^3 + 4x^2 + 3$. If the width is x and the depth is 2, find an algebraic expression to represent the height. (*Hint: $V = lwh$; $h = \frac{V}{lw}$.*)

In Exercises 65–67, use the STOre feature of your calculator for the arithmetic computation.

 65. The perimeter of a square is $x^3 - 12x^2$. Is $x = 2$ a valid value of x? Why or why not? Justify your response.

 66. The area of a rectangle is $4x^2 - 2x - 6$. If the length is $2x$, determine if 1 is a valid value of x for the width. Why or why not? Justify your response.

 67. The perimeter of a square is $8x^3 - 12x^2$. Is $x = 3$ a valid value for x? Why or why not? Justify your response.

REVIEW

68. Translate each phrase to an algebraic expression.

   a. 4 less than x

   b. 2 more than three times x

   c. half x increased by 5

69. The polynomial $P(x)$ has been evaluated for $x = 3$ by using the calculator. Based on the calculator screen, **a.** what was the polynomial and **b.** what was the value of the polynomial when $x = 3$?

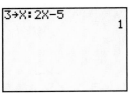

70. If $P(x) = 3x^2 + 5x - 1$ and $Q(x) = 2x^2 - 6x - 3$, find

   a. $P(x) - Q(x)$

   b. $Q(x) - P(x)$

2.7   DIVIDING POLYNOMIALS BY POLYNOMIALS

- DIVIDING A POLYNOMIAL BY A POLYNOMIAL

DIVIDING A POLYNOMIAL BY A POLYNOMIAL

To divide one polynomial by another, we can use a process similar to long division. We will begin the examples using division by a monomial, with the arithmetic division paralleled alongside.

EXAMPLE 1

Divide $6x^3 + 8x^2 - 4x$ by $2x$

SOLUTION

Arithmetic division

Divide 684 by 2.

$$\begin{array}{r} 3 \\ 2\overline{)684} \end{array}$$

How many times does 2 divide 6? $\frac{6}{2} = 3$

$$\begin{array}{r} 3 \\ 2\overline{)684} \\ \underline{6} \end{array}$$

Multiply the divisor, 2, by the partial quotient, 3.

$$\begin{array}{r} 3 \\ 2\overline{)684} \\ \underline{-6} \\ 8 \end{array}$$

Subtract; bring down the next term.

$$\begin{array}{r} 34 \\ 2\overline{)684} \\ \underline{-6} \\ 8 \end{array}$$

How many times does 2 divide 8? $\frac{8}{2} = 4$.
Place a 4 above the 8.

$$\begin{array}{r} 34 \\ 2\overline{)684} \\ \underline{-6} \\ 8 \\ \underline{8} \end{array}$$

Multiply the divisor of 2 by the 4 in the quotient.

$$\begin{array}{r} 34 \\ 2\overline{)684} \\ \underline{-6} \\ 8 \\ \underline{-8} \\ 4 \end{array}$$

Subtract; bring down the next term.

$$\begin{array}{r} 342 \\ 2\overline{)684} \\ \underline{-6} \\ 8 \\ \underline{-8} \\ 4 \end{array}$$

How many times does 2 divide 4? $\frac{4}{2} = 2$.
Place a 2 above the 4.

$$\begin{array}{r} 342 \\ 2\overline{)684} \\ \underline{-6} \\ 8 \\ \underline{-8} \\ 4 \\ \underline{4} \end{array}$$

Multiply 2 by 2.

Polynomial division

Divide $6x^3 + 8x^2 - 4x$ by $2x$

$$\begin{array}{r} 3x^2 \\ 2x\overline{)6x^3 + 8x^2 - 4x} \end{array}$$

How many times does $2x$ divide $6x^3$?
$\dfrac{6x^3}{2x} = 3x^2$

$$\begin{array}{r} 3x^2 \\ 2x\overline{)6x^3 + 8x^2 - 4x} \\ \underline{6x^3} \end{array}$$

Multiply the divisor, $2x$, by the partial quotient, $3x^2$.

$$\begin{array}{r} 3x^2 \\ 2x\overline{)6x^3 + 8x^2 - 4x} \\ \underline{-6x^3} \\ 8x^2 \end{array}$$

Subtract; bring down the next term.

$$\begin{array}{r} 3x^2 + 4x \\ 2x\overline{)6x^3 + 8x^2 - 4x} \\ \underline{-6x^3} \\ 8x^2 \end{array}$$

How many times does $2x$ divide $8x^2$?
$\dfrac{8x^2}{2x} = 4x$. Place $4x$ above the $8x^2$.

Remember: We are dividing *terms*;
the appropriate sign must be placed
between the terms.

$$\begin{array}{r} 3x^2 + 4x \\ 2x\overline{)6x^3 + 8x^2 - 4x} \\ \underline{-6x^3} \\ 8x^2 \\ \underline{8x^2} \end{array}$$

Multiply the divisor of $2x$ by the $4x$
in the quotient.

$$\begin{array}{r} 3x^2 + 4x \\ 2x\overline{)6x^3 + 8x^2 - 4x} \\ \underline{-6x^3} \\ 8x^2 \\ \underline{-8x^2} \\ -4x \end{array}$$

Subtract; bring down the next term.

$$\begin{array}{r} 3x^2 + 4x \\ 2x\overline{)6x^3 + 8x^2 - 4x} \\ \underline{-6x^3} \\ 8x^2 \\ \underline{-8x^2} \\ -4x \end{array}$$

How many times does $2x$ divide $-4x$?
$\dfrac{-4x}{2x} = -2$. Place (-2) above the $-4x$.

$$\begin{array}{r} 3x^2 + 4x - 2 \\ 2x\overline{)6x^3 + 8x^2 - 4x} \\ \underline{-6x^3} \\ 8x^2 \\ \underline{-8x^2} \\ -4x \\ \underline{-4x} \end{array}$$

Multiply $2x$ by (-2).

$$\begin{array}{r} 342 \\ 2\overline{)684} \\ \underline{-6} \\ 8 \\ \underline{-8} \\ 4 \\ \underline{-4} \\ 0 \end{array}$$ Subtract.

$684 \;\div\; 2 \;=\; 342$ The remainder is 0.

dividend $\div$ divisor $=$ quotient

Check by multiplication:

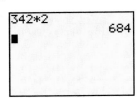

The solution checks.

$$\begin{array}{r} 3x^2 + 4x - 2 \\ 2x\overline{)6x^3 + 8x^2 - 4x} \\ \underline{-6x^3} \\ 8x^2 \\ \underline{-8x^2} \\ -4x \\ \underline{-(-4x)} \\ 0 \end{array}$$ Subtract.

$(6x^3 + 8x^2 - 4x) \div (2x) = 3x^2 + 4x - 2$ The remainder is 0.

dividend $\div$ divisor $=$ quotient

Check by multiplication:

$$2x(3x^2 + 4x - 2) = 2x(3x^2) + 2x(4x) - 2x(2)$$
$$= 6x^3 + 8x^2 - 4x$$

The solution checks.

You should make note of the pattern that occurs: divide, multiply, subtract, bring down the next term. We will now extend the concept to the division of a polynomial by a polynomial.

EXAMPLE 2

Divide $x^2 + 5x + 6$ by $x + 2$

SOLUTION   **Step 1:**

$$\begin{array}{r} x \\ x + 2\overline{)x^2 + 5x + 6} \end{array}$$

How many times does x divide x^2? $\frac{x^2}{x} = x$. Place the x above the division symbol.

Step 2:

$$\begin{array}{r} x \\ x + 2\overline{)x^2 + 5x + 6} \\ x^2 + 2x \end{array}$$

Multiply each term in the divisor by x. Place the product under $x^2 + 5x$ as indicated and draw a line.

Step 3:

$$\begin{array}{r} x \\ x + 2\overline{)x^2 + 5x + 6} \\ \underline{x^2 + 2x} \\ 3x + 6 \end{array}$$

Subtract $x^2 + 2x$ from $x^2 + 5x$ by adding the negative of $x^2 + 2x$ to $x^2 + 5x$.

Bring down the next term.

Step 4:

$$\begin{array}{r} x + 3 \\ x + 2\overline{)x^2 + 5x + 6} \\ \underline{x^2 + 2x} \\ 3x + 6 \end{array}$$

How many times does x divide $3x$? $\frac{3x}{x} = +3$. Place the $+3$ above the division symbol.

Step 5:

$$\begin{array}{r} x \ + 3 \\ x + 2 \overline{)\smash{)}\, x^2 + 5x + 6} \\ \underline{x^2 + 2x} \\ 3x + 6 \\ \underline{3x + 6} \end{array}$$

Multiply each term in the divisor by 3. Place the product under the $3x + 6$ as indicated and draw a line.

Step 6:

$$\begin{array}{r} x \ + 3 \\ x + 2 \overline{)\smash{)}\, x^2 + 5x + 6} \\ \underline{x^2 + 2x} \\ 3x + 6 \\ \underline{3x + 6} \\ 0 \end{array}$$

Subtract $3x + 6$ from $3x + 6$ by adding the negative of $3x + 6$.

The quotient is $x + 3$ and the remainder is 0.

Step 7: Check the work by verifying that the product of $x + 2$ and $x + 3$ is $x^2 + 5x + 6$.

$$(x + 2)(x + 3) = x^2 + 3x + 2x + 6$$
$$= x^2 + 5x + 6$$

The answer checks.

EXAMPLE 3

Divide: $\dfrac{6x^2 - 7x - 2}{2x - 1}$

SOLUTION   **Step 1:**

$$\begin{array}{r} 3x \\ 2x - 1 \overline{)\smash{)}\, 6x^2 - 7x - 2} \end{array}$$

How many times does $2x$ divide $6x^2$? $\frac{6x^2}{2x} = 3x$. Place the $3x$ above the division symbol.

Step 2:

$$\begin{array}{r} 3x \\ 2x - 1 \overline{)\smash{)}\, 6x^2 - 7x - 2} \\ \underline{6x^2 - 3x} \end{array}$$

Multiply each term in the divisor by $3x$. Place the product under $6x^2 - 7x$ as indicated and draw a line.

Step 3:

$$\begin{array}{r} 3x \\ 2x - 1 \overline{)\smash{)}\, 6x^2 - 7x - 2} \\ \underline{6x^2 - 3x} \\ - 4x - 2 \end{array}$$

Subtract $6x^2 - 3x$ from $6x^2 - 7x$ by adding the negative of $6x^2 - 3x$ to $6x^2 - 7x$.

Bring down the next term.

Step 4:

$$
\begin{array}{r}
3x - 2 \\
2x - 1{\overline{\smash{\big)}\,6x^2 - 7x - 2}} \\
\underline{6x^2 - 3x} \\
-\,4x - 2
\end{array}
$$

How many times does $2x$ divide $-4x$? $\frac{-4x}{2x} = -2$.
Place the -2 above the division symbol.

Step 5:

$$
\begin{array}{r}
3x - 2 \\
2x - 1{\overline{\smash{\big)}\,6x^2 - 7x - 2}} \\
\underline{6x^2 - 3x} \\
-\,4x - 2 \\
\underline{-\,4x + 2}
\end{array}
$$

Multiply each term in the divisor by -2. Place the product under the $-4x - 2$ as indicated and draw a line.

Step 6:

$$
\begin{array}{r}
3x - 2 \\
2x - 1{\overline{\smash{\big)}\,6x^2 - 7x - 2}} \\
\underline{6x^2 - 3x} \\
-\,4x - 2 \\
\underline{-\,4x + 2} \\
-\,4
\end{array}
$$

Subtract $-4x + 2$ from $-4x - 2$ by adding the negative of $-4x + 2$.

In this example, the quotient is $3x - 2$ and the remainder is -4. It is common to write the answer in quotient $+\ \frac{\text{remainder}}{\text{divisor}}$ form:

$$
3x - 2 + \frac{-4}{2x - 1}
$$

where the fraction $\dfrac{-4}{2x - 1}$ is formed by dividing the remainder by the divisor.

Step 7: To check the answer, we multiply $3x - 2 + \dfrac{-4}{2x - 1}$ by $2x - 1$. The product should be the dividend.

$$
\begin{aligned}
(2x - 1)\left(3x - 2 + \frac{-4}{2x - 1}\right) &= (2x - 1)(3x - 2) + (2x - 1)\left(\frac{-4}{2x - 1}\right) \\
&= (2x - 1)(3x - 2) - 4 \\
&= 6x^2 - 4x - 3x + 2 - 4 \\
&= 6x^2 - 7x - 2
\end{aligned}
$$

Because the result is the dividend, the answer checks.

EXAMPLE 4

Divide $4x^2 + 2x^3 + 12 - 2x$ by $x + 3$

SOLUTION   The division process works most efficiently if the terms of both the divisor and the dividend are written in descending order. This means that the term involving the highest power of x appears first, the term involving the second-highest power of x appears second, and so on. We can use the commutative property to rearrange the terms of the dividend in descending order and divide as follows:

$$
\begin{array}{r}
2x^2 - 2x + 4 \\
x + 3\overline{)2x^3 + 4x^2 - 2x + 12} \\
\underline{2x^3 + 6x^2} \\
-2x^2 - 2x \\
\underline{-2x^2 - 6x} \\
+4x + 12 \\
\underline{+4x + 12}
\end{array}
$$

Check: $(x + 3)(2x^2 - 2x + 4) = 2x^3 - 2x^2 + 4x + 6x^2 - 6x + 12$
$$= 2x^3 + 4x^2 - 2x + 12$$

EXAMPLE 5

Divide $\dfrac{x^2 - 4}{x + 2}$

SOLUTION   Since the binomial $x^2 - 4$ does not have a term involving x, we must either include the term $0x$ or leave a space for it. After this adjustment, the division is routine.

$$
\begin{array}{r}
x - 2 \\
x + 2\overline{)x^2 + 0x - 4} \\
\underline{x^2 + 2x} \\
-2x - 4 \\
\underline{-2x - 4}
\end{array}
$$

Check: $(x + 2)(x - 2) = x^2 - 2x + 2x - 4$
$$= x^2 - 4$$

EXAMPLE 6

A fish tank has a volume of $x^3 + 3x^2 + 2x$ cubic feet. Find an algebraic expression for the length if the width is $x + 1$ feet and the height is x feet. (*Hint:* The formula for volume is $V = lwh$. It follows that $l = \frac{v}{wh}$.)

length

Figure 2-2

SOLUTION

$$\dfrac{v}{wh} = \dfrac{x^3 + 3x^2 + 2x}{(x + 1)(x)}$$ Substitution

$$= \dfrac{x^3 + 3x^2 + 2x}{x^2 + x}$$ Simplify the denominator.

$$
= x^2 + x\overline{)x^3 + 3x^2 + 2x}
$$
$$
= \underline{x^3 + x^2}
$$
$$
= 2x^2 + 2x
$$
$$
= \underline{2x^2 + 2x}
$$
$$
0
$$

The length can be expressed as $x + 2$ feet.

EXERCISE 2.7

CONCEPTS

1. Divide each of the following:

   a. $\dfrac{6x^3}{2x}$

   b. $\dfrac{8x^2}{2x}$

   c. $\dfrac{-4x}{2x}$

   d. $\dfrac{6x^3 + 8x^2 - 4x}{2x}$

2. Arrange the terms in descending order by the powers of x.

   a. $11x + 10x^2 + 3$

   b. $-x - 21 + 2x^2$

   c. $xy - 2y^2 + 6x^2$

   d. $-10y^2 + 13xy + 3x^2$

3. Fill in the missing terms using 0 coefficients.

   a. $x^2 - 9$

   b. $x^3 - 8$

PRACTICE

4. Divide $x^2 + 4x + 4$ by $x + 2$

5. Divide $x^2 - 5x + 6$ by $x - 2$

6. Divide $y^2 + 13y + 12$ by $y + 1$

7. Divide $z^2 - 7z + 12$ by $z - 3$

8. Divide $a^2 + 2ab + b^2$ by $a + b$

9. Divide $a^2 - 2ab + b^2$ by $a - b$

In Exercises 10–15, divide.

10. $\dfrac{6a^2 + 5a - 6}{2a + 3}$

11. $\dfrac{8a^2 + 2a - 3}{2a - 1}$

12. $\dfrac{3b^2 + 11b + 6}{3b + 2}$

13. $\dfrac{3b^2 - 5b + 2}{3b - 2}$

14. $\dfrac{2x^2 - 7xy + 3y^2}{2x - y}$

15. $\dfrac{3x^2 + 5xy - 2y^2}{x + 2y}$

In Exercises 16–24, rearrange the terms so that the powers of x are in descending order, then divide.

16. $5x + 3 \overline{)11x + 10x^2 + 3}$

17. $2x - 7 \overline{)-x - 21 + 2x^2}$

18. $4 + 2x \overline{)-10x - 28 + 2x^2}$

19. $1 + 3x \overline{)9x^2 + 1 + 6x}$

20. $2x - y \overline{)xy - 2y^2 + 6x^2}$

21. $2y + x \overline{)3xy + 2x^2 - 2y^2}$

22. $3x - 2y \overline{)-10y^2 + 13xy + 3x^2}$

23. $2x + 3y \overline{)-12y^2 + 10x^2 + 7xy}$

24. $4x + y \overline{)-19xy + 4x^2 - 5y^2}$

In Exercises 25–30, divide.

25. $2x + 3 \overline{)2x^3 + 7x^2 + 4x - 3}$

26. $2x - 1 \overline{)2x^3 - 3x^2 + 5x - 2}$

27. $3x + 2 \overline{)6x^3 + 10x^2 + 7x + 2}$

28. $4x + 3 \overline{)4x^3 - 5x^2 - 2x + 3}$

29. $2x + y \overline{)2x^3 + 3x^2y + 3xy^2 + y^3}$

30. $3x - 2y \overline{)6x^3 - x^2y + 4xy^2 - 4y^3}$

In Exercises 31–40, divide. If there is a remainder, leave the answer in quotient + $\frac{remainder}{divisor}$ form.

31. $\dfrac{2x^2 + 5x + 2}{2x + 3}$

32. $\dfrac{3x^2 - 8x + 3}{3x - 2}$

33. $\dfrac{4x^2 + 6x - 1}{2x + 1}$

34. $\dfrac{6x^2 - 11x + 2}{3x - 1}$

35. $\dfrac{x^3 + 3x^2 + 3x + 1}{x + 1}$

36. $\dfrac{x^3 + 6x^2 + 12x + 8}{x + 2}$

37. $\dfrac{2x^3 + 7x^2 + 4x + 3}{2x + 3}$

38. $\dfrac{6x^3 + x^2 + 2x + 1}{3x - 1}$

39. $\dfrac{2x^3 + 4x^2 - 2x + 3}{x - 2}$

40. $\dfrac{3y^3 - 4y^2 + 2y + 3}{y + 3}$

In Exercises 41–50, divide.

41. $\dfrac{x^2 - 1}{x - 1}$

42. $\dfrac{x^2 - 9}{x + 3}$

43. $\dfrac{4x^2 - 9}{2x + 3}$

44. $\dfrac{25x^2 - 16}{5x - 4}$

45. $\dfrac{x^3 + 1}{x + 1}$

46. $\dfrac{x^3 - 8}{x - 2}$

47. $\dfrac{a^3 + a}{a + 3}$

48. $\dfrac{y^3 - 50}{y - 5}$

49. $3x - 4\overline{)15x^3 - 23x^2 + 16x}$

50. $2y + 3\overline{)21y^2 + 6y^3 - 20}$

APPLICATIONS *In Exercises 51–58, write an expression that includes polynomial division to model each of the problems below. Compute the division to respond to the question.*

51. Perimeter The perimeter of a square is $4x^2 - 20x + 16$. What algebraic expression represents the length of each side?

52. Using the STOre feature of your calculator, determine if $x = 2$ is a valid value for the variable in Exercise 51.

53. Perimeter The perimeter of an equilateral triangle is $27x^3 - 6x^2 + 9$. What algebraic expression represents the length of a side?

54. Using the STOre feature of your calculator, determine if $x = 2$ is a valid value for the variable in Exercise 53.

55. Perimeter a. What algebraic expression represents the width of a rectangle whose length is $x + 5$ and whose area is $x^2 + 12x + 35$? (*Hint:* If $A = lw$, then $w = \frac{A}{l}$.) **b.** What is the perimeter?

56. Perimeter a. Find the length of a rectangular garden whose area is $3x^2 - x - 10$ and whose width is $x - 2$. (*Hint:* If $A = lw$, then $w = \frac{A}{l}$.) **b.** What is the perimeter?

57. Volume The volume occupied by a swimming pool is $12x^6 + 54x^5 + 60x^4$ cu. ft. Find the depth if the width is $6x^4$ and the length is $2x + 5$. (*Hint:* If $V = lwh$, then $h = \frac{V}{lw}$.)

58. Volume The volume of cereal occupied by the box in Illustration 1 is $12x^3 + 2x^2 - 2x$. Find the width of the box. (*Hint:* If $V = lwh$, then $h = \frac{V}{lw}$.)

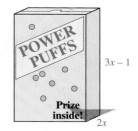

Illustration 1

REVIEW

59. If the dimensions of a crate are $2x - 3$ units high by $x + 2$ units wide by $3x$ units long, find the polynomial expression that represents the volume.

60. Use the STOre feature of your calculator to compute the volume of the crate in Exercise 59 if $x = 2.5$ ft.

61. Simplify:

a. $(4x^2)^3$

b. $\dfrac{4x^3y}{8x^2y^2}$

c. $\dfrac{(2x)^2}{(6x)^3}$

62. The dimensions of the crate in Exercise 59 are $2x - 3$, $x + 2$, and $3x$. Translate each of these dimensions to an English phrase.

63. Translate each English phrase to an algebraic expression and then find the sum of the expressions: 4 more than x, twice x decreased by 5, 4 times x.

CHAPTER SUMMARY

CONCEPTS

REVIEW EXERCISES

Section 2.1

Properties of exponents. If n is a natural number,

$$\text{then } x^n = \overbrace{x \cdot x \cdot x \cdot \ldots \cdot x}^{n \text{ factors of } x}$$

If m and n are integers, then

$x^m x^n = x^{m+n}$
$(x^m)^n = x^{mn}$
$(xy)^n = x^n y^n$
$\left(\dfrac{x}{y}\right)^n = \dfrac{x^n}{y^n}$ provided $y \neq 0$
$\dfrac{x^m}{x^n} = x^{m-n}$ provided $x \neq 0$
$x^0 = 1$ provided $x \neq 0$

Remember that most graphing calculators cannot perform symbolic operations. Expressions entered with variables will be evaluated, not simplified, at whatever value is stored for that specific variable.

Whole Number Exponents

1. Simplify each expression. Confirm results with the calculator, copying the screen display to justify your work.

   a. 5^3
   b. 3^5
   c. $(-8)^2$
   d. -8^2
   e. $3^2 + 2^2$
   f. $(3+2)^2$
   g. $3(3^3 + 3^3)$
   h. $1^{17} + 17^1$

2. Simplify each expression.

   a. $x^3 x^2$
   b. $x(x^2 y)$
   c. $y^7 y^3$
   d. $x^0 y^5$
   e. $2b^3 b^4 b^5$
   f. $(-z^2)(z^3 y^2)$
   g. $(4^4 x)s^2$
   h. $-3y(y^5)$

3. Simplify each expression.

   a. $(x^2 x^3)^3$
   b. $(2x^2 y)^2$
   c. $(3x^0)^2$
   d. $(3x^2 y^2)^0$
   e. $\dfrac{x^7}{x^3}$
   f. $\left(\dfrac{x^2 y^3}{xy^2}\right)^2$
   g. $\dfrac{8(y^2 x)^2}{2^3(yx)^2}$
   h. $\dfrac{(5x^0 y^2 z^3)^3}{25(yz)^5}$

Section 2.2

If x is any nonzero number and n is a natural number, then

$$x^{-n} = \dfrac{1}{x^n}.$$

Negative Integer Exponents and Scientific Notation

4. Simplify each expression, leaving no negative exponents or parentheses.

   a. $(-z^2)(z^3 y^2)$
   b. $(4x)x^2$
   c. $(x^2 x^3)^3$
   d. $(2x^2 y)^2$

5. Simplify each expression, leaving no negative exponents or parentheses.

   a. $\left(\dfrac{x^2 y}{xy^2}\right)^2$
   b. $\left(\dfrac{3x}{6x^2}\right)^3$
   c. $x^{-2} x^3$
   d. $y^4 y^{-3}$

6. Simplify each expression, leaving no negative exponents or parentheses.

   a. $\dfrac{x^3}{x^{-7}}$
   b. $(x^{-3} x^{-4})^{-2}$
   c. $\left(\dfrac{x^2}{x}\right)^{-5}$
   d. $\left(\dfrac{15x^4}{5y^3}\right)^{-2}$

A number is written in scientific notation if it is written as the product of a number between 1 (including 1) and 10 and an integer power of 10.

7. Write each number in scientific notation.

   a. 728   **b.** 0.00942

   c. 0.018×10^{-2} **d.** 600×10^{-3}

8. Write each number in standard notation.

   a. 7.26×10^5 **b.** 3.91×10^{-4} **c.** $\dfrac{(4800)(20,000)}{600,000}$

9. The calculator displays

$4E-7$

Write this number in standard notation.

10. The calculator displays

$2.1721E3$

Write this number in standard notation.

Section 2.3

Algebraic Expressions and Polynomials

The degree of a monomial with several variables is the sum of the exponents of those variables.

The degree of a polynomial is the same as the degree of its term with the highest degree.

$P(x)$ is read "P of x" and defines a polynomial P in a variable x.

11. Give the degree of each polynomial below and classify it as a monomial, a binomial, or a trinomial.

   a. $13x^7$ **b.** $5^3x + x^2$

   c. $-3x^5 + x - 1$ **d.** $9x + 21x^3$

12. Let $P(x) = 5x^4 - x$. Find each value. Confirm your result with the calculator, copying the screen display exactly as it appears to justify your results.

   a. $P(3)$ **b.** $P(0)$

   c. $P(-2)$ **d.** $P(2t)$

13. Let a, b, and c represent three real numbers. Write an algebraic expression to represent each quantity.

   a. a decreased by c **b.** the sum of b and c

   c. the product of a and b

14. Investing George has $10,000 to invest. If he invests b dollars in a money market account, how much money is left to invest in stocks?

15. Mileage A car averages 55 miles an hour on a trip. Write expressions to represent the number of miles traveled in 2 hours, 5 hours, and b hours.

Section 2.4

When adding polynomials, remove parentheses and combine like terms by adding or subtracting the numerical coefficients and using the same variables and the same exponents.

When subtracting polynomials, be careful to distribute the operation of subtraction when removing the parentheses.

Adding and Subtracting Polynomials

16. Simplify each expression.

    a. $3x + 5x - x$ **b.** $3x + 2y$

    c. $(xy)^2 + 3x^2y^2$ **d.** $-2x^2yz + 3yx^2z$

    e. $3x^2y^0 + 2x^2$ **f.** $2(x + 7) + 3(x + 7)$

    g. $(7a^2 + 2a - 5) - (3a^2 - 2a + 1)$

    h. $4(4x^3 + 2x^2 - 3x - 8) - 5(2x^3 - 3x + 8)$

17. Find the polynomial that represents the perimeter of the pictured figure.

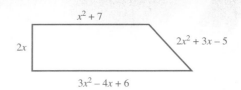

18. Find the difference when $4x^2 - 5x$ is subtracted from the sum of $3x^2 + 2x - 9$ and $5x^2 + 6x + 1$.

Section 2.5

To multiply two monomials, first multiply the numerical factors and then multiply the variable factors.

To multiply a polynomial with more than one term by a monomial, multiply each term of the polynomial by the monomial and simplify.

To multiply one polynomial by another, multiply each term of one polynomial by each term of the other polynomial and simplify.

To multiply two binomials, use the FOIL method, if desired.

Special products:
$(x + y)^2 = x^2 + 2xy + y^2$
$(x - y)^2 = x^2 - 2xy + y^2$
$(x + y)(x - y) = x^2 - y^2$

Multiplying Polynomials

19. Find each product.

    a. $(2x^2y^3)(5xy^2)$ **b.** $5(x + 3)$

    c. $x^2(3x^2 - 5)$ **d.** $-3xy(xy - x)$

    e. $(x + 3)(x + 2)$

20. Find each product.

    a. $(3a - 3)(2a + 2)$ **b.** $(a - 1)(a + 1)$

    c. $6(y - 2)(y + 2)$ **d.** $(x + 4)^2$

    e. $(x - 3)^2$

21. Find the polynomial that represents the area of the pictured rectangle.

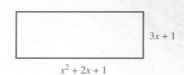

22. Find the product of $y^2 + 1$ and its conjugate.

Section 2.6

To simplify a fraction, divide out all factors common to the numerator and the denominator of the fraction.

To divide a polynomial by a monomial, divide each term of the polynomial by the monomial.

Dividing Polynomials by Monomials

23. Divide each polynomial by the given monomial.

a. $\dfrac{3x + 6y}{2xy}$

b. $\dfrac{14xy - 21x}{7xy}$

c. $\dfrac{15a^2bc + 20ab^2c - 25abc^2}{-5abc}$

d. $\dfrac{(x + y)^2 + (x - y)^2}{-2xy}$

Section 2.7

Use long division to divide one polynomial by another.

Dividing Polynomials by Polynomials

24. Perform the divisions.

a. $x - 1\overline{)x^2 - 6x + 5}$

b. $x + 2\overline{)x^2 + 3x + 5}$

c. $3x - 1\overline{)3x^2 + 14x - 2}$

d. $3x + 1\overline{)-13x - 4 + 9x^3}$

CHAPTER TEST

1. Use exponents to rewrite $2\,x\,x\,x\,x\,y\,y\,y\,y$

In Problems 2–5, simplify each expression. Write results with positive exponents.

2. $y^2(y\,y^3)$

3. $(2x^3)^5(x^{-2})^3$

4. $(2r\,r^2\,r^3)^{-3}$

5. a. $3x^0$ **b.** $(3x)^0$

6. Identify $3x^2 + 2$ as a monomial, a binomial, or a trinomial.

7. Find the degree of the polynomial
$$3x^2y^3z^4 + 2x^3y^2z - 5x^2y^3z^5$$

8. If $P(x) = x^2 + x - 2$, find $P(-2)$.

9. Simplify: $(xy)^2 + 5x^2y^2 - (3x)^2y^2$.

10. Add: $(3x^3 + 4x^2 - x - 7) + (2x^3 - 2x^2 + 3x + 2)$.

11. Subtract: $(2x^2 - 7x + 3) - (3x^2 - 2x - 1)$.

12. Simplify: $-6(x - y) + 2(x + y) - 3(x + 2y)$.

13. Simplify: $-2(x^2 + 3x - 1) - 3(x^2 - x + 2) + 5(x^2 + 2)$.

In Problems 14–17, find each product.

14. $(-2x^3)(2x^2y)$

15. $3y^2(y^2 - 2y + 3)$

16. $(2x - 5)(3x + 4)$

17. $(2x - 3)(x^2 - 2x + 4)$

18. Simplify the fraction: $\dfrac{8x^2y^3z^4}{16x^3y^2z^4}$.

19. Do the division: $\dfrac{6a^2 - 12b^2}{24ab}$.

20. Divide: $2x + 3\overline{)2x^2 - x - 6}$.

CUMULATIVE REVIEW

CHAPTERS 1 AND 2
Vocabulary / Concepts

1. A number greater than one, whose only factors are one and itself, is called a _____ number.

2. Explain, in a step-by-step format, how to find the least common denominator for two fractions.

   Following your step-by-step directions, find the least common denominator for $\frac{5}{24}$ and $\frac{7}{18}$.

3. Explain the difference between the expressions $4x$ and x^4.

4. Explain the difference between *evaluate* and *simplify* when applied to the expression $3x + 5x$.

5. Determine the absolute value of each expression and verify with your calculator.

   a. $|-3|$ b. $-|-3|$ c. $-|3|$

6. What rule supports the statement $-(-5) = 5$?

7. Explain why $\frac{x}{0}$ is undefined and $\frac{0}{x} = 0$ provided $x \neq 0$.

8. Which of the following represent a product?

   a. $(-4) - (4)$ b. $5(-3)$

   c. $-4(-3)$ d. $6 - (-6)$

9. List the numbers in the set $\{-5, -\frac{7}{2}, -1, 0, \sqrt{2}, 4.2, 6\}$ that are

   a. whole numbers

   b. integers

   c. rational numbers

   d. irrational numbers

   e. nonnegative integers

10. Determine whether perimeter, area, or volume would be used to determine the measure of the indicated quantity.

    a. molding for a window frame

    b. water in a hot tub

    c. sand on a barge

    d. carpeting for a living room

11. Complete the Laws of Exponents. Assume m and n are natural numbers.

    a. $x^m \cdot x^n =$ _____   b. $(x^m)^n =$ _____

    c. $(xy)^n =$ _____   d. $\left(\dfrac{x}{y}\right)^n =$ _____ $(y \neq 0)$

    e. $\dfrac{x^m}{x^n} =$ _____ $(x \neq 0)$

    f. $x^0 =$ ___ provided $x \neq 0$.

12. In the expression $3x^5$, identify the base and the exponent.

13. In the expression $3x^5$, 3 is called the _____ _____ of the term.

14. Polynomials consisting of two terms are called _____.

15. If George lives 15 blocks from his office in downtown Chicago and Kevin lives x blocks further from the same office building, write an expression to describe how far Kevin lives from the office building.

16. If $4(x + y) = 4x + 4y$, then the _____ property has been applied.

17. Which of the following is equal to $(x + 2)^3$?

    a. $x^3 + 8$ b. $(x + 2)(x + 2)(x + 2)$

18. The binomials $(4 + x)$ and $(4 - x)$ are called _____ binomials and have a product of _____.

19. Divide by using the distributive property: $\dfrac{4x + 12}{3}$.

20. Arrange in descending order by the powers of x: $4 - 3x - x^2$

Practice *In Exercises 21–28, perform the indicated operation and verify with your calculator.*

21. $\dfrac{5}{6} \cdot \dfrac{3}{20}$

22. $\dfrac{2}{3}\left(\dfrac{4}{5} \div \dfrac{28}{35}\right)$

23. $5^3 - (5 - 1)^3$

24. $4 + 2 \cdot 3 - [6 - (5 - 1)]$

25. $-|8 - 12| - 5$

26. $-\dfrac{7}{9} - \dfrac{7}{3}$

27. $\dfrac{-4 - 11}{4 - 7}$

28. $\dfrac{3(2^2) - 3}{-2^3 - 1}$

29. Arithmetically evaluate $zx + x^2 - z$ for $x = 3$ and $z = 2$ and confirm with the STOre feature of your calculator.

30. Arithmetically evaluate $\dfrac{x^2 - y}{x + y}$ for $x = -2$ and $y = -3$ and confirm with the STOre feature of your calculator.

In Exercises 31–32, write an expression involving only one positive exponent.

31. a. $-3x(x^4)$ **b.** $\left(\dfrac{x^8}{x^2}\right)^3$

32. a. $(y^{-2})^3$ **b.** $\left(\dfrac{x^{-2}x^3}{x^{-4}x^6}\right)^{-2}$

33. Translate each English phrase to an algebraic expression.

a. the product of x and y

b. twice x increased by 4

34. If $P(x) = 3x^2 - 4x + 2$, find

a. the degree of $P(x)$, and

b. $P(-2)$. Verify with your calculator.

35. Simplify: $(4x^2 - 2x + 6) - 2(x^2 - 2x - 3)$.

36. Find the difference when $2x + 4y$ is subtracted from the sum of $3x + 6y$ and $x - 4y$.

37. Multiply: $(3x + 5)(x^2 - 3x + 1)$.

38. Simplify: $(x + 2)^2 - (x + 3)^2$

39. Divide: $\dfrac{-30x^4 + 25xy^2 - 5xy}{-5xy}$.

40. Divide: $\dfrac{8x^3 - 6x^2 + 3x - 2}{2x + 1}$. If there is a remainder, express the answer in the form: *quotient* $+ \frac{remainder}{divisor}$.

41. Use the STOre feature to evaluate $(3x^2y)^2$ when $x = -3$ and $y = \frac{1}{3}$.

42. If $P(x) = -4x^2 - 5x + 2$, circle the line in the TABLE that indicates the value of $P(-1)$.

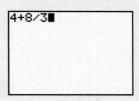

43. You are performing the computation $\dfrac{4 + 8}{3}$. Make the appropriate corrections on the displayed screen to ensure the correct result.

44. Simplify with your calculator. Copy your complete screen display to justify your result.
$10^2 + [10^3 \cdot (18 \div 2) - 3]$

45. Simplify with your calculator. Copy your complete screen display to justify your result.

$\dfrac{|4.02 - 0.001|}{|0.003 - 0.01|}$

46. An arithmetic expression is displayed on the screen at the right. What would this expression look like if written in this textbook?

Applications

47. Course enrollment   In the fall semester, 24 students were enrolled in the 12:30 P.M. Tuesday-Thursday Intermediate Algebra Course. Five-sixths of these students graduated from high school the previous spring.

  a. Write an arithmetic expression to determine how many students in the course graduated from high school in the previous spring. Simplify your expression.

  b. Of this group of recent high school graduates, $\frac{3}{4}$ had taken Algebra I and Algebra II in high school. Write an arithmetic expression to determine how many students had not taken Algebra I and Algebra II. Simplify your expression.

48. Telephone solicitor   A telephone solicitor earns $2.50 per hour plus 10 cents per call. Write an arithmetic expression to determine how much money is earned by the solicitor in one hour if 20 calls are made. Simplify your expression.

49. Room dimensions   Find the length of a room whose width is $x + 4$ and whose area is $x^3 + 5x^2 + 6x + 8$.

50. Grade distribution   At the end of the semester an instructor's grade distributions are as indicated in the chart below. Find an expression to model the total number of grades and thus the total number of students. Express your result as a simplified polynomial expression.

As	x grades
Bs	double the number of As
Cs	3 times as many As
Ds	4 more than the As
Fs	1 more than the As
Ws	2 less than the As

CAREERS & MATHEMATICS

Photodisk Collection/Getty Images

Actuary

Why do young people pay more for automobile insurance than older persons? How much should an insurance policy cost? How much should an organization contribute each year to its pension fund? Answers to these and similar questions are provided by actuaries, who design insurance and pension plans and follow their performance to make sure that the plans are maintained on a sound financial basis.

Qualifications

A good educational background for a beginning job in a large life insurance or casualty insurance company is a bachelor's degree in mathematics or statistics; a degree in actuarial science is even better. Companies prefer well-rounded individuals with a liberal arts background, including social science and communications courses, as well as a good technical background.

Job Outlook

According to the U.S. Department of Labor, Bureau of Labor Statistics, the outlook for employment for actuaries through the year 2010 is expected to grow more slowly than the average for all occupations. New employment opportunities should become available in medical and health insurance industries and there will continue to be a need for actuaries to evaluate risks associated with catastrophes such as tornadoes, floods, and other natural disasters. Job opportunities will be best for new college graduates who have passed at least two actuarial examinations while still in school and who have strong mathematical and statistical backgrounds.

Example Application 1

An **annuity** is a sequence of equal payments made periodically over a length of time. The sum of the payments and the interest earned during the **term** of the annuity is called the **amount** of the annuity.

After a sales clerk works 6 months, her employer will begin an annuity for her and will contribute $500 semiannually to a fund that pays 8% annual interest. After she has been employed two years, what will be the amount of her annuity?

Solution

Because the payments are to be made semiannually, there will be four payments of $500, each earning a rate of 4% per 6-month period. These payments will occur at the end of 6 months, 12 months, 18 months, and 24 months. The first payment, to be made after 6 months, will earn interest for three interest periods. Thus, the amount of the first payment after two years is $500(1.04)^3$. The amounts of each of the four payments after two years are shown in Table 1.

TABLE 1

Payment (at the end of period)	Amount of Payment at the End of 2 Years
1	$500(1.04)^3 = \$$ 562.43
2	$500(1.04)^2 = \$$ 540.80
3	$500(1.04)^1 = \$$ 520.00
4	$500 = \$$ 500.00
	$A_n = \$2{,}123.23$

The amount of the annuity is the sum of the amounts of the individual payments. This sum is $2,123.23.

Example Application 2   The **present value** of the above annuity is the lump-sum principal that must be invested on the first day of her employment at 8% annual interest, compounded semiannually, to grow to $2,123.23 at the end of two years. Find the present value of the annuity.

Solution   A sum of $500(1.04)^{-1}$ invested at 4% interest per 6-month period will grow to be the first payment of $500, to be made after 6 months. A sum of $500(1.04)^{-2}$ will grow to be the second payment of $500, to be made after 12 months. The present values of each of the four payments are shown in Table 2.

TABLE 2

Payment	Present Value of Each Payment
1	$500(1.04)^{-1} = \$\ \ 480.77$
2	$500(1.04)^{-2} = \$\ \ 462.28$
3	$500(1.04)^{-3} = \$\ \ 444.50$
4	$500(1.04)^{-4} = \$\ \ 427.40$
	Present value $= \$1,814.95$

The present value of the annuity is $1,814.95. Verify that $1,814.95 will grow to $2,123.23 in two years at 8% annual interest, compounded semiannually.

EXERCISES

1. Find the amount of an annuity if $1,000 is paid semiannually for two years at 6% annual interest. Assume that the first of the four payments is made immediately.

2. Find the present value of the annuity of Exercise 1.

3. Note that the amounts for the payments in Table 1 form a geometric progression. Verify the answer for the first example by using the formula for the sum of a geometric progression.

4. Note that the present values for the payments in Table 2 form a geometric progression. Verify the answer for the second example by using the formula for the sum of a geometric progression.

Solutions of Equations in One and Two Variables

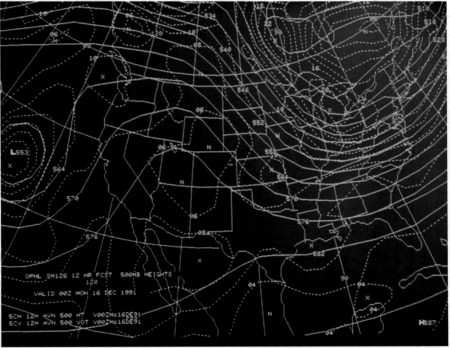

© Royalty-Free/CORBIS

 InfoTrac Project

Do a subject guide search on "crime and weather" and find a periodical article titled "The effect of temperature on crime." Write a summary of the article.

Use the following information to develop an equation relating temperature and the number of burglaries. When the high for the day was 78°, 24 burglaries were reported. The high for a second day was 96° and 60 burglaries were reported. Use your equation to predict the number of burglaries for a temperature of 85°. What information does the slope of your equation give you? If such an equation could be developed, how might this type of information be useful to the government of a town?

Complete this project after studying Section 3.4.

PERSPECTIVE

The concept of equations has a long history, and the techniques we will study in this chapter have been developed over many centuries.

The mathematical notation that we use today is the result of thousands of years of development. The ancient Egyptians used a word for variables, best translated as *heap*. Others used the word *res*, which is Latin for *thing*. In the 15th century, the letters *p:* and *m:* were used for *plus* and *minus*. What we would now write as $2x + 3 = 5$ might have appeared to those early mathematicians as

> 2 *res p:* 3 *aequalis* 5

Chapter 1 focused on the simplification of arithmetic expressions and the evaluation of algebraic expressions. When specified values are substituted for respective variables, an algebraic expression is converted to an arithmetic expression. The order of operations must be followed to correctly simplify/evaluate these expressions.

Chapter 2 focused on the simplification of algebraic expressions by removing grouping symbols and combining like terms. Properties of exponents were also introduced and used.

This chapter will focus on equations and their solutions. The main focus is on algebraic solutions of linear equations in one variable and absolute-value equations. However, equations in two variables, their graphs, and the basic concept of a function also will be introduced.

Remember, you are building a logic system — **expressions** may be **simplified** or **evaluated** and **equations** are **solved**. **Equations** may also be **graphed** to provide a visual mathematical model of given information. You will be writing equations to provide an algebraic model of given information.

3.1 INTRODUCTION TO LINEAR EQUATIONS IN ONE VARIABLE

- EQUATIONS AND THEIR SOLUTIONS
- PROPERTIES OF EQUALITY
- ALGEBRAIC SOLUTIONS OF LINEAR EQUATIONS IN ONE VARIABLE

- TYPES OF EQUATIONS AND THEIR SOLUTIONS
- INTRODUCTION TO PROBLEM SOLVING

EQUATIONS AND THEIR SOLUTIONS

An **equation** is a statement indicating that two expressions are equal. These are a combination of algebraic and/or arithmetic expressions separated by an equal sign. Below are some examples of equations.

$$x + 5 = 21 \qquad 3(3x + 4) = 3 + x \qquad 3x^2 - 4x + 5 = 0$$

In the equation $x + 5 = 21$, $x + 5$ is called the **left-hand side**, and 21 is called the **right-hand side**. The letter x is called a **variable**. The initial discussion will focus on **linear equations in one variable**. A linear equation in one variable, x, is any equation that *can be* written in the form $ax + c = 0$, where a and c are real numbers and $a \neq 0$.

An equation can be true or false. It is true when the left side is equivalent to the right side. The equation $16 + 5 = 21$ is true, whereas the equation $10 + 5 = 21$ is false. An equation that contains a variable may be true or false, depending upon the value of that variable. Consider the equation $2x - 5 = 11$. If $x = 8$, the equation is true because

$$2x - 5 = 11$$
$$2(8) - 5 \stackrel{?}{=} 11$$
$$16 - 5 \stackrel{?}{=} 11$$
$$11 = 11$$

However, this equation is false for all other values of x. Any number that makes an equation true when substituted for its variable is said to satisfy the equation. All numbers that satisfy an equation are called its **solution(s)** or **root(s)**. Because the number 8 is the only number that satisfies the equation $2x - 5 = 11$, it is its only solution.

Often solutions to equations are expressed as **solution sets**. In the example above the solution set of the equation is $\{8\}$. A solution set, therefore, is simply a set that contains the solution(s) of an equation.

EXAMPLE 1

Is 5 a solution of $3x - 5 = 2x$? The substitution property of equality allows us to replace x with any value and evaluate the resulting expressions. Only the solution will yield a true equation. Verify using the graphing calculator.

SOLUTION   We substitute 5 for x in the equation and simplify.

$$3x - 5 = 2x$$
$$3 \cdot 5 - 5 \stackrel{?}{=} 2 \cdot 5 \qquad \text{Replace } x \text{ with 5}$$
$$15 - 5 \stackrel{?}{=} 10$$
$$10 = 10$$

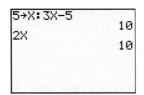

Store the value of 5 for the variable x and confirm that the left side of the equation is equivalent to the right side.

Because the left side is equivalent to the right side when x is replaced by 5, 5 is the solution or root of the equation.

EXAMPLE 2

Which of the following numbers, 2, 4, or $-\dfrac{3}{2}$, are solutions of $2x^2 = 5x + 12$? Verify using your graphing calculator.

SOLUTION
$$2(2)^2 \stackrel{?}{=} 5(2) + 12$$
$$2(4) \stackrel{?}{=} 10 + 12$$
$$8 \neq 22$$

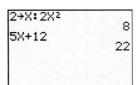

Because $8 \neq 22$, 2 is not a solution.

$$2(4)^2 \stackrel{?}{=} 5(4) + 12$$

$$2(16) \stackrel{?}{=} 20 + 12$$

$$32 = 32$$

Because $32 = 32$, 4 is a solution.

$$2\left(-\frac{3}{2}\right)^2 \stackrel{?}{=} 5\left(-\frac{3}{2}\right) + 12$$

$$\frac{9}{2} \stackrel{?}{=} \left(-\frac{15}{2}\right) + 12$$

$$\frac{9}{2} \stackrel{?}{=} -\frac{15}{2} + \frac{24}{2}$$

$$\frac{9}{2} = \frac{9}{2}$$

Because $\dfrac{9}{2} = \dfrac{9}{2}$, $-\dfrac{3}{2}$ is a solution.

```
4→X:2X²
              32
5X+12
              32
```

```
-3/2→X:2X²
            4.5
5X+12
            4.5
```

TECHNOLOGY TIP

If desired, the FRAC option of the graphing calculator can be used to perform arithmetic operations with fractions.

Example 1 is a first-degree equation in one variable, whereas Example 2 is a second-degree equation. In these examples we were merely checking solutions; we now turn our attention to algebraically solving first-degree equations in one variable.

▌▌▌ PROPERTIES OF EQUALITY

The four properties of equality are used to algebraically solve first-degree linear equations. Recall than an equation has a left side and a right side and is a statement that these two sides are equivalent. Thus, relationships between constants and variables must be preserved. The properties of equality are the rules that must be followed to preserve equality.

Addition Property of Equality

Suppose that a, b, and c are real numbers. It follows that

if $a = b$, then $a + c = b + c$.

Subtraction Property of Equality

Suppose that a, b, and c are real numbers. It follows that

if $a = b$, then $a - c = b - c$.

Multiplication Property of Equality

Suppose that a, b, and c are real numbers. It follows that

if $a = b$, then $a \cdot c = b \cdot c$.

Division Property of Equality

Suppose that a, b, and c are real numbers and $c \neq 0$. It follows that

$$\text{if } a = b, \text{ then } \frac{a}{c} = \frac{b}{c}.$$

Stated informally, the properties allow you to add, subtract, or multiply the same quantity to both *sides* of an equation and divide both *sides* of an equation by the same *nonzero* quantity. Careful application of these properties is used to algebraically solve linear equations.

Linear equations are solved by writing a series of **equivalent equations**. Equivalent equations are equations that have the same solution set.

EXPLORING THE CONCEPT

Consider the mathematical statement $x = 12$. This is an equation because it contains an equal sign and is true because replacement of x by 12 yields the statement $12 = 12$. Adding the number 2 to both sides of the equation results in a new equation, $x + 2 = 12 + 2$ or $x + 2 = 14$. The new equation is equivalent to the original equation because, again, when the variable is replaced by 12 the new equation is true. The original equation ($x = 12$) and the new equation ($x + 2 = 14$) have the same solution. Therefore, they are **equivalent equations**.

Create three equivalent equations by

a. Subtracting 4 from each side of the original equation ($x = 12$).

b. Multiplying each side of the original equation ($x = 12$) by 4.

c. Dividing each side of the original equation ($x = 12$) by 4.

CONCLUSION

a. $x - 4 = 8$; this is an equivalent equation to $x = 12$ because replacing the variable x with 12 results in the true statement $12 - 4 = 8$.

b. $4x = 48$; this is an equivalent equation to $x = 12$ because replacing the variable x with 12 results in the true statement $4 \cdot 12 = 48$.

c. $\frac{x}{4} = 3$; this is an equivalent equation to $x = 12$ because replacing the variable x with 12 results in the true statement $\frac{12}{4} = 3$.

Adding, subtracting, multiplying, or dividing both sides of an equation by the same nonzero constant results in a series of equivalent equations.

We will now examine the use of the properties of equality in solving an equation by transformation.

ALGEBRAIC SOLUTIONS OF LINEAR EQUATIONS IN ONE VARIABLE

To **solve** an equation means to find the value(s) of the variable(s) that will make the equation a true statement. The equations in this chapter will be solved by writing a series of equivalent equations.

When solving equations it is best to first simplify both sides of the equation and then begin transforming the equation by using the properties of equality to write a series of equivalent equations. To solve any linear equation, we must isolate the variable on one side. This can be a multistep process that may require combining like terms. As we solve equations, we will follow the steps below.

Steps for Solving a Linear Equation in One Variable

1. Use the distributive property to remove grouping symbols.
2. Combine like terms on each **side** of the equation.
3. Clear fractions, if desired, by using the multiplication property of equality and multiplying each **term** on both sides of the equation by the Lowest Common Denominator (LCD).
4. Use the addition or subtraction property of equality to collect all terms containing the variable on one side of the equation and all constants on the other side.
5. Use the multiplication or division property of equality to isolate the variable. The equation should be in the form:

$$\textbf{variable} = \textbf{constant}$$

6. Check by ensuring that the proposed solution produces a true statement when the variable is replaced by the constant. Always use the original equation when checking.

EXAMPLE 3

Solve $\dfrac{5x}{4} = 9$.

SOLUTION

$$4 \cdot \frac{5x}{4} = 4 \cdot 9 \qquad \text{Use the multiplication property of equality to clear the equation of fractions.}$$

$$5x = 36 \qquad \text{Simplify both sides.}$$

$$\frac{5x}{5} = \frac{36}{5} \qquad \text{Use the division property of equality to isolate the variable.}$$

$$x = \frac{36}{5} \qquad \text{Simplify.}$$

Therefore, $\frac{36}{5}$ is the solution of the original equation; the solution set is $\{\frac{36}{5}\}$. The check is shown below, both analytically (paper-and-pencil computation) and by using the STOre feature of the calculator:

$$\frac{5 \cdot \dfrac{36}{5}}{4} \overset{?}{=} 9$$

$$\frac{36}{4} \overset{?}{=} 9$$

$$9 = 9$$

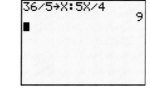

EXAMPLE 4

Solve $\dfrac{x}{3} + 4 = 5$.

SOLUTION

$$3 \cdot \dfrac{x}{3} + 3 \cdot 4 = 3 \cdot 5$$ Use the multiplication property of equality to clear the equation of fractions. Each **term** of both **sides** of the equation must be multiplied by the LCD.

$$x + 12 = 15$$ Simplify both sides of the equation.

$$x + 12 - \mathbf{12} = 15 - \mathbf{12}$$ Subtract 12 from both **sides** of the equation to place all constants on the right (notice the variable is on the left).

$$x + 0 = 3$$ Simplify both sides of the equation.

$$x = 3$$ Because zero is the *additive identity*, $x + 0$ is equivalent to x.

The solution, or root, is 3; the solution set is {3}. Verify with the calculator.

```
3→X:X/3+4
              5
■
```

EXAMPLE 5

Solve $-4x - 8x + 5 = 17$.

SOLUTION

$$-12x + 5 = 17$$ Combine like terms.

$$-12x + 5 - \mathbf{5} = 17 - \mathbf{5}$$ Use the subtraction property of equality to place constants together on one side of the equation.

$$-12x = 12$$ Simplify.

$$\dfrac{-12x}{\mathbf{-12}} = \dfrac{12}{\mathbf{-12}}$$ Use the division property of equality to isolate the variable.

$$x = -1$$

The solution (root) is -1; the solution set is $\{-1\}$. Verify with the calculator.

```
-1→X: -4X-8X+5
                17
■
```

EXAMPLE 6

Solve $2x - 5 = x + 4$.

SOLUTION

$$2x - 5 = x + 4$$

$$2x - 5 - x = x + 4 - x$$ The subtraction property of equality allows us to subtract x from both sides of the equation.

$$x - 5 = 4$$ Simplify both sides.

$$x - 5 + 5 = 4 + 5$$ The addition property of equality allows us to add 5 to both sides of the equation.

$$x = 9$$ Simplify both sides.

```
9→X:2X-5
              13
X+4
              13
```

The equation $2x - 5 = x + 4$ is true when x has a value of 9. The solution (root) is 9; the solution set is {9}. Verify with the calculator.

To solve these equations, a series of equivalent equations was written using the properties of equality, which guarantee preservation of the relationship of equality.

EXAMPLE 7   Solve $3(x + 2) - 2x = 4x$.

SOLUTION $3x + 6 - 2x = 4x$ Use the distributive property to remove grouping.

$x + 6 = 4x$ Combine like terms.

$x + 6 - x = 4x - x$ Use the subtraction property of equality to get all variable terms together.

$6 = 3x$ Simplify each side.

$\dfrac{6}{3} = \dfrac{3x}{3}$ Use the division property of equality to isolate the variable.

$2 = x$ Simplify each side.

Some students do not like having the variable on the right. However, the relationship of equality is **symmetric**. This property ensures that entire *sides* of equations may be transposed. Therefore, $2 = x$ is an equivalent mathematical statement to $x = 2$. The variable terms could have been placed on the left side of the equation. This is illustrated below, beginning with the step in bold print. The steps prior to that would remain the same regardless of where you prefer the variable to be.

$x + 6 = 4x$

$x + 6 - 4x = 4x - 4x$ Use the subtraction property of equality.

$-3x + 6 = 0$ Simplify both sides.

$-3x + 6 - 6 = 6 - 6$ Use the subtraction property of equality to place the constant on the right side of the equation.

$-3x = -6$ Simplify.

$\dfrac{-3x}{-3} = \dfrac{-6}{-3}$ Use the division property of equality to isolate the variable.

$x = 2$

```
2→X:3(X+2)-2X
              8
4X
              8
```

Because the properties of equality have been carefully applied, placement of the variable does not affect the solution to the equation. Verify with the calculator.

TYPES OF EQUATIONS AND THEIR SOLUTIONS

The equations in Examples 1–7 are true for some real numbers. These equations are called **conditional equations** because they are true for some, but not all, real numbers.

An equation that is true for all real numbers is called an **identity**. Regardless of the value used to replace the variable, the equation will be true. For example, the equation $x + x = 2x$ is an identity because it is true regardless of the real number substituted in for the variable x. Its solution is **all real numbers**. This can be expressed with the symbol $\mathbb{R}$, which means "the set of all real numbers."

An equation that has no solution is called a **contradiction**. The equation $x = x + 1$ is a contradiction because no number is equal to itself plus 1. Its solution is **no solution**, which can be represented with the symbol $\varnothing$ representing the **null set** or the set that contains no elements.

WARNING!   Do not confuse the **type** of equation with its **solution**. The terms *conditional*, *identity*, and *contradiction* refer to the type. Remember the solution(s) or root (s)are the value(s) of the variable(s) that make the equation a true statement.

Examples 8 and 9 illustrate the solutions of equations that represent a contradiction and an identity, respectively.

EXAMPLE 8

Solve $6(x + 1) - 4x = 2(x + 1) + 1$.

SOLUTION

$6x + 6 - 4x = 2x + 2 + 1$	Use the distributive property to clear grouping.
$2x + 6 = 2x + 3$	Combine like terms on each side of the equation.
$2x + 6 - 2x = 2x + 3 - 2x$	Use the subtraction property of equality to put all the variables on one side of the equation.
$6 = 3$	The result is a false statement.

The equation is a contradiction and has **no solution**. This is often denoted with the symbol $\varnothing$ or { } to denote an empty set.

STUDY TIP

When designating an *empty set* use either the notation ϕ (the Greek letter phi) or { }. Do not use the notation $\{\phi\}$ as this set is not empty — it contains one element.

Think about the equation printed in red in Example 8. At that point in the solving process it is obvious that the equation has no solution. There is no real number that you can double and add six to that will equal itself doubled plus three. As soon as you realize the equation is a contradiction, you may state its solution as $\varnothing$.

EXAMPLE 9

Solve $x + 6 = 2\left(3 + \dfrac{x}{2}\right)$.

SOLUTION

Use the distributive property to clear grouping symbols and yield

$$x + 6 = 6 + x$$

Because the left side of the equation is equivalent to the right side, the equation is an identity. There is no need to proceed further. Therefore the solution is **all real numbers**, which is often symbolized as $\mathbb{R}$.

STUDY TIP

When learning to solve equations students often check solutions only by referring to the back of the text or comparing answers with a study partner. It is important to check all solutions so that the process becomes a natural part of problem solving.

Carefully look back over the examples in this section. As you work through the exercises for this section, keep in mind that **not** all the equations you will be asked to solve will be conditional equations. Therefore, be very careful.

▮ INTRODUCTION TO PROBLEM SOLVING

Application problems illustrate the power of algebra in solving practical problems, or ones that occur every day. Several types of application problems will be considered in the rest of this section and throughout the text. However, rarely do real-life applications fall into these categories. The emphasis, therefore, will be on developing a repertoire of problem-solving *tools*. The strategies below will help you begin. It is suggested that these be copied on a note card for quick reference when working application problems.

Problem Solving Strategies

1. Read the problem several times and organize the facts. What is known? What are you asked to find? It may help you to organize the facts under the two headings, **known** and **unknown**.
2. a. If there is only one unknown quantity, choose a variable to represent that quantity. Clearly define what the variable is and what it represents with a "Let" statement.
   b. When there is more than one unknown, choose a variable to represent one of the quantities and express all other important quantities in terms of the chosen variable.
3. Translate the information expressed in the words of the problem to an algebraic equation.
4. Solve the equation.
5. Answer the question posed in the problem.
6. Check your results in terms of both the equation and with the words of the problem. Make sure your solution makes sense within the context of the problem.

Recall the set of integers $\{\ldots, -3, -2, -1, 0, 1, 2, 3, \ldots\}$. If x represented an integer, $x + 1$ would represent the next consecutive integer. If x represented an even or an odd integer, the next consecutive even or odd integer would be represented as $x + 2$.

EXAMPLE 10   The sum of three consecutive even integers is 84. What is the largest integer?

SOLUTION   Let x = the first even integer
$x + 2$ = the next even integer
$x + 4$ = the third even integer

Because the sum of all three integers is 84, the algebraic equation is

$$\underset{x}{\underbrace{\text{1st even integer}}} + \underset{x + 2}{\underbrace{\text{2nd even integer}}} + \underset{x + 4}{\underbrace{\text{3rd even integer}}} = 84$$

Now solve the equation

$$3x + 6 = 84$$
$$3x + 6 - 6 = 84 - 6$$
$$3x = 78$$
$$\frac{3x}{3} = \frac{78}{3}$$
$$x = 26$$

Because the largest integer was represented by the expression $x + 4$ and x has a value of 26, the largest integer is 30.

EXAMPLE 11

Five more than double a number is 19. What is the number?

SOLUTION Because there is only one unknown, x will represent the unknown number. This is expressed by using a "Let" statement.

Let $x =$ the number   Clearly define the variable and what it represents.

Five **more than double** a number is 19:

$$5 \quad + \quad 2 \cdot \quad x \quad = 19$$

The words of the problem are translated to an algebraic equation: $5 + 2x = 19$.

$$5 + 2x = 19 \qquad \text{Solve the equation.}$$
$$5 + 2x - 5 = 19 - 5$$
$$2x = 14$$
$$\frac{2x}{2} = \frac{14}{2}$$
$$x = 7$$

The number is 7.   Answer the question.

Make sure your result makes sense in **both** the equation and in the context of the problem. Checking the equation can be done by the tools previously introduced: analytically (substitution) or by using the STOre feature of the calculator.

EXAMPLE 12

A rental car agency charges a flat fee of $37 a day plus $0.21 per mile driven. If the rental charge for one day is $61.15, how many miles were driven?

SOLUTION Let $x =$ number of miles driven.   Clearly identify a variable and what it represents.

$37 a day **plus $0.21 per mile** translates to the expression

$$37 \quad + \quad .21 \quad \cdot \quad x$$

The *total* rental charge is $61.15, so the resulting equation is

$$37 + .21x = 61.15$$

Now solve the equation:

$$37 + .21x = 61.15$$
$$37 + .21x - 37 = 61.15 - 37$$
$$.21x = 24.15$$
$$\frac{.21x}{.21} = \frac{24.15}{.21}$$
$$x = 115$$

The car was driven 115 miles.   Answer the question.

EXERCISE 3.1

VOCABULARY AND NOTATION   *In Exercises 1–10, fill in the blank to make a true statement.*

1. A _____ equation in one variable is any equation in one variable, x, that can be written in the form $ax + b = 0$.

2. All numbers that satisfy an equation are called _____ or _____ .

3. Equations with the same solution are called _____ equations.

4. To _____ an equation means to find the value(s) for the variable(s) that will make the equation a true statement.

5. The statement that $x = 2$ is equivalent to the statement $2 = x$ is an example of the _____ property.

6. An equation that is true for *all* acceptable real numbers is called an _____ .

7. A contradiction is an equation that is true for ____ real numbers.

8. An equation that is true for some real numbers is called a _____ .

9. If an equation has no solution, then the solution set is represented by the symbol ___ .

10. When the solution to an equation is the set of all real numbers, we represent it by the symbol ___ .

CONCEPTS

11. In Exercises a–f, the calculator screens pictured show the verification of a solution to an equation. Write the equation that was being checked and express its solution as a solution set.

a.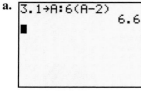

b.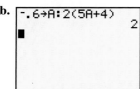

c.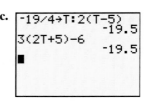

d.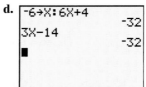

e.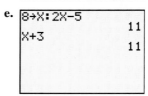

f.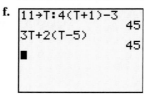

12. If x is the first integer, write expressions for each of the next two consecutive integers.

13. If x is the first even integer, write expressions for each of the next two consecutive even integers.

14. If x is the first odd integer, write expressions for each of the next two consecutive odd integers.

15. If x is the number, write an expression for each of the following:
    a. a number increased by 4
    b. a number decreased by 7

16. If x is the number, write an expression for each of the following:
    a. twice a number decreased by 3
    b. half a number increased by 7

PRACTICE   *In Exercises 17–28, state whether the given number is a solution of the equation. Justify your response.*

17. $x + 2 = 3$; 1

18. $x - 2 = 4$; 6

19. $2x = 4; 2$

20. $3x = 6; 3$

21. $\dfrac{x}{5} = x; 0$

22. $\dfrac{x}{7} = 7x; 0$

23. $3k + 5 = 5k - 1; 3$

24. $2s - 1 = s + 7; 6$

25. $2(t + 5) - 3t = -2t + 1; -9$

26. $3(a - 5) + 2 = 3a - 13; 6$

27. $\dfrac{5 + x}{10} - x = \dfrac{1}{2}; 0$

28. $\dfrac{x - 5}{6} = 12 - x; 11$

*In Exercises 29–68, solve each equation algebraically, showing **all** steps in the solution process. Solutions should be expressed as sets. Check all problems, either analytically or by using the STOre feature of the calculator, to verify roots.*

29. $\dfrac{b}{3} = \dfrac{1}{3}$ **30.** $\dfrac{a}{13} = \dfrac{1}{26}$

31. $3x + 2 = 29$ **32.** $8x + 6 = 70$

33. $6x - 1 = 35$ **34.** $3x - 5 = 10$

35. $-5x - 24 = 31$ **36.** $-2x - 18 = 22$

37. $3x - 6x = 18$ **38.** $5x + 7x = 36$

39. $5x - 6x = -(x + 2)$

40. $-2x + 4 + 3x = x - 3$

41. $6x - 5 = 7 + 2x$ **42.** $4x + 2 = 9 - 3x$

43. $3x + 2 - 5x = -2x + 6 - 4$

44. $2x + 6 - 1 - 8x = -6x - 10 + 15$

45. $2v - \dfrac{7}{3} = -\dfrac{5}{6}$ **46.** $z + \dfrac{7}{9} = \dfrac{2}{9}$

47. $5(b + 7) = 6b - 2$

48. $8(b + 2) = 9b$

49. $2 + 3(x - 5) - 4(x - 1)$

50. $10x + 3(2 - x) = 5(x + 2) - 4$

51. $11x + 6(3 - x) = 3 + x$

52. $9(t - 1) = 6(t + 2) - t$

53. $9(x + 11) + 5(13 - x) = 2x$

54. $3(x + 15) + 4(11 - x) = 5x$

55. $\dfrac{3(t - 7)}{2} = t - 6$ **56.** $\dfrac{2(t + 9)}{3} = t - 8$

57. $8x + 3(2 - x) = 5(x + 2) - 4$

58. $4(a + 2) = 2(a - 4) + 16$

59. $3(b - 1) + 18 = 4(b + 5)$

60. $x + 7 = \dfrac{2x + 6}{2} + 4$

61. $\dfrac{2(t - 1)}{6} - 2 = \dfrac{t + 2}{6}$

62. $2(3z + 4) = 2(3z - 2) + 13$

Recall that the decimal, 0.1, is equivalent to the fraction $\frac{1}{10}$. Therefore, if a problem contains decimals, you may clear them in exactly the same way as you clear fractions — by multiplying each term on both sides by a number that will clear the decimal.

63. $4.8x + 6 = 0.8x - 2$

64. $6.5x - 5.2 = 2.5x + 6.8$

65. $3.1(x - 2) = 1.3x + 2.8$

66. $0.6x - 0.8 = 0.8(2x - 1) - 0.7$

67. $3.2x + 5.64 = 5.6x + x - 1.16$

68. $0.65x + 2.7 = 0.05x + 2(x + 1)$

APPLICATIONS

69. Perimeter   The perimeter of a square picture that is to be framed is 28 inches and each side is $(2x + 3)$ inches. Find the length of each side.

    a. Write an equation to model the perimeter.

    b. Simplify the equation.

    c. Solve the equation.

    d. Determine the length of each side.

70. Perimeter   The perimeter of a square picture that is to be framed is 32 inches and each side is $(3x - 1)$ inches in length. Find the length of the sides of the picture.

    a. Write an equation to model the perimeter.

    b. Simplify the equation.

    c. Solve the equation.

    d. Determine the length of the sides.

71. Perimeter   The perimeter of a rectangular window is 20 feet. The length is $(x - 4)$ feet and the width is x feet. Find the dimensions.

  a. Write an equation to model the perimeter.

  b. Simplify the equation.

  c. Solve the equation.

  d. Determine the dimensions of the window.

72. Perimeter   The perimeter of a rectangular window is 250 feet. The length is $(2x + 5)$ feet and the width is $4x$ feet. Find the dimensions of the rectangle.

  a. Write an equation to model the perimeter.

  b. Simplify the equation.

  c. Solve the equation.

  d. Determine the length of each side.

In Exercises 73–80, (a) write an equation to model the problem, (b) simplify the equation, (c) solve the equation, and (d) answer the question posed.

73. Integer   The sum of two consecutive even integers is 54. Find the integers.

74. Integer   The sum of two consecutive odd integers is 88. Find the integers.

75. Number   Twice a number increased by 5 is 15. Find the number.

76. Number   Four less than three times a number is 17, find the number.

77. Cable charges   The basic cable rate is $29.95 a month, plus $3.75 for each pay-per-view movie. If a subscriber's total bill is $52.45, how many movies were viewed?

78. Catering charges   A wedding consultant charges $1250 plus $12.75 per guest for a wedding reception. How many guests attended the reception if the bill was $3940.25?

79. Kennel charges   A sportsman boarded his dog at the kennel for $16, plus $12 a day. If the stay cost $100, how many days was the owner gone?

80. Phone rates   A call to Tucson from a pay phone in Chicago cost $0.85 for the first minute and $0.27 for each additional minute or portion of a minute. If a student has $8.50 in change, how long can she talk?

REVIEW

81. Simplify each of the following:

  a. $\dfrac{4}{5} - \dfrac{1}{3}$ **b.** $\left(3\dfrac{1}{2}\right) \cdot \left(-\dfrac{3}{14}\right)$

  c. $-\dfrac{4}{5} \div \left(1\dfrac{1}{2}\right)$

82. Evaluate each of the following for $x = -1$ and $y = 2$:

  a. $(xx^7y^3)^2$ **b.** $2x^2(x^3 - 4xy)$

  c. $(2x - y)^2$

83. Simplify: $(xx^7y^3)^2$

84. If you write $a(b + c) = ab + ac$, then the _____ property has been applied.

85. Simplify: $2x^2(x^3 - 4xy)$

86. Simplify: $(2x - y)^2$

3.2   LITERAL EQUATIONS AND FORMULAS

  • SOLVING LITERAL EQUATIONS   • FORMULAS

SOLVING LITERAL EQUATIONS

Equations with several variables are called **literal equations**. Often these equations are **formulas**, such as $A = lw$, the formula for finding the area of a rectangle. Suppose that we wish to find the lengths of several rectangles whose areas and widths are known. It would be tedious to substitute values for A and w into the formula and then repeatedly solve the formula for l. It would be better to solve the formula $A = lw$ for l first and then substitute values for A and w and compute l directly.

EXAMPLE 1

Solve $A = lw$ for l.

SOLUTION   To isolate l on the right-hand side, we eliminate the variable w by dividing both sides of the equation by w.

$$A = lw$$

$$\frac{A}{w} = \frac{\overset{1}{l\cancel{w}}}{\underset{1}{\cancel{w}}} \qquad \text{Divide both sides by } w.$$

$$\frac{A}{w} = l \qquad \text{Simplify.}$$

$$l = \frac{A}{w} \qquad \text{Symmetric property of equality.}$$

STUDY TIP

As much as possible, study when you are most alert. Beginning a two-hour study session at 11:00 P.M. is often nonproductive.

EXAMPLE 2

The formula $A = \frac{1}{2}bh$ gives the area of a triangle with base b and height h. Solve this formula for b.

SOLUTION

$$A = \frac{1}{2}bh$$

$$2A = 2 \cdot \frac{1}{2}bh \qquad \text{Multiply both sides by 2.}$$

$$2A = bh \qquad \text{Simplify.}$$

$$\frac{2A}{h} = \frac{b\overset{1}{\cancel{h}}}{\underset{1}{\cancel{h}}} \qquad \text{Divide both sides by } h.$$

$$\frac{2A}{h} = b \qquad \text{Simplify.}$$

$$b = \frac{2A}{h} \qquad \text{Symmetric property of equality.}$$

If the area A and the height h of a triangle are known, the base b is given by the formula $b = \frac{2A}{h}$.

EXAMPLE 3

Solve $3x + 2y = 8$, for y.

SOLUTION

$$3x + 2y - 3x = -3x + 8 \qquad \text{Subtract } 3x \text{ from both sides.}$$

$$2y = -3x + 8 \qquad \text{Simplify both sides.}$$

Method 1:

$$\frac{2y}{2} = \frac{-3x + 8}{2}$$

$$y = \frac{-3x + 8}{2}$$

Method 2:

$$\frac{2y}{2} = \frac{-3x}{2} + \frac{8}{2} \qquad \text{Divide both sides by 2. This is equivalent to dividing each term by 2.}$$

$$y = -\frac{3}{2}x + 4 \qquad \text{Simplify.}$$

EXAMPLE 4

The formula $C = \frac{5}{9}(F - 32)$ is used to convert Fahrenheit temperature readings into their Celsius equivalents. Solve the formula for F.

SOLUTION

$$C = \frac{5}{9}(F - 32)$$

$$C = \frac{5}{9}F - \frac{160}{9} \qquad \text{Remove parentheses.}$$

$$9 \cdot C = 9 \cdot \frac{5}{9}F - 9 \cdot \frac{160}{9} \qquad \text{Multiply each term by the LCD of 9.}$$

$$9C = 5F - 160 \qquad \text{Simplify.}$$

$$9C + 160 = 5F - 160 + 160 \qquad \text{Add 160 to both sides.}$$

$$9C + 160 = 5F \qquad \text{Simplify.}$$

$$\frac{9C}{5} + \frac{160}{5} = F \qquad \text{Divide each term by 5 to isolate the variable, } F.$$

$$\frac{9}{5}C + 32 = F \qquad \text{Simplify.}$$

$$F = \frac{9}{5}C + 32 \qquad \text{Symmetric property of equality.}$$

The formula $F = \frac{9}{5}C + 32$ is used to convert degrees Celsius to degrees Fahrenheit.

EXAMPLE 5

The area A of the trapezoid shown in Figure 3-1 is given by the formula

$$A = \frac{1}{2}(B + b)h$$

where B and b are its bases and h is its height. Solve the formula for b.

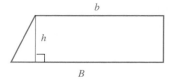

Figure 3-1

SOLUTION   *Method 1:*

$$A = \frac{1}{2}(B + b)h$$

$$2A = 2 \cdot \frac{1}{2}(B + b)h \qquad \text{Multiply both sides by 2.}$$

$$2A = Bh + bh \qquad \text{Simplify and remove parentheses.}$$

$$2A - Bh = Bh + bh - Bh \qquad \text{Subtract } Bh \text{ from both sides.}$$

$$2A - Bh = bh \qquad \text{Combine like terms.}$$

$$\frac{2A - Bh}{h} = \frac{\overset{1}{\cancel{b}h}}{\underset{1}{\cancel{h}}} \qquad \text{Divide both sides by } h.$$

$$\frac{2A - Bh}{h} = b \qquad \text{Simplify.}$$

Method 2: $A = \frac{1}{2}(B + b)h$

$2A = 2 \cdot \frac{1}{2}(B + b)h$ Multiply both sides by 2.

$2A = (B + b)h$ Simplify.

$\dfrac{2A}{h} = \dfrac{(B + b)h}{h}$ Divide both sides by *h*.

$\dfrac{2A}{h} = B + b$ Simplify.

$\dfrac{2A}{h} - B = B + b - B$ Subtract *B* from both sides.

$\dfrac{2A}{h} - B = b$ Combine like terms.

Although they look different, the results of Methods 1 and 2 are equivalent.

▍▍▍ FORMULAS

In Section 1.5, many formulas were given. It is very important that before using a formula you understand what each variable represents. Example 6 illustrates this idea.

EXAMPLE 6

The volume *V* of the right-circular cone shown in Figure 3-2 is given by the formula

$$V = \frac{1}{3}Bh$$

where *B* is the area of its circular base and *h* is its height. Solve the formula for *h* and find the height of a right-circular cone with a volume of 64 cubic centimeters and a base area of 16 square centimeters.

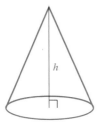

Figure 3-2

SOLUTION   We first solve the formula for *h*.

$V = \frac{1}{3}Bh$

$3V = 3 \cdot \frac{1}{3}Bh$ Multiply both sides by 3.

$3V = Bh$ Simplify.

$\dfrac{3V}{B} = \dfrac{Bh}{\frac{B}{1}}$ Divide both sides by *B*.

$\dfrac{3V}{B} = h$ Simplify.

$h = \dfrac{3V}{B}$ Symmetric property of equality.

We then substitute 64 for V and 16 for B and simplify.

$$h = \frac{3V}{B}$$

$$h = \frac{3(64)}{16}$$

$$= 3(4)$$

$$= 12$$

The height of the right-circular cone is 12 centimeters.

EXERCISE 3.2

VOCABULARY AND NOTATION *Fill in the blank to make a true statement.*

1. Equations with several variables are called

_____.

CONCEPTS *In Exercises 2–14, solve for the indicated variable. Each exercise requires the use of exactly one property of equality from page 138 of Section 3.1. Identify the property used.*

2. $K = A + 32$; for A

3. $P = a + b + c$; for c

4. $P = 4s$; for s

5. $d = rt$; for t

6. $E = IR$; for I

7. $i = prt$; for r

8. $V = lwh$; for h

9. $C = 2\pi r$; for r

10. $P = I^2R$; for R

11. $y + 2 = x - 8$; for y

12. $\dfrac{y}{2} = x$; for y

13. $3y = 8x$; for y

14. $5y = 16x$; for y

PRACTICE *In Exercises 15–34, solve each formula for the variable indicated.*

15. $P = 2l + 2w$; for w

16. $A = P + Prt$; for t

17. $A = \dfrac{1}{2}(B + b)h$; for h

18. $K = \dfrac{wv^2}{2g}$; for w

19. $I = \dfrac{E}{R}$; for R

20. $V = \dfrac{1}{3}\pi r^2 h$; for h

21. $2x + y = 8$; for y

22. $3x + y = 2$; for y

23. $x - y = 6$; for y

24. $2x - y = 8$; for y

25. $3x + 5y = 15$; for y

26. $6x + 3y = 18$; for y

27. $\dfrac{1}{2}x + 3y = 10$; for y

28. $\dfrac{2}{3}x + 4y = 12$; for y

29. $\dfrac{1}{2}x + \dfrac{2}{3}y = 5$; for y

30. $x + 4y = \dfrac{1}{2}$; for y

31. $\dfrac{2}{3}(x + 4) + y = -1$; for y

32. $\dfrac{1}{2}(x - 5) + y = 6$; for y

33. $2(x - 4) + 3(y + 2) = 6$; for y

34. $3(x + 2) - 4(y - 1) = 5$; for y

In Exercises 35–42, solve each formula for the variable indicated. Then evaluate to find that variable's value.

35. $d = rt$; find t if $d = 135$ and $r = 45$.

36. $d = rt$; find r if $d = 275$ and $t = 5$.

37. $i = prt$; find t if $i = 12$, $p = 100$, and $r = 0.06$.

38. $i = prt$; find r if $i = 120$, $p = 500$, and $t = 6$.

39. $P = a + b + c$; find c if $P = 37$, $a = 15$, and $b = 19$.

40. $y = mx + b$; find x if $y = 30$, $m = 3$, and $b = 5$.

41. $K = \frac{1}{2}h(a + b)$; find h if $K = 48$, $a = 7$, and $b = 5$.

42. $\frac{x}{2} + y = z^2$; find x if $y = 3$ and $z = 3$.

APPLICATIONS

43. Ohm's law   The formula $E = IR$, called **Ohm's law**, is used in electronics. Solve for I and then calculate the current I if the voltage E is 48 volts and the resistance R is 12 ohms. Current has units of *amperes*.

44. Volume of a cone   The volume V of a cone is given by the formula $V = \frac{1}{3}\pi r^2 h$. Solve the formula for h and then calculate the height h if V is 36π cubic inches and the radius r is 6 inches.

45. Circumference of a circle   The circumference C of a circle is given by $C = 2\pi r$, where r is the radius of the circle. Solve the formula for r and then calculate the radius of a circle with a circumference of 14.32 feet. Round your answer to the nearest hundredth of a foot.

46. Growth of money   At a simple interest rate r, an amount of money P grows to an amount A in t years according to the formula $A = P(1 + rt)$. Solve the formula for P. After $t = 3$ years, a girl has an amount $A = \$4357$ on deposit. What amount P did she start with? Assume an interest rate of 2%.

47. Power loss   The power P lost when an electric current I passes through a resistance R is given by the formula $P = I^2 R$. Solve for R. If P is 2700 watts and I is 14 amperes, calculate R to the nearest hundredth of an ohm.

48. Geometry   The perimeter P of a rectangle with length l and width w is given by the formula $P = 2l + 2w$. Solve this formula for w. If the perimeter of a certain rectangle is 58.37 meters and its length is 17.23 meters, find its width. Round the answer to two decimal places.

49. Force of gravity   The masses of the two objects in Illustration 1 are m and M. The force of gravitation, F, between the masses is given by

$$F = \frac{GmM}{d^2}$$

where G is a constant and d is the distance between them. Solve for m.

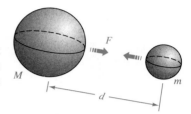

Illustration 1

50. Thermodynamics   In thermodynamics, the Gibbs free-energy function is given by

$$G = U - TS + pV$$

Solve this equation for the pressure, p.

51. Pulleys   The approximate length L of a belt joining two pulleys of radii r and R feet with centers D feet apart is given by the formula $L = 2D + 3.25(r + R)$. (See Illustration 2.) Solve the formula for D. If a 25-foot belt joins pulleys of radius 1 foot and 3 feet, how far apart are the centers of the pulleys?

Illustration 2

52. Geometry   The measure a of an interior angle of a regular polygon with n sides is given by the formula $a = 180°(1 - \frac{2}{n})$. (See Illustration 3.) Solve the formula for n. How many sides does a regular polygon have if an interior angle is 108°? (*Hint:* Distribute first.)

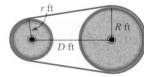

Illustration 3

REVIEW

53. Evaluate the expression $2(x - 1) + 2$ for $x = -3$.

54. Find $P(4)$ when $P(x) = 2(x - 1) + 2$.

55. Simplify the expression $2(x - 1) + 2$.

56. Solve for x: $2(x - 1) + 2 = 3x - 5$.

3.3 INTRODUCTION TO EQUATIONS IN TWO VARIABLES

- VERIFYING SOLUTIONS OF EQUATIONS IN TWO VARIABLES
- SOLUTIONS TO EQUATIONS IN TWO VARIABLES
- FUNCTIONS

▓ VERIFYING SOLUTIONS OF EQUATIONS IN TWO VARIABLES

Section 1.5 addressed several types of everyday applications that involved two related quantities. For example, the relationship of temperature measurement in Fahrenheit versus Celsius can be expressed as $C = \frac{5}{9}(F - 32)$. There are other similar relationships that can be modeled using two variables. In order to represent these relationships graphically with our calculator, we will use the variables x and y. Thus $y = \frac{5}{9}(x - 32)$, where $x =$ Fahrenheit temperature and $y =$ Celsius temperature. Using the distributive property, we can simplify this equation to $y = \frac{5}{9}x - \frac{160}{9}$.

Solutions to equations in two unknowns must be expressed as **ordered pairs**. The pair is called an ordered pair because the x value is always written first while the y value is always written second: (x, y). Ordered pairs are *always* enclosed in parentheses.

EXAMPLE 1

Verify that the following ordered pairs are solutions of the equation $4x - 5y = 2$.

a. $(-7, -6)$ **b.** $(3, 2)$

SOLUTION   To verify that each ordered pair is a solution, we will substitute the given values for x and y into the equation.

a.
$$4x - 5y = 2$$
$$4(-7) - 5(-6) \overset{?}{=} 2$$
$$-28 + 30 \overset{?}{=} 2$$
$$2 = 2$$

The ordered pair $(-7, -6)$ is a solution.

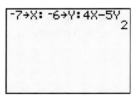

b.
$$4x - 5y = 2$$
$$4(3) - 5(+2) \overset{?}{=} 2$$
$$12 - 10 \overset{?}{=} 2$$
$$2 = 2$$

The ordered pair $(3, 2)$ is a solution.

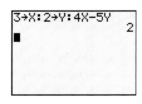

EXAMPLE 2

Verify that the following ordered pairs are solutions of the equation $y = x^2 + 3$.

a. $(-3, 12)$ **b.** $(5, 28)$

SOLUTION   To verify that each ordered pair is a solution, we will substitute the given values for x and y into the equation.

a. $y = x^2 + 3$

$12 \stackrel{?}{=} (-3)^2 + 3$

$12 \stackrel{?}{=} 9 + 3$

$12 = 12$

The ordered pair $(-3, 12)$ is a solution.

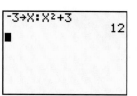

b. $y = x^2 + 3$

$28 \stackrel{?}{=} (5)^2 + 3$

$28 \stackrel{?}{=} 25 + 3$

$28 = 28$

The ordered pair $(5, 28)$ is a solution.

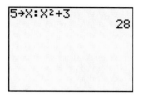

▐ SOLUTIONS TO EQUATIONS IN TWO VARIABLES

When finding solutions to equations in two variables, we are trying to find the x and y values that make the equation a true statement when the variables are replaced with the given values. The examples above demonstrated this relationship. We will now determine the value of one variable when given the value of the other variable.

EXAMPLE 3

Find the ordered pair solution to $y = 5x - 7$ when $x = 3$.

SOLUTION   Substitute **3** for x and evaluate for y.

$y = 5(3) - 7$

$= 15 - 7$

$= 8$

The ordered pair solution is $(3, 8)$.

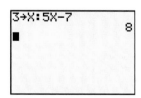

EXAMPLE 4

Find the ordered pair solution to $y = 5x - 7$ when $y = -4$.

SOLUTION   Substitute **−4** for y and solve for x.

$-4 = 5x - 7$

$-4 + 7 = 5x - 7 + 7$ Add 7 to both sides.

$3 = 5x$

$\dfrac{3}{5} = \dfrac{5x}{5}$ Divide both sides by 5.

$\dfrac{3}{5} = x$

The ordered pair solution is $\left(\tfrac{3}{5}, -4\right)$.

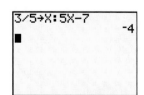

When solving a linear equation in x and y, you may select any value for x and solve for the related y value, or select any value of y and solve for the related x value. There are an infinite number of ordered pair solutions to an equation in two variables.

Many different ordered pair solutions can be viewed in the TABLE of your calculator. The expression $5x - 7$ must be entered on the Y= screen thus producing the equation $y = 5x - 7$. Set the TABLE (TBLSET) to begin at $x = -3$ (any value can be selected for the table start value) and to be incremented by 1 (ΔTbl = 1).

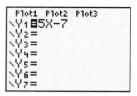 

From the TABLE we can see that the equation $y = 5x - 7$ has the ordered pairs $(-3, -22), (-2, -17), (-1, -12), (0, -7), (1, -2), (2, 3),$ and $(3, 8)$ as solutions. In fact, the equation has infinitely many solutions; you can see many of them if you scroll through the table. Because the table was incremented by units of 1, we will not be able to see ordered pair solutions that involve rational numbers such as $\frac{3}{5}$ unless we change the table increment or the table start value. The screens below indicate a table increment of $\frac{1}{5}$.

Observe that the table increment of $\frac{1}{5}$ is converted to .2 when displayed in the table. The ordered pair solution of $(\frac{3}{5}, -4)$ occurs in the table as $(.6, -4)$.

EXAMPLE 5

The following table of values gives selected values for x and y. Complete the table by substituting the given value into the equation and solving for the remaining variable.

$$3x - 2y = 6$$

x	y
4	
-2	
	0
	1
0	

SOLUTION If $x = 4$, then $3(4) - 2y = 6$
$$12 - 2y = 6$$
$$-2y = -6$$
$$y = 3 \qquad \text{The ordered pair (4, 3) is a solution.}$$

If $x = -2$, then $3(\mathbf{-2}) - 2y = 6$

$$-6 - 2y = 6$$

$$-2y = 12$$

$$y = -6 \qquad \text{The ordered pair } (-2, -6) \text{ is a solution.}$$

If $y = 0$, then $3x - 2(\mathbf{0}) = 6$

$$3x = 6$$

$$x = 2 \qquad \text{The ordered pair } (2, 0) \text{ is a solution.}$$

If $y = 1$, then $3x - 2(\mathbf{1}) = 6$

$$3x - 2 = 6$$

$$3x = 8$$

$$x = \frac{8}{3} \qquad \text{The ordered pair } (\tfrac{8}{3}, 1) \text{ is a solution.}$$

If $x = 0$, then $3(\mathbf{0}) - 2y = 6$

$$-2y = 6$$

$$y = -3 \qquad \text{The ordered pair } (0, -3) \text{ is a solution.}$$

The completed chart looks like this:

x	y
4	3
-2	-6
2	0
$\frac{8}{3}$	1
0	-3

The ordered pairs that result from this chart are $(4, 3)$, $(-2, -6)$, $(2, 0)$, $(\tfrac{8}{3}, 1)$, and $(0, -3)$.

To verify with the TABLE on your calculator, you will need to solve the equation $3x - 2y = 6$ for the variable y in order to enter the appropriate expression on the Y= screen.

$$3x - 2y = 6$$

$$3x - \mathbf{3x} - 2y = 6 - \mathbf{3x} \qquad \text{Subtract } 3x \text{ from both sides of the equation.}$$

$$-2y = 6 - 3x \qquad \text{Simplify.}$$

$$\frac{-2y}{\mathbf{-2}} = \frac{6 - 3x}{\mathbf{-2}} \qquad \text{Divide both sides of the equation by } -2.$$

$$y = \frac{6 - 3x}{-2} \qquad \text{Simplify.}$$

$$y = \frac{6}{-2} - \frac{3x}{-2}$$

$$y = -3 + \frac{3}{2}x \qquad \text{or} \qquad y = \frac{3}{2}x - 3$$

```
Plot1 Plot2 Plot3
\Y1◘(3/2)X-3
\Y2=
\Y3=
\Y4=
\Y5=
\Y6=
\Y7=
```

```
  X   │ Y1  │
━━━━━━┿━━━━━┿━━━━
 -2   │ -6  │
 -1   │ -4.5│
  0   │ -3  │
  1   │ -1.5│
  2   │  0  │
  3   │  1.5│
  4   │  3  │
X= -2
```

In order to verify the ordered pair $(\frac{8}{3}, 1)$, we could either set the table increment to $\frac{1}{3}$ or merely set the table to start at $\frac{8}{3}$ and verify the single ordered pair.

EXAMPLE 6

The following table of values gives selected values for x. Complete the table by substituting the given value into the equation and solving for the remaining variable.

$$y = x^3 - 2$$

x	y
-4	
0	
2	

SOLUTION

If $x = -4$, then $y = (-4)^3 - 2$ The ordered pair $(-4, -66)$ is a solution.

$$y = -64 - 2$$
$$y = -66$$

If $x = 0$, then $y = (0)^3 - 2$ The ordered pair $(0, -2)$ is a solution.

$$y = 0 - 2$$
$$y = -2$$

If $x = 2$, then $y = (2)^3 - 2$ The ordered pair $(2, 6)$ is a solution.

$$y = 8 - 2$$
$$y = 6$$

The completed chart looks like this:

x	y
-4	-66
0	-2
2	6

The ordered pairs that result from this chart are $(-4, -66)$, $(0, -2)$, and $(2, 6)$. Verified with the TABLE we have the following:

```
Plot1 Plot2 Plot3
\Y1▤X^3-2
\Y2=
\Y3=
\Y4=
\Y5=
\Y6=
\Y7=
```

```
 X    Y1
-4    -66
-3    -29
-2    -10
-1    -3
 0    -2
 1    -1
 2    6
X=-4
```

This view of the table provides several additional ordered pair solutions to the equation $y = x^3 - 2$.

▌ FUNCTIONS

All of the equations we have examined have formed a relationship between a specific value for *x* and a specific value for *y*. We call the equations *relations*. A **relation** is merely a set of ordered pairs. When each *x* value is related to exactly one *y* value, we call the relation a *function*.

Function
A **function** is a collection of ordered pairs in which each *x* value of the ordered pair is paired with a unique *y* value.

EXAMPLE 7

The table of values represents the conversion of the measurement of cups to pints. Does the relationship represent a function?

x = number of cups	$\frac{1}{2}$	1	$1\frac{1}{2}$	2	$2\frac{1}{2}$
y = number of pints	$\frac{1}{4}$	$\frac{1}{2}$	$\frac{3}{4}$	1	$1\frac{1}{4}$

SOLUTION

Because each *x* value is paired with a unique (one and only one) *y* value, we have a function. In other words, $\frac{1}{2}$ cup is always equal to $\frac{1}{4}$ of a pint. We do not have to worry about $\frac{1}{2}$ cup sometimes equaling a different amount. The ordered pairs for this function would be $\{(\frac{1}{2}, \frac{1}{4}), (1, \frac{1}{2}), (1\frac{1}{2}, \frac{3}{4}), (2, 1), (2\frac{1}{2}, 1\frac{1}{4})\}$.

EXAMPLE 8

Is the formula for converting Fahrenheit temperature to Celsius a function? The formula is

$$y = \frac{5}{9}(x - 32)$$

where *y* = degrees Celsius and *x* = degrees Fahrenheit.

SOLUTION

We need to answer the question "does each *x* produce a unique *y*?" Select a value for *x*, say 41°. Then

$$y = \frac{5}{9}(41 - 32)$$

$$y = \frac{5}{9}(9) \qquad \text{Subtraction produces a unique result.}$$

$$y = 5 \qquad \text{Multiplication produces a unique result.}$$

Because the operations of addition, subtraction, multiplication, and division produce unique results, any equation that applies these operations to *x* will produce unique results for *y* and is thus a function.

Domain of a Function
The domain of a function is the set of all x values of the ordered pairs of the function.

Range of a Function
The range of a function is the set of all y values of the ordered pairs of the function.

EXAMPLE 9

In the conversion chart, identify the domain and range.

x = number of cups	$\frac{1}{2}$	1	$1\frac{1}{2}$	2	$2\frac{1}{2}$
y = number of pints	$\frac{1}{4}$	$\frac{1}{2}$	$\frac{3}{4}$	1	$1\frac{1}{4}$

SOLUTION   The domain is $\{\frac{1}{2}, 1, 1\frac{1}{2}, 2, 2\frac{1}{2}\}$. The braces indicate we have a **set**, a collection of objects. The range is $\{\frac{1}{4}, \frac{1}{2}, \frac{3}{4}, 1, 1\frac{1}{4}\}$. The ordered pairs for this function would be $\{(\frac{1}{2}, \frac{1}{4}), (1, \frac{1}{2}), (1\frac{1}{2}, \frac{3}{4}), (2, 1), (2\frac{1}{2}, 1\frac{1}{4})\}$.

EXAMPLE 10

An arching stream of water in a botanical garden fountain can be modeled by the function $y = -0.05x^2 + 2x$, where x represents the horizontal distance and y the vertical height. Complete the table of values and determine the domain and range.

x = horizontal distance	10	15	20
y = vertical height			

Height

Horizontal distance

Figure 3-3

SOLUTION

$y = -0.05x^2 + 2x$
$y = -0.05(10)^2 + 2(10)$
$y = -0.05(100) + 20$
$y = -5 + 20$
$y = 15$

$y = -0.05x^2 + 2x$
$y = -0.05(15)^2 + 2(15)$
$y = -0.05(225) + 30$
$y = -11.25 + 30$
$y = 18.75$

$y = -0.05x^2 + 2x$
$y = -0.05(20)^2 + 2(20)$
$y = -0.05(400) + 40$
$y = -20 + 40$
$y = 20$

Set the TABLE to begin at $x = 10$ and to be incremented by 5 to verify your results.

```
TABLE SETUP
 TblStart=10
 ΔTbl=5
Indpnt: Auto Ask
Depend: Auto Ask
```

X	Y1
10	15
15	18.75
20	20
25	18.75
30	15
35	8.75
40	0

X=10

x = horizontal distance	10	15	20
y = vertical height	15	18.75	20

This table translates to the ordered pairs (10, 15), (15, 18.75), and (20, 20), which means that the arch of water reaches a height of 15 feet when the arch travels 10 feet, a height of 18.75 feet when the arch travels 15 feet, and a height of 20 feet when the arch travels 20 feet.

The domain of the function, as indicated by the hand-constructed table, is {10, 15, 20} and the range is {15, 18.75, 20}. Remember, the domain and range contain many other values than those that are represented in our table.

EXAMPLE 11

A parking garage charges the building tenants $5.00 for valet service and $1.50 for each hour (or part of an hour) that they are parked.

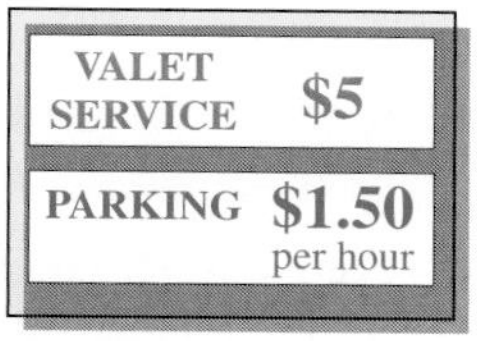

a. Write an equation where y is the total cost and x is the number of hours to model this relationship.

b. Is this relationship a function?

c. During any one 24-hour period, what is the domain and range?

| VALET SERVICE | $5 |
| PARKING | $1.50 per hour |

Figure 3-5

SOLUTION   **a.**   The unknown amount is the number of hours, thus

$$x = \text{number of hours parked}$$

We know that $1.50 is charged for each hour, thus

$$1.50x = \text{cost for parking}$$

and a flat fee of $5 is charged for valet service. The equation that models the total cost of parking is

$$y = 1.5x + 5$$

b. This equation is a function because the operations of multiplication (1.5 times x) and addition (plus 5) yield unique results. A table is helpful in picturing this relationship.

x = no. of hours	1	2	3	4	5	6	7	...	23	24
y = total cost	6.5	8	9.5	11	12.5	14	15.5	...	39.5	41

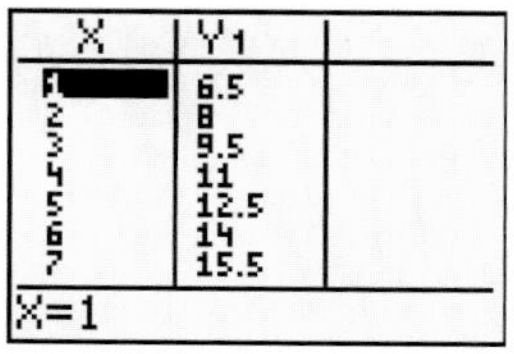

X	Y₁
1	6.5
2	8
3	9.5
4	11
5	12.5
6	14
7	15.5

X=1

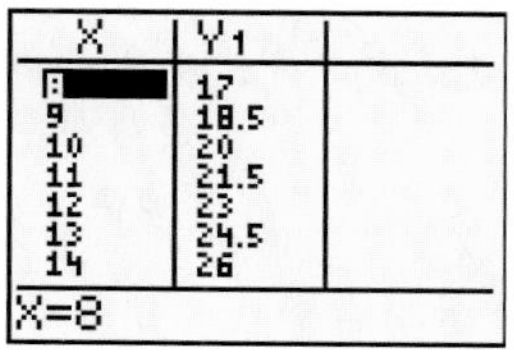

X	Y₁
8	17
9	18.5
10	20
11	21.5
12	23
13	24.5
14	26

X=8

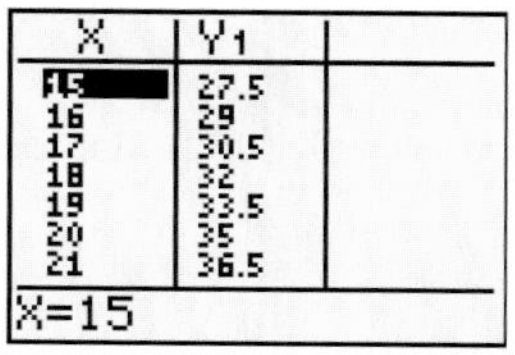

X	Y₁
15	27.5
16	29
17	30.5
18	32
19	33.5
20	35
21	36.5

X=15

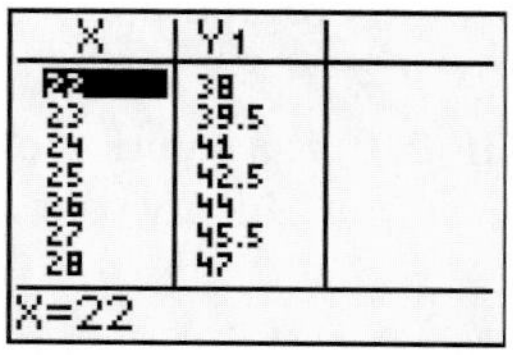

X	Y₁
22	38
23	39.5
24	41
25	42.5
26	44
27	45.5
28	47

X=22

c. The domain is {1, 2, 3, 4, ... , 23, 24} because there are 24 hours in a day. The range is {6.5, 8, 9.5, 11, 12.5, 14, 15.5, 17, 18.5, 20, 21.5, 23, 24.5, 26, 27.5, 29, 30.5, 32, 33.5, 35, 36.5, 38, 39.5, 41} because you will pay $5 plus $1.50 per hour for any given number of hours parked during this 24-hour period.

The equation in Example 10, $y = -0.05x^2 + 2x$, is a second-degree (also called *quadratic*) equation/function, and the equation in Example 6, $y = x^3 - 2$, is a third-degree equation/function. Throughout this text we will examine functions of different degrees and their behavior. Initially, however, our focus will be on first-degree (also called *linear*) equations/functions. The last example demonstrated a linear function.

EXERCISE 3.3

VOCABULARY AND NOTATION   *In Exercises 1–7, fill in the blanks to make a true statement.*

1. The equation $y = \frac{5}{9}(x - 32)$ is an equation in _____ variables.

2. Solutions to equations in two unknowns must be expressed as _____.

3. ΔTbl determines how the TABLE of the calculator is to be _____.

4. A ____ is a collection of objects.

5. A _____ is a collection of ordered pairs where each x value is paired with a unique y value.

6. The _____ is the set of all x coordinates in a collection of ordered pairs.

7. The _____ is the set of all y coordinates in a collection of ordered pairs.

CONCEPTS

8. Is the ordered pair $(-2, -2)$ a solution to the equation $y = 2x^3 - 5$?

9. Is the ordered pair $(2, 2)$ a solution to the equation $3x - 4y = 2$?

10. Is the ordered pair $(3, -5)$ a solution of the equation $3x - y = 14$?

11. Is the ordered pair $(-2, 6)$ a solution of the equation $5x + 2y = 2$?

12. If $(2, 0)$ is a solution of $2x + y = 4$, will $(0, 2)$ also be a solution? Justify your answer.

13. In general, if (a, b) is a solution to a linear equation in two variables, will (b, a) be a solution? Justify your answer.

14. If an equation produces the table of values displayed at the right, write all the displayed ordered pairs that are solutions to the equation.

X	Y₁
-2	-14
-1.5	-13
-1	-12
-.5	-11
0	-10
.5	-9
1	-8

X= -2

15. Does the table of values in Exercise 14 represent a function? Explain in complete sentences.

16. What is the pictured domain of the set of ordered pairs in Exercise 14?

17. What is the pictured range of the set of ordered pairs in Exercise 14?

PRACTICE   *In Exercises 18–33, find the corresponding y value for the given x value. Write the result as an ordered pair.*

18. $x + y = 2$; $x = 2$, $x = -3$

19. $x - y = 3$; $x = -1$, $x = 5$

20. $2x + y = -1$; $x = -3$, $x = 4$

21. $x + 3y = -4$; $x = -5$, $x = -8$

22. $y = 3x + 2$; $x = \frac{2}{3}$, $x = -\frac{5}{3}$

23. $y = -4x + 1$; $x = \frac{1}{4}$, $x = -\frac{3}{4}$

24. $x^2 = y$; $x = -3$, $x = 9$

25. $x^2 - 20 = y$; $x = 10$, $x = -5$

26. $y = 2(x - 1)^2$; $x = -3$, $x = 4$

27. $y = 3(x + 2)^2$; $x = 2$, $x = -6$

28. $y = \frac{1}{4}(x + 2)^2$; $x = 0$, $x = 3$

29. $y = \frac{2}{3}(x + 3)^2 - 2$; $x = -1$, $x = 2$

30. $y = x^3$; $x = 4$, $x = -5$

31. $y = -x^3$; $x = -2$, $x = 10$

32. $y = -\frac{2}{3}x^3 + 5$; $x = 3$, $x = -6$

33. $y = \frac{4}{5}x^3 - 1$; $x = 1$, $x = -1$

In Exercises 34–37, complete the table of values for the given equation.

34. $5x + y = 7$

x	y
-2	
0	
2	

35. $5x + y = 7$

x	y
	-3
	0
	3

36. $\frac{1}{2}x + \frac{2}{3}y = 7$

x	y
-4	
2	
6	

37. $\frac{1}{2}x + \frac{2}{3}y = 7$

x	y
	-6
	0
	3

In Exercises 38–47, determine if each relationship or equation defines a function and determine the domain and range.

38. $\{(2, 1), (3, -2), (4, 5)\}$

39. $\{(-4, 5), (5, -4), (3, -4)\}$

40. $\{(4, 3), (4, 5), (4, 7)\}$

41. $\{(2, 5), (5, 2), (2, 0)\}$

42.

X	Y1
-8	-26
-7	-24
-6	-22
-5	-20
-4	-18
-3	-16
-2	-14

X= -8

43.

X	Y1
-1	-8
0	-5
1	-2
2	1
3	4
4	7
5	10

X= -1

44. $y = 2x + 3$

45. $2x + y = 1$

46. $y = x^2 + 5$

47. $y = x^3 - 1$

APPLICATIONS

48. Rental car   A rental company charges $25 plus $0.21 per mile to rent a car. The linear equation $y = .21x + 25$ models the problem, where y represents the total cost and x is the number of miles traveled. Fill in the table of values to determine the cost, y, for each of the given mileages.

x	5	25	100	150
y				

Answer each of the following questions in a complete sentence:

a. What is the cost of traveling 25 miles?

b. What is the cost of traveling 150 miles?

c. If the company increased the cost per mile to $0.25, what would the new equation be?

d. What would the cost be for traveling 150 miles with the new price?

49. Entertainment   The county fair charges $5 gate admission and $0.75 per carnival ride. The linear equation that models the relationship between the total cost, y, and the number of rides, x, is $y = .75x + 5$. Fill in the table of values to determine the cost of attending the fair if riding the indicated number of rides.

x	0	5	10	15
y				

Answer each of the following questions in a complete sentence:

a. What is the cost of riding 5 rides?

b. What is the cost of riding 15 rides?

c. If the carnival decreased the cost per ride to $0.50, what would the new equation be?

d. What would the cost be for riding 15 rides with the new price?

50. Depreciation   A car that originally costs $18,000 depreciates $1,200 per year. The linear equation that models this depreciation is $y = -1,200x + 18,000$, where x represents the number of years and y represents the value of the car.

a. Find the value of the car after 8 years.

b. After how many years will the car have no value?

51. Appreciation   A house purchased in 1987 for $82,000 increases in value at a rate of $1,250 each year. If y represents the value of the house and x represents the number of years, the linear equation that models this appreciation is $y = 1,250x + 82,000$.

a. Find the value of the house in 12 years.

b. When will the house have a value of $92,000?

52. Tuition   The cost of tuition at a local university is $112 per credit hour. The linear equation that models this is $y = 112x$, where x represents the number of course hours taken and y is the cost.

   a. Find the cost of taking 6 credit hours.

   b. Find the cost of taking 8 credit hours.

   c. If a student could only afford to pay $500 for the semester, what is the maximum number of credit hours that could be taken?

53. Tuition   The cost of tuition at a local community college is $75 per credit hour. The linear equation that models this is $y = 75x$, where x represents the number of course hours taken and y is the cost.

   a. Find the cost of taking 4 credit hours.

   b. Find the cost of taking 10 credit hours.

   c. If a student could only afford to pay $500 for the semester, what is the maximum number of credit hours that could be taken?

In Exercises 54–59, write a linear equation in the form $y = mx + b$ to model the problem. Use your model to complete the given tables.

54. Tutoring   A tutor charges $15 for a study guide and $12 per hour for tutoring. After determining the equation to model this relationship, fill in the table of values to determine the tutoring costs for the indicated hours.

hours	2	6	$8\frac{1}{2}$
cost			

55. Auto repairs   A repair service charges a $20 "diagnostic fee" and $37.50 per hour for repairs. After determining the equation to model this relationship, fill in the table of values to determine the total repair costs.

hours	1 hour	1.5 hours	5 hours	5.75 hours
cost				

56. Appreciation   An antique gold necklace purchased in 1950 originally cost $12. Its current appraised value indicates it has been increasing in value at the rate of $8.20 per year.

   a. Write an equation to model this appreciation.

   b. What is the necklace's value in 50 years?

   c. If it continues to appreciate at its current rate, after how many years will it have a value of $627?

57. Depreciation   A computer system purchased for $4,000 depreciates at a rate of $800 per year. Write an

equation to model this relation and determine the value of the computer after 4 years.

58. Distance   The distance, y, that a vehicle travels can be determined by multiplying the rate of speed times the time, x.

   a. Write an equation in x and y to model the distance traveled by a vehicle that averages 55 mph.

   b. Fill in the table to determine the designated distance.

x	2	3.5	$4\frac{3}{4}$
y			

59. Child care   If the campus day care charges $5.50 per hour for "drop-ins," write an equation to model the cost, y for x number of hours.
Use your model to determine the cost for 3.5 hours of drop-in day care.

REVIEW

60. Evaluate $x^4 y^3$ for $x = -2$ and $y = 3$.

61. Let $P(x) = 3x^2 - 5x + 2$.

   a. State the degree of the polynomial.

   b. Find the value of $P(-3)$.

62. Simplify: $(2x^2 - 3x + 5) - (3x^2 - 5x + 4)$.

63. Find the volume of the figure in Illustration 1.

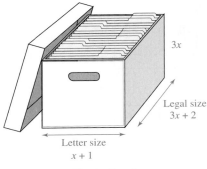

3x

Legal size
3x + 2

Letter size
x + 1

Illustration 1

64. Perimeter   The perimeter of a square is $12 + 16x$. Write an expression involving division to determine the length of a side. Simplify your expression.

65. Class grades   An algebra class has 36 students. One-third of the class made an A. How many students did not make an A?

66. Interest   Jorge signs a one-year note to borrow $9,000 for tuition. If the rate of interest is $7\frac{1}{4}\%$, how much interest must he pay?

3.4 THE RECTANGULAR COORDINATE SYSTEM

- REVIEW OF ORDERED PAIR SOLUTIONS
- THE RECTANGULAR COORDINATE SYSTEM
- GRAPHING LINEAR EQUATIONS
- GRAPHING WITH A GRAPHING CALCULATOR
- GRAPHS OF NONLINEAR FUNCTIONS

▓ REVIEW OF ORDERED PAIR SOLUTIONS

In Chapter 1 we drew one-dimensional graphs on a number line. We now want to draw two-dimensional graphs in a number plane. Just as each point on a number line is associated with a real number, each point in a number plane is associated with an ordered pair of real numbers.

The equation $3x + 2 = 5$ has the single variable x, and its only solution is 1. This solution can be graphed (or plotted) on the number line, as in the figure below.

−4 −3 −2 −1 0 1 2 3 4 5 6

The equation $x + 2y = 5$ contains the two variables x and y. The solutions of such equations are ordered pairs of numbers. For example, the pair $x = 1$ and $y = 2$, $(1, 2)$, is a solution, because the equation is satisfied when $x = 1$ and $y = 2$.

$$x + 2y = 5$$
$$1 + 2(2) = 5 \qquad \text{Substitute 1 for } x \text{ and 2 for } y.$$
$$1 + 4 = 5$$
$$5 = 5$$

The pair $x = 5$ and $y = 0$ $(5, 0)$, is also a solution, because the pair satisfies the equation:

$$x + 2y = 5$$
$$5 + 2(0) = 5 \qquad \text{Substitute 5 for } x \text{ and 0 for } y.$$
$$5 + 0 = 5$$
$$5 = 5$$

▓ THE RECTANGULAR COORDINATE SYSTEM

Solutions of equations with two variables can be plotted on a **rectangular coordinate system**, sometimes called a **Cartesian coordinate system** after the 17th-century French mathematician René Descartes. To design the McDonald's arches a mathematical model had to be created on just such a coordinate system.

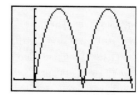

The rectangular coordinate system consists of two number lines, called the **x-axis** and the **y-axis**, drawn at right angles to each other, as shown in Figure 3-6. The two axes intersect at a point called the **origin**, which is the 0 point on each axis. The positive direction on the x-axis is to the right, and the positive direction on the y-axis is upward. The two axes divide the coordinate system into four regions, called **quadrants**, which are numbered as shown in Figure 3-6.

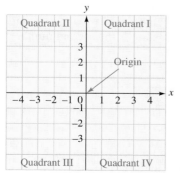

Figure 3-6

Each point P is associated with a unique ordered pair (x, y) of real numbers called the **coordinates of the point**.

EXPLORING THE CONCEPT

Use your graphing calculator to determine the sign of the x- and y-coordinates in each quadrant. Display the coordinate axes on your calculator graph screen by using the GRAPH feature. The directional arrow keys, ▶ ◀ ▲ ▼, will allow you to move the cursor (cross hairs) around the various quadrants. Observe the x- and y-coordinates as they are displayed at the bottom of the screen and record the sign of the coordinates in the table below.

Quadrant	Sign of x-coordinate	Sign of y-coordinate
I		
II		
III		
IV		

CONCLUSION

Quadrant I: $(+, +)$ Quadrant III: $(-, -)$

Quadrant II: $(-, +)$ Quadrant IV: $(+, -)$

To plot the ordered pair $(3, 2)$, where $x = 3$ and $y = 2$, we start at the origin and move 3 units to the right along the x-axis and then 2 units up in the positive y direction, as in Figure 3-7. This locates point A. Point A has an **x-coordinate** or **abscissa** of 3 and a **y-coordinate** or **ordinate** of 2. This information is denoted by the pair of numbers $(3, 2)$, called the **coordinates** of point A. Point A is the **graph** of the pair $(3, 2)$.

To plot the pair $(-4, -3)$, we start at the origin, move 4 units to the left along the x-axis, and then move 3 units down to point B in the figure. Point C has coordinates of $(2, 3)$.

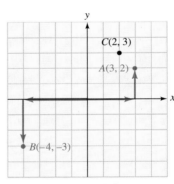

Figure 3-7

WARNING! The order of the coordinates of a point is important. Point A with coordinates of $(3, 2)$ is not the same as point C with coordinates $(2, 3)$. For this reason, the pair of coordinates (x, y) of a point is called an **ordered pair.** The x-coordinate is the first number in the pair, and the y-coordinate is the second number.

EXAMPLE 1

On a rectangular coordinate system, plot the points   **a.** $A(-1, 2)$,   **b.** $B(0, 0)$, **c.** $C(5, 0)$,   **d.** $D(-\frac{5}{2}, -3)$,   and   **e.** $E(3, -2)$.

SOLUTION   **a.** The point $A(-1, 2)$ has an x-coordinate of -1 and a y-coordinate of 2. To plot point A, we start at the origin and move 1 unit to the left and then 2 units up. (See Figure 3-8.) Point A lies in quadrant II.

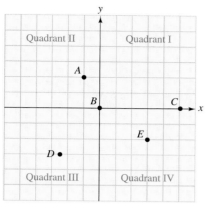

Figure 3-8

b. To plot point $B(0, 0)$, we start at the origin and move 0 units to the right and 0 units up. Because there was no movement, point B is the origin.

c. To plot point $C(5, 0)$, we start at the origin and move 5 units to the right and 0 units up. Point C lies on the x-axis, 5 units to the right of the origin.

d. To plot point $D(-\frac{5}{2}, -3)$, we start at the origin and move $\frac{5}{2}$ units ($\frac{5}{2}$ units = $2\frac{1}{2}$ units) to the left and then 3 units down. Point D lies in quadrant III.

e. To plot point $E(3, -2)$, we start at the origin and move 3 units to the right and then 2 units down. Point E lies in quadrant IV.

EXAMPLE 2

Braking distances Approximate braking distances for a car in good condition on dry pavement are given in Table 3-1.

TABLE 3-1

Speed (in miles per hour)	20	40	55	65
Braking distance (in meters)	6	22	46	60

Plot these points on a rectangular coordinate system.

SOLUTION If we plot the speeds along the x-axis and the braking distances along the y-axis, we obtain the graph shown in Figure 3-9. Notice that each axis can be incremented as desired. We have chosen increments of 10 units on both axes.

From the position of the points plotted in the figure, we can see that the braking distances are much longer at greater speeds.

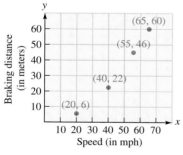

Figure 3-9

GRAPHING LINEAR EQUATIONS

We have seen that solutions to equations in x and y are ordered pairs of numbers. To find some ordered pairs (x, y) that satisfy $y = 5 - x$, we pick numbers at random, substitute them for x, and find the corresponding values of y. If we pick $x = 1$, we have

$$y = 5 - x$$
$$y = 5 - 1 \quad \text{Substitute 1 for } x.$$
$$y = 4$$

The ordered pair $(1, 4)$ satisfies the equation. If we let $x = 2$, we have

$$y = 5 - x$$
$$y = 5 - 2 \quad \text{Substitute 2 for } x.$$
$$y = 3$$

A second solution of the equation is $(2, 3)$. If we let $x = 5$, we have

$$y = 5 - x$$
$$y = 5 - 5 \qquad \text{Substitute 5 for } x.$$
$$y = 0$$

A third solution of the equation is $(5, 0)$. If we let $x = -1$, we have

$$y = 5 - x$$
$$y = 5 - (-1) \qquad \text{Substitute } -1 \text{ for } x.$$
$$y = 6$$

A fourth solution is $(-1, 6)$. If we let $x = 6$, we have

$$y = 5 - x$$
$$y = 5 - 6 \qquad \text{Substitute 6 for } x.$$
$$y = -1$$

A fifth solution is $(6, -1)$.

We list the ordered pairs $(1, 4)$, $(2, 3)$, $(5, 0)$, $(-1, 6)$, and $(6, -1)$ in the table shown in Figure 3-10. Connecting these points forms the **graph** of the equation, which is a straight line.

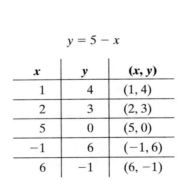

$y = 5 - x$

x	y	(x, y)
1	4	$(1, 4)$
2	3	$(2, 3)$
5	0	$(5, 0)$
-1	6	$(-1, 6)$
6	-1	$(6, -1)$

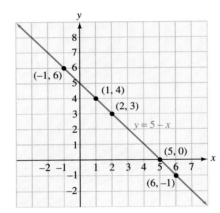

Figure 3-10

Any equation of the form $y = mx + b$ whose graph is a line is called a *linear equation in two variables*. Any point on the line has coordinates that satisfy the equation, and the graph of any pair (x, y) that satisfies the equation is a point on the line.

The graph is a way to picture all solutions of the equation. Because the equation has an infinite number of ordered pair solutions, it would be impossible to list them all.

Linear Function

An equation that can be written in the form $y = mx + b$, where m and b are real numbers, is called a *linear function*.

Although only two points are needed to graph a linear equation, we often plot a third point as a check. If the three points do not lie on a line, at least one of them is in error.

Procedure for Graphing Linear Equations in *x* and *y*

1. Find two ordered pairs (x, y) that satisfy the equation by selecting a value for x and solving for the corresponding value of y. A third point provides a check.
2. Plot each resulting pair (x, y) on a rectangular coordinate system. If they do not lie on a line, check your calculations.
3. Draw the line passing through the points.

EXAMPLE 3

Graph the ordered pair solutions to the equation $y = 3x - 4$.

SOLUTION

If we substitute 1 for x, we get $y = -1$, and the ordered pair $(1, -1)$ is a solution. If we substitute 2 for x, we get $y = 2$, and the ordered pair $(2, 2)$ is a solution. If we substitute 3 for x, we get $y = 5$, and the ordered pair $(3, 5)$ is a solution. These ordered pairs appear in the table accompanying Figure 3-11. We plot these points and join them with a line, which is the graph of the solutions of the equation $y = 3x - 4$. Every point that lies on this line is a solution to the equation $y = 3x - 4$, and every ordered pair whose coordinates satisfy the equation $y = 3x - 4$ lies on the line.

$y = 3x - 4$

x	*y*	*(x, y)*
1	−1	(1, −1)
2	2	(2, 2)
3	5	(3, 5)

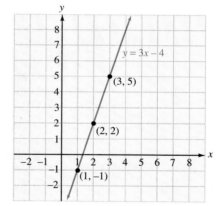

Figure 3-11

Scroll through the TABLE on your calculator to view other ordered pairs that are solutions to the equation. A few of the possible pairs are displayed below.

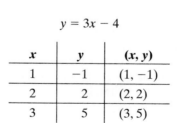

▮ GRAPHING WITH A GRAPHING CALCULATOR

TECHNOLOGY TIP

TI-83 Plus: Set your viewing window to ZStandard by pressing [ZOOM] [6]. TI-86: Set your viewing window to ZStandard by pressing [GRAPH] [F3](ZOOM) [F4](ZSTD).

When you sketch a graph on your graph paper, the size of your graph is limited by the size of your paper space. Theoretically, each axis extends indefinitely, which we represent by arrows at the end of each axis. However, the number of tic marks you place on each axis limits its visual size. Similar limitations occur when using a graphing calculator to display a coordinate system. You can limit the visual size of your calculator graph by designating a minimum and maximum value for the x-axis and a minimum and maximum value for the y-axis.

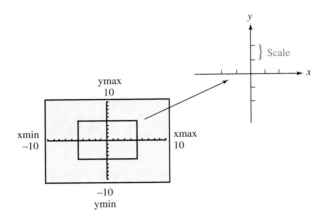

To condense this information, we will use the following notation:

$$\underset{[-10, 10]}{\overset{\text{xmin xmax}}{}} \quad \text{by} \quad \underset{[-10, 10]}{\overset{\text{ymin ymax}}{}}$$

Observe that there are 10 tic marks to the left, right, above, and below the origin on the calculator graph. The distance between the tic marks is called the *scale*. On this graph the scale for both the x and y axes is 1. We would write this as xscl = 1 and yscl = 1. Unless otherwise noted, it should be assumed xscl = 1 and yscl = 1 on all graphs displayed in this text.

In order to set the limitations on your calculator graph, you should be familiar with the key that accesses this feature. It may be denoted as "WINDOW," "RANGE," or "WIND" on your calculator. The viewing window above $[-10, 10]$ by $[-10, 10]$ is called the **standard viewing window**. It can be set manually by entering the information or automatically using the ZOOM feature.

Before proceeding to the examples, set your calculator viewing window to Zstandard either manually by entering the window values of $[-10, 10]$ by $[-10, 10]$ or automatically by using the ZOOM menu of your calculator.

EXAMPLE 4

To use the graphing calculator to graph the function $y = 3x - 4$ the expression $3x - 4$ should be entered at the y1= prompt on the y= screen. Display the graph of the equation.

SOLUTION

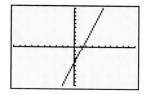

TECHNOLOGY TIP

Lines and curves that are graphed with the graphing calculator are actually straight lines and smooth curves. It is the resolution of the calculator screen that makes them appear "jagged."

$y = 3x - 4$

x	y
1	-1
2	2
3	5

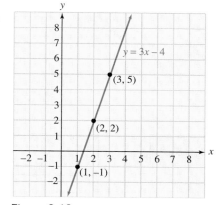

Figure 3-12

TECHNOLOGY TIP

Both the standard and integer viewing windows can be automatically set by accessing them from the ZOOM menu. See "Technology 3.4: Graphing Basics" on page 182 for details.

You may now access the TABLE feature of the calculator to compare your table of values, as shown in Figure 3-12, to those of the calculator.

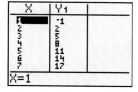

If you access the TRACE feature you can use the left and right cursor keys to trace along the line. You will notice that the coordinate values are cumbersome decimal values.

For this reason, we will *not* use the standard viewing window when we want to trace. The calculator has a preset window called ZInteger that yields integer values when tracing (see "Technology 3.4: Graphing Basics"). To set these values manually, enter $[-47, 47]$ by $[-31, 31]$ on your WINDOW screen now (TI-86 users set your WIND values as $[-63, 63]$ by $[-31, 31]$), with x and y scales equal to 10. Trace and observe the x and y coordinates.

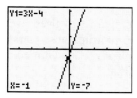 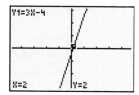 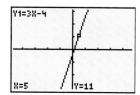

EXAMPLE 5

Graph $3x + 2y = 6$ in the ZInteger window: $[-47, 47]$ by $[-31, 31]$ (TI-86: $[-63, 63]$ by $[-31, 31]$), with x and y scales equal to 10. Trace to confirm the values displayed in the table in Figure 3-13.

SOLUTION First, solve the equation for y in order to enter it on the y= screen.

$$3x + 2y = 6$$
$$3x - 3x + 2y = 6 - 3x$$
$$2y = 6 - 3x$$
$$\frac{2y}{2} = \frac{6 - 3x}{2}$$
$$y = 3 - \frac{3}{2}x$$
$$y = -\frac{3}{2}x + 3$$

$3x + 2y = 6$

x	y	(x, y)
0	3	$(0, 3)$
2	0	$(2, 0)$
4	−3	$(4, -3)$

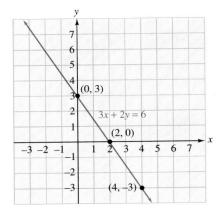

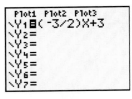

ZInteger

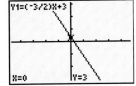

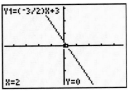

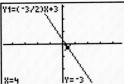

Figure 3-13

EXAMPLE 6

Graph the equations **a.** $y = 3$ and **b.** $x = -2$.

SOLUTION **a.** We can write the equation $y = 3$ in standard form as $0x + y = 3$. Since the coefficient of x is 0, the numbers assigned to x have no effect on y. The value of y is always 3. For example, if we replace x with -3, we get

$$0x + y = 3$$
$$0(-3) + y = 3$$
$$0 + y = 3$$
$$y = 3$$

The table in Figure 3-14 gives several ordered pairs that satisfy the equation $y = 3$. After plotting these pairs $(x, 3)$ and joining them with a line, we see that the graph of $y = 3$ is a horizontal line, parallel to the x-axis and intersecting the y-axis at 3.

$y = 3$

x	y	(x, y)
−3	3	$(-3, 3)$
0	3	$(0, 3)$
2	3	$(2, 3)$
4	3	$(4, 3)$

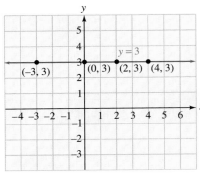

Figure 3-14

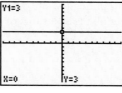
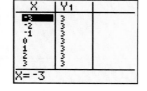

[−10, 10] by [−10, 10]

b. We can write the equation $x = -2$ in standard form as $x + 0y = -2$. Since the coefficient of y is 0, the values of y have no effect on x. The number x is always negative 2. A table of values and the graph are shown in Figure 3-15.

The graph of $x = -2$ is a vertical line, parallel to the y-axis and intersecting the x-axis at negative 2.

$x = -2$

x	y	(x, y)
-2	-2	$(-2, -2)$
-2	0	$(-2, 0)$
-2	2	$(-2, 2)$
-2	3	$(-2, 3)$

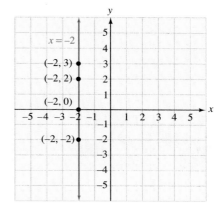

Figure 3-15

The graphing calculator will graph linear functions in the form $y = mx + b$ and linear functions that are horizontal lines, $y = b$, as in Example 6a. However, the calculator cannot graph the equation of vertical lines, $x = a$, as in Example 6b, because these lines are not functions. If you examine the table of ordered pairs for the graph of the line $x = -2$, you will discover that each x value **does not** have a *unique* y value. Thus, $x = -2$ is *not* a function and cannot be graphed with your calculator.

From the results of Example 6, we have the following facts.

Equations of Lines Parallel to the Coordinate Axes

The equation $y = b$ represents a **horizontal line** that intersects the y-axis at $(0, b)$. If $b = 0$, the line is the x-axis.

The equation $x = a$ represents a **vertical line** that intersects the x-axis at $(a, 0)$. If $a = 0$, the line is the y-axis.

▌ GRAPHS OF NONLINEAR FUNCTIONS

Not all equations are linear; therefore, not all graphs are straight lines. The next two examples graph the solutions of equations that are not linear. Notice that the number of coordinates used to determine the graph is **not** three. It is important that you find enough x and y values to be able to connect points to form an accurate graph of all the solutions of the given equation.

EXAMPLE 7

Graph the ordered pair solutions to $y = x^2$. Use the specified x values to determine your graph.

x	y
-3	
-2	
-1	
0	
1	
2	
3	

SOLUTION   The equation $y = x^2$ is a second degree equation and is therefore not linear. Complete the table of values by substituting each value for x into the equation and solving for y.

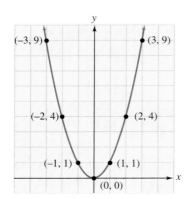

Figure 3-16

x	y
-3	9
-2	4
-1	1
0	0
1	1
2	4
3	9

Verify your table of values by using the TABLE feature of your calculator.

Plot these points and connect them with a smooth curve.

EXAMPLE 8

Graph the ordered pair solutions to $y = x^3$ by using the selected x values to determine your graph.

x	y
-2	
-1	
0	
1	
2	

SOLUTION   The equation $y = x^3$ is a third degree equation and is therefore not linear. Complete the table of values by substituting each value for x into the equation and solving for y.

x	y
−2	−8
−1	−1
0	0
1	1
2	8

Verify your table of values by using the TABLE feature of your calculator.

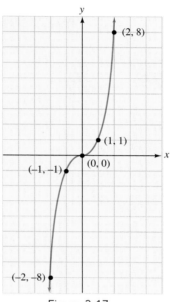

Plot these points and connect them with a smooth curve.

Figure 3-17

EXERCISE 3.4

VOCABULARY AND NOTATION   *In Exercises 1–10, fill in the blanks to make a true statement.*

1. Solutions of equations in two variables can be plotted on a _____.

2. The ___ axis is the horizontal axis and the ___ axis is the vertical axis.

3. The two axes intersect at the _____.

4. The two axes divide the coordinate system into four regions called _____.

5. Another name for the *x*-coordinate of the ordered pair (x, y) is the _____.

6. Another name for the *y*-coordinate of the ordered pair (x, y) is the _____.

7. An equation in the form $Ax + By = C$ is called the _____ form for the equation of a line.

8. The calculator viewing window indicated at the right is called the _____ viewing window.

```
WINDOW
Xmin=-10
Xmax=10
Xscl=1
Ymin=-10
Ymax=10
Yscl=1
Xres=1
```

9. The space between the tic marks on the axes is called the _____.

10. The TI-83 Plus viewing window that yields integer values when tracing has xmin = _____, xmax = ___, ymin = _____, and ymax = ___. It also has both scales set to 10.

CONCEPTS

11. In a–d below, plot each point on a rectangular coordinate system. Indicate in which quadrant each point lies.

a. $E(2, 5)$ **b.** $F(-3, 1)$

c. $G(-2, -3)$ **d.** $H(3, -2)$

12. In a–f below, plot each point on a rectangular coordinate system. Indicate on which axis each point lies.

    a. $F(0, 5)$ **b.** $E(0, -2)$

    c. $D(2, 0)$ **d.** $C(-4, 0)$

    e. $B(0, 0)$ **f.** $A(-5, 0)$

13. In a–h below, refer to Illustration 1 and find the coordinates of each point.

    a. H **b.** G **c.** F

    d. E **e.** B **f.** C

    g. D **h.** A

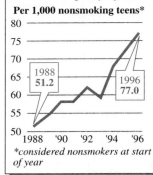

Illustration 1

14. Data analysis   Based on the graph in Illustration 2, express the information in the boxes, ┃1988┃ and ┃1996┃, as ordered pairs.
Explain the meaning of the ordered pairs as they relate to the graph.

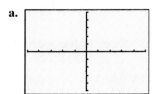

Rise in teen smoking

The incidents of teens becoming regular smokers jumped 50 percent between 1988 and 1996. A look at the rate of teens who started smoking daily, age 12-17:

Per 1,000 nonsmoking teens*

1988 51.2

1996 77.0

considered nonsmokers at start of year

SOURCE: Centers for Disease Control and Prevention

Illustration 2

15. Data analysis
Write an ordered pair that relates the information that "there were 1035 Hispanic students enrolled in Clarksville-Montgomery County Schools in 2001" based on the graph in Illustration 3.

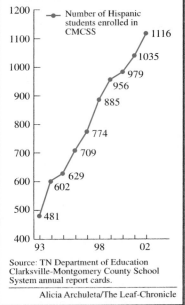

Hispanic student enrollment increases

The number of Hispanic students enrolled in Clarksville-Montgomery County schools has increased 132 percent from 1993-2002. The increase has put the county in need of translators and impacted programs, teachers and ESOL classes.

Number of Hispanic students enrolled in CMCSS

Source: TN Department of Education Clarksville-Montgomery County School System annual report cards.

Alicia Archuleta/The Leaf-Chronicle

Illustration 3

16. In a–d, which of the ordered pairs would be displayed in the viewing window that is indicated at the right?

    a. $(2, 5)$

    b. $(-3.7, 2.9)$

    c. $(4.8, -3.2)$

    d. $(-3.1, -4.7)$

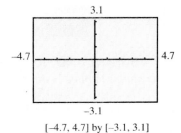

$[-4.7, 4.7]$ by $[-3.1, 3.1]$

17. In a–d, determine the viewing window and indicate it in the form [xmin, xmax] by [ymin, ymax]. Both the x and y scales are equal to 1.

    a.   **b.**

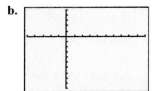

c. 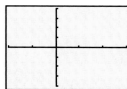   **d.**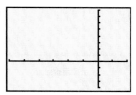

18. In a–d, sketch the set of coordinate axes indicated by the given viewing window.

a. $[-6, 6]$ by $[-6, 6]$
xscl = 1, yscl = 1

b. $[-4, 4]$ by $[-4, 4]$
xscl = 1, yscl = 1

c. $[-8, 8]$ by $[-8, 8]$
xscl = 2, yscl = 1

d. $[-8, 8]$ by $[-8, 8]$
xscl = 4, yscl = 4

19. In order to more fully understand why lines and curves on the graphing calculator appear jagged, shade every box that the curve on the graph in Illustration 4 passes through. This simulates the calculator action of lighting up the pixel squares that display a graph.

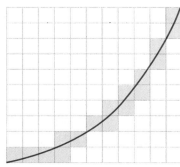

Illustration 4

20. Explain why the ZInteger viewing window is a better choice of viewing window than ZStandard when using the TRACE feature.

21. Discuss the disadvantages to displaying a graph in the ZInteger viewing window.

22. Which of the following are linear functions? In your own words, justify your answer.

a. $y = x^2 - 5x$ **b.** $2x - y = 6$
c. $y = x^4 + 2$ **d.** $y = x + 2$

23. Explain why you would want to graph three ordered pairs when graphing the equation of a line when it only takes two points to determine a line.

24. What is the equation of the vertical line pictured in Illustration 5? Is this the equation of a function? Explain.

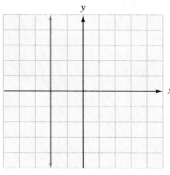

Illustration 5

25. What is the equation of the horizontal line pictured in Illustration 6? Is this the equation of a function? Explain.

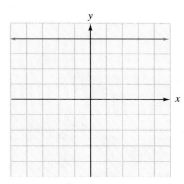

Illustration 6

26. Find the equation of the *x*-axis.

27. Find the equation of the *y*-axis.

PRACTICE *In Exercises 28–41, complete the table of solutions for each equation and graph the equation on graph paper. Confirm both the table and the graph with your calculator. (Hint: Be sure the calculator is in the ZInteger viewing window so you can TRACE.)*

28. $y = x + 2$

x	y
3	
1	
−2	

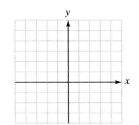

29. $y = x - 4$

x	y
5	
4	
−1	

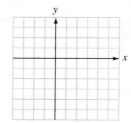

30. $y = -2x$

x	y
2	
1	
−3	

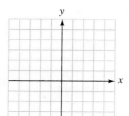

31. $y = \dfrac{x}{2}$

x	y
1	
−1	
−4	

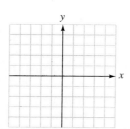

32. $y = 2x - 1$

x	y
3	
−1	
−2	

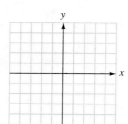

33. $y = 3x + 1$

x	y
−2	
0	
1	

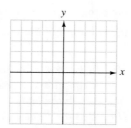

34. $y = \dfrac{x}{2} - 2$

x	y
8	
0	
−2	

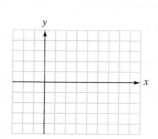

35. $y = \dfrac{x}{3} - 3$

x	y
6	
0	
−3	

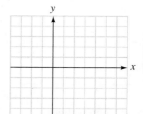

36. $y = x^2 + 2$

x	y
−2	
−1	
0	
1	
2	

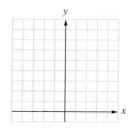

37. $y = x^2 - 3$

x	y
−2	
−1	
0	
1	
2	

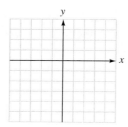

38. $y = (x - 5)^2$

x	y
3	
4	
5	
6	
7	

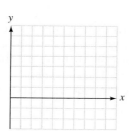

39. $y = (x + 9)^2 - 7$

x	y
−11	
−10	
−9	
−8	
−7	

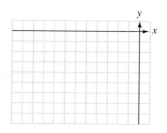

40. $y = x^3 + 4$

x	y
-2	
-1	
0	
1	
2	

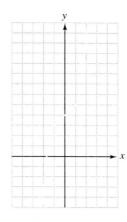

41. $y = x^3 - 4$

x	y
-2	
-1	
0	
1	
2	

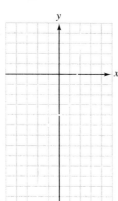

In Exercises 42–53, graph each equation on graph paper and verify your solution graphically when possible.

42. $y = -5$

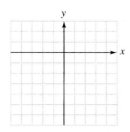

43. $x = 4$

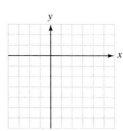

44. $x = 5$

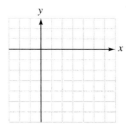

45. $y = 4$

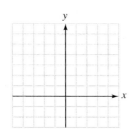

46. $y = 0$

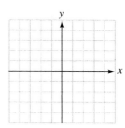

47. $x = 0$

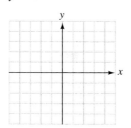

48. $2x = 5$

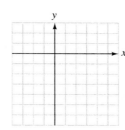

49. $3y = 7$

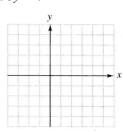

50. $3(x + 2) + x = 4$

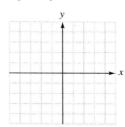

51. $2(y + 3) - y = 5$

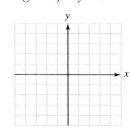

52. $3(y - 2) + 2 = y$

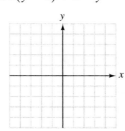

53. $4(x + 2) - 2(x - 1) = 6$

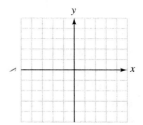

APPLICATIONS

54. Stopping distances Approximate stopping distances for a car in good condition on dry pavement, including the normal $\frac{3}{4}$-second reaction time, are as shown in Table 1. Plot the points and draw any appropriate conclusions.

Speed (in miles per hour)	20	40	55	65
Stopping distance (in meters)	12	35	64	80

TABLE 1

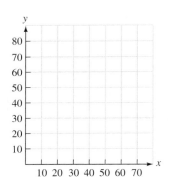

55. Tutoring A tutor charges $12 an hour for tutoring mathematics. Complete Table 2 showing the tutor's fee for working 1, 2, 3, and 5 hours.

Hours worked	1	2	3	5
Tutor's fee				

TABLE 2

56. Tutoring Let y represent the tutor's fee for working x hours. Plot the pairs found in Exercise 55. Do the points lie on a line?

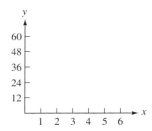

57. Stopping distances Consuming two or three drinks can easily increase the reaction time to put on the brakes by 1 second. Approximate stopping distances for a car in good condition on dry pavement, including $1\frac{3}{4}$ seconds for reaction time, are as shown in Table 3. Plot the points and draw any appropriate conclusions.

Speed (in miles per hour)	20	40	55	65
Stopping distance (in meters)	22	53	87	109

TABLE 3

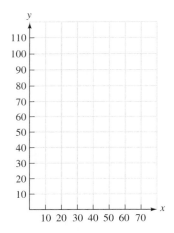

58. Maximum heart rate A rule of thumb for a person's maximum heart rate for safe aerobic exercise is 220 minus the person's age. Complete Table 4 showing the maximum heart rate for persons 20, 30, 40, and 60 years old.

Age	20	30	40	60
Heart rate				

TABLE 4

59. Let y represent the maximum heart rate for a person x years old. (See Exercise 58.) Write an equation that relates x and y and graph it.

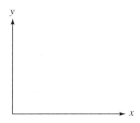

60. Installing computer systems A computer system installer charges $10 for materials and $40 per hour. Complete Table 5 showing the charge for jobs taking 1, 2, 3, and $5\frac{1}{2}$ hours.

Hours worked	1	2	3	$5\frac{1}{2}$
Charge				

TABLE 5

61. Let y represent the charge for working x hours. Plot the pairs found in Exercise 60. Do the points lie on a line?

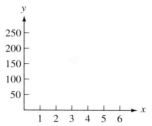

REVIEW

62. The area of a kitchen floor is $12x^2 + 5x - 2$. If the length of one side of the room is $3x + 2$ feet, find the polynomial that represents the length of the adjacent side.

63. A circle has a radius of $2x^2$. Find the area, leaving π as part of your answer.

64. The calculator screen indicates that an equation with a supposed solution of $x = 4$ is being checked. Does the display verify that 4 is a solution or does it dispute it? Explain.

```
4→X:3(X-1)
                    9
2X²-5X-2
                   10
```

65. Verify that $F = 95$ is a solution to $35 = \dfrac{5}{9}(F - 32)$.

Use your calculator for the verification.

66. Classify these equations as an identity, contradiction, or conditional equation.

  a. $2x - 1 = 5$

  b. $2x + 4 = 2(x - 2)$

  c. $2(x + 1) = 2(x + 3) - 4$

TECHNOLOGY 3.4

Graphing Basics

Read the headings carefully to be sure that you are following the directions for your specific calculator model. All the TI-86 information begins on page 185.

Setting Up the Graph Display: TI-83 Plus

The calculator's display is controlled through the **MODE** and **WINDOW FORMAT** screens.

1. Press the **[MODE]** key. **MODE** controls how numbers and graphs are displayed and interpreted. The current settings on each row should be highlighted as displayed. The blinking rectangle can be moved using the four **cursor** (arrow) keys. To change the setting on a particular row, move the blinking rectangle to the desired setting and press **[ENTER]**.

Note: Items must be highlighted to be activated.

Normal vs. scientific notation

Floating decimal vs. fixed to 9 places

Type of angle measurement

Type of graphing: function, parametric, polar, sequence

Graphed points connected or dotted

Functions graphed sequentially or simulatneously

Results displayed with real or complex numbers

Full screen, horizontally split screen, or a vertically split screen that displays a graph and table simultaneously

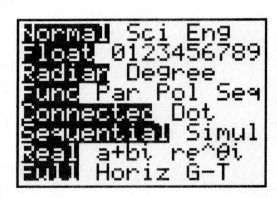

2. To return to the home screen at this point, press **[CLEAR]** or **[2nd] [QUIT]**.

3. Press **[y=]**. The calculator can graph up to ten different equations at the same time. Because **MODE** is in the sequential setting, the graphs will be displayed sequentially. Note that cursoring down accesses additional **y=** prompts. The display of the equations entered on the **y=** screen is controlled by the size of the viewing window. The dimensions of the viewing window are determined by the values entered on the **WINDOW** screen.

This calculator has a feature called the *graph style icon* that allows you to distinguish between the graphs of equations. Press **[y=]** and observe the "\" in front of each y. Use the left arrow to cursor over to this "\". The diagonal should now be moving up and down. Pressing **[ENTER]** once changes the diagonal from thin to thick. The graph of y1 will now be displayed as a thick line. Repeatedly pressing **[ENTER]** displays the following:

  ◥: shades above y1
  ◣: shades below y1
 −o: traces the leading edge of the graph followed by the graph
  o: traces the path but does not plot
 ⋰: displays graph in dot, not connected, MODE

It is important to remember that the graphing icon takes precedence over the **MODE** screen. If the icon is set for a solid line and the **MODE** screen is set for DOT and not solid, the graphing icon will determine how the graph is displayed. Pressing [**CLEAR**] to delete an entry at the **y=** prompt will automatically reset the graphing icon to default, a solid line.

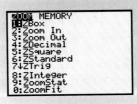

4. Press [**ZOOM**] [**6:ZStandard**] [**WINDOW**]. This is called the standard viewing window. The information on this screen indicates that in a rectangular coordinate system the *x*-values will range from −10 to 10 and the *y*-values will range from −10 to 10. The interval notation for this is [−10, 10] by [−10, 10]. The xscl=1 and yscl=1 settings indicate that the tic marks on the axes are one unit apart. The values entered on this screen may be changed by using the cursor arrows to move to the desired line and typing over the existing entry. When drawing a graph, you may set the desired viewing rectangle on the calculator as well as scale the *x*-axis and *y*-axis. This calculator has a row labeled Xres=. This determines the screen resolution. It should be set equal to 1, which means that each pixel on the *x*-axis will be evaluated and graphed.

5. Access **FORMAT** (located above the ZOOM key) by pressing [**2nd**] [**FORMAT**]. The following settings should be highlighted:

Graphs on a rectangular coordinate system

Cursor location is displayed on screen

Graphing grid is not displayed

Axes are visible

Axes are not labeled with an *x* and *y*

If **ExprOn** is highlighted, the equation is displayed on the graph screen when the TRACE feature is activated.

Altering the Viewing Window: TI-83 Plus

The calculator viewing window is 95 pixel points wide by 63 pixel points high; there are 94 horizontal spaces and 62 vertical spaces to light up. When tracing on a graph, the readout changes according to the size of the space. The size of the space can be controlled by the following formulas:

$$\frac{\text{Xmax} - \text{Xmin}}{94} = \text{horizontal space width}, \quad \frac{\text{Ymax} - \text{Ymin}}{62} = \text{vertical space height}$$

We will now examine some preset viewing windows and how they affect the pixel space size.

6. Press [**ZOOM**]. There are 10 entries on this screen. The down arrow key can be used to view remaining entries.

1: Boxes in and enlarges a designated area.
2: Acts like a telephoto lens and "zooms in."
3: Acts like a wide-angle lens and "zooms out."
4: Moves cursor $\frac{1}{10}$ of a unit per move.
5: "Squares up" the previously used viewing window.
6: Sets axes to [−10, 10] by [−10, 10].
7: Used when graphing trigonometric functions.

8: Moves cursor ONE integer unit per move.
9: Used when graphing statistics.
0: Adjusts the viewing window to fit the graph.

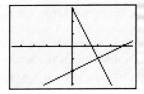

7. a. **ZDecimal** is useful for graphs that require the use of the calculator's TRACE feature. Enter Y1 = −2X + 3 and Y2 = 1/2X − 2 on the **Y=** screen. Press **[ZOOM] [4:ZDecimal]** to display the graph of these two lines in the ZDecimal viewing window. TRACE along the graph of one of the lines and observe the changes in x-values. The change should be $\frac{1}{10}$ of a unit. (Remember, the y-values are dependent on the values selected for x.)

b. **ZInteger** is useful for application problems where the x-value is valid only if represented as an integer (such as when x equals the number of tickets sold, number of passengers in a vehicle, etc.). ZInteger is the only preset viewing window that allows you to set the cursor to recenter the axes and the only preset window that *requires* you to press **[ENTER]** to activate the viewing window.

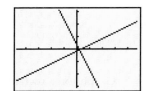

Press **[ZOOM] [8:ZInteger]**, move the cursor to the origin of the graph (x = 0 and y = 0), and press **[ENTER]** to display the graph of these two lines, as pictured. TRACE along the graph of one of the lines and observe the x-values changing by 1 unit.

c. **ZStandard** provides a good visual comparison between hand-sketched graphs (or textbook graphs) that are approximately [−10, 10] by [−10, 10]. The x-values will change by .212765974 each time the TRACE cursor is moved. If you are using the TRACE feature, you will usually want a screen with "friendlier" x-values than this one provides.

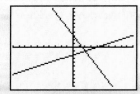

Press **[ZOOM] [6:ZStandard]** to display the graph of these two lines as pictured. TRACE along the graph of one of the lines and observe the changes in x-values.

8. **ZDecimal** frequently does not provide a large-enough viewing window. When this is the case, you may multiply the Xmin and Xmax by the same constant and the Ymin and Ymax by the same constant to produce a larger viewing rectangle that still provides cursor moves in tenths of units. Multiplying Xmin and Xmax by 2 would mean cursor moves of $\frac{2}{10}$ of a unit, whereas multiplying by 3 would mean cursor moves of $\frac{3}{10}$ of a unit. The screen pictured is the **ZDecimal** screen with the max and min values multiplied by 2. This WINDOW will be referred to in the future as ZDecimal × 2.

Setting Up the Graph Display: TI-86

The calculator's display is controlled through the **MODE** and **GRAPH/FORMAT** screens.

1. Press **[2nd] [MODE]**. The current settings on each row should be highlighted as displayed. The blinking rectangle can be moved using the four **cursor** (arrow) keys. To change the setting on a particular row, move the blinking rectangle to the desired setting and press **[ENTER]**.

Note: Items must be highlighted to be activated.

Normal vs. scientific notation

Floating decimal vs. fixed to 11 places

Type of angle measurement

Complex number display

Type of graphing: function, polar, parametric, differential equation

Computations performed in bases other than base 10

Format of vector display

Type of differentiation

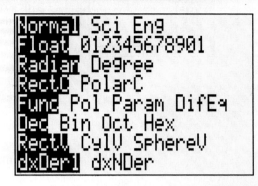

2. The **FORMAT** screen is accessed by pressing **[GRAPH] [MORE] [F3](FORMAT)**. The following settings should be highlighted:

Graphs on both rectangular and polar coordinate system

Cursor location is displayed on the screen

Graphed points are connected or discrete

Functions displayed sequentially or simultaneously

Graphing grid is not displayed

Axes are visible

Axes are not labeled with an *x* and *y*

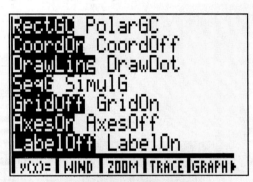

3. To return to the home screen at this point, press **[CLEAR]** or **[EXIT]**.

4. Press **[GRAPH] [F1](y(x)=)**. The calculator can graph up to 99 different equations at the same time. Because **MODE** is in the sequential setting, the graphs will be displayed sequentially. The display of the equation(s) entered on the **y(x)=** screen is controlled by the size of the viewing window. The dimensions of the viewing window are determined by the values entered on the **WIND** screen, which is referred to as the **WINDOW** screen in the text.

This calculator has a feature called the "graph style icon" that allows you to distinguish between the graphs of equations. Press **[GRAPH] [F1](y(x)=)** and observe the "\" in front of each y. Press **[MORE]** followed by **[F3](STYLE)**. The diagonal should now have changed from thin to thick. The graph of y1 will now be displayed as a thick line. Repeatedly pressing **[STYLE]** displays the following:

◥: shades above y1

◣: shades below y1

−o: traces the leading edge of the graph followed by the graph

o: traces the path but does not plot

∴: displays graph in dot, not connected, MODE

It is important to remember that the graphing icon takes precedence over the **MODE** screen. If the icon is set for a solid line and the **MODE** screen is set for DOT and not solid, the graphing icon will determine how the graph is displayed. Pressing **[CLEAR]** to delete an entry at the **y(x)=** prompt will automatically reset the graphing icon to default, a solid line.

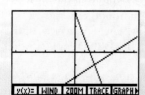

5. Press **[GRAPH] [F3](ZOOM) [F4](ZSTD) [2nd] [M2](WIND)**. This is called the standard viewing window. The information on this screen indicates that in a rectangular coordinate system the x-values will range from −10 to 10 and the y-values will range from −10 to 10. The interval notation for this is [−10, 10] by [−10, 10]. The xScl=1 and yScl=1 settings indicate that the tic marks on the axes are one unit apart. The values entered on this screen may be changed by using the cursor arrows to move to the desired line and typing over the existing entry. When drawing a graph, you may set the desired viewing window on the calculator as well as scale the x-axis and y-axis.

Remember: Every time the text indicates "press **[WINDOW]**" you will need to press **[GRAPH] [F2](WIND)**. When the text indicates "press **[GRAPH]**", you should press **[GRAPH] [F5](GRAPH)**.

Altering the Viewing Window: TI-86

The calculator viewing window is 127 pixel points wide by 63 pixel points high; there are 126 horizontal spaces and 62 vertical spaces to light up. When tracing on a graph, the readout changes according to the size of the space. The size of the space can be controlled by the following formulas:

$$\frac{xMax - xMin}{126} = \text{horizontal space width}, \quad \frac{yMax - yMin}{62} = \text{vertical space height}$$

We will now examine some preset viewing windows and how they affect the pixel space size. Press **[GRAPH] [F3](ZOOM)** to display a partial zoom menu.

6. a. **ZDecimal (ZDECM)** is useful for graphs that require the use of the calculator's TRACE feature. ZDecimal can be located on the ZOOM menu by pressing **[MORE]**. Enter y1 = −2x + 3 and y2 = (1/2)x − 2 on the y(x)= screen. Press **[2nd] [M3](ZOOM) [MORE] [F4](ZDECM)** to display the graph of these two lines in the ZDecimal viewing window. TRACE along the graph of one of the lines and observe the changes in x-values. The change should be $\frac{1}{10}$ of a unit. (Remember, the y-values are dependent on the values selected for x.)

b. **ZInteger** is useful for application problems where the x-value is valid only if represented as an integer (such as when x equals the number of tickets sold, number of passengers in a vehicle, etc.). Press **[F3](ZOOM) [MORE] [MORE] [F5](ZINT)**, move the cursor to the origin of the graph (x = 0 and y = 0), and press **[ENTER]** to display the graph of these two lines as pictured. TRACE along the graph of one of the lines and observe the changes in x-values. The change should be one unit.

c. **ZStandard** provides a good visual comparison between hand-sketched graphs (or textbook graphs) that are approximately $[-10, 10]$ by $[-10, 10]$. The x-values will change by 0.15873015873 each time the TRACE cursor is moved. If you are using the TRACE feature, you will usually want a screen with "friendlier" x-values than this one provides.

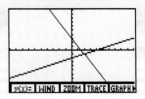

Press **[GRAPH] [F3](ZOOM) [F4](ZSTD)** to display the graph of these two lines as pictured. TRACE along the graph of one of the lines and observe the changes in x-values.

7. **ZDecimal** frequently does not provide a large-enough viewing window. When this is the case, you may multiply the xMin and xMax by the same constant and the yMin and yMax by the same constant to produce a larger viewing rectangle that still provides cursor moves in tenths of units. Multiplying xMin and xMax by 2 would mean cursor moves of $\frac{2}{10}$ of a unit, whereas multiplying by 3 would mean cursor moves of $\frac{3}{10}$ of a unit. The screen at the left is the ZDecimal screen with the max and min values multiplied by 2. This WINDOW will be referred to in the future as ZDecimal $\times$ 2.

8. Press **[GRAPH] [F3](ZOOM)**. Pressing **[MORE]** repeatedly will display the remaining menu selections and then return the display to the original screen.

ZBOX	Boxes in and enlarges a designated area.
ZIN	Acts like a telephoto lens and "zooms in."
ZOUT	Acts like a wide-angle lens and "zooms out."
ZSTD	Automatically sets standard viewing window to $[-10, 10]$ by $[-10, 10]$.
ZPREV	Resets **WIND** values to values used prior to the previous ZOOM operation.
ZFIT	Resets yMin and yMax on the WIND screen to include the minimum and maximum y-values that occur between the current xMin and xMax settings.
ZSQR	"Squares up" the previously used viewing window.
ZTRIG	Set the WIND to built-in trig values.
ZDECM	Sets cursor moves to $\frac{1}{10}$ of a unit per move.
ZDATA	Automatically sets the viewing window to accommodate statistical data.
ZRCL	Sets **WIND** values to those stored by the user (see ZSTO).
ZFACT	Sets the zoom factors used in **ZIN** and **ZOUT**.
ZOOMX	Bases graph display on xFact only when zooming in or out.
ZOOMY	Bases graph display on yFact only when zooming in or out.
ZINT	Moves cursor ONE integer unit per move.
ZSTO	Stores current WIND values for future use; values are recalled by **ZRCL**.

3.5   LINEAR APPLICATIONS AND MODELS

- x- AND y-INTERCEPTS
- SETTING GOOD VIEWING WINDOWS
- LINEAR APPLICATIONS
- LINEAR MODELS

x-AND y-INTERCEPTS

The points where a graph intersects the x- and y-axes are called **intercepts** of the graph.

x- and y-Intercepts

An **x-intercept** of a graph is a point $(a, 0)$ where the graph intersects the x-axis. To find a, substitute 0 for y in the equation of the graph and solve for x.

The **y-intercept** of a graph is the point $(0, b)$ where the graph intersects the y-axis. To find b, substitute 0 for x in the equation of the graph and solve for y.

Plotting the y- and x-intercepts to graph a first-degree equation is called the **intercept method of graphing**. This method is useful for graphing equations written in **standard form**.

Standard Form of the Equation of a Line

If A, B, and C are real numbers and A and B are not both 0, then the equation

$$Ax + By = C$$

is called the **standard form** of the equation of a line.

Whenever possible, we will write the standard form $Ax + By = C$ so that A, B, and C are integers and $A \geq 0$.

EXAMPLE 1

Graph the solutions to the equation $3x + 2y = 6$ by using the intercept method of graphing.

SOLUTION   To find the y-intercept, we let $x = 0$ and solve for y.

$$3x + 2y = 6$$
$$3(\mathbf{0}) + 2y = 6 \qquad \text{Substitute 0 for } x.$$
$$2y = 6 \qquad \text{Simplify.}$$
$$y = 3 \qquad \text{Divide both sides by 2.}$$

The y-intercept is the ordered pair $(0, 3)$.
To find the x-intercept, we let $y = 0$ and solve for x.

$$3x + 2y = 6$$
$$3x + 2(\mathbf{0}) = 6 \qquad \text{Substitute 0 for } y.$$
$$3x = 6 \qquad \text{Simplify.}$$
$$x = 2 \qquad \text{Divide both sides by 3.}$$

The x-intercept is the pair $(2, 0)$.

As a check, we plot one more point. If $x = 4$, then

$$3x + 2y = 6$$
$$3(4) + 2y = 6 \qquad \text{Substitute 4 for } x.$$
$$12 + 2y = 6 \qquad \text{Simplify.}$$
$$2y = -6 \qquad \text{Add } -12 \text{ to both sides.}$$
$$y = -3 \qquad \text{Divide both sides by 2.}$$

The point $(4, -3)$ lies on the graph. We plot these three points and join them with a line. The graph of $3x + 2y = 6$ is shown in Figure 3-18.

$$3x + 2y = 6$$

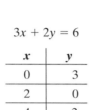

x	y
0	3
2	0
4	-3

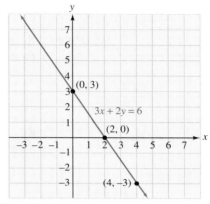

Figure 3-8

Verify the solutions and the graph with your calculator. To do this, $3x + 2y = 6$ must be solved for y.

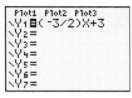

ZInteger

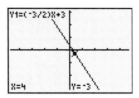

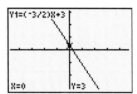

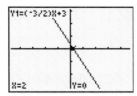

The graphs of nonhorizontal and nonvertical lines will have one x-intercept and one y-intercept.

▍▍ SETTING GOOD VIEWING WINDOWS

We will now focus on efficiently using our calculators to aid us in interpreting mathematical models represented by a graphical display. We must, therefore, obtain a graph that displays all necessary information. We will begin by displaying the x- and y-intercepts of the graph.

To find the *y*-intercept, simply press the TRACE key. The coordinates of the *y*-intercept will be displayed. You can then adjust your window (by changing either *y*-min or *y*-max) so that the intercept is displayed on the graph screen. We suggest that you choose a value that is either higher (for a positive *y*-intercept) or lower (for a negative *y*-intercept) for clarity. Once the *y*-intercept is displayed, you may either trace or make educated guesses to locate the *x*-intercept, adjusting the *x*-values on the window screen as appropriate. These techniques are illustrated in Example 2.

EXAMPLE 2

Graph the linear function $y = \dfrac{3}{4}x - 18$.

SOLUTION

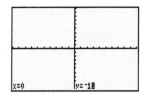

Begin by displaying the graph in the standard viewing window. (Use the appropriate ZOOM feature to obtain this window, or manually set your values for the *x*- and *y*-axes.) No part of the graph is visible.

Press TRACE to see the coordinates of the *y*-intercept, as pictured.

Readjust your window values so that the *y*-intercept is visible. (We chose a value of -25, which is smaller than the *y*-intercept.) Then view the graph in this window:

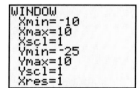
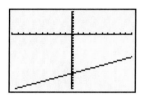

Because of the inclination of the line, we can tell that the *x*-intercept is positive. Changing our *x*-max value will allow us to see this intercept. The screens below show an *x*-max of 20, 30, 40, and 50, respectively. This illustrates the fact that many different values are acceptable.

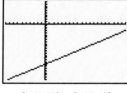
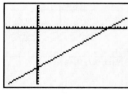
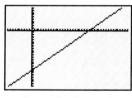
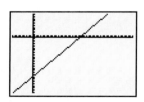

$[-10, 20]$ by $[-25, 10]$ $[-10, 30]$ by $[-25, 10]$ $[-10, 40]$ by $[-25, 10]$ $[-10, 50]$ by $[-25, 10]$

In all of the graphs pictured, the scale on both axes was equal to 1. Notice how difficult it becomes to see the tic marks as we increased our *x*-maximum. Remember, you may set the scales to any value you desire to include 0. A scale of 0 removes all tic marks from the axes.

EXAMPLE 3

Suppose $12,000 is invested in two accounts, a money market account earning $3\frac{1}{2}\%$ interest and a six-month certificate of deposit earning **5%** interest. If *x* dollars is invested in the money market account then there will be **$12,000 − x** left to invest in the CD. The equation $y = .035x + .05(12,000 - x)$ models the total interest earned by the two accounts. Graph this model in an appropriate viewing window.

SOLUTION   Upon entering the equation on the y-edit screen and displaying the graph in the standard viewing window, you discover that none of the graph is visible.

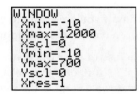

Pressing TRACE reveals that the y-intercept is located at $(0, 600)$. This means that the y-axis must be adjusted to display $(0, 600)$. We will set the ymax to 700.

We know that $y = .035x + .05(12{,}000 - x)$ *is not* a horizontal line because it is not in the form $y = b$. It is, however, linear because the equation $y = .035x + .05(12{,}000 - x)$ simplifies to $y = -.015x + 600$. For application models it is important to take into consideration what your variables represent as you are adjusting the viewing window. In this problem, x represents the amount invested in the $3\frac{1}{2}\%$ account. The *total* investment is $12,000, so it is safe to assume that x will not exceed $12,000. Thus we set the xmax to $12,000. The pictured graph has a viewing window of $[-10, 12{,}000]$ by $[-10, 700]$. The x- and y-scales of the pictured graph are set equal to 1. The axes appear thickened because of the larger number of tic marks on each axis. Setting the x- and y-scales equal to zero returns them to a more normal appearance.

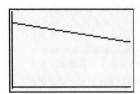

```
WINDOW
 Xmin=-10
 Xmax=12000
 Xscl=0
 Ymin=-10
 Ymax=700
 Yscl=0
 Xres=1
```

You can now TRACE or make a reasonable guess for the value of xmax to determine the approximate x-intercept and complete the graph as displayed below.

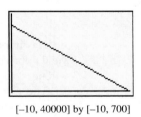

$[-10, 40000]$ by $[-10, 700]$

▐▐▐ LINEAR APPLICATIONS

Let us continue to investigate the investment of $12,000 in the two different interest-bearing accounts.

EXAMPLE 4   Suppose $12,000 is invested in two accounts, a money market account earning $3\frac{1}{2}\%$ interest and a six-month certificate of deposit earning **5%** interest. If x dollars is invested in the money market account then there will be **$12,000 − x** left to invest in the CD. The equation $y = .035x + .05(12{,}000 - x)$ models the total interest earned by the two accounts. Use the graphical display to determine the amount of interest in the account when $4246 is deposited in the $3\frac{1}{2}\%$ account.

SOLUTION   **a.** We could TRACE to *approximately* x = 4246 and determine that the amount of interest, y, is approximately $536.

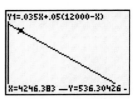

The STOre feature indicates that the exact amount of interest when $4246 is invested in the $3\frac{1}{2}$% account is $536.31.

The TABLE feature supports the STOre feature.

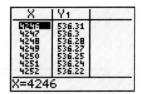

Thus, if you are asked to graphically determine a specific value, TRACE is not the most accurate approach.

b. The use of your calculator's VALUE feature (or eval feature on some calculators) provides an accurate value *and* at the same time displays the information graphically. This feature is found under the CALC menu.

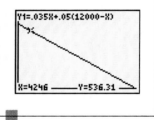

TECHNOLOGY TIP

TI-83 Plus: To access the value feature, press [2nd] [CALC] [1:VALUE]. At the blinking cursor enter the desired *x*-value and press [ENTER]. TI-86: To access the eval *x* feature press [GRAPH], with only one line of menu displayed, press [MORE] [MORE], [FI] (EVAL). At the blinking cursor, enter the desired *x*-value and press [ENTER]. When using the value or eval x feature, your *x*-value must be between the values defined for xmin and xmax on the Window screen.

It is important to remember that in most viewing windows, TRACE will only yield approximate values. Unless you can TRACE to the exact value required, you will need to use the value/eval feature.

The next example uses a preset viewing window (ZInteger) because the unknown quantity, x, represents the number of rides at a fair.

EXAMPLE 5

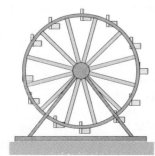

The first day of the fair, admission is $5, and it costs $0.75 to ride each ride. We can represent this algebraically with the equation y = 5 + .75x where x represents the number of rides and y represents the total amount of money spent.

a. Graph the equation in the ZInteger viewing window [−47, 47] by [−31, 31] (TI-86: [−63, 63] by [−31, 31]) with both x and y scales equal to 10. Trace along the graph of the line. How would you interpret the information displayed at the bottom of the screen in the context of the state fair?

b. Continuing to trace, answer each question: How much would you spend if you rode 5 rides? 15 rides? If you cursor left past the y-axis, you will get negative values of x and y. These points are meaningless in the context of our problem. Why?

SOLUTION   a. The *x*-value represents the number of tickets purchased and the *y*-value represents the total amount of money spent. In fact, the ZInteger screen must be used in order for the *x*-value displayed to be an integer because *x* represents the quantity of tickets purchased.

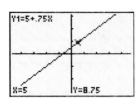

b. The total cost for 5 rides would be $8.75 (the interpretation of the graph pictured at the left). Trace to $x = 15$. The interpretation of your graphical display is that to ride 15 rides, the cost would be $16.25.

The *x* values that are less than zero are meaningless because *x* represents the quantity of tickets purchased.

EXAMPLE 6

TECHNOLOGY TIP

To deselect an equation so it is not displayed on the graph, place the cursor over the equal sign and press ENTER. To select the equation to be graphed, place the cursor over the equal sign and press ENTER to highlight the equal sign. When expressions are entered on the y= screen, the equal sign will always be highlighted (the equation is automatically selected) unless **you** turn it off.

The normal price of the fair is $1 admission and $1.25 per ride. The second night that the fair is in town, it will run its normal rates. The equation that would represent cost in terms of number of rides would be $y = 1 + 1.25x$. Graph the equation of this line at the y2 prompt. To view *only* the new graph, deselect the equation entered at y1.

a. Trace along the graph of the line to determine how much you would spend if you rode 5 rides, 15 rides.

b. Reselect the graph of $y = 5 + .75x$ and display both graphs, representing the cost and number of rides for both nights. How many rides would you have to ride for the cost to be the same no matter which night you attended the fair?

SOLUTION   a. The total cost for 5 rides would be $7.25 and for 15 rides would be $19.75.

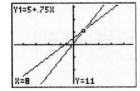

b. Trace to the point of intersection of the two graphs. You may use the up and down arrow keys to move from the graph of one equation to the graph of the other equation. Be sure that the point of intersection is the **one** point that the two lines have in common. The common ordered pair is (8, 11). Translated to the context of the problem, the ordered pair means that when 8 tickets have been purchased, the total cost will be $11 regardless of which night you attend the fair.

EXAMPLE 7

The third night that the fair is in town, the managers decide to run a special: $10 admission and no additional cost per ride. We would represent this with the equation $y = 10$. Enter this equation at the y3= prompt and display the graph of all three equations.

How many rides would you have to ride for the $10 price to be a better deal than the first night? second night? (*Hint*: You might want to turn the graphs ON and OFF to make comparisons.)

SOLUTION   After six rides the cost of attending the first night exceeds the cost of attending the third night.

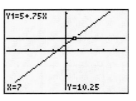

After seven rides the cost of attending the second night exceeds the cost of attending the third night.

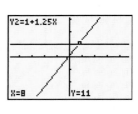

 EXAMPLE 8

Intelix Home Video produces videotapes of weddings, baptisms, etc. For each tape produced there is a basic fixed cost of $100 and a variable cost of $2 per tape. The tapes then sell for $20 each. Write an equation to represent the production costs y for x tapes and an equation to represent the revenues y for x tapes. Explore these graphs and their meaning in the context of the problem.

a. What is the cost of producing 23 tapes?

b. What is the revenue for producing 23 tapes?

c. Does the company sustain a loss or a profit when 23 tapes are sold?

d. How many tapes must be produced and sold for the company to break even (cost = revenue)?

SOLUTION   **a.** The cost equation would be $y = 100 + 2x$. Graph this equation in the viewing window $[-20, 74]$ by $[-5, 181]$ with x- and y-scales equal to 10. *Note:* We did not display the x-intercept because it is a negative number, which is not applicable to the context of the problem. Trace to find the production cost for 23 tapes. At the point $(23, 146)$ we can see that the cost for producing 23 tapes is $146.

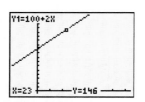

b. The revenue equation would be $y = 20x$. Graph this equation on the same screen as the cost equation and trace on the **revenue** equation to determine the revenues produced for 23 tapes. The point $(23, 460)$ indicates that 23 tapes produce a revenue of $460.

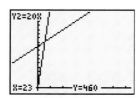

c. Does the company sustain a loss or make a profit when 23 tapes are sold? Because the production cost is $146 and the revenues are $460, the company makes a profit of $314.

d. How many tapes must be produced and sold for the company to break even (cost = revenue)? Attempting to trace to the point of intersection is futile because the x value at the point of intersection is not an integer. (The window

TECHNOLOGY TIP

See "Technology 3.5: Intersect Method for Solving Equations" on page 200 for specific help in using the INTERSECT feature of your calculator.

you were given was designed to produce integer values for x when tracing.) The x-value of the point of intersection of the two equations represents the solution to the equation cost = revenue (i.e., $100 + 2x = 20x$). Use the INTERSECT feature of your calculator to locate the point of intersection displayed at the right. The company would need to produce approximately 5.6 tapes to break even.

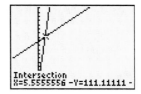

▌ LINEAR MODELS

Many times the data that we are given to work with do not produce a straight-line graph. When this happens, we use a **line of best fit** to predict trends in the data. We will not discuss the methods used to determine the equation as this is a topic for statistics; however, you may want to refer to your calculator manual to learn how your calculator can be used to determine these types of equations.

EXAMPLE 9

The following table gives the winning time (in seconds) for the men's 400-meter freestyle swimming event in the Olympics from 1904 to 2000. The Olympics were not held during the years 1916, 1940, and 1944 due to the two world wars.

YEAR	1904	1908	1912	1920	1924	1928	1932	1936	1948	1952
TIME	376.2	336.8	324.4	326.8	304.2	301.6	288.4	284.5	281	270.7

1956	1960	1964	1968	1972	1976	1980	1984	1988	1992	1996	2000
267.3	258.3	252.2	249	240.3	231.9	231.3	231.2	227	225	227.97	213.6

SOURCE: *The New York Times 2002 Almanac.*

The linear equation that is the best fit for the set of data points is

$$y = -1.40x + 3006.02$$

Use the graph to predict the winning times for the missing years during the two world wars.

SOLUTION   The data points are graphed as (year, time) at the right in the viewing window [1895, 2004] by [199, 401] with both x- and y-scales set to zero.

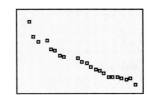

The graph at the right demonstrates the fit between the graph of the equation and the data points.

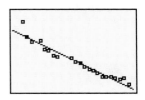

TECHNOLOGY TIP

When using the value option, the value chosen for x **must** be between the xMax and xMin as defined in the window.

Access the value option on your calculator to enter specific x-values and have the calculator plot the location of the point and give the coordinates. Entering 1914 at the $x =$ prompt computes a winning time of 326.42 seconds, as displayed on the graph at the right. Repeating the procedure for the years 1940 and 1944 produces winning times of 290.02 seconds and 284.42 seconds, respectively.

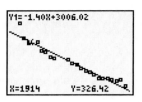

EXERCISE 3.5

Note: Read the information in Technology 3.5 on page 200 before beginning these Exercises.

VOCABULARY AND NOTATION   *In Exercises 1–3, fill in the blank to make a true statement.*

1. A line of _____ is used to predict trends in data when the data are not linear.

2. The point $(0, b)$ of a line represents the ___ intercept.

3. The point $(a, 0)$ of a line represents the ___ intercept.

CONCEPTS

4. Give the **coordinates** of the x- and y-intercepts of the graph in Illustration 7.

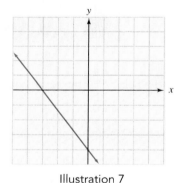

Illustration 7

5. Give the **coordinates** of the x- and y-intercepts of the graph of $y = |x| - 4$ in Illustration 8.

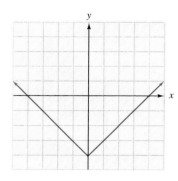

Illustration 8

6. You have been asked to determine the x- and y-intercepts of the equation $2x - 4y = 7$. Explain the process in detail.

7. Explain what constitutes a "good" or "appropriate" graphical display.

8. Allison's mother pays her babysitter, Sherah, $5.00 per hour to babysit Allison and an additional $5 for gas used in driving to and from her home. If $y = 5x + 5$ models the cost of babysitting for x hours, interpret the meaning of the graphical display.

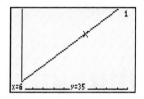

9. The cost for using Fast-Copy printers is $10 plus 9¢ a copy for bulk printing, and the cost for using Quick-Copy is $20 plus 5¢ a copy for bulk printing. If $y =$ cost and $x =$ number of copies, then $y1 = 10 + .09x$ would model the cost for using Fast-Copy and $y2 = 20 + .05x$ would model the cost at Quick-Copy. Interpret the meaning of the graphical display.

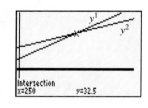

10. In Exercise 9, if more than 250 copies are printed, which company offers a "better deal"?

PRACTICE *In Exercises 11–22, graph the solution to each equation by using the intercept method of graphing. Verify the intercepts and the graph with your calculator. Remember, to graph with the calculator you will need to solve each equation for y.*

11. $x + y = 7$

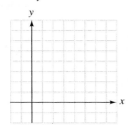

12. $x + y = -2$

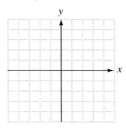

13. $x - y = 7$

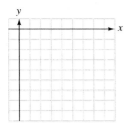

14. $x - y = -2$

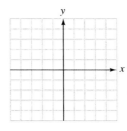

15. $2x + y = 5$

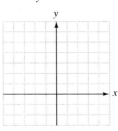

16. $3x + y = -1$

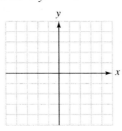

17. $2x + 3y = 12$

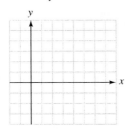

18. $3x - 2y = 6$

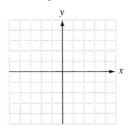

19. $3x + 12 = 4y$

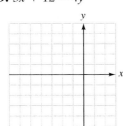

20. $2x + 12 = 9y$

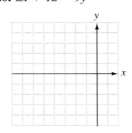

21. $2(x + 2) - y = 4$

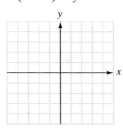

22. $3(y + 1) - x = 4$

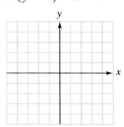

In Exercises 23–30, graph each function in the ZInteger viewing window and use TRACE to determine the x and y intercepts.

23. $y = 2x + 8$

24. $y = -3x - 12$

25. $2x - 3y = 10$

26. $4x - y = 8$

27. $y = x^2 - 3x - 28$

28. $y = -x^2 + 2x - 15$

29. $y = -0.01(x - 10)^3$

30. $y = (x - 3)^3$

In Exercises 31–40, graph each function in an appropriate viewing window. Begin each graph in the standard viewing window and adjust the window so the x- and y-intercepts are visible, as well as any interesting features of the graph.

31. $y = 3x + 15$

32. $y = 5x + 18$

33. $y = -\frac{1}{2}x - 8$

34. $y = -\frac{3}{5}x - 9$

35. $y = \frac{2}{5}x - 12$

36. $y = \frac{1}{5}x - 14$

37. $y = -2x + 15$

38. $y = -5x + 20$

39. $y = \frac{1}{3}x - 8$

40. $y = \frac{3}{8}x - 6$

APPLICATIONS

41. Stopping distance The approximate stopping distance of a car in good conditions on dry pavement includes a $\frac{3}{4}$-second reaction time. Two or three drinks can increase the reaction time to $1\frac{3}{4}$ seconds. The equation that models this is $y = \frac{87}{45}x - \frac{50}{3}$, where x is the speed of the vehicle and y is the stopping distance in meters. Graph this model in an appropriate viewing window.

42. College fees   If a local community college charges an activity fee of $50 per student and $25 per credit hour for a technology fee, the equation $y = 25x + 50$ models the cost y for x credit hours taken. Graph this equation in an appropriate viewing window.

43. Gratuity   George and Annette eat out every Friday night at their favorite restaurant and leave the waitress a 15% tip.

  a. Write an equation where y is the total cost of the meal plus tip and x is the cost of their meal.

  b. Graph the equation in the ZInteger viewing window.

  c. Trace to determine the amount of money they can spend on food if they only have $20 to spend tonight.

  d. Use the value feature to determine the total bill if they order the $35.25 steak dinner for two.

44. Sales tax   The combined state and local sales tax for Clarksville, TN is $9\frac{3}{4}\%$.

  a. Write an equation where y is the amount of tax and x is the amount of the purchase.

  b. Graph the equation in the ZInteger viewing window.

  c. Trace to determine the amount of the purchase if the tax is 2.73.

45. Interest   Jean received gifts of money from her grandparents for graduation. The money she received in May was deposited in a certificate of deposit earning 4% simple interest. She received twice as much money from her parents and deposited this money in a CD earning 5% simple interest.

  a. Write an equation in x and y where y represents the total amount of interest earned in one year on the two accounts and x represents the amount invested at 4%.

  b. Graph this equation in the viewing window [0, 141] by [0, 30] where the x- and y-scales are 10.

  c. Trace to determine the amount of interest earned in the two accounts if her grandparents gave her $108.

  d. How much money would be in the 5% account if her total interest for the year is $31.50?

46. Ticket sales   The local elementary school sells tickets to a play at a cost of $0.50 per ticket for faculty and stu-

dents (to cover the cost of printing the programs) and $2.50 for everyone else.

  a. Write an equation to model the income y from the sale of 25 total tickets.

  b. Graph this equation in the viewing window [0, 94] by [0, 62] where the x- and y-scales equal 10.

  c. Trace to determine the number of faculty/student tickets sold if the total income for the event is $36.50

  d. Trace to determine the number of $2.50 tickets sold if the income is $20.50

47. Grade averages   A student makes a 75 and a 65 on the first two tests.

  a. Write an equation where y represents the average grade after three tests.

  b. Graph in the viewing window designated in Exercise 46.

  c. Trace to the point on the graph that indicates the student would have an average of 72 if he made a 76 on the third test.

  d. Is it possible for the student to have an average of 81 after the third test? If so, what grade would he have to make on the third test?

48. Markup   The campus bookstore purchases books at the retail price from a textbook dealer and marks the cost up by 40% before selling to the student.

  a. Write an equation to represent the cost y (to the student) of a text purchased for x dollars.

  b. Graph this equation in the viewing window [0, 94] by [0, 75], with x- and y-scales equal to 10.

  c. Trace to determine the retail cost of a textbook that sells for $46.20.

  d. Use the value feature to determine the cost of a textbook to the student if the bookstore purchases the text for $48.

49. Printing costs   The Mathematics Department uses the campus media center to print flyers advertising the sale of $3\frac{1}{2}$-inch computer disks. The cost of printing the flyers is $15 and the disks are purchased from a wholesale catalogue for $0.75 each. The disks are then sold to the student at a cost of $1.10 each.

  a. Write an equation to represent the cost y to the department.

  b. Graph the cost equation in the ZInteger viewing window.

  c. Trace to determine the number of disks purchased by the department if the total cost is $81.

d. Write an equation to represent the revenue y produced by the sale of x disks.

e. Reset the viewing window to ZInteger and display both the cost equation and the revenue equation.

f. Adjust the size of the viewing window to display the point of intersection and use the INTERSECT feature to determine how many disks must be sold to break even.

g. How many disks must be sold in order for the department to make a profit?

50. Data analysis A survey (see Table 1) was taken of notably tall buildings on the east coast, midwest, and west coast. The survey compares height, in feet from sidewalk to roof, to number of stories (beginning at street level). The linear equation that models this data is $y = 10.69x + 191.11$ where x represents the number of stories and y represents the height. Graph this equation in the viewing window [26, 120] by [348, 1615] with both x- and y-scales set to zero. If the Empire State building has 102 stories, use the given information to predict the height of the building.

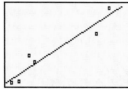

TABLE 1

Building	Stories	Height (ft)
Baltimore U.S. Fidelity and Guaranty Co.	40	529
Maryland National Bank (Baltimore, MD)	34	509
Sears Tower (Chicago, IL)	110	1454
John Hancock Building (Chicago, IL)	100	1127
Transamerica Pyramid (San Francisco, CA)	48	853
Bank of America (San Francisco, CA)	52	778

SOURCE: *The World Almanac and Book of Facts 1995*; actual height of the Empire State building is 1250 ft.

51. Statistics The statistics in Table 2 represent dollars per 100 pounds for cattle. The data is plotted in the viewing window [1934, 2010] by [−5, 100] and displayed below. The linear equation that models this data is $y = 1.19x - 2309.35$, where x is the year and y is the price.

a. Graph the linear equation in the given viewing window and predict the price per 100 pounds of cattle for the years 2005–2007.

b. Are these predictions a "sure thing"? Discuss the factors that might affect your predictions.

TABLE 2

Year	1940	1950	1960	1970	1975	1979
Price	7.56	23.30	20.40	27.10	32.20	66.10
1980	1984	1985	1986	1987	1988	1989
62.40	57.30	53.70	52.60	61.10	66.60	69.50
1990	1991	1992	1993	1994	1995	1996
74.60	72.70	71.30	72.60	66.70	61.80	58.70

SOURCE: *The World Almanac and Book of Facts 1998.*

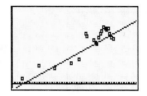

REVIEW

52. Determine the dimensions of the viewing window that could simultaneously display the following groups of points.

a. $(2, 3)$, $(-11, 3)$, $(-15, -4)$, and $(9, -20)$

b. $(-18, 13.4)$, $(16.5, -4)$ and $(25, 30)$

c. $(50, -50)$, $(-75, 15)$, and $(25, -80)$

53. Graph the linear equation $y = x - 15$ in both the standard (ZStandard) viewing window and in the integer (ZInteger) viewing window and explain the difference in what you see.

54. Graph each of the following functions in the integer viewing window and TRACE to determine the coordinates of the x-intercepts.

a. $y = x^2 - 14x + 45$

b. $y = x^2 - x - 56$

c. $y = x^2 + 11x + 24$

55. Trace to find the y-intercept of each of the functions defined in Exercise 54. Compare each of the y-values of the intercepts with the constant value of the function. What do you observe?

TECHNOLOGY 3.5

INTERSECT Method for Solving Equations

The INTERSECT features on the TI-86 and the TI-83Plus work similarly. If using the TI-86, watch for parenthetical information that specifies the differences for your calculator.

To graphically solve an equation, the graphical representation of the algebraic expression on each side of the equation will be examined. Consider the equation $3x + 4 = 2 - x$. The left side of the equation, $3x + 4$, will be designated as y1. Enter $3x + 4$ after **y1=** (remember to use the [X,T,θ,n] key on the TI-83 Plus and the [x-VAR] key on the TI-86). The right side of the equation, $2 - x$, will be designated as y2. Enter $2 - x$ after **y2=**. We want to determine graphically *where* y1 = y2.

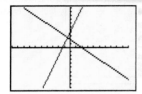

Press **[ZOOM] [6:ZStandard] [GRAPH]** (TI-86 users press **[2nd] [M3](ZOOM) [F4](ZSTD)**) to view the graphical representation of each of the algebraic expressions graphed in the standard viewing window. The graphical solution to the equation is the intersection of y1 and y2. At the intersection point, y1 is equal to y2. Circle the intersection point.

To interpret this graph, we want the *x-value* that is the solution to the equation. The *x*-value can be found by using the intersect feature (denoted as ISECT on the TI-86). The only requirement of the intersect feature is that the point of intersection be visible on the screen.

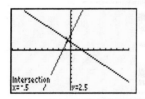

To access the INTERSECT option, press **[2nd] [CALC] [5:Intersect]** to select INTERSECT (TI-86 users press **[MORE] [MATH] [MORE] [F3](ISECT)**). Move the cursor along the first curve to the approximate point of intersection and press **[ENTER]**. At the *second curve* prompt, press **[ENTER]** again because the cursor will still be close to the point of intersection. At the *guess* prompt, press **[ENTER]**. This is instructing the calculator that our *guess* is the approximate point of intersection that was designated at the *first curve* prompt. The solution for *x* is -0.5. Your screen should look like the one pictured.

Check the solution. Recall the two options that have been introduced thus far for checking solutions, the **STO**re key and TABLE feature of the calculator. Comparing the results of your check work to the information displayed on the INTERSECT screen shows that when -0.5 is substituted for *x*, both sides of the equation evaluate to 2.5. This is the *y*-value displayed on the intersection screen. Although the ordered pair $(-0.5, 2.5)$ is the coordinate of the intersection of the two lines displayed, ONLY -0.5 is the solution to the one variable equation, $3x + 4 = 2 - x$. The value -0.5 is stored in *x*. You may return to the home screen and press **[X,T,θ,n] [MATH] [1:▶Frac]** (TI-86 users press **[x-var]** and access your ▶Frac feature) to convert the decimal to its equivalent fraction.

3.6 ALGEBRAIC AND GRAPHICAL SOLUTIONS TO ONE-VARIABLE LINEAR EQUATIONS

- THE THEORY BEHIND EQUATION SOLVING
- USING THE TABLE TO CONFIRM ALGEBRAIC SOLUTIONS

- GRAPHICAL SOLUTIONS OF LINEAR EQUATIONS
- PROBLEM SOLVING REVISITED

▌▌▌ THE THEORY BEHIND EQUATION SOLVING

The steps for equation solving presented in Section 3.1 can be used to solve any linear equation in one variable.

Linear Equation

A linear equation in one variable, x, is any equation that *can be* written in the form

$$ax + c = 0$$

where a and c are real numbers, $a \neq 0$.

We suggest that the steps be carefully followed until you are comfortable with the process.

The theory behind solving linear equations by transformation (writing a series of equivalent equations) is simple: Using inverse operations, arithmetic processes are reversed or "undone." This is usually done in the reverse order of the order of operations. For example, additions and subtractions of terms are done prior to multiplications and divisions.

When the solving procedure is examined, the first two steps involve the simplification of the expressions on each side of the equal sign. The third step suggests multiplying each term on both sides of the equation by the LCD to eliminate fractions. This step is a convenience rather than a necessity. The example below illustrates two valid algebraic solutions of the same equation. Study them carefully.

EXAMPLE 1

Solve $\dfrac{2}{3}x + 5 = \dfrac{1}{2}x - 2$.

Method 1:

$$\frac{2}{3}x + 5 = \frac{1}{2}x - 2$$

$$6 \cdot \frac{2}{3}x + 6 \cdot 5 = 6 \cdot \frac{1}{2}x - 6 \cdot 2$$

$$4x + 30 - 3x = 3x - 12 - 3x$$

$$x + 30 = -12$$

$$x + 30 - 30 = -12 - 30$$

$$x = -42$$

Method 2:

$$\frac{2}{3}x + 5 = \frac{1}{2}x - 2$$

$$\frac{2}{3}x + 5 - \frac{1}{2}x = \frac{1}{2}x - 2 - \frac{1}{2}x$$

$$\frac{1}{6}x + 5 = -2$$

$$\frac{1}{6}x + 5 - 5 = -2 - 5$$

$$\frac{1}{6}x = -7$$

$$x = -42$$

TECHNOLOGY TIP

The graphing calculator can be used to simplify $\frac{2}{3}x - \frac{1}{2}x$. Recall that the distributive property allows you to rewrite the expression as $\left(\frac{2}{3} - \frac{1}{2}\right)x$. Enter $\frac{2}{3} - \frac{1}{2}$ on your calculator. The fraction option returns the correct result of $\frac{1}{6}$.

The equation on the left was solved by clearing the fraction first, whereas the equation on the right kept the fraction throughout the solution process. Make sure you can justify each step in the solving process for **both** the solutions in Example 1.

Example 1 illustrates the fact that clearing an equation of fractions is a solving *option* rather than a *requirement*. Most students choose to eliminate fractions before beginning the solving process. Care must be taken, however, if you choose to eliminate the fraction(s), that each *term* on both sides of the equation be multiplied by the LCD. Consider the two solutions presented in Example 2. In the solution on the left, the fraction is eliminated before anything else is done. The solution on the right retains the fraction. Both solutions are correct.

EXAMPLE 2

Solve $\frac{1}{2}(x + 4) + 5 = x - 3$.

Method 1:

$$\frac{1}{2}(x + 4) + 5 = x - 3$$

$$2 \cdot \frac{1}{2}(x + 4) + 2 \cdot 5 = 2 \cdot x - 2 \cdot 3$$

$$(x + 4) + 10 = 2x - 6$$

$$x + 14 = 2x - 6$$

$$x + 14 - 2x = 2x - 6 - 2x$$

$$-x + 14 = -6$$

$$-x + 14 - 14 = -6 - 14$$

$$-x = -20$$

$$\frac{-x}{-1} = \frac{-20}{-1}$$

$$x = 20$$

Method 2:

$$\frac{1}{2}(x + 4) + 5 = x - 3$$

$$\frac{1}{2}x + \frac{1}{2} \cdot 4 + 5 = x - 3$$

$$\frac{1}{2}x + 2 + 5 = x - 3$$

$$\frac{1}{2}x + 7 - x = x - 3 - x$$

$$-\frac{1}{2}x + 7 = -3$$

$$-\frac{1}{2}x + 7 - 7 = -3 - 7$$

$$-\frac{1}{2}x = -10$$

$$(-2)\left(-\frac{1}{2}x\right) = (-2)(-10)$$

$$x = 20$$

Care must be taken to justify steps when solving. **Think** about and mentally justify each step in the solving process.

▌▌▌ USING THE TABLE TO CONFIRM ALGEBRAIC SOLUTIONS

The STOre feature of the graphing calculator was used previously to check solutions. Another way to verify solutions is by constructing a table. Consider the equation $5x + 4 = 2x + 10$ and the following chart in which different values of x were substituted into the left and right side of the equation. Each side was simplified and then compared. Notice that the number 2 is the solution to the equation because both sides of the equation evaluate to 14 when x has a value of 2.

x	$5x + 4$	$2x + 10$
1	$5(1) + 4 = 9$	$2(1) + 10 = 12$
2	$5(2) + 4 = 14$	$2(2) + 10 = 14$
3	$5(3) + 4 = 19$	$2(3) + 10 = 16$
4	$5(4) + 4 = 24$	$2(4) + 10 = 18$

The TABLE feature of the graphing calculator can be used to generate a table similar to the preceding one. This is illustrated in the checking process of the following example.

EXAMPLE 3

Solve $3(x - 5) = 4(x + 9)$.

SOLUTION

$$3x - 15 = 4x + 36 \qquad \text{Distributive property.}$$
$$3x - 15 - \mathbf{4x} = 4x + 36 - \mathbf{4x} \qquad \text{Subtraction property of equality.}$$
$$-x - 15 = 36 \qquad \text{Simplify each side.}$$
$$-x - 15 + \mathbf{15} = 36 + \mathbf{15} \qquad \text{Addition property of equality.}$$
$$-x = 51 \qquad \text{Simplify each side.}$$
$$\frac{-x}{\mathbf{-1}} = \frac{51}{\mathbf{-1}} \qquad \text{Division property of equality.}$$
$$x = -51 \qquad \text{Simplify both sides.}$$

Check:

a. Using the TABLE feature of the graphing calculator: To use the calculator's TABLE feature to check, first enter the left side of the equation at the Y1= prompt and the right side at the Y2= prompt. Access the TABLE setup and have the TABLE start at the value being checked. Access the TABLE. If the solution checks, the value in the Y1 column will be the same as the value in the Y2 column.

```
Plot1 Plot2 Plot3
\Y1■3(X-5)
\Y2■4(X+9)
\Y3=
\Y4=
\Y5=
\Y6=
\Y7=
```

```
TABLE SETUP
 TblStart=-51
 △Tbl=1
Indent: Auto Ask
Depend: Auto Ask
```

X	Y₁	Y₂
-51	-168	-168
-50	-165	-164
-49	-162	-160
-48	-159	-156
-47	-156	-152
-46	-153	-148
-45	-150	-144

X=-51

b. Analytically,
$$3x - 15 = 4x + 36$$
$$3(-51) - 15 \overset{?}{=} 4(-51) + 36$$
$$-153 - 15 \overset{?}{=} -204 + 36$$
$$-168 = -168$$

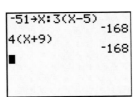

```
-51→X:3(X-5)
              -168
4(X+9)
              -168
■
```

c. Using the STOre feature of the graphing calculator.

EXAMPLE 4

Solve $\dfrac{3x + 11}{5} = x + 3$.

$$5 \cdot \left(\dfrac{3x + 11}{5} \right) = 5 \cdot x + 5 \cdot 3$$

$$3x + 11 = 5x + 15$$

$$3x + 11 - 3x = 5x + 15 - 3x$$

$$11 = 2x + 15$$

$$11 - 15 = 2x + 15 - 15$$

$$-4 = 2x$$

$$\dfrac{-4}{2} = \dfrac{2x}{2}$$

$$-2 = x$$

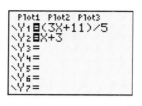

TECHNOLOGY TIP

Recall the entire numerator, $3x + 11$, **must** be enclosed in parentheses when entering the fraction into the calculator.

Study the algebraic solution and make sure you can justify each step in the solution process.

GRAPHICAL SOLUTIONS OF LINEAR EQUATIONS

In each of the examples below, the equations are solved both algebraically and with the graphing calculator. For complete keystroking information, see "Technology 3.5: INTERSECT Method for Solving Equations" on page 200.

EXAMPLE 5

Solve $2(x + 2) - x = 10 - x$.

SOLUTION   Algebraic solution:

$$2x + 4 - x = 10 - x$$
$$x + 4 = 10 - x$$
$$2x + 4 = 10$$
$$2x = 6$$
$$x = 3$$

Graphic solution:
$$y1 = 2(x + 2) - x$$
$$y2 = 10 - x$$

Notice the display at the bottom of the screen. This means that when the variable x has a value of 3, the expressions entered at the y= prompts evaluate to 7. The solution is 3, not 7, since we began with an equation written in "x" only.

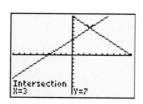

Check: $2(3) + 4 - 3 \stackrel{?}{=} 10 - 3$

$6 + 4 - 3 \stackrel{?}{=} 7$

$7 = 7$

EXAMPLE 6   Solve $-\dfrac{3}{5}x + \dfrac{21}{5} = -\dfrac{4}{3}x + \dfrac{17}{3}$.

SOLUTION   Algebraic solution: $\mathbf{15} \cdot -\dfrac{3}{5}x + \mathbf{15} \cdot \dfrac{21}{5} = \mathbf{15} \cdot -\dfrac{4}{3}x + \mathbf{15} \cdot \dfrac{17}{3}$

$$-9x + 63 = -20x + 85$$

$$11x = 22$$

$$x = 2$$

Graphic solution: $y1 = (-3/5)x + 21/5$

$y2 = (-4/3)x + 17/3$

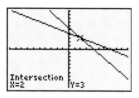

The check is left to the reader. We suggest that the check be done analytically, using the TABLE feature and the STOre feature of the calculator.

▌▌▌ PROBLEM SOLVING REVISITED

The strategies below were introduced in Section 3.1. These should have been copied on a note card for quick reference. This card will be used throughout the text when working application problems.

Problem-Solving Strategies

1. Read the problem several times and organize the facts. What is known? What are you asked to find? It may help you to organize the facts under the two headings, **known** and **unknown**.
2. a. If there is only one unknown quantity, choose a variable to represent that quantity. Clearly define what the variable is and what it represents with a "Let" statement.
   b. When there is more than one unknown, choose a variable to represent one of the quantities and express all other important quantities in terms of the chosen variable.
3. Translate the information expressed in the words of the problem into an algebraic equation.
4. Solve the equation.
5. Answer the question posed in the problem.
6. Check your results both in terms of the equation and the words of the problem. Make sure your solution makes sense within the context of the problem.

EXAMPLE 7

Three more than double a number is 3 times the number plus 15. What is the number?

SOLUTION   Because there is only one unknown, x will represent the unknown number. This is expressed by using a "Let" statement.

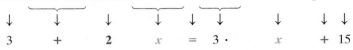

Let x = the number.   Clearly define the variable and what it represents.

The words of the problem are translated to an algebraic equation:

$$3 + 2x = 3x + 15$$

Solve the equation

$$3 + 2x = 3x + 15$$
$$3 + 2x - 3x = 3x + 15 - 3x$$
$$3 - x = 15$$
$$3 - x - 3 = 15 - 3$$
$$-x = 12$$
$$x = -12$$

The number is -12.   Answer the question.

Make sure your result makes sense **both** in the equation and in the context of the problem. Checking the equation can be done by the tools previously introduced: analytically (substitution), using the STOre feature of the calculator, or using the TABLE feature of the calculator. Another helpful strategy is to draw a picture or diagram whenever possible. You should add this to your list of problem-solving strategies. This is illustrated in Example 8.

EXAMPLE 8

Plumbing   A plumber wants to cut a 17-foot pipe into three sections. The longest section is to be 3 times as long as the shortest, and the middle-sized section is to be 2 feet longer than the shortest. How long should each section be?

SOLUTION   Known:
        a 17-foot pipe will be cut into three sections
        the longest section is 3 times as long as the **shortest**
        the middle section is 2 feet longer than the **shortest**

Unknown: the length of each section.

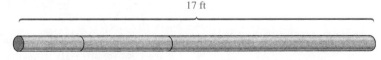

17 ft

Because there is more than one unknown, we must first determine which piece of the pipe will be represented by the variable x. Because the lengths of the longest section and the middle section are given *in terms of* the length of the shortest section,

> Let x = the length of the shortest section.

We can now translate our known phrases into algebraic statements:

> longest section is 3 **times as long** as **the shortest**
> longest section = 3   · x
> longest section = $3x$

> middle section is 2 feet **longer than the shortest**
> middle section = 2   + x
> middle section = $2 + x$

Couple this with a labeled drawing:

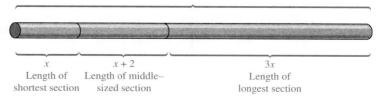

17 ft = total length

| x | $x + 2$ | $3x$ |
| Length of shortest section | Length of middle-sized section | Length of longest section |

Because the pipe is 17 feet long, the sum of the 3 pieces will equal the total length. This translates to the equation below.

$$x + (x + 2) + 3x = 17$$
$$x + x + 2 + 3x = 17 \qquad \text{Remove grouping.}$$
$$5x + 2 = 17 \qquad \text{Combine like terms.}$$
$$5x = 15 \qquad \text{Subtract 2 from both sides.}$$
$$x = 3 \qquad \text{Divide both sides by 5.}$$

The equation could have been solved with the calculator.

   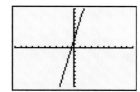

Accessing the graph in the ZStandard viewing window does not display the intersection of the y1 and y2 equations. Because y2 = 17 is the horizontal line whose *y*-intercept is 17, the ymax on the WINDOW screen must be adjusted. Let the ymax equal 20. The point of intersection is clearly displayed and the INTERSECT option under the CALC menu is used to find the intersection of the two graphs.

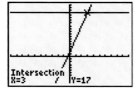

This confirms our analytic solution that x must have a value of 3 for the equation to be a true statement. We then substitute in 3 for x and find the length of our pieces of pipe. The shortest piece is 3 feet long, the middle section is 5 feet long, and the longest section is 9 feet.

EXAMPLE 9

Road construction   A truck made five trips hauling asphalt to a road construction site. A larger truck made eight trips. When fully loaded, the larger truck carries 2 tons more asphalt than the smaller truck. If the two trucks hauled a total of 55 tons, how much asphalt can the smaller truck carry in one load?

SOLUTION

Known:
    there are two trucks, one large and one small
    the small truck made five trips hauling asphalt
    the larger truck made eight trips hauling asphalt
    the large truck carries 2 tons more asphalt than the smaller
    the two trucks hauled a total of 55 tons

Unknown: how much asphalt can the smaller truck carry in one load?

Because we're only asked for the amount of asphalt carried by one truck (the smaller), our let statement becomes

Let x = amount of asphalt carried by the smaller truck.

We know there are two trucks, so we look in our known information for a phrase that relates the amounts hauled by each truck. Once we find the phrase, we translate it.

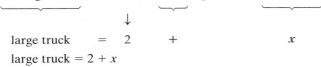

the large truck carries 2 tons **more** asphalt than **the smaller**

large truck   =   2   + x

large truck = 2 + x

Our variable expressions, x and $x + 2$, represent the amounts carried by each truck per trip. We know the small truck made five trips and the larger truck made eight trips. Therefore, the amount hauled by each truck is

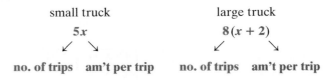

small truck   large truck
    $5x$ $8(x + 2)$

no. of trips   am't per trip   no. of trips   am't per trip

Because the two trucks hauled a total amount of 55 tons of asphalt, the key is the word *total*, which implies the arithmetic operation of addition. Again, we translate to an equation, solve, and check:

Total tonnage carried by small truck	plus	total tonnage carried by large truck	equals	total tonnage carried by both trucks
$5x$	$+$	$8(x + 2)$	$=$	55

$5x + 8(x + 2) = 55$ The equation to solve.

$5x + 8x + 16 = 55$ Remove parentheses.

$13x + 16 = 55$ Combine like terms.

$13x = 39$ Subtract 16 from both sides.

$x = 3$ Divide both sides by 13.

The smaller truck can carry 3 tons of asphalt.

This equation could have been solved with the calculator.

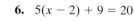

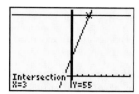

Because the point of intersection is not displayed, the WINDOW screen is accessed and the ymax is adjusted. Because the number 55 was entered at the y2= prompt, the ymax is set equal to 60 to accommodate the value. The INTERSECT option under the CALC menu is used to find the intersection. The intersect screen on the left confirms our algebraic value of x.

EXERCISE 3.6

VOCABULARY AND NOTATION   *In Exercises 1–2, fill in the blanks to make a true statement.*

1. A linear equation in **one** variable, x, is any equation that can be written in the form _____, where a and c are real numbers and $a \neq 0$.

2. Two features of the calculator that can be used to verify solutions are the _____ and _____ features.

CONCEPTS

3. The graphical solution to an equation whose left side has been graphed at y1 and whose right side has been graphed at y2 is displayed at the right. What is the solution?

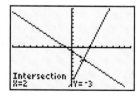

4. The solution to an equation whose left side has been entered at y1 and whose right side has been entered at y2 is verified in the table at the right. What is the solution?

In Exercises 5–14, the left side of each equation was entered at the y1= prompt and the right side of the equation was entered at the y2= prompt. The pictured display is in the

ZStandard viewing window, and the point of intersection is not visible. To display the point of intersection, a new viewing window must be set. Specify an appropriate viewing window for each of the following.

5. $3(x - 4) - 6 = 16$

6. $5(x - 2) + 9 = 20$

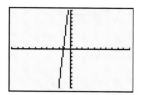

7. $8x - 2 = 3(x - 8)$

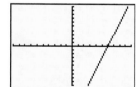

8. $\dfrac{8x + 3}{5} + 6 = 22$

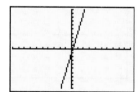

9. $\dfrac{5x - 4}{2} - 8 = x - 15$

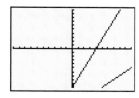

10. $\dfrac{1}{2}(x + 4) - \dfrac{1}{2} = -\dfrac{23}{2} - \dfrac{1}{2}x$

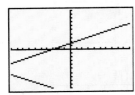

11. $[3(x + 1) - 2] - 5 = -16$

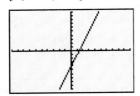

12. $\dfrac{3x + 10}{5} + 7 = 12$ **13.** $\dfrac{x + 3}{5} + 12 = 13$

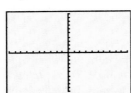

14. $2[4(x - 1) + 3] - 6 = -12$

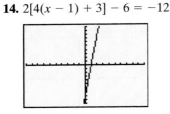

PRACTICE *In Exercises 15–30, solve each equation algebraically, showing all steps of the solving process. Check solutions using the STOre and/or TABLE feature(s) of the graphing calculator.*

15. $\dfrac{5x + 10}{7} = 0$ **16.** $\dfrac{5k - 8}{9} = 1$

17. $\dfrac{17k - 28}{21} + \dfrac{4}{3} = 0$ **18.** $\dfrac{3t - 5}{5} + \dfrac{1}{2} = -\dfrac{19}{2}$

19. $\dfrac{11(x - 12)}{2} = 9 - 2x$

20. $\dfrac{8(5 - s)}{5} = -2s$ **21.** $\dfrac{5z}{3} + 3 = -2$

22. $\dfrac{3y}{2} + 3 = -2$

23. $19.1x - 4(x + 0.3) = -46.5$

24. $14.3(x + 2) + 13.7(x - 3) = 15.5$

25. $\dfrac{2}{3}(x + 6) + \dfrac{5}{6}(x - 6) = x + 3$

26. $\dfrac{1}{8}(x - 8) + \dfrac{1}{4}(x - 4) = x + 2$

27. $\dfrac{2x + 3}{5} = \dfrac{3x - 5}{3}$ **28.** $\dfrac{6a - 1}{8} = \dfrac{5a + 2}{3}$

29. $-[2x - (5x - 1) + 3] = -4(x + 3)$

30. $3[x - (3x - 1)] = 3(x + 4)$

In Exercises 31–40, solve each equation using the INTER-SECT feature of the graphing calculator. Copy the graphing calculator screen display exactly as it appears, recording all displayed information. If the WINDOW values were adjusted from ZStandard, specify the viewing window used to obtain a graph that clearly displayed the point of intersection.

31. $3(x + 4) - 8 = 5(x - 1) + 16$

32. $4(x - 1) + 2 = 3x - 1$

33. $\dfrac{2x + 4}{2} = \dfrac{5x - 2}{8}$ **34.** $\dfrac{3x + 1}{5} = \dfrac{4x - 1}{6}$

35. $\dfrac{1}{5}(x - 10) - 8 = -11 + 2x$

36. $3x + 2(x - 4) = -2x - 22$

37. $2(x + 6) - 10 = 20$

38. $3(x - 6) = 2(x - 5)$

39. $\dfrac{x}{5} - 3 = 15$

40. $\dfrac{x}{3} - 2 = -10$

In Exercises 41–50, translate each statement below into an algebraic equation and solve. Let x represent the variable.

41. Three more than twice a number is 5 less than the number. What is the number?

42. Six less than 3 times a number is 8 more than the number. What is the number?

43. Twice the sum of a number and 3 is 5 times the number. What is the number?

44. Triple the difference of a number and 2 to get an expression equal to twice the number plus 3. What is the number?

45. A number and three times its additive inverse is equal to 4. What is the number?

46. The additive inverse and 5 times a number is equal to 6 less than the number. What is the number?

47. Four times the quantity of a number and 6 is 3 less than the number. What is the number?

48. Eight times the quantity of a number decreased by 2 is equal to 8. What is the number?

49. The quotient of a number and 2 is equivalent to the number plus 3. What is the number?

50. The quotient of a number and 3 is 14 less than 5 times the number. What is the number?

APPLICATIONS *Solve each of the application problems below. Make sure that all four component parts of a complete solution are present. This includes the appropriate "Let" statement(s), an equation, the complete solution to the equation, and a response to the question or problem posed.*

51. Shares of stock At a recent stockholders' meeting, 4.5 million shares were voted in favor of a proposal for a mandatory retirement age for the board of directors. This represented 75% of the total number of shares, so the proposal passed. How many shares were there?

52. Buying real estate The cost of a condominium is $57,595 less than the cost of a house. If the house costs $102,744, find the cost of the condominium.

53. Buying paint After reading the ad in Illustration 1, a decorator bought one gallon of primer, one gallon of paint, and a brush. The total cost was $30.44. Find the cost of the brush.

Illustration 1

54. Customer satisfaction Two-thirds of the movie audience left the theater is disgust. If 78 angry patrons walked out, how many were there originally?

55. Stock split After a three-for-two stock split, a shareholder will own 1.5 times as many shares as before. If 555 shares are owned after the split, how many were owned before?

56. Off-campus housing Four-sevenths of the senior class is living in off-campus housing. If 868 students live off campus, how large is the senior class?

57. Union membership The 2,484 union members represent 90% of a factory's workforce. How many employees are there?

58. Hospital occupancy Thirty-six percent of hospital patients stay for less than 2 days. If 1,008 patients in January stayed for less than 2 days, what total number of patients did the hospital treat in January?

59. Advertising The manager of a supermarket hires a woman to distribute advertising circulars door to door. She will be paid $5 a day plus $0.05 for every advertisement distributed. How many circulars must she distribute to earn $42.50?

60. Integer problem Six less than 3 times a certain integer is 9. Find the integer.

61. Integer problem If a certain integer is increased by 7 and that result is divided by 2, the integer 5 is obtained. Find the original integer.

62. Apartment rental A student moves into a bigger apartment, which rents for $600 per month. That rent is $200 less than twice what she is now paying. Find the current rate.

63. Auto repair A mechanic charged $50 an hour to repair the water pump on a car, plus $95 for parts. If the total bill was $245, how many hours did the repair take?

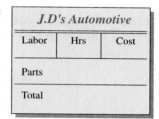

J.D's Automotive		
Labor	Hrs	Cost
Parts		
Total		

64. Water billing The city's water department charges $7 per month plus 42 cents for every 100 gallons of water used. Last month, one homeowner used 1900 gallons and received a bill for $17.98. Was the billing correct? Justify your response.

65. Monthly sales A clerk's sales in February were $2,000 less than three times her sales in January. If her February sales were $7,000, by what amount did her sales increase?

66. Ticket sales A music group charges $1,500 for each performance plus 20% of the total ticket sales. After a concert, the group received $2,980. How much money did the ticket sales raise?

67. Carpentry The 12-foot board in Illustration 2 has been cut into two sections, one twice as long as the other. How long is each section?

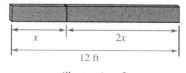

Illustration 2

68. Plumbing The 20-foot pipe in Illustration 3 has been cut into two sections, one 3 times as long as the other. How long is each section?

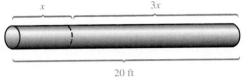

Illustration 3

69. Discount shopping A stereo system discounted 20% is selling for $969.20. What was its original price?

STEREO $969.20
SAVE $$$

70. Installing solar heating One solar panel in Illustration 4 is 3.4 feet wider than the other. Find the width of each.

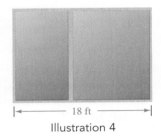

|← 18 ft →|

Illustration 4

71. Waste disposal Two tanks hold a total of 45 gallons of a toxic solvent. One tank holds 6 gallons more than twice the amount in the other. Can the smaller tank be emptied into a 10-gallon waste disposal canister?

72. Buying vitamins If you buy one bottle of vitamins, you can get a second bottle for half price. Two bottles cost $2.25. Find the usual price for a single bottle of vitamins.

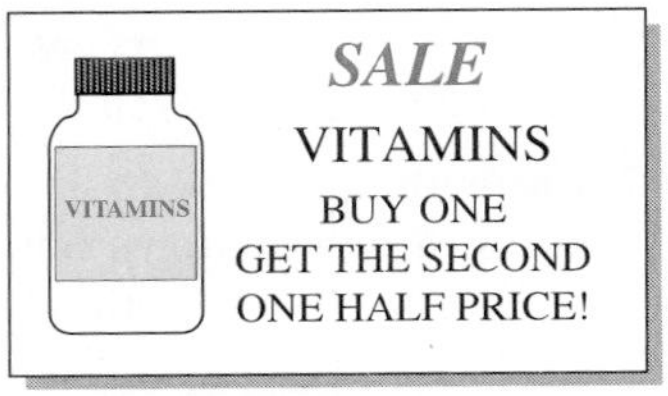

SALE
VITAMINS
BUY ONE
GET THE SECOND
ONE HALF PRICE!

VITAMINS

73. Atomic weights Water is made up of hydrogen and oxygen. In water, the mass of the oxygen is 8 times the mass of the hydrogen. How much hydrogen is in 2,700 grams of water?

74. Choosing salary options Carrie has her choice of two salary options. The first plan pays $600 per month plus a 2% commission on sales. The second plan is straight commission—5% of sales. To decide which plan to choose, Carrie needs to know the monthly sales that will produce equal income. Find the monthly sales.

75. Sales quotas The salesforce is hoping that last month's sales will exceed their one-million-dollar quota. Jennifer was last month's top sales representative, responsible for 45% of total sales. Jim came in second at 23%, with the remaining $317,920 credited to others. Did the sales team make their quota?

76. Publisher's inventories A novel can be purchased in a hardcover edition for $15.95 or in paperback for $4.95. The publisher printed 11 times as many paperbacks as hardcover books, a total of 114,000 copies. How many hardcover books were printed?

77. Manufacturing concrete Concrete contains 3 times as much gravel as cement. How much cement is in 500 pounds of dry concrete mix?

REVIEW

78. Dosing guidelines The dosing guidelines for synthetic thyroid replacement hormone are given in Illustration 5. What is the **minimum** dosage for a child of 7 years of age weighing 55 pounds (25 kilograms)?

Dosing Guidelines for Pediatric Hypothyroidism	
Age	Daily dose per kg body weight*
0-3 mos	10-15 mcg
3-6 mos	8-10 mcg
6-12 mos	6-8 mcg
1-5 yrs	5-6 mcg
6-12 yrs	4-5 mcg
>12 years	2-3 mcg
Growth & puberty complete	1.6 mcg

*To be adjusted on the basis of clinical response and laboratory tests (see **Laboratory Tests**).

SOURCE: Knoll Pharmaceutical Company

Illustration 5

79. Sports A baseball diamond is 90 feet on each side as shown in Illustration 6. Find the area.

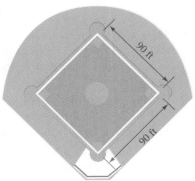

90 ft

90 ft

Illustration 6

80. Cricket chirps   The number of cricket chirps, N, per minute is related to the temperature, T, by the formula

$$T = 50 + \frac{N - 92}{4.7}$$

a. Fill in the following table to determine the relationship between temperature and chirps. Round all results to the nearest whole number.

N		235		0
T	90°		40°	

 b. What is the relationship between the number of chirps and the temperature?

 c. Is this a linear relationship? (That is, if T were replaced with y and N with x, would the equation define a linear function and would its graph be a straight line?)

d. Graph the equation in an appropriate viewing window.

81. Write as an algebraic phrase: "the product of x and y divided by 3."

82. Write as an English phrase: $2x - 5$.

83. In order to get to work every day, a college student walks $x + 2$ miles to the bus stop, rides the bus $2x - 1$ miles, and then walks another $x - 1$ miles to campus. What algebraic expression represents the total distance walked?

3.7 EQUATIONS CONTAINING ABSOLUTE VALUE

- ABSOLUTE VALUE
- EQUATIONS WITH ONE ABSOLUTE VALUE
- EQUATIONS WITH TWO ABSOLUTE VALUES

ABSOLUTE VALUE

We begin this section by defining the absolute value of x.

Absolute Value
$$\text{If } x \geq 0, \text{ then } |x| = x$$
$$\text{If } x < 0, \text{ then } |x| = -x$$

This definition provides a way to associate a nonnegative real number with any real number:

$\quad$ If $x \geq 0$, then x is its own absolute value.

$\quad$ If $x < 0$, then $-x$ (which is positive) is the absolute value.

Either way, $|x|$ is positive or 0:

$|x| \geq 0$ **for all real numbers x**

Because absolute value is always positive and distance is always positive, absolute value may also be thought of as a number's distance from 0 on the number line.

EXAMPLE 1

Find   **a.** $|9|$,   **b.** $|-5|$,   **c.** $|0|$.

SOLUTION   **a.**   Since $9 \geq 0$, 9 is its own absolute value: $|9| = 9$.

    b.   Since $-5 < 0$, the negative of -5 is the absolute value: $|-5| = -(-5) = 5$.

    c.   Since $0 \geq 0$, 0 is its own absolute value: $|0| = 0$.

Reconsidering the three parts of Example 1 within the context of distance reveals

a. $|9| = 9$ because the number 9 is located 9 units from zero on the number line:

b. $|-5| = 5$ because the number -5 is located 5 units from zero on the number line.

c. $|0| = 0$ because the number 0 is 0 units from zero on the number line.

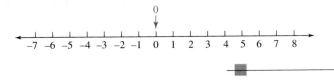

▮ EQUATIONS WITH ONE ABSOLUTE VALUE

Consider the absolute value equation $|x| = 5$. The variable x can have a value of either 5 or -5 because both of these numbers are a distance of 5 units from 0 on the number line.

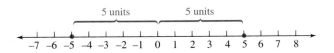

Therefore, the solutions to the equation are $x = 5$ or $x = -5$.
    This leads to the following generalization:

Absolute Value Equations
If $k > 0$, then $|x| = k$ is equivalent to $x = k$ or $x = -k$

It is important to note two things about this statement:

1.   Absolute value is isolated on one side of the equation.
2. k is a positive constant.

EXAMPLE 2

Solve the equation $|x - 3| = 7$.

SOLUTION The equation $|x - 3| = 7$ indicates that a point on the number line with a coordinate of $x - 3$ is 7 units from the origin. Thus, $x - 3$ can be either 7 or -7.

$$x - 3 = 7 \quad \text{or} \quad x - 3 = -7$$
$$x = 10 \quad | \quad x = -4$$

The solutions of the equation $|x - 3| = 7$ are 10 and -4. (See Figure 3-19.) If either of these numbers is substituted for x in the equation $|x - 3| = 7$, the equation is satisfied:

$$|x - 3| = 7 \qquad \qquad |x - 3| = 7$$
$$|10 - 3| \overset{?}{=} 7 \qquad |-4 - 3| \overset{?}{=} 7$$
$$|7| \overset{?}{=} 7 \qquad \qquad |-7| \overset{?}{=} 7$$
$$7 = 7 \qquad \qquad 7 = 7$$

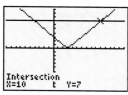

Figure 3-19

Solution set: $\{10, -4\}$

Calculator solution:

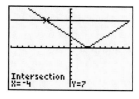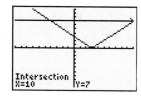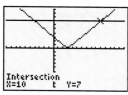

$[-10, 15]$ by $[-10, 10]$

Notice on the second screen that the point of intersection is right at the edge of the screen. It would be clearer had the xmax been increased to provide a better window. The third screen has an adjusted xmax value of 15.

TECHNOLOGY TIP

Recall that to solve an equation graphically, the left side is entered at the y1= prompt and the right side at the y2= prompt. Once the point(s) of intersection are displayed on the screen, the INTERSECT/ISECT feature is accessed. Because there are obviously two points of intersection in the given absolute value equation, the INTERSECT/ISECT feature must be accessed twice. It is important that the cursor be moved close to the desired point of intersection at the "Guess" prompt.

EXAMPLE 3

Solve the equation $|3x - 2| = 5$.

SOLUTION We can write $|3x - 2| = 5$ as

$$3x - 2 = 5 \quad \text{or} \quad 3x - 2 = -5$$

and solve each equation for x:

$$3x - 2 = 5 \qquad \text{or} \qquad 3x - 2 = -5$$
$$3x = 7 \qquad \qquad 3x = -3$$
$$x = \frac{7}{3} \qquad \qquad x = -1$$

Solution set: $\{\frac{7}{3}, -1\}$

Verify that both solutions check.

Calculator solution:

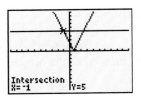 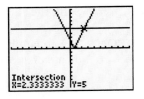 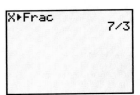

Solution set: $\{-1, \frac{7}{3}\}$

EXAMPLE 4

Solve the equation $\left|\frac{2}{3}x + 3\right| + 4 = 10$

SOLUTION   Careful examination of the algorithm reveals that it is valid if and only if the absolute value is isolated on one side and a constant is on the other side.
    We first isolate the absolute value on the left-hand side.

$$\left|\frac{2}{3}x + 3\right| + 4 = 10$$

STUDY TIP

Be sure to isolate the quantity in the absolute value before applying the definition of absolute value to solve the equation.

1. $\left|\frac{2}{3}x + 3\right| = 6$ Subtract 4 from both sides.

We can now write Equation 1 as

$$\frac{2}{3}x + 3 = 6 \quad \text{or} \quad \frac{2}{3}x + 3 = -6$$

and solve each equation for x:

$$\frac{2}{3}x + 3 = 6 \quad \text{or} \quad \frac{2}{3}x + 3 = -6$$

$$\frac{2}{3}x = 3 \qquad\qquad \frac{2}{3}x = -9$$

$$2x = 9 \qquad\qquad 2x = -27$$

$$x = \frac{9}{2} \qquad\qquad x = \frac{-27}{2}$$

Solution set: $\{\frac{9}{2}, -\frac{27}{2}\}$
Verify that both solutions check.

Calculator solution:

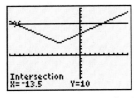 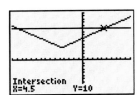

$[-15, 10]$ by $[-10, 20]$

Solution set: $\{-13.5, 4.5\}$

EXAMPLE 5

Solve the equation $\left| 7x + \dfrac{1}{2} \right| = -4$.

SOLUTION

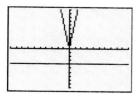

Again, careful examination of the algorithm reveals that our constant, k, must be positive in order to apply the algorithm. In this example, $k = -4$. Therefore, the algorithm cannot be applied. Relating absolute value to distance on the number line translates to "what number(s) are a *negative* four units from 0 on the number line?" This does not make sense bcause we know that distance is always positive. We also know that absolute value is always equal to 0 or some positive number. We therefore conclude that the given equation is a contradiction, and has no solution.

The calculator display confirms our conclusion because there are no points of intersection.

EXAMPLE 6

Solve the equation $\left| \dfrac{1}{2}x - 5 \right| - 4 = -4$.

SOLUTION

We first isolate the absolute value on the left-hand side.

$$\left| \frac{1}{2}x - 5 \right| - 4 = -4$$

$$\left| \frac{1}{2}x - 5 \right| = 0 \qquad \text{Add 4 to both sides.}$$

TECHNOLOGY TIP

Notice that when confirming our algebraic solution with the graphing calculator we enter the original expressions on each side of the equation at the appropriate prompt. This helps ensure that our conclusion is not based on a careless arithmetic error.

Because 0 is the only number whose absolute value is 0, the binomial $\frac{1}{2}x - 5$ must be equal to 0, and we have

$$\frac{1}{2}x - 5 = 0$$

$$\frac{1}{2}x = 5 \qquad \text{Add 5 to both sides.}$$

$$x = 10 \qquad \text{Multiply both sides by 2.}$$

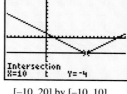

[−10, 20] by [−10, 10]

The calculator confirms our algebraic solution. Solution set: {10}

▍ EQUATIONS WITH TWO ABSOLUTE VALUES

The equation $|a| = |b|$ is true when $a = b$ or when $a = -b$. For example,

$$|3| = |3| \qquad \text{or} \qquad |3| = |-3|$$
$$3 = 3 \qquad \quad | \qquad \quad 3 = 3$$

Thus, we have the following result.

Equations with Two Absolute Values

If a and b represent algebraic expressions, the equation $|a| = |b|$ is equivalent to the pair of equations

$$a = b \qquad \text{or} \qquad a = -b$$

EXAMPLE 7

Solve the equation $|5x + 3| = |3x + 25|$.

SOLUTION   This equation is true when $5x + 3 = 3x + 25$, or when $5x + 3 = -(3x + 25)$. We solve each equation for x.

$$
\begin{array}{c|c}
5x + 3 = 3x + 25 & 5x + 3 = -(3x + 25) \\
2x = 22 & 5x + 3 = -3x - 25 \\
x = 11 & 8x = -28 \\
& x = -\dfrac{28}{8} \\
& x = -\dfrac{7}{2}
\end{array}
$$

Solution set: $\{11, -\frac{7}{2}\}$
The graphing calculator can be used to confirm the solutions.

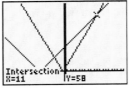

$[-20, 20]$ by $[-10, 70]$

Solution set: $\{11, -3.5\}$

EXAMPLE 8

Solve the equation $|x + 4| = |x - 4|$.

SOLUTION   This equation is true when $x + 4 = x - 4$ or when $x + 4 = -(x - 4)$. We solve each equation for x.

$$
\begin{array}{c|c}
x + 4 = x - 4 & x + 4 = -(x - 4) \\
4 = -4 & x + 4 = -x + 4 \\
\varnothing & 2x + 4 = 4 \\
& 2x = 0 \\
& x = 0
\end{array}
$$

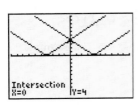

The graphing calculator can be used to confirm the solution. Solution set: $\{0\}$

EXERCISE 3.7

VOCABULARY AND NOTATION   *In Exercises 1–2, fill in the blanks to make a true statement.*

1. If $x \geq 0$, then $|x| = $ ___ . If $x < 0$, then $|x| = $ ____ .

2. Absolute value can be thought of as a number's _____ from zero on the number line.

CONCEPTS

3. Determine the indicated absolute value.

    a. $|4|$ **b.** $|-4|$

    c. $-|-4|$ **d.** $|a|$

    e. $|3 - \pi|$ (Do not approximate!)

4. If $|x| = 8$, then $x = $ ___ or $x = $ ____

5. Explain why the statement $|x| = -2$ makes no sense.

6. Is it possible to have a solution for $|x| + 7 = 2$? Explain why or why not.

PRACTICE   *In Exercises 7–40, solve each equation algebraically. Confirm your solutions graphically.*

7. $|x| = 8$ **8.** $|x| = 9$

9. $|x - 3| = 6$ **10.** $|x + 4| = 8$

11. $|2x - 3| = 5$ **12.** $|4x - 4| = 20$

13. $|3x + 2| = 16$ **14.** $|5x - 3| = 22$

15. $\left|\dfrac{7}{2}x + 3\right| = -5$ **16.** $|2x + 10| = 0$

17. $\left|\dfrac{x}{2} - 1\right| = 3$ **18.** $\left|\dfrac{4x - 64}{4}\right| = 32$

19. $|3 - 4x| = 5$ **20.** $|8 - 5x| = 18$

21. $|3x + 24| = 0$ **22.** $|x - 21| = -8$

23. $\left|\dfrac{3x + 48}{3}\right| = 12$ **24.** $\left|\dfrac{x}{2} + 2\right| = 4$

25. $|x + 3| + 7 = 10$ **26.** $|2 - x| + 3 = 5$

27. $\left|\dfrac{3}{5}x - 4\right| - 2 = -2$ **28.** $\left|\dfrac{3}{4}x + 2\right| + 4 = 4$

29. $|2x + 1| = |3x + 3|$

30. $|5x - 7| = |4x + 1|$

31. $|3x - 1| = |x + 5|$

32. $|3x + 1| = |x - 5|$

33. $|2 - x| = |3x + 2|$

34. $|4x + 3| = |9 - 2x|$

35. $\left|\dfrac{x}{2} + 2\right| = \left|\dfrac{x}{2} - 2\right|$

36. $|7x + 12| = |x - 6|$

37. $\left|x + \dfrac{1}{3}\right| = |x - 3|$

38. $\left|x - \dfrac{1}{4}\right| = |x + 4|$

39. $|3x + 7| = -|8x - 2|$

40. $-|17x + 13| = |3x - 14|$

41. Construct several examples to show that $|a \cdot b| = |a| \cdot |b|$

42. Construct several examples to show that $\left|\dfrac{a}{b}\right| = \dfrac{|a|}{|b|}$

43. Construct several examples to show that $|a + b| \neq |a| + |b|$

44. Construct several examples to show that $|a - b| \neq |a| - |b|$

REVIEW

45. Simplify and verify with your calculator.

    a. $\dfrac{4}{5} + \dfrac{-8}{9}$ **b.** $\left(-\dfrac{7}{8}\right)\left(-\dfrac{3}{5}\right) \div \left(\dfrac{-9}{20}\right)$

46. Identify the property that is exemplified below:

    a. $-(-x) = x$

    b. $a(b + c) = ab + ac$

    c. $xy = yx$

47. Find $P(-1)$ if $P(x) = -2x^2 - x + 2$. Verify with your calculator.

48. If each tic mark is equal to 1 unit, give the dimensions of each viewing window.

    a.    **b.**

49. Graph $y = \frac{1}{2}x - 12$ in a good viewing window and record the viewing window's dimensions in the following form: [xmin, xmax] by [ymin, ymax]. Larger viewing windows are acceptable.

CHAPTER SUMMARY

CONCEPTS

REVIEW EXERCISES

Section 3.1

Any real number can be added to (or subtracted from) both sides of an equation to form another equation with the same solution(s) as the original equation.

Both sides of an equation can be multiplied by (or divided by) any *nonzero* real number to form another equation with the same solution(s) as the original equation.

Conditional equations are true for some, but not all, real numbers.

Identities are equations that are true for **all** real numbers. Their solutions are represented by $\mathbb{R}$, a symbol for "the set of all real numbers."

Contradictions are equations that are **never** true for any real number. Their solution is represented by $\varnothing$, the null set.

This section includes "Steps for Solving a Linear Equation in One Variable" as well as "Problem-Solving Strategies."

Introduction to Linear Equations in One Variable

1. Tell whether the indicated number is a solution of the given equation.

    a. $3x + 7 = 1; -2$

    b. $5 - 2x = 3; -1$

    c. $2(x + 3) = x; -3$

    d. $5(3 - x) = 2 - 4x; 13$

2. Solve each equation algebraically, showing all steps of your work. Check all results in the original equation.

    a. $3x + 7 = 1$

    b. $7 - 9x = 8$

    c. $\dfrac{x + 3}{4} = 2$

    d. $\dfrac{x - 7}{2} = -2$

    e. $\dfrac{x}{2} + \dfrac{7}{2} = 3$

    f. $\dfrac{x}{3} - 3 = 2$

3. Solve each equation algebraically, showing all steps of your work. Check all results in the original equation.

    a. $2x - 19 = 2 - x$

    b. $5x - 19 = 20 - 2x$

    c. $10(x - 3) = 3(x + 11)$

    d. $2(5x - 7) = 2(x - 16)$

    e. $\dfrac{2}{3}(5x - 3) = 38$

    f. $\dfrac{3}{2}(3x - 12) = 9$

4. Write an equation in one variable to model the problem and solve the equation algebraically. Make sure your result makes sense within the context of the problem.

    a. If twice an integer is decreased by 7, the result is 9. Find the original integer.

    b. Four less than 3 times a number is 77. Find the number.

    c. A contractor charges $35 for the installation of rain gutters plus $1.50 per foot. One installation costs $162.50. How many feet of gutter is required?

    d. The electric company charges $17.50 per month plus 18 cents for every kilowatt hour of energy used. One resident's bill was $43.96. How many kilowatt hours were used?

    e. A 45-foot rope is to be cut into three sections. One section is to be 15 feet long. Of the remaining sections, one must be 2 feet less than 3 times the length of the other. Find the length of the shortest section.

Section 3.2

A literal equation, or formula, can often be solved for any of its variables.

Literal Equations and Formulas

5. Solve each equation for the indicated variable.

    a. $3x + y = 12; y$

    b. $4x - 3y = 8; y$

    c. $d = rt; t$

    d. $P = 2l + 2w; w$

e. $A = 2\pi rh; h$ **f.** $A = P + Prt; r$

g. $P = \dfrac{RT}{V};$

Section 3.3

Linear equations in two variables have an infinite number of ordered pair solutions, (x, y).

A function is a collection of ordered pairs in which each x-value is paired with a unique y-value.

The domain of a function is the set of all x-values of the ordered pairs of the function.

The range of a function is the set of all y-values of the ordered pairs of the function.

A linear equation in Standard Form is of the form $Ax + By = C$, where A, B, and C are real numbers.

Introduction to Equations in Two Variables

6. Tell whether the indicated ordered pair is a solution to the linear equation in two unknowns. Do this both analytically (by substitution and simplifying) and using the STOre feature of the calculator.

  a. $4x + y = 7; (2, -1)$ **b.** $x - 3y = 5; (11, 2)$

  c. $3x - 2y = -1; (3, -4)$ **d.** $2x + 3y = 3; (-2, -3)$

7. Find the corresponding y-value for each given x-value. Write the result as an ordered pair.

  a. $x - 2y = 6; x = 2$ **b.** $5x + 2y = 9; x = 3$

  c. $x^2 + 3x + 2 = y; x = -1$ **d.** $x^2 - 9 = y; x = 3$

8. Complete the table of values for the equation $3x + y = 6$.

x	y
-2	
-1	
0	
1	
2	

9. Determine if each relation or equation specified below defines a function. Determine both the domain and range.

  a. $\{(2, 3), (4, 6), (2, -3)\}$ **b.** $\{(1, 5), (2, 8), (3, 11)\}$

  c. $y = 4x + 5$ **d.** $y = x^2 - 1$

Section 3.4

Ordered pairs of numbers (x, y) are associated with points in a rectangular coordinate system.

To graph a linear equation in the variables x and y, choose at least three values for x, use substitution to find the corresponding values of y, plot the points (x, y), and finally draw the line that passes through the points.

The Rectangular Coordinate System

10. Graph the equation by hand on graph paper.

  a. $y = 3x - 1$ **b.** $y = -4x + 2$ **c.** $3x + y = 8$

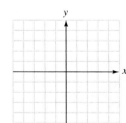
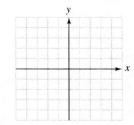
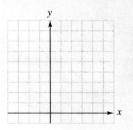

Equations of nonvertical and nonhorizontal lines can be written either in the form $y = mx + b$ or the standard form $Ax + By = C$.

The equation $x = a$ is the equation of a vertical line. If $a = 0$ then $x = a$ represents the y-axis. If $a \neq 0$ then $x = a$ represents a line parallel to the y-axis.

The equation $y = b$ is the equation of a horizontal line. If $b = 0$ then $y = b$ represents the x-axis. If $b \neq 0$ then $y = b$ represents a line parallel to the x-axis.

The Standard (ZStandard) viewing window for a graphing calculator is $[-10, 10]$ by $[-10, 10]$ with x- and y-scales set equal to 1.

Graphing Basics are explained in "Technology 3.4" on page 182.

d. $x - 4y = 6$ **e.** $y = 3$ **f.** $x = -5$

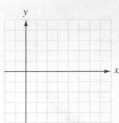 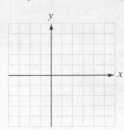 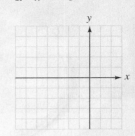

g. $y = -5x - 2$ **h.** $2x - y = 3$ **i.** $x + 2y = 4$

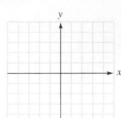 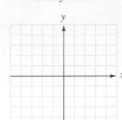 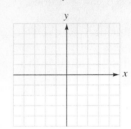

j. $2x - 5 = 7$ **k.** $3(y - 4) = 0$ **l.** $2x + 4y = 8$

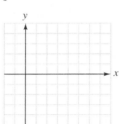 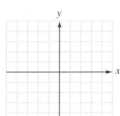 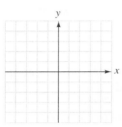

Linear Applications and Models

To find the x-intercept of a line, let $y = 0$ and solve for the value of x.

To find the y-intercept of a line, let $x = 0$ and solve for the value of y.

A good viewing window is one that displays x-intercepts, y-intercepts, and any high or low points of the graph.

Both graphs and tables of the graphing calculator can be used to explore application problems in two unknowns.

11. Algebraically, find the x- and y-intercepts of the equations in Exercise 10 (if they exist). Identify those equations that are horizontal and vertical lines.

12. Graph each equation in the ZInteger viewing window and TRACE to determine x- and y-intercepts.

   a. $y = 3x - 6$ **b.** $y = x^2 - 4x - 5$

   c. $y = (x - 4)^3$ **d.** $4x - y = 12$

13. Determine an appropriate viewing window for each function specified below. Begin in ZStandard.

   a. $y = 3x - 20$ **b.** $y = 5x - 100$

   c. $y = \dfrac{2}{5}x + 12$

Solutions to linear equations in one variable can be confirmed via the STOre feature or by using the TABLE feature of the graphing calculator.

Equations may be solved graphically by graphing each side of the equation and then using the INTERSECT feature to determine the point(s) of intersection. See "Technology 3.5."

In Review Exercises 14–16, use only your graphing calculator to respond to the questions.

14. A machine shop buys a lathe for $1970. Its expected life is 10 years. If the linear equation that models this is $y = -170x + 1970$, then

    a. graph the equation in the viewing window of [0, 11.75] by [0, 1974].

    b. trace to determine the value of the lathe in 8 years.

    c. use the VALUE feature to determine the value of the lathe in 6 years 9 months.

15. An antique clock was purchased for $125. It has been increasing in value in a linear pattern at a rate of $35 per year.

    a. Write an equation to model its value y as related to years x.

    b. Graph this equation in the viewing window of [0, 9.4] by [−10, 500].

    c. TRACE to determine the value in 3 years.

    d. TRACE to determine how many years it will take to have a value of $405.

16. A community theater has receipts of $1,069 after selling adult tickets for $12 and student/child tickets for $5. Let x represent the number of adult tickets sold and y the number of student tickets sold.

    a. Write an equation in x and y to model the total sales.

    b. Solve the equation for y and graph it in a [0, 94] by [0, 215] viewing window.

    c. Using TRACE or VALUE, determine the number of student tickets if 42 adult tickets are sold.

    d. Is it possible for 60 adult tickets to be sold? Based on the graph, why or why not?

    e. Is it possible for 90 adult tickets to be sold? Why or why not?

Section 3.6

The INTERSECT feature of the calculator, the TABLE, and the STOre feature can be used to confirm algebraic solutions to equations.

Algebraic and Graphical Solutions to One-Variable Linear Equations

17. Using the equations in Review Exercises 2 and 3,

    a. solve each equation graphically using the INTERSECT feature.

    b. confirm solutions in the TABLE.

    c. confirm solutions using the STOre feature.

18. Using Review Exercise 4,

    a. solve each equation graphically.

    b. confirm solutions in the TABLE.

    c. confirm solutions using the STOre feature.

Section 3.7   Equations Containing Absolute Value

Absolute Value: If $x \geq 0$, then $|x| = x$.

If $x < 0$, then $|x| = -x$.

If $k > 0$, then $|x| = k$ is equivalent to $x = k$ or $x = -k$.

If a and b represent algebraic expressions, then the equation $|a| = |b|$ is equivalent to the pair of equations $a = b$ or $a = -b$.

19. Solve each equation algebraically and confirm your solution graphically.

   a. $|3x + 1| = 10$
   b. $\left|\dfrac{3}{2}x - 4\right| = 9$

   c. $|x - 5| = -3$
   d. $|2x - 3| + 2 = 5$

   e. $|4x - 1| - 7 = -3$

20. Solve each equation algebraically and confirm your solution graphically.

   a. $|3x + 2| = |2x - 3|$
   b. $|5x - 4| = |4x - 5|$

CHAPTER TEST

In Problems 1 and 2, tell whether the indicated number is a solution of the given equation.

1. $-3(2 - x) = 0;\ -2$

2. $x(x + 1) = x^2 + 1;\ -1$

In Problems 3 and 4, solve each equation for the indicated variable.

3. $4x + 3y = 8$; solve for y

4. $\dfrac{1}{2}x + 2y = \dfrac{3}{5}$; solve for y

In Problems 5 and 6, tell whether the indicated ordered pair is a solution of the given equation.

5. $2x - 5y = 7;\ (-1, 2)$

6. $x + 3y = -2;\ (1, -1)$

In Problems 7–14, solve for x both algebraically and graphically.

7. $\dfrac{x}{7} = 1$
         8. $8x + 2 = -14$

9. $23 - 5(x + 10) = -12$

10. $\dfrac{5}{3}(x - 7) = 15(x + 1)$

11. $-(x + 1) - 2x = -3x + 5 - 8$

12. $\left|\dfrac{2}{3}x - 5\right| = 6$
         13. $|3x - 2| + 4 = 2$

14. $|4x + 1| = |2x - 3|$

In Problems 15–18, graph the equation on graph paper and with your graphing calculator. Indicate the x and y intercepts on your hand-sketched graph. Copy the screen display from your graphing calculator and specify the window values.

15. $y = 4x - 1$
       16. $y = -\dfrac{1}{2}x + 5$

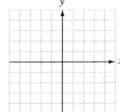
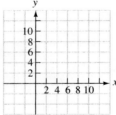

17. $2x + y = 1$
       18. $x + 3y = -2$

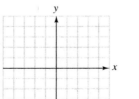
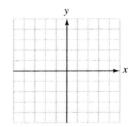

19. The number of television sets that consumers buy depends on the price. The higher the price, the fewer TVs people will buy. The equation that relates the price, y, to the number, x, of TVs sold at that price is the demand equation. The demand equation for a 13-inch TV is

$$y = -\frac{1}{10}x + 170$$

a. Graph the demand equation in a [0, 188] by [0, 200] viewing window and sketch the graph.

b. Use the graph to determine how many televisions will be sold at a price of $150.

In Problems 20–22, (a) write an equation to model the problem, (b) solve the equation, (c) answer the question in a complete sentence.

20. The sum of two consecutive odd integers is 36. Find the integers.

21. A Turkish rug was purchased for $560. It has increased 96% in value. Find its current value.

22. A real estate agent earns $600 plus a 6% commission for the sale of each house. If she earns $5700 on the sale of a house, what was the value of the house?

CAREERS & MATHEMATICS

© Royalty-Free/CORBIS

Pilot   Pilots are highly trained professionals who fly airplanes and helicopters to carry out a wide variety of tasks. Most pilots transport passengers and cargo, but some are involved in crop dusting, spreading seed, testing aircraft, firefighting, tracking criminals, traffic reporting, and rescue and evacuation.

Qualifications   All pilots paid to transport passengers or cargo must have a commercial pilot's license with an instrument rating. They must also have good health, vision correctable to 20/20, good hearing, and no physical handicaps that would interfere with safety. Most employers also require at least 2 years of college, and many prefer college graduates. Requirements for a private pilot's license are less restrictive.

Job Outlook   Pilots are expected to face strong competition for jobs through the year 2010. Due to high earnings and free to low-cost travel benefits, many qualified applicants seek these jobs. Relatively few jobs will be created from the rising demand, even though employment is expected to increase about as fast as for all occupations through 2010.

Example Application   The air passing over the curved top of the wing in Illustration 1 moves faster than the air beneath, because it has a greater distance to travel. This faster-moving air exerts less pressure on the wing than the air beneath, causing *lift*.

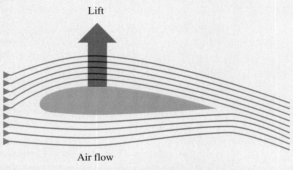

Lift

Air flow

Illustration 1

Two factors that determine lift are controlled by the pilot. One is the speed or *velocity* of the plane. Another, the *angle of attack*, is the angle between the direction the plane is aimed and the direction it is actually moving, as shown in Illustration 2.

For one particular light aircraft, the lift, velocity, and angle of attack are related by the equation

$$L = (0.017a + 0.023)V^2$$

where L is the lift in pounds, a is the angle of attack in degrees, and V is the velocity in feet per second. To support this aircraft, the lift must equal the plane's weight, 2050 pounds. If the velocity is 130 feet per second, find the correct angle of attack.

Solution   First, the pilot substitutes the values for lift and velocity into the given equation and solves for a, the angle of attack.

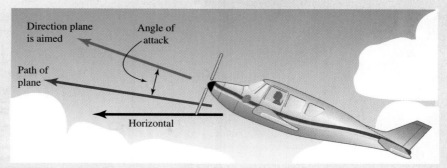

Illustration 2

$$L = (0.017a + 0.023)V^2$$

$$\mathbf{2050} = (0.017a + 0.023)(\mathbf{130})^2 \qquad \text{Substitute 2050 for } L \text{ and 130 for } V.$$

$$\frac{2050}{130^2} = 0.017a + 0.023 \qquad \begin{array}{l}\text{Undo the multiplication on the right by} \\ \text{dividing both sides by } 130^2.\end{array}$$

$$\frac{2050}{130^2} - 0.023 = 0.017a \qquad \begin{array}{l}\text{Undo the addition on the right by} \\ \text{subtracting 0.023 from both sides.}\end{array}$$

$$0.0983017751 = 0.017a \qquad \begin{array}{l}\text{Use a calculator to simplify the} \\ \text{left-hand side.}\end{array}$$

$$\frac{0.0983017751}{0.017} = a \qquad \begin{array}{l}\text{Undo the multiplication on the right by} \\ \text{dividing both sides by 0.017.}\end{array}$$

$$5.782457362 = a \qquad \text{Use a calculator.}$$

To provide proper lift, the pilot must adjust the climb to maintain an angle of attack of about $5\frac{3}{4}°$.

EXERCISES   As the angle of attack approaches about 17°, the plane begins to stall. With more cargo on the return trip, the same plane weighs 2325 pounds. If the pilot allows the velocity to drop to 80 feet per second (about 55 miles per hour), will the plane stall?

(*Answer:* The calculated angle of attack is 20°. The plane will stall unless the pilot increases the speed.)

An Extended Look at Applications and Inequalities

© Photodisc Collection/Getty Images

InfoTrac Project

Do a keyword search on "Pythagoras" and find the article "Number secrets: In the world of Pythagoras, 'all is number.'" Write a summary of the article and include one of the discoveries Pythagoras made concerning certain numbers.

The length of the diagonal of the screen is used to determine the size of a television. Using the Pythagorean theorem, the sum of the square of the legs of a right triangle equals the square of the hypotenuse, find the size of a television with a screen length of 22 inches and a screen width of 16 inches. How much would the diagonal increase by if the length and the width of the screen were each increased by 5 inches?

Complete this project after studying Section 4.1.

PERSPECTIVE

How to Solve It

As a young student, George Polya (1888–1985) enjoyed his studies of mathematics and understood the solutions presented by his teachers. However, Polya had questions still asked by mathematics students today: "Yes, the solution works, but how is it possible to come up with such a solution? How could I discover such things by myself?" These questions still concerned him years later when, as Professor of Mathematics at Stanford University, he developed an approach to teaching mathematics that was very popular with faculty and students. His book, *How to Solve It*, became a bestseller.

Polya's problem-solving approach involves four steps.

- *Understand the problem*. What is the unknown? What information is known? What are the conditions?
- *Devise a plan*. Have you seen anything like it before? Do you know any related problems you have solved before? If you can't solve the pro-

posed problem, can you solve a similar but easier problem?
- *Carry out the plan*. Check each step. Can you explain why each step is correct?
- *Look back*. Examine the solution. Can you check the result? Can you use the result, or the method, to solve any other problem?

George Polya (1888–1985)

4.1 APPLICATIONS—A NUMERIC APPROACH

- CONSTRUCTING TABLES
- GEOMETRY PROBLEMS
- SOLVING PROPORTIONS
- RATIO AND PROPORTION
- SIMILAR TRIANGLES

▮ CONSTRUCTING TABLES

When investigating problem-solving strategies, it becomes critical to first organize the given data. In the previous chapter, the strategy of writing down the known information and the unknown information provided a method for organization. Tables are useful for both organizing data and helping generalize from the specific to the general. Consider the following example.

EXAMPLE 1

Blaire is offered a summer job for 3 months at a popular teen clothing store in the local mall. She has a choice of salary — either a set salary of $1,040 per month or $500 per month plus 10% of all sales. How much would she have to sell each month for the salaried offer to equal the commission offer?

SOLUTION   She constructs the following table to organize her information. She begins computing sales of $400 per month and moves up by increments of $200.

Sales	$400	$600	$800	x
Pay	$.10(400) + 500 = 540$	$.10(600) + 500 = 560$	$.10(800) + 500 = 580$	$.10x + 500$

The table of specific values illustrates a pattern. She can now move from the specifics of a set amount of sales to a generalized expression for the amount of money earned from sales. This allows her to set up an equation that can be solved and she can compute the amount of sales necessary to equal the proposed salary of $1,040.

Let x = amount of sales per month

$.10x$ = amount of commission

$$.10x + 500 = 1,040$$
$$.10x + 500 - 500 = 1,040 - 500$$
$$.10x = 540$$
$$x = 5,400$$

She would have to sell $5,400 worth of clothing each month to equal the fixed salary of $1,040.

Calculator solution: Once Blaire constructed her equation, she entered the expression $.10x + 500$ at the y1= prompt. She set the table to start at 400 (to represent $400 in clothing sales) and incremented the table by 200. Scrolling down the table allows her to compare the amount of clothing sales with her income from those sales.

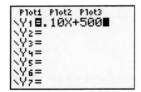

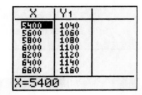

Recall the set of integers is $\{\ldots, -5, -4, -3, -2, -1, 0, 1, 2, 3, 4, 5, \ldots\}$. This example illustrates a consecutive integer problem where x is the first integer and $x + 1$ represents the second integer. If x represents an *odd* or an *even* integer, the expression $x + 2$ represents the *next* odd (or even) integer.

EXAMPLE 2

Bonnie is a mathematics teacher who loves to read. She's currently reading *War and Peace*. While placing her bookmark in the book, she notices that the sum of the page numbers is 2,065. Between which two pages is the bookmark?

SOLUTION   Again, a table can help us organize the information. In this problem, we begin with the pages 1,000 and 1,001 (this is simply a guess). The table allows us to generalize from the specific to the general and write an algebraic equation that we can solve to answer the question posed.

Left page number	1,000	1,002	x
Right page number	1,001	1,003	$x + 1$
Sum	2,001	2,005	$x + (x + 1)$

Let x = left page number

$x + 1$ = right page number

$$x + (x + 1) = 2{,}065$$
$$2x + 1 = 2{,}065$$
$$2x + 1 - 1 = 2{,}065 - 1$$
$$2x = 2{,}064$$
$$x = 1{,}032$$

The bookmark is between pages 1,032 and 1,033.

Calculator solution: Once the equation is constructed, enter the left side at the y1= prompt. Set the table to start at 1,000 and to increment by 1. Enter 2,065 at the y2= prompt. Scroll down until the value in the Y1 column equals the value in the Y2 column.

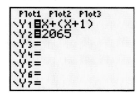

EXAMPLE 3

Marie wants to attend the state fair, yet cannot decide on which night she should go. On Friday night, the price of admission is $3.00 and it costs $0.75 per ride. Saturday night is armband night and the admission price of $15 allows you to purchase an armband and ride all the rides you desire. Help her decide which is the best night to go.

Number of rides	1	2	x
Cost for Friday	$3 + 1(.75)$	$3 + 2(.75)$	$3 + x(.75)$

Again, the table allows Marie to move from the specific to the general and construct the following equation:

Let x = number of rides

$3 + x(.75)$ = cost on Friday night

$$3 + x(.75) = 15.00$$
$$3 + x(.75) - 3 = 15.00 - 3$$
$$.75x = 12.00$$
$$x = 16$$

Marie concludes she would have to ride 16 rides on Friday night for the cost of attendance to equal the cost on Saturday.

The calculator solution is left up to the student.

▎▎ GEOMETRY PROBLEMS

The technique of using a table introduced previously in this section coupled with the problem-solving strategies introduced in the previous chapter can be used as you work through the remaining examples in the section. The solutions provided illustrate only one algebraic method of obtaining the equation to be solved. It is often useful to draw a picture when working with geometry problems.

EXAMPLE 4

The length of a rectangle is 4 meters longer than twice its width. If the perimeter of the rectangle is 26 meters, find its dimensions.

SOLUTION   We can draw a sketch as illustrated. The facts are organized into known facts and those that are unknown.

Known:

the length is 4 meters longer than twice the width
the perimeter is 26 meters
the formula for the perimeter of a rectangle is $p = 2l + 2w$

Unknown:

the dimensions (length and width)

Algebraic solution: Let w represent the width of the rectangle. Then $4 + 2w$ represents the length of the rectangle.
We can form the equation

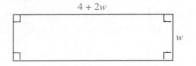

| $2 \cdot$ | the length | plus $2 \cdot$ | the width | equals | the perimeter |

$2(4 + 2w) + 2w = 26$	The equation to solve.
$8 + 4w + 2w = 26$	Remove parentheses.
$6w + 8 = 26$	Combine like terms.
$6w = 18$	Subtract 8 from both sides.
$w = 3$	Divide both sides by 6.

The width of the rectangle is 3 meters, and the length, $4 + 2w$, is 10 meters.

Graphic solution:

$$y1 = 2(4 + 2x) + 2x$$
$$y2 = 26$$

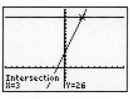

[−10, 10] by [−10, 30]

Check: If a rectangle has a width of 3 meters and a length of 10 meters, then the length is 4 meters longer than twice the width ($4 + 2 \cdot 3 = 10$). Furthermore, the perimeter is $2 \cdot 10 + 2 \cdot 3$, or 26 meters. The solution checks.

▉ RATIO AND PROPORTION

An indicated quotient of two numbers is often called a **ratio**.

Ratio
A **ratio** is the comparison of two numbers by their indicated quotient.

The definition implies that a ratio is a fraction. Some examples of ratios are

$$\frac{7}{8}, \quad \frac{21}{24}, \quad \text{and} \quad \frac{117}{223}$$

Recall, from Chapter 1, that fractions can represent parts of a whole or an indicated division. Another interpretation of a fraction is that it represents the **ratio** of one number to another. The fraction $\frac{7}{8}$ can be read as "the ratio of 7 to 8," the fraction $\frac{21}{24}$ can be read as "the ratio of 21 to 24," and the fraction $\frac{117}{223}$ can be read as "the ratio of 117 to 223." Because the fractions $\frac{7}{8}$ and $\frac{21}{24}$ represent equal numbers, they are called **equal ratios**. Ratios should be expressed in lowest terms.

Proportion
A **proportion** is a statement indicating that two ratios are equal.

Some examples of proportions are

$$\frac{1}{2} = \frac{3}{6}, \quad \frac{3}{7} = \frac{9}{21}, \quad \text{and} \quad \frac{8}{1} = \frac{40}{5}$$

The proportion $\frac{1}{2} = \frac{3}{6}$ can be read as "1 is to 2 as 3 is to 6," the proportion $\frac{3}{7} = \frac{9}{21}$ can be read as "3 is to 7 as 9 is to 21," and the proportion $\frac{8}{1} = \frac{40}{5}$ can be read as "8 is to 1 as 40 is to 5."

In the proportion $\frac{1}{2} = \frac{3}{6}$, the numbers 1 and 6 are called the **extremes** of the proportion, and the numbers 2 and 3 are called the **means**. If we find the product of the extremes and the product of the means in this proportion, we see that the products are equal:

$$1 \cdot 6 = 6 \quad \text{and} \quad 3 \cdot 2 = 6$$

Theorem

In any proportion, the product of the extremes is equal to the product of the means.

EXAMPLE 5

Determine if the equation $\dfrac{x}{3y} = \dfrac{xy + 3x}{3y^2 + 9y}$ is a proportion.

SOLUTION   We check to see if the product of the means is equal to the product of the extremes.

$$3y(xy + 3x) = 3xy^2 + 9xy \qquad \text{The product of the means.}$$
$$x(3y^2 + 9y) = 3xy^2 + 9xy \qquad \text{The product of the extremes.}$$

Because the products are equal, the equation is a proportion.

▌▌ SOLVING PROPORTIONS

EXAMPLE 6   Solve the proportion $\dfrac{12}{18} = \dfrac{3}{x}$ for x.

SOLUTION   We proceed as follows:

$$\frac{12}{18} = \frac{3}{x}$$

$12x = 54$ The product of the extremes equals the product of the means.

$x = \dfrac{54}{12}$ Divide both sides by 12.

$x = \dfrac{9}{2}$ Simplify.

Thus, x represents the fraction $\frac{9}{2}$.

TECHNOLOGY TIP

When a variable appears in the denominator of a fraction, the student should be advised to *not* solve graphically at this point in time. Use the STOre or TABLE feature to check solutions.

EXAMPLE 7   Solve the proportion $\dfrac{y+1}{y} = \dfrac{y}{y+2}$ for y.

SOLUTION $$\frac{y+1}{y} = \frac{y}{y+2}$$

$y^2 = (y+1)(y+2)$ The product of the means is equal to the product of the extremes.

$y^2 = y^2 + 3y + 2$ Remove parentheses.

$0 = 3y + 2$ Add $-y^2$ to both sides.

$-3y = 2$ Add $-3y$ to both sides.

$y = -\dfrac{2}{3}$ Divide both sides by -3.

EXAMPLE 8   **Grocery shopping**   If 5 tomatoes cost \$1.15, how much will 16 tomatoes cost?

SOLUTION   We can let c represent the cost of 16 tomatoes. Because the ratio of the number of tomatoes is the same as the ratio of their costs, we can express this relationship as a proportion and find c.

$$\frac{5}{16} = \frac{1.15}{c}$$

$5c = 1.15(16)$ The product of the extremes is equal to the product of the means.

$5c = 18.4$ Do the multiplication.

$c = \dfrac{18.4}{5}$ Divide both sides by 5.

$c = 3.68$ Simplify.

Sixteen tomatoes will cost \$3.68.

SIMILAR TRIANGLES

Two triangles are **similar** when corresponding angles have the same measure and the lengths of corresponding sides are in proportion. The following theorem enables us to measure sides of triangles indirectly.

Theorem

If two triangles are similar, then all pairs of corresponding sides are in proportion.

Using this theorem, we can find the height of a tree on a sunny day and stay safely on the ground.

EXAMPLE 9

A tree casts a shadow 18 feet long at the same time a woman 5 feet tall casts a shadow that is 1.5 feet long. Find the height of the tree.

SOLUTION   Figure 4-1 shows the triangles determined by the tree and its shadow and the woman and her shadow.

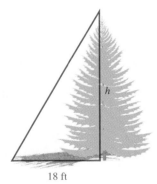

5 ft

1.5 ft

h

18 ft

Figure 4-1

Because the triangles have the same shape, they are similar, and the lengths of their corresponding sides are in proportion. If we let h represent the height of the tree, we can find h by solving the following proportion.

$$\frac{h}{5} = \frac{18}{1.5}$$ $\dfrac{\text{Height of the tree}}{\text{Height of the woman}} = \dfrac{\text{Shadow of the tree}}{\text{Shadow of the woman}}$

$$1.5h = 90$$ In a proportion, the product of the extremes is equal to the product of the means.

$$h = 60$$ To undo the multiplication by 1.5, divide both sides by 1.5 and simplify.

The tree is 60 feet tall.

EXERCISE 4.1

VOCABULARY AND NOTATION   *In Exercises 1–6, fill in the blanks to make a true statement.*

1. A _____ is a comparison of two numbers by their indicated quotient.

2. _____ ratios represent _____ numbers.

3. In any proportion, the product of the _____ is equal to the product of the _____.

4. Two triangles with the same shape are called _____ triangles.

5. The statement "4 is to 5 as 8 is to 10" is the proportion _____.

6. In the proportion $\dfrac{4xy}{x^2} = \dfrac{z}{z^2}$, the product of the means is _____.

CONCEPTS

7. Express each phrase as a ratio in lowest terms.
   a. 5 to 7   b. 3 to 5
   c. 17 to 34   d. 19 to 38
   e. 22 to 33   f. 14 to 21

8. Tell whether each statement is a proportion.
   a. $-\dfrac{7}{3} = \dfrac{14}{-6}$ b. $\dfrac{13}{-19} = \dfrac{-65}{95}$
   c. $\dfrac{25}{2} = \dfrac{5}{4}$ d. $\dfrac{50}{25} = \dfrac{2}{1}$
   e. $\dfrac{5}{35} = \dfrac{1}{7}$ f. $\dfrac{7}{3} = -\dfrac{7}{3}$

9. Identify the pairs that represent similar triangles, assuming corresponding angles have the same measure as indicated.
   a.
   b.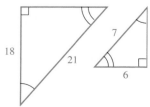

   c.

10. In two similar triangles, all pairs of corresponding sides are in proportion. Set up a proportion for each set of similar triangles.

    a.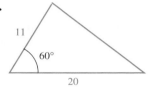

    b.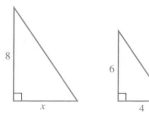

PRACTICE   *In Exercises 11–16, solve for the variable in each proportion.*

11. $\dfrac{x + 3}{12} = \dfrac{-7}{6}$ 12. $\dfrac{x + 7}{-4} = \dfrac{3}{12}$

13. $\dfrac{3x - 2}{7} = \dfrac{x}{28}$ 14. $\dfrac{2x + 1}{9} = \dfrac{x}{27}$

15. $\dfrac{7(x + 6)}{6} = \dfrac{6(x + 3)}{5}$ 16. $\dfrac{3(x + 5)}{2} = \dfrac{5(x - 2)}{3}$

APPLICATIONS   *In Exercises 17–30, use the problem-solving strategies you have learned thus far to solve each of the given problems. Experiment with setting up tables and separating the known information from what is unknown. For each problem, make sure you clearly define your variable and what it represents, write an equation, provide a complete solution to that equation, and answer the question(s) posed. Be sure and check all results both in the equation as well as in the context of the problem.*

17. **Finding consecutive even integers**   The sum of two consecutive even integers is 54. Find the integers.

18. **Finding consecutive odd integers**   The sum of two consecutive odd integers is 88. Find the integers.

19. Finding consecutive integers The sum of three consecutive integers is 120. Find the integers.

20. Finding consecutive even integers The sum of three consecutive even integers is 72. Find the three even integers.

21. Finding integers The sum of an integer and twice the next integer is 23. Find the smaller integer.

22. Finding integers If 4 times the smallest of three consecutive integers is added to the largest, the result is 112. Find the three integers.

23. Finding integers The larger of two integers is 10 greater than the smaller. The larger is 3 less than twice the smaller. Find the smaller integer.

24. Finding integers The smaller of two integers is one-half of the larger, and 11 greater than one-third of the larger. Find the smaller integer.

25. Triangular bracing The outside perimeter of the triangular brace in Illustration 1 is 57 feet. If all three sides are equal, find the length of each side.

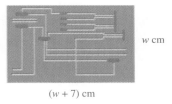

Illustration 1

26. Circuit boards The perimeter of the circuit board in Illustration 2 is 90 centimeters. Find the dimensions of the board.

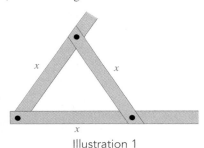

w cm

(*w* + 7) cm

Illustration 2

27. Swimming pools The width of a rectangular swimming pool is 11 meters less than the length. The perimeter is 94 meters. Find the dimensions of the pool.

28. Wooden truss The truss in Illustration 3 is in the form of an isosceles triangle. Each of the two equal sides is 4 feet less than the third side. If the perimeter is 25 feet, find the lengths of the sides.

Illustration 3

29. Framing pictures The length of a rectangular picture frame is 5 inches greater than twice the width. If the perimeter is 112 inches, find the dimensions of the frame.

30. Land areas The perimeter of a square parcel of land is twice the perimeter of an equilateral (equal-sided) triangular plot. If one side of the square is 60 meters, find the length of a side of the triangle.

In Exercises 31–42, set up and solve the required proportion.

31. Gardening Garden seeds are on sale at 3 packets for 50 cents. How much will 39 packets cost?

32. Increasing a recipe A recipe for spaghetti sauce requires four 16-ounce bottles of ketchup to make 2 gallons of sauce. How many bottles of ketchup are needed to make 10 gallons of sauce?

33. Model railroading An HO-scale model railroad engine is 9 inches long. The HO scale is 87 feet to 1 foot. How long is a real engine?

34. Model railroading An N-scale model railroad caboose is 3.5 inches long. The N scale is 169 feet to 1 foot. How long is a real caboose to the nearest tenth of a foot?

35. Mixing fuel The instructions on a can of oil intended to be added to lawn mower gasoline read as follows:

Recommended	Gasoline	Oil
50 to 1	6 gal	16 oz

Are these instructions correct? (*Hint*: There are 128 ounces in 1 gallon.)

36. Height of a tree A tree casts a shadow of 26 feet at the same time as a 6-foot man casts a shadow of 4 feet. (See Illustration 4.) Find the height of the tree.

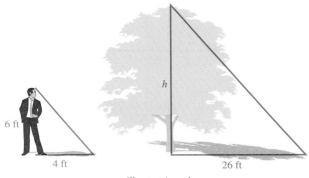

6 ft

4 ft

h

26 ft

Illustration 4

37. Height of a flagpole A man places a mirror on the ground and sees the reflection of the top of a flagpole, as in Illustration 5. The two triangles in the illustration are similar. Find the height, h, of the flagpole rounded to the nearest tenth.

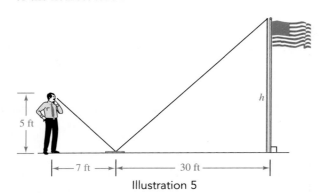

Illustration 5

38. Width of a river Use the dimensions in Illustration 6 to find w, the width of the river. The two triangles in the illustration are similar.

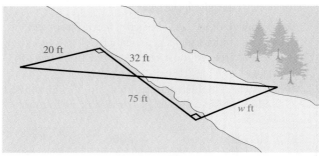

Illustration 6

39. Flight path An airplane ascends 100 feet as it flies a horizontal distance of 1,000 feet. How much altitude will it gain as it flies a horizontal distance of 1 mile? (See Illustration 7.) (*Hint:* 5,280 feet = 1 mile.)

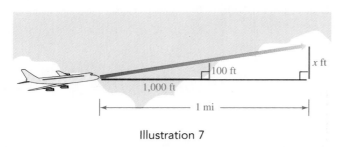

Illustration 7

40. Flight path An airplane descends 1,350 feet as it flies a horizontal distance of 1 mile. How much altitude is lost as it flies a horizontal distance of 5 miles? (*Hint:* 5,280 feet = 1 mile.)

41. Ski runs A half-mile long ski course falls 100 feet in every 300 feet of horizontal run. Find the height of the hill.

42. Mountain travel A mountain road ascends 375 feet in every 2,000 feet of travel. By how much will the road rise in a trip of 10 miles? (*Hint:* 5,280 feet = 1 mile)

REVIEW

43. Simplify each expression by performing the indicated operation.

   a. $(4x^2 - 5x + 2) - (6x^2 - 3x - 4)$

   b. $(2x + 5)^3$

   c. $(x^3 + 27)$ divided by $(x + 3)$

44. If $P(x) = 4x^5 - 5x^2 + 6x$, find $P(-1)$ and verify with your calculator.

45. The left side of an equation has been graphed at y1 and the right side at y2.

   a. Based on the calculator display, what is the solution of the equation y1 = y2?

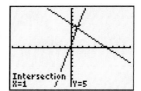

   b. Based on the table, what is the solution of the equation y1 = y2?

X	Y₁	Y₂
-3	-11	9
-2	-7	8
-1	-3	7
0	1	6
1	5	5
2	9	4
3	13	3

X= -3

46. Algebraically solve the equation $\frac{2}{3}(x - 6) + 4 = 2x + 1$. Determine the **three** ways that you can use the graphing calculator to check your results and then apply one of these methods as your check.

4.2 BUSINESS APPLICATIONS

- BREAK-EVEN ANALYSIS
- QUANTITY AND VALUE PROBLEMS
- INVESTMENT PROBLEMS

▌ BREAK-EVEN ANALYSIS

In manufacturing, there are two types of costs — **fixed costs** and **unit costs**. Fixed costs do not depend on the amount of product manufactured. Fixed costs would include the cost of plant rental, insurance, and machinery. Unit costs depend on the amount of product manufactured. Unit costs would include the cost of raw materials and labor.

Break-even analysis is used to find the production level where revenue will just offset the cost of production. When production exceeds the break-even point, the company will make a profit.

EXAMPLE 1

An electronics company has fixed costs of $6,405 a week and a unit cost of $75 for each compact disk player manufactured. If the company can sell all the CD players it can make at a wholesale price of $90, find the company's break-even point.

SOLUTION   *Known:*

  fixed cost of $6,405
  unit cost of $75 for each CD player produced
  selling price of $90 per CD player

Unknown:

  break-even point (the number of CD players produced that would make cost equal revenue)

Suppose the company manufactures x CD players each week. The cost of manufacturing these players is the sum of the fixed and unit costs. We are given that the fixed costs are $6,405 each week. The unit cost is the product of x, the number of CD players manufactured each week, and $75, the cost of manufacturing a single player. Thus, the weekly cost is

  Total weekly cost = Weekly unit cost + weekly fixed cost

  $$\text{Cost} = 75x + 6{,}405$$

The company can sell all the machines it can make, so the weekly revenue is the product of x, the number of players manufactured (and sold) each week, and $90, the wholesale price of each CD player. Thus, the weekly revenue is

  $$\text{Revenue} = 90x$$

The break-even point is the value of x for which the weekly revenue is equal to the weekly cost.

Because x represents the number of CD players manufactured each week,

  $90x$ represents the weekly revenue, and

  $75x + 6{,}405$ represents the weekly cost.

Because the break-even point occurs when revenue equals cost, we set up and solve the following equation:

| Total weekly revenue | = | Total weekly cost |

$$90x = 75x + 6{,}405$$

$$15x = 6{,}405 \qquad \text{Subtract } 75x \text{ from both sides.}$$

$$x = 427 \qquad \text{Divide both sides by 15.}$$

If the company manufactures 427 CD players, the revenue will equal the cost, and the company will break even.

Graphic solution:

$$y1 = 90x$$
$$y2 = 75x + 6{,}405$$

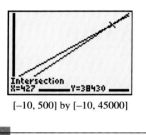

[−10, 500] by [−10, 45000]

It is often convenient to practice solving application problems graphically. The values represented by x and the expressions denoted at the y1 and y2 prompts often aid in setting accurate viewing windows.

▌▌ QUANTITY AND VALUE PROBLEMS

EXAMPLE 2

An investor has stock worth $240,000 in three segments of the economy. He owns five times as many shares in transportation as he owns in utilities, and 2,000 more shares in pharmaceuticals than in transportation. Each share in transportation is worth $55, each utility share is worth $75, and each share in pharmaceuticals is worth $30. How many shares does he own in each segment?

SOLUTION   *Known:*

the total worth of the stock is $240,000
he owns five times as many shares in transportation as in utilities
he owns 2,000 more shares in pharmaceuticals than in transportation
each share in transportation is worth $55
each share in utilities is worth $75
each share in pharmaceuticals is worth $30

Unknown:

the number of shares in each segment

Upon examination of the known information, it becomes apparent that the number of shares in transportation is given in terms of the number of shares in utilities, therefore,

Let n = number of shares in utilities

Translating, we see that

$5n$ = number of shares in transportation and

$5n + 2{,}000$ = number of shares in pharmaceuticals

It is important to distinguish between the *number* of shares and the *value* of those shares. Because the investor owns n shares of utility stock worth $75 each, the value of the stock is $75n$. Because he owns $5n$ shares of transportation stock worth $55 each, their value is $55(5n)$. Finally, his $(5n + 2,000)$ shares of pharmaceutical stock, at $30 each, are worth $30(5n + 2,000)$. The total value can be expressed in two ways: as

$$\$[75n + 55(5n) + 30(5n + 2,000)] \quad \text{and as} \quad \$240,000$$

The value of the utility stock	+	the value of the transportation stock	+	the value of the pharmaceutical stock	=	the total value
$75n$	+	$55(5n)$	+	$30(5n + 2,000)$	=	$240,000$

$75n + 275n + 150n + 60,000 = 240,000$	Remove parentheses.
$500n + 60,000 = 240,000$	Combine like terms.
$500n = 180,000$	Subtract 60,000 from both sides.
$n = 360$	Divide both sides by 500.

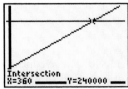

Intersection
X=360   Y=240000
[−10, 500] by [−10, 300000]

Graphic solution:

$$y1 = 75x + 55(5x) + 30(5x + 2,000)$$
$$y2 = 240,000$$

Summarizing the results in a table is a helpful way to check:

Segment	Shares	Value
Utilities	$n = 360$	$27,000
Transportation	$5n = 1,800$	$99,000
Pharmaceuticals	$5n + 2,000 = 3,800$	$114,000
Total		$240,000

▌ INVESTMENT PROBLEMS

EXAMPLE 3

Investing money   A retired teacher invested part of $12,000 at 4% annual interest and the rest at 3%. If the annual income from these investments was $435, how much did the teacher invest at each rate?

ANALYSIS   The interest i earned by an amount p invested at an annual rate r for t years is given by the formula $i = prt$. In this example, $t = 1$ year. Hence, if x dollars were invested at 4%, the interest earned would be $0.04x$. If x dollars were invested at 4%, then the rest of the money, $(12,000 − x)$, would be invested at 3%. The interest earned on that money would be $.03(12,000 − x)$ dollars. The total interest earned in dollars can be expressed in two ways: as 435 and as the sum $.04x + .03(12,000 − x)$.

SOLUTION   Let x represent the amount of money invested at 4%. Then $12,000 - x$ represents the amount of money invested at 3%.
We can form an equation as follows:

The interest earned at 4%	plus	the interest earned at 3%	equals	the total interest
$0.04x$	$+$	$0.03(12,000 - x)$	$=$	435

$$0.04x + .03(12,000 - x) = 435 \quad \text{The equation to solve.}$$
$$4x + 3(12,000 - x) = 43,500 \quad \text{Multiply both sides by 100 to clear the equation of decimals.}$$
$$4x + 36,000 - 3x = 43,500 \quad \text{Remove parentheses.}$$
$$x + 36,000 = 43,500 \quad \text{Combine like terms.}$$
$$x = 7,500 \quad \text{Subtract 36,000 from both sides.}$$

The teacher invested $7,500 at 4% and $12,000 - $7,500$, or $4,500 at 3%.

Graphic solution:

$$y1 = .04x + .03(12000 - x)$$
$$y2 = 435$$

Check: The first investment yielded 4% of $7,500, or $300. The second investment yielded 3% of $4,500, or $135. Because the total return was $300 + $135, or $435, the answers check.

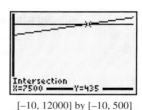

Intersection
X=7500 ————— Y=435

[−10, 12000] by [−10, 500]

EXERCISE 4.2

VOCABULARY AND NOTATION   *In Exercises 1–4, fill in the blanks to make a true statement.*

1. Costs that do not depend on the quantity of a product manufactured are called _____ costs.

2. Costs that depend on the quantity of a product manufactured are called _____ costs.

3. We use _____ analysis to find the production level where the revenue will just offset the cost of production.

4. To determine the amount of interest earned in one year on $500 at 4%, compute $500(4\%) = 500($____$)$.

CONCEPTS

5. If paint sells for $21 per gallon and x represents the number of gallons sold, is $21x$ a representation of cost of production or revenue?

6. If the cost of production is represented by $C(x) = \$4,000 + \$25x$, identify the unit cost and the fixed cost.

APPLICATIONS   *In Exercises 7–28, pick a variable to represent the unknown quantity, set up an equation involving the variable, solve the equation, and check it.*

7. A paint manufacturer can choose between two processes for manufacturing house paint, with monthly costs shown in Table 1. The paint can be sold for $21 per gallon.

a. Find the break-even point for process A.

b. Find the break-even point for process B.

Process	Fixed costs	Unit cost (per gallon)
A	$ 75,000	$11
B	$128,000	$ 5

TABLE 1

c. If expected sales will be 8,800 gallons per month, which process should the company choose?

d. If expected sales will be 9,000 gallons per month, which process should the company choose?

8. Performance bond A construction company has agreed to finish a highway repair in 60 days, and pay a $1,000 fine for each day the work takes beyond 60 days. To finish by the deadline, management expects to spend $12,000 per day. By cutting the work crew, expenses would be $9,400 per day, but the job would take longer. How many days beyond deadline can the project last to make expenditures equal?

9. Manufacturing shoes A shoe company has fixed costs of $9,600 per month and a unit cost of $20 per pair of shoes. The company can sell all the shoes it can make at a wholesale price of $30 per pair. Find the break-even point.

10. Manufacturing insulators A manufacturer of high-voltage insulators has fixed costs of $5,400 per month and a unit cost of $12 per insulator. It can sell all the units it can make at a wholesale price of $15. Find the break-even point.

11. Scheduling machine usage A machine shop has two machines that can mill a certain brass plate. One machine has a setup cost of $500 and a cost of $2 per plate, while the other machine has a setup cost of $800 and a cost of $1 per plate. How many plates should be manufactured if the cost is to be the same using either machine?

12. Oriental rugs A rug manufacturer has two looms for weaving Oriental-style rugs. One loom has a setup cost of $750 and can produce a rug for $115. The other loom has a setup cost of $950 and can produce a rug for $95.

How many rugs can be manufactured if the cost is to be the same using either loom?

13. Software sales The three best-selling software applications at one computer store are priced as in Table 2.

Spreadsheet	$150
Database	$195
Word processing	$210

TABLE 2

Spreadsheet and database programs sold in equal numbers, but 15 more word processing applications were sold than the other two combined. The three applications generated sales of $72,000. How many spreadsheets were sold?

14. Appliance store inventory Stoves and refrigerators are expected to sell in equal numbers. With summer approaching, the number of air conditioners sold is expected to be double that of stoves and refrigerators combined. Stoves sell for $350, refrigerators for $450, and air conditioners for $500, and sales of $56,000 are expected. How many of each should be stocked?

15. Sorting hardware Liz has an assortment of bolts, washers, and locknuts — 99 pieces in all. She has twice as many washers as bolts and one less locknut than washers. How many bolts does she have?

16. CD collections A musician has a collection of 178 CDs. He has twice as many jazz as rock and 23 more classical than jazz. How many of each does he have?

17. Stock performance An investor owns twice as many shares of stock with average performance as he owns of shares that have far outpaced the market. He owns 200 shares of a stock that is now worthless. If he owns 3,500 shares in all, how many shares have performed better than average?

18. Raising capital 30% of a corporation's stock offering is preferred stock worth $25 per share. The rest is common stock, selling at $15 per share. If the entire offering raised $306 million, how many common shares were sold?

19. Warehousing costs Monthly storage costs are $1.50 for each portable television, $4.00 for each wide screen television, and $7.50 for each flat screen television. An appliance store warehouses 40 more portables than flat-screen sets, and 15 fewer wide screens than portables. Television storage costs $277 for the month. How many flat-screen sets are in stock?

20. Apartment rental   The designers of a new apartment complex will include one-, two-, and three-bedroom apartments. They feel they can rent equal numbers of each, with the monthly rents given in Table 3.

One bedroom	$550
Two bedroom	$700
Three bedroom	$900

TABLE 3

If the total monthly rental income will be $36,550, how many units of each kind are planned?

21. Investment problem   A broker has invested $24,000 in two accounts, one earning 5% annual interest and the other earning 4%. After 1 year, his combined interest is $1,155. How much was invested at each rate?

22. Investment problem   A woman's rollover IRA of $18,750 is invested in two accounts, one earning 7% interest and the other earning 5%. After 1 year, the combined interest income is $1,179.50. How much has she invested at each rate?

23. Investment problem   One investment pays 5% and another pays 5.5%. If equal amounts are invested in each, the combined interest income for 1 year is $702.03. How much is invested at each rate?

24. Investment problem   When equal amounts are invested in each of three accounts paying 3%, 4%, and 5%, one year's combined interest income is $882. How much is invested in each account?

25. Investment problem   A college professor wants to supplement her retirement income with investment interest. If she invests $15,000 at 4% annual interest, how much more would she have to invest at 6% to achieve a goal of $1,200 in supplemental income?

26. Investment problem   A retired teacher has a choice of two investment plans: an insured fund that pays 6% interest or a riskier investment that promises an 8% return. The same amount invested at the higher rate would generate an extra $150 per year. How much does the teacher have to invest?

27. Investment problem   A financial adviser recommends investing twice as much in CDs as in a bond fund. A client follows his advice and invests $21,000 in CDs paying 1% more interest than the fund. The CDs generate $840 more interest than the fund. Find the two rates. (*Hint:* 1% = 0.01)

28. Investment problem   The amount of annual interest earned by $8,000 invested at a certain rate is $200 less than $12,000 would earn at a 1% lower rate. At what rate is the $8,000 invested?

REVIEW

29. Simplify.

    a. $(4x)^2$ **b.** $(4x + 3)^2$

30. Evaluate for $x = -2$ and verify with your calculator.

    a. $(4x)^2$ **b.** $(4x + 3)^2$

31. Algebraically solve $4\left(\dfrac{2}{3}x - 1\right) = (2x + 5) - (x - 1)$ and verify your results with

    a. the STOre feature

    b. a graphical solution

    c. the TABLE

32. Algebraically solve $\dfrac{5x - 1}{3} = \dfrac{6x + 2}{4}$

4.3   MORE APPLICATIONS OF EQUATIONS

- LIQUID MIXTURE PROBLEMS
- DRY MIXTURE PROBLEMS
- UNIFORM MOTION PROBLEMS

This section examines several different types of problems that are commonly encountered in algebra courses. An analysis of each problem is provided that offers another problem-solving strategy. Remember the tools available for problem solving: writing the known and the unknown facts, translating from English words to algebraic symbols, using a table to organize information, and moving from the specific to the general. These are valuable tools that you must experiment with in order to become a successful problem solver. You will probably discover that you use a combination of tools rather than any single problem-solving strategy. As you study and work through the examples in the section, think about ways to organize the information other than those illustrated.

▐ LIQUID MIXTURE PROBLEMS

EXAMPLE 1

Mixing acids   A chemistry instructor has one solution that is 50% sulfuric acid and another that is 20% sulfuric acid. How much of each should she use to make 12 liters of a solution that is 30% acid?

ANALYSIS   The sulfuric acid present in the final mixture comes from the two solutions to be mixed. If x represents the number of liters of the 50% solution required for the mixture, then the rest of the mixture, $(12 - x)$ liters, must be the 20% solution. (See Figure 4-2.) Only 50% of the x liters and only 20% of the $(12 - x)$ liters are pure sulfuric acid. The total of these amounts is also the amount of acid in the final mixture, which is 30% of 12 liters.

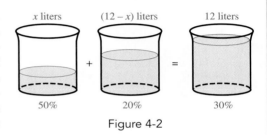

Figure 4-2

SOLUTION   Since x represents the required number of liters of the 50% solution, $12 - x$ represents the required number of liters of the 20% solution.

We can form the equation

The acid in the 50% solution	plus	the acid in the 20% solution	equals	the acid in the final mixture
50% of x	+	20% of $(12 - x)$	=	30% of 12

$$0.50x + 0.20(12 - x) = 0.30(12)$$ The equation to solve.

$$5x + 2(12 - x) = 3(12)$$ Multiply both sides by 100 to clear the equation of decimals.

$$5x + 24 - 2x = 36$$ Remove parentheses.

$$3x + 24 = 36$$ Combine like terms.

$$3x = 12$$ Subtract 24 from both sides.

$$x = 4$$ Divide both sides by 3.

The chemist must mix 4 liters of the 50% solution and 8 liters $((12 - 4)$ liters$)$ of the 20% solution.

Graphic solution:

$$y1 = .50x + .20(12 - x)$$
$$y2 = .30(12)$$

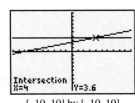

[−10, 10] by [−10, 10]

▌▌▌ DRY MIXTURE PROBLEMS

EXAMPLE 2

Mixing nuts   Fancy cashews are not selling at $9 per pound because they are too expensive. Filberts are selling at $6 per pound. How many pounds of filberts should be combined with 50 pounds of cashews to obtain a mixture that can be sold at $7 per pound?

ANALYSIS   Dry mixture problems are based on the formula $v = pn$, where v is the value of the mixture, p is the price per pound, and n is the number of pounds. Suppose x pounds of filberts are used in the mixture. At $6 per pound, they are worth $6x. At $9 per pound, the 50 pounds of cashews are worth $9 · 50, or $450. The mixture will weigh $(50 + x)$ pounds, and at $7 per pound, it will be worth $7(50 + x). The value of the ingredients, $(6x + 450)$, is equal to the value of the mixture, $7(50 + x)$. (See Figure 4-3.)

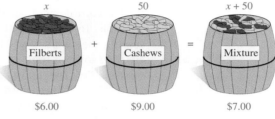

Figure 4-3

SOLUTION   Let x represent the number of pounds of filberts in the mixture. We can form the equation

The value of the filberts	plus	the value of the cashews	equals	the value of the mixture
$6x	+	$9 · 50	=	$7(50 + x)

$$6x + 9 \cdot 50 = 7(50 + x) \quad \text{The equation to solve.}$$
$$6x + 450 = 350 + 7x \quad \text{Remove parentheses and simplify.}$$
$$100 = x \quad \text{Subtract } 6x \text{ and } 350 \text{ from both sides.}$$

The storekeeper should use 100 pounds of filberts in the mixture.

Check:

The value of 100 pounds of filberts at $6 per pound is $600
The value of 50 pounds of cashews at $9 per pound is   450
The value of the mixture is $1,050

The value of 150 pounds of mixture at $7 per pound is also $1,050.

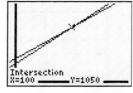

[−10, 200] by [−10, 1500]

Graphic solution:

$$y1 = 6x + 9(50)$$
$$y2 = 7(50 + x)$$

▮ UNIFORM MOTION PROBLEMS

EXAMPLE 3   **Driving times**   Chicago, Illinois, and Green Bay, Wisconsin, are about 200 miles apart. A car leaves Chicago traveling toward Green Bay at 55 miles per hour. At the same time, a truck leaves Green Bay bound for Chicago at 45 miles per hour. How long will it take them to meet?

ANALYSIS   Uniform motion problems are based on the formula $d = rt$, where d is the distance traveled, r is the rate, and t is the time. We can organize the information of this problem in chart form, as in Figure 4-4.

	r	$\cdot$	t	$=$	d
Car	55		t		$55t$
Truck	45		t		$45t$

Figure 4-4

We know that the two vehicles travel for the same amount of time; say, t hours. The faster car travels $55t$ miles, and the slower truck travels $45t$ miles. The total distance can be expressed in two ways: as the sum $55t + 45t$ and as 200 miles.

SOLUTION   Let t represent the time that each vehicle travels until they meet. Then

$55t$ represents the distance traveled by the car, and
$45t$ represents the distance traveled by the truck.

After referring to Figure 4-4, we form the equation

The distance the car goes	plus	the distance the truck goes	equals	the total distance
$55t$	$+$	$45t$	$=$	200

$$55t + 45t = 200 \qquad \text{The equation to solve.}$$
$$100t = 200 \qquad \text{Combine like terms.}$$
$$t = 2 \qquad \text{Divide both sides by 100.}$$

The vehicles will meet after 2 hours.

Graphic solution:

$$y1 = 55x + 45x$$
$$y2 = 200$$

Intersection
X=2   Y=200
[–10, 10] by [–10, 250]

Check: During those 2 hours, the car travels $55 \cdot 2$, or 110 miles, while the truck travels $45 \cdot 2$, or 90 miles. The total distance traveled is $110 + 90$, or 200 miles. This is the total distance between Chicago and Green Bay. The answer checks.

EXERCISE 4.3

CONCEPTS

1. Express each percentage as a decimal number *and* as a fraction.

   a. 3%   **b.** 30%

   c. $3\frac{1}{2}$%   **d.** 3.5%

2. If a 20-ounce bottle of saline solution for contacts is 5% salt, how much salt is in the bottle?

3. If the total liquid mixture is 30 grams and the liquid was originally divided between two bottles, how much was in each of the original bottles? (*Hint:* Your response will be in terms of polynomials, not specific quantities.)

4. If one beaker contains 25 ounces of liquid and a second beaker contains an unknown amount, x, how much liquid will there be if the liquid of the two beakers is combined into one larger beaker?

APPLICATIONS   *In Exercises 5–24, use the problem-solving strategies you have learned thus far to solve each of the given problems. For each problem, make sure you clearly define your variable and what it represents, write an equation, provide a complete solution to that equation, and answer the question(s) posed. Be sure and check all results both in the equation as well as in the context of the problem.*

5. **Mixing fuels**   How many gallons of fuel costing $1.15 per gallon must be mixed with 20 gallons of a fuel costing $0.85 per gallon to obtain a mixture costing $1 per gallon? (See Illustration 1.)

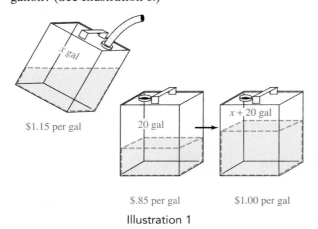

$1.15 per gal

20 gal $x + 20$ gal

$.85 per gal $1.00 per gal

Illustration 1

6. **Mixing paint**   Paint costing $19 per gallon is to be mixed with 5 gallons of a $3-per-gallon thinner to make

a paint that can be sold for $14 per gallon. How much paint will be produced?

7. **Brine solution**   How many gallons of a 3% salt solution must be mixed with 50 gallons of a 7% solution to obtain a 5% solution?

8. **Making cottage cheese**   To make low-fat cottage cheese, milk containing 4% butterfat is mixed with 10 gallons of milk containing 1% butterfat to obtain a mixture containing 2% butterfat. How many gallons of the richer milk must be used?

9. **Antiseptic solutions**   A nurse wishes to add water to 30 ounces of a 10% solution of benzalkonium chloride to dilute it to an 8% solution. How much water must she add?

10. **Mixing photographic chemicals**   A photographer wishes to mix 2 liters of a 5% acetic acid solution with a 10% solution to get a 7% solution. How many liters of 10% solution must be added?

11. **Mixing candy**   Lemon drops worth $1.90 per pound are to be mixed with jelly beans that cost $1.20 per pound to make 100 pounds of a mixture worth $1.48 per pound. How many pounds of each candy should be used?

12. **Blending gourmet tea**   One grade of tea worth $3.20 per pound is to be mixed with another grade worth $2 per pound to make 20 pounds that will sell for $2.72 per pound. How much of each grade of tea must be used?

13. **Mixing nuts**   A pound of peanuts is worth $0.30 less than a pound of cashews. Equal amounts of peanuts and cashews are used to make 40 pounds of a mixture that sells for $1.05 per pound. How much is a pound of cashews worth?

14. **Mixing candy**   Twenty pounds of lemon drops are to be mixed with cherry chews to make a mixture that will sell for $1.80 per pound. How much of the more expensive candy should be used? (See Table 1.)

	Price per pound
Peppermint patties	$1.35
Lemon drops	$1.70
Licorice lumps	$1.95
Cherry chews	$2.00

TABLE 1

15. Coffee blends A store sells regular coffee for $6 a pound and a gourmet coffee for $10 a pound. To get rid of 40 pounds of the gourmet coffee, the shopkeeper makes a gourmet blend that he will put on sale for $7.25 a pound. How many pounds of regular coffee should be used?

16. Lawn seed blends A garden store sells Kentucky bluegrass seed for $6 per pound and ryegrass seed for $3 per pound. How much rye must be mixed with 100 pounds of bluegrass to obtain a blend that will sell for $5 per pound?

17. Travel time Ashford and Bartlett are 315 miles apart. A car leaves Ashford bound for Bartlett at 50 miles per hour. At the same time, another car leaves Bartlett and heads toward Ashford at 55 miles per hour. In how many hours will the two cars meet?

18. Travel time Granville and Preston are 535 miles apart. A car leaves Preston bound for Granville at 47 miles per hour. At the same time, another car leaves Granville and heads toward Preston at 60 miles per hour. How long will it take them to meet?

19. Travel time Two cars leave Peoria at the same time, one heading east at 60 miles per hour and the other west at 50 miles per hour. (See Illustration 2.) How long will it take them to be 715 miles apart?

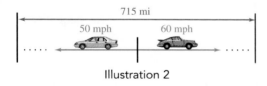

Illustration 2

20. Boating Two boats leave port at the same time, one heading north at 35 knots (nautical miles per hour), the other south at 47 knots. How long will it take them to be 738 nautical miles apart?

21. Travel time Two cars start together and head east, one at 42 miles per hour and the other at 53 miles per hour. (See Illustration 3.) In how many hours will the cars be 82.5 miles apart?

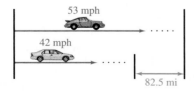

Illustration 3

22. Speed of trains Two trains are 330 miles apart, and their speeds differ by 20 miles per hour. They travel toward each other and meet in 3 hours. Find the speed of each train.

23. Speed of an airplane Two planes are 6,000 miles apart, and their speeds differ by 200 miles per hour. They travel toward each other and meet in 5 hours. Find the speed of the slower plane.

24. Average speed An automobile averaged 40 miles per hour for part of a trip and 50 miles per hour for the remainder. If the 5-hour trip covered 210 miles, for how long did the car average 40 miles per hour?

REVIEW

25. Simplify:
   a. $[4 - (2 - 3)^2 + 5] - (6 - 2)^3$
   b. $(3x + 5)(2x^2 - x + 1)$
   c. $\dfrac{4x^2y^3 - 6xy^2 + 8x^3}{4xy}$

26. Evaluate $(3x + 5)(2x^2 - x + 1)$ for $x = -2$. Verify with your calculator.

27. Solve graphically $6x + 2 = 4 - 3x$ and express the solution as a fraction.

28. Finding consecutive even integers The sum of two consecutive even integers is 206. Find the numbers.

29. Finding integers Four more than 3 times the quantity of an integer plus 2 is equal to 13. Find the integer.

4.4 SOLVING SIMPLE LINEAR INEQUALITIES

- GRAPHING INEQUALITIES
- INTERVAL NOTATION
- SET NOTATION

- SOLVING LINEAR INEQUALITIES
- GRAPHICAL SOLUTIONS
- PROBLEM SOLVING

GRAPHING INEQUALITIES

The solution of a conditional linear equation in one variable is a single value. For example, the solution of the equation $2x + 5 = 11$ is 3. We could graph the solution on a number line as pictured below:

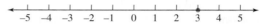

Suppose, however, that our solution had been all numbers *less than or equal to* three. First, we would need the following symbols to translate our English expression to the mathematical format of $x \leq 3$.

Inequality Symbols

$<$ means	"is less than"
$>$ means	"is greater than"
$\leq$ means	"is less than or equal to"
$\geq$ means	"is greater than or equal to"

An **inequality** is a mathematical statement that indicates that two quantities are not necessarily equal. A **solution of an inequality** is any number that makes the inequality a true statement. The number 2 is a solution of the inequality

$$x \leq 3$$

because $2 \leq 3$.

The inequality $x \leq 3$ has many more solutions, because any real number that is less than or equal to 3 will satisfy the inequality. We can use a graph on the number line to illustrate the solutions of the inequality $x \leq 3$. The colored arrow in Figure 4-5 indicates all those points with coordinates that satisfy the inequality $x \leq 3$. The solid circle at the point with coordinate 3 indicates that the number 3 is a solution of the inequality $x \leq 3$. Because the value of 3 divides the number line into the set of points that **are** solutions and the set of points that **are not** solutions, it is called the **critical point**.

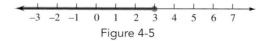

Figure 4-5

The graph of the inequality $x > 1$ appears in Figure 4-6. The colored arrow indicates all those points whose coordinates satisfy the inequality $x > 1$. The open circle at the point with coordinate 1 indicates that 1 is not a solution of the inequality $x > 1$; however, 1 is a critical point.

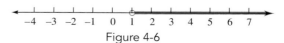

Figure 4-6

EXAMPLE 1

Graph each inequality on a number line:

a. $x > 5$ **b.** $x \geq 5$

Critical point   5 is not part of the solution   Critical point   5 is part of the solution

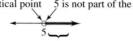

All numbers greater than 5   All numbers greater than 5
are part of the solution   are part of the solution

c. $x < 5$ **d.** $x \leq 5$

Critical point   5 is not part of the solution   Critical point   5 is part of the solution

All numbers less than 5   All numbers less than 5
are part of the solution   are part of the solution

▮▮ INTERVAL NOTATION

Inequalities are graphed as regions on a number line. The regions are called *intervals*. **Interval notation** is a concise way to describe the region that has been graphed on the number line. A parenthesis is used to enclose infinity, ∞, and for critical points that are not a part of the solution. A bracket is used for critical points that **are** a part of the solution.

EXAMPLE 2

Describe each inequality using interval notation.

a. $x > 5$ **b.** $x \geq 5$ **c.** $x < 5$ **d.** $x \leq 5$

SOLUTION   It will be easier to express or describe each inequality using interval notation if you have a number line graph to look at. Your interval should correspond to the number line with the smaller value designating the left-hand end of the interval and the larger value designating the larger, right-hand end of the interval.

a. $x > 5$ **b.** $x \geq 5$

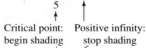

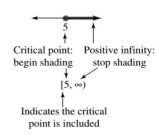

Critical point: Positive infinity: Critical point: Positive infinity:
begin shading   stop shading   begin shading   stop shading

Interval notation: $(5, \infty)$ Interval notation: $[5, \infty)$

Indicates the critical   Indicates the critical
point is not included   point is included

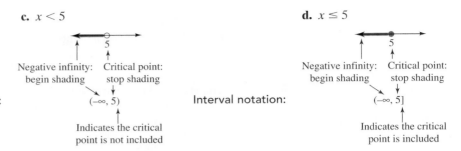

c. $x < 5$

Negative infinity: Critical point:
begin shading   stop shading

Interval notation: $(-\infty, 5)$

Indicates the critical
point is not included

d. $x \leq 5$

Negative infinity: Critical point:
begin shading   stop shading

Interval notation: $(-\infty, 5]$

Indicates the critical
point is included

The infinity symbols (∞ and $-\infty$) indicate that the interval has no bound. For example, $x > 5$ and $x \geq 5$ have no right-hand boundary, whereas $x < 5$ and $x \leq 5$ have no left-hand boundary. When an interval has no left-hand boundary or no right-hand boundary, we say the graph is an **unbounded interval**. The symbols ∞ and $-\infty$ are not real numbers and therefore cannot be included in the interval. Thus, a parenthesis will always be used when an infinity symbol is involved.

▌▌ SET NOTATION

In Chapter 1 we defined a set as a collection of objects called elements. When the elements can be listed we can write the set in **roster notation**. For example, the set of natural numbers is the set $N = \{1, 2, 3, 4, \ldots\}$, which is in roster notation. The ellipses, $\ldots$, imply that this set continues to infinity in some pattern established by the first four elements that are listed.

Set-builder notation is a notation that is used when roster notation is cumbersome or impossible, describing the set by indicating the conditions it must satisfy. It is a symbolic way of saying "the set of all numbers, x, such that x has this property . . . ." In symbols, we would write $\{x \mid \text{description of the variable } x\}$. The vertical bar, "$\mid$," is read as "such that." In set-builder notation, a rule for determining what is in the set must follow this bar. A comparison between roster notation and set-builder notation is given below:

Roster Notation	Set-Builder Notation
$\{2, 1, 0, -1, -2, \ldots\}$	$\{x \mid x \text{ is an integer less than 3}\}$

We will use inequality symbols to further simplify the description of the sets, as seen in the example that follows.

EXAMPLE 3

Describe using set-builder notation:

a. $x < 5$ **b.** $[-4, \infty)$ **c.**
3

SOLUTION   **a.** To write in set-builder notation (set notation, for short), we need a description of the inequality $x < 5$. Thus, the set notation would be $\{x \mid x < 5\}$.

b. To write in set notation, we need a description of the inequality $[-4, \infty)$. This will be graphed first to make the inequality description more apparent. See Figure 4-7.

−4

Figure 4-7

This translates to $x \geq -4$, which becomes our description. The set notation is $\{x \mid x \geq -4\}$.

c. Translate the graphical display to the inequality $x \leq 3$ and use this as the set description. The set notation is $\{x \mid x \leq 3\}$.

Summary of Unbounded Intervals			
Inequality	**Graph**	**Interval Notation**	**Set Notation**
$x > a$		(a, ∞)	$\{x \mid x > a\}$
$x \geq a$		$[a, \infty)$	$\{x \mid x \geq a\}$
$x < a$		$(-\infty, a)$	$\{x \mid x < a\}$
$x \leq a$		$(-\infty, a]$	$\{x \mid x \leq a\}$

▌▐ SOLVING LINEAR INEQUALITIES

A **linear inequality** is any inequality that can be expressed in the form

$$ax + c < 0 \qquad ax + c > 0 \qquad ax + c \leq 0 \qquad ax + c \geq 0 \quad (a \neq 0)$$

To solve these inequalities we use the following properties of inequalities to transform our inequality into equivalent inequalities, much the same way that we used the property of equality to transform linear equations into equivalent linear equations.

Addition Property of Inequality

If a, b, and c are real numbers, and

if $a < b$,   then $a + c < b + c$.

Similar statements can be made for the symbols $>$, $\leq$, and $\geq$.

The **addition property of inequality** can be stated this way: *If any quantity is added to both sides of an inequality, the resulting inequality has the same direction as the original inequality.*

Subtraction Property of Inequality

If a, b, and c are real numbers, and

if $a < b$,   then $a - c < b - c$.

Similar statements can be made for the symbols $>$, $\leq$, and $\geq$.

The **subtraction property of inequality** can be stated this way: *If any quantity is subtracted from both sides of an inequality, the resulting inequality has the same direction as the original inequality.*

The subtraction property of inequality is included in the addition property: To *subtract* a number c from both sides of an inequality, we could instead *add* the opposite of c to both sides.

EXAMPLE 4

Solve the inequality $2x + 5 > x - 4$. Graph the solution on the number line and translate to interval and set notation.

SOLUTION   To isolate the x on the left-hand side of the $>$ sign, we proceed as we would when solving equations.

$$2x + 5 > x - 4$$

$$2x + 5 - 5 > x - 4 - 5 \qquad \text{Subtract 5 from both sides.}$$

$$2x > x - 9 \qquad \text{Combine like terms.}$$

$$2x - x > x - 9 - x \qquad \text{Subtract } x \text{ from both sides.}$$

$$x > -9 \qquad \text{Combine like terms.}$$

The graph of this solution (see Figure 4-8) includes all points to the right of -9 but does not include -9 itself. For that reason, we use an open circle at -9.

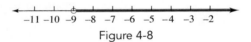

Figure 4-8

Because the critical point -9 is not included in the graph of the solution, we will use a parenthesis in the interval notation: $(-9, \infty)$. In set notation this would be $\{x \mid x > 9\}$.

If both sides of the true inequality $-2 < 5$ are multiplied by a *positive* number, such as 3, another true inequality results.

$$-2 < 5$$

$$\mathbf{3 \cdot (-2) < 3 \cdot 5} \qquad \text{Multiply both sides by 3.}$$

$$-6 < 15$$

The inequality $-6 < 15$ is a true inequality. However, if both sides of the inequality $-2 < 5$ are multiplied by a negative number, such as -3, the direction of the inequality symbol must be reversed to produce another true inequality.

$$-2 < 5$$

$$\mathbf{-3 \cdot (-2) > -3 \cdot 5} \qquad \text{Multiply both sides by the } \textit{negative} \text{ number } -3 \text{ and reverse the direction of the inequality.}$$

$$6 > -15$$

The inequality $6 > -15$ is a true inequality because 6 lies to the right of -15 on the number line.

Multiplication Property of Inequality

If a, b, and c are real numbers, and

$$\text{If } a < b \quad \text{and} \quad c > 0, \quad \text{then} \quad ac < bc.$$
$$\text{If } a < b \quad \text{and} \quad c < 0, \quad \text{then} \quad ac > bc.$$

There is a similar property for division.

Division Property of Inequality

If a, b, and c are real numbers, $c \neq 0$ and

$$\text{If } a < b \quad \text{and} \quad c > 0, \quad \text{then} \quad \frac{a}{c} < \frac{b}{c}.$$

$$\text{If } a < b \quad \text{and} \quad c < 0, \quad \text{then} \quad \frac{a}{c} > \frac{b}{c}.$$

To *divide* both sides of an inequality by a nonzero number c, we could instead *multiply* both sides by $\frac{1}{c}$. The multiplication and division properties of inequality are also true for $\leq$, $>$, and $\geq$.

EXAMPLE 5

Solve $3x + 7 \leq -5$ and express the solution both as a number line graph and in interval and set notation.

SOLUTION

$$3x + 7 \leq -5$$
$$3x + 7 - 7 \leq -5 - 7 \qquad \text{Subtract 7 from both sides.}$$
$$3x \leq -12 \qquad \text{Combine like terms.}$$
$$\frac{3x}{3} \leq \frac{-12}{3} \qquad \text{Divide both sides by 3.}$$
$$x \leq -4$$

The solution of $3x + 7 \leq -5$ consists of all real numbers less than, and also including, -4. The solid circle at -4 in the graph of the solution in Figure 4-9 indicates that -4 is one of the solutions of the given inequality.

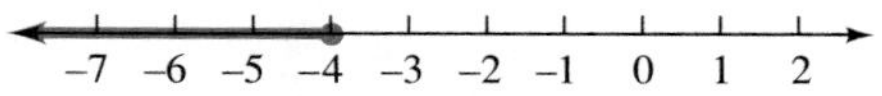

$$\overset{\longleftarrow}{\underset{-7 \;\; -6 \;\; -5 \;\; -4 \;\; -3 \;\; -2 \;\; -1 \;\;\; 0 \;\;\; 1 \;\;\; 2}{\vrule height 6pt}}$$

Figure 4-9

A square bracket will be used in the interval notation to indicate that the critical point of -4 is part of the solution: $(-\infty, -4]$. In set notation this would be $\{x \mid x \leq -4\}$.

EXAMPLE 6

Solve $5 - 3x \leq 14$ and express the solution both as a number line graph and in interval and set notation.

SOLUTION

$$5 - 3x \leq 14$$

$5 - 3x - \mathbf{5} \leq 14 - \mathbf{5}$ Subtract 5 from both sides.

$-3x \leq 9$ Combine like terms.

$\dfrac{-3x}{-3} \geq \dfrac{9}{-3}$ Divide both sides by -3 and reverse the direction of the $\leq$ symbol.

$x \geq -3$

In the last step, both sides of the inequality were divided by -3. Because -3 is negative, the direction of the inequality was *reversed*. The graph of the solution appears in Figure 4-10. The solid circle at -3 indicates that -3 is one of the solutions.

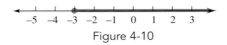

Figure 4-10

The square bracket indicates that the critical point of -3 is one of the solutions to the inequality: $[-3, \infty)$. In set notation this would be $\{x \mid x \geq -3\}$.

WARNING! Do not forget to reverse the direction of the inequality symbol when multiplying or dividing both sides of the inequality by a negative number.

EXPLORING THE CONCEPT

Solve each equation/inequality below and graph the solution set on a number line graph.

a. $4x - 5 = 2 - 3x$ **b.** $4x - 5 \leq 2 - 3x$

c. $4x - 5 < 2 - 3x$ **d.** $4x - 5 \geq 2 - 3x$

e. $4x - 5 > 2 - 3x$

f. What is the relationship between the critical points in **b–e** and the solution to the linear equation in **a**?

g. If $ax + b > cx + d$ has $x > -2$ as the solution, what is the solution to each of the following:

$$ax + b = cx + d \qquad ax + b \geq cx + d \qquad ax + b < cx + d$$

SOLUTION   **a.** $x = 1$    **b.** $x \leq 1$

c. $x < 1$ **d.** $x \geq 1$

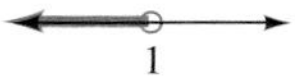

e. $x > 1$

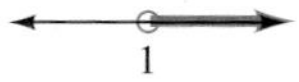

f. The critical points are all the same regardless of the direction or type of the inequality symbol, and they are the same as the solution to the equation.

g. $ax + b = cx + d$ would have $x = -2$ as its solution.
$ax + b \geq cx + d$ would have $x \geq -2$ as its solution.
$ax + b < cx + d$ would have $x < -2$ as its solution.

▓ GRAPHICAL SOLUTIONS

Consider the inequality in Example 5, $5 - 3x \leq 14$. The solution of the inequality is the set of all values for x such that $x \geq -3$.

The calculator can be used to *confirm* that $x \geq -3$ is a solution. Enter the left side of the original inequality at y1 and the right side of the inequality at y2. Set the TABLE to begin at the critical point of $x = -3$ and scroll through the TABLE to examine the x-values that are less than -3, greater than -3, and equal to -3. Compare the y1 and y2 columns that represent the left and right sides of the inequality.

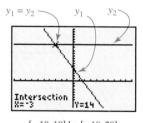

- At all values of $x < -3$, y1 > y2 (the expression $5 - 3x$, designated as y1, has values greater than 14).
- At $x = -3$, y1 = y2 (the expression $5 - 3x$ is equal to 14).
- At all values of $x > -3$, y1 < y2 (the expression $5 - 3x$, designated as y1, has values less than 14).

The inequality is true for values of x that satify y1 ≤ y2, so our solution of $x \geq -3$ is confirmed.

To graphically *solve* an inequality, we will use both the TABLE *and* a graphical display to interpret the solution. If you are graphically solving an inequality, you should graph both the left and right sides of the inequality and determine the point of intersection just as you do when graphically solving equations. (See Figure 4-11.)

$[-10, 10]$ by $[-10, 20]$
Figure 4-11

The critical point of the inequality occurs at the point of intersection. As we discovered in the Exploring the Concept section, the critical point of the inequality and the solution to the corresponding equation are the same. Thus, we can now conclude that our number line will be divided by the value -3 (Figure 4-12).

Figure 4-12

The graph screen should be read from left to right. It is a visual representation of the numerical information that was previously displayed in the TABLE. For what x-values is y1 less than y2? Remember, the INTERSECT feature has already shown us the value for which y1 = y2. It should be the section that is highlighted on the graph displayed. (See Figure 4-13.) In general, y1 < y2 where y1 is **below** the graph of y2.

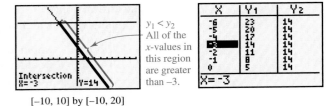

[−10, 10] by [−10, 20]

Figure 4-13

The x-values greater than -3 produce the graph of the highlighted section and satisfy the requirement that y1 $<$ y2. We can combine this information with the critical point on the number line to produce the final solution graphed on a number line (Figure 4-14).

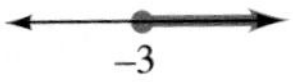

−3

Figure 4-14

Translated to interval notation: $[-3, \infty)$. In set notation: $\{x \mid x \geq -3\}$.

 EXAMPLE 7

Graphically solve $10 - 3x < 2x + 5$.

SOLUTION

$$y1 = 10 - 3x$$
$$y1 = 2x + 5$$

From the INTERSECT screen (Figure 4-15) we can determine that the critical point of the inequality is 1. The value of 1 will not be included as part of the solution because the original inequality is strictly "less than." On your number line graph place an open circle at the coordinate 1. (See Figure 4-16.)

Now determine where y1 $<$ y2 by determining which x-values on the graph will occur where y1 $<$ y2. Confirm this with the TABLE. (See Figure 4-17.)

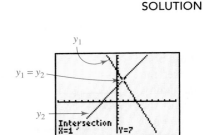

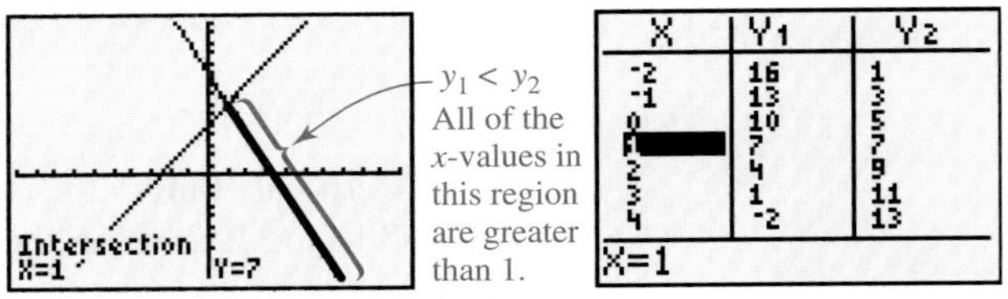

[−10, 10] by [−10, 15]

Figure 4-15

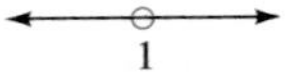

1

Figure 4-16

Figure 4-17

Combine the information determined from interpreting the graph with the critical point determined through the INTERSECT feature to produce a number line graph of the solution, as shown in Figure 4-18. Translated to interval notation: $(1, \infty)$. In set notation: $\{x \mid x > 1\}$.

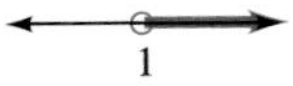

Figure 4-18

▐▐ PROBLEM SOLVING

EXAMPLE 8

Averaging grades   A student has scores of 72%, 74%, and 78% on three mathematics examinations. What score does he need on the last exam to earn a grade of at least a B (80% or better)?

SOLUTION

We can let x represent the score on the fourth (and last) exam. To find the average grade, we add the four scores and divide by 4. To earn a B, this average must be greater than or equal to 80%.

$$\boxed{\begin{array}{c}\text{The averages of}\\\text{the four grades}\end{array}} \quad \geq \quad \boxed{80}$$

$$\frac{72 + 74 + 78 + x}{4} \quad \geq \quad 80$$

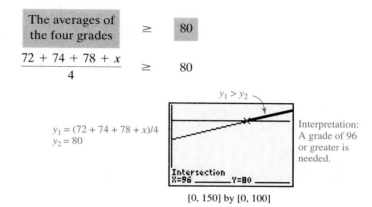

$y_1 = (72 + 74 + 78 + x)/4$
$y_2 = 80$

Interpretation:
A grade of 96
or greater is
needed.

[0, 150] by [0, 100]

We can solve this inequality for x.

$$\frac{224 + x}{4} \geq 80 \qquad 72 + 74 + 78 = 224.$$

$$224 + x \geq 320 \qquad \text{Multiply both sides by 4.}$$

$$x \geq 96 \qquad \text{Subtract 224 from both sides.}$$

To earn a B, the student must score 96% or better on the last exam. The graph of this solution appears in Figure 4-19.

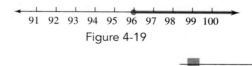

Figure 4-19

EXAMPLE 9

Geometry   If the perimeter of an equilateral triangle is greater than 15 feet, how long could a side be?

SOLUTION

Recall that each side of an equilateral triangle is the same length and that the perimeter of a triangle is the sum of the lengths of its three sides.

Let x represent the length of one side of the triangle.
Then $x + x + x$ represents the perimeter.

The perimeter is to be greater than 15 feet. We indicate this fact with the following inequality.

$$x + x + x > 15$$
$$3x > 15 \qquad \text{Combine like terms.}$$
$$x > 5 \qquad \text{Divide both sides by 3.}$$

Each side of the triangle must be more than 5 feet long.

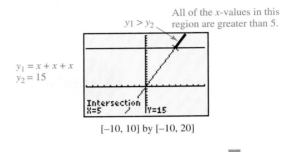

$y_1 = x + x + x$
$y_2 = 15$

$y_1 > y_2$

All of the x-values in this region are greater than 5.

Intersection
X=5   Y=15

$[-10, 10]$ by $[-10, 20]$

EXERCISE 4.4

VOCABULARY AND NOTATION *In Exercises 1–10, fill in the blanks to make a true statement.*

1. An _____ is a mathematical expression that indicates that two quantities are not necessarily equal.

2. The _____ divides the number line into the set of points that are the solution and the set of points that are not the solution.

3. _____ notation is a concise method used to describe the region that has been graphed on the number line.

4. An _____ interval has no left- or right-hand boundary.

5. Any number that makes the inequality a true statement is a _____ of the inequality.

6. The symbol, $>$, means _____, whereas the symbol $\geq$ means _____.

7. The symbol, $<$, means _____, whereas the symbol, $\leq$ means _____.

8. In interval notation, the use of the symbol "["or"]" would indicate that the point is to be _____ (included/not included) on the number line graph.

9. In interval notation, the use of the symbol "("or")" would indicate that the point is to be _____ (included/not included) on the number line graph.

10. The symbol, ∞, means _____.

CONCEPTS

11. If the number 5 is added to both sides of the inequality $-4 < 6$, the resulting inequality is _____.

12. If the number 5 is subtracted from both sides of the inequality $-4 < 6$, the resulting inequality is _____.

13. If both sides of the inequality $-4 < 6$ are multiplied by 2 the resulting inequality is _____.

14. If both sides of the inequality $-4 < 6$ are multiplied by -2 the resulting inequality is _____.

15. If both sides of the inequality $-4 < 6$ are divided by 2 the resulting inequality is _____.

16. If both sides of the inequality $-4 < 6$ are divided by -2 the resulting inequality is _____.

17. Restate the inequality $x \geq 3$ using interval notation _____.

18. Restate the inequality $x < 2$ using interval notation _____.

19. Restate the inequality $x \leq 5$ using interval notation _____.

20. Restate the inequality $x > 4$ using interval notation _____.

21. The left side of an equation has been stored at y1 and the right side at y2. Based on the TABLE displayed,

determine the x-value that is the solution to y1 = y2.

X	Y1	Y2
-1	-7	-2
0	-5	-1
1	-3	0
2	-1	1
3	1	2
4	3	3
5	5	4

X = -1

CONCEPTS

22. The left side of an equation has been stored at y1 and the right side at y2. Based on the TABLE shown, determine the x-value that is the solution to y1 = y2.

X	Y1	Y2
-12	-32	-29
-11	-29	-27
-10	-26	-25
-9	-23	-23
-8	-20	-21
-7	-17	-19
-6	-14	-17

X = -12

23. The solution to y1 = y2 can be determined from the TABLE shown. Based on the information displayed in the TABLE, determine a number line graph of the solution to the following inequalities.

X	Y1	Y2
-4	1	4
-3	3	5
-2	5	6
-1	7	7
0	9	8
1	11	9
2	13	10

X = -1

  a. y1 < y2

  b. y1 ≤ y2

  c. y1 > y2

  d. y1 ≥ y2

24. The solution to y1 = y2 can be determined from the TABLE shown. Based on the information displayed in the TABLE, determine a number line graph of the solution to the following inequalities.

X	Y1	Y2
-5	-29	-20
-4	-22	-16
-3	-15	-12
-2	-8	-8
-1	-1	-4
0	6	0
1	13	4

X = -2

  a. y1 < y2

  b. y1 ≤ y2

  c. y1 > y2

  d. y1 ≥ y2

25. The solution to y1 = y2 can be determined from the accompanying graph. Based on the information displayed, determine a number line graph of the solution to the following inequalities.

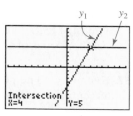

  a. y1 < y2

  b. y1 ≤ y2

  c. y1 > y2

  d. y1 ≥ y2

26. The solution to y1 = y2 can be determined from the accompanying graph. Based on the information displayed, determine a number line graph of the solution to the following inequalities.

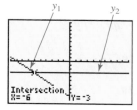

  a. y1 < y2

  b. y1 ≤ y2

  c. y1 > y2

  d. y1 ≥ y2

27. The solution to y1 = y2, y1 < y2, y1 ≤ y2, y1 > y2, and y1 ≥ y2 can be determined from the pictured graph. Based on the information displayed, determine a number line graph of the solution to the following equation and inequalities.

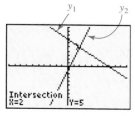

  a. y1 = y2

  b. y1 < y2

  c. y1 ≤ y2

  d. y1 > y2

  e. y1 ≥ y2

28. The solution to y1 = y2, y1 < y2, y1 ≤ y2, y1 > y2, and y1 ≥ y2 can be determined from the pictured graph. Based on the information displayed, determine a number line graph of the solution of each of the following.

X	Y1	Y2
-4	8	-4
-3	7	-1
-2	6	2
-1	5	5
0	4	8
1	3	11
2	2	14

X = -4

  a. y1 = y2

  b. y1 < y2

  c. y1 ≤ y2

d. $y1 > y2$

e. $y1 \geq y2$

PRACTICE *In Exercises 29–48, solve each inequality and express the solution using interval notation, a number line graph, and set notation. Verify your solution by solving graphically.*

29. $2x - 3 \leq 5$

30. $-3x - 5 < 4$

31. $-3x - 7 > -1$

32. $-5x + 7 \leq 12$

33. $2x + 9 \leq x + 8$

34. $3x + 7 \leq 4x - 2$

35. $9x + 13 \geq 8x$

36. $7x - 16 < 6x$

37. $7 - x \leq 3x - 1$

38. $2 - 3x \geq 6 + x$

39. $3(x - 8) < 5x + 6$

40. $9(x - 11) > 13 + 7x$

41. $8(5 - x) \leq 10(8 - x)$

42. $17(3 - x) \geq 3 - 13x$

43. $\dfrac{5}{2}(7x - 15) + x \geq \dfrac{13}{2}x - \dfrac{3}{2}$

44. $\dfrac{5}{3}(x + 1) \leq -x + \dfrac{2}{3}$

45. $\dfrac{3x - 3}{2} < 2x + 2$

46. $\dfrac{x + 7}{3} \geq x - 3$

47. $\dfrac{2(x + 5)}{3} \leq 3x - 6$

48. $\dfrac{3(x - 1)}{4} > x + 1$

APPLICATIONS *In Exercises 49–57, express each solution as a statement of inequality.*

49. Calculating grades A student has test scores of 68%, 75%, and 79%. What must she score on the last exam to earn a B (80% or better)?

50. Calculating grades A student has test scores of 70%, 74%, and 84%. What does he need to maintain a C (70% or better)?

51. Fleet averages An auto manufacturer produces three sedan models in equal quantities. One model has an economy rating of 17 miles per gallon, and the second model is rated at 19 mpg. The manufacturer is required to have a fleet average of at least 21 mpg. What economy rating is required for the third model car?

52. Avoiding a service charge When the average daily balance of a customer's checking account falls below $500 in any week, the bank assesses a $5 service charge. Bill's account balances for the week were

Monday	$540.00
Tuesday	435.50
Wednesday	345.30
Thursday	310.00

What must Friday's balance be to avoid the service charge?

53. FDA guidelines The food and Drug Administration recommends that an individual eat at least five servings of fruit each day. If for the first six days of the week, you have five, four, three, six, seven, and four servings, how many servings of fruit should you eat on the seventh day of the week in order to average at least five servings per day?

54. Car rental A rental car agency charges a daily rate of $37.50 per day plus $0.21 per mile. Sheila has budgeted $150 per day for car rental during her vacation. To the nearest tenth of a mile, how far can she travel per day to stay within her budget?

55. Rent-to-own The local Rent-to-Own store rents washing machines for a mere $0.67 a day. If a new washer costs $385, after how many days will it be cheaper to buy a machine instead of renting it?

56. Geometry The perimeter of an equilateral triangle is at most 57 feet. What could be the length of a side? (*Hint:* All three sides of an equilateral triangle are equal.)

57. Geometry The perimeter of a square is no less than 68 centimeters. How long can a side be?

REVIEW

58. Determine if each of the following represents a function:

   a. $\{(-3, 1), (-2, 0), (0.4, 7)\ (-3, 0)\}$

   b.

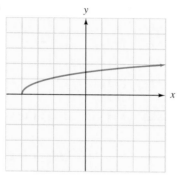

   c. $y = 4x - 5$

   d.

59. Solve each of the equations both analytically (algebraically) and graphically. You may use either the Root/Zero or INTERSECT features. Copy your complete screen display to justify your work.

   a. $2[3 - 2(3 + x)] = 2(3 + x)$

   b. $|4x - 3| + 2 = 0$

   c. $3x^2 + 19x = -28$

60. Verify that $-3/4$ is a solution of $4x^2 + 11x + 6 = 0$ using both the STOre and TABLE features of your calculator.

4.5 COMPOUND INEQUALITIES

- AND STATEMENTS
- OR STATEMENTS
- GRAPHICAL SOLUTIONS

Compound inequalities are inequalities that are linked by the words **and** or **or**. When solving compound inequalities you will be required to determine either the **intersection** (common values) or the **union** (all represented values) of solutions to simple inequalities in order to determine the solution to the compound inequality statement.

▐ AND STATEMENTS

Compound inequalities linked by the word "and" are called **conjunctions**. The solution to a conjunction is all values of the variable that make **both** simple inequalities true.

    The set of all values that are both greater than two *and* less than 5,

$$x > 2 \quad and \quad x < 5,$$

represents a compound inequality. The graph of each separate inequality is shown below.

$x > 2$

$x < 5$

The compound inequality of $x > 2$ *and* $x < 5$ is represented by the intersection of the two inequalities. The intersection of the two inequalities is the set of all values that satisfy (are common to) *both* inequality statements.

$x > 2$ *and* $x < 5$

This inequality of $x > 2$ *and* $x < 5$ can be written in the abbreviated form of $2 < x < 5$. It is read as "x is greater than 2 and x is less than 5." This is often called a **continued inequality**.

When written in the interval form, $(2, 5)$, we say that the interval is an **open interval** because neither endpoint is included in the solution.

EXAMPLE 1

Express each compound inequality as a number line graph, in interval notation, and in set notation.

a. $-1 < x < 3$ $(-1, 3)$ $\{x \mid -1 < x < 3\}$

b. $-1 \le x < 3$ $[-1, 3)$ $\{x \mid -1 \le x < 3\}$

c. $-1 < x \le 3$ $(-1, 3]$ $\{x \mid -1 < x \le 3\}$

d. $-1 \le x \le 3$ $[-1, 3]$ $\{x \mid -1 \le x \le 3\}$

Because there is a finite endpoint on each of these intervals, we say the intervals are **bounded**. The chart below summarizes bounded intervals.

Summary of Bounded Intervals			
Inequality	**Number Line Graph**	**Interval Notation**	**Set Notation**
$a < x < b$		(a, b) **open interval**	$\{x \mid a < x < b\}$
$a \le x \le b$		$[a, b]$ **closed interval**	$\{x \mid a \le x \le b\}$
$a < x \le b$		$(a, b]$ **half-open interval**	$\{x \mid a < x \le b\}$
$a \le x < b$		$[a, b)$ **half-open interval**	$\{x \mid a \le x < b\}$

EXAMPLE 2

Solve the compound inequality.

$x - 4 < 5$ *and* $x + 2 \ge 5$

SOLUTION   Solve each inequality separately.

$x - 4 < 5$ *and* $x + 2 \ge 5$
$x < 9$ *and* $x \ge 3$

Graph each inequality. The compound inequality is true for all values of x that are less than 9 *and* greater than or equal to 3.

$x < 9$

$x \geq 3$

The intersection of these two graphs is

The solution to the inequality is $3 \leq x < 9$, which becomes $[3, 9)$ when written in interval notation. The set notation would be $\{x \mid 3 \leq x < 9\}$.

WARNING!   When writing compound inequalities be sure that the inequality makes sense when read from left to right. For example, $4 \leq x < -8$ does not make sense because $4 \not\leq -8$.

EXAMPLE 3   Solve the compound inequality

$$4x - 5 > 3 \quad and \quad 4 - \frac{1}{2}x < 1$$

SOLUTION   Solve each simple inequality.

$$4x - 5 > 3 \quad and \quad 4 - \frac{1}{2}x < 1$$

$$4x > 8 \quad and \quad -\frac{1}{2}x < -3$$

$$x > 2 \quad and \quad x > 6$$

The graphs are shown below.

$x > 2$

$x > 6$

The solution to the compound inequality is the set of all values that satisfy both inequalities: all real numbers greater than 2 and also greater than 6.

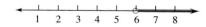

Expressed in interval notation this is $(6, \infty)$. Expressed in set notation, it is $\{x \mid x > 6\}$.

EXAMPLE 4

Solve the compound inequality

$$2(x - 5) \geq 2 \quad and \quad 4 - (3 - x) < 3$$

SOLUTION   Solve each simple inequality.

$$2(x - 5) \geq 2 \quad and \quad 4 - (3 - x) < 3$$
$$2x - 10 \geq 2 \quad and \quad 4 - 3 + x < 3$$
$$2x \geq 12 \quad and \quad 1 + x < 3$$
$$x \geq 6 \quad and \quad x < 2$$

Express each inequality as a graph.

$x \geq 6$

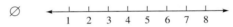

$x < 2$

These two graphs have **no** common points of intersection. There are no values for x that are both less than 2 and simultaneously greater than 6. The solution is the empty set, $\varnothing$.

$\varnothing$

The number line would not be shaded.

EXAMPLE 5

Solve the compound inequality

$$2(x - 1) > -4 \quad and \quad 2(x - 1) \leq 4$$

SOLUTION   We can solve each inequality separately as in the previous examples, or we can combine the two inequalities into a continued inequality because both inequalities share a common side of $2(x - 1)$.

$$-4 < 2(x - 1) \leq 4$$ Be sure that the double inequality makes sense. Is $-4 \leq 4$?

$$-4 < 2x - 2 \leq 4$$ Remove grouping symbols.

$$-4 + 2 < 2x - 2 + 2 \leq 4 + 2$$ Add 2 to *all three parts of the inequality*.

$$-2 < 2x \leq 6$$

$$-\frac{2}{2} < \frac{2x}{2} \leq \frac{6}{2}$$ Divide all three parts of the inequality by 2.

$$-1 < x \leq 3$$

The solution expressed as a number line graph is

Expressed in interval notation, the solution is $(-1, 3]$ and as a set it is $\{x \mid -1 < x \leq 3\}$.

▮▮▮ OR STATEMENTS

Inequalities that involve the use of the word **or** are also compound inequalities. They are called **disjunctions**. When the inequality is an *or* statement, we are looking for values of the variable that will satisfy either inequality. These values are represented by the **union** of all possible solution values. For example, the solution to

$$x < 5 \qquad or \qquad x > 9$$

is graphed as

$x < 5$ (number line graph: open circle at 5, shaded left; marks 4 5 6 7 8 9 10)

$x > 9$ (number line graph: open circle at 9, shaded right; marks 4 5 6 7 8 9 10)

which is the combined set of values (number line graph: open circle at 5 shaded left, open circle at 9 shaded right; marks 4 5 6 7 8 9 10)

The word *or* indicates that only one of the inequalities needs to be true to make the statement true. To express an *or* statement in interval notation, we will use an interval to describe each shaded section of the number line graph. The interval $(-\infty, 5)$ describes the portion of the number line graph represented by

(number line graph: open circle at 5, shaded left; marks 4 5 6 7 8 9 10 11)

and the interval $(9, \infty)$ describes (number line graph: open circle at 9, shaded right; marks 4 5 6 7 8 9 10 11)

It is the union of these two intervals that produces the solution set. Thus we use the symbol ∪ (the symbol for union) to indicate that union.

$$(-\infty, 5) \cup (9, \infty)$$

The set would be $\{x \mid x < 5 \text{ or } x > 9\}$.

EXAMPLE 6

Solve the compound inequality

$$5x + 3 > 4x \qquad or \qquad 3(x + 2) - 1 < 2x$$

SOLUTION   Solve each inequality separately.

$$
\begin{array}{lll}
5x + 3 > 4x & or & 3(x + 2) - 1 < 2x \\
3 > -x & or & 3x + 6 - 1 < 2x \\
-3 < x & or & 3x + 5 < 2x \\
& & 5 < -x \\
& & -5 > x
\end{array}
$$

Graph each inequality separately.

$x > -3$ (number line graph: open circle at −3, shaded right; marks −7 −6 −5 −4 −3 −2 −1 0)

$x < -5$ (number line graph: open circle at −5, shaded left; marks −7 −6 −5 −4 −3 −2 −1 0)

The original inequality is an *or* statement, so we want the union of these two sets of values for the solution.

(number line graph: open circle at −5 shaded left, open circle at −3 shaded right; marks −7 −6 −5 −4 −3 −2 −1 0)

The interval notation would be

$$(-\infty, -5) \cup (-3, \infty).$$

The set notation would be $\{x \mid x < -5 \text{ or } x > -3\}$.

EXAMPLE 7

Solve the compound inequality

$$\frac{1}{2}(4x - 2) + 1 > 5 \qquad or \qquad 3\left(\frac{1}{3}x - 1\right) < 2$$

SOLUTION   Solve each inequality.

$$\frac{1}{2}(4x - 2) + 1 > 5 \qquad or \qquad 3\left(\frac{1}{3}x - 1\right) < 2$$

$$2x - 1 + 1 > 5 \qquad or \qquad x - 3 < 2$$

$$2x > 5 \qquad or \qquad x < 5$$

$$x > \frac{5}{2}$$

The graph of each inequality would be

$x > \dfrac{5}{2}$

$x < 5$

The union of these two inequalities would be every real number: $\mathbb{R}$.

Expressed as an interval, this would be $(-\infty, \infty)$.

▎▎ GRAPHICAL SOLUTIONS

EXPLORING THE CONCEPT

To graphically solve the continued inequality $-4 \le 2(x - 1) \le 4$ in Example 5, graph y1 $= -4$, y2 $= 2(x - 1)$, and y3 $= 4$. We can now abbreviate our inequality statement to y1 $\le$ y2 $\le$ y3.

a. Sketch the graph in the standard viewing window.

b. Use the INTERSECT feature to find the intersection of y1 and y2.

c. Use the INTERSECT feature to find the intersection of y2 and y3.

d. On your paper, highlight the section of y2 that is **between** the graph of y1 and y3.

e. Determine the x-values for which y1 $\le$ y2 $\le$ y3.

f. Write the solution set of y1 $\le$ y2 $\le$ y3 in interval notation.

SOLUTION **a.**

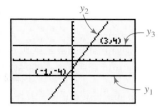

NOTE: Because the INTERSECT feature is used more than once, points of intersection are labeled with the ordered pairs instead of repeatedly displaying the INTERSECT screen "$x=$ $y=$" in the text. Begin using this notation now to condense your graphs.

b. The intersection of y1 and y2 is the coordinate $(-1, -4)$. The x-value of -1 is the critical point for the inequality.

c. The intersection of y2 and y3 is the coordinate $(3, 4)$. The x-value of 3 is the critical point for the inequality.

d.

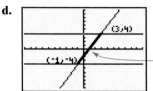

y_1 y_2 y_3
All of the x-values in this region are between $x = -1$ and $x = 3$.

e. When x is between -1 and 3, then y1 $\leq$ y2 $\leq$ y3.

f. The number line graph of the solution to the one-variable inequality would be

-1 3

Converted to interval notation, this would be $[-1, 3]$ and in set notation, it would be $\{x \mid -1 \leq x \leq 3\}$.

EXAMPLE 8

Graphically solve $-1 \leq \dfrac{4}{5}x + 2 < 6$.

SOLUTION Graph y1 = -1:

$$y2 = \frac{4}{5}x + 2$$

$$y3 = 6$$

and use the INTERSECT feature to determine the critical points.

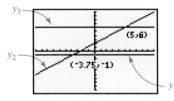

The critical points for the compound inequality are -3.75 and 5.

Finally, interpret the graphical display.

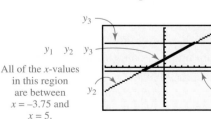

y_1 y_2 y_3
All of the x-values in this region are between $x = -3.75$ and $x = 5$.

The solution consists of the critical point -3.75 and all x-values *between* the critical points of -3.75 and 5.

-3.75 5 $[-3.75, 5)$

EXAMPLE 9

To hold the temperature of a room between 19° and 22° Celsius, what Fahrenheit temperatures must be maintained? (Recall that the Fahrenheit temperature, F, and Celsius temperature, C, are related by the formula $C = \frac{5}{9}(F - 32)$.)

SOLUTION   Write the compound inequality to model the problem:

$$19 \leq \frac{5}{9}(x - 32) \leq 22$$

Graph the inequality where y1 = 19, y2 = $\frac{5}{9}(x - 32)$, and y3 = 22, and use the INTERSECT feature to determine the critical points.

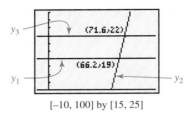

[−10, 100] by [15, 25]

The critical points are the x-values of 71.6 and 66.2. Interpret the graph to determine y1 ≤ y2 ≤ y3.

66.2   71.6

The Fahrenheit temperature must be between 66.2° and 71.6°.

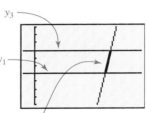

y_1 y_2 y_3
All of the x-values
in this region
are between
$x = 66.2$ and
$x = 71.6$.

EXERCISE 4.5

VOCABULARY AND NOTATION   *In Exercises 1–8, fill in the blanks to make a true statement.*

1. _____ inequalities are inequalities that are linked by the words *and* or *or*.

2. A compound inequality such as $2 < x < 5$ is also called a _____ inequality.

3. *And* statements represent the _____ of the set of values that satisfy both inequalities.

4. *Or* statements represent the _____ of the set of values that satisfy both inequalities.

5. Two simple inequalities joined by the word *and* are called _____.

6. Two simple inequalities joined by the word *or* are called _____.

7. In a compound inequality the notation (2, 3) is the inequality _____ but on the rectangular coordinate system (2,3) locates a point in Quadrant __.

8. In a compound inequality, the notation $(-2, -1)$ is the inequality _____ but on the rectangular coordinate system $(-2, -1)$ locates a point in Quadrant ____.

CONCEPTS   *In Exercises 9–14, solve each compound inequality by interpreting the information displayed on the screen. Express the solution in interval notation.*

9.

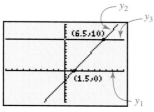

[−10, 10] by [−10, 15]

a. y1 < y2 < y3

b. y2 < y1   or   y2 > y3

10.

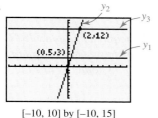

[−10, 10] by [−10, 15]

a. y1 < y2 < y3

b. y2 < y1   or   y2 > y3

11.

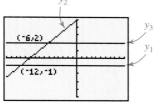

[−15, 10] by [−5, 5]

a. y1 < y2 < y3

b. y2 < y1   or   y2 > y3

12.

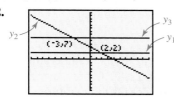

[−10, 10] by [−10, 10]

a. y1 < y2 < y3

b. y2 < y1   or   y2 > y3

13. Match each compound inequality with the appropriate number line graph.

a. $-1 < x < 4$

b. $x < -1$ or $x > 4$

c. $-1 < x \le 4$

d. $x \le -1$ or $x > 4$

i.

ii.

iii.

iv.

14. Match each number line graph with the appropriate interval notation.

a.

b.

c.

d.

i. $(2, 3)$

ii. $(-\infty, 2] \cup [3, \infty)$

iii. $[2, 3]$

iv. $(-\infty, 2) \cup (3, \infty)$

In Exercises 15–22, classify each statement as either true or false.

15. The number 8 is a solution of the conjunction $x < -1$ and $x > 4$.

16. The number 8 is a solution of the disjunction $x < -1$ or $x > 4$.

17. The number 4 is a solution of the conjunction $-2 < x \le 4$.

18. The number 4 is a solution of the conjunction $-2 < x < 4$.

19. The solution of the conjunction $x < 2$ and $x > 5$ is $\varnothing$.

20. The solution of the disjunction $x < 2$ or $x > 5$ is $\varnothing$.

21. The solution of $x > -2$ and $x < 3$ is $\mathbb{R}$.

22. The solution of $x > -2$ or $x < 3$ is $\mathbb{R}$.

PRACTICE   *In Exercises 23–34, express each compound inequality as a number line graph, in interval notation, and in set notation.*

23. $x \le 5$ and $x > -2$

24. $x \ge 3$ and $x < 11$

25. $x < 10$ and $x \le 4$

26. $x \ge -5$ and $x \ge -1$

27. $x \le 6$ and $x \ge 12$

28. $x < -3$ and $x > 1$

29. $x > 5$ or $x < -2$

30. $x \le 3$ or $x > 11$

31. $x > 10$ or $x \geq 4$

32. $x > 5$ or $x > 8$

33. $x < 6$ or $x \geq -2$

34. $x > 4$ or $x \leq 5$

In Exercises 35–54, solve each compound inequality and express the solution using a number line graph, interval notation, and set notation.

35. $5x - 3 > 12$ and $7x - 23 < 5$

36. $-3x - 1 \geq 5$ and $-2x + 6 \leq 16$

37. $-9x \geq -x$ and $4 - 3x \leq x$

38. $8 - 5x \geq -x$ and $4 - 2x \leq 2x$

39. $7 - 4x \geq 8$ and $2 + 3x \leq 8$

40. $8 - 9x \geq 4x$ and $4 - 3x \geq x$

41. $-3(x + 2) > 2(x + 1)$ and $-4(x - 1) < x + 8$

42. $\frac{1}{2}x + 2 \leq \frac{1}{3}x - 4$ and $\frac{1}{4}x - \frac{1}{3} \leq x + 2$

43. $-2 < -x + 3 < 5$

44. $4 < -x - 2 < 9$

45. $15 > 2x - 7 > 9$

46. $25 > 3x - 2 > 7$

47. $0 \leq \dfrac{4 - x}{3} \leq 2$

48. $-2 \leq \dfrac{5 - 3x}{2} \leq 2$

49. $3x + 2 < 8$ or $2x - 3 > 11$

50. $3x + 4 < -2$ or $3x + 4 > 10$

51. $-4(x + 2) \geq 12$ or $3x + 8 < 11$

52. $5(x - 2) \geq 0$ or $-3x < 9$

53. $4x < -24$ or $-3x > 18$

54. $1 + 2x \leq 13$ or $x - 2 > 4$

In Exercises 55–62, solve each inequality graphically. Copy your screen display and in complete sentences explain how you interpreted the graphical display to determine the solution. Solutions should be expressed in interval notation.

55. $-2 < 3x + 4 \leq 7$

56. $0 \leq 5 - 2x \leq 8$

57. $0 \leq 1 - \frac{2}{3}x < 5$

58. $2 < x - 5 < 7$

59. $3x + 4 < -2$ or $3x + 4 \geq 7$

60. $5 - 2x \leq 0$ or $5 - 2x \geq 8$

61. $1 - \frac{2}{3}x \leq 0$ or $1 - \frac{2}{3}x > 5$

62. $x - 5 < 2$ or $x - 5 > 7$

APPLICATIONS   *In Exercises 63–68, express each solution as an inequality. Solve algebraically and confirm graphically.*

63. Melting iron   To melt iron, the temperature of a furnace must be at least 1,540° C but no more than 1,650° C. What range of Celsius temperatures must be maintained?

64. Phonograph records   The radii of phonograph records must lie between 5.9 and 6.1 inches. What variation in circumference can occur? (*Hint:* The circumference of a circle is given by the formula $C = 2\pi r$, where r is the radius. Let $\pi = 3.14$.)

65. Pythons   A large snake, the African Rock Python, can grow to a length of 25 feet. To the nearest hundredth, what is the range of lengths in meters? (*Hint:* There are about 3.281 feet in one meter.)

66. Comparing weights   The normal weight of a 6-foot-2-inch man is between 150 and 190 pounds. To the nearest hundredth, what would such a person weigh in kilograms? (*Hint:* There are 2.2 pounds in one kilogram.)

67. Geometry   A rectangle's length is 3 feet less than twice its width, and its perimeter is between 24 and 48 feet. What might be its width?

68. Geometry   A rectangle's width is 8 feet less than 3 times its length, and its perimeter is between 8 and 16 feet. What might be its length?

REVIEW

69. The solution to y1 = y2 can be determined from the accompanying graph. Based on the information displayed, determine a number line graph of the solution to the following inequalities.

a. y1 < y2

b. y1 ≤ y2

c. y1 > y2

d. y1 ≥ y2

70. Algebraically solve $\frac{2}{3}(x - 5) \le 6 - 12x$ and record the solution as a number line graph and in interval notation.

71. Construct a hand-drawn graph of $y = \frac{2}{3}(x - 5)$, designating the *x*- and *y*-intercepts.

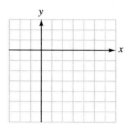

72. Simplify
   a. $5(x^2 - 5y)(x^2 + 5y)$
   b. $(3x - 4)(2x - 3)$
   c. $x(4x + 1)(x - 5)$

4.6   ABSOLUTE VALUE INEQUALITIES

- INEQUALITIES OF THE FORM |x| < k
- INEQUALITIES OF THE FORM |x| > k

Recall the definition of absolute value.

Absolute Value

$|x| = x$ when $x \ge 0$

$|x| = -x$ when $x < 0$

We can relate this to the number line by translating the expression $|x|$ to mean "find the number(s) that are a distance of *x* units from zero on the number line." Thus $|x| = 5$ translates to

We conclude that *x* must have a value of 5 or −5 in order for $|x| = 5$.

▮▮ INEQUALITIES OF THE FORM |x| < k

Now consider $|x| < 5$. We can translate this expression to mean "find all numbers x whose distance is 5 or less units from zero. Thus, x is between -5 and 5, and

$$|x| < 5 \qquad \text{is equivalent to} \qquad -5 < x < 5$$

EXPLORING THE CONCEPT

Graphically solve the absolute value inequality $|x| < 5$.

a. Graph both sides of the inequality in the standard viewing window.

b. Determine the value of the critical points by using the INTERSECT feature.

c. Interpret the graphical display to determine the x-values that satisfy the inequality.

d. Express the solution as a number line graph and in interval notation.

SOLUTION

a. Let $y1 = \text{abs}(x)$
$\qquad y2 = 5$

b.

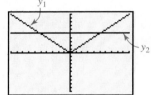

c.

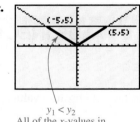

Critical points: the x-values of -5 and 5.

$y_1 < y_2$
All of the x-values in this region are between $x = -5$ and $x = 5$.

d.

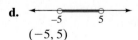

$(-5, 5)$

In general, we have the following property:

Solving Absolute Value Inequalities

If k is a positive number then $|x| < k$ is equivalent to $-k < x < k$.

EXAMPLE 1

Solve: $|2x - 3| < 9$.

SOLUTION   **Analytical solution:**

We write the inequality as a continued inequality

$|2x - 3| < 9$ is equivalent to $-9 < 2x - 3 < 9$

and solve for x:

$$-9 < 2x - 3 < 9$$
$$-6 < 2x < 12 \qquad \text{Add 3 to all three parts.}$$
$$-3 < x < 6 \qquad \text{Divide all parts by 2.}$$

Any number between -3 and 6, not including either -3 or 6, is in the solution set. This is the interval $(-3, 6)$, whose set is $\{x \mid -3 < x < 6\}$. The graph is shown in Figure 4-20.

Graphical solution:

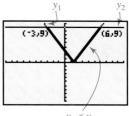

$y_1 < y_2$
All of the x-values in this region are between $x = -3$ and $x = 6$.

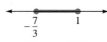

Figure 4-20

EXAMPLE 2

Solve: $|3x + 2| \leq 5$.

SOLUTION   **Analytical solution:**

We write the expression as a continued inequality:

$|3x + 2| \leq 5$ is equivalent to $-5 \leq 3x + 2 \leq 5$

and solve for x:

$$-5 \leq 3x + 2 \leq 5$$
$$-7 \leq 3x \leq 3 \qquad \text{Subtract 2 from all three parts.}$$
$$-\frac{7}{3} \leq x \leq 1 \qquad \text{Divide all three parts by 3.}$$

The solution set is $\{x \mid -\frac{7}{3} \leq x \leq 1\}$ and the interval is $[-\frac{7}{3}, 1]$, whose graph is shown in Figure 4-21.

Graphical solution:

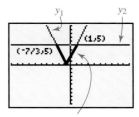

$y_1 < y_2$
All of the x-values in this region are between $x = -\frac{7}{3}$ and $x = 1$.

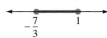

Figure 4-21

▮ INEQUALITIES OF THE FORM |x| > k

The inequality $|x| > 5$ can be interpreted to mean that a point with coordinate x is greater than 5 units from the origin. (See Figure 4-22.)

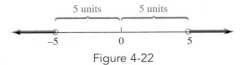

Figure 4-22

Thus, $x < -5$ or $x > 5$.

EXPLORING THE CONCEPT   Graphically solve $|x| > 5$.

a. Graph both sides of the inequality in the standard viewing window.

b. Determine the value of the critical points through the use of the INTERSECT feature.

c. Interpret the graphical display to determine the x-values that satisfy the inequality.

d. Express the solution as a number line graph and in interval notation.

SOLUTION   a. Let $y1 = \text{abs}(x)$
$\quad\quad\quad y2 = 5$

b.
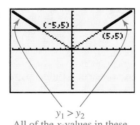

c.
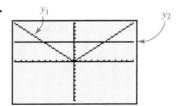

Critical points: the x-values of -5 and 5.

$y_1 > y_2$
All of the x-values in these regions are less than $x = -5$ or greater than $x = 5$.

d.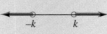

$(-\infty, -5) \cup (5, \infty)$

Solving Absolute Value Inequalities

If k is a positive number then $|x| > k$ is equivalent to $x < -k$ or $x > k$.

EXAMPLE 3   Solve the inequality $|5x - 10| > 20$.

SOLUTION   Analytical solution:

We write the inequality as two separate inequalities

$|5x - 10| > 20$ is equivalent to $5x - 10 < -20$ or $5x - 10 > 20$

and solve each one for x:

$$5x - 10 < -20 \quad \text{or} \quad 5x - 10 > 20$$
$$5x < -10 \qquad\qquad\qquad 5x > 30 \qquad \text{Add 10 to both sides.}$$
$$x < -2 \qquad\qquad\qquad x > 6 \qquad \text{Divide both sides by 5.}$$

Thus, x is either less than -2 or greater than 6:

$$x < -2 \quad \text{or} \quad x > 6$$

This is the interval $(-\infty, -2) \cup (6, \infty)$. The solution set is $\{x \mid x < -2 \text{ or } x > 6\}$. The graph appears in Figure 4-23.

Graphical solution:

$[-10, 10]$ by $[-10, 25]$

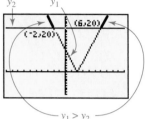

$y_1 > y_2$
All of the x-values in these regions are less than $x = -2$ or greater than $x = 6$.

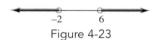

Figure 4-23

EXAMPLE 4   Solve the inequality $\left|\dfrac{3 - x}{5}\right| \geq 6$.

SOLUTION   Analytical solution:

We write the inequality as two separate inequalities

$\left|\dfrac{3 - x}{5}\right| \geq 6$ is equivalent to $\dfrac{3 - x}{5} \leq -6$ or $\dfrac{3 - x}{5} \geq 6$

and solve each one for x:

$$\frac{3 - x}{5} \leq -6 \quad \text{or} \quad \frac{3 - x}{5} \geq 6$$

$$3 - x \leq -30 \qquad\qquad 3 - x \geq 30 \qquad \text{Multiply both sides by 5.}$$

$$-x \leq -33 \qquad\qquad -x \geq 27 \qquad \text{Subtract 3 from both sides.}$$

$$x \geq 33 \qquad\qquad\quad x \leq -27 \qquad \begin{array}{l}\text{Divide both sides by } -1 \text{ and}\\ \text{reverse the direction of the}\\ \text{inequality symbol.}\end{array}$$

The solution set is $\{x \mid x \leq -27 \text{ or } x \geq 33\}$ and the interval is $(-\infty, -27] \cup [33, \infty)$, whose graph appears in Figure 4-24.

Graphical solution:

$[-40, 45]$ by $[-10, 10]$

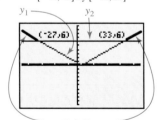

$y_1 > y_2$
All of the x-values in these regions are less than $x = -27$ or greater than $x = 33$.

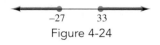

Figure 4-24

EXAMPLE 5

Solve the inequality $\left|\dfrac{2}{3}x - 2\right| - 3 > 6$.

SOLUTION

Analytical solution:

We begin by adding 3 to both sides to isolate the absolute value on the left-hand side. We then proceed as follows:

$$\left|\frac{2}{3}x - 2\right| - 3 > 6$$

$$\left|\frac{2}{3}x - 2\right| > 9 \qquad \text{Add 3 to both sides.}$$

$$\frac{2}{3}x - 2 < -9 \qquad \text{or} \qquad \frac{2}{3}x - 2 > 9$$

$$\frac{2}{3}x < -7 \qquad\qquad \frac{2}{3}x > 11 \qquad \text{Add 2 to both sides.}$$

$$2x < -21 \qquad\qquad 2x > 33 \qquad \text{Multiply both sides by 3.}$$

$$x < -\frac{21}{2} \qquad\qquad x > \frac{33}{2} \qquad \text{Divide both sides by 2.}$$

The solution set is $\{x \mid x < -\frac{21}{2} \text{ or } x > \frac{33}{2}\}$ and the interval is $(-\infty, -\frac{21}{2}) \cup (\frac{33}{2}, \infty)$, whose graph appears in Figure 4-25.

Figure 4-25

Graphical solution:

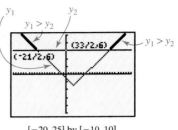

$[-20, 25]$ by $[-10, 10]$

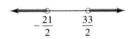

 WARNING! The expression involving the absolute value, $|ax + b|$, must be isolated before applying either of the properties for solving absolute value inequalities.

EXAMPLE 6

Solve the inequality $|3x - 5| + 4 \geq 2$.

SOLUTION

Analytical solution:

Isolate the absolute value by subtracting 4 from both sides of the inequality:

$$|3x - 5| + 4 - 4 \geq 2 - 4$$

$$|3x - 5| \geq -2$$

Because the absolute value of any number is non-negative and because any nonnegative number is larger than -2, the inequality is true for all x.

Graphical solution:

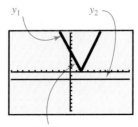

All of the x-values occur on the graph of y1, thus every real number is an acceptable solution.

The solution set, $\mathbb{R}$, is the interval $(-\infty, \infty)$, whose graph appears in Figure 4-26.

Every portion of the graph of y1 is greater than (above) the graph of y2.

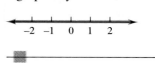

Figure 4-26

EXERCISE 4.6

VOCABULARY AND NOTATION *Fill in the blank to make a true statement.*

1. The _____ of a number can be defined as the distance between a number and zero on the number line.

2. Absolute value can never be less than ___.

CONCEPTS *In Exercises 3–6, use the graphical display to determine the solution to the given inequality or equation.*

3. **a.** $|2x - 4| - 6 = 0$

   b. $|2x - 4| - 6 < 0$

   c. $|2x - 4| - 6 > 0$

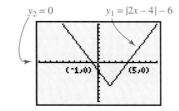

4. **a.** $-|3x - 2| + 4 = -3$

   b. $-|3x - 2| + 4 < -3$

   c. $-|3x - 2| + 4 > -3$

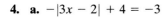
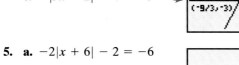

5. **a.** $-2|x + 6| - 2 = -6$

   b. $-2|x + 6| - 2 \le -6$

   c. $-2|x + 6| - 2 \ge -6$

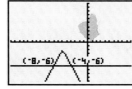

6. **a.** $3|x - 2| - 6 = 3$

   b. $3|x - 2| - 6 \le 3$

   c. $3|x - 2| - 6 \ge 3$

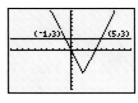

In Exercises 7–12, use the graphical display to determine the solution to the given inequality or equation.

7. **a.** y1 = y2

   b. y1 < y2

   c. y1 > y2

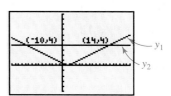

8. **a.** y1 = y2

   b. y1 < y2

   c. y1 > y2

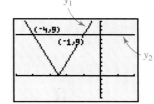

9. **a.** y1 = y2

   b. y1 < y2

   c. y1 > y2

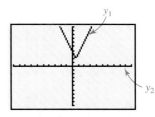

10. **a.** y1 = y2

    b. y1 < y2

    c. y1 > y2

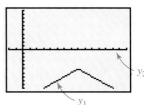

11. **a.** y1 = y2

    b. y1 < y2

    c. y1 > y2

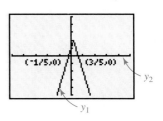

12. a. $y1 = y2$

  b. $y1 < y2$

  c. $y1 > y2$

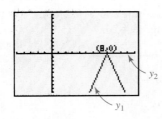

28. $|2x - 5| > 25$

29. $|4x + 3| > -5$

30. $|4x + 3| > 0$

PRACTICE *In Exercises 13–52 solve each inequality and confirm graphically. Write the solution using interval notation, a number line graph, and set notation.*

31. $-|3x + 1| < -8$

32. $|8x - 3| > 0$

13. $|2x| < 8$

33. $|7x + 2| > -8$

14. $|3x| < 27$

15. $|x + 9| \leq 12$

34. $\left|\dfrac{x - 2}{3}\right| \leq 4$

16. $|x - 8| \leq 12$

35. $\left|\dfrac{x - 2}{3}\right| > 4$

17. $|3x + 2| \leq -3$

18. $|3x - 2| < 10$

36. $|3x + 1| + 2 < 6$

19. $|4x - 1| \leq 7$

37. $|3x - 2| + 2 \geq 0$

20. $|5x - 12| < -5$

38. $3|2x + 5| \geq 9$

21. $|3 - 2x| < 7$

39. $-2|3x - 4| < 16$

22. $|4 - 3x| \leq 13$

40. $|5x - 1| + 4 \leq 0$

41. $-|5x - 1| + 2 < 0$

23. $|5x| > 5$

24. $|7x| > 7$

42. $\left|\dfrac{1}{3}x + 7\right| - 5 > 6$

25. $|x - 12| \geq 24$

43. $\left|\dfrac{1}{2}x - 3\right| - 4 < 2$

26. $|x + 5| \geq 7$

44. $\left|\dfrac{1}{5}x - 5\right| + 4 > 4$

27. $|3x + 2| > 14$

45. $\left|\dfrac{1}{6}x + 6\right| + 2 < 2$

46. $\left|\dfrac{3}{5}x + \dfrac{7}{3}\right| < 2$

47. $\left|\dfrac{7}{3}x - \dfrac{5}{3}\right| \geq 1$

48. $\left|3\left(\dfrac{x + 4}{4}\right)\right| > 0$

49. $3\left|\dfrac{1}{3}(x - 2)\right| + 2 \leq 3$

50. $|2x + 1| + 2 \leq 2$

51. $\left|\dfrac{x - 5}{10}\right| < 0$

52. $\left|\dfrac{3}{5}x - 2\right| + 4 \leq 3$

In Exercises 53 and 54 experiment with various values of x and y with your calculator to determine a response to the questions. Be specific in explaining your conclusions.

 **53.** Under what condtions is $|x| + |y| > |x + y|$?

54. Under what conditions is $|x| + |y| = |x + y|$?

 55. If $k > 0$, explain the differences between the solution sets of $|x| < k$ and $|x| > k$.

 56. If $k > 0$, explain why $|x| + k < 0$ would have $\varnothing$ as its solution.

 57. If $k > 0$, explain why $|x| + k > 0$ would have the solution $(-\infty, \infty)$.

REVIEW

58. Solve $p = 2l + 2w$ for w.

59. Solve $|3x - 5| + 3 = 2$.

60. Determine the x-intercepts of the accompanying graph.

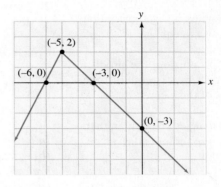

61. Is the graph in Exercise 60 a function? Explain why.

CHAPTER SUMMARY

CONCEPTS

Section 4.1

Constructing tables is a valuable way to organize information in application problems.

A ratio is a comparison of two numbers by their indicated quotient.

A proportion is a statement that two ratios are equal.

In any proportion, the product of the extremes is equal to the product of the means.

If two triangles are similar, then the corresponding sides are in proportion.

REVIEW EXERCISES

Applications—A Numeric Approach

1. In each of the problems below, determine if each statement is a proportion.

   a. $-\dfrac{4}{5} = \dfrac{28}{-35}$

   b. $\dfrac{14}{19} = \dfrac{-42}{57}$

   c. $\dfrac{9}{3} = \dfrac{15}{6}$

   d. $\dfrac{6}{30} = \dfrac{2}{10}$

2. Solve each proportion. Check your results both graphically (using the INTERSECT feature) and by using the STOre feature of the calculator.

   a. $\dfrac{x-4}{4} = \dfrac{x+4}{8}$

   b. $\dfrac{x+3}{2} = \dfrac{x-3}{3}$

   c. $\dfrac{x-5}{3} = \dfrac{2x+1}{2}$

   d. $\dfrac{3x-1}{2} = \dfrac{4x+5}{3}$

In Review Exercises 3–6, (a) clearly identify the variable and what it represents, (b) write an equation that models the problem, (c) solve the equation either algebraically or graphically, and (d) answer the question posed in the problem with a complete sentence.

3. If the length of the rectangular painting in Illustration 1 is 3 inches more than twice the width, how wide is the rectangle?

84 in.

Illustration 1

4. If 5 tons of iron ore yields 3 tons of pig iron, how much iron ore is needed to make 18 tons of pig iron?

5. A pharmacist mixes 3 grams of medicine with 300 milliliters of sugar syrup. Find out how much medicine is in a single dose of 5 milliliters.

6. How tall is the building if a 5-foot-tall woman casts a shadow of 2 feet (Illustration 2)?

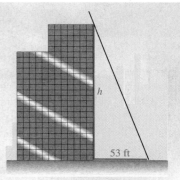

h

53 ft

Illustration 2

Section 4.2

The break-even point occurs when the costs equal the revenue.

It is important to distinguish between *number* and *value* when translating written expressions to algebraic equations.

Business Applications

In Review Exercises 7–10, (a) clearly identify the variable and what it represents, (b) write an equation that models the problem, (c) solve the equation either algebraically or graphically, and (d) answer the question posed in the problem with a complete sentence.

7. Costs for maintaining automotive parts run $668.50 per week plus variable costs of $1.25 per unit. The company can sell as many as it can make for $3 each. Find the break-even point.

8. A company can manufacture baseball caps on either of two machines, with costs shown in Table 1. At the projected sales level, they find the costs of the two machines equal. Find the expected sales.

Machine	Startup cost	Unit cost (per cap)
1	$ 85	$3
2	$105	$2.50

TABLE 1

9. A gumball machine takes nickels, dimes, and quarters. When the vendor empties the machine and separates the coins, he notices that there are twice the number of dimes as nickels and 5 more quarters than dimes. The value of the coins is $27.50. How many of each type of coin was in the machine?

With investment problems remember that $i = prt$, where i represents interest earned, p represents amount invested, r is annual rate, and t is time. Rate must be expressed as a decimal or fraction for computation purposes.

10. Part of $13,750 is invested at $4\frac{1}{2}\%$ and the rest at 5%. After one year, the account paid $647.50. How much was invested at the lower rate?

Section 4.3

More Applications of Equations

When working liquid mixture problems it is important that the concentrations represented in equations are consistent in that they all represent the same material, whether it be salt, alcohol, etc.

Dry mixture problems are based on the formula $v = pn$, where v is the value of the mixture, p is the price per pound, and n is the number of pounds.

Uniform motion problems are based on the formula $d = rt$, where d is the distance traveled, r is the rate, and t is the time.

In Review Exercises 11–13, (a) clearly identify the variable and what it represents, (b) write an equation that models the problem, (c) solve the equation either algebraically or graphically, and (d) answer the question posed in the problem with a complete sentence.

11. How many liters of water must be added to 30 liters of a 10% brine solution to dilute it to an 8% solution?

12. A store manager mixes candy worth $0.90 per pound with candy worth $1.65 per pound to make 20 lb of a mixture worth $1.20 per pound. How many pounds of each must be mixed?

13. A bicycle path is 5 miles long. A man walks from one end at the rate of 3 mph. At the same time a friend bicycles from the other end, traveling 12 mph. In how many minutes will they meet?

Section 4.4

Solving Simple Linear Inequalities

The critical point of an inequality is the solution to the corresponding equation and divides the number line into regions that contain solution points versus the regions that contain no solution points.

Interval notation is a means of symbolically describing where the shading begins and ends on a number line graph.

Always reverse the direction of the inequality symbol when multiplying or dividing by a negative value.

The graphical display for solving an inequality graphically looks the same as the graphical display for solving the corresponding equation graphically. It is your *interpretation* that makes the difference.

14. Solve each linear inequality and express the solution in interval notation, as a number line graph, and in set notation. Verify results with the graphing calculator.

    a. $3x - 4 < 5$ **b.** $5 - 2x < 6$

    c. $2(x - 3) \leq 7$ **d.** $-3(x - 1) + 5 \geq 6$

In Review Exercises 16–17, solve the indicated equation or inequality based on the displayed graph.

15. a. $y_1 = y_2$ **16. a.** $y_1 = y_2$

    b. $y_1 < y_2$ **b.** $y_1 < y_2$

    c. $y_1 > y_2$ **c.** $y_1 > y_2$

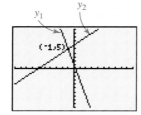

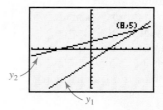

Section 4.5	Compound Inequalities

When solving compound inequalities linked by the word *and* (conjunctions), the intersection of the component simple inequalities comprises the solution.

When solving compound inequalities linked by the word *or* (disjunctions), the union of the component simple inequalities comprises the solution.

Compound inequalities can be solved graphically by graphing the *equations* corresponding to the simple inequalities and interpreting the results. Use of the INTERSECT feature of the calculator is helpful in locating critical points.

17. Solve each compound inequality and express the solution using interval notation, as a number line graph, and in set notation.

  a. $3 < 3x + 4 < 10$ **b.** $4 > 3x + 2 > -3$

  c. $6x + 1 < 5$ or $3(x - 9) > -2$

  d. $5 - \dfrac{1}{2}x \geq -1$ or $6 - x < -1$

In Review Exercises 18–19, solve the compound inequality by interpreting the information displayed on the screen.

18. a. $y_1 < y_2 < y_3$ **19. a.** $y_1 < y_2 < y_3$

  b. $y_2 < y_1$ or $y_2 > y_3$ **b.** $y_2 < y_1$ or $y_2 > y_3$

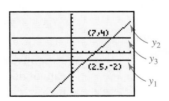   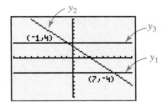

Section 4.6	Absolute Value Inequalities

If k is a positive number, then $|x| < k$ is equivalent to the compound inequality $-k < x < k$.

If k is a positive number, then $|x| > k$ is equivalent to $x > k$ or $x < -k$.

20. Solve each absolute value inequality and confirm graphically. Write the solution using interval notation, as a number line graph, and in set notation.

  a. $|x + 3| \leq 4$ **b.** $|2x + 3| < 11$

  c. $|2x - 4| > 22$ **d.** $|3x + 1| \geq 5$

  e. $|4x - 1| + 7 < 2$ **f.** $\left|\dfrac{1}{2}x + 3\right| - 8 > -11$

In Review Exercises 21–22, use the graphical display to determine the solution to the given inequality or equation.

21. a. $y_1 = y_2$

  b. $y_1 < y_2$

  c. $y_1 > y_2$

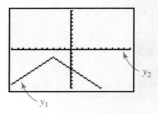

22. a. $y_1 = y_2$

  b. $y_1 < y_2$

  c. $y_1 > y_2$

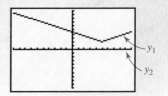

CHAPTER TEST

1. Is $-\dfrac{42}{22} = \dfrac{14}{-11}$ a proportion?

2. Is $\dfrac{2x - 1}{3} = \dfrac{4x - 1}{6}$ a proportion?

3. Solve $\dfrac{4 - x}{13} = \dfrac{11}{26}$.

4. Solve $\dfrac{2x + 5}{4} = \dfrac{x - 1}{6}$.

5. A plane drops 575 feet as it flies a horizontal distance of $\frac{1}{2}$ mile. How much altitude will it lose as it flies a horizontal distance of 7 miles (Illustration 1)?

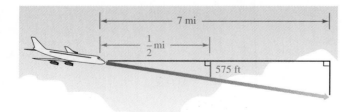

Illustration 1

6. A car leaves Rockford, Illinois at a rate of 65 mph, bound for Madison, Wisconsin. At the same time, a truck leaves

Madison at a rate of 55 mph, bound for Rockford. If the cities are 72 miles apart, how long will it take the truck to meet the car?

7. Rose invests part of a $50,000 inheritance in a certificate of deposit at a rate of 4%. The rest is invested in a high-interest savings account earning $2\frac{1}{2}\%$ to be accessible for emergencies that her $1,140-a-month Social Security check cannot cover. In one year, the two accounts earn $1,925. How much is in the certificate of deposit?

8. A leading juice brand advertises that its pink grapefruit juice cocktail is 35% juice. How many ounces of juice must be mixed with 13 ounces of other ingredients (e.g., filtered water, corn syrup, preservatives, etc.) to produce a bottle of pink grapefruit juice cocktail that is 35% juice?

9. One side of a triangle is 4 more than twice the shortest side. If the third side is twice the shortest side, find the short side if the perimeter is 24.

10. A high school student invests $100 in supplies for an "earn at home" envelope-stuffing job. It costs her $0.37 to mail each envelope, but the company pays her $0.45 per envelope mailed. How many envelopes will she need to stuff and mail to break even?

In Problems 11 and 12, use the graphical display to determine the solution to the equation or inequality.

11. a. $y_1 = y_2$
    b. $y_1 < y_2$
    c. $y_1 > y_2$

12. a. $y_1 = y_2$
    b. $y_1 < y_2$
    c. $y_1 > y_2$

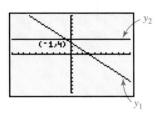

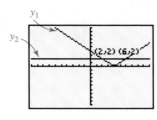

In Problems 13–20, solve each inequality and express the solution using interval notation, as a number line graph, and in set notation.

13. $\frac{1}{4}x - 2 \le -2$

14. $-2 \le 4 - x < 5$

15. $5 - 2x > 6$ or $x + 2 > 1$

16. $4x + 1 < x$ and $2x - 5 < 1$

17. $|2x - 5| < 4$

18. $\left|\frac{1}{3}x - 1\right| \ge 5$

19. $\left|\frac{1}{2}x + 3\right| + 4 \le 2$

20. $|2x + 4| + 4 > 1$

CUMULATIVE REVIEW

CHAPTERS 3 AND 4
Vocabulary/Concepts

1. An identity is an equation that is true _____ values of the variable, and the symbol ____ represents the solution.

2. A contradiction is an equation that is true for ____ real numbers and the symbol ____ represents the solution.

3. Solutions to equations in two variables are expressed as _____.

4. A relation in which each x-value is paired with a single y-value is called a _____.

5. _____ inequalities are inequalities that are linked by the words *and* or *or*.

6. Determine the quadrant location of each ordered pair.

    a. $(-2, 3)$ **b.** $(-4, -1)$

    c. $(3, -4)$ **d.** $(5, -2)$

7. Sketch the set of coordinate axes that is indicated by the given viewing window: $[-5, 5]$ by $[-10, 10]$ with xscl = 1 and yscl = 2.

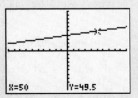

8. Explain the difference between tracing on a graph in the standard (ZStandard) viewing window and tracing on a graph in an integer viewing window (ZInteger).

9. The display at the right is a graphical model of the linear equation $y = \$39 + \$0.21x$, which represents the cost, y, of renting a car for one day at a fixed rate of \$39 for x miles at 21¢ a mile. Interpret the meaning of the graphical display.

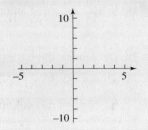

10. Two features of the calculator that can be used to verify solutions of linear equations in one variable are the _____ and _____ features.

11. A ratio is a _____ of two numbers by their indicated quotient.

12. Express the phrase "18 to 24" as a ratio in lowest terms.

13. Two triangles with the same shape are called _____ triangles, and their corresponding sides are in _____ .

14. If a CD sells for $16.99 and x represents the quantity sold, write an expression to represent the revenue from the sale of x number of CDs.

15. Express each percentage as a decimal number.

a. 6%   **b.** $7\frac{3}{4}\%$

16. If 15 liters of a solution is added to a quantity of alcohol to produce a total liquid solution of $15 + x$ liters, how much alcohol is there?

17. If $a \geq 0$ then $|a| =$ ___ . If $a < 0$ then $|a| =$ ___ .

18. Why is $|x| = -4$ an invalid statement?

19. Determine the indicated absolute value.

a. $-|-3|$

b. $|2 - \pi|$ (do not approximate)

20. Solve for p: $i = prt$.

Practice

21. Algebraically solve $3x + 4(2 - x) = 3(x + 4) - 5$ and verify with the STOre feature of your calculator.

22. Algebraically solve $\dfrac{3(x - 1)}{6} - 4 = \dfrac{x + 3}{6}$ and verify with the STOre feature of your calculator.

23. Which ordered pair(s) is *not* a solution of $2x + y = 5$?

a. $(2, 1)$ **b.** $(-2, -1)$ **c.** $(-3, 11)$ **d.** $(5, -5)$

24. Which of the following are *not* functions?

a. $\{(2, 4), (-2, 4), (3, 1), (-3, 1)\}$

b. $\{(2, 4), (2, -4), (3, 1), (-3, -1)\}$

c. $y = 2x - 5$

25. Give the coordinate of the x- and y-intercepts in Illustration 1.

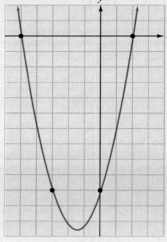

Illustration 1

26. Graph the equation of the line $y = 4x - 1$.

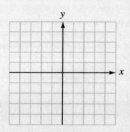

27. a. Graph $y = x^2 - 8x + 7$ in the ZInteger viewing window and use the TRACE feature to determine the x- and y-intercepts.

b. Explain why ZInteger is a better viewing window for part **a** than ZStandard would be.

28. Why is the standard viewing window not a good viewing window for solving the equation $3 + [x - (2 - x)] = -16$? What must be done to have a good viewing window?

29. Solve for x in the proportion $\dfrac{4(x + 1)}{3} = \dfrac{x - 2}{4}$.

In Exercises 30–32, algebraically solve each equation and verify by solving graphically.

30. $|6x - 5| = 2$

31. $\left|\dfrac{x}{2} + 1\right| - 5 = 3$

32. $|2x - 1| = |3x + 4|$

In Exercises 33–34, solve for the indicated variable.

33. $\dfrac{1}{2}x - \dfrac{1}{3}y = 6$; for y

34. $Ax + By = C$; for y

In Exercises 35–41, algebraically solve each inequality and express the solution using interval notation, as number line graph, and in set notation.

35. $\dfrac{2}{3}x - 4 \geq 6$

36. $2 < 6 - x \leq 8$

37. $3x + 5 < 8$ or $2x - 5 > 11$

38. $2x + 6 < 8$ and $6 - 5x > 11$

39. $\left|\dfrac{1}{2}x - 1\right| < 6$

40. $-|x + 2| \leq -3$

41. $|x + 2| \leq -3$

Applications

42. Mixing solutions   A veterinarian recommends a solution of 3 parts vinegar to 1 part alcohol as ear drops for a dog that is prone to ear infections. If the solution is to be stored in a 5-ounce bottle, will 3.75 ounces of vinegar and 1.25 ounces of alcohol be the correct amounts?

43. Finding integers   The sum of an integer and three times the next integer is 171. Find the smaller integer.

44. Investing money $3,000 earns 3% simple interest in a savings acount, and a $4,000 investment earns $1\frac{3}{4}$% simple interest in another savings account. How long will it take for the accounts to have the same amount of money?

45. Buying stamps   Sherah purchased 37-cent stamps for her Christmas cards and 34-cent stamps for post-cards. If she spent $27.30 and purchased four times as many 37-cent stamps as 34-cent stamps, how many 37-cent stamps did she purchase?

46. Computing travel time   A postal transport truck leaves town traveling at 55 miles per hour. Unfortunately, one hour after the truck left the main post office, it was discovered that a large bag of priority mail had not been loaded. A vehicle was immediately dispatched to carry the priority mail to the transport truck. If the vehicle can travel 70 miles per hour, how long will it take to overtake the transport?

47. Mixing solutions   A fruit juice beverage containing 10% juice is mixed with 64 ounces of another fruit juice beverage containing 50% juice. How many ounces of the 10% juice beverage will be needed to produce a beverage that is 35% juice?

48. Volume   The volume of a sphere is given by the formula $V = \frac{4}{3}\pi r^3$. How many cubic inches of air are in a beach ball having a radius of 9 inches (to the nearest tenth of a cubic inch)?

49. Investments   At a simple interest rate r, an amount of money P grows to an amount A in t years according to the formula $A = P(1 + rt)$. If $2,500 grows to $2,937.50 when invested at a $3\frac{1}{2}$% interest rate, find the length of time the money was invested.

In Section 4.2, Example 1, the equation for the break-even point for an electronics company was determined to be $90x = 75x + 6{,}405$. Review the example and then graph the left and right sides of the equation in the window $[-10, 930]$ by $[-10, 60{,}000]$ before completing Exercises 50–51.

50. Trace along the graph of $y1 = 90x$ to complete the following.

   a. When the number of CD players is 230, the revenue is _____ .

   b. When the revenue is $15,300, the number of CD players produced is _____ .

51. Trace along the graph of $y2 = 75x + 6405$ to complete the following.

   a. When 170 CD players are produced the cost is _____ .

   b. How does this compare to the revenue earned (Exercise 50) from this same quantity of CD players?

   c. If the cost is $52,905, how many CD players were produced? Is the company making a profit or loss at this point?

CAREERS & MATHEMATICS

© Royalty-Free/CORBIS

Banker   Practically every bank has a group of officers who make decisions affecting bank operations: the president, who directs overall operations; one or more vice-presidents, who act as general managers or are in charge of bank departments, such as trust or credit; a comptroller, or cashier, who (as an executive officer) is generally responsible for all bank property; and treasurers and other senior officers as well as junior officers, who supervise sections within departments.

These officers make decisions within a framework of policy set by a board of directors and existing laws and regulations. They must have a broad knowledge of business activities to relate to the operations of their departments, because their customers will include a variety of individuals and businesses applying for loans, seeking investment advice, organizing trusts, setting up pensions, and so on. Besides supervising these financial services, officers advise individuals and businesses and participate in community projects.

Qualifications   Bank officer and management positions are filled by management trainees and by promoting outstanding bank clerks or tellers. A college degree in finance or liberal arts, including accounting, economics, commercial law, political science, and statistics, is necessary. A Master of Business Administration degree (MBA) is preferred by some banks, although people with backgrounds as diverse as nuclear physics and forestry are hired by some banks to meet the needs of the complex, high-technology industries with which they deal.

Job Outlook   Through the year 2010, employment of bank officers is expected to increase faster than the average for other occupations, due to expanding bank services, both domestic and international, and the increasing dependence on computers.

Example Application   In banking, the calculation of compound interest uses exponential expressions. At the end of each period, a savings account is credited with a dividend, calculated as a fixed percentage of the savings balance. The money will grow to an amount FV, called the **future value**, according to the formula

$$FV = PV(1 + i)^n$$

where PV, called the **present value**, is the initial deposit. The percent of the current balance credited as interest is called the **periodic interest rate**, denoted by i. The **number of periods** that the money is left on deposit is n.

If \$1,000 is deposited in an account that earns 6% annual interest compounded quarterly, find the amount in the account after 3 years.

Solution   Interest paid quarterly is paid four times each year. The periodic interest rate is one-fourth of 6%, or 1.5%. Express the periodic interest rate as a decimal: $i = 0.015$. In 3 years, there will be $3 \cdot 4$, or 12 periods. Thus, $n = 12$. The present value is the initial deposit: $PV = \$1,000$. Substitute these values into the equation and calculate FV.

$$FV = \mathbf{PV}(1 + \mathbf{i})^n$$
$$= \mathbf{1,000}(1 + \mathbf{0.015})^{12} \quad \text{Substitute 1,000 for } PV, 0.015 \text{ for } i, \text{ and 12 for } n.$$
$$= 1,000(1.015)^{12}$$
$$= 1,195.618171 \quad \text{Use a calculator.}$$

After 3 years, the account will contain \$1,195.62.

EXERCISES

1. If $10,000 is deposited in an account earning 6% annual interest, compounded once a year, how much will be in the account at the end of 2 years?

2. If $10,000 is deposited in an account earning 6% annual interest, compounded twice a year, how much will be in the account at the end of 2 years?

3. If $10,000 is deposited in an account earning 6% annual interest, compounded four times a year, how much will be in the account at the end of 2 years?

4. Refer to Exercises 1–3. How much more is earned by compounding twice a year, instead of once a year? How much more is earned by compounding four times a year, instead of twice a year?

(*Answers:* 1. $11,236   2. $11,255.09   3. $11,264.93   4. $19.09; $9.84)

Factoring

© AFP/CORBIS

InfoTrac Project

Do a subject guide search on "arches," choose "View periodical references," and find the article "Another grand arch for Paris." Write a summary of the article.

The area of a cross section of the arch is 7,700 square meters. In the following equation, x represents 1/10 of the width of the arch. Using factoring, solve the equation for x.

$$2x^2 - 3x - 77 = 0$$

Use your answer to find the width of the arch and then find its height.
Complete this project after studying Section 5.5.

PERSPECTIVE

No End of Primes

A composite number always has prime factors that are smaller than the number itself. The composite number 30, for example, has the prime factors 2, 3, and 5—all less than 30. We might think that the larger a number is, the more likely it is to be composite, because there are so many smaller prime numbers that could be possible factors. Maybe all numbers, if large enough, would be composite, and therefore there would be a largest prime number.

That is not what happens, however. In about 300 B.C. the Greek mathematician Euclid proved that the number of primes is unlimited—that there are infinitely many prime numbers. This is an important fact of a branch of mathematics called number theory. Although Euclid is best known for his study of geometry, many of his writings deal with number theory.

Who could possibly care that prime numbers can be very large? Cryptographers working for the federal government care, because several codes that are very difficult to crack rely on the properties of large primes.

Euclid (about 300 B.C.)

In this chapter we shall reverse the operation of multiplication and find the factors of a polynomial. In Chapter 2, polynomials were discussed. Polynomials are comprised of terms; ungrouped addition and subtraction signs are term indicators. The process of factoring a polynomial enables us to write polynomials as a product of **factors** rather than a sum/difference of terms. Factoring is the reverse of multiplication.

5.1   USING THE DISTRIBUTIVE PROPERTY TO FACTOR

- IDENTIFYING THE GREATEST COMMON FACTOR
- FACTORING OUT THE GREATEST COMMON FACTOR
- FACTORING OUT A NEGATIVE FACTOR
- POLYNOMIALS AS GREATEST COMMON FACTORS
- FACTORING BY GROUPING

▚ IDENTIFYING THE GREATEST COMMON FACTOR

The **factors** of an expression (either a constant or a polynomial) are those expressions (a) that evenly divide the given expression or (b) whose product is equivalent to the given expression. To **factor** a number or a polynomial simply means to write the given quantity as a product of factors. Because the number 2 divides the number 6 evenly, 2 is a **factor** of 6. Three is also a factor of 6 because it divides 6 evenly. Therefore,

$$6 = 2 \cdot 3$$

factor   factor

The numbers 2 and 3 are each factors of 6 because (a) they both evenly divide 6 and (b) their product, $2 \cdot 3$, is equivalent to 6. The expression $2 \cdot 3$ is the factored form of 6. To factor a natural number means to write the number as a product of prime numbers. Recall that a prime number is a natural number *greater than 1* whose only factors are one and the number itself.

Prime Factored Form
A natural number is said to be in **prime factored form** if it is written as the product of factors that are prime numbers.

The Fundamental Theorem of Arithmetic states that there is *exactly one* prime factorization for every natural number greater than 1. When asked to **factor** in algebra, it is implicitly stated that the prime factorization is required.

EXAMPLE 1

Factor each of the following: **a.** 42   **b.** 60   **c.** 90.

SOLUTION   **a.** $42 = 2 \cdot 21$ **b.** $60 = 2 \cdot 30$ **c.** $90 = 2 \cdot 45$
$\quad\quad\quad\quad = 2 \cdot 3 \cdot 7 \quad\quad\quad = 2 \cdot 2 \cdot 15 \quad\quad\quad = 2 \cdot 3 \cdot 15$
$\quad\quad\quad\quad\quad\quad\quad\quad\quad\quad\quad\; = 2 \cdot 2 \cdot 3 \cdot 5 \quad\quad = 2 \cdot 3 \cdot 3 \cdot 5$
$\quad\quad\quad\quad\quad\quad\quad\quad\quad\quad\quad\; = 2^2 \cdot 3 \cdot 5 \quad\quad\;\; = 2 \cdot 3^2 \cdot 5$

It is a good habit to check factorization by multiplying the factors together to ensure the product of the factors is equivalent to the original number.

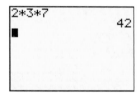
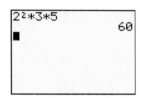
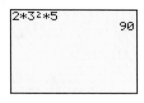

Examining the prime factorizations of 42, 60, and 90 reveals that the three numbers have several common factors. We are interested in the **greatest common factor** of the three numbers. This is defined as the largest natural number that divides each of several given numbers. In Example 1, 6 is the GCF because it is the product of the two common factors of 2 and 3.

Determining the Greatest Common Factor

1. Determine the prime factorization of each constant. Use exponents rather than repeated factors to write the number in prime factored form.
2. List each prime number that is a factor of all the numbers in the list. When exponents appear, choose the common factor with the *smallest* exponent.
3. Multiply the listed primes from the previous step. This is the greatest common factor, or GCF.

EXAMPLE 2

State the GCF for the numbers

20, 40, 80

SOLUTION

$$20 = 2 \cdot 10 \qquad 40 = 4 \cdot 10 \qquad 80 = 2 \cdot 40$$
$$= 2 \cdot 2 \cdot 5 \qquad = 2 \cdot 2 \cdot 2 \cdot 5 \qquad = 2 \cdot 2^3 \cdot 5$$
$$= 2^2 \cdot 5 \qquad = 2^3 \cdot 5 \qquad = 2^4 \cdot 5$$

It is a good idea to first double-check your factorization by confirming that the product of the factors is equivalent to the original number before determining the GCF.

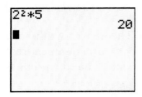 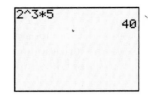 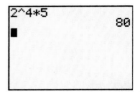

The common factors are 2^2 and 5 (notice that 2^2 was chosen because it has the smallest exponent of all factors with a base of 2). The GCF is the *product* of the common factors: $2^2 \cdot 5 = 20$. The greatest common factor of 20, 40, and 80 is 20.

EXAMPLE 3

Find the GCF of $6x^2y^3$, $4x^3y^2$, and $18x^2y$.

SOLUTION

$$6x^2y^3 = 2 \cdot 3 \cdot x^2 \cdot y^3 \qquad \text{Write each constant factor in prime factored form using exponents.}$$

$$4x^3y^2 = 2^2 \cdot x^3 \cdot y^2$$
$$18x^2y = 2 \cdot 3^2 \cdot x^2 \cdot y$$
$$2, x^2, y \qquad \text{List the common factors, specifying the factor with the smallest exponent (this applies to bases that are constants or variables).}$$

The greatest common factor is $2x^2y$, the product of the common factors.
    You can check your choice of GCF by ensuring that it evenly divides each monomial **and** that the quotients have no common factors.

Check:

$$\frac{6x^2y^3}{2x^2y} = 3y^2 \qquad \frac{4x^3y^2}{2x^2y} = 2xy \qquad \frac{18x^2y}{2x^2y} = 9$$

The quotients $3y^2$, $2xy$, and 9 are **relatively prime**, meaning that they have no common factors.

▌ FACTORING OUT THE GREATEST COMMON FACTOR

Recall the distributive property: $a(b + c) = ab + ac$. This property allows multiplication over the operation of addition (and subtraction, because subtraction is defined as adding the opposite of a number). Consider the use of the distributive property in the following expression:

$$3x^2(2x - 3y) = 3x^2 \cdot 2x - 3x^2 \cdot 3y$$
$$= 6x^3 - 9x^2y$$

To factor the polynomial $6x^3 - 9x^2y$, we find the greatest common factor of each *term* and then rewrite each *term* as the product of the GCF and another factor.

The GCF of the two terms in the polynomial is $3x^2$. Rewriting each term of the polynomial $6x^3 - 9x^2y$ as the indicated product of the GCF and another factor gives us the expression

$$3x^2 \cdot 2x - 3x^2 \cdot 3y$$

We then use the distributive property to rewrite the polynomial as the *product of factors* rather than the *sum of terms*:

$$3x^2(2x - 3y)$$

The steps for factoring out a GCF are summarized as follows:

Factoring Out the Greatest Common Factor

1. Determine the GCF of the terms of the polynomial.
2. Rewrite each *term* as the product of the GCF and another factor.
3. Use the distributive property to rewrite the polynomial as the *product* of the GCF and a polynomial.

EXAMPLE 4

Factor: $12a^3b^2 + 20a^2b^3$.

SOLUTION

$$12a^3b^2 = 2^2 \cdot 3 \cdot a^3 \cdot b^2$$

Factor each term of the polynomial to determine the GCF.

$$20a^2b^3 = 2^2 \cdot 5 \cdot a^2 \cdot b^3$$

$$\text{GCF} = 2^2 a^2 \, b^2$$

The GCF is the product of the common factors.

$$12a^3b^2 + 20a^2b^3 = 4a^2b^2 \cdot 3a + 4a^2b^2 \cdot 5b$$

Rewrite each term as the product of the GCF and another factor.

$$= 4a^2b^2(3a + 5b)$$

Use the distributive property.

Check the result by verifying that $4a^2b^2(3a + 5b) = 12a^3b^2 + 20a^2b^3$.

EXAMPLE 5

Factor: $a^2b^2 - ab$.

SOLUTION The GCF is ab.

$$a^2b^2 - ab = ab \cdot ab - ab \cdot 1$$

Rewrite each term as the product of the GCF and another factor.

$$= ab(ab - 1)$$

Use the distributive property.

Check the result by verifying that $ab(ab - 1) = a^2b^2 - ab$.

WARNING!   It is important to include the number 1 in the binomial factor in Example 5. It serves as a place holder.

▍▍ FACTORING OUT A NEGATIVE FACTOR

It is often convenient when working with polynomials for the leading coefficient to be a positive number. For this reason, a negative one (-1) is usually factored out of a polynomial when the leading coefficient is negative.

EXAMPLE 6

Factor: $-18a^2b + 6ab^2 - 12a^2b^2$.

SOLUTION

$$-18a^2b = -2 \cdot 3^2 \cdot a^2 \cdot b$$ Factor each term of the polynomial to determine the GCF.

$$6ab^2 = 2 \cdot 3 \cdot a \cdot b^2$$
$$-12a^2b^2 = -2^2 \cdot 3 \cdot a^2 \cdot b^2$$

The GCF is $-2 \cdot 3ab$, or $-6ab$

Because the leading coefficient is negative, we will use negative one as part of our greatest common factor.

$$-18a^2b + 6ab^2 - 12a^2b^2 = (-6ab)3a - (-6ab)b + (-6ab)2ab$$
$$= -6ab(3a - b + 2ab)$$

Check by verifying that $-6ab(3a - b + 2ab) = -18a^2b + 6ab^2 - 12a^2b^2$.

As you progress through this chapter on factoring, **always** look for and factor out the GCF first.

▍▍ POLYNOMIALS AS GREATEST COMMON FACTORS

Sometimes the greatest common factor of several terms is a polynomial, and we can factor out that polynomial as a common factor. It is helpful to carefully identify the number of terms in a polynomial before beginning the process.

EXAMPLE 7

Factor $a(x + 2y) - b(x + 2y)$.

SOLUTION

The polynomial has two terms, $a(x + 2y)$ and $-b(x + 2y)$. The quantity, $x + 2y$, is a common factor of both terms. We may then rewrite the polynomial in factored form as

$$(x + 2y)(a - b)$$ Use the distributive property to factor out the quantity $(x + 2y)$ from each term.

EXAMPLE 8

Factor $a + 3$ out of the expression $(a + 3) + (a + 3)^2$.

SOLUTION   The polynomial has two terms, $(a + 3)$ and $(a + 3)^2$. We recall that $a + 3$ is equal to $(a + 3)1$ and that $(a + 3)^2$ is equal to $(a + 3)(a + 3)$. We can then factor out the GCF of $(a + 3)$ and simplify:

$$(a + 3) + (a + 3)^2 = (a + 3)1 + (a + 3)(a + 3)$$

$$= (a + 3)[1 + (a + 3)] \qquad$$ Use the distributive property to factor out the quantity $(a + 3)$ from each term of the polynomial.

$$= (a + 3)(a + 4) \qquad$$ Simplify the second factor.

EXAMPLE 9

Factor: $6a^2b^2(x + 2y) - 9ab(x + 2y)$.

SOLUTION $6a^2b^2(x + 2y) - 9ab(x + 2y) = (x + 2y)(6a^2b^2 - 9ab) \qquad$ Factor out the common factor $(x + 2y)$.

$$= (x + 2y)(3ab)(2ab - 3) \qquad$$ Factor out the common factor of $(3ab)$ from $(6a^2b^2 - 9ab)$.

$$= 3ab(x + 2y)(2ab - 3) \qquad$$ Use the commutative property of multiplication.

It is important that you develop the habit of always checking each polynomial factor to ensure it is prime before concluding that the factoring process is complete.

FACTORING BY GROUPING

Suppose we wish to factor the expression

$$ax + ay + cx + cy$$

Although no factor is common to all four terms, there is a common factor of a in $ax + ay$ and a common factor of c in $cx + cy$. We can factor out the a and c as follows:

$$ax + ay + cx + cy = (ax + ay) + (cx + cy) \qquad$$ Associative property of addition.

$$= a(x + y) + c(x + y) \qquad$$ Distributive property (factoring the GCF of a out of the first two terms and the GCF of c out of the last two terms).

$$= (x + y)(a + c) \qquad$$ Distributive property (factoring the GCF of the quantity $(x + y)$ out of each term).

Therefore, $ax + ay + cx + cy = (x + y)(a + c)$. This type of factoring is called **factoring by grouping**, and applies to polynomials containing four terms.

EXAMPLE 10

Factor: $2c + 2d - cd - d^2$.

SOLUTION $2c + 2d - cd - d^2 = (2c + 2d) + (-cd - d^2)$ Associative property and definition of subtraction.

$$= 2(c + d) - d(c + d)$$ Distributive property. Notice that after grouping $(-cd - d^2)$, the leading coefficient of this group was negative. Therefore, a negative one became part of the GCF.

$$= (c + d)(2 - d)$$ Distributive property.

WARNING!   When factoring expressions by grouping make sure that the process of taking out a GCF is continued until the result is a *product of factors*. There should be no ungrouped addition or subtraction signs in the final result of any factoring exercise.

EXAMPLE 11

Factor: **a.** $a(c - d) + b(d - c)$ and   **b.** $ac + bd - ad - bc$.

SOLUTION   **a.** $a(c - d) + b(d - c) = a(c - d) - b(-d + c)$ Use the distributive property to factor a negative one from $(d - c)$.

$$= a(c - d) - b(c - d)$$ The commutative property assures us that $-d + c = c - d$.

$$= (c - d)(a - b)$$ Factor out $(c - d)$ using the distributive property.

b. $ac + bd - ad - bc$

In this example, we cannot factor anything (other than 1) from the first two terms or the last two terms. However, if we use the commutative property to rearrange the terms, the factoring is routine.

$$ac + bd - ad - bc = ac - ad + bd - bc$$ By the commutative property, $bd - ad = -ad + bd$.

$$= a(c - d) - b(-d + c)$$ Factor a from $ac - ad$ and $-b$ from $bd - bc$ using the distributive property.

$$= a(c - d) - b(c - d)$$ $-d + c = c - d$

$$= (c - d)(a - b)$$ Factor out $(c - d)$ using the distributive property.

EXERCISE 5.1

VOCABULARY AND NOTATION   *In Exercises 1–9, fill in each blank to make a true statement.*

1. The _____ of an expression evenly divides the expression.

2. Two or more expressions whose product is equivalent to a given expression are called _____ of the original expression.

3. A _____ number is a number whose *only* factors are 1 and itself.

4. The expression $3 + x$ is an indicated *sum* of _____.

5. The _____ property of multiplication over addition enables us to factor polynomials.

6. Factorization of a polynomial requires that a _____ or _____ of terms be rewritten as a _____ of factors.

7. The greatest common factor of a polynomial can be a monomial or a _____.

8. The Fundamental Theorem of Arithmetic states there is _____ prime factorization for every natural number greater than 1.

9. If a polynomial contains four terms, it may be possible to factor the polynomial by _____.

CONCEPTS   *In Exercises 10–19, classify each of the statements as true or false.*

10. The factors of 12 are 1, 2, 3, 4, 6, and 12.

11. The prime factors of 12 are 1, 2, and 3.

12. The numbers 10 and 17 have no common factors (other than 1).

13. The greatest common factor of 12 and 16 is 4.

14. The polynomial $3x(x + 3) + 4(x - 5)$ is completely factored.

15. The polynomial $3x^2(x + 3)(x - 2)$ is completely factored.

16. $6(x^2 - y^3)$ is equivalent to the polynomial $6x^2 - 6y^3$.

17. $6x^2y - 3x^2 = 3x(2xy - x)$ illustrates a prime factorization of a polynomial.

18. The GCF of the polynomial $3x^2y^3 + 6xy^4 - 9x^3y^5$ is $3x$.

19. The GCF of the polynomial $15a^2b^3c - 20a^3b^2c^2 + 25a^2b^4c^2$ is $5a^2b^2c$.

In Exercises 20–25, determine the GCF of each polynomial.

20. $24 + 12x$

21. $36y + 6$

22. $2x^2y^4 + 6x^2y^5 - 2x^2y^3$

23. $4x^3y^5 - 2x^2y^4 + 6x^3y^3$

24. $16a^3b^2c - 4a^2bc^3 + 8a^3b^2c^2$

25. $20t^3v - 15t^2v^3 + 3tv$

In Exercises 26–31, complete the factorization.

26. $2x^2 - 6x = 2x(\quad)$

27. $6x^3 + 3x = 3x(\quad)$

28. $3x^2y + 2x^3y^2 = x^2y(\quad)$

29. $5a^3b^2 + 7a^2b = a^2b(\quad)$

30. $-6x^2y^3 - 3xy = -3xy(\quad)$

31. $-4x^3y - 2xy = -2xy(\quad)$

PRACTICE   *In Exercises 32–53, factor each polynomial. When appropriate, factor out negative one as part of the GCF.*

32. $3x + 6$

33. $xy - xz$

34. $-a - b$

35. $-2x + 5y$

36. $t^3 + 2t^2$

37. $r^4 + r^2$

38. $-2a + 3b$

39. $-3m - 4n + 1$

40. $a^3b^3z^3 - a^2b^3z^2$

41. $24x^2y^3z^4 + 8xy^2z^3$

42. $12uvw^3 - 18uv^2w^2$

43. $3x + 3y - 6z$

44. $-3xy + 2z + 5w$

45. $-3ab - 5ac + 9bc$

46. $ab + ac - ad$

47. $4y^2 + 8y - 2xy$

48. $-3x^2y - 6xy^2$

49. $-4a^2b^3 + 12a^3b^2$

50. $-4a^2b^2c^2 + 14a^2b^2c - 10ab^2c^2$

51. $-14a^6b^6 + 49a^2b^3 - 21ab$

52. $abx - ab^2x + abx^2$

53. $4x^2y^2z^2 - 6xy^2z^2 - 12xyz^2$

In Exercises 54–61, factor each polynomial.

54. $3(r - 2s) - x(r - 2s)$

55. $x(a + 2b) + y(a + 2b)$

56. $(x - 3)^2 + (x - 3)$

57. $(3t + 5)^2 - (3t + 5)$

58. $3x^2(r + 3s) - 6y^2(r + 3s)$

59. $9a^2b^2(3x - 2y) - 6ab(3x - 2y)$

60. $(3x - y)(x^2 - 2) + (x^2 - 2)$

61. $(x - 5y)(a + 2) - (x - 5y)$

In Exercises 62–71, factor each polynomial.

62. $xr + xs + yr + ys$

63. $pm - pn + qm - qn$

64. $2ab + 2ac + 3b + 3c$

65. $3ac + a + 3bc + b$

66. $2x^2 + 2xy - 3x - 3y$

67. $3ab + 9a - 2b - 6$

68. $mp - np - m + n$

69. $6x^2u - 3x^2v + 2yu - yv$

70. $x(a - b) + y(b - a)$

71. $p(m - n) - q(n - m)$

In Exercises 72–77, factor each polynomial completely. You may have to rearrange some terms.

72. $2r - bs - 2s + br$

73. $5x + ry + rx + 5y$

74. $ax + by + bx + ay$

75. $mr + ns + ms + nr$

76. $ac + bd - ad - bc$

77. $sx - ry + rx - sy$

78. When we add $5x$ and $7x$, we combine like terms to simplify the expression: $5x + 7x = 12x$. Explain how this is related to factoring.

79. A student factors the polynomial $3x^2y + 6xy^2$ as follows: $3x^2y + 6xy^2 = 3xy(x + 2)$. To check his work, he used his graphing calculator, storing 8 for x and 6 for y. His screen display is

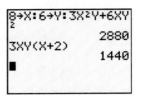

Is his factorization correct? Why or why not?

Another student factors the same polynomial as follows: $3x^2y + 6xy^2 = 3y(x^2 + 2xy)$. To check his work, he used his graphing calculator, storing 5 for x and 6 for y. His screen display is

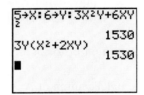

Is his factorization correct? Why or why not?

REVIEW

80. Use the distributive property to multiply the two polynomials $(3x + 4)$ and $(5x - 2)$. Show each step of the process.

81. Simplify the expression $3x + 4y - 2(x - 2y)$.

82. a. Evaluate the expression $4x^2 + 5x^2y$ both analytically and by using the STOre feature of the graphing calculator when $x = -2$ and $y = 3$.

  b. Evaluate the expression $x^2(4 + 5y)$ both analytically and by using the STOre feature of the graphing calculator using the same values given for x and y in part (a).

  c. Should the results of the evaluations in part (a) and part (b) be the same? Why or why not?

83. a. Solve the equation $3(x + 2) - 4x = \frac{1}{2}(x - 3) + 2$ analytically.

  b. Solve the same equation graphically.

  c. Is the solution a rational or an irrational number? Justify your conclusion.

5.2 FACTORING TRINOMIALS

- METHOD I: FACTORING TRINOMIALS BY GROUPING
- METHOD II: FACTORING TRINOMIALS BY TRIAL AND ERROR
- PERFECT-SQUARE TRINOMIALS
- PRIME POLYNOMIALS

▎ METHOD I: FACTORING TRINOMIALS BY GROUPING

Recall that trinomials are polynomials that contain three terms. The method of factoring by grouping can be used to factor trinomials of the form $ax^2 + bx + c$, where $a > 0$. The following steps will be helpful:

Factoring a Trinomial by Grouping

1. Write the trinomial in descending powers of one variable.
2. Factor out any GCF (including -1 if that is necessary to make the coefficient of the first term positive).
3. Identify a, b, and c, where a is the coefficient of the first term ($a > 0$), b is the coefficient of the middle term, and c is the coefficient of the third term (or the constant term).
4. Find the product $a \cdot c$.
5. Find two numbers whose **product** is ac and whose **sum** is b.
6. Replace the middle term of the original trinomial with two terms, using the numbers found in step 5 as coefficients. You now have a four-term polynomial.
7. Factor by grouping.
8. Verify the results by multiplication.

EXAMPLE 1

Factor: $4x^2 - 4x - 3$.

SOLUTION The trinomial is in descending powers of the variable, x.

Identify a, b, and c: $a = 4$, $b = -4$, $c = -3$.
The product $a \cdot c = 4 \cdot (-3) = -12$.
The factors of -12 are $-1(12)$, $-2(6)$, $-3(4)$, and $1(-12)$, $2(-6)$, $3(-4)$.

Because the sum of the factors must equal -4 (the value of b), the only pair of factors that will work are 2 and -6: $2 + -6 = -4$.

Replace the middle term with two terms using 2 and -6 as coefficients:

$$4x^2 - 4x - 3 = 4x^2 + 2x - 6x - 3$$

Factor by grouping:

$$4x^2 + 2x - 6x - 3 = (4x^2 + 2x) + (-6x - 3)$$

Use the associative property and the definition of subtraction to group terms.

$$= 2x(2x + 1) - 3(2x + 1)$$

Use the distributive property to factor the GCF out of each grouped pair.

$$= (2x + 1)(2x - 3)$$

Use the distributive property to factor out the GCF of $(2x + 1)$.

Verify the results by multiplication:

$$(2x + 1)(2x - 3) = 2x(2x) + 2x(-3) + 1(2x) + 1(-3)$$ Use the distributive property.

$$= 4x^2 - 6x + 2x - 3$$

$$= 4x^2 - 4x - 3$$ Combine like terms.

EXAMPLE 2

Factor: $x^2 - 5x - 24$.

SOLUTION   The trinomial is in descending powers of the variable x.

Identify a, b, and c: $a = 1$, $b = -5$, $c = -24$.
The product $a \cdot c = 1(-24) = -24$.
The factors of -24 are $-1(24)$, $-2(12)$, $-3(8)$, $-4(6)$, and $4(-6)$, $3(-8)$, $2(-12)$, $1(-24)$.

The only pair of factors that will have a sum of -5 are $+3$ and -8 **and** $3 + (-8) = -5$.

Replace the middle term using 3 and -8 as the coefficients.

$$x^2 + 3x - 8x - 24 = (x^2 + 3x) + (-8x - 24)$$ Definition of subtraction. And use of the associative property.

$$= x(x + 3) - 8(x + 3)$$ Use of the distributive property to factor the GCF out of each grouped pair.

$$= (x + 3)(x - 8)$$

Verify the result by multiplication.

$$(x + 3)(x - 8) = x \cdot x + x(-8) + 3 \cdot x + 3(-8)$$ Use of the distributive property.

$$= x^2 - 8x + 3x - 24$$ Combine like terms.

$$= x^2 - 5x - 24$$

EXAMPLE 3

Factor: $x^2 - 7x + 12$.

SOLUTION   The trinomial is in descending powers of the variable x. In addition, the leading coefficient on the x^2 is 1 so we can apply the observation made in the Study Tip from Example 2.

We need factors of $+12$ that could replace the middle term of $-7x$ if we desired to proceed to factor by grouping:

$$(-3)(-4) = +12 \quad \text{and} \quad (-3) + (-4) = -7$$

Using the observation from the Study Tip, we see that -3 and -4 will be the terms in the factors of the trinomial:

$$x^2 - 7x + 12 = (x - 3)(x - 4)$$

Verify the result by multiplication.

$$(x - 3)(x - 4) = x \cdot x + x(-4) - 3(x) - 3(-4)$$ Use of the distributive property.

$$= x^2 - 4x - 3x + 12$$ Combine like terms.

$$= x^2 - 7x + 12$$

Remember, this technique is valid **only** for trinomials with a lead coefficient of 1.

It is often helpful to chart your information when trying to find two numbers whose product is ac and sum is b. This is illustrated in the next example.

EXAMPLE 4

Factor: $6x^3 + 3x - 11x^2$.

SOLUTION $6x^3 + 3x - 11x^2 = 6x^3 - 11x^2 + 3x$ Write the polynomial in descending powers of a variable.

$$= x(6x^2 - 11x + 3)$$ Factor out the GCF of x.

To factor the trinomial, identify a, b, and c: $a = 6$, $b = -11$, $c = 3$.
Find the product, $a \cdot c$: $a \cdot c = 6(3) = 18$.

Because the product of the two numbers is positive and the sum of the same two numbers must equal b, which is negative, the desired factors must both be negative.

Two numbers	Product is 18	Sum is -11
$-1, -18$	$(-1)(-18) = 18$	$(-1) + (-18) = -19$
$-2, -9$	$(-2)(-9) = 18$	$(-2) + (-9) = -11$✓

The required numbers are -2 and -9.
Replace the middle term of the trinomial with two terms, using -2 and -9 as coefficients. Do not forget the GCF of x when working with the trinomial. Factor by grouping.

$$x(6x^2 - 11x + 3) =$$
$$x[6x^2 - 2x - 9x + 3] = x[2x(3x - 1) - 3(3x - 1)]$$
$$= x(3x - 1)(2x - 3)$$

Verify the result by multiplication.

EXAMPLE 5

Factor: $2x^2y - 8x^3 + 3xy^2$.

SOLUTION $2x^2y - 8x^3 + 3xy^2 = -8x^3 + 2x^2y + 3xy^2$ Write the polynomial in descending powers of a variable.

$$= -x(8x^2 - 2xy - 3y^2)$$ Factor out the GCF of $-x$.

To factor the trinomial, identify a, b, and c: $a = 8$, $b = -2$, $c = -3$.
Find the product $a \cdot c$: $a \cdot c = 8(-3) = -24$.

Because the product of the two numbers is negative and the sum of the same two numbers must equal b, which is negative, the desired factors must have opposite signs. Moreover, the number with the larger absolute value will be negative.

Two numbers	Product is -24	Sum is -2
$1, -24$	$(1)(-24) = -24$	$1 + (-24) = -23$
$2, -12$	$(2)(-12) = -24$	$2 + (-12) = -10$
$3, -8$	$(3)(-8) = -24$	$3 + (-8) = -5$
$4, -6$	$(4)(-6) = -24$	$4 + (-6) = -2$ ✓

The required numbers are 4 and -6.

Replace the middle term of the trinomial with two terms, using 4 and -6 as coefficients. Do not forget the GCF of $-x$ when working with the trinomial.

Factor by grouping:

$$
\begin{aligned}
-8x^3 + 2x^2y + 3xy^2 &= -x(8x^2 - 2xy - 3y^2) \\
&= -x(8x^2 + 4xy - 6xy - 3y^2) \\
&= -x[4x(2x + y) - 3y(2x + y)] \\
&= -x(2x + y)(4x - 3y)
\end{aligned}
$$

Verify the result by multiplication.

▮ METHOD II: FACTORING TRINOMIALS BY TRIAL AND ERROR

The previous examples illustrate the fact that the product of two binomials is often a trinomial. Many polynomials, therefore, can be factored by simply testing reasonable products of binomial factors (trial and error). When the leading coefficient is 1, this is often not difficult. Consider factoring the trinomial

$$x^2 + 5x + 6$$

To find its binomial factors, note that the product of the first two terms must be x^2. Thus, the first term of each binomial must be x. Recalling the mnemonic FOIL (first, outer, inner, last) used for multiplying two binomials may be helpful when factoring by trial and error.

$$
\overset{\overparen{x^2}}{(x \quad)(x \quad)}
$$

Because the product of their last terms must be 6, and the sum of the products of the outer and inner terms must be $5x$, we must find two numbers whose product is 6 and whose sum is 5.

$$
\overset{\overparen{6}}{(x + ?)(x + ?)}
$$
$$
\underset{O + I = 5x}{}
$$

Two such numbers are $+3$ and $+2$. Thus, we have

$$x^2 + 5x + 6 = (x + 3)(x + 2)$$

We can verify this factorization by multiplying $x + 3$ and $x + 2$ and observing that the product is $x^2 + 5x + 6$.

$$(x + 3)(x + 2) = x^2 + 2x + 3x + 6$$
$$= x^2 + 5x + 6 \qquad \text{Combine like terms.}$$

Because of the commutative property of multiplication, the order of the factors is not important. Therefore,

$$x^2 + 5x + 6 = (x + 2)(x + 3)$$

We must consider more combinations when we factor trinomials with lead coefficients other than 1. This is illustrated in the next example.

EXAMPLE 6

Factor: $3y^2 - 4y - 4$.

SOLUTION Because the first term of this trinomial is $3y^2$, the first terms of the binomial factors must be $3y$ and y.

$$\overset{3y^2}{(3y \quad)(y \quad)}$$

The product of the last terms must be -4, and the sum of the products of the outer terms and inner terms must be $-4y$.

$$\overset{-4}{(3y \quad ?)(y \quad ?)}$$
$$O + I = -4y$$

Because $(1)(-4)$, $(-1)(4)$, and $(-2)(2)$ all give a product of -4, there are six possible combinations to consider:

$$(3y + 1)(y - 4) \qquad (3y - 4)(y + 1)$$
$$(3y - 1)(y + 4) \qquad (3y + 4)(y - 1)$$
$$(3y - 2)(y + 2) \qquad (3y + 2)(y - 2)$$

Again, only the last possibility gives the required middle term of $-4y$. Thus,

$$3y^2 - 4y - 4 = (3y + 2)(y - 2)$$

We can check by multiplication.

$$(3y + 2)(y - 2) = 3y^2 - 6y + 2y - 4$$
$$= 3y^2 - 4y - 4$$

EXAMPLE 7   Factor: $-8x^3 + 22x^2 - 12x$.

SOLUTION   We factor out the common monomial factor of $-2x$.

$$-8x^3 + 22x^2 - 12x = -2x(4x^2 - 11x + 6)$$

Then we find the binomial factors of $4x^2 - 11x + 6$.

$$-8x^3 + 22x^2 - 12x = -2x(x - 2)(4x - 3)$$

We can check by multiplication.

$$-2x(x - 2)(4x - 3) = -2x(4x^2 - 3x - 8x + 6)$$
$$= -2x(4x^2 - 11x + 6)$$
$$= -8x^3 + 22x^2 - 12x$$

It is not easy to give specific rules for factoring trinomials by trial and error because guesswork is often necessary. However, the following hints are often helpful:

Factoring a Trinomial by Trial and Error

1. Write the trinomial in descending powers of one variable.
2. Factor out any greatest common factor (including -1 if that is necessary to make the coefficient of the first term positive).
3. If the sign of the third term is positive, the signs between the terms of the binomial factors are the same as the sign of the middle term of the trinomial. If the sign of the third term is negative, the signs between the terms of the binomial factors are opposite.
4. Mentally try various combinations of first terms and last terms until you find one that works, or until you exhaust all the possibilities. In that case, the trinomial does not factor using only integer coefficients.
5. Check the factorization by multiplication.

PERFECT-SQUARE TRINOMIALS

Remember, when we factor a polynomial we are writing it as a product of factors; factoring is the reverse of multiplication. For example, we can multiply $(3x - 2)(x + 1)$ to get the result $3x^2 + x - 2$. Reversing the process means that if we factor $3x^2 + x - 2$ we get the result $(3x - 2)(x + 1)$.

Notice the pattern when a binomial is squared:

$$(2x + 3)^2 = (2x + 3)(2x + 3) = 4x^2 + 12x + 16$$
$$(4x + 5)^2 = (4x + 5)(4x + 5) = 16x^2 + 40x + 25$$
$$(3x - 2)^2 = (3x - 2)(3x - 2) = 9x^2 - 12x + 4$$
$$(5x - 3)^2 = (5x - 3)(5x - 3) = 25x^2 - 30x + 4$$

In each trinomial result, the first term is the *square* of the first term of the binomial. The middle term of the trinomial is *twice* the product of the two terms in the binomial. The last term of the trinomial is the *square* of the second term of the binomial.

Let's follow the pattern to square the binomial $(5x - 2)^2$.

First term: $(5x)^2 = 25x^2$
Middle term: $2(5x)(-2) = -20x$
Last term: $(-2)^2 = 4$

Thus the trinomial is $25x^2 - 20x + 4$.

We can observe that both the first and the last terms are positive and that the middle term retains the sign of the original binomial. Therefore, to factor a trinomial that is a perfect-square trinomial, we reverse this process.

Consider factoring the trinomial $4x^2 + 12x + 9$. Observe that the trinomial could also be expressed as $(2x)^2 + 2(2x)(3) + (3)^2$. The middle term, $12x$, is twice the product of the bases of the first and last terms.

$$12x = 2(2x)(3)$$

When this pattern can be observed, our trinomial is a perfect-square trinomial and its factored form is

$$(2x + 3)^2.$$

This leads to the special product formulas specified below.

Special Product Formulas

$$x^2 + 2xy + y^2 = (x + y)^2$$
$$x^2 - 2xy + y^2 = (x - y)^2$$

EXAMPLE 8

Factor $36x^2 + 84xy + 49y^2$.

SOLUTION   We notice that the trinomial can be expressed as

$$(6x)^2 + 2(6x)(7y) + (7y)^2$$

Therefore, it is a perfect-square trinomial and its factored form is

$$(6x + 7y)^2$$

Please note that all factorable trinomials can be factored either by the grouping method or by the trial-and-error method. However, identifying a perfect-square trinomial will make the factoring process quicker.

▐ PRIME POLYNOMIALS

Not all polynomials are factorable — some are prime. The trinomial $x^2 + 2x + 3$ is an example of a polynomial that is prime. It cannot be factored using only integer coefficients by either method presented in this section.

EXAMPLE 9

Factor: $x^2 + 2x + 3$.

SOLUTION   **By grouping:**
Identify a, b, and c: $a = 1$, $b = 2$, $c = 3$.
Find two numbers whose product is $a \cdot c = (1)(3) = 3$ **and** whose sum is 2. The only factors of 3 are 1 and 3, and the sum of these two numbers is 4 (which does not equal the value of b).
Therefore, the polynomial is prime.

By trial and error:

Because the last term of the trinomial is 3 and the middle term is $2x$, we must find two factors of 3 whose sum is 2 so that

Because 3 factors only as $(1)(3)$ and $(-1)(-3)$, it has no factors whose sum is 2. Thus, $x^2 + 2x + 3$ cannot be factored over the integers. It is a prime polynomial.

When factoring trinomials by grouping, the chart becomes a valuable tool for identifying prime polynomials. This is illustrated in the following example.

EXAMPLE 10

Factor: $6x^2 + 3x + 2$.

SOLUTION   Identify a, b, and c: $a = 6$, $b = 3$, $c = 2$.
The product, $a \cdot c$, is 12, therefore, we must find two numbers whose product is 12 and whose sum is 3.

Two numbers	Product is 12	Sum is 3
1, 12	$1 \cdot 12 = 12$	$1 + 12 = 13$
2, 6	$2 \cdot 6 = 12$	$2 + 6 = 8$
3, 4	$3 \cdot 4 = 12$	$3 + 4 = 7$
4, 3	$4 \cdot 3 = 12$	$4 + 3 = 7$
6, 2	$6 \cdot 2 = 12$	$6 + 2 = 8$
12, 1	$12 \cdot 1 = 12$	$12 + 1 = 13$

There are no combinations of factors of 12 whose sum is 3. Observe the pattern in the sum column: 13, 8, **7, 7**, 8, 13. A sum of 3 would have had to occur between the two 7s. This is impossible when using integer factors of 12. The rows in bold are actually a repeat of the first three rows. Because multiplication is commutative, they were not required.

EXERCISE 5.2

VOCABULARY AND NOTATION   *Fill in each blank to make a true statement.*

1. _____ are polynomials that contain three terms.

2. When factoring three-term polynomials, begin by writing the terms in _____ order of one of the variables.

3. A correct factorization can always be justified by _____.

4. Polynomials that are not factorable are called _____ polynomials.

5. A factorable trinomial can be expressed as the product of two (or more) _____.

CONCEPTS

6. To factor $10x^2 - 27 - 3x$, we must first rewrite the trinomial as _____.

7. Given the trinomial $x^2 + 5x + 6$,
   a. What is the leading coefficient?
   b. What is the coefficient of the second term?
   c. What is the coefficient of the first term?
   d. What is the constant term?

8. Complete each factorization:
   a. $2x^2 + x - 15 = (2x - 5)(x + \underline{?})$.
   b. $10x^2 + 41x - 18 = (5x - \underline{?})(2x + 9)$
   c. $3x^2 + x - 10 = (\underline{?} - 5)(x + \underline{?})$

In Exercises 9–14, decide which of the three lettered choices is the correct factored form of the given polynomial.

9. $x^2 - 2x - 15$
   (a) $(x + 3)(x - 5)$ **(b)** $(x - 3)(x + 5)$
   (c) $(x + 15)(x - 1)$

10. $x^2 + 4x - 12$
   (a) $(x + 4)(x - 3)$ **(b)** $(x - 6)(x + 2)$
   (c) $(x + 6)(x - 2)$

11. $6x^2 - x - 15$
   (a) $(2x - 3)(3x + 5)$ **(b)** $(3x + 3)(2x - 5)$
   (c) $(2x + 3)(3x - 5)$

12. $10x^2 - 19x + 6$
   (a) $(2x - 6)(5x - 1)$ **(b)** $(5x - 2)(2x - 3)$
   (c) $(2x + 3)(5x + 2)$

13. $2x^2 + 2x - 4$
   (a) $(2x + 4)(x - 1)$ **(b)** $2(x + 2)(x - 1)$
   (c) $(2x - 4)(x + 1)$

14. $3x^2 - 6x - 24$
   (a) $3(x + 2)(x - 4)$ **(b)** $(3x + 6)(x + 4)$
   (c) $(3x + 6)(x + 1)$

15. When factoring $x^2 + 5x + 6$, one student wrote $(x + 3)(x + 2)$, while another wrote $(x + 2)(x + 3)$. Are they both correct? Why or why not?

16. When factoring $x^2 + 10x + 9$, one student wrote $(x + 9)(x + 1)$, while another student wrote $(x + 1)(x + 9)$. Are they both correct? Why or why not?

PRACTICE: *In Exercises 17–57, factor each polynomial completely. If it is prime, so state. Do not forget to factor out a -1 if the leading coefficient is negative and to **always** look for and factor out a GCF first.*

17. $x^2 - 9x - 12$

18. $y^2 + 11y - 26$

19. $4 - 5x + x^2$

20. $y^2 + 5 + 6y$

21. $x^2 + 3x + 2$

22. $y^2 + 4y + 3$

23. $3x^2 + x - 4$

24. $5x^2 - 14x - 24$

25. $6x^2 + 13x - 5$

26. $6x^2 + 13x - 28$

27. $6x^2 + 11x - 10$

28. $20x^2 + 21x - 5$

29. $6x^2 - 22x - 8$

30. $8x^2 - 23x - 3$

31. $6x^2 + 11x - 7$

32. $20x^2 + 9x - 18$

33. $15x^2 - 17x + 4$

34. $12x^2 - 25x + 12$

35. $8x^2 - 26x + 15$

36. $5x^2 + x + 4$

37. $x^2 - 2x - 1$

38. $2x^2 + 10x + 12$

39. $3y^2 - 21y + 18$

40. $-5t^2 + 25t - 30$

41. $-2x^2 + 20x - 18$

42. $x^2 + 9x + 5$

43. $y^2 - 8y + 6$

44. $u^2 + 2uv - 15v^2$

45. $m^2 + 3mn - 10n^2$

46. $-3rs + r^2 + 2s^2$

47. $-13yz + y^2 - 14z^2$

48. $-4x^2y - 4x^3 + 24xy^2$

49. $8x^2 - 12xy - 8y^2$

50. $24y^2 + 14xy + 2x^2$

51. $9y^3 + 3y^2 - 6y$

52. $x^2 + 9x + 20$

53. $y^2 - 8y + 15$

54. $6x^2 - 7x - 1$

55. $2y^2 + 5y - 2$

56. $4r^2 - 4rs - 8s^2$

57. $6x^2 + 3xy - 18y^2$

In Exercises 58–63, factor each of the polynomials by grouping.

58. $6x^2 - 15x + 4x - 10$

59. $35x^2 - 63x + 10x - 18$

60. $2y^2 + 16y - 5y - 40$

61. $4x^2 + 12x - x - 3$

62. $42x^2 + 48x - 7x - 8$

63. $5y^2 - 5y - 2y + 2$

*In Exercises 64–73, each of the trinomials is a **perfect-square trinomial**. Factor each, using the method presented in this section.*

64. $x^2 + 6x + 9$

65. $x^2 + 10x + 25$

66. $y^2 - 8y + 16$

67. $z^2 - 2z + 1$

68. $t^2 + 20t + 100$

69. $r^2 + 24r + 144$

70. $u^2 - 18u + 81$

71. $v^2 - 14v + 49$

72. $x^2 + 4xy + 4y^2$

73. $x^2 + 6xy + 9y^2$

APPLICATIONS

74. Volume The volume of a rectangular solid is represented by the polynomial $h^3 + 7h^2 + 12h$, where h represents its height. Recall that the volume of a rectangular solid is the product of its length, width, and height ($V = lwh$).

    a. How much longer are the length and width than the height? (assume the length is the longest dimension)

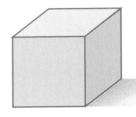

    b. State the length and the width if the height is 6 inches.

75. Volume The volume of a rectangular solid is represented by the polynomial $h^3 + 7h^2 + 10h$, whre h represents its height. Recall that the volume of a rectangular solid is the product of its length, width, and height ($V = lwh$).

    a. How much longer are the length and width than the height? (assume the length is the longest dimension)

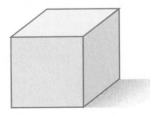

    b. State the length and the width if the height is 4 inches.

76. For what values of b will the polynomial $6x^2 + bx + 6$ be factorable?

77. For what values of b will the polynomial $5x^2 + bx + 4$ be factorable?

REVIEW

78. Simplify: $(x + 3)^2$.

79. a. Simplify the expression $(5x^2y^3)^2$.

  b. Evaluate the original expression in part (a) if $x = -2$ and $y = -4$.

  c. Evaluate your simplified expression from part (a) when $x = -2$ and $y = -4$.

  d. Are the evaluations equal? Why or why not?

80. Rewrite $\dfrac{a^{-4}b^2}{c^{-5}}$ without negative exponents.

81. If $P(x) = 2x^2 + 3x - 4$, find each of the following:

  a. $P(-5)$

  b. $P(6)$

  c. $P(-2.3)$ (use the STOre feature of your calculator)

82. Solve the equation

  $$x^2 + 5x - 3 = x(x - 4) + 3(2x - 6)$$

  a. algebraically.

  b. graphically (specify a window that shows a complete graph).

5.3 FACTORING BY SPECIAL PRODUCTS

- FACTORING THE DIFFERENCE OF TWO SQUARES
- FACTORING THE SUM AND DIFFERENCE OF TWO CUBES

This section will focus on the factoring of binomials that fit a pattern. Much of the section will be devoted to the discovery of those patterns.

FACTORING THE DIFFERENCE OF TWO SQUARES

EXPLORING THE CONCEPT

Using the distributive property, multiply each of the following binomials together.

  a. $(x + 2)(x - 2) = ?$ **b.** $(x + 2)(x + 2) = ?$

  c. $(ab - 1)(ab + 1) = ?$ **d.** $(ab - 1)(ab + 2) = ?$

  e. $(2x + 3)(2x - 3) = ?$ **f.** $(2x + 3)(3x - 2) = ?$

CONCLUSION   ***a.** $x^2 - 4$ **b.** $x^2 + 4x + 4$

  *c. $a^2b^2 - 1$ **d.** $a^2b^2 + ab - 2$

  *e. $4x^2 - 9$ **f.** $6x^2 + 5x - 6$

  *Whenever we multiply a binomial of the form $x + y$ by a binomial of the form $x - y$, we obtain another binomial:

  $$(x + y)(x - y) = x^2 - y^2$$

  The binomials $(x + y)$ and $(x - y)$ are called **conjugate binomials**.

The binomial $x^2 - y^2$ is called the **difference of two squares**, because x^2 is the square of x and y^2 is the square of y. The difference of the squares of two quantities such as x and y always factors into the *sum* of those two quantities multiplied by the *difference* of those two quantities. The product of conjugate binomials is *always* the difference of squares.

Factoring the Difference of Two Squares

$$x^2 - y^2 = (x + y)(x - y)$$

To factor $x^2 - 9$, for example, we note that $x^2 - 9$ can be written in the form $(x)^2 - (3)^2$ and that this is the *difference* between the square of x and the square of 3. Thus, it factors into the product of the sum of the bases (x plus 3) and the difference between the bases (x minus 3).

$$x^2 - 9 = (x)^2 - (3)^2$$
$$= (x + 3)(x - 3)$$

As in all of our factoring problems, we can check the result by verifying that $(x + 3)(x - 3) = x^2 - 9$.

A perfect square is any quantity that can be written as an exact square of a rational quantity. Thus, the square root key, $\sqrt{}$, can be used to determine if a constant is a perfect square. Expressions containing variables such as x^4y^2 are also perfect squares because they can be written as the square of a quantity:

$$x^4y^2 = (x^2y)^2$$

EXAMPLE 1

Factor: $4y^4 - 25z^2$.

SOLUTION   Because the binomial $4y^4 - 25z^2$ can be written in the form $(2y^2)^2 - (5z)^2$, it represents the difference of the squares of $2y^2$ and $5z$. Thus, it factors into the sum of these two quantities times the difference of these two quantities.

$$4y^4 - 25z^2 = (2y^2)^2 - (5z)^2$$
$$= (2y^2 + 5z)(2y^2 - 5z)$$

Check this result by multiplication.

We can often factor out a greatest common factor before factoring the difference of two squares. For example, to factor $8x^2 - 32$, we begin by factoring out the greatest common factor of 8 and then factor the resulting difference of two squares.

$$8x^2 - 32 = 8(x^2 - 4)$$ Factor out 8.
$$= 8(x^2 - 2^2)$$ Write 4 as 2^2.
$$= 8(x + 2)(x - 2)$$ Factor the difference of two squares.

We can verify this result by multiplication:

$$8(x + 2)(x - 2) = 8(x^2 - 4)$$
$$= 8x^2 - 32$$

EXAMPLE 2

Factor: $2a^2x^3y - 8b^2xy$.

SOLUTION

$$2a^2x^3y - 8b^2xy = 2xy(a^2x^2 - 4b^2)$$ Factor out $2xy$.

$$= 2xy[(\mathbf{ax})^2 - (\mathbf{2b})^2]$$

$$= 2xy(\mathbf{ax} + \mathbf{2b})(\mathbf{ax} - \mathbf{2b})$$ Factor the difference of two squares.

Check this result by multiplication.

Sometimes we must factor a difference of two squares more than once to completely factor a polynomial. For example, the binomial $625a^4 - 81b^4$ can be written in the form $(25a^2)^2 - (9b^2)^2$. The difference of the squares of $25a^2$ and $9b^2$ factors as

$$625a^4 - 81b^4 = (\mathbf{25a^2})^2 - (\mathbf{9b^2})^2$$
$$= (\mathbf{25a^2} + \mathbf{9b^2})(\mathbf{25a^2} - \mathbf{9b^2}).$$

The factor $25a^2 - 9b^2$, however, can be written in the form $(\mathbf{5a})^2 - (\mathbf{3b})^2$ and can be factored as $(\mathbf{5a} + \mathbf{3b})(\mathbf{5a} - \mathbf{3b})$. Thus, the complete factorization of $625a^4 - 81b^4$ is

$$625a^4 - 81b^4 = (25a^2 + 9b^2)(5a + 3b)(5a - 3b)$$

WARNING!   The binomial $x^2 + 9$ is called the **sum of two squares** because it can be written in the form $(x)^2 + (3)^2$. Such binomials cannot be factored if we are limited to integer coefficients. Polynomials that do not factor over the integers are called **irreducible** or **prime polynomials**.

EXAMPLE 3

Factor: $2x^4y - 32y$.

SOLUTION

$$2x^4y - 32y = 2y \cdot x^4 - 2y \cdot 16$$

$$= 2y(x^4 - 16)$$ Factor out $2y$.

$$= 2y(x^2 + 4)(x^2 - 4)$$ Factor $x^4 - 16$.

$$= 2y(x^2 + 4)(x + 2)(x - 2)$$ Factor $x^2 - 4$. Note that $x^2 + 4$ does not factor.

Example 4 requires the techniques of factoring out a common factor, factoring by grouping, and factoring the difference of two squares.

EXAMPLE 4   Factor: $2x^3 - 8x + 2yx^2 - 8y$.

SOLUTION $2x^3 - 8x + 2yx^2 - 8y = 2(x^3 - 4x + yx^2 - 4y)$ Factor out 2.

$$= 2[x(x^2 - 4) + y(x^2 - 4)]$$ Factor out x from $x^3 - 4x$ and factor out y from $yx^2 - 4y$.

$$= 2(x^2 - 4)(x + y)$$ Factor out $x^2 - 4$.

$$= 2(x + 2)(x - 2)(x + y)$$ Factor $x^2 - 4$.

Check by multiplication.

WARNING!   To *factor* an expression means to factor the expression *completely*. Each polynomial factor must be prime.

FACTORING THE SUM AND DIFFERENCE OF TWO CUBES

There are formulas similar to the difference-of-two-squares formula for factoring the sum of the cubes of two quantities and the difference of the cubes of two quantities. To discover these formulas, we need to find the following two products:

$$(a - b)(a^2 + ab + b^2) \quad \text{and} \quad (a + b)(a^2 - ab + b^2)$$

Because we are multiplying two polynomials, we use the distributive property to multiply each term of one polynomial by each term of the other.

$$(a - b)(a^2 + ab + b^2) = a(a^2) + a(ab) + a(b^2) + (-b)(a^2) + (-b)(ab) + (-b)(b^2)$$
$$= a^3 + a^2b + ab^2 - a^2b - ab^2 - b^3$$
$$= a^3 - b^3$$

and

$$(a + b)(a^2 - ab + b^2) = a(a^2) + a(-ab) + a(b^2) + b(a^2) + b(-ab) + b(b^2)$$
$$= a^3 - a^2b + ab^2 + a^2b - ab^2 + b^3$$
$$= a^3 + b^3$$

Notice the pattern of signs in the polynomial factors.

The first binomial factor has the same sign as the original binomial:

$$a^3 - b^3 = (a - b)(a^2 + ab + b^2)$$
$$\uparrow \qquad\qquad \uparrow$$

$$a^3 + b^3 = (a + b)(a^2 - ab + b^2)$$
$$\uparrow \qquad\qquad \uparrow$$

The second factor is always a trinomial and the middle sign is the opposite of the sign of the original binomial:

$$a^3 - b^3 = (a - b)(a^2 + ab + b^2)$$
$$\uparrow \qquad\qquad\qquad \uparrow$$

TECHNOLOGY TIP

Determining whether or not a number is a perfect cube can be done by accessing the cube root function under the MATH menu on the TI-83 Plus graphing calculator. TI-86 users may customize this feature. See Technology 1.1.

$$a^3 + b^3 = (a + b)(a^2 - ab + b^2)$$

Finally, notice the pattern of bases in the polynomial factors:

$$a^3 - b^3 = (a \quad - \quad b)(\quad a^2 \quad + \quad ab \quad + \quad b^2)$$

base of 1st term	base of 2nd term	square of first base	product of bases	square of second base

$$a^3 + b^3 = (a \quad + \quad b)(\quad a^2 \quad - \quad ab \quad + \quad b^2)$$

Factoring the Difference of Two Cubes

$$a^3 - b^3 = (a - b)(a^2 + ab + b^2)$$

Factoring the Sum of Two Cubes

$$a^3 + b^3 = (a + b)(a^2 - ab + b^2)$$

EXAMPLE 5

Factor: $x^3 + 8$.

SOLUTION Since $x^3 + 8$ can be written as $x^3 + 2^3$, we have the sum of two cubes, which factors as follows:

$$F^3 + L^3 = (F + L)(F^2 - FL + L^2) \qquad \text{F for first term, L for last term.}$$
$$x^3 + 2^3 = (x + 2)(x^2 - x2 + 2^2) \qquad \text{Substitute } x \text{ for F and 2 for L.}$$
$$= (x + 2)(x^2 - 2x + 4)$$

Thus, $x^3 + 8 = (x + 2)(x^2 - 2x + 4)$.
Verify this result by multiplication.

EXAMPLE 6

Factor: $27a^3 - 64b^3$.

SOLUTION Since $27a^3 - 64b^3$ can be written as $(3a)^3 - (4b)^3$, we have the difference of two cubes, which factors as follows:

$$F^3 - L^3 = (F - L)(F^2 + FL + L^2)$$
$$(3a)^3 - (4b)^3 = (3a - 4b)[(3a)^2 + (3a)(4b) + (4b)^2] \qquad \text{Substitute } 3a \text{ for F and } 4b \text{ for L.}$$
$$= (3a - 4b)(9a^2 + 12ab + 16b^2)$$

Thus, $27a^3 - 64b^3 = (3a - 4b)(9a^2 + 12ab + 16b^2)$.
Verify this result by multiplication.

EXAMPLE 7

Factor: $a^3 - (c + d)^3$.

SOLUTION

$$a^3 - (c + d)^3 = [a - (c + d)][a^2 + a(c + d) + (c + d)^2]$$
$$= (a - c - d)(a^2 + ac + ad + c^2 + 2cd + d^2)$$

Remove internal grouping symbols.

EXAMPLE 8

Factor: $x^6 - 64$.

SOLUTION

This expression can be factored as either the difference of two squares or the difference of two cubes.

$$(x^3)^2 - 8^2 \qquad \text{or} \qquad (x^2)^3 - 4^3$$

However, it is much easier to factor the difference of two squares first.

$$x^6 - 64 = (x^3)^2 - 8^2$$
$$= (x^3 + 8)(x^3 - 8)$$

Each of these factors further, however, for one is the sum of two cubes and the other is the difference of two cubes:

$$x^6 - 64 = (x^3 + 2^3)(x^3 - 2^3)$$
$$= (x + 2)(x^2 - 2x + 4)(x - 2)(x^2 + 2x + 4)$$

Try to factor $x^6 - 64$ as the difference of two cubes first and see what difficulty you encounter.

EXAMPLE 9

Factor: $2a^5 + 128a^2$.

SOLUTION

We first factor out the greatest common monomial factor of $2a^2$ to obtain

$$2a^5 + 128a^2 = 2a^2(a^3 + 64)$$

Then we factor $a^3 + 64$ as the sum of two cubes to obtain

$$2a^5 + 128a^2 = 2a^2(a^3 + 4^3)$$
$$= 2a^2(a + 4)(a^2 - 4a + 16)$$

Verify this result by multiplication.

EXERCISE 5.3

VOCABULARY AND NOTATION

1. A binomial of the form $x^2 - y^2$ is called the _____ of two squares.

2. A binomial of the form $x^2 + y^2$ is called the _____ of two squares.

3. The binomial $x^2 - y^2$ factors into the product of _____.

4. A binomial of the form $x^3 - y^3$ is called the _____ of two _____.

5. A trinomial of the form $x^3 + y^3$ is called the _____ of two _____.

CONCEPTS   *In Exercises 6–8, classify each of the statements as true or false. If a statement is false, justify your response.*

6. The binomial $x^2 - 2y^2$ is an example of the difference of two squares.

7. The binomial $x^2 - 25$ is equivalent to the product of the two binomials $(x - 5)(x + 5)$.

8. To factor $16x^2 - 25y^2$, first write the polynomial as $(4x)^2 - (5y)^2$.

PRACTICE   *In Exercises 9–32, factor each expression completely. It is suggested that you check your factoring by multiplying factors so as to gain proficiency in moving from factoring to multiplication. If a polynomial is prime, so state.*

9. $x^2 - 4$

10. $y^2 - 9$

11. $4y^2 - 25$

12. $9x^2 - 1$

13. $49x^2 - 81y^2$

14. $9x^2 - 16y^2$

15. $81a^4 - 16b^2$

16. $16x^4 - 81y^2$

17. $x^2 + 25$

18. $144a^2 - b^2$

19. $625a^2 - 169b^4$

20. $4y^2 + 9z^4$

21. $81a^4 - 49b^2$

22. $64r^6 - 121s^2$

23. $36x^4y^2 - 49z^4$

24. $100a^2b^4c^6 - 225d^8$

25. $(x + y)^2 - z^2$

26. $a^2 - (b - c)^2$

27. $(a - b)^2 - c^2$

28. $(m + n)^2 - p^2$

29. $a^4 - b^4$

30. $x^4 - y^4$

31. $y^4 - 256$

32. $x^4 - 81$

In Exercises 33–44, factor each expression.

33. $2x^2 - 2$

34. $4y^2 - 4$

35. $2x^2 - 288$

36. $8x^2 - 72$

37. $2x^3 - 32x$

38. $3x^3 - 243x$

39. $5x^3 - 125x$

40. $6x^4 - 216x^2$

41. $4x^2y - 9xy^2$

42. $4x^2y - 16y^2z$

43. $r^2s^2t^2 - t^2x^4y^2$

44. $16a^4b^3c^4 - 64a^2bc^6$

In Exercises 45–64, factor each expression.

45. $y^3 + 1$

46. $x^3 - 8$

47. $a^3 - 27$

48. $b^3 + 125$

49. $8 + x^3$

50. $27 - y^3$

51. $r^3 + s^3$

52. $t^3 - v^3$

53. $x^3 - 8y^3$

54. $a^3 + 8b^3$

55. $x^3 - 27y^3$

56. $27a^3 + b^3$

57. $27a^3 - b^3$

58. $64x^3 - 27$

59. $27x^3 + 125$

60. $27x^3 - 125y^3$

61. $64x^3 + 27y^3$

62. $64a^3 - 125b^6$

63. $x^6 + y^6$

64. $x^9 + y^9$

In Exercises 65–74, factor each expression.

65. $5x^3 + 625$

66. $2x^3 - 128$

67. $4x^5 - 256x^2$

68. $2x^6 + 54x^3$

69. $128u^2v^3 - 2t^3u^2$

70. $56rs^2t^3 + 7rs^2v^6$

71. $(a + b)x^3 + 27(a + b)$

72. $(c - d)r^3 - (c - d)s^3$

73. $6a^3b^3 - 6z^3$

74. $18x^3y^3 + 18c^3d^3$

5.4 SUMMARY OF FACTORING TECHNIQUES

To completely factor a polynomial means to write the given polynomial as a *product* of prime polynomial factors. Factoring can always be checked by multiplying factors and ensuring that the product of factors is equivalent to the original simplified polynomial. Remember to check that each polynomial factor is prime before stopping. Also, recall that there should be no *ungrouped* addition or subtraction signs in your final result. Following is a summary of techniques that will help you in your factoring.

Factoring Polynomials

1. Always check for and factor out the greatest common factor (GCF) first. If the leading coefficient is negative, factor out a -1 with/as the GCF. Refer to Section 5.1 if you need more practice in factoring out the GCF.
2. Count the number of terms in the polynomial and then follow the given procedures for that specific type of polynomial:

Binomial

If it is the difference of two squares: $a^2 - b^2 = (a + b)(a - b)$

If it is the difference of two cubes: $a^3 - b^3 = (a - b)(a^2 + ab + b^2)$

If it is the sum of two cubes: $a^3 + b^3 = (a + b)(a^2 - ab + b^2)$

If it is *not* one of these problem types, it is prime. Refer to Section 5.3 if you need more practice in factoring binomials.

Trinomial

Make sure the trinomial is in descending powers of a variable and then attempt to factor it by either grouping or trial and error.

The special product formulas below may be used where appropriate, but are not required:

$$x^2 + 2xy + y^2 = (x + y)^2$$
$$x^2 - 2xy + y^2 = (x - y)^2$$

Refer to Section 5.2 if you need more practice in factoring trinomials.

Four-Term Polynomials

Try to factor by grouping. Refer to Section 5.1 if you need more practice in factoring polynomials that contain 4 terms.

Always cycle back after each factorization and check to ensure that each factor is prime before stopping.

EXAMPLE 1

Factor: $3x^3 - 12xy^2$.

SOLUTION   We begin by factoring out the greatest common factor of $3x$.

$$3x^3 - 12xy^2 = 3x(x^2 - 4y^2)$$

Since the expression $x^2 - 4y^2$ is a binomial that is a difference of squares, it can be factored further.

$$3x^3 - 12xy^2 = 3x[(x)^2 - (2y)^2]$$
$$= 3x(x + 2y)(x - 2y)$$

Because each of these factors is prime, the process is complete. Verify your factorization by multiplication.

EXAMPLE 2

Factor: $6x^2 + 7xy - 3y^2$.

SOLUTION   There are no common factors and the trinomial is in descending order so we will factor by grouping.

Identify a, b, and c: $a = 6$, $b = 7$, $c = -3$.
Find the product $a \cdot c = 6(-3) = -18$ with factors whose sum is 7.

Two Numbers	Product Is -18	Sum Is 7
-1 and 18	$-1(18) = -18$	$-1 + 18 = 17$
-2 and 9	$-2(9) = -18$	$-2 + 9 = 7$

Replace the middle term of the trinomial using -2 and 9 as the coefficients.

$6x^2 - 2xy + 9xy - 3y^2$ Remember, addition is commutative. The polynomial $6x^2 + 9xy - 2xy - 3y^2$ will work also.

Factor by grouping.

$$6x^2 - 2xy + 9xy - 3y^2 = (6x^2 - 2xy) + (9xy - 3y^2)$$
$$= 2x(3x - y) + 3y(3x - y)$$
$$= (3x - y)(2x + 3y) \qquad \text{Factor out } (3x - y).$$

Because each of the binomial factors is prime (there is no GCF and neither fit any of the binomial patterns), the expression is completely factored. Verify your factorization by multiplication.

EXAMPLE 3

Factor $x^4y + 7x^3y - 18x^2y$.

SOLUTION   We begin by factoring out the greatest common factor of x^2y.

$$x^4y + 7x^3y - 18x^2y = x^2y(x^2 + 7x - 18)$$

We now examine the polynomial *factor* and attempt to factor it further. The trinomial factor, $x^2 + 7x - 18$, can be factored using either grouping or trial and error. Continuing the process,

$$x^4y + 7x^3y - 18x^2y = x^2y(x^2 + 7x - 18)$$
$$= x^2y(x + 9)(x - 2)$$

Because the binomial factors of $(x + 9)$ and $(x - 2)$ do not fit any of the patterns for factoring binomials, the process is complete. Verify your factorization by multiplication.

EXAMPLE 4

Factor $x^4 - y^4$.

SOLUTION   Because there is no common factor, we begin by noting that our polynomial is a binomial difference, and we see that it can be written as the difference of two squares:

$$x^4 - y^4 = (x^2)^2 - (y^2)^2$$
$$= (x^2 + y^2)(x^2 - y^2)$$

Examining each of the binomial factors $(x^2 - y^2)$ and $(x^2 + y^2)$, we see that $(x^2 - y^2)$ is the difference of two squares, which can be factored further. The first factor, $(x^2 + y^2)$, is prime because it does not fit any binomial pattern. Therefore, continuing the factorization we have:

$$x^4 - y^4 = (x^2)^2 - (y^2)^2$$
$$= (x^2 + y^2)(x^2 - y^2)$$
$$= (x^2 + y^2)(x + y)(x - y)$$

Because each of these binomial factors is prime, the process is complete. Verify your factorization by multiplication.

EXAMPLE 5

Factor $x^6 - x^4y^2 - 2x^3y^3 + 2xy^5$.

SOLUTION   We begin by factoring out the greatest common factor of x.

$$x^6 - x^4y^2 - 2x^3y^3 + 2xy^5 = x(x^5 - x^3y^2 - 2x^2y^3 + 2y^5)$$

Because the expression $x^5 - x^3y^2 - 2x^2y^3 + 2y^5$ has four terms, we try factoring by grouping:

$$x^6 - x^4y^2 - 2x^3y^3 + 2xy^5 = x(x^5 - x^3y^2 - 2x^2y^3 + 2y^5)$$
$$= x[x^3(x^2 - y^2) - 2y^3(x^2 - y^2)]$$
$$= x(x^2 - y^2)(x^3 - 2y^3) \qquad \text{Factor out } x^2 - y^2.$$

Finally, we factor the difference of two squares.

$$x^6 - x^4y^2 - 2x^3y^3 + 2xy^5 = x(x^2 - y^2)(x^3 - 2y^3)$$
$$= x(x + y)(x - y)(x^3 - 2y^3)$$

Because each of the polynomial factors is prime, the given expression is completely factored. Verify by multiplication.

EXAMPLE 6   Factor: $x^8 + 8x^2y^3$.

SOLUTION   Begin by factoring out the greatest common factor of x^2.

$$x^8 + 8x^2y^3 = x^2(x^6 + 8y^3)$$
$$= x^2[(x^2)^3 + (2y)^3] \qquad \text{Write as a sum of cubes.}$$
$$= x^2(x^2 + 2y)[(x^2)^2 - (x^2)(2y) + (2y)^2]$$
$$= x^2(x^2 + 2y)(x^4 - 2x^2y + 4y^2)$$

Because each of these polynomial factors is prime, the given expression is completely factored. Verify by multiplication.

EXERCISE 5.4

VOCABULARY AND NOTATION

1. The process of factoring polynomials is not complete until each polynomial factor is _____ .

2. Always look for a _____ first when factoring polynomials.

3. A polynomial containing four terms often can be factored by _____ .

CONCEPTS   *In Exercises 4–6, classify each of the statements as true or false. If a statement is false, justify your response.*

4. Correct factorization of polynomials can be confirmed by multiplying factors and comparing the result to the original simplified polynomial.

5. A polynomial cannot be prime.

6. Factor the polynomial $6x^2 + 14x - 15x - 35$
   a. by grouping
   b. after combining like terms
   c. Explain why the results are the same.

PRACTICE   *In Exercises 7–66, factor each polynomial completely. If a polynomial is prime, say so. Verify by multiplication.*

7. $6x + 3$

8. $x^2 - 9$

9. $x^3 - 27$

10. $x^3 + 8$

11. $x^2 - 11x + 10$

12. $x^2 + 5x + 6$

13. $6t^2 + 7t - 3$

14. $2t^2 + 15t + 18$

15. $3rs^2 - 6r^2st$

16. $32a^3 - 4ab^9$

17. $4x^2 - 25$

18. $36 - x^2$

19. $ac + ad + bc + bd$

20. $6ax + 4bx - 3ay - 2by$

21. $t^2 - 2t + 1$

22. $6p^2 - 3p - 2$

23. $3a^2 - 12$

24. $2x^2 - 32$

25. $2xy^2 + 8xy - 24x$

26. $-3x^2 + 3x + 60$

27. $-5x^2 - 10x + 40$

28. $5x^2 + 16x + 12$

29. $t^4 - 16$

30. $-8p^3q^7 - 4p^2q^3$

31. $27x^3 - 8y^3$

32. $64y^3 - z^3$

33. $3a^2 + 13a + 4$

34. $7x^2 + 36x + 5$

35. $6x^2 - x - 12$

36. $8x^2 - 14x + 5$

37. $3rs + 6r^2 - 18s^2$

38. $x^2 + 7x + 1$

39. $25x^2 + 16y^2$

40. $3a^2 + 27b^2$

41. $16 - 40z + 25z^2$

42. $25 - 30x + 9x^2$

43. $80a^5 - 5a$

44. $-34t^3 + 40t^2 + 6t^4$

45. $6x^2 + 7x - 20$

46. $-6x^2 - 7x + 20$

47. $-20x^2 - 11x + 3$

48. $-6x^2 - 7xy + 3y^2$

49. $16x^3 - 54y^3$

50. $64a^3 - 4b^3$

51. $a^2(x - a) - b^2(x - a)$

52. $x^2y^2 - 2x^2 - y^2 + 2$

53. $8a^2x^3y - 2b^2xy$

54. $81p^4 - 16q^4$

55. $6x^2 - x - 16$

56. $4x^2 + 9y^2$

57. $30x^4 + 5x^3 - 200x^2$

58. $10r^2 - 13r - 4$

59. $a^2x^2 + b^2y^2 + b^2x^2 + a^2y^2$

60. $ae + bf + af + be$

61. $625x^4 - 256y^4$

62. $36x^4 - 36$

63. $a^4 - 13a^2 + 36$

64. $x^4 - 17x^2 + 16$

65. $x^4 - 18x^2 + 81$

66. $x^4 - 16x^2 + 64$

REVIEW

In Exercises 67–70, solve each of the following linear equations algebraically.

67. $-2(x + 5) = 30 - x$

68. $5(x + 4) = -2(x - 3)$

69. $3(x + 2) - 2 = -(5 + x) + x$

70. $2x + 3 = \dfrac{2}{3}x - 1$

5.5 USING FACTORING TO SOLVE EQUATIONS

- REVIEW OF EQUATION SOLVING AND BASIC DEFINITIONS
- SOLVING QUADRATIC EQUATIONS BY FACTORING
- SOLVING HIGHER-DEGREE EQUATIONS BY FACTORING
- GRAPHICAL SOLUTIONS
- APPLICATIONS

REVIEW OF EQUATION SOLVING AND BASIC DEFINITIONS

Factoring can be used to solve equations that cannot be solved by methods introduced previously in Chapter 3. Recall that we solved equations by a process called *transformation*. After simplifying each side of the equation, we used the addition, subtraction, multiplication, and/or division properties of equality to rewrite (transform) the original equation into a series of equivalent equations until the variable was isolated on one side of the equation. This process is used to solve linear (first-degree) equations.

The degree of a polynomial equation (when simplified) is the same as the highest degree of any of its terms. Equations that are of degree two (second-degree equations) are called **quadratic equations**.

Quadratic Equation

A quadratic equation is any equation that can be written in the form $ax^2 + bx + c = 0$, where a, b, and c are real numbers, $a \neq 0$.

The solution (or root) of an equation is the value (or values) of the variable that make the equation a true statement. To find the solutions of an equation that is quadratic or of higher degree, the **zero factor property** is used.

Zero Factor Property

If $a \cdot b = 0$ and a and b are real numbers, it follows that $a = 0$ or $b = 0$.

This property simply states that if a product of factors is zero, *at least one* of the factors must be equal to zero. To apply this property, the polynomial equations must be set equal to zero. A polynomial equation is said to be in **standard form** when one side of the equation is zero.

▐▌ SOLVING QUADRATIC EQUATIONS BY FACTORING

EXAMPLE 1

Solve: $(x - 4)(x + 5) = 0$.

SOLUTION

Because the product of two factors equals 0, the zero factor property assures us that one of the factors must be equivalent to zero.

$$\text{Either } x - 4 = 0 \quad \text{ or } \quad x + 5 = 0$$

We can solve each of these linear equations to get

$$x = 4 \quad \text{ or } \quad x = -5$$

Therefore, the equation $(x - 4)(x + 5) = 0$ has two solutions, 4 and -5. These could be expressed as a solution set: $\{4, -5\}$.

These solutions can be verified in any one of the following ways:

Analytically: By substituting the solutions into the original equation:

Using the STOre feature:

When $x = 4$

$$(x - 4)(x + 5) = 0$$
$$(4 - 4)(4 + 5) = 0$$
$$(0)(9) = 0$$
$$0 = 0 \checkmark$$

When $x = -5$

$$(x - 4)(x + 5) = 0$$
$$(-5 - 4)(-5 + 5) = 0$$
$$(-9)(0) = 0$$
$$0 = 0 \checkmark$$

```
4→X: (X-4)(X+5)
                    0
-5→X: (X-4)(X+5)
                    0
```

Using the TABLE feature: The left side of the equation is entered at the y1 prompt and the right side is entered at the y2 prompt. Initially, the table is set up to start with x having a value of -5. The second screen verifies the solution of -5 because the value in the y1 column is equal to the value in the y2 column. Scrolling down the table, the third screen shows the verification of the solution 4.

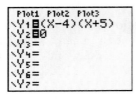

Solving Equations by Factoring

1. Simplify both sides of the equation by removing grouping symbols and combining like terms. Fractions may be eliminated by multiplying each term on both sides of the equation by the LCD.
2. Put the equation in standard form (set one side equal to zero).
3. Factor the polynomial.
4. Set each factor containing a variable equal to zero.
5. Solve the resulting equation.
6. Check solutions either analytically or with the graphing calculator.

EXAMPLE 2

Solve: $9x^2 - 6x = 0$.

SOLUTION Because the polynomial equation is in standard form, we begin by factoring the polynomial.

$$9x^2 - 6x = 0$$

$$3x(3x - 2) = 0 \qquad \text{Factor out the GCF of } 3x.$$

$$3x = 0 \quad \text{or} \quad 3x - 2 = 0 \qquad \text{Use the zero factor property to set each factor equal to zero.}$$

$$\frac{3x}{3} = \frac{0}{3} \qquad\qquad 3x = 2 \qquad \text{Solve each linear equation for the variable.}$$

$$\qquad\qquad\qquad \frac{3x}{3} = \frac{2}{3}$$

$$x = 0 \qquad\qquad\qquad x = \frac{2}{3}$$

TECHNOLOGY TIP

Verify solutions to equations using the method you like best and with which you feel most comfortable. However, experiment with both methods presented in Example 1 and link the calculator verification to the analytical check.

Solutions: $0, \frac{2}{3}$; Solution set: $\{0, \frac{2}{3}\}$.

The verification using the STOre feature is shown; however, remember that the TABLE feature or an analytical verification could also have been used. Also, remember that solutions should be verified using the original equation.

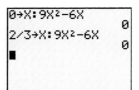

EXAMPLE 3

Solve: $\dfrac{3}{4}(6x^2 - x) = \dfrac{3}{2}$.

SOLUTION

$$\frac{18}{4}x^2 - \frac{3}{4}x = \frac{3}{2}$$ Use the distributive property to eliminate grouping.

$$4 \cdot \frac{18}{4}x^2 - 4 \cdot \frac{3}{4}x = 4 \cdot \frac{3}{2}$$ Multiply through the equation by the LCD of 4 to eliminate fractions.

$$18x^2 - 3x = 6$$ Simplify.

$$18x^2 - 3x - 6 = 0$$ Put the equation in standard form.

$$3(6x^2 - x - 2) = 0$$ Factor the GCF of 3 from the polynomial.

$$3(3x - 2)(2x + 1) = 0$$ Factor the trinomial.

$3x - 2 = 0$ | $2x + 1 = 0$ Using the zero factor property, set each factor con-**taining a variable** equal to zero. It does not make sense to set the factor of 3 equal to 0.

$3x = 2$ | $2x = -1$ Solve each linear equation.

$x = \dfrac{2}{3}$ | $x = -\dfrac{1}{2}$

Solutions: $\frac{2}{3}, -\frac{1}{2}$; Solution set: $\{\frac{2}{3}, -\frac{1}{2}\}$.

The verification using the STOre feature is shown:

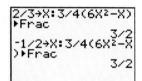

```
2/3→X:3/4(6X²-X)
▶Frac
                3/2
-1/2→X:3/4(6X²-X
)▶Frac
                3/2
```

WARNING! In the previous example, factoring the polynomial produced the equation $3(3x - 2)(2x + 1) = 0$. The product of three factors equaled zero. The zero could only have occurred in the factor $(3x - 2)$ or the factor $(2x + 1)$ because the third factor was the constant, 3.

▍ SOLVING HIGHER-DEGREE EQUATIONS BY FACTORING

EXAMPLE 4

Solve: $x^3 - 2x^2 = 63x$.

SOLUTION

This equation is *not* a quadratic equation (it is a third-degree equation), but it can be solved by factoring.

$$x^3 - 2x^2 - 63x = 0$$ Put the equation in standard form.

$$x(x^2 - 2x - 63) = 0$$ Begin factoring the polynomial by factoring out the GCF of x.

$$x(x - 9)(x + 7) = 0$$ Factor the trinomial.

$$x = 0 \qquad x - 9 = 0 \qquad x + 7 = 0 \qquad \text{Use the zero factor property to set each factor equal to 0.}$$

$$x = 0 \qquad x = 9 \qquad x = -7 \qquad \text{Solve each linear equation.}$$

Solutions: $0, 9, -7$; Solution set: $\{0, 9, -7\}$.

The verification using the TABLE feature is illustrated below:

The left side of the original equation is entered at the y1 prompt and the right side is entered at the the y2 prompt.

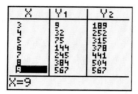

The TABLE is set up to begin with an x value of -7. Because the values in the y1 and y2 columns are equivalent, -7 is verified as a solution. Scrolling through the TABLE verifies the other roots of 0 and 9.

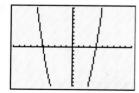

GRAPHICAL SOLUTIONS

Quadratic and higher-degree equations can also be solved graphically, using the INTERSECT feature of the calculator. Each equation that has been solved algebraically in this section will now be solved graphically, using this feature.

EXAMPLE 5

Solve $(x - 4)(x + 5) = 0$ graphically.

SOLUTION   Enter the left side of the equation at the y1 prompt and the right side at the y2 prompt and view the graph in the standard viewing window $[-10, 10]$ by $[-10, 10]$:

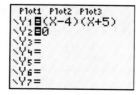

TECHNOLOGY TIP

The right side of the equation, 0, was entered at the y2 prompt. The graph of this line is on top of the x-axis and is not distinguishable as a separate line. Turning your axes off will enable you to view the line.

To get a complete graph, reset the ymin to -25.

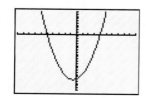

Using the INTERSECT feature twice gives the roots of the equation:

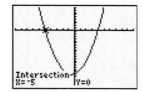 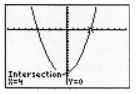

This verifies the solutions found in Example 1.

EXAMPLE 6

Solve $9x^2 - 6x = 0$ graphically.

SOLUTION   Enter the left side of the equation at the y1 prompt and the right side at the y2 prompt and view the graph in the standard viewing window. To get a better look at the points of intersection, the window values are set to $[-2, 2]$ by $[-2, 2]$.

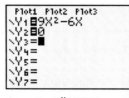 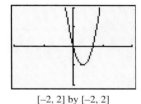

| y = edit screen | ZStandard | $[-2, 2]$ by $[-2, 2]$ |

Using the INTERSECT feature twice gives the roots of the equation:

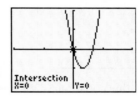 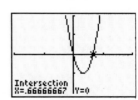 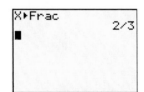

Remember, the value for x that appears in the middle screen is automatically stored in the "x" slot in the calculator when the INTERSECT feature is used. Returning to the home screen allows us to convert this value to its equivalent fraction form. This verifies the solution found in Example 2.

 ## EXAMPLE 7

Solve $\dfrac{3}{4}(6x^2 - x) = \dfrac{3}{2}$ graphically.

SOLUTION   Enter the left side of the equation at the y1 prompt and the right side at the y2 prompt and view the graph in the standard viewing window. To get a better view of the points of intersection, the window values are reset to $[-2, 2]$ by $[-3, 3]$. This was not necessary; the points of intersection could have been found when the graph was in the ZStandard viewing window; however, we suggest that you practice getting a good, clear graph of desired points.

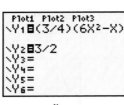
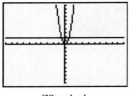
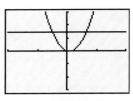

| y = edit screen | ZStandard | [–2, 2] by [–3, 3] |

The solutions, or roots, are found using the INTERSECT feature.

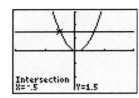
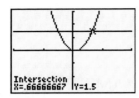

This verifies the solutions found in Example 3.

EXAMPLE 8

Solve $x^3 - 2x^2 = 63x$ graphically.

SOLUTION   Enter the left side of the equation at the y1 prompt and the right side at the y2 prompt and view the graph in the standard viewing window.

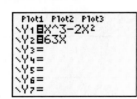
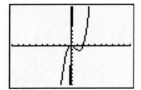

TECHNOLOGY TIP

Tracing from left to right on the curve, you will discover that the y-coordinate peaks at approximately $y = 156$ and it has a valley at approximately $y = -241$. The viewing window for the y-axis needs to be at least as big as $[-241, 156]$. We opted for $[-250, 200]$.

The right side of the equation ($63x$) has graphed almost on top of the y-axis. To get a better graph, we'll rewrite the original equation in standard form, $x^3 - 2x^2 - 63x = 0$, and graph it in the ZStandard viewing window (enter the left side at y1 and the right side at y2).

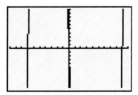

Because this is not a good window, we TRACE to find the peaks and valleys of the graph so that we can reset our window values. The graph is pictured below in a window of $[-10, 10]$ by $[-250, 200]$, with all intersections displayed.

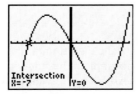
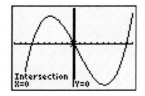
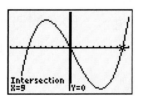

This verifies the solution found in Example 4.

▮▮ APPLICATIONS

EXAMPLE 9

Integer problem   One negative integer is 5 less than another, and their product is 84. Find the integers.

SOLUTION   Let x represent the larger number. Then $x - 5$ represents the smaller number. Because their product is 84, we form the equation $x(x - 5) = 84$ and solve it.

$$x(x - 5) = 84$$
$$x^2 - 5x = 84 \qquad \text{Simplify by removing parentheses.}$$
$$x^2 - 5x - 84 = 0 \qquad \text{Put the equation in standard form.}$$
$$(x - 12)(x + 7) = 0 \qquad \text{Factor.}$$
$$x - 12 = 0 \quad \text{or} \quad x + 7 = 0 \qquad \text{Set each factor containing a variable equal to 0.}$$
$$x = 12 \quad \text{or} \quad x = -7 \qquad \text{Solve each linear equation.}$$

Because we need two negative numbers, we discard the result $x = 12$. The two integers are

$$x = -7 \quad \text{and} \quad x - 5 = -7 - 5$$
$$= -12$$

Check: The number -12 is 5 less than -7, and $(-12)(-7) = 84$.
The graphical solutions are pictured below.

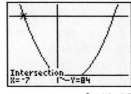 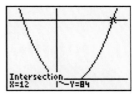

$$[-10, 15] \text{ by } [-10, 100]$$

EXAMPLE 10

Ballistics   If an object is thrown straight up into the air with an initial velocity of 112 feet per second, its height after t seconds is given by the formula

$$h = 112t - 16t^2$$

where h represents the height of the object in feet. After this object has been thrown, in how many seconds will it hit the ground?

SOLUTION   When the object hits the ground, its height will be 0. Thus, we set h equal to 0 and solve for t.

$$h = 112t - 16t^2$$
$$0 = 112t - 16t^2$$
$$0 = 16t(7 - t) \qquad \text{Factor.}$$
$$16t = 0 \quad \text{or} \quad 7 - t = 0 \qquad \text{Set each factor equal to zero.}$$
$$t = 0 \quad \text{or} \quad t = 7 \qquad \text{Solve each linear equation.}$$

When $t = 0$, the object's height above the ground is 0 feet, because it has not been released. When $t = 7$, the height is again 0 feet. The object has hit the ground. The solution is 7 seconds.

The graphical solutions are pictured below.

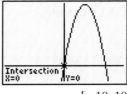
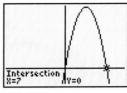

$[-10, 10]$ by $[-50, 200]$

Recall that the area of a rectangle is given by the formula $A = lw$, where A represents the area, l the length, and w the width of the rectangle. The perimeter of a rectangle is given by the formula $P = 2l + 2w$, where P represents the perimeter, l the length, and w the width of the rectangle.

EXAMPLE 11

Perimeter of a rectangle   Assume that the rectangle in Figure 5-1 has an area of 52 square centimeters and that its length is 1 centimeter more than 3 times its width. Find the perimeter of the rectangle.

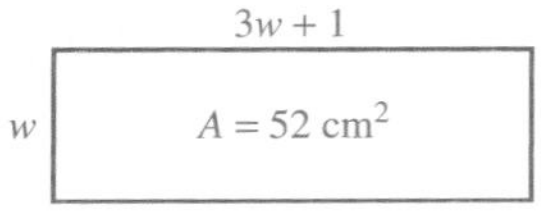

$3w + 1$

w | $A = 52$ cm^2

Figure 5-1

SOLUTION   Let w represent the width of the rectangle. Then $3w + 1$ represents its length. Because the area is 52 square centimeters, we substitute 52 for A and $3w + 1$ for l in the formula $A = lw$ and solve for w.

$$A = lw$$
$$52 = (3w + 1)w$$
$$52 = 3w^2 + w \qquad \text{Simplify by removing parentheses.}$$
$$0 = 3w^2 + w - 52 \qquad \text{Put the equation in standard form by subtracting 52 from both sides.}$$
$$0 = (3w + 13)(w - 4) \qquad \text{Factor.}$$

$$3w + 13 = 0 \qquad \text{or} \qquad w - 4 = 0 \qquad \text{Set each factor equal to 0.}$$
$$3w = -13 \qquad \text{or} \qquad w = 4 \qquad \text{Solve each linear equation.}$$
$$w = -\frac{13}{3}$$

Because the width of a rectangle cannot be negative, we discard the result $w = -\frac{13}{3}$. Thus, the width of the rectangle is 4, and the length is given by

$$3w + 1 = 3(4) + 1$$
$$= 12 + 1$$
$$= 13$$

The dimensions of the rectangle are 4 centimeters by 13 centimeters. We find the perimeter by substituting 13 for l and 4 for w in the formula for the perimeter.

$$P = 2l + 2w$$
$$= 2(13) + 2(4)$$
$$= 26 + 8$$
$$= 34$$

The perimeter of the rectangle is 34 centimeters.

Check: A rectangle with dimensions of 13 centimeters by 4 centimeters does have an area of 52 square centimeters, and the length is 1 centimeter more than 3 times the width. A rectangle with these dimensions has a perimeter of 34 centimeters.

The graphical solutions are pictured below.

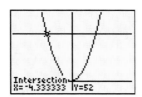
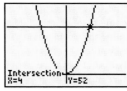

$$[-10, 10] \text{ by } [-10, 75]$$

EXAMPLE 12

Dimensions of a triangle   The triangle in Figure 5-2 has an area of 10 square centimeters and a height that is 3 centimeters less than twice the length of its base. Find the length of the base and the height of the triangle.

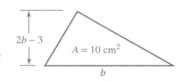

Figure 5-2

SOLUTION   Let b represent the length of the base of the triangle. Then $2b - 3$ represents the height. Because the area is 10 square centimeters, we substitute 10 for A and $2b - 3$ for h in the formula $A = \frac{1}{2}bh$ and solve for b.

$$A = \frac{1}{2}bh$$

$$10 = \frac{1}{2}b(2b - 3)$$

$$10 = b^2 - \frac{3}{2}b \qquad \text{Remove grouping symbols by use of the distributive property.}$$

$$20 = 2b^2 - 3b \qquad \text{Multiply both sides by 2 to clear fractions.}$$

$$0 = 2b^2 - 3b - 20 \qquad \text{Put the equation in standard form.}$$

$$0 = (2b + 5)(b - 4) \qquad \text{Factor.}$$

$$2b + 5 = 0 \qquad \text{or} \qquad b - 4 = 0 \qquad \text{Set each factor equal to 0.}$$

$$2b = -5 \qquad\qquad b = 4 \qquad \text{Solve each linear equation.}$$

$$b = -\frac{5}{2}$$

Because a triangle cannot have a negative number for the length of its base, we discard the result $b = -\frac{5}{2}$. The length of the base of the triangle is 4 centimeters. Its height is $2(4) - 3$, or 5 centimeters.

Check: If the base of the triangle has a length of 4 centimeters and the height of the triangle is 5 centimeters, its height is 3 centimeters less than twice the length of its base. Its area is 10 centimeters.

$$A = \frac{1}{2}bh$$
$$= \frac{1}{2}(4)(5)$$
$$= 2(5)$$
$$= 10$$

The graphical solution is pictured below.

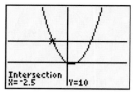 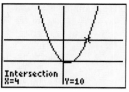

$$[-10, 10] \text{ by } [-10, 25]$$

EXERCISE 5.5

VOCABULARY AND NOTATION   *Fill in each blank to make a true statement.*

1. The process used to solve linear equations in one variable is called _____ because the addition, subtraction, multiplication, and division properties of equality are used to rewrite the original equation into a series of equivalent equations.

2. Quadratic equations are of degree _____.

3. Another word for the solution of an equation is its _____.

4. The _____ property is used to find solutions of quadratic and higher-degree equations when factoring is used.

5. The property referred to in Exercise 4 states that if a product of factors is zero, _____ of its factors must actually be zero.

6. A polynomial equation is said to be in _____ when one side of the equation is equal to zero.

7. The _____ feature of the graphing calculator can be used to solve equations graphically.

8. Solutions to equations can be checked by using the _____ or _____ features of the calculator.

9. A complete graphical display should show both ___ and ___ intercepts as well as any peaks and valleys of the graph.

CONCEPTS

10. Simplify $2x^2 - 10x + 3x - 15$.

11. Factor $2x^2 - 10x + 3x - 15$.

12. Evaluate $2x^2 - 10x + 3x - 15$ when x has a value of 6.

13. Solve $2x^2 - 7x - 15 = 0$.

14. Check the solutions obtained in Exercise 13. Does this relate to the process of simplifying, evaluating, or factoring? Explain.

15. Classify each of the following statements as an *equation* or an *expression*.

  a. $x^2 - 25$ **b.** $x^2 = -3x + 10$

  c. $x^2 - 25 = 0$ **d.** $x^2 + 3x - 10$

16. Classify each of the following equations as linear, quadratic, or higher degree.

  a. $3x(x + 4) = 9$

  b. $5x - 6 = 2x - 8(x - 3)$

  c. $x^2(x + 1) = 10 + 20\left(x - \dfrac{1}{2}\right)$

Solve each equation and state your solution as a set.

17. $(x + 2)(x - 5) = 0$

18. $(x + 5)(x - 2) = 0$

19. $(x - 6)(x + 8) = 0$

20. $3x(x + 2)(x - 1) = 0$

21. $3(x + 2)(x - 1) = 0$

22. $(x - 1)(x + 3)(x - 4) = 0$

PRACTICE   *In Exercises 23–55, solve each equation by factoring. Verify your results by using either the STOre or TABLE feature of your calculator.*

23. $x^2 - 9 = 0$ **24.** $y^2 - 16 = 0$

25. $5y^2 - 10y = 0$ **26.** $x^2 + x = 0$

27. $x^2 - 3x = 0$ **28.** $5y^2 - 25y = 0$

29. $y^2 - 36 = 0$ **30.** $z^2 + 8z + 15 = 0$

31. $w^2 + 7w + 12 = 0$

32. $y^2 - 7y + 6 = 0$ **33.** $n^2 - 5n + 6 = 0$

34. $y^2 - 7y + 12 = 0$ **35.** $x^2 - 3x + 2 = 0$

36. $15x^2 - 20x = 0$ **37.** $10x^2 - 2x = 0$

38. $2x^2 - 5x + 2 = 0$ **39.** $2x^2 + x - 3 = 0$

40. $5x^2 - 6x + 1 = 0$ **41.** $6x^2 - 5x + 1 = 0$

42. $x(6x + 5) = 6$ **43.** $x(2x - 3) = 14$

44. $(x + 1)(8x + 1) = 18x$

45. $4x(3x + 2) = x + 12$

46. $x^2(x + 10) = 2x(x - 8)$

47. $\dfrac{1}{2}x^2 = 6 - 2x$

48. $\dfrac{3}{2}x^2 + \dfrac{27}{2}x = -30$

49. $6x^3 - 8x^2 = 8x$

50. $4x^2 = 81$ **51.** $9y^2 = 64$

52. $(x - 2)^2 - 3 = -3$

53. $(x + 3)^2 - 4 = 0$

54. $\dfrac{1}{8}x^2 - \dfrac{1}{4}x - 6 = 0$

55. $\dfrac{1}{3}x^2 + x - 6 = 0$

In Exercises 56–63, (a) match each graphical display with its equation; (b) state the solutions of the equation; (c) write the equation in standard form (multiply the factors so the left side is a polynomial).

56. $(x - 2)(x + 3) = 0$

57. $x(x - 5) = 0$

58. $x(x - 3)(x + 3) = 0$

59. $x(x - 2)(x + 2) = 0$

60. $x(x - 3)(x + 2) = 0$

61. $x(x - 2)(x + 3) = 0$

62. $(x + 4)(x - 2) = 0$

63. $(x + 4)(x + 2) = 0$

A.
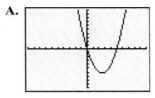

B.
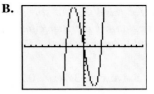

C.
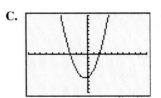

D.
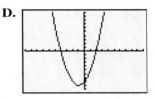

E. 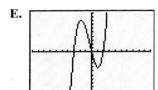   **F.**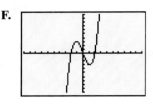

G. 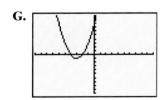   **H.**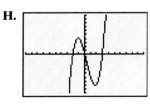

In Exercises 64–75, solve each higher-degree equation.

64. $x^3 + x^2 = 0$

65. $2x^4 + 8x^3 = 0$

66. $y^3 - 49y = 0$

67. $2z^3 - 200z = 0$

68. $x^3 - 4x^2 - 21x = 0$

69. $x^3 + 8x^2 - 9x = 0$

70. $z^4 - 13z^2 + 36 = 0$

71. $y^4 - 10y^2 + 9 = 0$

72. $3a(a^2 + 5a) = -18a$

73. $7t^3 = 2t\left(t + \dfrac{5}{2}\right)$

74. $\dfrac{x^2(6x + 37)}{35} = x$

75. $x^2 = -\dfrac{4x^3(3x + 5)}{3}$

In Exercises 76–81, solve each equation using the graphing calculator. Unless specified otherwise, the standard viewing window may be used.

76. Solve $x^2 + 3.44x - 0.21 = 0$.

77. Solve $x^2 - 3.88x - 1.806 = 0$.

78. Solve $x^2 + 4.9x - 1.56 = 0$.

79. Solve $x^2 + 4.08x - 2.392 = 0$.

80. Solve $x^3 + 1.2x^2 - 15x + 2.944 = 0$.
(Window: $[-10, 10]$ by $[-15, 35]$)

81. Solve $x^3 + x^2 - 7.79x - 2.4 = 0$.

APPLICATIONS *In Exercises 82–87, an object has been thrown straight up into the air. The formula $h = vt - 16t^2$ gives the height, h, of the object above the ground after t seconds when it is thrown upward with an initial velocity, v.*

82. Time of flight After how many seconds will an object hit the ground if it was thrown with a velocity of 144 feet per second?

83. Time of flight After how many seconds will an object hit the ground if it was thrown with a velocity of 160 feet per second?

84. Ballistics If a cannonball is fired with an upward velocity of 220 feet per second, at what times will it be at a height of 600 feet?

85. Ballistics A cannonball's initial upward velocity is 128 feet per second. At what times will it be 192 feet above the ground?

86. Exhibition diving A popular tourist attraction consists of swimmers diving from a cliff to the water 64 feet below. A diver's height, h, above the water t seconds after diving is given by $h = -16t^2 + 64$. How long does a dive last?

87. Forensic medicine The kinetic energy, E, of a moving object is given by $E = \frac{1}{2}mv^2$, where m is the mass of the object (in kilograms) and v is the object's velocity (in meters per second). Kinetic energy is measured in joules. By measuring the damage done to a victim who has been struck by a 3-kilogram club, a police pathologist finds that the energy at impact was 54 joules. Find the velocity of the club at impact.

In Exercises 88–106, write an equation to model each problem. Solve the equation and state the answer to the question posed in the problem with a complete sentence. Make sure you check results in the words of the problem.

88. Integer problem One positive integer is 2 more than another. Their product is 35. Find the integers.

89. Integer problem One positive integer is 5 less than 4 times another. Their product is 21. Find the integers.

90. Integer problem If 4 is added to the square of a composite integer, the result is 5 less than 10 times that integer. Find the integer. (*Hint:* A composite number is one that is *not* prime or equal to 1.)

91. Number problem If 3 times the square of a certain natural number is added to the number itself, the result is 14. Find the number.

92. Insulation The area of the rectangular slab of foam insulation in Illustration 1 is 36 square meters. Find the dimensions of the slab.

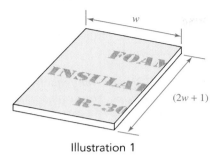

Illustration 1

93. Shipping pallet The length of a rectangular shipping pallet is 2 feet less than 3 times its width. Its area is 21 square feet. Find the dimensions of the pallet.

94. Carpentry A room containing 143 square feet is 2 feet longer than it is wide. How long a crown molding is needed to trim the perimeter of the ceiling?

95. Designing solar panels The length of a standard rectangular solar heat exchange panel is 2 meters longer than the width. If the length remained the same but the width were doubled, the area would be 48 square meters. Find the perimeter of a standard panel.

96. Designing a tent The length of the base of the triangular sheet of canvas above the door of the tent shown in Illustration 2 is 2 feet more than twice its height. The area is 30 square feet. Find the height and the length of the base of the triangle.

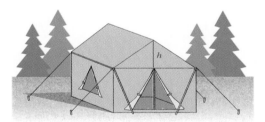

Illustration 2

97. Dimensions of a triangle The height of a triangle is 2 inches less than 5 times the length of its base. The area is 36 square inches. Find the length of the base and the height of the triangle.

98. Area of a triangle The base of a triangle is numerically 3 less than its area, and the height is numerically 6 less than its area. Find the area of the triangle.

99. Area of a triangle The length of the base and the height of a triangle are numerically equal. Their sum is 6 less than the number of units in the area of the triangle. Find the area of the triangle.

100. Dimensions of a parallelogram The formula for the area of a parallelogram is $A = bh$. The area of the parallelogram in Illustration 3 is 200 square centimeters. If its base is twice its height, how long is the base?

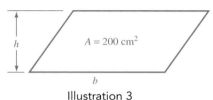

Illustration 3

101. Swimming pool border The owners of the rectangular swimming pool in Illustration 4 want to surround the pool with a crushed-stone border of uniform width. They have enough stone to cover 74 square meters. How wide should they make the border? (*Hint:* The area of the larger rectangle minus the area of the smaller is the area of the border.)

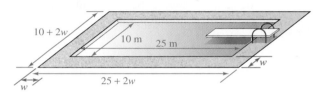

Illustration 4

102. House construction The formula for the area of a trapezoid is $A = \dfrac{h(B + b)}{2}$. The area of the trapezoidal truss in Illustration 5 is 24 square meters. Find the height of the trapezoid if one base is 8 meters and the other base is the same as the height.

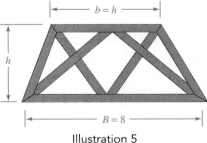

Illustration 5

103. Volume The volume of a rectangular solid is given by the formula $V = lwh$, where l is the length, w is the width, and h is the height. The volume of the rectangular solid in Illustration 6 is 210 cubic centimeters.

Find the width of the rectangular solid if its length is 10 centimeters and its height is 1 centimeter longer than twice its width.

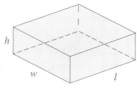

Illustration 6

104. Volume of a pyramid The volume of a pyramid is given by the formula $V = \dfrac{Bh}{3}$, where B is the area of its base and h is its height. The volume of the pyramid in Illustration 7 is 192 cubic centimeters. Find the dimensions of its rectangular base if one edge of the base is 2 centimeters longer than the other and the height of the pyramid is 12 centimeters.

105. Volume of a pyramid The volume of a pyramid is 84 cubic centimeters. Its height is 9 centimeters, and one side of its rectangular base is 3 centimeters shorter than the other. Find the dimensions of its base. (See Exercise 104).

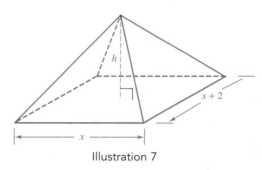

Illustration 7

106. Volume of a solid The volume of a rectangular solid is 72 cubic centimeters. Its height is 4 centimeters, and its width is 3 centimeters shorter than its length. Find the sum of its length and width. (See Exercise 103).

107. Here is an easy-sounding word problem: The length of a rectangle is 2 feet greater than the width, and the area is 18 square feet. Find the width of the rectangle.

    a. Write an equation to model the data.

    b. Can you solve the given equation? How did you solve it? Explain fully.

In Exercises 108–112, write an equation that has the specified values as solutions. Your equation should contain only integral coefficients.

108. $3, -4$

109. $-2, 5$

110. $-3, 0, \dfrac{7}{3}$

111. $-5, -1, \dfrac{1}{2}$

112. $\dfrac{3}{4}, \dfrac{2}{3}$

REVIEW *Solve each of the following equations.*

113. $\dfrac{2}{3}(5x - 3) = 38$

114. $2q^2 - 9 = q(q + 3) + q^2$

115. $\dfrac{4x + 1}{15} = \dfrac{3x - 3}{6} - 2$

116. Graph $y = 2x + 1$.

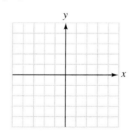

117. Is $(6, -2)$ a solution of $2x + 3y = 6$?

CHAPTER SUMMARY

CONCEPTS

REVIEW EXERCISES

Section 5.1

Using the Distributive Property to Factor

To factor a polynomial means to write the indicated sum and/or difference of terms as a product of factors.

Identifying the Greatest Common Factor
1. Determine the prime factorization of each constant. Use exponents rather than repeated factors to write the number in prime factored form.
2. List each prime number that is a factor of all the numbers in the list. When exponents appear, choose the common factor with the *smallest* exponent.
3. Multiply the listed primes from the previous step. This is the greatest common factor, or GCF.

Factoring Out the Greatest Common Factor
1. Determine the GCF of the terms of the polynomial.
2. Rewrite each *term* as the product of the GCF and another factor.
3. Use the distributive property to rewrite the polynomial as the *product* of the GCF and a polynomial.

When the leading coefficient is negative, a -1 is usually factored out of the polynomial.

In Review Exercises 1–20, factor each polynomial completely.

1. $3x + 9y$

2. $5ax^2 + 15a$

3. $7x^2 + 14x$

4. $3x^2 - 3x$

5. $2x^3 + 4x^2 - 8x$

6. $ax + ay + az$

7. $ax + ay - a$

8. $x^2yz + xy^2z$

9. $5a^2 + 6ab^2 + 10acd - 15a$

10. $7axy + 21x^2y - 35x^3y + 7xy^2$

11. $(x + y)a + (x + y)b$

12. $(x + y)^2 + (x + y)$

13. $2x^2(x + 2) + 6x(x + 2)$

14. $3x(y + z) - 9x(y + z)^2$

15. $3p + 9q + ap + 3aq$

16. $ar - 2as + 7r - 14s$

17. $x^2 + ax + bx + ab$

18. $xy + 2x - 2y - 4$

19. $3x^2y - xy^2 - 6xy + 2y^2$

20. $5x^2 + 10x - 15xy - 30y$

Section 5.2

Factoring Trinomials

Factoring a Trinomial by Grouping

1. Write the trinomial in descending powers of one variable.
2. Factor out any GCF (including -1 if that is necessary to make the coefficients of the first term positive).
3. Identify a, b, and c, where a is the coefficient of the first term ($a > 0$), b is the coefficient of the middle term, and c is the coefficient of the third term (or the constant term).
4. Find the product, $a \cdot c$.
5. Find two numbers whose **product** is ac and whose **sum** is b.
6. Replace the middle term of the original trinomial with two terms, using the numbers found in step 5 as coefficients. You now have a four-term polynomial.
7. Factor by grouping.
8. Verify the results by multiplication.

Factoring a Trinomial by Trial and Error

1. Write the trinomial in descending powers of one variable.
2. Factor out any greatest common factor (including -1 if that is necessary to make the coefficient of the first term positive).
3. If the sign of the third term is positive, the signs between the terms of the binomial factors are the same as the sign of the middle term of the trinomial. If the sign of the third term is negative, the signs between the terms of the binomial factors are opposite.

In Review Exercises 21–36, factor each polynomial completely.

21. $x^2 + 10x + 21$

22. $x^2 + 4x - 21$

23. $x^2 + 2x - 24$

24. $x^2 - 4x - 12$

25. $2x^2 - 5x - 3$

26. $3x^2 - 14x - 5$

27. $6x^2 + 7x - 3$

28. $6x^2 + 3x - 3$

29. $6x^3 + 17x^2 - 3x$

30. $4x^3 - 5x^2 - 6x$

31. $x^2 - 2xy - 3y^2$

32. $4x^2 + 4x + 1$

33. $-8x^2 + 14x - 5$

34. $2x^2 - 13xy + 18y^2$

35. $5x^2 + 28x + 15$

36. $3x^2 + 10xy - 8y^2$

4. Mentally, try various combinations of first terms and last terms until you find one that works, or until you exhaust all the possibilities. In that case, the trinomial does not factor using only integer coefficients.
5. Check the factorization by multiplication.

Section 5.3

To factor the difference of two squares, use the pattern
$x^2 - y^2 = (x - y)(x + y)$.

The binomials $(x + y)$ and $(x - y)$ are called *conjugate binomials*.

The sum of two squares is **not** factorable over the set of real numbers.

To factor the difference of two cubes, use the pattern
$x^3 - y^3 = (x - y)(x^2 + xy + y^2)$.

To factor the sum of two cubes, use the pattern
$x^3 + y^3 = (x + y)(x^2 - xy + y^2)$.

Factoring by Special Products

In Review Exercises 37–50, factor each polynomial completely.

37. $x^2 - 9$

38. $64 - x^2$

39. $25x^2 - 16y^2$

40. $6x^2y - 24y^3$

41. $x^4 - 16$

42. $162 - 2x^4$

43. $x^3 + 27$

44. $27 - x^3$

45. $x^3 + 8y^3$

46. $64x^3 + y^3$

47. $5 - 40y^3$

48. $64x^3 - 1$

49. $x^3 + 125$

50. $x^5 + 216x^2y^3$

Section 5.4

Factoring Polynomials
1. Always check for and factor out the greatest common factor (GCF) first. If the leading coefficient is negative, factor out a -1 with/as the GCF.
2. Count the number of terms in the polynomial, and then follow the given procedures for that specific type of polynomial.

Summary of Factoring Techniques

In Review Exercises 51–68, factor each polynomial completely.

51. $24xy^2 - 6xy$

52. $x^3 + 27$

53. $6x^2 + x - 2$

54. $9x^2 - 49$

55. $16x^2 - 25$

56. $2x^2 - 10x + 3x - 15$

57. $4x^2 + 8x - 12$

58. $4x^2 + 12x + 9$

59. $x^3 - 1$

60. $2x^3 - 16$

61. $2x^2 - 7xy + 6y^2$

62. $x^2y^2 - 25$

63. $3x^2 + 2x - 1$

64. $54x^3 + 2y^3$

65. $8x^2 - 32y^2$

66. $12x^2 - 5x - 3$

67. $8x^2 - 6x - 27$

68. $3x + 81x^4$

Section 5.5

The degree of a polynomial equation (when simplified) is the same as the highest degree of any of its terms.

A quadratic equation is an equation of degree two that can be written in the form $ax^2 + bx + c = 0$.

Zero factor property
If $a \cdot b = 0$ and a and b are real numbers, it follows that $a = 0$ or $b = 0$.

Solving Equations by Factoring
1. Simplify both sides of the equation by removing grouping symbols and combining like terms. Fractions may be eliminated by multiplying each term on both sides of the equation by the LCD.
2. Put the equation in standard form (set one side equal to zero).
3. Factor the polynomial.
4. Set each factor containing a variable equal to zero.
5. Solve the resulting equation.
6. Check solutions either analytically or with the graphing calculator.

Using Factoring to Solve Equations

Solve each equation below by factoring. Check roots using the STOre feature of the calculator.

69. $x^2 + 2x = 0$

70. $2x^2 - 6x = 0$

71. $x^2 - 9 = 0$

72. $x^2 - 25 = 0$

73. $a^2 - 7a + 12 = 0$

74. $x^2 - 2x - 15 = 0$

75. $2x - x^2 + 24 = 0$

76. $16 + x^2 - 10x = 0$

77. $2x^2 - 5x - 3 = 0$

78. $2x^2 + x - 3 = 0$

79. $4x^2 = 1$

80. $9x^2 = 4$

81. $x^3 - 7x^2 + 12x = 0$

82. $x^3 + 5x^2 + 6x = 0$

83. $2x^3 + 5x^2 = 3x$

84. $3x^3 - 2x = x^2$

Solve each of the following problems. Make sure you clearly identify what the variable represents, write an equation, provide a complete solution to the equation (either algebraically or graphically), and answer the question posed. Be sure to check your solution in the words of the problem.

85. Integer problem   The sum of two integers is 12, and their product is 35. Find the integers.

86. Number problem   If 3 times the square of a positive number is added to 5 times the number, the result is 2. Find the number.

87. Construction   The base of the triangular preformed concrete panel in Illustration 1 is 3 feet longer than twice its height, and its area is 45 square feet. How long is the base?

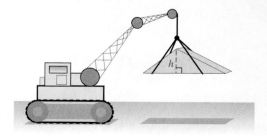

Illustration 1

88. Bombing run   A pilot releases a bomb from an altitude of 3000 feet. The bomb's height, h, above the target t seconds after its release is given by the formula $h = 3000 + 40t - 16t^2$. How long will it be before the bomb hits the target?

89. Gardening   A rectangular flower bed is 3 feet longer than twice its width, and its area is 27 square feet. Find its dimensions.

90. Geometry   A rectangle is 3 feet longer than it is wide. Its area is numerically equal to its perimeter. Find its dimensions.

CHAPTER TEST

In Problems 1 and 2, factor out the greatest common factor.

1. $4xy^2 + 12x^2y^3 - 6xz$

2. $3x^2(a + b) - 6xy(a + b)$

In Problems 3–14, factor each expression.

3. $ax + ay + bx + by$

4. $x^2 - 25$

5. $3a^2 - 27b^2$

6. $16x^4 - 81y^4$

7. $x^2 + 4x + 3$

8. $x^2 - 9x - 22$

9. $27 - x^3$

10. $3x^2 + 13x + 4$

11. $2y^2 + 5y - 12$

12. $2x^2 + 3xy - 2y^2$

13. $12 - 25x + 12x^2$

14. $2y^4 + 16y$

In Problems 15–20, solve each equation.

15. $(x - 5)(x + 5) = 0$

16. $2x^2 + 5x + 3 = 0$

17. $10x^2 - 13x = 9$

18. $-3(y - 6) + 2 = y^2 + 2$

19. Cannon fire   A cannonball is fired into the air with a velocity of 192 feet per second. In how many seconds will it hit the ground? (Its height above the ground is given by the formula $h = vt - 16t^2$, where v is the velocity and t is the time.)

20. Base of a triangle   The base of a triangle with an area of 40 square meters is 2 meters longer than its height. Find the base of the triangle.

CAREERS & MATHEMATICS

© Royalty-Free/CORBIS

Computer Programmer

Computers process vast quantities of information rapidly and accurately when they are given programs to follow. Computer programmers write those programs, which logically list the steps the machine must follow to organize data, solve a problem, or do some other task.

Programmers work from descriptions (prepared by systems analysts) of the task that the computer system should perform. The programmer then writes the specific program from the description, by breaking down each step into a series of coded instructions using one of several possible computer languages.

Application programmers are usually oriented toward business, engineering, or science. System programmers maintain the **software** that controls the computer system. Because of their knowledge of operating systems, system programmers often help application programmers determine the source of problems that may occur with their programs.

Qualifications

Most programmers have taken special courses in computer programming or have degrees in computer or information science, mathematics, engineering, or the physical sciences. Graduate degrees are required for some jobs.

Some employers who use computers for business applications do not require college degrees but prefer applicants who have had college courses in data processing.

Employers look for people who can think logically, who are capable of exacting, analytical work, and who demonstrate ingenuity and imagination when solving problems.

Job Outlook

The need for computer programmers will increase as business, government, schools, and scientific organizations develop new applications and require improvements in the software they already use. System programmers will be needed to develop and maintain the complex operating systems required by new, more powerful computer languages and by the networking of desktop computers. Job prospects are excellent into the next century for college graduates who have had computer-related courses — especially for those who have a major in computer science or a related field. Graduates of two-year programs in data-processing technologies also have good prospects, primarily in business applications.

Example Application

The polynomial

$$3x^4 + 2x^3 + 5x^2 + 7x + 1$$

can be written as

$$3 \cdot x \cdot x \cdot x \cdot x + 2 \cdot x \cdot x \cdot x + 5 \cdot x \cdot x + 7 \cdot x + 1$$

to illustrate that it involves ten multiplications and four additions. Multiplications are more time-consuming on a computer than are additions. Rewrite the polynomial to require fewer multiplications.

Solution Factor the common factor of x from the first four terms of the given polynomial as follows.

$$3x^4 + 2x^3 + 5x^2 + 7x + 1$$
$$x(3x^3 + 2x^2 + 5x + 7) + 1$$

Now factor the common x from the first three terms of the polynomial appearing within the parentheses.

$$x[x(3x^2 + 2x + 5) + 7] + 1$$

Again, factor an x from the first two terms within the set of parentheses to get

$$x\{x[x(3x + 2) + 5] + 7\} + 1$$

To emphasize the number of multiplications, we rewrite the previous expression as

$$x \cdot \{x \cdot [x \cdot (3 \cdot x + 2) + 5] + 7\} + 1$$

Each colored dot represents one multiplication. There are now only four multiplications, although there are still four additions. If the polynomial were to be evaluated for many different values of x, the savings in computer time afforded by using the revised form of the polynomial would be substantial.

EXERCISES

1. Evaluate the polynomial in the example at $x = -1$. Do this twice — once by direct substitution into the original polynomial, and then by substitution into the revised form.

2. Write $5x^7 + 3x^4 + 9x + 2$ in revised form.

3. Evaluate the polynomial of Exercise 2 at $x = 1$ by using each method.

4. Evaluate the polynomial of Exercise 2 at $x = -2$ by using each method.

Answers: **1.** 0 **2.** $x \cdot \{x \cdot [x \cdot x \cdot (5 \cdot x \cdot x \cdot x + 3)] + 9\} + 2$ **3.** 19 **4.** -608

Rational Expressions and Equations

© Royalty-Free/Getty Images

InfoTrac Project

Do a subject guide search on "volume," and under "volume (cubic content)," choose "periodicals." Find the article "Rules of thumb: oil flow, line volume, line-fill in barrels, small hole leakage." In the formula for V (speed), "d sup 2" means d^2. What is the speed of oil moving through a 17-inch pipe with a throughput of 225,000 barrels of oil per day? Using the formula for V (speed), solve for Q.

Two pipes are emptying into a large oil tanker at the same time. One pipe has an inner diameter of 14 inches and the other has an inner diameter of 15 inches. The oil flows through the 15-inch pipe 1.5 mph faster than through the 14-inch pipe. If the two pipes together have a throughput of 1,006,111 barrels per day, find the speed of the oil flowing through the 14-inch pipe.

Complete this project after studying Section 6.6.

PERSPECTIVE

Some Complex Fractions in Architecture

Each of the complex factions in the list

$$1 + \frac{1}{2}, \; 1 + \frac{1}{1 + \frac{1}{2}}, \; 1 + \frac{1}{1 + \frac{1}{1 + \frac{1}{2}}}, \; 1 + \frac{1}{1 + \frac{1}{1 + \frac{1}{1 + \frac{1}{2}}}}, \; \ldots$$

can be simplified by using the value of the expression preceding it. For example, to simplify the second expression in the list, replace $1 + \frac{1}{2}$ with $\frac{3}{2}$:

$$1 + \frac{1}{1 + \frac{1}{2}} = 1 + \frac{1}{\frac{3}{2}} = 1 + \frac{2}{3} = \frac{5}{3}$$

To simplify the third expression, replace

$$1 + \frac{1}{1 + \frac{1}{2}} \text{ with } \frac{5}{3}:$$

$$1 + \frac{1}{1 + \frac{1}{1 + \frac{1}{2}}} = 1 + \frac{1}{\frac{5}{3}} = 1 + \frac{3}{5} = \frac{8}{5}$$

Can you show that the expressions in the list simplify to the fractions $\frac{3}{2}, \frac{5}{3}, \frac{8}{5}, \frac{13}{8}, \frac{21}{13}, \frac{34}{21}, \ldots$? Do you see a pattern, and can you predict the next fraction?

Use a calculator to write each of these fractions as a decimal. The values produced get closer and closer to the irrational number $1.61803398875\ldots$, which is known as the golden ratio. This number often appears in the architecture of the ancient Greeks and Egyptians. The width of the stairs in front of the Greek Parthenon (Illustration 1), divided by the building's height, is the golden ratio. The height of the triangular face of the Great Pyramid of Cheops (Illustration 2), divided by the pyramid's width, is also the golden ratio.

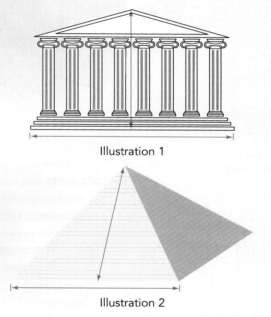

Illustration 1

Illustration 2

6.1 SIMPLIFYING RATIONAL EXPRESSIONS

- BASIC DEFINITIONS AND EXCLUDED VALUES
- SIMPLIFYING RATIONAL EXPRESSIONS
- EQUIVALENT RATIONAL EXPRESSIONS

▐ BASIC DEFINITIONS AND EXCLUDED VALUES

Just as the quotient of two integers (where the denominator is not equal to zero) is called a rational number, the quotient of two polynomials is called a **rational expression**. The denominator of this expression cannot equal zero since division by zero is undefined.

Rational Expression

A rational expression is any expression that can be written as $\dfrac{P}{Q}$, where P and Q are polynomials, $Q \neq 0$.

Examples of rational expressions are $\dfrac{3}{x+5}$, $\dfrac{2x^2}{3y}$, $\dfrac{2x^2+3x+4}{x^2-4}$, and $\dfrac{x^2+3}{x^2+1}$. An **excluded value** of a rational expression is the value(s) of the variable that would make the denominator equal 0. When the denominator of a fraction is equal to 0, the fraction is undefined. This is also true of a rational expression. To find the excluded values, set the denominator equal to zero and solve the resulting equation for the variable.

EXAMPLE 1

State the excluded values of **a.** $\dfrac{3}{x+5}$, **b.** $\dfrac{2x^2}{x^2+5x+6}$, **c.** $\dfrac{2x^2+3x+4}{x^2-4}$, and **d.** $\dfrac{x^2+3}{x^2+1}$.

SOLUTION

a. The fraction $\dfrac{3}{x+5}$ is undefined when its denominator equals zero.

$$x + 5 = 0 \qquad \text{Set the denominator equal to zero.}$$
$$x = -5 \qquad \text{Solve for the variable.}$$

The number -5 is an excluded value for the expression $\dfrac{3}{x+5}$. Another way to state this is $x \neq -5$, or $\{x \mid x \neq -5\}$.

b. The fraction $\dfrac{2x^2}{x^2+5x+6}$ is undefined when its denominator equals zero.

$$x^2 + 5x + 6 = 0 \qquad \text{Set the denominator equal to zero.}$$
$$(x + 3)(x + 2) = 0 \qquad \text{Because this is a quadratic (second-degree) equation, we factor the polynomial side.}$$
$$x + 3 = 0 \quad \text{or} \quad x + 2 = 0 \qquad \text{Use the zero factor property and set each factor equal to zero.}$$
$$x = -3 \quad \text{or} \quad x = -2 \qquad \text{Solve for the variable.}$$

The numbers -3 and -2 are excluded values for the rational expression $\dfrac{2x^2}{x^2+5x+6}$. Another way to state this is $x \neq -3$ or -2, or $\{x \mid x \neq -3 \text{ or } -2\}$.

c. The fraction $\dfrac{2x^2+3x+4}{x^2-4}$ is undefined when its denominator equals zero.

$$x^2 - 4 = 0 \qquad \text{Set the denominator equal to zero.}$$
$$(x - 2)(x + 2) = 0 \qquad \text{Because this is a quadratic (second-degree) equation, factor the polynomial side.}$$
$$x - 2 = 0 \quad \text{or} \quad x + 2 = 0 \qquad \text{Use the zero factor property and set each factor equal to zero.}$$
$$x = 2 \quad \text{or} \quad x = -2 \qquad \text{Solve for the variable.}$$

TECHNOLOGY TIP

Enter the expression $3/(x + 5)$ at the y1 prompt. Set the TABLE to begin at -5 and then access the TABLE.

X	Y1
-5	ERROR
-4	3
-3	1.5
-2	1
-1	.75
0	.6
1	.5

X= -5

This shows us that when $x = -5$, the expression entered at the y1 prompt, $\dfrac{3}{x+5}$, is undefined. This is the significance of the ERROR message that is displayed.

The numbers 2 and -2 are excluded values for the expression $\dfrac{2x^2 + 3x + 4}{x^2 - 4}$. Another way to state this is $x \neq 2$ or -2, or $\{x \mid x \neq 2 \text{ or } -2\}$.

d. The fraction $\dfrac{x^2 + 3}{x^2 + 1}$ is undefined when its denominator is equal to zero.

$$x^2 + 1 = 0 \qquad \text{Set the denominator equal to zero.}$$

There are *no* excluded values for the given rational expression because squaring a real number always yields a result that is either 0 or positive. Adding the constant 1 makes the denominator larger than 0.

▌ SIMPLIFYING RATIONAL EXPRESSIONS

A rational number is in lowest terms (simplified) when the numerator and the denominator have no common *factors* other than 1. The same is true of a rational expression. Recall the fundamental property of fractions from Chapter 1.

The Fundamental Property of Fractions

If a represents a real number, and b and x represent nonzero real numbers, then

$$\frac{a \cdot x}{b \cdot x} = \frac{a}{b}.$$

This property also is true for rational expressions. It allows the process of dividing out common factors in the numerator and the denominator of a rational expression (often called canceling by students). It is imperative to remember that only common *factors* can be divided out and that no factors of the denominator are equal to zero.

EXAMPLE 2

Simplify: **a.** $\dfrac{18}{30}$ and **b.** $\dfrac{21x^2y}{14xy^2}$.

SOLUTION

The simplification of a rational number and a rational expression will be paralleled so that the similarities are apparent.

Begin by writing the numerator and the denominator as a product of factors:

a. $\dfrac{18}{30} = \dfrac{2 \cdot 3 \cdot 3}{2 \cdot 3 \cdot 5}$

b. $\dfrac{21x^2y}{14xy^2} = \dfrac{3 \cdot 7 \cdot x \cdot x \cdot y}{2 \cdot 7 \cdot x \cdot y \cdot y}$

Divide out common factors (apply the fundamental property of fractions):

$$= \dfrac{\overset{1}{\cancel{2}} \cdot \overset{1}{\cancel{3}} \cdot 3}{\underset{1}{\cancel{2}} \cdot \underset{1}{\cancel{3}} \cdot 5}$$

$$= \dfrac{\overset{1}{\cancel{3}} \cdot \overset{1}{\cancel{7}} \cdot \cancel{x} \cdot x \cdot \overset{1}{\cancel{y}}}{\underset{1}{\cancel{2}} \cdot \underset{1}{\cancel{7}} \cdot \cancel{x} \cdot y \cdot \underset{1}{\cancel{y}}}$$

Express the numerator and the denominator as the product of the resulting factors:

$$= \dfrac{3}{5}$$

$$= \dfrac{3x}{2y}$$

Because the numerator and denominator have no common factors (other than 1) the fractions are simplified.

Just as the fraction $\dfrac{18}{30}$ is equivalent to $\dfrac{3}{5}$, the rational expression $\dfrac{21x^2y}{14xy^2}$ is equivalent to $\dfrac{3x}{2y}$ for all values of x and y except 0. The number zero is an excluded value for both variables x and y.

The rational expression could also have been simplified by using the rules of exponents:

$$\frac{21x^2y}{14xy^2} = \frac{3\cdot 7}{2\cdot 7}\cdot x^{2-1}\cdot y^{1-2}$$

$$= \frac{3}{2}xy^{-1}$$

$$= \frac{3}{2}\cdot\frac{x}{1}\cdot\frac{1}{y} \qquad \text{Recall } y^{-1}=\frac{1}{y}.$$

$$= \frac{3x}{2y}$$

The following steps can be used when simplifying rational expressions:

Simplifying a Rational Expression

1. Factor the numerator and the denominator completely.
2. Divide out common factors.

EXAMPLE 3   Simplify the rational expression $\dfrac{x^2+3x}{3x+9}$.

SOLUTION
$$\frac{x(x+3)}{3(x+3)} \qquad \text{Factor the numerator and the denominator.}$$

$$= \frac{x}{3} \qquad \text{Divide out the common factor of } (x+3).$$

The rational expression $\dfrac{x^2+3x}{3x+9}$ is equivalent to $\dfrac{x}{3}$ for all values of x *except* -3. Notice that the excluded value of -3 must be found by setting the denominator of the *original* expression equal to zero rather the denominator of the reduced expression.

From now on we will assume that our simplified rational expression is equivalent to the original rational expression for all values of the variable that *do not* make a denominator equal to zero.

EXAMPLE 4   Simplify: **a.** $\dfrac{x - y}{y - x}$ and   **b.** $\dfrac{2a - 1}{1 - 2a}$.

SOLUTION   We can rearrange terms in each numerator, factor out -1, and proceed as follows:

a. $\dfrac{x - y}{y - x} = \dfrac{-y + x}{y - x}$

$= \dfrac{-(y - x)}{y - x}$

$= \dfrac{\overset{1}{-\cancel{(y - x)}}}{\underset{1}{\cancel{y - x}}}$

$= -1$

b. $\dfrac{2a - 1}{1 - 2a} = \dfrac{-1 + 2a}{1 - 2a}$ Commutative property

$= \dfrac{-(1 - 2a)}{1 - 2a}$ Distributive property

$= \dfrac{\overset{1}{-\cancel{(1 - 2a)}}}{\underset{1}{\cancel{1 - 2a}}}$

$= -1$

The results of Example 4 illustrate the following fact.

Division of Negatives

The quotient of any nonzero expression and its opposite is -1.

 WARNING!   It is important to remember that only *factors* that are common to the *entire numerator* and the *entire denominator* can be divided out. Consider the arithmetic fraction $\dfrac{5 + 8}{5}$. It simplifies to $\dfrac{13}{5}$. It would be incorrect to simplify it by dividing out the common *term* of 5.

EXAMPLE 5   Simplify: $\dfrac{5(x + 3) - 5}{7(x + 3) - 7}$.

SOLUTION   Resist the urge to divide out the binomial $(x + 3)$. It is *not* a common factor of the *entire numerator* and the *entire denominator*. Instead, we simplify both the numerator and denominator, factor, and divide out any common factors.

$\dfrac{5(x + 3) - 5}{7(x + 3) - 7} = \dfrac{5x + 15 - 5}{7x + 21 - 7}$ Remove parentheses.

$= \dfrac{5x + 10}{7x + 14}$ Combine like terms.

$= \dfrac{5(x + 2)}{7(x + 2)}$ Factor the numerator and denominator.

$= \dfrac{5\overset{1}{\cancel{(x + 2)}}}{7\underset{1}{\cancel{(x + 2)}}}$ Divide out the common factor of $x + 2$.

$= \dfrac{5}{7}$

EXAMPLE 6

Simplify: $\dfrac{xy + 2x + 3y + 6}{x^2 + x - 6}$.

SOLUTION   We can factor the numerator by grouping and factor the denominator as a general trinomial.

$$\frac{xy + 2x + 3y + 6}{x^2 + x - 6} = \frac{x(y + 2) + 3(y + 2)}{(x - 2)(x + 3)}$$ Begin to factor the numerator and factor the denominator.

$$= \frac{(y + 2)(x + 3)}{(x - 2)(x + 3)}$$ Complete the factorization of the numerator.

$$= \frac{(y + 2)\overset{1}{\cancel{(x + 3)}}}{(x - 2)\underset{1}{\cancel{(x + 3)}}}$$ Divide out the common factor of $x + 3$.

$$= \frac{y + 2}{x - 2}$$

Do not assume every rational expression will simplify. Some are already in lowest terms.

EXAMPLE 7

Simplify: $\dfrac{x^2 + x - 2}{x^2 + x}$.

SOLUTION $\dfrac{x^2 + x - 2}{x^2 + x} = \dfrac{(x + 2)(x - 1)}{x(x + 1)}$ Factor the numerator and the denominator.

Because there are no factors common to the numerator and denominator, this fraction is already simplified.

The following property holds for both rational numbers (fractions) and rational expressions:

Properties of Fractions

$$\frac{a}{b} = \frac{-a}{-b} = -\frac{a}{-b} = -\frac{-a}{b}, \qquad (b \neq 0)$$

and

$$-\frac{a}{b} = \frac{-a}{b} = \frac{a}{-b}, \qquad (b \neq 0)$$

The second property of fractions is particularly useful when a result involves a negative. That negative can be placed with the numerator, with the denominator, or in front of the entire rational expression. The rational expression below can be written in several different ways, all of which are equivalent:

$$-\frac{3x+4}{2x-1} = \frac{-(3x+4)}{2x-1} = \frac{-3x-4}{2x-1} \quad \text{or} \quad -\frac{3x+4}{2x-1} = \frac{3x+4}{-(2x-1)} = \frac{3x+4}{-2x+1} = \frac{3x+4}{1-2x}$$

▍ EQUIVALENT RATIONAL EXPRESSIONS

When adding and subtracting fractions, it is often necessary to rewrite a given fraction as an equivalent fraction with a different denominator. The fundamental property of fractions allows us to divide out common *factors* **and** to multiply the numerator and denominator of a given fraction by the same quantity. This is equivalent to multiplying by the number 1, which is the identity for multiplication.

EXAMPLE 8

a. Rewrite the fraction $\frac{2}{3}$ with a denominator of 6.

b. Rewrite the fraction $\frac{3}{x-2}$ with a denominator of $x^2 - 4$.

SOLUTION   **a.** $\dfrac{2}{3} = \dfrac{?}{6}$ **b.** $\dfrac{3}{x-2} = \dfrac{?}{x^2-4}$

Begin by factoring denominators:

$$\frac{2}{3} = \frac{?}{2\cdot 3} \qquad\qquad \frac{3}{x-2} = \frac{?}{(x-2)(x+2)}$$

Comparing denominators (old to new), we see that in part (a) the new denominator has the extra factor of 2, and in part (b) the new denominator has the extra factor of $(x+2)$. We then multiply the original fraction by a fraction that has the appropriate "missing factor" as its numerator and denominator.

$$\frac{2}{3}\cdot\frac{2}{2} = \frac{4}{6} \qquad \frac{3}{x-2}\cdot\frac{x+2}{x+2} = \frac{3(x+2)}{(x-2)(x+2)} = \frac{3x+6}{(x-2)(x+2)}$$

For now, simplify numerators but leave denominators in factored form. The reason for this will become apparent when we add and subtract fractions in the next section.

EXAMPLE 9   Rewrite the fraction $\dfrac{4x+1}{2x+3}$ with a denominator of $2x^2 - 7x - 15$.

SOLUTION

$$\frac{4x+1}{2x+3} = \frac{?}{2x^2 - 7x - 15}$$

Factor denominators so that a comparison of factors can be made between the original denominator and the desired denominator. Notice that the original denominator is a prime polynomial. It is enclosed in parentheses so that it is treated as a single factor.

$$\frac{4x+1}{(2x+3)} = \frac{?}{(2x+3)(x-5)}$$

The missing factor is the quantity $(x-5)$.

Multiply the numerator and denominator of the original rational expression by the missing factor, $(x-5)$.

$$\frac{4x+1}{2x+3} \cdot \frac{x-5}{x-5}$$
$$= \frac{(4x+1)(x-5)}{(2x+3)(x-5)}$$
$$= \frac{4x^2 - 20x + x - 5}{(2x+3)(x-5)}$$ Simplify the numerator.
$$= \frac{4x^2 - 19x - 5}{(2x+3)(x-5)}$$

Therefore, $\dfrac{4x+1}{2x+3} = \dfrac{4x^2 - 19x - 5}{(2x+3)(x-5)}$.

EXERCISE 6.1

VOCABULARY AND NOTATION *In Exercises 1–8, fill in the blanks to make a true statement.*

1. The quotient of two polynomials is called a _____ expression.

2. The denominator of a fraction cannot equal _____.

3. The _____ is the value(s) of the variable that will make the denominator equal to 0.

4. The algebraic expressions $x - 3$ and $3 - x$ are _____ of one another.

5. To _____ a fraction means to reduce it to lowest terms.

6. A fraction is simplified when the numerator and the denominator have no common _____ (other than ± 1).

7. When simplifying a fraction, we factor the _____ and the _____ to divide out common factors.

8. The Fundamental Property of Fractions states that $\dfrac{a \cdot x}{b \cdot x} = $ ____.

CONCEPTS

9. State the excluded values of each rational expression.
   a. $\dfrac{3+x}{x}$ b. $\dfrac{5+x}{2x}$
   c. $\dfrac{x+2}{x-5}$ d. $\dfrac{2(x+4)}{x+6}$

10. One by one, enter each of the rational expressions in Exercises 9(a) and 9(d) at the y1 prompt and record the y1 value displayed in the TABLE for each restricted value. Explain why the "values" are the same and their meaning.

11. State the excluded value(s) of each rational expression.

a. $\dfrac{6 + x}{x - 2}$ **b.** $\dfrac{3x + 5}{x}$

c. $\dfrac{x + 2}{3x}$ **d.** $\dfrac{3(x - 1)}{x + 5}$

e. $\dfrac{x + 1}{x^2 - 5x - 6}$

12. For each TABLE below, a rational expression has been entered at the y1 prompt. Based on the TABLE, determine the excluded value(s) for each rational expression.

a.
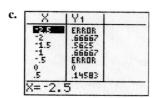

X	Y1
0	0
1	-.25
2	-.6667
3	-1.5
4	-4
5	ERROR
6	6

X=0

b.

X	Y1
-3	-.5
-2	ERROR
-1	.25
0	0
1	-.1667
2	-.5
3	ERROR

X=-3

c.

X	Y1
-2.5	ERROR
-2	.66667
-1.5	.5625
-1	.66667
-.5	ERROR
0	0
.5	.14583

X=-2.5

13. Simplify the following. Assume that no variable has a value that would make the denominator 0.

a. $\dfrac{x + y}{y + x}$ **b.** $\dfrac{x - y}{y - x}$

c. $\dfrac{a - 5}{5 - a}$ **d.** $\dfrac{a + 5}{5 + a}$

14. Simplify each rational expression. Assume that no variable has a value that would make the denominator 0.

a. $\dfrac{8}{52}$ **b.** $\dfrac{15}{21}$

c. $\dfrac{-6x}{18}$ **d.** $\dfrac{-25y}{5}$

e. $\dfrac{x + 3}{3(x + 3)}$ **f.** $\dfrac{2(x + 7)}{(x + 7)}$

15. Which of the following expressions is/are equivalent to $-\dfrac{x - y}{x + y}$?

a. $\dfrac{-x - y}{x + y}$ **b.** $\dfrac{y - x}{x + y}$

c. $\dfrac{x - y}{-x + y}$ **d.** $\dfrac{-x + y}{-x - y}$

16. Which of the following expressions is/are equivalent to $-\dfrac{x + y}{x - y}$?

a. $\dfrac{-(x + y)}{-(x - y)}$ **b.** $\dfrac{-x - y}{x - y}$

c. 1 **d.** $\dfrac{-x - y}{x + y}$

17. Rewrite each fraction as an equivalent fraction with the given denominator.

a. $\dfrac{25}{4} = \dfrac{?}{20}$ **b.** $\dfrac{19}{21} = \dfrac{?}{42}$

c. $\dfrac{8}{x} = \dfrac{?}{x^2 y}$ **d.** $\dfrac{7}{y} = \dfrac{?}{xy^2}$

PRACTICE *In Exercises 18–23, state the excluded value(s) of each rational expression (if any exist).*

18. $\dfrac{x^2 + 9}{x^2 + 5}$

19. $\dfrac{x - 4}{x^2 + 4}$

20. $\dfrac{3x + 2}{2x - 3}$ **21.** $\dfrac{5x - 4}{4x + 3}$

22. $\dfrac{x - 3}{x^2 - 2x - 24}$ **23.** $\dfrac{x + 5}{x^2 - x - 30}$

In Exercises 24–57, simplify each rational expression. If a fraction is already simplified, so state. Assume that no variable has a value that would make a denominator equal to zero.

24. $\dfrac{15x^2 y^2}{5xy^2}$ **25.** $\dfrac{12xz^3}{4xz^2}$

26. $\dfrac{5x + 35}{x + 7}$ **27.** $\dfrac{x - 9}{3x - 27}$

28. $\dfrac{x^2 + 3x}{2x + 6}$ **29.** $\dfrac{xz - 2x}{yz - 2y}$

30. $\dfrac{15x - 3x^2}{25y - 5xy}$ **31.** $\dfrac{3y + xy}{3x + xy}$

32. $\dfrac{x - 7}{7 - x}$ **33.** $\dfrac{d - c}{c - d}$

34. $\dfrac{6x - 3y}{3y - 6x}$

35. $\dfrac{3c - 4d}{4d - 3c}$

36. $\dfrac{a + b - c}{c - a - b}$

37. $\dfrac{x - y - z}{z + y - x}$

38. $\dfrac{x^2 + 3x + 2}{x^2 + x - 2}$

39. $\dfrac{x^2 + x - 6}{x^2 - x - 2}$

40. $\dfrac{x^2 - 8x + 15}{x^2 - x - 6}$

41. $\dfrac{x^2 - 6x - 7}{x^2 + 8x + 7}$

42. $\dfrac{2x^2 - 8x}{x^2 - 6x + 8}$

43. $\dfrac{3y^2 - 15y}{y^2 - 3y - 10}$

44. $\dfrac{xy + 2x^2}{2xy + y^2}$

45. $\dfrac{3x + 3y}{x^2 + xy}$

46. $\dfrac{x^2 + 3x + 2}{x^3 + x^2}$

47. $\dfrac{6x^2 - 13x + 6}{3x^2 + x - 2}$

48. $\dfrac{x^2 - 8x + 16}{x^2 - 16}$

49. $\dfrac{3x + 15}{x^2 - 25}$

50. $\dfrac{x^2 - 2x - 15}{x^2 + 2x - 15}$

51. $\dfrac{x^2 + 4x - 77}{x^2 - 4x - 21}$

52. $\dfrac{x^2 - 3(2x - 3)}{9 - x^2}$

53. $\dfrac{x(x - 8) + 16}{16 - x^2}$

54. $\dfrac{4(x + 3) + 4}{3(x + 2) + 6}$

55. $\dfrac{4 + 2(x - 5)}{3x - 5(x - 2)}$

56. $\dfrac{x^2 - 9}{(2x + 3) - (x + 6)}$

57. $\dfrac{x^2 + 5x + 4}{2(x + 3) - (x + 2)}$

In Exercises 58–63, (a) Using the STOre feature of your calculator, evaluate the given rational expressions when $x = 2$, $y = -3$, and $z = 5$. Copy the screen displays exactly as they appear. (b) Algebraically simplify the given expression. (c) Evaluate the reduced form of the rational expression for the given values of the variable. The results from parts (a) and (c) should be equal.

58. $\dfrac{28x}{32y}$

59. $\dfrac{14xz^2}{7x^2z^2}$

60. $\dfrac{6x - 6y + 6z}{9x - 9y + 9z}$

61. $\dfrac{3x - 3y - 6}{2x - 2y - 4}$

62. $\dfrac{2x^2 - 8}{x^2 - 3x + 2}$

63. $\dfrac{3x^2 - 27}{x^2 + 3x - 18}$

In Exercises 64–73, rewrite each fraction as an equivalent fraction with the given denominator.

64. $\dfrac{3x}{x + 1}; (x + 1)^2$

65. $\dfrac{5y}{y - 2}; (y - 2)^2$

66. $\dfrac{2y}{x}; x^2 + x$

67. $\dfrac{3x}{y}; y^2 - y$

68. $\dfrac{z}{z - 1}; z^2 - 1$

69. $\dfrac{y}{y + 2}; y^2 - 4$

70. $\dfrac{x + 2}{x - 2}; x^2 - 4$

71. $\dfrac{x - 3}{x + 3}; x^2 - 9$

72. $\dfrac{2}{x + 1}; x^2 + 3x + 2$

73. $\dfrac{3}{x - 1}; x^2 + x - 2$

74. Graph the rational expression $\dfrac{x + 3}{x - 3}$ on your graphing calculator.

    a. Sketch the graph (in the standard viewing window).

    b. Reset your calculator in dot mode and graph the same expression. Sketch the graph. (*Note:* See Technology 3.4 for instructions on the use of the graph style icon for your particular calculator.)

    c. Explain (reconcile) the differences between the two graphs. (*Hint:* Think about the excluded values of the rational expression.)

75. Graph the rational expression $\dfrac{x - 4}{x + 4}$ on your graphing calculator.

    a. Sketch the graph (in the standard viewing window).

    b. Reset your calculator in dot mode and graph the same expression. Sketch the graph. (*Note:* See Technology 3.4 for instructions on the use of the graph style icon for your particular calculator.)

    c. Explain (reconcile) the differences between the two graphs. (*Hint:* Think about the excluded values of the rational expression.)

76. a. Simplify the rational expression

$$\dfrac{x(x + 3) - 3(x - 1)}{x^2 + 3} \text{ algebraically.}$$

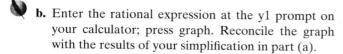

b. Enter the rational expression at the y1 prompt on your calculator; press graph. Reconcile the graph with the results of your simplification in part (a).

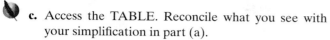

c. Access the TABLE. Reconcile what you see with your simplification in part (a).

77. a. Simplify the rational expression

$$\frac{x(x-5) + 2(x-5)}{x^2 - 3x - 10}$$ algebraically.

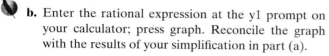

b. Enter the rational expression at the y1 prompt on your calculator; press graph. Reconcile the graph with the results of your simplification in part (a).

c. Access the TABLE. Reconcile what you see with your simplification in part (a).

REVIEW

78. Factor completely:

  a. $4x^3 - 10x^2 + 6x$

  b. $y^2 + 2xy - 63x^2$

  c. $15 + 3b - 5a^2 - a^2 b$

79. The perimeter of a triangle is 18 cm. If the third side is 3 cm less than the first side and the second side is 5 cm more than twice the first side, find the measure of the first side.

80. Graph $y = \frac{2}{3}x - 5$ by plotting the x- and y-intercepts.

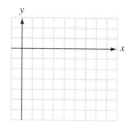

81. Graph $y = \frac{2}{3}x - 5$ in the standard viewing window and sketch the graph.

82. Solve $\frac{1}{2}(x - 1) + 2 = 3 - 4x$ graphically. Copy your screen display to justify your work.

6.2 MULTIPLYING AND DIVIDING RATIONAL EXPRESSIONS

- MULTIPLYING RATIONAL EXPRESSIONS
- DIVIDING RATIONAL EXPRESSIONS
- COMBINED OPERATIONS

MULTIPLYING RATIONAL EXPRESSIONS

Recall that to multiply rational numbers (fractions) we multiply the numerators and multiply the denominators. The same is true for rational expressions.

Multiplying Rational Expressions

If $\dfrac{P}{Q}$ and $\dfrac{R}{S}$ are rational expressions, then $\dfrac{P}{Q} \cdot \dfrac{R}{S} = \dfrac{PR}{QS}$.

Because all rational expressions should be ultimately expressed in simplified form, it is suggested that you *express* the product of the numerators over the product of the denominators rather than actually multiplying the polynomials in the numerators and denominators. The reason for this is because simplification occurs when common factors are divided.

EXAMPLE 1

Multiply: $\dfrac{35x^2y}{7y^2z} \cdot \dfrac{z}{5xy}$.

SOLUTION

$$\dfrac{35x^2y}{7y^2z} \cdot \dfrac{z}{5xy} = \dfrac{35x^2y \cdot z}{7y^2z \cdot 5xy} = \dfrac{7 \cdot 5 \cdot x \cdot x \cdot y \cdot z}{7 \cdot y \cdot y \cdot z \cdot 5 \cdot x \cdot y}$$

$$= \dfrac{\overset{1}{7} \cdot \overset{1}{5} \cdot \overset{1}{x} \cdot x \cdot \overset{1}{y} \cdot \overset{1}{z}}{\underset{1}{7} \cdot \underset{1}{y} \cdot y \cdot \underset{1}{z} \cdot \underset{1}{5} \cdot \underset{1}{x} \cdot y}$$

Divide out common factors to simplify.

$$= \dfrac{x}{y^2}$$

EXAMPLE 2

Find the product: $\dfrac{x^2 - x}{2x + 4} \cdot \dfrac{x + 2}{x}$.

SOLUTION

$$\dfrac{x^2 - x}{2x + 4} \cdot \dfrac{x + 2}{x} = \dfrac{(x^2 - x)(x + 2)}{(2x + 4)(x)}$$

Multiply rational expressions by *expressing* the product of the numerators over the product of the denominators.

$$= \dfrac{x(x - 1)(x + 2)}{2(x + 2)(x)}$$

Factor the numerator and denominator completely.

$$= \dfrac{\overset{1}{x}(x - 1)\overset{1}{(x + 2)}}{2\underset{1}{(x + 2)}\underset{1}{(x)}}$$

Divide out common factors to simplify.

$$\dfrac{x - 1}{2}$$

STUDY TIP

Place a piece of paper over the solution to an example in your text. Read the example and try to work the problem on your paper. If you get stuck, slide the paper down and jump-start yourself.

EXAMPLE 3

Find the product of $\dfrac{x^2 - 3x}{x^2 - x - 6}$ and $\dfrac{x^2 + x - 2}{x^2 - x}$.

SOLUTION:

$$\dfrac{x^2 - 3x}{x^2 - x - 6} \cdot \dfrac{x^2 + x - 2}{x^2 - x} = \dfrac{(x^2 - 3x)(x^2 + x - 2)}{(x^2 - x - 6)(x^2 - x)}$$

$$= \dfrac{x(x - 3)(x + 2)(x - 1)}{(x - 3)(x + 2)x(x - 1)}$$

$$= \dfrac{\overset{1}{x}\overset{1}{(x - 3)}\overset{1}{(x + 2)}\overset{1}{(x - 1)}}{\underset{1}{(x - 3)}\underset{1}{(x + 2)}\underset{1}{x}\underset{1}{(x - 1)}}$$

Divide out common factors.

$$= 1$$

The product, $\dfrac{x^2 - 3x}{x^2 - x - 6} \cdot \dfrac{x^2 + x - 2}{x^2 - x}$, is equal to the constant 1 for all values of x except the excluded values of 3, −2, 0, and 1. This is verified next using different options available on the graphing calculator.

First, store any value for x (other than an excluded value) and evaluate the original expression:

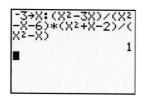

Another option is to enter the expression at y1 and access the TABLE. The advantage of this method is that it allows you to evaluate the expression for many values of the variable at one time.

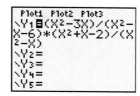 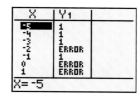

Notice that $-2, 0$, and 1 yield an error message because they are the excluded values. If we scroll down the table to where $x = 3$ we would also see an error message in the y1 column.

EXAMPLE 4

Multiply: $\dfrac{x^2 + 2x}{xy - 2y} \cdot \dfrac{x + 1}{x^2 - 4} \cdot \dfrac{x - 2}{x^2 + x}$.

SOLUTION

$\dfrac{x^2 + 2x}{xy - 2y} \cdot \dfrac{x + 1}{x^2 - 4} \cdot \dfrac{x - 2}{x^2 + x}$

Express the product of the numerators over the product of the denominators.

$= \dfrac{x(x + 2)(x + 1)(x - 2)}{y(x - 2)(x + 2)(x - 2)x(x + 1)}$

Factor the numerator and the denominator completely.

$= \dfrac{\overset{1}{\cancel{x}}\overset{1}{\cancel{(x + 2)}}\overset{1}{\cancel{(x + 1)}}\overset{1}{\cancel{(x - 2)}}}{y\underset{1}{\cancel{(x - 2)}}\underset{1}{\cancel{(x + 2)}}(x - 2)\underset{1}{\cancel{x}}\underset{1}{\cancel{(x + 1)}}}$

Divide out common factors.

$= \dfrac{1}{y(x - 2)}$

▌▌▌ DIVIDING RATIONAL EXPRESSIONS

Recall that when dividing rational numbers we multiply by the *reciprocal* (multiplicative inverse) of the divisor. This is stated formally below:

Dividing Rational Expressions

If $\dfrac{P}{Q}$ and $\dfrac{R}{S}$ are rational expressions, then $\dfrac{P}{Q} \div \dfrac{R}{S} = \dfrac{P}{Q} \cdot \dfrac{S}{R} = \dfrac{PS}{QR}$.

EXAMPLE 5   Divide: $\dfrac{x^2 + x}{3x - 15} \div \dfrac{x^2 + 2x + 1}{6x - 30}$.

SOLUTION $\dfrac{x^2 + x}{3x - 15} \div \dfrac{x^2 + 2x + 1}{6x - 30}$

$= \dfrac{x^2 + x}{3x - 15} \cdot \dfrac{6x - 30}{x^2 + 2x + 1}$ Invert the divisor and multiply.

$= \dfrac{x(x + 1) \cdot 2 \cdot 3(x - 5)}{3(x - 5)(x + 1)(x + 1)}$ Express the product of the numerators over the product of the denominators. Factor the numerator and the denominator completely.

$= \dfrac{x \overset{1}{\cancel{(x + 1)}} \cdot 2 \cdot \overset{1}{\cancel{3}} \overset{1}{\cancel{(x - 5)}}}{\underset{1}{\cancel{3}} \, \underset{1}{\cancel{(x - 5)}} \, \underset{1}{\cancel{(x + 1)}}(x + 1)}$ Divide out common factors.

$= \dfrac{2x}{x + 1}$

EXAMPLE 6   Divide: $\dfrac{2x^2 - 3x - 2}{2x + 1} \div (4 - x^2)$.

SOLUTION $\dfrac{2x^2 - 3x - 2}{2x + 1} \div (4 - x^2)$

$= \dfrac{2x^2 - 3x - 2}{2x + 1} \cdot \dfrac{1}{(4 - x^2)}$ Invert the divisor and multiply.

$\dfrac{(2x + 1)(x - 2) \cdot 1}{(2x + 1)(2 + x)(2 - x)}$ Express the product of the numerators over the product of the denominators. Factor the numerator and the denominator completely.

$= \dfrac{\overset{1}{\cancel{(2x + 1)}} \overset{-1}{\cancel{(x - 2)}} \cdot 1}{\underset{1}{\cancel{(2x + 1)}}(2 + x)\underset{1}{\cancel{(2 - x)}}}$ Divide out common factors. Because $(x - 2)$ and $(2 - x)$ are opposites, their quotient is -1.

$= \dfrac{-1}{2 + x}$

which could be written as $-\dfrac{1}{2 + x}$ or $\dfrac{1}{-(2 + x)}$ or $\dfrac{1}{-2 - x}$.

▮ COMBINED OPERATIONS

The order of operations holds for the addition, subtraction, multiplication, and division of rational expressions. Unless parentheses indicate otherwise, perform multiplications and divisions in order from left to right.

EXAMPLE 7

Simplify the expression $\dfrac{x^2 - x - 6}{x - 2} \div \dfrac{x^2 - 4x}{x^2 - x - 2} \cdot \dfrac{x - 4}{x^2 + x}$.

SOLUTION

$$\dfrac{x^2 - x - 6}{x - 2} \div \dfrac{x^2 - 4x}{x^2 - x - 2} \cdot \dfrac{x - 4}{x^2 + x}$$

$$= \dfrac{x^2 - x - 6}{x - 2} \cdot \dfrac{x^2 - x - 2}{x^2 - 4x} \cdot \dfrac{x - 4}{x^2 + x}$$
Invert the divisor and multiply.

$$= \dfrac{(x + 2)(x - 3)(x + 1)(x - 2)(x - 4)}{(x - 2)x(x - 4)x(x + 1)}$$
Express the product of the numerators over the product of the denominators. Factor the numerator and the denominator completely.

$$= \dfrac{(x + 2)(x - 3)\overset{1}{\cancel{(x + 1)}}\overset{1}{\cancel{(x - 2)}}\overset{1}{\cancel{(x - 4)}}}{\underset{1}{\cancel{(x - 2)}}x\underset{1}{\cancel{(x - 4)}}x\underset{1}{\cancel{(x + 1)}}}$$
Divide out common factors.

$$= \dfrac{(x + 2)(x - 3)}{x^2}$$

EXERCISE 6.2

VOCABULARY AND NOTATION *In Exercises 1–4, fill in the blanks to make a true statement.*

1. If $\dfrac{P}{Q}$ and $\dfrac{R}{S}$ are rational expressions, then $\dfrac{P}{Q} \cdot \dfrac{R}{S} =$

_____ .

2. If $\dfrac{P}{Q}$ and $\dfrac{R}{S}$ are rational expressions, then

$\dfrac{P}{Q} \div \dfrac{R}{S} = \dfrac{P}{Q} \cdot \dfrac{}{} = $.

3. When dividing rational expressions, two changes are made: we find the _____ of the divisor and change the operation from division to _____ .

4. We can always write a polynomial with a denominator of ___ .

In Exercises 5–14, perform the indicated multiplications and divisions. Simplify answers if possible.

5. $\dfrac{25}{35} \cdot \dfrac{21}{55}$

6. $\dfrac{27}{24} \cdot \dfrac{56}{35}$

7. $\dfrac{2}{3} \cdot \dfrac{15}{2} \cdot \dfrac{1}{7}$

8. $\dfrac{2}{5} \cdot \dfrac{10}{9} \cdot \dfrac{3}{2}$

9. $\dfrac{5y}{7} \cdot \dfrac{7x}{5z}$

10. $\dfrac{4x}{3y} \cdot \dfrac{3y}{7x}$

11. $\dfrac{35}{2} \div \dfrac{15}{2}$

12. $\dfrac{6}{14} \div \dfrac{10}{35}$

13. $\dfrac{x^2}{3} \div \dfrac{2x}{4}$

14. $\dfrac{z^2}{z} \div \dfrac{z}{3z}$

PRACTICE *In Exercises 15–34, perform the indicated multiplications. Simplify answers if possible.*

15. $\dfrac{8x^2y^2}{4x^2} \cdot \dfrac{2xy}{2y}$

16. $\dfrac{9x^2y}{3x} \cdot \dfrac{3xy}{3y}$

17. $\dfrac{-2xy}{x^2} \cdot \dfrac{3xy}{2}$

18. $\dfrac{-3x}{x^2} \cdot \dfrac{2xz}{3}$

19. $\dfrac{z + 7}{7} \cdot \dfrac{z + 2}{z}$

20. $\dfrac{a - 3}{a} \cdot \dfrac{a + 3}{5}$

21. $\dfrac{x-2}{2} \cdot \dfrac{2x}{x-2}$

22. $\dfrac{y+3}{y} \cdot \dfrac{3y}{y+3}$

23. $\dfrac{(x+1)^2}{x+1} \cdot \dfrac{x+2}{x+1}$

24. $\dfrac{(y-3)^2}{y-3} \cdot \dfrac{y-3}{y-3}$

25. $\dfrac{2x+6}{x+3} \cdot \dfrac{3}{4x}$

26. $\dfrac{3y-9}{y-3} \cdot \dfrac{y}{3y^2}$

27. $\dfrac{x^2+x-6}{5x} \cdot \dfrac{5x-10}{x+3}$

28. $\dfrac{z^2+4z-5}{5z-5} \cdot \dfrac{5z}{z+5}$

29. $\dfrac{x^2+7xy+12y^2}{x^2+2xy-8y^2} \cdot \dfrac{x^2-xy-2y^2}{x^2+4xy+3y^2}$

30. $\dfrac{m^2+9mn+20n^2}{m^2-25n^2} \cdot \dfrac{m^2-9mn+20n^2}{m^2-16n^2}$

31. $\dfrac{3r^2+15rs+18s^2}{6r^2-24s^2} \cdot \dfrac{2r-4s}{3r+9s}$

32. $\dfrac{2u^2+8u}{2u+8} \cdot \dfrac{4u^2+8uv+4v^2}{u^2+5uv+4v^2}$

33. $\dfrac{x^2+5x+6}{x^2} \cdot \dfrac{x^2-2x}{x^2-9} \cdot \dfrac{x^2-3x}{x^2-4}$

34. $\dfrac{x^2-1}{1-x} \cdot \dfrac{4x}{x+x^2} \cdot \dfrac{x^2+2x+1}{2x+2}$

In Exercise 35–48, perform the indicated divisions. Simplify answers when possible.

35. $\dfrac{3x}{2} \div \dfrac{x}{2}$

36. $\dfrac{y}{6} \div \dfrac{2}{3y}$

37. $\dfrac{3x}{y} \div \dfrac{2x}{4}$

38. $\dfrac{12x}{y} \div 2x$

39. $\dfrac{4x}{3x} \div \dfrac{2y}{9y}$

40. $\dfrac{14}{7y} \div \dfrac{10}{5z}$

41. $\dfrac{x^2y}{3xy} \div \dfrac{xy^2}{6y}$

42. $\dfrac{2xz}{z} \div \dfrac{4x^2}{z^2}$

43. $\dfrac{x+2}{3x} \div \dfrac{x+2}{2}$

44. $\dfrac{z-3}{3z} \div \dfrac{z+3}{z}$

45. $\dfrac{(z-2)^2}{3z^2} \div \dfrac{z-2}{6z}$

46. $\dfrac{(x+7)^2}{x+7} \div \dfrac{(x-3)^2}{x+7}$

47. $\dfrac{5x^2+13x-6}{x+3} \div \dfrac{5x^2-17x+6}{x-2}$

48. $\dfrac{x^2-x-6}{2x^2+9x+10} \div \dfrac{x^2-25}{2x^2+15x+25}$

In Exercises 49–58, perform the indicated operations. The order of operations is valid for rational expressions and must be followed.

49. $\dfrac{2}{3} \cdot \dfrac{15}{5} \div \dfrac{10}{5}$

50. $\dfrac{6}{5} \div \dfrac{3}{5} \cdot \dfrac{5}{15}$

51. $\dfrac{x^2}{18} \div \dfrac{x^3}{6} \div \dfrac{12}{x^2}$

52. $\dfrac{y^3}{3y} \cdot \dfrac{3y^2}{4} \div \dfrac{15}{20}$

53. $\dfrac{x^2-1}{x^2-9} \cdot \dfrac{x+3}{x+2} \div \dfrac{5}{x+2}$

54. $\dfrac{2}{3x-3} \div \dfrac{2x+2}{x-1} \cdot \dfrac{5}{x+1}$

55. $\dfrac{x-x^2}{x^2-4}\left(\dfrac{2x+4}{x+2} \div \dfrac{5}{x+2}\right)$

56. $\dfrac{2}{3x-3} \div \left(\dfrac{2x+2}{x-1} \cdot \dfrac{5}{x+1}\right)$

57. $\dfrac{y^2}{x+1} \cdot \dfrac{x^2+2x+1}{x^2-1} \div \dfrac{3y}{xy-y}$

58. $\dfrac{x^2-y^2}{x^4-x^3} \div \dfrac{x-y}{x^2} \div \dfrac{x^2+2xy+y^2}{x+y}$

REVIEW

59. Simplify: $\dfrac{x^3+2x^2-25x-50}{x^2-10x+25}$.

60. Evaluate $\dfrac{x^3+2x^2-25x-50}{x^2-10x+25}$ when

    a. $x=3$

    b. $x=5$

     c. Explain the significance of the result of part (b).

61. Solve $4x-3y=7$ for y.

62. Solve:

    a. $|3x-1|=2$

    b. $|x+5|-4=-1$

    c. $|2x+1|=|x-5|$

6.3 ADDING AND SUBTRACTING RATIONAL EXPRESSIONS

- ADDING AND SUBTRACTING RATIONAL EXPRESSIONS WITH LIKE DENOMINATORS
- FINDING THE LCD

- ADDING AND SUBTRACTING RATIONAL EXPRESSIONS WITH UNLIKE DENOMINATORS

▓ ADDING AND SUBTRACTING RATIONAL EXPRESSIONS WITH LIKE DENOMINATORS

Rational expressions can be added and subtracted in exactly the same manner in which rational numbers are added or subtracted. As long as the denominators are the same, keep the common denominator, add (or subtract) the numerators, and then simplify the rational expressions.

Adding or Subtracting Rational Expressions with Common Denominators

If $\dfrac{P}{Q}$ and $\dfrac{R}{Q}$ are rational expressions, then $\dfrac{P}{Q} + \dfrac{R}{Q} = \dfrac{P + R}{Q}$ and $\dfrac{P}{Q} - \dfrac{R}{Q} = \dfrac{P - R}{Q}$.

EXAMPLE 1

Add or subtract, as the signs indicate. Simplify results, if possible.

a. $\dfrac{x}{7} + \dfrac{3x}{7}$ **b.** $\dfrac{5x}{3} - \dfrac{2x}{3}$ **c.** $\dfrac{3x + y}{5x} + \dfrac{x + y}{5x}$

SOLUTION

a. $\dfrac{x}{7} + \dfrac{3x}{7} = \dfrac{x + 3x}{7}$

$= \dfrac{4x}{7}$

b. $\dfrac{5x}{3} - \dfrac{2x}{3} = \dfrac{5x - 2x}{3}$

$= \dfrac{3x}{3}$ Combine like terms.

$= \dfrac{x}{1}$ Simplify.

$= x$ Denominators of 1 need not be written.

c. $\dfrac{3x + y}{5x} + \dfrac{x + y}{5x} = \dfrac{3x + y + x + y}{5x}$

$= \dfrac{4x + 2y}{5x}$ Combine like terms.

Factoring the numerator, $\dfrac{2(2x + y)}{5x}$, shows that the rational expression does not simplify. The final result could be left in factored form, if desired.

WARNING! When subtracting fractions with polynomial numerators, care must be taken to use the distributive property to subtract the entire numerator rather than just a single term from the numerator.

EXAMPLE 2

Subtract and simplify: $\dfrac{5x + 1}{x - 3} - \dfrac{4x - 2}{x - 3}$.

SOLUTION

$$\frac{5x + 1}{x - 3} - \frac{4x - 2}{x - 3} = \frac{(5x + 1) - (4x - 2)}{x - 3}$$

Express the subtraction of the numerators over the common denominator of $(x - 3)$.

$$= \frac{5x + 1 - 4x + 2}{x - 3}$$

Use the distributive property to remove grouping in the numerator.

$$= \frac{x + 3}{x - 3}$$

Combine like terms.

EXAMPLE 3

Perform the indicated operations: $\dfrac{3xy}{x - y} - \dfrac{x(3y - x)}{x - y} - \dfrac{x(x - y)}{x - y}$.

SOLUTION

$$\frac{3xy}{x - y} - \frac{x(3y - x)}{x - y} - \frac{x(x - y)}{x - y} = \frac{3xy - x(3y - x) - x(x - y)}{x - y}$$

Express the subtraction of the numerators over the common denominator.

$$= \frac{3xy - 3xy + x^2 - x^2 + xy}{x - y}$$

Use the distributive property to remove parentheses.

$$= \frac{xy}{x - y}$$

Combine like terms.

▏▏▏ FINDING THE LCD

To add rational numbers, they must first be converted to fractions with equivalent denominators (if they do not have like denominators originally). The same is true for adding and subtracting rational expressions. To find the least (lowest) common denominator, use the following steps:

Finding the Least Common Denominator (LCD)

1. Write down each of the different denominators that appear in the given fractions.
2. Factor each of these denominators completely; use exponential form.
3. Form a product using each of the different factors obtained in step 2. Use each different factor the *greatest* number of times it appears in any *single* factorization. This product is the least common denominator.

EXAMPLE 4

Find the LCD of the given pairs of rational expressions:

a. $\dfrac{5}{4x^4y^2}; \dfrac{3}{6x^3y^3}$ **b.** $\dfrac{3x}{x+2}; \dfrac{5x}{x^2-4}$ **c.** $\dfrac{x-2}{x^2+2x-15}; \dfrac{x+2}{x^2-6x+9}$

SOLUTION

a. $4x^4y^2; 6x^3y^3$

$4x^4y^2 = 2^2 \cdot x^4 \cdot y^2$

$6x^3y^3 = 2 \cdot 3 \cdot x^3 \cdot y^3$

$\text{LCD} = 2^2 \cdot 3 \cdot x^4 \cdot y^3$

$\quad = 12x^4y^3$

Write down each denominator.

Factor each denominator.

The LCD is the product of the different factors. The highest power of any common factor is chosen.

b. $x+2; x^2-4$

$x^2 - 4 = (x+2)(x-2)$

$\text{LCD} = (x+2)(x-2)$

Write down each denominator.

Factor each denominator completely.

The LCD is the product of the different factors.

c. $x^2 + 2x - 15; x^2 - 6x + 9$

$x^2 + 2x - 15 = (x+5)(x-3)$

$x^2 - 6x + 9 = (x-3)(x-3) = (x-3)^2$

$\text{LCD} = (x+5)(x-3)^2$

Write down each denominator.

Factor each denominator completely.

The LCD is the product of the different factors. The highest power of any common factor is chosen.

▌ ADDING AND SUBTRACTING RATIONAL EXPRESSIONS WITH UNLIKE DENOMINATORS

To add or subtract rational expressions with unlike denominators, use the following five-step process:

Adding/Subtracting with Unlike Denominators

1. Identify the LCD of the rational expressions. Use the process outlined earlier in this section.
2. Convert each rational expression to an equivalent rational expression having the LCD as denominator. This process was demonstrated in Section 6.1.
3. Add or subtract the rational expressions by *expressing* the addition or subtraction over the common denominator.
4. Simplify the numerators, but leave the denominators in factored form.
5. Simplify the rational expression. Recall that this requires you to divide out common factors in the numerator and the denominator. Because the denominator is in factored form, you need only factor the numerator.

EXAMPLE 5

Add or subtract as the signs indicate: $\dfrac{3}{x^2y} + \dfrac{2}{xy} - \dfrac{1}{xy^2}$.

SOLUTION   Find the least common denominator.

$$
\left.\begin{array}{l} x^2y \\ xy \\ xy^2 \end{array}\right\} \quad \text{Write down each denominator.}
$$

$$
\text{LCD} = x^2y^2 \qquad \begin{array}{l}\text{The LCD is the product of the different factors. The highest power}\\ \text{of any common base is chosen.}\end{array}
$$

We rewrite each original fraction as an equivalent fraction with a denominator of x^2y^2.

$$
\frac{3}{x^2y} + \frac{2}{xy} - \frac{1}{xy^2}
$$

$$
= \frac{3 \cdot y}{x^2y \cdot y} + \frac{2 \cdot xy}{xy \cdot xy} - \frac{1 \cdot x}{xy^2 \cdot x} \qquad \begin{array}{l}\text{Multiply each numerator and denominator of the}\\ \text{original fraction by any factor(s) in the LCD that}\\ \text{were not in the denominator of the original fraction.}\end{array}
$$

$$
= \frac{3y}{x^2y^2} + \frac{2xy}{x^2y^2} - \frac{x}{x^2y^2} \qquad \begin{array}{l}\text{Simplify numerators and denominators by performing}\\ \text{the multiplication.}\end{array}
$$

$$
= \frac{3y + 2xy - x}{x^2y^2} \qquad \begin{array}{l}\text{Express the sum/difference of the numerators over}\\ \text{the common denominator.}\end{array}
$$

EXAMPLE 6

Subtract: $\dfrac{x}{x + 1} - \dfrac{3}{x}$.

SOLUTION   The least common denominator is $(x + 1)x$.

$$
\frac{x}{x + 1} - \frac{3}{x} = \frac{x(x)}{(x + 1)x} - \frac{3(x + 1)}{x(x + 1)} \qquad \begin{array}{l}\text{Multiply each numerator and denominator}\\ \text{of the original fraction by any factor(s) in}\\ \text{the LCD that were not in the denominator}\\ \text{of the original fraction.}\end{array}
$$

$$
= \frac{x^2}{(x + 1)x} - \frac{3x + 3}{x(x + 1)} \qquad \text{Perform the multiplication in the numerator.}
$$

$$
= \frac{x^2 - 3x - 3}{x(x + 1)} \qquad \begin{array}{l}\text{Subtract the numerators and keep the}\\ \text{common denominator.}\end{array}
$$

EXAMPLE 7

Perform the indicated operations: $\dfrac{3}{x^2 - y^2} + \dfrac{2}{x - y} - \dfrac{1}{x + y}$.

SOLUTION   Find the least common denominator.

$$
\left.\begin{array}{l} x^2 - y^2 = (x - y)(x + y) \\ x - y = x - y \\ x + y = x + y \end{array}\right\} \quad \text{Factor each denominator where possible.}
$$

Because the least common denominator is $(x - y)(x + y)$, we convert each given fraction into an equivalent fraction with that common denominator.

$$\frac{3}{x^2 - y^2} + \frac{2}{x - y} - \frac{1}{x + y} = \frac{3}{(x - y)(x + y)} + \frac{2}{x - y} - \frac{1}{x + y}$$

Factor $x^2 - y^2$.

$$= \frac{3}{(x - y)(x + y)} + \frac{2(x + y)}{(x - y)(x + y)} - \frac{1(x - y)}{(x + y)(x - y)}$$

Multiply each numerator and denominator of the original fraction by any factor(s) in the LCD that were not in the denominator of the original fraction.

$$= \frac{3}{(x - y)(x + y)} + \frac{2x + 2y}{(x - y)(x + y)} - \frac{x - y}{(x + y)(x - y)}$$

Remove parentheses.

$$= \frac{3 + 2x + 2y - x + y}{(x - y)(x + y)}$$

Add and subtract the numerators of the fractions and keep the common denominators.

$$= \frac{3 + x + 3y}{(x - y)(x + y)}$$

Combine like terms.

EXAMPLE 8   Perform the subtraction, $\dfrac{a}{a - 1} - \dfrac{2}{a^2 - 1}$, and simplify.

SOLUTION   Factor $a^2 - 1$ and write each fraction as an equivalent fraction with a denominator of $(a + 1)(a - 1)$.

$$\frac{a}{a - 1} - \frac{2}{a^2 - 1} = \frac{a(a + 1)}{(a - 1)(a + 1)} - \frac{2}{(a + 1)(a - 1)}$$

$$= \frac{a^2 + a}{(a - 1)(a + 1)} - \frac{2}{(a + 1)(a - 1)}$$

Remove parentheses in the numerator.

$$= \frac{a^2 + a - 2}{(a - 1)(a + 1)}$$

Subtract the numerators and keep the common denominator.

$$= \frac{(a + 2)\overset{1}{\cancel{(a - 1)}}}{\underset{1}{\cancel{(a - 1)}}(a + 1)}$$

Factor the numerator and divide out the common factor of $a - 1$.

$$= \frac{a + 2}{a + 1}$$

EXAMPLE 9   Perform the addition: $\dfrac{m + 1}{2m + 6} + \dfrac{4 - m^2}{2m^2 + 2m - 12}$.

SOLUTION   Factor $2m + 6$ and $2m^2 + 2m - 12$ to determine the LCD and write each fraction with the common denominator of $2(m + 3)(m - 2)$:

$$\frac{m+1}{2m+6} + \frac{4-m^2}{2m^2+2m-12} = \frac{m+1}{2(m+3)} + \frac{4-m^2}{2(m+3)(m-2)}$$

$$= \frac{(m+1)(m-2)}{2(m+3)(m-2)} + \frac{4-m^2}{2(m+3)(m-2)}$$

Now add the fractions by adding the numerators and keeping the common denominator, and simplify.

$$= \frac{(m+1)(m-2)+4-m^2}{2(m+3)(m-2)}$$

$$= \frac{m^2-m-2+4-m^2}{2(m+3)(m-2)} \qquad \text{Multiply } (m+1)(m-2).$$

$$= \frac{-m+2}{2(m+3)(m-2)} \qquad \text{Combine like terms.}$$

$$= \frac{-\overset{1}{\cancel{(m-2)}}}{2(m+3)\underset{1}{\cancel{(m-2)}}} \qquad \text{Factor } -1 \text{ from } -m+2.$$

$$= \frac{-1}{2(m+3)} \qquad \text{Divide out the common factor of } m-2.$$

EXERCISE 6.3

VOCABULARY AND NOTATION *In Exercises 1–3, fill in the blanks to make a true statement.*

1. To add or subtract rational expressions they must have a common _____ .

2. To find the LCD, we first _____ all denominators.

3. The LCD is the _____ (largest/smallest) number that each denominator will divide evenly.

CONCEPTS *In Exercises 4–9, add or subtract the given rational expressions as the sign indicates. All results should be expressed as simplified rational expressions.*

4. $\dfrac{8}{7} + \dfrac{6}{7}$

5. $\dfrac{9}{11} + \dfrac{2}{11}$

6. $\dfrac{9y}{3x} - \dfrac{6y}{3x}$

7. $\dfrac{24}{7y} - \dfrac{10}{7y}$

8. $\dfrac{15z}{22} - \dfrac{-15z}{22}$

9. $\dfrac{-30rs}{21} + \dfrac{30rs}{21}$

In Exercises 10–19, find the least common denominator for the pair of fractions.

10. $\dfrac{2}{21} ; \dfrac{5}{42}$

11. $\dfrac{4}{15} ; \dfrac{3}{10}$

12. $\dfrac{4}{3x} ; \dfrac{5}{9xy}$

13. $\dfrac{3}{x^2} ; \dfrac{8}{3xy}$

14. $\dfrac{5}{x-2} ; \dfrac{8}{x^2-4}$

15. $\dfrac{6}{x+3} ; \dfrac{9}{x^2-9}$

16. $\dfrac{5}{x^2+2x+1} ; \dfrac{8}{x+1}$

17. $\dfrac{7}{x^2+6x+9} ; \dfrac{5}{x+3}$

18. $\dfrac{2}{x^2 + 5x + 6}; \dfrac{3}{x + 2}$

19. $\dfrac{5}{x^2 + 6x + 5}; \dfrac{2}{x + 1}$

46. $\dfrac{x + 1}{x - 2} - \dfrac{2(x - 3)}{x - 2} + \dfrac{3(x + 1)}{x - 2}$

47. $\dfrac{x^2 - 4}{x + 2} + \dfrac{2(x^2 - 4)}{x + 2} - \dfrac{3(x^2 - 5)}{x + 2}$

In Exercises 20–29, supply the missing numerator.

20. $\dfrac{8}{x} = \dfrac{?}{x^2 y}$

21. $\dfrac{7}{y} = \dfrac{?}{xy^2}$

22. $\dfrac{3x}{x + 1} = \dfrac{?}{(x + 1)^2}$

23. $\dfrac{5y}{y - 2} = \dfrac{?}{(y - 2)^2}$

24. $\dfrac{2y}{x} = \dfrac{?}{x^2 + x}$

25. $\dfrac{3x}{y} = \dfrac{?}{y^2 - y}$

26. $\dfrac{z}{z - 1} = \dfrac{?}{z^2 - 1}$

27. $\dfrac{y}{y + 2} = \dfrac{?}{y^2 - 4}$

28. $\dfrac{2}{x + 1} = \dfrac{?}{x^2 + 3x + 2}$

29. $\dfrac{3}{x - 1} = \dfrac{?}{x^2 + x - 2}$

In Exercises 48–55, find the LCD for the given pairs of fractions.

48. $\dfrac{x + 3}{x^2 - 1}; \dfrac{x - 2}{x + 1}$

49. $\dfrac{y - 2}{y^2 - 9}; \dfrac{y + 3}{y - 3}$

50. $\dfrac{6}{x^2 + 6x}; \dfrac{9}{x + 6}$

51. $\dfrac{x + 4}{xy^2 - xy}; \dfrac{x - 2}{xy}$

52. $\dfrac{x + 5}{x^2 - x - 2}; \dfrac{x + 6}{(x - 2)^2}$

53. $\dfrac{x + 4}{x^2 + 2x - 3}; \dfrac{x + 5}{(x + 3)^2}$

54. $\dfrac{x + 2}{x^2 - 4x - 5}; \dfrac{x + 3}{x^2 - 25}$

55. $\dfrac{x + 4}{x^2 - x - 6}; \dfrac{x + 2}{x^2 - 9}$

In Exercises 30–35, simplify the sum or difference of the rational expressions. Results should be expressed in simplified form.

30. $\dfrac{1}{2} + \dfrac{2}{3}$

31. $\dfrac{2}{3} - \dfrac{5}{6}$

32. $\dfrac{2y}{9} + \dfrac{y}{3}$

33. $\dfrac{8a}{15} - \dfrac{5a}{12}$

34. $\dfrac{21x}{14} - \dfrac{5x}{21}$

35. $\dfrac{4x}{3} + \dfrac{2x}{y}$

In Exercises 56–71, simplify the sum and/or difference of the given rational expressions. All results should be expressed in simplified form.

56. $\dfrac{y + 2}{5y} + \dfrac{y + 4}{15y}$

57. $\dfrac{x + 3}{x^2} + \dfrac{x + 5}{2x}$

58. $\dfrac{x + 5}{xy} - \dfrac{x - 1}{x^2 y}$

59. $\dfrac{y - 7}{y^2} - \dfrac{y + 7}{2y}$

60. $\dfrac{3}{x - 2} - (x - 1)$

61. $a + 1 - \dfrac{3}{a + 3}$

PRACTICE *In Exercises 36–47, add or subtract the given rational expressions as the sign indicates. All results should be expressed as simplified rational expressions.*

36. $\dfrac{y + 2}{2z} - \dfrac{y - 4}{2z}$

37. $\dfrac{2x - 3}{x^2} - \dfrac{x - 3}{x^2}$

38. $\dfrac{y + 2}{5z} + \dfrac{y + 4}{5z}$

39. $\dfrac{x + 3}{x^2} + \dfrac{x + 5}{x^2}$

40. $\dfrac{3x - 5}{x - 2} + \dfrac{6x - 13}{x - 2}$

41. $\dfrac{8x - 7}{x + 3} + \dfrac{2x + 37}{x + 3}$

42. $\dfrac{3y - 2}{y + 3} - \dfrac{2y - 5}{y + 3}$

43. $\dfrac{5x + 8}{x + 2} - \dfrac{3x - 2}{x + 2}$

44. $\dfrac{3x}{y + 2} - \dfrac{3y}{y + 2} + \dfrac{x + y}{y + 2}$

45. $\dfrac{3y}{x - 5} - \dfrac{x}{x - 5} - \dfrac{y - x}{x - 5}$

62. $\dfrac{2x + 2}{x - 2} - \dfrac{2x}{x + 2}$

63. $\dfrac{y + 3}{y - 1} - \dfrac{y + 4}{y + 1}$

64. $\dfrac{x}{(x - 2)^2} + \dfrac{x - 4}{(x + 2)(x - 2)}$

65. $\dfrac{a - 2}{(a + 3)^2} - \dfrac{a}{a - 3}$

66. $\dfrac{2x}{x^2 - 3x + 2} + \dfrac{2x}{x - 1} - \dfrac{x}{x - 2}$

67. $\dfrac{4x}{x - 2} - \dfrac{3x}{x - 3} + \dfrac{4x}{x^2 - 5x + 6}$

68. $\dfrac{x+1}{2x+4} - \dfrac{x^2}{2x^2-8}$

69. $\dfrac{x+1}{x+2} - \dfrac{x^2+1}{x^2-x-6}$

70. $\dfrac{x-1}{x+2} + \dfrac{x}{3-x} + \dfrac{9x+3}{x^2-x-6}$

71. $\dfrac{x+1}{x-3} - \dfrac{x^2+9x}{x^2-2x-3} - \dfrac{5}{3-x}$

In Exercises 72–77, simplify each expression. Be sure to follow the order of operations.

72. $\dfrac{x+3}{x-2} \div \dfrac{x+1}{x-2} + \dfrac{3}{x+1}$

73. $\dfrac{x-5}{x+4} \div \dfrac{x+6}{x+4} + \dfrac{8}{x+6}$

74. $\dfrac{x+2}{x^2-4} + \dfrac{x+5}{x-2} \cdot \dfrac{x+2}{x+5}$

75. $\dfrac{x+3}{x^2-9} + \dfrac{x-2}{x-3} \cdot \dfrac{x+3}{x-2}$

76. $\dfrac{x+2}{x+5} - \dfrac{x-5}{x+4} \div \dfrac{x^2-25}{x^2-16}$

77. $\dfrac{x+3}{x+7} - \dfrac{x-7}{x+2} \div \dfrac{x^2-49}{x^2-4}$

78. Perimeter
   a. Find a rational expression that models the perimeter of the pictured square.

$\dfrac{x}{x-6}$ ft

b. If x has a value of 8, what would be the perimeter of the square?

 c. Could x have a value of 6 feet? Why or why not?

79. Perimeter
   a. Find a rational expression that models the perimeter of the pictured square.

$\dfrac{x}{x-5}$ ft

b. If x has a value of 10, what would be the perimeter of the square?

 c. Could x have a value of 5 feet? Why or why not?

REVIEW

80. Factor $2x^3 - 7x^2 - 15x$.

81. If $P(x) = 2x^3 - 7x^2 - 15x$, find

   a. $P(5)$ **b.** $P\left(-\dfrac{3}{2}\right)$ **c.** $P(0)$

82. Solve $2x^3 - 7x^2 - 15x = 0$ graphically. Copy your screen display to justify your work.

83. Solve $2x^3 - 7x^2 - 15x = 0$ analytically and verify your solution with the calculator using the STOre feature.

6.4 COMPLEX FRACTIONS

- COMPLEX FRACTIONS AND ORDER OF OPERATIONS
- COMPLEX FRACTIONS AND LCD

COMPLEX FRACTIONS AND ORDER OF OPERATIONS

Consider the arithmetic expression $2\dfrac{2}{3} \div 3\dfrac{3}{4}$. Because a fraction bar also can be

used to indicate division, the expressions could have been written as $\dfrac{2\frac{2}{3}}{3\frac{3}{4}}$ or $\dfrac{\frac{8}{3}}{\frac{15}{4}}$.

This is an example of a **complex fraction**. A **complex fraction** is a fraction that contains a fraction in the numerator, the denominator, or both the numerator and denominator. One way to simplify a complex fraction is to replace the *main* fraction bar with a division sign. The expression is then simplified following the normal order of operations.

To simplify the complex fraction $\dfrac{\dfrac{3}{5} + 1}{2 - \dfrac{1}{5}}$ we can rewrite it as a division problem:

$\left(\dfrac{3}{5} + 1\right) \div \left(2 - \dfrac{1}{5}\right)$. The parentheses are required because the numerator of the complex fraction contains more than one term, as does the denominator. Continuing the simplification process:

$$\left(\dfrac{3}{5} + \dfrac{5}{5}\right) \div \left(\dfrac{10}{5} - \dfrac{1}{5}\right)$$

Simplify within grouping first by finding the common denominator and adding/subtracting the fractions as the signs indicate.

$$= \dfrac{8}{5} \div \dfrac{9}{5}$$

Continue to simplify grouping.

$$= \dfrac{8}{5} \cdot \dfrac{5}{9}$$

Invert the divisor and change the operation to multiplication.

$$= \dfrac{8 \cdot 5}{5 \cdot 9}$$

Express the multiplication of the fractions.

$$= \dfrac{8 \cdot \cancel{5}}{\cancel{5} \cdot 9}$$

Divide out common factors to simplify.

$$= \dfrac{8}{9}$$

Our analytical work is confirmed with the graphing calculator.

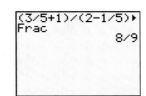

▮ COMPLEX FRACTIONS AND LCD

A second method for simplifying complex fractions uses the fundamental property of fractions. Because any expression can be multiplied by 1 to obtain an equivalent expression, we may multiply the numerator and the denominator of the complex fraction by the LCD of its component fractions. This makes the complex fraction a simple fraction.

Consider $\dfrac{\dfrac{3}{5} + 1}{2 - \dfrac{1}{5}}$. We may multiply the numerator and denominator of the complex fraction by the LCD of its component fractions ($\frac{3}{5}$ and $\frac{1}{5}$). We are actually multiplying the fraction by the multiplicative identity of 1, yet it is expressed as $\frac{5}{5}$.

$$\frac{\dfrac{3}{5}+1}{2-\dfrac{1}{5}}\cdot\frac{5}{5}$$

$$=\frac{5\cdot\left(\dfrac{3}{5}+1\right)}{5\cdot\left(2-\dfrac{1}{5}\right)}$$ Express the multiplication.

$$=\frac{5\cdot\dfrac{3}{5}+5\cdot1}{5\cdot2-5\cdot\dfrac{1}{5}}$$ Use the distributive property to remove parentheses.

$$=\frac{3+5}{10-1}$$ Use the order of operations to simplify.

$$=\frac{8}{9}$$

This process is verified with the graphing calculator, as displayed by the following screen:

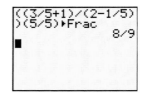

Either of the two methods illustrated and summarized previously can be used to simplify a complex fraction.

Method 1: Simplifying a Complex Fraction Using the Order of Operations

1. Replace the main fraction bar with a division sign, grouping the numerator and denominator appropriately.
2. Simplify, using the order of operations.

Method 2: Simplifying a Complex Fraction by Multiplying by the LCD

1. Determine the LCD of the component fractions of the complex fraction.
2. Multiply the numerator and denominator of the complex fraction by the LCD. Be sure to distribute when the numerator and/or denominator contain more than one term.

The examples that follow illustrate both methods of simplification.

EXAMPLE 1   Simplify: $\dfrac{\dfrac{x}{3}}{\dfrac{y}{3}}$.

SOLUTION   **Method 1**

$$\dfrac{\dfrac{x}{3}}{\dfrac{y}{3}} = \dfrac{x}{3} \div \dfrac{y}{3}$$

$$= \dfrac{x}{3} \cdot \dfrac{3}{y}$$

$$= \dfrac{3x}{3y}$$

$$= \dfrac{x}{y}$$

Method 2

$$\dfrac{\dfrac{x}{3}}{\dfrac{y}{3}} = \dfrac{\left(\dfrac{x}{3}\right)3}{\left(\dfrac{y}{3}\right)3}$$

$$= \dfrac{\dfrac{x}{1}}{\dfrac{y}{1}}$$

$$= \dfrac{x}{y}$$

EXAMPLE 2   Simplify: $\dfrac{\dfrac{x}{x + 1}}{\dfrac{y}{x}}$.

SOLUTION   **Method 1**

$$\dfrac{\dfrac{x}{x + 1}}{\dfrac{y}{x}} = \dfrac{x}{x + 1} \div \dfrac{y}{x}$$

$$= \dfrac{x}{x + 1} \cdot \dfrac{x}{y}$$

$$= \dfrac{x^2}{y(x + 1)}$$

Method 2

$$\dfrac{\dfrac{x}{x + 1}}{\dfrac{y}{x}} = \dfrac{\left(\dfrac{x}{x + 1}\right)x(x + 1)}{\left(\dfrac{y}{x}\right)x(x + 1)}$$

$$= \dfrac{\dfrac{x^2}{1}}{\dfrac{y(x + 1)}{1}}$$

$$= \dfrac{x^2}{y(x + 1)}$$

EXAMPLE 3

Simplify: $\dfrac{1 + \dfrac{1}{x}}{1 - \dfrac{1}{x}}$.

SOLUTION

Method 1

$$\dfrac{1 + \dfrac{1}{x}}{1 - \dfrac{1}{x}} = \left(1 + \dfrac{1}{x}\right) \div \left(1 - \dfrac{1}{x}\right)$$

$$= \left(\dfrac{x}{x} + \dfrac{1}{x}\right) \div \left(\dfrac{x}{x} - \dfrac{1}{x}\right)$$

$$= \dfrac{x + 1}{x} \div \dfrac{x - 1}{x}$$

$$= \dfrac{(x + 1)}{x} \cdot \dfrac{x}{(x - 1)}$$

$$= \dfrac{(x + 1) \cdot \overset{1}{\cancel{x}}}{\underset{1}{\cancel{x}} \cdot (x - 1)}$$

$$= \dfrac{x + 1}{x - 1}$$

Method 2

$$\dfrac{1 + \dfrac{1}{x}}{1 - \dfrac{1}{x}} = \dfrac{\left(1 + \dfrac{1}{x}\right)x}{\left(1 - \dfrac{1}{x}\right)x}$$

$$= \dfrac{1 \cdot x + \dfrac{1}{x} \cdot x}{1 \cdot x - \dfrac{1}{x} \cdot x}$$

$$= \dfrac{x + 1}{x - 1}$$

EXAMPLE 4

Simplify the complex fraction $\dfrac{1}{1 + \dfrac{1}{x + 1}}$.

SOLUTION   We use Method 2.

$$\dfrac{1}{1 + \dfrac{1}{x + 1}} = \dfrac{1(x + 1)}{\left(1 + \dfrac{1}{x + 1}\right)(x + 1)} \qquad \text{Multiply numerator and denominator by } x + 1.$$

$$= \dfrac{x + 1}{1(x + 1) + 1} \qquad \text{Remove parentheses and simplify.}$$

$$= \dfrac{x + 1}{x + 2} \qquad \text{Simplify.}$$

EXERCISE 6.4

VOCABULARY AND NOTATION *Fill in the blank to make a true statement.*

1. A _____ fraction is a fraction that contains a fraction in the numerator, the denominator, or both.

2. The denominator of the complex fraction $\dfrac{\dfrac{3}{x} + \dfrac{x}{y}}{\dfrac{1}{x} + 2}$ is _____ .

3. The component fractions of the complex fraction $\dfrac{\dfrac{3}{x} + \dfrac{x}{y}}{\dfrac{1}{x} + 2}$ are _____ .

4. When using Method 1 to simplify a complex fraction, we first rewrite the problem as a _____ problem, taking care to follow the _____ .

5. When using Method 2 to simplify a complex fraction, we multiply each term of the numerator and the denominator of the complex fraction by the LCD of the _____ fractions. If the numerator and/or the denominator contain more than one term, we must use the _____ property.

CONCEPTS

6. Is the expression $\dfrac{\dfrac{1}{2}}{4}$ equal to the expression $\dfrac{1}{\dfrac{2}{4}}$? Why or why not?

7. Is the expression $\dfrac{\dfrac{3}{2}}{5}$ equal to the expression $\dfrac{3}{\dfrac{2}{5}}$? Why or why not?

8. **a.** Perform the indicated division: $\dfrac{x^2}{3} \div \dfrac{2x}{4}$.

   b. Rewrite the expression in part (a) as a complex fraction.

   c. Would the complex fraction written in part (b) (when simplified) be equivalent to the quotient in part (a)? Why or why not?

9. **a.** Perform the indicated division: $\dfrac{x+2}{3x} \div \dfrac{x+2}{2}$.

   b. Rewrite the expression in part (a) as a complex fraction.

   c. Would the complex fraction written in part (b) (when simplified) be equivalent to the quotient in part (a)? Why or why not?

10. **a.** Evaluate the expression $\dfrac{\dfrac{1}{2} + \dfrac{2}{3}}{4}$ using the graphing calculator. Copy your screen display exactly as it appears.

    b. Rewrite the complex fraction in part (a) as a division problem.

    c. Evaluate the expression you wrote in part (b). Is the result equal to the result from part (a)? Should it be? Why or why not?

11. **a.** Evaluate the expression $\dfrac{\dfrac{3}{4} + \dfrac{1}{5}}{6}$ using the graphing calculator. Copy your screen display exactly as it appears.

    b. Rewrite the complex fraction in part (a) as a division problem.

    c. Evaluate the expression you wrote in part (b). Is the result equal to the result from part (a)? Should it be? Why or why not?

PRACTICE *In Exercises 12–53, simplify each complex fraction. Check results of 12–19 using the graphing calculator.*

12. $\dfrac{\dfrac{2}{3}}{\dfrac{3}{4}}$

13. $\dfrac{\dfrac{3}{5}}{\dfrac{2}{7}}$

14. $\dfrac{\dfrac{4}{5}}{\dfrac{32}{15}}$

15. $\dfrac{\dfrac{7}{8}}{\dfrac{49}{4}}$

16. $\dfrac{\dfrac{2}{3} + 1}{\dfrac{1}{3} + 1}$

17. $\dfrac{\dfrac{3}{5} - 2}{\dfrac{2}{5} - 2}$

18. $\dfrac{\dfrac{1}{2} + \dfrac{3}{4}}{\dfrac{3}{2} + \dfrac{1}{4}}$

19. $\dfrac{\dfrac{2}{3} - \dfrac{5}{2}}{\dfrac{2}{3} - \dfrac{3}{2}}$

20. $\dfrac{\dfrac{x}{y}}{\dfrac{1}{x}}$

21. $\dfrac{\dfrac{y}{x}}{xy}$

22. $\dfrac{\dfrac{5t^2}{9x^2}}{\dfrac{3t}{x^2t}}$

23. $\dfrac{\dfrac{5w^2}{4tz}}{\dfrac{15wt}{z^2}}$

24. $\dfrac{\dfrac{1}{x} - 3}{\dfrac{5}{x} + 2}$

25. $\dfrac{\dfrac{1}{y} + 3}{\dfrac{3}{y} - 2}$

26. $\dfrac{\dfrac{2}{x} + 2}{\dfrac{4}{x} + 2}$

27. $\dfrac{\dfrac{3}{x} - 3}{\dfrac{9}{x} - 3}$

28. $\dfrac{\dfrac{3y}{x} - y}{y - \dfrac{y}{x}}$

29. $\dfrac{\dfrac{y}{x} + 3y}{y + \dfrac{2y}{x}}$

30. $\dfrac{\dfrac{1}{x+1}}{1 + \dfrac{1}{x+1}}$

31. $\dfrac{\dfrac{1}{x-1}}{1 - \dfrac{1}{x-1}}$

32. $\dfrac{\dfrac{x}{x+2}}{\dfrac{x}{x+2} + x}$

33. $\dfrac{\dfrac{2}{x-2}}{\dfrac{2}{x-2} - 1}$

34. $\dfrac{1}{\dfrac{1}{x} + \dfrac{1}{y}}$

35. $\dfrac{1}{\dfrac{b}{a} - \dfrac{a}{b}}$

36. $\dfrac{\dfrac{2}{x}}{\dfrac{2}{y} - \dfrac{4}{x}}$

37. $\dfrac{\dfrac{2y}{3}}{\dfrac{2y}{3} - \dfrac{8}{y}}$

38. $\dfrac{3 + \dfrac{3}{x-1}}{3 - \dfrac{3}{x}}$

39. $\dfrac{2 - \dfrac{2}{x+1}}{2 + \dfrac{2}{x}}$

40. $\dfrac{\dfrac{3}{x} + \dfrac{4}{x+1}}{\dfrac{2}{x+1} - \dfrac{3}{x}}$

41. $\dfrac{\dfrac{5}{y-3} - \dfrac{2}{y}}{\dfrac{1}{y} + \dfrac{2}{y-3}}$

42. $\dfrac{\dfrac{2}{x} - \dfrac{3}{x+1}}{\dfrac{2}{x+1} - \dfrac{3}{x}}$

43. $\dfrac{\dfrac{5}{y} + \dfrac{4}{y+1}}{\dfrac{4}{y} - \dfrac{5}{y+1}}$

44. $\dfrac{\dfrac{1}{y^2+y} - \dfrac{1}{xy+x}}{\dfrac{1}{xy+x} - \dfrac{1}{y^2+y}}$

45. $\dfrac{\dfrac{2}{b^2-1} - \dfrac{3}{ab-a}}{\dfrac{3}{ab-a} - \dfrac{2}{b^2-1}}$

46. $\dfrac{1 + x^{-1}}{x^{-1} - 1}$

47. $\dfrac{y^{-2} + 1}{y^{-2} - 1}$

48. $\dfrac{a^{-2} + a}{a + 1}$

49. $\dfrac{t - t^{-2}}{1 - t^{-1}}$

50. $\dfrac{2x^{-1} + 4x^{-2}}{2x^{-2} + x^{-1}}$

51. $\dfrac{x^{-2} - 3x^{-3}}{3x^{-2} - 9x^{-3}}$

52. $\dfrac{1 - 25y^{-2}}{1 + 10y^{-1} + 25y^{-2}}$

53. $\dfrac{1 - 9x^{-2}}{1 + 6x^{-1} + 9x^{-2}}$

54. Simplify the four complex fractions below:

$$\dfrac{1}{1+1}, \quad \dfrac{1}{1+\dfrac{1}{2}}, \quad \dfrac{1}{1+\dfrac{1}{1+\dfrac{1}{2}}}, \quad \text{and} \quad \dfrac{1}{1+\dfrac{1}{1+\dfrac{1}{1+\dfrac{1}{2}}}}$$

(*Hint:* Use Method 2, and work from the bottom up.)

REVIEW

55. Solve: $\dfrac{x+5}{x-3} = \dfrac{4}{5}$.

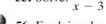

56. Explain why $x = 3$ would not be an acceptable solution to the equation in Exercise 55.

57. Samuel holds a yardstick perpendicular to the sidewalk and measures a 4-foot shadow. He then measures the shadow of the bell tower on campus and determines it to be 50 feet long. Using what he knows about similar triangles, he will be able to determine the height of the bell tower. What height does he find?

58. Simplify each of the following. Assume that all variables represent positive numbers.

   a. $\dfrac{(x^2y)^3}{x^2}$

   b. $(2x^4)^3$

   c. $(2x + 3)^2$

6.5 ALGEBRAIC AND GRAPHICAL SOLUTIONS OF RATIONAL EQUATIONS

- EXTRANEOUS SOLUTIONS
- ALGEBRAIC SOLUTIONS OF RATIONAL EQUATIONS
- GRAPHICAL SOLUTIONS OF RATIONAL EQUATIONS

▌▌ EXTRANEOUS SOLUTIONS

Sections 6.1 through 6.4 of this chapter focused on *simplifying* rational expressions. We were required to perform the arithmetic operations of addition, subtraction, multiplication, and division. We now turn our attention to *solving* equations that contain rational expressions. The differences between the *simplification of expressions* and the *solving of equations* are reiterated below.

When simplifying an *expression*, follow the order of operations to remove groupings, raise to powers, and perform all arithmetic operations to rewrite the original *expression* as an equivalent *expression* in which all like terms have been combined. All fractions (rational numbers) and rational expressions must be expressed in lowest terms.

When solving an *equation* that is linear, we use the addition, subtraction, multiplication, and division properties of equality to rewrite the original *equation* as a series of equivalent *equations* until the variable is isolated on one side of the equation. For equations of degree two or higher, factoring and the zero factor property are applied. The solutions, or roots, of the equation are the value(s) of the variable that make the equation a true statement.

Care must be taken when solving equations containing rational expressions with variables in the denominator. This is illustrated below.

EXAMPLE 1

Solve the equation $\dfrac{x + 3}{x - 1} = \dfrac{4}{x - 1}$.

SOLUTION

Generalizing the method learned previously, begin by multiplying through the equation by the LCD, which is the quantity $(x - 1)$.

$$\frac{x + 3}{x - 1} = \frac{4}{x - 1}$$

$$(x - 1) \cdot \frac{x + 3}{x - 1} = (x - 1) \cdot \frac{4}{x - 1}$$
Multiply both sides of the equation by the LCD.

$$\frac{(x - 1) \cdot (x + 3)}{(x - 1)} = \frac{(x - 1) \cdot 4}{x - 1}$$
Express the multiplication.

$$\frac{\overset{1}{\cancel{(x - 1)}} \cdot (x + 3)}{\underset{1}{\cancel{x - 1}}} = \frac{\overset{1}{\cancel{(x - 1)}} \cdot 4}{\underset{1}{\cancel{x - 1}}}$$
Simplify, by dividing out common factors.

$$x + 3 = 4$$
Solve the resulting linear equation by subtracting 3 from both sides.

$$x = 1$$

Check: To check analytically, replace the variable, x, with its value, 1, in the original equation and simplify.

$$\frac{x+3}{x-1} = \frac{4}{x-1}$$

$$\frac{1+3}{1-1} = \frac{4}{1-1} \qquad \text{Simplify.}$$

$$\frac{4}{0} = \frac{4}{0}$$

Because zero appears in the denominator, the fractions are undefined. Therefore, the number 1 *cannot* be a solution of the equation. The original equation was a contradiction and has *no solution*.

The check can also be done with the graphing calculator using either the STOre feature or the TABLE. This is illustrated in the screens below.

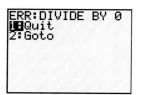

Before ENTER is
pressed.

After ENTER is
pressed.

The number 1 is stored for the variable x and the left side of the equation is entered for evaluation. When ENTER is pressed we immediately see that the fraction is undefined. There is no need to repeat the process with the right side of the equation.

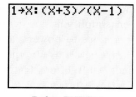

The left side of the equation is entered at the y1 prompt and the right side is entered at the y2 prompt. The TABLE is set to begin at 0 and to be incremented by 1. The error messages confirm the fact that 1 cannot be a solution to the equation.

Looking back at the solution of the equation it does not appear that we have used any algebraic algorithms or properties incorrectly. The fallacy lies in the fact that the LCD used $(x-1)$ contained a variable. Recall that the multiplication property of equality states multiplying through an equation by a *nonzero* quantity produces an equivalent equation. Because a variable appeared in our LCD, we were *not* assured that the multiplication produced an equivalent equation. In this case, it did not. The equation $x+3=4$ is **not** equivalent to the original equation, $\frac{x+3}{x-1} = \frac{4}{x-1}$.

Sometimes when solving rational equations apparent false solutions result. These are called **extraneous solutions** and are rejected. Any apparent solution that makes a denominator equal to zero is an extraneous solution.

▮ ALGEBRAIC SOLUTIONS OF RATIONAL EQUATIONS

When algebraically solving equations that contain variables in the denominators, solutions **must** be verified through a checking process. Extraneous solutions **must** be rejected. For this reason, it is suggested that excluded values be determined first. The following steps can be used:

Solving Rational Equations Algebraically

1. Determine all excluded values by setting each denominator containing a variable equal to zero and solving for the variable.
2. Determine the LCD.
3. Multiply each *term* on both sides of the equation by the LCD to clear fractions.
4. Simplify both sides of the equation.
5. Solve the equation. Remember, linear (first-degree) equations are solved by transformation, whereas quadratic (second-degree) and higher-order equations require the additional process of applying the zero factor property after the equation is put in standard form and the polynomial is factored.
6. Reject any extraneous solutions (those values that equal the excluded values found in step 1).
7. Check solutions in the original equation to verify solutions.

EXAMPLE 2   Solve the equation $\dfrac{3x + 1}{x + 1} - 2 = \dfrac{3(x - 3)}{x + 1}$.

SOLUTION   Determine the excluded values by setting each denominator that contains a variable equal to zero and solving.

$$x + 1 = 0$$ Set the denominator equal to zero. Solve by subtracting 1 from both sides of the equation.

$$x = -1$$

The variable, x, **cannot** have a value of -1. We state this as $x \neq -1$.

We begin solving by multiplying both sides of the equation by $x + 1$, the least common denominator of the fractions contained in the equation.

$$(x + 1) \cdot \left(\frac{3x + 1}{x + 1} - 2 \right) = (x + 1) \left[\frac{3(x - 3)}{x + 1} \right]$$

$$(x + 1) \cdot \frac{(3x + 1)}{(x + 1)} - 2(x + 1) = \frac{3(x - 3)(x + 1)}{x + 1}$$ Use the distributive property to multiply each term on both sides by the LCD.

$$3x + 1 - 2x - 2 = 3x - 9$$ Use the distributive property to remove parentheses.

$$x - 1 = 3x - 9$$ Combine like terms.

$$-2x - 1 = -9$$ Subtract $3x$ from both sides.

$$-2x = -8$$ Add 1 to both sides.

$$x = 4$$ Divide both sides by -2.

Because our solution of 4 is not equal to the excluded value of -1, 4 is a valid (rather than an extraneous) solution.

Check:

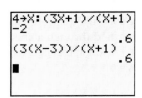

The STOre feature confirms the left side of the equation equals the right side when x is replaced by 4.

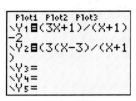

The pictured TABLE confirms both the solution and the excluded value.

EXAMPLE 3   Solve the equation $\dfrac{x+2}{x+3} + \dfrac{1}{x^2+2x-3} = 1$.

SOLUTION   Determine the excluded values by setting each denominator equal to zero and solving for the variable.

$$x + 3 = 0 \quad \text{and} \quad x^2 + 2x - 3 = 0$$

Because $x^2 + 2x - 3 = 0$ is a quadratic equation, we factor the polynomial side and use the zero factor property to find the solutions.

$$x = -3 \qquad (x+3)(x-1) = 0$$

$$x + 3 = 0 \qquad \text{or} \qquad x - 1 = 0$$

$$x = -3 \qquad \text{or} \qquad x = 1$$

The excluded values are -3 and 1. Therefore, $x \neq -3, x \neq 1$.

Now that the excluded values have been found, we begin solving the equation.

$$\frac{x+2}{x+3} + \frac{1}{x^2+2x-3} = 1$$

$$\frac{x+2}{x+3} + \frac{1}{(x+3)(x-1)} = 1 \qquad \text{To determine the LCD, factor the denominators.}$$

The LCD is $(x+3)(x-1)$. We multiply each term on both sides of the equation to eliminate fractions.

$$(x+3)(x-1) \cdot \frac{x+2}{x+3} + (x+3)(x-1) \cdot \frac{1}{(x+3)(x-1)} = (x+3)(x-1) \cdot 1$$

$$(x-1)(x+2) + 1 = (x+3)(x-1) \qquad \text{Simplify by dividing out common factors.}$$

$$x^2 + 2x - x - 2 + 1 = x^2 - x + 3x - 3 \qquad \text{Remove parentheses.}$$

$$x^2 + x - 1 = x^2 + 2x - 3 \qquad \text{Combine like terms.}$$

$$x - 1 = 2x - 3 \qquad \text{Subtract } x^2 \text{ from both sides.}$$

$$-x - 1 = -3 \qquad \text{Subtract } 2x \text{ from both sides.}$$

$$-x = -2 \qquad \text{Add 1 to both sides.}$$

$$x = 2 \qquad \text{Divide both sides by } -1.$$

Because 2 is *not* an excluded value, it *is* a solution to the original equation.

The verification using the STOre feature of the calculator is pictured below:

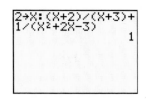

Only the left side was evaluated for the value of $x = 2$ because the right side was a constant.

EXAMPLE 4

Solve the equation $\dfrac{2}{x-1} + 4 = \dfrac{4}{x^2 - 1} + 3$.

SOLUTION   Determine the excluded values by setting each denominator equal to zero and solving for the variable.

$$x - 1 = 0 \qquad \text{and} \qquad x^2 - 1 = 0$$
$$x = 1 \qquad\qquad\qquad (x-1)(x+1) = 0$$

Because $x^2 - 1 = 0$ is a quadratic equation, we factor the polynomial side and use the zero factor property to find the solutions.

$$x - 1 = 0 \qquad \text{or} \qquad x + 1 = 0$$
$$x = 1 \qquad \text{or} \qquad x = -1$$

The excluded values are 1 and -1. Therefore, $x \neq 1$, $x \neq -1$.

We now begin solving the rational equation. To determine the LCD, we factor denominators that are not prime.

$$\frac{2}{x-1} + 4 = \frac{4}{x^2 - 1} + 3$$

$$\frac{2}{x-1} + 4 = \frac{4}{(x-1)(x+1)} + 3$$

The LCD is $(x-1)(x+1)$.

$$(x-1)(x+1) \cdot \frac{2}{x-1} + (x-1)(x+1) \cdot 4 = (x-1)(x+1) \cdot \frac{4}{(x-1)(x+1)} + (x-1)(x+1) \cdot 3$$

$$2(x+1) + 4(x^2 - 1) = 4 + 3(x^2 - 1)$$

Simplify by dividing out common factors.

$$2x + 2 + 4x^2 - 4 = 4 + 3x^2 - 3$$

Simplify by removing grouping.

$$4x^2 + 2x - 2 = 3x^2 + 1$$

Combine like terms.

$$x^2 + 2x - 3 = 0$$

Because the equation is quadratic, use the subtraction property of equality to put it in standard form.

$$(x + 3)(x - 1) = 0$$

Factor the polynomial side.

$$x + 3 = 0 \qquad \text{or} \qquad x - 1 = 0$$

Use the zero factor property to set each factor equal to 0 and solve for the variable.

$$x = -3 \qquad \text{or} \qquad x = 1$$

Because the number 1 is an excluded value, it is rejected as a root. The solution is -3.

The solution of -3 is verified below:

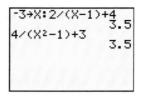

The fact that 1 is an extraneous root is also verified below:

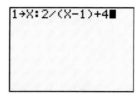   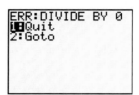

Before ENTER is   After ENTER is
pressed.   pressed.

The TABLE feature demonstrates that -3 is a solution because the expressions entered at the y1 and y2 prompts evaluate to the same constant (3.5) for $x = -3$. It also confirms the fact that 1 cannot be a solution. The ERROR message displayed indicates that **you** have made an error. In this case, you are "asking" the calculator to evaluate expressions for a value of the variable that makes those expressions undefined.

▓ GRAPHICAL SOLUTIONS OF RATIONAL EQUATIONS

The INTERSECT option of the graphing calculator can also be used to solve rational equations. The INTERSECT feature is used in the examples that follow.

EXAMPLE 5   Solve $\dfrac{2}{x} + 6 = 8$ graphically.

SOLUTION   Enter the left side of the equation at the y1 prompt and the right side at the y2 prompt. Access the graph.

   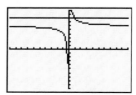

When we access the INTERSECT feature, we see the first curve prompt. Notice that no value is displayed for y at the bottom of the screen. This is because when $x = 0$, the fraction $\frac{2}{x}$ is undefined. Therefore, cursor either right or left until a value is displayed for both x and y at the bottom of the screen. You may then respond to the prompts as usual.

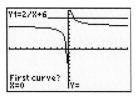

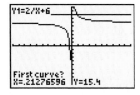
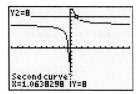
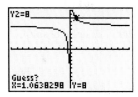
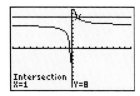

The solution is 1. This can be verified algebraically or by using the STOre or TABLE feature of the calculator.

EXAMPLE 6

Solve $\dfrac{x^2 - 2x - 8}{x + 2} - 2x = x - 5$ both graphically and algebraically.

SOLUTION   **Graphical solution:**   Enter the left side of the equation at the y1 prompt and the right side at the y2 prompt. Use the INTERSECT feature to find the solution to the equation.

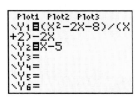
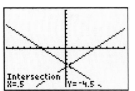

The solution is $\frac{1}{2}$ (remember, because the original equation contained rational expressions, the solution should be expressed as a fraction rather than in its decimal form).

Algebraic solution: The graphical display shows us that *both* the left and the right sides of the equation are linear. However, the left side of the original equation does not appear linear. Therefore, as we begin solving we will start by simplifying the left side rather than multiplying through the entire equation by the LCD.

$$\frac{x^2 - 2x - 8}{x + 2} - 2x = x - 5$$

$$\frac{(x - 4)(x + 2)}{x + 2} - 2x = x - 5 \qquad \text{Simplify the rational expression by factoring and dividing out common factors.}$$

$$x - 4 - 2x = x - 5$$

$$-x - 4 = x - 5 \qquad \text{Combine like terms.}$$

$$-4 = 2x - 5 \qquad \text{Add } x \text{ to both sides of the equation.}$$

$$1 = 2x \qquad \text{Add 5 to both sides of the equation.}$$

$$\frac{1}{2} = x \qquad \text{Divide both sides of the equation by 2.}$$

This confirms the graphical solution.

Notice that looking at the graphical display first helped us in our algebraic solution. Had we solved the equation by first multiplying through by the LCD of $(x + 2)$ we would have obtained the extraneous root of -2 as well as the actual root of $\frac{1}{2}$. The graphical displays only show actual roots.

EXAMPLE 7

Solve $\dfrac{5}{x} = -3x + 4$ both graphically and algebraically.

SOLUTION   **Graphical solution:**   Enter the left side of the equation at the y1 prompt and the right side at the y2 prompt. The display is pictured below.

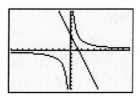

Because there are no points of intersection, the equation is a contradiction and has *no solution*.

Algebraic solution:

$$\frac{5}{x} = -3x + 4 \qquad \text{The excluded value is 0. Therefore, } x \neq 0.$$

$$x \cdot \frac{5}{x} = -x \cdot 3x + x \cdot 4 \qquad \text{Multiply through the equation by the LCD of } x.$$

$$5 = -3x^2 + 4x \qquad \text{Simplify by performing multiplications.}$$

$$3x^2 - 4x + 5 = 0 \qquad \text{Because the equation is quadratic, put it into standard form.}$$

Attempts to factor the quadratic polynomial $(3x^2 - 4x + 5)$ are unsuccessful. It is prime and has no rational-number solutions.

EXAMPLE 8

Solve $\dfrac{1}{x - 1} + \dfrac{3}{x - 1} = 5$ graphically.

SOLUTION   When we first look at the graphical display, we see a vertical line that we have not seen before. However if we change our calculator mode from CONNECTED to DOT (or change the graph style icon), this vertical line disappears.

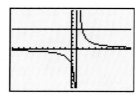   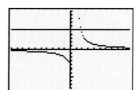

The reason the line disappears is because it is not a part of the graph of the expression $\dfrac{1}{x - 1} + \dfrac{3}{x - 1}$. When in connected mode, the grapher has connected the two branches that are the graph of the side of the equation containing rational expressions. In dot mode this does not happen.

Using the INTERSECT feature we find the solution is 1.8, or, as a fraction, $\frac{9}{5}$.

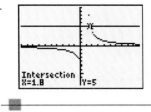

Equations containing rational expressions are sometimes difficult to solve graphically. However, the graphical display can often provide insights into the solving process. The following suggestions may prove helpful. Remember, explore the options available to you with the graphing calculator.

Graphical Displays of Rational Expressions

1. View the graph in *both* connected and dot mode.
2. Use the TABLE to help you see, numerically, what is happening. Reconcile the numerical information with the graphical display.
3. TRACE along the graph watching for patterns.
4. Use the INTERSECT feature to find/confirm solutions graphically.
5. For clarity, you may want to turn the *x* and *y* axes off to unclutter the screen. To do this, access the FORMAT menu highlight AxesOff and press ENTER.
6. Even when solving equations algebraically, look at the graphical display and practice reconciling your algebraic solution with the graphical display.

EXERCISE 6.5

VOCABULARY AND NOTATION *Fill in the blank to make a true statement.*

1. False solutions are called _____ solutions.

2. If 3 and 2 are excluded values of a rational expression that is contained in an equation, we may write that $x \neq$ ___ and $x \neq$ ___.

3. To clear an equation of fractions we multiply each term on both sides by the _____ of the fractions in the equation.

4. _____ equations are solved by writing a series of equivalent equations until the variable is isolated on one side of the equal sign.

5. _____ and _____ equations are solved by factoring and the application of the zero factor property.

CONCEPTS

6. Determine the restricted values for each rational expression.

   a. $\dfrac{x}{x^2 - 1}$ b. $\dfrac{2}{4x}$

   c. $\dfrac{3}{x^2 - 2x - 8}$

7. Identify each item as an expression or an equation. If it is an expression, simplify it. If it is an equation, solve it.

   a. $\dfrac{x}{3} + \dfrac{x}{2}$ b. $\dfrac{x}{3} + \dfrac{x}{2} = 6$

   c. $\dfrac{3x}{x - 1} - \dfrac{x}{x^2 - 1}$ d. $\dfrac{3}{x} + \dfrac{2}{4x}$

e. $\dfrac{3x}{x-1} - \dfrac{x}{x^2-1} = 3$

f. $\dfrac{3}{x} + \dfrac{2}{4x} = 7$

PRACTICE *In Exercises 8–23, solve each equation algebraically. Check solutions using either the STOre or TABLE feature of the calculator:*

8. $\dfrac{2y}{5} - 8 = \dfrac{4y}{5}$

9. $\dfrac{3x}{4} - 6 = \dfrac{x}{4}$

10. $\dfrac{x}{5} - \dfrac{x}{3} = -8$

11. $\dfrac{2}{3} + \dfrac{x}{4} = 7$

12. $\dfrac{3a}{2} + \dfrac{a}{3} = -22$

13. $\dfrac{x}{2} + x = \dfrac{9}{2}$

14. $\dfrac{x}{2} + 4 = \dfrac{3x}{2}$

15. $\dfrac{y}{3} + 6 = \dfrac{4y}{3}$

16. $\dfrac{x}{3} + 1 = \dfrac{x}{2}$

17. $\dfrac{x}{2} - 3 = \dfrac{x}{5}$

18. $\dfrac{x-3}{3} + 2x = -1$

19. $\dfrac{x+2}{2} - 3x = x + 8$

20. $\dfrac{z-3}{2} = z + 2$

21. $\dfrac{b+2}{3} = b - 2$

22. $\dfrac{5(x+1)}{8} = x + 1$

23. $\dfrac{3(x-1)}{2} + 2 = x$

In Exercises 24–45, solve each rational equation algebraically. Check solutions using either the STOre or TABLE feature of the calculator.

24. $\dfrac{5}{x} - \dfrac{4}{x} = 8 + \dfrac{1}{x}$

25. $\dfrac{11}{x} + \dfrac{13}{x} = 12$

26. $\dfrac{1}{x-1} + \dfrac{3}{x-1} = 1$

27. $\dfrac{3}{p+6} - 2 = \dfrac{7}{p+6}$

28. $\dfrac{x^2}{x+2} - \dfrac{4}{x+2} = x$

29. $\dfrac{x}{x-2} + \dfrac{2}{3} = \dfrac{2}{x-2}$

30. $\dfrac{x}{x-5} - \dfrac{5}{x-5} = 3$

31. $\dfrac{3}{y-2} + 1 = \dfrac{3}{y-2}$

32. $\dfrac{3r}{2} - \dfrac{3}{r} = \dfrac{3r}{2} + 3$

33. $\dfrac{2p}{3} - \dfrac{1}{p} = \dfrac{2p-1}{3}$

34. $\dfrac{1}{3} + \dfrac{2}{x-3} = 1$

35. $\dfrac{3}{5} + \dfrac{7}{x+2} = 2$

36. $\dfrac{7}{q^2-q-2} + \dfrac{1}{q+1} = \dfrac{3}{q-2}$

37. $\dfrac{-5}{x^2+x-2} + \dfrac{3}{x+2} = \dfrac{1}{x-1}$

38. $\dfrac{3x}{3x-6} + \dfrac{8}{x^2-4} = \dfrac{2x}{2x+4}$

39. $\dfrac{x-3}{4x-4} + \dfrac{1}{9} = \dfrac{x-5}{6x-6}$

40. $\dfrac{5}{4y+12} - \dfrac{3}{4} = \dfrac{5}{4y+12} - \dfrac{y}{4}$

41. $\dfrac{3}{5x-20} - \dfrac{4}{5} = \dfrac{3}{5x-20} - \dfrac{x}{5}$

42. $\dfrac{x}{x-1} - \dfrac{12}{x^2-x} = \dfrac{-1}{x-1}$

43. $1 - \dfrac{3}{x} = \dfrac{-8x}{x^2+3x}$

44. $\dfrac{x}{x^2-9} + \dfrac{x+8}{x+3} = \dfrac{x-8}{x-3}$

45. $\dfrac{x-3}{x-2} = \dfrac{1}{x} + \dfrac{x-3}{x}$

In Exercises 46–51, solve each rational equation graphically. Copy the INTERSECT screen exactly as it appears, including all displayed information, to justify your work. If a window other than Zstandard is used, specify it in the form [xmin, xmax] by [ymin, ymax].

46. $\dfrac{2x}{5} - 8 = \dfrac{4x}{5}$

47. $\dfrac{3x}{4} - 6 = \dfrac{x}{4}$

48. $\dfrac{2}{x+1} - \dfrac{12}{x+1} = -5$

49. $\dfrac{1}{x-3} + \dfrac{2}{x-3} = 1$

50. $x + \dfrac{2}{3} = \dfrac{2x-12}{3x-9}$ (*Hint:* Use a window in which you can clearly see both roots.)

51. $x + \dfrac{3}{4} = \dfrac{3x-50}{4x-24}$ (*Hint:* Use a window in which you can clearly see both roots.)

52. Consider the equation $\dfrac{x-4}{x-3} + \dfrac{x-2}{x-3} = x - 3$.

    a. Solve the given equation by first multiplying through by the LCD.

    b. Solve the given equation by first simplifying the left side.

    c. Solve the given equation graphically.

    d. Which solution method did you like best? Why?

    e. Would looking at the graphical display *first* have helped you in your algebraic solution? Why or why not?

53. Consider the equation $\dfrac{x-5}{x-4} + \dfrac{x-3}{x-4} = x - 4$.

    a. Solve the given equation by first multiplying through by the LCD.

    b. Solve the given equation by first simplifying the left side.

    c. Solve the given equation graphically.

    d. Which solution method did you like best? Why?

    e. Would looking at the graphical display *first* have helped you in your algebraic solution? Why or why not?

REVIEW

54. Simplify: $2(3x^2 - 4x + 5) - (x^2 + 25)$

55. Divide: $\dfrac{4x^2 + 6xy - 12x^3}{4x}$

56. Uniform motion Two cars leave from Ashland City driving in opposite directions on Highway 41A. If one car is traveling at 55 mph and the other at 65 mph, how long before they are 60 miles apart?

57. Auto repair The service department at the local car dealership charges a flat rate of $75 for diagnosing problems and then $15 per hour for labor.

    a. Write an equation to model the cost where x represents the number of hours and y the total cost.

    b. Graph the equation in an appropriate viewing window on your calculator.

    c. Explain how to use the graph on the calculator (and the appropriate feature(s) of the calculator) to determine the cost for 3 hours of work.

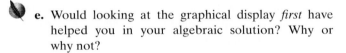

6.6 APPLICATIONS OF RATIONAL EQUATIONS

- NUMBER PROBLEMS
- UNIFORM MOTION PROBLEMS
- WORK PROBLEMS
- INVESTMENT PROBLEMS
- AVERAGE COST

▍ NUMBER PROBLEMS

The ability to solve equations containing rational expressions allows us to expand our knowledge of applications. We will begin with number problems. You may want to review the problem-solving strategies introduced in Section 3.1.

EXAMPLE 1

If the same number is added to both the numerator and the denominator of the fraction $\frac{3}{5}$, the result is $\frac{4}{5}$. Find the number.

SOLUTION

Because there is only one unknown, we will let $x =$ the number. Translate the words of the problem to an algebraic equation: "The *same number* is added to both the numerator and denominator . . . the result is $\frac{4}{5}$":

$$\frac{3+x}{5+x} = \frac{4}{5}$$

To solve the equation, you may multiply through by the LCD of $5(5 + x)$ *OR*, because the equation is a proportion, cross-multiply, as shown here:

$$5(3 + x) = 4(5 + x)$$ Cross-multiply (the product of the means equals the product of the extremes).

$$15 + 5x = 20 + 4x$$ Use the distributive property to remove parentheses.

$$15 + x = 20$$ Subtract $4x$ from both sides.

$$x = 5$$ Subtract 15 from both sides.

The number is 5.

The check is most easily done analytically, substituting 5 for x in the expression $\dfrac{3 + x}{5 + x}$ to obtain the simplified result of $\dfrac{4}{5}$:

$$\frac{3 + 5}{5 + 5} = \frac{8}{10} = \frac{4}{5}$$

UNIFORM MOTION PROBLEMS

Earlier, when solving uniform motion problems, we discovered two important tools for solving these types of problems, the formula $d = r \cdot t$ and organizing given information into a chart. This is particularly helpful when considering the movement of more than one object.

EXAMPLE 2

A coach can run 10 miles in the same amount of time that his best student-athlete can run 12 miles. If the student can run 1 mile per hour faster than the coach, how fast does each run?

SOLUTION

Because the student's rate is given in terms of the coach's rate, we begin by writing our "Let" statements.

Let x = the coach's rate

$x + 1$ = the student's rate

Organize the given information into a chart:

	d	r	t
Coach	10	x	
Student	12	$x + 1$	

We do not know the times of either the coach or the student. However, we *do* know that if $d = r \cdot t$, then $t = \frac{d}{r}$. We can represent the times for each of the runners by dividing *their* distance by *their* rate. The last *column* of the table is determined by the information in the given row.

Completing the table, we have

	d	r	t
Coach	10	x	$\dfrac{10}{x}$
Student	12	$x+1$	$\dfrac{12}{x+1}$

Now that the information is organized in a table, return to the words of the problem to write the equation.

The first sentence tells us that their times are equal. Our equation can now be written and solved.

$$\frac{10}{x} = \frac{12}{x+1} \qquad \text{Equate the runners' times.}$$

$$10(x+1) = 12x \qquad \text{Cross-multiply.}$$

$$10x + 10 = 12x \qquad \text{Distribute.}$$

$$10 = 2x \qquad \text{Subtract } 10x \text{ from both sides of the equation.}$$

$$5 = x \qquad \text{Divide both sides by 2.}$$

Make sure you answer the question posed in the problem with a complete sentence: The coach runs 5 miles per hour and the student runs 6 miles per hour.

▎▎ WORK PROBLEMS

Work problems often involve people or machines completing a job together. These problems are closely related to uniform motion problems, as will be illustrated in the following example. In each problem you must consider the *rate* at which each individual or machine works, the *time* actually worked, and the *part of the task* that is *completed* by that individual or machine. A table often proves helpful in organizing the information.

EXAMPLE 3

Mary can clean the house completely in 4 hours, whereas it takes her husband 6 hours to do the same job. How long will it take them to completely clean the house if they work together?

SOLUTION

Let x = the time to complete the task when working together. Organize the information into a chart. Since it takes Mary 4 hours to complete the task, she is doing $\frac{1}{4}$ of the task per hour. Similarly, her husband is doing $\frac{1}{6}$ of the task per hour since it takes him 6 hours to complete the task.

	Time worked	Rate of work
Mary	4 hours	$\dfrac{1}{4}$
Husband	6 hours	$\dfrac{1}{6}$
Time Together	x hours	$\dfrac{1}{x}$

Because they are working together to complete the same job, the individual work rates should produce a total that is equal to the rate of time worked together to produce the completed task. We may then solve the problem.

$$\frac{1}{4} + \frac{1}{6} = \frac{1}{x}$$

$3x + 2x = 12$ Multiply through the equation by the LCD of 12x.

$5x = 12$ Combine *like* terms.

$x = \frac{12}{5}$ Divide both sides by 5.

They can complete the task together in $\frac{12}{5}$ hours. This is equivalent to 2.4 hours, which is better expressed as 2 hours and 24 minutes (four-tenths of an hour is $0.4 \cdot 60 = 24$ minutes).

The equation can be solved graphically, as illustrated below. The window used is $[-5, 5]$ by $[-5, 5]$.

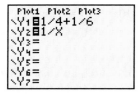

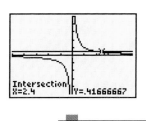

▓ INVESTMENT PROBLEMS

Recall the formula $i = prt$, where i represents the interest, p the principal, r the rate (expressed as a decimal rather than a percentage), and t the time. Previously, we worked problems in which the principal amount was usually the unknown quantity. The example below illustrates a type of investment problem that results in a rational equation. Again, a chart is often helpful.

EXAMPLE 4

At a bank, a sum of money invested for one year will earn $150 interest. If invested in bonds, that same money would earn $200, because the interest rate paid by the bonds is 1% greater than that paid by the bank. Find the bank's rate.

SOLUTION   Let $x =$ the bank's rate.

$x + .01 =$ the bond's rate

Organize the known information in a table:

	Interest	Principal	Rate	Time
Bank	150		x	1
Bonds	200		$x + .01$	1

The principal for each investment can be expressed in terms of the interest, rate, and time. $i = prt$; therefore, $p = \frac{i}{rt}$. In each row we divide the interest earned by the product of the rate and time. We can now complete the table.

	Interest	Principal	Rate	Time
Bank	150	$\dfrac{150}{x}$	x	1
Bonds	200	$\dfrac{200}{x + .01}$	$x + .01$	1

We must now write an equation. Referring back to the original problem, we notice the phrase "... that same money." This means that the principal values are equal. We can now write an equation and solve it.

$$\frac{150}{x} = \frac{200}{x + .01}$$ Equate the principal amounts from the two investments.

$$150(x + .01) = 200x$$ Because the equation is a proportion, cross-multiply.

$$150x + 1.5 = 200x$$ Distribute.

$$1.5 = 50x$$ Subtract $150x$ from both sides of the equation.

$$.03 = x$$ Divide both sides by 50 to isolate the variable.

The bank's interest rate is 0.03, which is expressed as 3%. The bonds pay 4%, a rate that is 1% greater than that paid by the bank.

▍▍▍ AVERAGE COST

Average cost is just what the term implies, the average cost to the consumer for the production of goods. Often, manufacturers must determine an average cost for each unit produced. Average cost is equal to the total cost divided by the number of units. The example below illustrates this concept.

EXAMPLE 5

A CD manufacturer can produce CDs at a cost of $8 per unit. It has operating costs of $24,000 per month. How many CDs must be produced monthly to have an average cost of $12 per unit?

SOLUTION   Let x = number of CDs produced.

$$\text{Total cost} = 24{,}000 + 8x$$

$$\text{Average cost} = \text{Total cost} \div \text{Number of units}$$

We substitute our information into this formula and solve the resulting equation:

$$12 = \frac{24{,}000 + 8x}{x}$$

$$12x = 24{,}000 + 8x$$ Multiply through the equation by the LCD of x.

$$4x = 24{,}000$$ Subtract $8x$ from both sides of the equation.

$$x = 6{,}000$$ Divide both sides by 4.

STUDY TIP

Always get a good night's rest before a test. This helps you avoid careless errors.

The company must produce 6,000 CDs monthly to keep the average cost of the CD at $12.

EXERCISE 6.6

CONCEPTS

1. If x and $x + 1$ are two consecutive integers, find their reciprocals.

2. What fraction do you have if the denominator of $\frac{7}{8}$ is doubled and the numerator is decreased by 2?

3. Write the rational expression determined by decreasing the numerator of $\frac{x + 2}{x}$ by 3 and increasing the denominator by 3.

APPLICATIONS *For each problem below, clearly identify the variable and what it represents, write an equation, provide a complete solution to that equation, and answer the question posed in a complete sentence.*

4. Number problem If the denominator of the fraction $\frac{3}{4}$ is increased by a number and the numerator of the fraction is doubled, the result is 1. Find the number.

5. Number problem If a number is added to the numerator of the fraction $\frac{7}{8}$ and the same number is subtracted from the denominator, the result is 2. Find the number.

6. Number problem If a number is added to the numerator of the fraction $\frac{3}{4}$ and twice as much is added to the denominator, the result is $\frac{4}{7}$. Find the number.

7. Number problem If a number is added to the numerator of the fraction $\frac{5}{7}$ and twice as much is subtracted from the denominator, the result is 8. Find the number.

8. Number problem The sum of a number and its reciprocal is $\frac{13}{6}$. Find the numbers.

9. Number problem The sum of the reciprocals of two consecutive even integers is $\frac{7}{24}$. Find the integers.

10. Sightseeing A tourist can bicycle 28 miles in the same time he can walk 8 miles. If he can ride 10 miles per hour faster than he can walk, how fast can he walk?

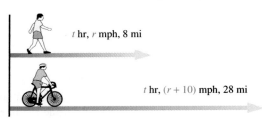

Illustration 1

11. Comparing travel A plane can fly 300 miles in the same time it takes a car to go 120 miles. If the car travels 90 miles per hour slower than the plane, how fast is the plane?

12. Boating in a river A boat that can travel 18 miles per hour in still water can travel 22 miles downstream in the same amount of time that it can travel 14 miles upstream. Find the speed of the current in the river. (See Illustration 2.)

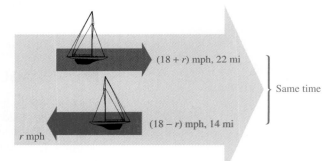

Illustration 2

13. Wind speed A plane can fly 300 miles downwind in the same amount of time it can travel 210 miles upwind. Find the velocity of the wind if the plane can fly at 255 miles per hour in still air.

14. River tours A riverboat tour begins by going 60 miles upstream against a 5 mph current. There, the boat turns around and returns with the current. What still-water speed should the captain use to complete the tour in 5 hours?

15. Travel time The company president flew 680 miles in the corporate jet and returned in a smaller plane that could fly only half as fast. If the total travel time was 6 hours, find the speeds of the planes.

16. Roofing a house A homeowner estimates that it will take 7 days to roof his house. A professional roofer estimates that he could roof the house in 4 days. How long will it take if the homeowner helps the roofer?

17. Filling a pool   An inlet pipe can fill an empty pool in 5 hours, and another inlet pipe can fill the pool in 4 hours. How long will it take both pipes to fill the pool?

18. Filling a pool   One inlet pipe can fill an empty pool in 4 hours, and a drain can empty the pool in 8 hours. How long will it take the pipe to fill the pool if the drain is left open? (*Hint:* Draining is a negative action.)

19. Mowing lawns   Blaine can mow his yard in 3 hours. When his teen-aged son helps, they can complete the lawn in 2 hours. How long would it take the son to mow the entire yard himself?

20. Painting a room   It takes Blaire 5 hours to paint her bedroom by herself. Her brother, Trey, can paint the entire room in 3 hours. How long will it take them to paint the room if they work together?

21. Processing forms   Becky, a secretary, can process 500 university forms in $2\frac{1}{2}$ days. Sherry can process the same 500 forms in 2 days. If they work together, how long would it take them to process 500 of the university's forms?

22. Comparing investments   Two certificates of deposit pay interest at rates that differ by 1%. Money invested for 1 year in the first CD earns $135 interest. The same principal invested in the other CD earns $180. Find the two rates of interest.

23. Comparing interest rates   Two bond funds pay interest at rates that differ by 2%. Money invested for 1 year in the first fund earns $200 interest. The same amount invested in the other fund earns $360. Find the lower rate of interest.

24. Average cost   A calculator manufacturing company has a yearly overhead of $120,000. It costs $ 84 to produce a single graphing calculator.

a. How many graphing calculators must be produced annually for the average production cost of a calculator to be $100?

 **b.** The left side of the equation is entered at the y1 prompt and the right side of the equation is entered at the y2 prompt (as shown below). The window values shown below were entered. The third graphical display shows *x* and *y* values. Interpret those values within the context of the problem.

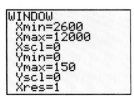

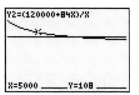

 c. Interpret the given display within the context of the problem.

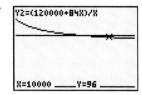

25. Average cost   A textbook company wants to keep the costs of its texts as low as possible.

a. If it costs them $55 to produce a single mathematics text with overhead costs of $50,000 annually, how many texts must be produced for the average production cost per text to be held at $75?

 b. The left side of the equation is entered at the y1 prompt and the right side of the equation is entered at the y2 prompt (as shown below). The window values shown below were entered. The third graphical display shows *x* and *y* values. Interpret those values within the context of the problem.

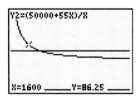

 c. Interpret the given display within the context of the problem.

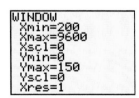

 d. Typically, the graphs of rational equations have two branches. In parts (b) and (c) above, only the right branch has been displayed for interpretation. The left branch of the graph of the expression entered at the y2 prompt follows. Interpret the displayed information, as best you can. Is it applicable to this problem? Why or why not?

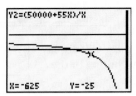

Simplify each expression. Write each result without negative exponents.

26. $(m^2n^{-3})^{-2}$

27. $\dfrac{a^{-1}}{a^{-1} + 1}$

28. $\dfrac{a^0 + 2a^0 - 3a^0}{(a - b)^0}$

29. $(4x^{-2} + 3)(2x^2 - 4)$

CHAPTER SUMMARY

CONCEPTS

REVIEW EXERCISES

Section 6.1

A rational expression is simplified when the numerator and the denominator have no common factors other than 1.

Simplifying (Reducing) a Rational Expression
1. Factor the numerator and the denominator completely.
2. Divide out common factors.

The fundamental property of fractions allows us to divide out common factors and to multiply the numerator and denominator of a given fraction by the same quantity. These processes produce equivalent fractions.

Simplifying Rational Expressions

In Review Exercises 1–3, simplify each rational expression. If a fraction is already in lowest terms, so indicate. State any excluded values of the variable.

1. a. $-\dfrac{51}{153}$ **b.** $-\dfrac{105}{45}$

   c. $\dfrac{3x^2}{6x^3}$ **d.** $\dfrac{5xy^2}{2x^2y^2}$

2. a. $\dfrac{x^2}{x^2 + x}$ **b.** $\dfrac{x + 2}{x^2 + 2x}$

   c. $\dfrac{x^2 + 4x + 3}{x^2 - 4x - 5}$ **d.** $\dfrac{x^2 - x - 56}{x^2 - 5x - 24}$

3. a. $\dfrac{2x^2 - 16x}{2x^2 - 18x + 16}$ **b.** $\dfrac{x^2 + x - 2}{x^2 - x - 2}$

Section 6.2

To multiply rational expressions, multiply the numerators and multiply the denominators. Divide out common factors to simplify (reduce).

When dividing rational expressions, multiply the dividend by the reciprocal of the divisor.

Multiplying and Dividing Rational Expressions

In Review Exercises 4–5, perform the indicated operations. Make sure all results are simplified.

4. a. $\dfrac{3xy}{2x} \cdot \dfrac{4x}{2y^2}$ **b.** $\dfrac{3x}{x^2 - x} \cdot \dfrac{2x - 2}{x^2}$

   c. $\dfrac{3x^2}{5x^2y} \div \dfrac{6x}{15xy^2}$ **d.** $\dfrac{x^2 + 5x}{x^2 + 4x - 5} \div \dfrac{x^2}{x - 1}$

5. a. $\dfrac{x^2 + 3x + 2}{x^2 + 2x} \cdot \dfrac{x}{x + 1}$ **b.** $\dfrac{x^2 + 4x + 4}{x^2 - x - 6} \cdot \dfrac{x^2 - 9}{x^2 + 5x + 6}$

   c. $\dfrac{x^2 - x - 6}{2x - 1} \div \dfrac{x^2 - 2x - 3}{2x^2 + x - 1}$ **d.** $\dfrac{3n^2 + 5n - 2}{12n^2 - 13n + 3} \div \dfrac{n^2 + 3n + 2}{4n^2 + 5n - 6}$

Section 6.3

To add rational expressions with common denominators, keep the common denominator, add (or subtract) the numerators as the signs indicate, and then reduce the rational expression.

Adding and Subtracting Rational Expressions

In Review Exercises 6–7, simplify each rational expression by adding or subtracting as the sign indicates. All results should be expressed in simplified form.

6. a. $\dfrac{x}{x + y} + \dfrac{y}{x + y}$ **b.** $\dfrac{3x}{x - 7} - \dfrac{x - 2}{x - 7}$

To add or subtract fractions with unlike denominators, you must first find the LCD. Then, convert each fraction to an equivalent fraction with the LCD as denominator.

Finding the Least Common Denominator (LCD)

1. Write down each of the different denominators that appear in the given fractions.
2. Factor each of these denominators completely (use exponential form).
3. Form a product using each of the different factors obtained in step 2. Use each different factor the *greatest* number of times it appears in any *single* factorization. The product is the least common denominator.

Adding/Subtracting with Unlike Denominators

1. Identify the LCD of the rational expressions.
2. Convert each rational expression to an equivalent rational expression having the LCD as denominator.
3. Add or subtract the rational expressions by *expressing* the addition or subtraction over the common denominator.
4. Simplify the numerators, but leave the denominators in factored form.
5. Simplify by reducing the rational expression to lowest terms. This requires you to divide out common factors in the numerator and the denominator. Because the denominator is in factored form, you need only factor the numerator.

c. $\dfrac{x}{x-1} + \dfrac{1}{x}$

d. $\dfrac{1}{7} - \dfrac{1}{x}$

7. a. $\dfrac{3}{x+1} - \dfrac{2}{x}$

b. $\dfrac{x+2}{2x} - \dfrac{2-x}{x^2}$

c. $\dfrac{x}{x+2} + \dfrac{3}{x} - \dfrac{4}{x^2+2x}$

d. $\dfrac{2}{x-1} - \dfrac{3}{x+1} + \dfrac{x-5}{x^2-1}$

Section 6.4

Complex fractions may be simplified by using either of the methods specified below. All complex fractions can be simplified using either method.

Method 1: Simplifying a Complex Fraction Using the Order of Operations
1. Replace the main fraction bar with a division sign, grouping the numerator and denominator appropriately.
2. Simplify, using the order of operations.

Method 2: Simplifying a Complex Fraction by Multiplying by the LCD
1. Determine the LCD of the component fractions of the complex fraction.
2. Multiply the numerator and denominator of the complex fraction by the LCD. Be sure to distribute when the numerator and/or denominator contain more than one term.

Complex Fractions

8. Simplify each complex fraction.

a. $\dfrac{\dfrac{1}{x} + 1}{\dfrac{1}{x} - 1}$

b. $\dfrac{1 + \dfrac{3}{x}}{2 - \dfrac{1}{x^2}}$

c. $\dfrac{\dfrac{2}{x-1} + \dfrac{x-1}{x+1}}{\dfrac{1}{x^2-1}}$

d. $\dfrac{\dfrac{a}{b} + c}{\dfrac{b}{a} + c}$

Section 6.5

Solving Rational Equations Algebraically
1. Determine any and all excluded values by setting each denominator containing a variable equal to zero and solving for the variable.
2. Determine the LCD.
3. Multiply each *term* on both sides of the equation by the LCD to clear fractions.
4. Simplify both sides of the equation.
5. Solve the equation.
6. Reject any extraneous solutions (those values that equal the excluded values found in step 1).

Algebraic and Graphical Solutions of Rational Equations

9. Solve each equation algebraically. Check all results using either the STOre or TABLE features of the graphing calculator.

a. $\dfrac{3}{x} = \dfrac{2}{x-1}$

b. $\dfrac{5}{x+4} = \dfrac{3}{x+2}$

c. $\dfrac{2}{3x} + \dfrac{1}{x} = \dfrac{5}{9}$

d. $\dfrac{2x}{x+4} = \dfrac{3}{x-1}$

e. $\dfrac{2}{x-1} + \dfrac{3}{x+4} = \dfrac{-5}{x^2+3x-4}$

f. $\dfrac{4}{x+2} - \dfrac{3}{x+3} = \dfrac{6}{x^2+5x+6}$

g. $\dfrac{5}{4x+12} - \dfrac{3}{4} = \dfrac{5}{4x+12} - \dfrac{x}{4}$

7. Check solutions in the original equation to verify solutions.

Graphical Displays of Rational Expressions
1. View the graph in *both* connected and dot mode.
2. Use the TABLE to help you see, numerically, what is happening. Reconcile the numerical information with the graphical display.
3. TRACE along the graph watching for patterns.
4. Use the INTERSECT feature to find/confirm solutions graphically.
5. For clarity, you may want to turn the *x*- and *y*-axes off to unclutter the screen. To do this, access the FORMAT menu, highlight AxesOff, and press ENTER.
6. Even when solving equations algebraically, look at the graphical display and practice reconciling your algebraic solution with the graphical display.

10. Solve the rational equation $\dfrac{1}{x-1} + \dfrac{3}{x-1} = 1$ graphically. Sketch your screen display and copy all displayed information.

| Section 6.6 | Applications of Rational Equations |

Applications of Rational Equations

Be familiar with the problem-solving strategies introduced in Section 3.1.

The rate of work can be determined by the fraction $\dfrac{1}{\text{time worked}}$.

In Review Exercises 11–16, solve each problem. Be sure to clearly identify what the variable represents, to write an equation and include a complete solution, and to answer the question posed with a sentence.

11. Pumping a basement If one pump can empty a flooded basement in 18 hours and a second pump can empty the basement in 20 hours, how long will it take to empty the basement if both pumps are used?

12. Painting houses If a homeowner can paint a house in 14 days and a professional painter can paint it in 10 days, how long will it take if they work together?

13. Jogging A jogger can bicycle 30 miles in the same time as he can jog 10 miles. If he can ride 10 miles per hour faster than he can jog, how fast can he jog?

14. Wind speed A plane can fly 400 miles downwind in the same amount of time as it can travel 320 miles upwind. If the plane can fly at 360 miles per hour in still air, find the velocity of the wind.

15. Comparing interest rates   Two savings accounts at a local bank pay interest rates that differ by $\frac{1}{2}$%. Money invested for 1 year in one of the accounts earns $108 in interest, whereas the same amount of money invested in the other account earns $121.50. Find the lower rate of interest.

16. Average cost   A new day care center is opening for small children. The overhead is $6,000 monthly. Expenditures per child (for supplies and lunches) will be approximately $150 per month. How many students must be enrolled to keep the average cost per month for each child at $400?

CHAPTER TEST

1. Simplify: $\dfrac{48x^2y}{54xy^2}$

2. Simplify $\dfrac{2x^2 - x - 3}{4x^2 - 9}$ and state excluded values for the variable. $; x \neq \dfrac{3}{2}, x \neq -\dfrac{3}{2}$

3. Simplify $\dfrac{3(x + 2) - 3}{2x - 4 - (x - 5)}$ and state the excluded values for the variable.

4. Multiply and simplify: $\dfrac{12x^2y}{15xyz} \cdot \dfrac{25y^2z}{16xt}$.

5. Multiply and simplify: $\dfrac{x^2 + 3x + 2}{3x + 9} \cdot \dfrac{x + 3}{x^2 - 4}$.

6. Divide and simplify: $\dfrac{8x^2y}{25xt} \div \dfrac{16x^2y^3}{30xyt^3}$.

7. Divide and simplify: $\dfrac{x^2 - x}{3x^2 + 6x} \div \dfrac{3x - 3}{3x^3 + 6x^2}$.

8. Add: $\dfrac{5x - 4}{x - 1} + \dfrac{5x + 3}{x - 1}$.

9. Subtract: $\dfrac{3y + 7}{2y + 3} - \dfrac{3(y - 2)}{2y + 3}$.

10. Add: $\dfrac{x + 1}{x} + \dfrac{x - 1}{x + 1}$.

11. Simplify: $\dfrac{\dfrac{8x^2}{xy^3}}{\dfrac{4y^3}{x^2y^3}}$.

12. Simplify: $\dfrac{1 + \dfrac{y}{x}}{\dfrac{y}{x} - 1}$.

13. Solve: $\dfrac{x}{10} - \dfrac{1}{2} = \dfrac{x}{5}$.

14. Solve: $3x - \dfrac{2(x + 3)}{3} = 16 - \dfrac{x + 2}{2}$.

15. Solve: $\dfrac{7}{x + 4} - \dfrac{1}{2} = \dfrac{3}{x + 4}$.

16. Solve: $\dfrac{y}{y - 1} = \dfrac{y - 2}{y}$.

17. Cleaning highways   One highway worker could pick up all the trash on a strip of highway in 7 hours, and his helper could pick up the trash in 9 hours. How long will it take them if they work together?

18. Boating   A boat can motor 28 miles downstream in the same amount of time as it can motor 18 miles upstream. Find the speed of the current if the boat can motor at 23 miles per hour in still water.

19. Cycling   A veteran cyclist can make a 35-mile trip in the same time that a novice cyclist can make a 20-mile trip. If the veteran is cycling 10 mph faster than the novice, how long will it take the novice to travel 20 miles?

20. Jogging   Betsy can jog the 5 miles to the softball field in the same amount of time that she can cycle 20 miles to work. If she is cycling 10 mph faster than she is jogging, how fast is she jogging?

CUMULATIVE REVIEW

CHAPTERS 5 AND 6
Vocabulary/Concepts

1. True or False: The polynomial $4(x - 2) + x(x - 2)$ is completely factored.

2. True or False: The greatest common factor of the polynomial $3x^2(x - 1)^2 + 9x^3(x - 1)$ is $3x^2(x - 1)$.

3. A polynomial that contains three terms is called a _____.

4. What is the leading coefficient of the polynomial $-x^2 - 5x + 6$?

5. Why is $x^3 + x^2 + 1$ not a factorable trinomial?

6. A binomial of the form _____ is called the difference of two squares.

7. In order for a polynomial to be completely factored, each polynomial factor must be _____.

8. Quadratic equations are equations of degree ____.

9. Explain the difference between factoring $x^2 + 5x + 6$ and solving $x^2 + 5x + 6 = 0$.

10. Classify each equation below as either linear or quadratic.
   a. $4x(x - 5) = 2x(x + 1) - 3$
   b. $x^2 - 5x = 2x(x + 1) - x^2$
   c. $(x + 2)(x - 5) = (2x + 1)(x - 4)$

11. The quotient of two polynomials is called a _____ expression.

12. State the excluded value for each expression.
   a. $\dfrac{x + 5}{x - 2}$ **b.** $\dfrac{3x}{4x + 1}$ **c.** $\dfrac{5}{6x}$

13. Perform the indicated operation and simplify:
$$\frac{14xy}{15} - \frac{-6xy}{15}.$$

14. Determine the least common denominator for $\dfrac{5z}{6x^2y}$ and $\dfrac{3}{9xy^2}$.

15. Perform the indicated operation and simplify the result:
   a. $\dfrac{3xy}{5x^2} \cdot \dfrac{10}{6xy^2}$ **b.** $\dfrac{-8y^2}{6x} \div \dfrac{2}{5y^2x}$

16. A fraction is defined to be a quotient of integers, and a _____ fraction is a quotient of fractions.

17. Is $\dfrac{\frac{1}{2} + \frac{2}{3}}{\frac{3}{4}}$ equivalent to $\dfrac{1}{2} + \dfrac{2}{3} \div \dfrac{3}{4}$? If not, explain why.

18. Apparent solutions that are not valid are called _____ solutions.

19. Identify the excluded values and either simplify the expression or solve the equation.
   a. $\dfrac{1}{x - 3} + \dfrac{2}{x - 3} = 1$
   b. $\dfrac{1}{x - 3} + \dfrac{2}{x - 3}$

20. If you have two consecutive odd integers, using the variable x, write the two rational expressions that represent their reciprocals.

Practice   *In Cumulative Review Exercises 21–28, factor the polynomials completely. If prime, so state.*

21. $x^2 - 49$

22. $x^2 - 8x - 9$

23. $18x^2 - 27x - 5$

24. $x^3 + 2x^2$

25. $x^2 - 18x + 81$

26. $5x^3 - 125x$

27. $ac - ad + bc - bd$

28. $a^3 - 8y^3$

In Cumulative Review Exercises 29–30, solve each equation by factoring. Verify your results using either the STOre or TABLE feature of the calculator.

29. $x(x + 8) - 5x = 5(1 - x) - 12$

30. $x^3 + x^2 - 12x = 0$

31. Solve graphically using the INTERSECT feature. Copy your complete screen display to justify your work. Solutions should be approximated to the nearest hundredth. $-0.65x^2 + 1.41x + 4 = 0$

32. The pictured screens display the solution to a quadratic equation. Write an equation in $ax^2 + bx + c = 0$ form that has these solutions.

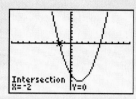 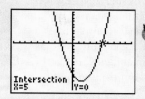

33. A student entered a polynomial at y1 and at y2 entered what he *believed* to be the correct factorization of the polynomial entered at y1. He then displayed the TABLE at the right to verify his factorization. What does the TABLE reveal about the factorization and why?

X	Y1	Y2
3	-10	-2
4	-6	-2
5	0	0
6	8	4
7	18	10
8	30	18
9	44	28

X=3

34. Simplify by reducing to lowest terms: $\dfrac{4a + 4b}{a^2 + ab}$.

35. Rewrite the fraction as an equivalent fraction with the given denominator: $\dfrac{x}{x + 5}$; $x^2 - 25$.

36. Find the least common denominator and then subtract the two fractions: $\dfrac{3}{y + 2} - \dfrac{2}{y - 2}$.

37. Explain why $\dfrac{a - b}{b - a}$ is equal to -1.

38. Perform the indicated operations and simplify:
$$\dfrac{(x + 3)^5}{x^2 + 4x + 3} \div \dfrac{x^2 + 6x + 9}{x^2 - 3x} \cdot \dfrac{2x^2 + 8x + 6}{x^2 - 9}.$$

39. Perform the indicated multiplication and simplify:
$$\dfrac{3 - a}{b^2 - b} \cdot \dfrac{b^2 - 2b + 1}{a^2 - 9}.$$

40. Simplify $\dfrac{\frac{2}{3} + \frac{4}{5}}{\frac{2}{3} - \frac{4}{5}}$ and verify with your calculator.

41. Simplify: $\dfrac{\frac{a}{b} - \frac{b}{a}}{\frac{a}{b^2} - \frac{b}{a^2}}$.

42. Algebraically solve $\dfrac{5}{a - 3} + \dfrac{30}{9 - a^2} = 1$.

43. Graphically solve $\dfrac{6}{5x} + \dfrac{4}{5} - 2x = 0$.

44. Explain the meaning of the indicated "line" in the graphical solution of the equation $\dfrac{x - 1}{x + 1} = 3$ that is displayed at the right.

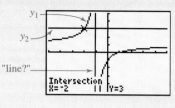

45. The TABLE at the right indicates that the left side of an equation has been entered at y1 and the right side at y2. Determine the excluded values, if any, and the solution of the equation.

X	Y1	Y2
1	2	-2
2	2	-1
3	ERROR	0
4	2	1
5	2	2
6	2	3
7	2	4

X=1

Applications

46. Time of flight An object has been thrown straight up into the air. The formula $h = vt - 16t^2$ gives the height, h, of the object above the ground after t seconds with an initial velocity of v. After how many seconds will an object hit the ground if the initial velocity was 80 feet per second?

47. Fencing A rancher wants to enclose a rectangular region for a garden because deer are eating the vegetables. If the length is to be 8 yards more than twice the width and the area is 10 square yards, find the measure of the width.

48. Data processing Becky can enter admissions data on her computer terminal in 4 hours. Sherry can do the same job from her computer terminal in 6 hours because she is frequently interrupted by the telephone. If they work together, how long will it take to enter the admissions data?

49. Travel speed The *Queen of the Cumberland* riverboat can travel 12 miles downstream in the same amount of time that it takes it to travel 8 miles upstream. If the current of the river is 1.5 miles per hour, what is the boat's rate in still water?

CAREERS & MATHEMATICS

© Georgina Bowater/CORBIS

Mechanical Engineer   Mechanical engineers design and develop power-producing machines such as internal combustion engines, steam and gas turbines, and jet and rocket engines, as well as power-using machines such as refrigeration and air-conditioning equipment, machine tools, printing presses, and steel rolling mills. Many mechanical engineers do research, test, and design work; others work in maintenance, technical sales, and production operations. Some teach in colleges and universities or work as consultants.

Qualifications   A bachelor's degree in engineering is generally required for beginning engineering jobs. College graduates with a degree in a natural science or mathematics also may qualify for some jobs.

Job Outlook   Employment of mechanical engineers is expected to increase through the year 2010 due to the growing demand for complex industrial machinery and processes. New job opportunities also will be created with emerging technologies in information technology, biotechnology, and nanotechnology.

Example Application   Mechanical power is often transmitted from one location to another by means of a shaft — the drive shaft in an automobile is a good example. It is one concern of the mechanical engineer that the shaft be strong enough that it will not twist, bend, or snap. One critical value in the required calculations is the shaft's **stiffness**, denoted here by the letter k. The stiffness of a shaft is measured in units of inch pounds per radian.

If a single shaft consists of two sections (see Illustration 1), then the shaft's overall stiffness k is given by the formula

$$k = \frac{1}{\dfrac{1}{k_1} + \dfrac{1}{k_2}}$$

Section 1   Section 2

Illustration 1

where k_1 and k_2 are the individual stiffnesses of Sections 1 and 2, respectively.

If the stiffness, k_2, of Section 2 of the shaft in Illustration 1 is 4,200,000 in. lb/rad, and the design specifies that the overall stiffness, k, of the entire shaft be 1,900,000 in. lb/rad, what must k_1, the stiffness of Section 1, be?

Solution   We begin by solving the given equation for k_1.

$$k = \frac{1}{\dfrac{1}{k_1} + \dfrac{1}{k_2}}$$

$$k = \frac{1 k_1 k_2}{\left(\dfrac{1}{k_1} + \dfrac{1}{k_2}\right) k_1 k_2}$$ Simplify the complex fraction by multiplying numerator and denominator by $k_1 k_2$.

$$k = \frac{k_1 k_2}{k_2 + k_1}$$

$$k(k_2 + k_1) = \frac{k_1 k_2}{k_2 + k_1}(k_2 + k_1)$$ Clear the equation of fractions.

$$kk_2 + kk_1 = k_1 k_2$$

$$kk_2 = k_1 k_2 - kk_1$$ Prepare to solve for k_1 by subtracting kk_1 from both sides.

$$kk_2 = k_1(k_2 - k)$$ Factor out k_1 from the right-hand side.

$$\frac{kk_2}{k_2 - k} = k_1$$ Solve for k_1 by dividing both sides by $k_2 - k$.

To find the stiffness required for Section 1, substitute 4,200,000 for k_2 and 1,900,000 for k into the equation and find k_1.

$$k_1 = \frac{kk_2}{(k_2 - k)}$$

$$k_1 = \frac{(1,900,000)(4,200,000)}{4,200,000 - 1,900,000}$$

$$k_1 \approx 3,500,000$$ Use a calculator.

The stiffness of Section 1 must be approximately 3,500,000 in. lb/rad.

EXERCISES

1. Find k_1, the stiffness of Section 1, if $k_2 = 3,700,000$ in. lb/rad and $k = 1,530,000$ in. lb/rad.

2. What is the stiffness of a shaft whose two sections have the same stiffness, 4,000,000 in. lb/rad?

3. Show that the formula of this example may also be written as

$$k = \frac{k_1 k_2}{k_1 + k_2}.$$

4. The **compliance** of a shaft is the reciprocal of the stiffness. Show that the compliance of the entire shaft is the sum of the compliances of each of its sections.

(*Answers:* **1.** approximately 2,600,000 in. lb/rad   **2.** 2,000,000 in. lb/rad)

CHAPTER

7

Making the Transition from Elementary to Intermediate Algebra

© Royalty-Free/CORBIS

 InfoTrac Project

Do a keyword search on "cholesterol" and find an article titled "Cooking oil to fight fat and cholesterol." Write a summary of the article.

If a container of "functional oil" contains 1.8 liters of flaxseed oil, how many liters of "functional oil" are in the container? Write your equation and solve. Use this information to determine the quantity of each of the other types of oil that will be in the mixture.

Complete this project after studying Section 7.3.

PERSPECTIVE

The Metric System

A common metric unit of length is the kilometer, which is 1,000 meters. Because 1,000 is 10^3, we can write 1 km = 10^3 m. Similarly, 1 centimeter is one-hundredth of a meter: 1 cm = 10^{-2} m. In the metric system, prefixes such as *kilo* and *centi* refer to powers of 10. Other prefixes are used in the metric system, as shown in the table.

Prefix	Symbol	Meaning	
peta	P	10^{15}	= 1,000,000,000,000,000.
tera	T	10^{12}	= 1,000,000,000,000.
giga	G	10^{9}	= 1,000,000,000.
mega	M	10^{6}	= 1,000,000.
kilo	k	10^{3}	= 1,000.
deci	d	10^{-1}	= 0.1
centi	c	10^{-2}	= 0.01
milli	m	10^{-3}	= 0.001
micro	μ	10^{-6}	= 0.000 001
nano	n	10^{-9}	= 0.000 000 001
pico	p	10^{-12}	= 0.000 000 000 001
femto	f	10^{-15}	= 0.000 000 000 000 001
atto	a	10^{-18}	= 0.000 000 000 000 000 001

To appreciate the magnitudes involved, consider these facts: Light, which travels 186,000 miles every second, will travel about 1 foot in one nanosecond. The distance to the nearest star is 43 petameters, and the diameter of an atom is about 10 nanometers. To measure some quantities, however, even these units are inadequate. The sun, for example, radiates 5×10^{26} watts. That's a lot of light bulbs!

Chapter 7 provides an overview of Chapters 1–6, which typically define an Elementary Algebra course. This chapter also provides the foundation for an Intermediate Algebra course. Students should work through Technology Sections 1.1 and 1.2 to become familiar with the capabilities and operation of their graphing calculators. Keystroking information is provided so that calculator support for arithmetic and algebraic processes may be integrated. The technology sections do not require faculty instruction and should be done independently by the student.

7.1 REVIEW OF EXPONENTS AND POLYNOMIALS

- BASIC PROPERTIES AND DEFINITIONS OF EXPONENTS
- BASIC DEFINITIONS AND THE ADDITION AND SUBTRACTION OF POLYNOMIALS
- EVALUATION OF POLYNOMIALS
- MULTIPLICATION AND DIVISION OF POLYNOMIALS

▮ BASIC PROPERTIES AND DEFINITIONS OF EXPONENTS

Points to Remember:

1. Natural numbers can be used as exponents to indicate repeated multiplications.

Exponential Form	*Expanded Form*
	n factors of x
x^n n is a natural number	$x \cdot x \cdot x \cdot \ldots \cdot x$

2. Exponents of 1 often are not expressed.

   $x^1 = x$

3. The exponent states the number of times the base is to be used as a factor.

   x^4 The exponent of 4 states that the base, x, is to be used as a factor 4 times: $x \cdot x \cdot x \cdot x$.

4. A negative is not a part of a base unless grouping is used. In the expression -4^2, the base is 4; the expanded form is $-4 \cdot 4$; the value is -16.
   In the expression $(-4)^2$, the base is -4; the expanded form is $(-4)(-4)$; the value is 16.

5. Any nonzero base raised to the zero power, x^0, is equal to 1.

   $5^0 = 1$ $(3x^2)^0 = 1$

6. Negative exponents are defined as follows:

   $x^{-n} = \dfrac{1}{x^n}$ and $\dfrac{1}{x^{-n}} = x^n$

7. The following rules define operations with expressions containing exponents.

   Product Rule: $x^m \cdot x^n = x^{m+n}$
   Power to a Power Rule: $(x^m)^n = x^{mn}$
   Product to a Power and Quotient to a Power:

   $(xy)^n = x^n y^n$ and $\left(\dfrac{x}{y}\right)^n = \dfrac{x^n}{y^n}$

   Quotient rule: $\dfrac{x^m}{x^n} = x^{m-n}$

   When simplifying expressions containing exponents, do not leave any negative exponents in a result.

EXAMPLE 1

Evaluate each expression. Confirm your results with the graphing calculator.
a. -3^2 **b.** $(-3)^2$ **c.** $(2^2)^3$ **d.** $(-3^2)^4$

SOLUTION   **a.** $-3^2 = -3 \cdot 3 = -9$

a.
```
-3²
        -9
```

b. $(-3)^2 = -3 \cdot -3 = 9$

b.
```
(-3)²
         9
```

c. $(2^2)^3 = (4)^3 = 64$

c.
```
(2²)^3
        64
```

d. $(-3^2)^4 = (-9)^4 = 6561$

d.
```
(-9)^4
      6561
```

EXAMPLE 2   Simplify each expression: **a.** $(x^3)(x^4)$ **b.** $(2x^3)(3xy^2)$ **c.** $(3x^2y^3)^4$
d. $(a^2b^3)^4(ab^2)^3$

SOLUTION   **a.** $(x^3)(x^4) = x^{3+4} = x^7$ Product rule for exponents.

b. $(2x^3)(3xy^2) = (2 \cdot 3)(x^3 \cdot x)(y^2)$ Commutative and associative properties.

$= 6x^4y^2$ Product rule for exponents.

c. $(3x^2y^3)^4 = (3)^4(x^2)^4(y^3)^4 = 81x^8y^{12}$ Product to a power rule and power to a power
rule for exponents.

d. $(a^2b^3)^4(ab^2)^3 = (a^8b^{12})(a^3b^6)$ Product to a power rule and power to a power
rule for exponents.

$= a^{11}b^{18}$ Product rule for exponents.

EXAMPLE 3   **a.** $\dfrac{6x^3}{3x^4}$ **b.** $\dfrac{x^2y^{-3}}{y^2x^{-1}}$ **c.** $\left(\dfrac{4}{x^2}\right)^3$ **d.** $\left(\dfrac{2x^3}{3y^2}\right)^3$

SOLUTION   **a.** $\dfrac{6x^3}{3x^4} = 2x^{3-4} = 2x^{-1}$ Quotient rule for exponents.

$= \dfrac{2}{x}$ Definition of a negative exponent.

b. $\dfrac{x^2y^{-3}}{y^2x^{-1}} = x^{2-(-1)}y^{-3-2} = x^3y^{-5}$ Quotient rule for exponents.

$= \dfrac{x^3}{y^5}$ Definition of a negative exponent.

c. $\left(\dfrac{4}{x^2}\right)^3 = \dfrac{4^3}{x^6} = \dfrac{64}{x^6}$ Quotient to a power rule for exponents and power
to a power rule.

d. $\left(\dfrac{2x^3}{3y^2}\right)^3 = \dfrac{(2)^3(x^3)^3}{(3)^3(y^2)^3} = \dfrac{8x^9}{27y^6}$ Quotient to a power rule and power to a power rule
for exponents.

EXAMPLE 4   Simplify: **a.** $6x^0y^5$ **b.** $\left(\dfrac{x^3}{y^2}\right)^0$ **c.** $(6x^5y^2)^0$ **d.** $\dfrac{x^8}{x^8}$

SOLUTION   **a.** $6x^0y^5 = 6(1)(y^5) = 6y^5$ Definition of a zero exponent.

b. $\left(\dfrac{x^3}{y^2}\right)^0 = 1$ Definition of a zero exponent.

c. $(6x^5y^2)^0 = 1$ Definition of a zero exponent.

d. $\dfrac{x^8}{x^8} = x^{8-8} = 1$ Quotient rule for exponents and definition of a zero exponent.

▮▮▮ BASIC DEFINITIONS AND THE ADDITION AND SUBTRACTION OF POLYNOMIALS

Points to remember:

1. Variables and constants combined with the operations of arithmetic produce **algebraic expressions**. The constant factor is the **numerical coefficient**, often simply referred to as the **coefficient**.

   > $8x$ is an algebraic expression; the coefficient is 8
   >
   > x is an algebraic expression; the coefficient is 1

2. Ungrouped addition and subtraction signs are term indicators.

   > $3xy^2 + 4x - 2$ is an algebraic expression with three terms: $3xy^2$, $4x$, and -2.
   >
   > $5x^2y - 2(x + y)$ is an algebraic expression with two terms: $5x^2y$ and $-2(x + y)$.

3. Algebraic expressions having only whole number exponents on the variables and no variables in the denominators are called **polynomials**.

4. Polynomials containing one term are called **monomials**, those with two terms, **binomials**, and those with three terms are **trinomials**. We may refer to these expressions as **polynomials**.

   > $8xy$ is a monomial
   >
   > $3x + 2$ and $2x - 3(x + 4)$ are binomials
   >
   > $2x^2 + 3x - 5$ is a trinomial
   >
   > $4x^3 + 3x^2 - 2x + 9$ is a polynomial

5. The **degree of a monomial** is the sum of the exponents on its variable factors.

   > The degree of the monomial $3x^3y^2$ is 5.
   >
   > The degree of the monomial 3^3x^4y is also 5.
   >
   > The degree of a constant term is 0.

6. The **degree of a polynomial** is the same as the term of highest degree.
   The degree of the polynomial $3x^3 - 2x^2 + 3x - 4$ is 3 because it is the highest of the degrees of the 4 monomial terms.

7. Terms with identical variable factors are called **like terms**. The distributive property allows us to combine like terms:

   > $3x^2y + 5x^2y = (3 + 5)x^2y = 8x^2y$
   >
   > $2x^2y + 3x$ are not like terms and cannot be combined.

8. To simplify an expression means to remove grouping symbols and to combine like terms.

EXAMPLE 5

Simplify each algebraic expression.

a. $5x^2 - 3x + 8 - 2x + 4x^2 - 2$ **b.** $-4x^2y + 3xy^2 + 5x^2y - 6xy^2$

c. $(2x^3)^2 + 5x^6$ **d.** $5(x^2 + x - 4) - 3(2x^2 + 4x - 2)$

SOLUTION   **a.** $5x^2 - 3x + 8 - 2x + 4x^2 - 2$

$$= 5x^2 + 4x^2 - 3x - 2x + 8 - 2 \qquad \text{Commutative property.}$$

$$= (5 + 4)x^2 - (3 + 2)x + (8 - 2) \qquad \text{Distributive property.}$$

$$= 9x^2 - 5x + 6 \qquad \text{Perform operations within grouping symbols.}$$

b. $-4x^2y + 3xy^2 + 5x^2y - 6xy^2$

$$= -4x^2y + 5x^2y + 3xy^2 - 6xy^2 \qquad \text{Commutative property.}$$

$$= x^2y - 3xy^2 \qquad \text{Combine like terms.}$$

c. $(2x^3)^2 + 5x^6 = 4x^6 + 5x^6 \qquad \text{Definition of an exponent and power rule.}$

$$= 9x^6 \qquad \text{Combine like terms.}$$

d. $5(x^2 + x - 4) - 3(2x^2 + 4x - 2)$

$$= 5x^2 + 5x - 20 - 6x^2 - 12x + 6 \qquad \text{Distributive property.}$$

$$= 5x^2 - 6x^2 + 5x - 12x - 20 + 6 \qquad \text{Commutative property.}$$

$$= -x^2 - 7x - 14 \qquad \text{Combine like terms.}$$

Because of the distributive property and the fact that the number 1 is the identity for the operation of multiplication, we can remove parentheses enclosing a polynomial when the sign preceding the parentheses is negative by changing the sign of each term of the polynomial.

$$-(3x^2 - 4x + 2) = -1(3x^2 - 4x + 2) = -3x^2 + 4x - 2$$

▉▉▉ EVALUATION OF POLYNOMIALS

The word **evaluate** means to "find the value of." When asked to evaluate a polynomial we substitute given values for the specified variable and then simplify the resulting arithmetic expression following the order of operations. We suggest that you first use parentheses for the variable factors and then substitute specified values. The STOre feature of the graphing calculator may be used to confirm your analytical work.

EXAMPLE 6

TECHNOLOGY TIP

You can review the use of the STOre feature in "Technology 1.2" and the use of the TABLE feature in Section 3.3.

Evaluate each polynomial for the given value(s) of the variables(s). Confirm your analytical work by using the STOre feature of the graphing calculator.

a. $3x^2 - 4x + 2$, when $x = 3$

b. $-x^2 - 4xy + y^2$, when $x = -1$ and $y = -2$

c. $2x^3 - 4x^2 + 3x + 6$, when $x = -4$

SOLUTION   **a.** $3x^2 - 4x + 2 = 3(\ )^2 - 4(\ ) + 2 \qquad \text{Insert parentheses for the variable.}$

```
3→X: 3X²-4X+2
              17
```

$$= 3(3)^2 - 4(3) + 2 \qquad \text{Substitute the value of the variable within the parentheses.}$$

$$= 3(9) - 4(3) + 2 \qquad \text{Following the order of operations, simplify exponents.}$$

$$= 27 - 12 + 2 \qquad \text{Perform multiplications.}$$

$$= 17 \qquad \text{Perform additions and subtractions.}$$

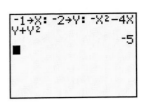

b. $-x^2 - 4xy + y^2$

$$= -(\ )^2 - 4(\ )(\ ) + (\ )^2$$ Insert parentheses for the variables.

$$= -(-1)^2 - 4(-1)(-2) + (-2)^2$$ Substitute the values of the variables within the parentheses.

$$= -1 - 4(-1)(-2) + 4$$ Following the order of operations, simplify exponents.

$$= -1 - 8 + 4$$ Perform multiplications.

$$= -5$$ Perform addition and subtraction.

c. $2x^3 - 4x^2 + 3x + 6$

$$= 2(\ )^3 - 4(\ )^2 + 3(\ ) + 6$$ Insert parentheses for the variables.

$$= 2(-4)^3 - 4(-4)^2 + 3(-4) + 6$$ Substitute the values of the variables within the parentheses.

$$= 2(-64) - 4(16) + 3(-4) + 6$$ Following the order of operations, simplify exponents.

$$= -128 - 64 - 12 + 6$$ Perform multiplications.

$$= -198$$ Perform addition and subtraction.

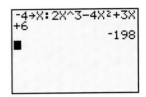

▮▮▮ MULTIPLICATION AND DIVISION OF POLYNOMIALS

Points to remember:

1. The commutative and associative properties coupled with the product rule for exponents are used to perform the multiplication of monomials.

    The product of $(2x^2y^3)(-3x^4y)$ could be rewritten as $(2 \cdot -3)(x^2 \cdot x^4)(y^3 \cdot y)$ to produce the result $-6x^6y^4$

2. The distributive property is used for the multiplication of a monomial times a polynomial. We remove grouping by distributing the multiplication over each term of the polynomial, thus producing several monomial products that can be simplified as above.

$$3x^2(2x^2 - 3x - 5) = 3x^2 \cdot 2x^2 + 3x^2 \cdot -3x + 3x^2 \cdot -5$$
$$= 6x^4 - 9x^3 - 15x^2$$

3. When multiplying two polynomials, each term of one polynomial is multiplied by each term of the other polynomial. Like terms are then combined.

$$(3x + 4)(2x - 5) = 3x \cdot 2x + 3x \cdot -5 + 4 \cdot 2x + 4 \cdot -5$$
$$= 6x^2 - 15x + 8x - 20$$
$$= 6x^2 - 7x - 20$$

4. The mnemonic **FOIL** may be used when multiplying two binomials. The letter **F** refers to the product of the first term of each binomial, **O** refers to the product of the outer terms of the original product, **I** refers to the product of the inner terms of the original product, and **L** refers to the product of the last term of each binomial factor.

    Thus, the product $(3x + 4)(2x - 5)$ could have been simplified using the mnemonic, however, the process is the same:

First   Inner

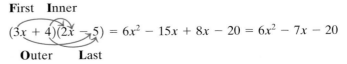

$$(3x + 4)(2x - 5) = 6x^2 - 15x + 8x - 20 = 6x^2 - 7x - 20$$

Outer   **Last**

5. When dividing polynomials, common factors may be divided out. Care must be taken to divide out common *factors* rather than common *terms*. The quotient rule is used to simplify exponential expressions with a common base. When dividing a polynomial by a monomial, divide each *term* of the polynomial by the monomial.

$$\frac{8x^2 - 4x + 2}{2x}$$

$$= \frac{8x^2}{2x} - \frac{4x}{2x} + \frac{2}{2x}$$ Divide each term of the polynomial by the monomial.

$$= 4x - 2 + \frac{1}{x}$$ Divide out common factors and use the quotient rule for exponents.

Examples 7 and 8 illustrate these points.

EXAMPLE 7

Simplify:

a. $(-5x^2y)(4x^3y^2)$

b. $3x^2y(-2x + 5xy - y)$

c. $(4x + 3)(3x - 2)$

d. $(3x - 1)(2x^2 + 4x - 7)$

SOLUTION

a. $(-5x^2y)(4x^3y^2) = (-5 \cdot 4)(x^2 \cdot x^3)(y \cdot y^2)$ Commutative and associative properties.

$$= -20x^5y^3$$ Product rule for exponents.

b. $3x^2y(-2x + 5xy - y)$

$$= 3x^2y(-2x) + 3x^2y(5xy) + 3x^2y(-y)$$ Distributive property.

$$= -6x^3y + 15x^3y^2 - 3x^2y^2$$ Commutative and associative properties and product rule for exponents.

c. $(4x + 3)(3x - 2)$

$$= 4x(3x) + 4x(-2) + 3(3x) + 3(-2)$$ Distributive property.

$$= 12x^2 - 8x + 9x - 6$$ Commutative and associative properties and product rule for exponents.

$$= 12x^2 + x - 6$$ Combine like terms.

d. $(3x - 1)(2x^2 + 4x - 7)$

$$= 3x(2x^2) + 3x(4x) + 3x(-7) - 1(2x^2) - 1(4x) - 1(-7)$$ Distributive property.

$$= 6x^3 + 12x^2 - 21x - 2x^2 - 4x + 7$$ Commutative and associative properties and product rule for exponents.

$$= 6x^3 + 10x^2 - 25x + 7$$ Combine like terms.

WARNING! In the expression $\frac{6x^2}{3}$, the 6 and the 3 are both divided by 3 to yield the equivalent expression $2x^2$. Because they are *factors*, they may both be divided by the same nonzero number. In the expression $\frac{6 + x^2}{3}$, the 6 and the 3 cannot both be divided by 3 because 6 is a *term* rather than a factor.

EXAMPLE 8   Simplify:

a. $\dfrac{-6a^4b^3}{3a^6b^2}$ b. $\dfrac{6x^2 + 3x^3y + 3x}{3x}$

c. $\dfrac{4x^2y^3 - 5xy^4 + 2y^3}{2x^2y^2}$ d. $\dfrac{2(x + y) - 5(x^2 + y)}{xy}$

SOLUTION   a. $\dfrac{-6a^4b^3}{3a^6b^2} = \dfrac{-6}{3}a^{4-6}b^{3-2}$ Quotient rule for exponents.

$= -2a^{-2}b^1$ Substitution.

$= -2 \cdot \dfrac{1}{a^2} \cdot b$ Definition of a negative exponent.

$= \dfrac{-2b}{a^2}$ Substitution.

b. $\dfrac{6x^2 + 3x^3y + 3x}{3x} = \dfrac{6x^2}{3x} + \dfrac{3x^3y}{3x} + \dfrac{3x}{3x}$ Express the division of each term of the polynomial by the monomial.

$= 2x + x^2y + 1$ Simplify each fraction.

c. $\dfrac{4x^2y^3 - 5xy^4 + 2y^3}{2x^2y^2}$

$= \dfrac{4x^2y^3}{2x^2y^2} - \dfrac{5xy^4}{2x^2y^2} + \dfrac{2y^3}{2x^2y^2}$ Express the division of each term of the polynomial by the monomial.

$= 2y - \dfrac{5y^2}{2x} + \dfrac{y}{x^2}$ Simplify each fraction.

d. $\dfrac{2(x + y) - 5(x^2 + y)}{xy} = \dfrac{2x + 2y - 5x^2 - 5y}{xy}$ Distributive property.

$= \dfrac{2x - 3y - 5x^2}{xy}$ Combine like terms.

$= \dfrac{2x}{xy} - \dfrac{3y}{xy} - \dfrac{5x^2}{xy}$ Express the division of each term of the polynomial by the monomial.

$= \dfrac{2}{y} - \dfrac{3}{x} - \dfrac{5x}{y}$ Simplify each fraction.

EXERCISE 7.1

Read the information in Technology 1.1 and Technology 1.2 for your particular calculator before beginning these exercises.

VOCABULARY AND NOTATION   *In Exercises 1–11, fill in the blank to make a true statement.*

1. In the exponential expression -4^3, the base is ___ and the exponent is ___.

2. In the exponential expression $(-4)^3$, the base is _____ and the exponent is ___.

3. Variables and constants combined with the operations of arithmetic produce _____ expressions.

4. Ungrouped addition and subtraction signs are called _____ indicators.

5. _____ terms are terms with identical variable factors.

6. In the expressions $4x^2$, 4 is the numerical _____.

7. A polynomial of one term is a _____.

8. A polynomial of two terms is a _____.

9. A polynomial of three terms is called a _____.

10. The degree of a monomial is the sum of the exponents on its _____ factors.

11. The degree of a polynomial is the same degree as its term of _____ degree.

CONCEPTS

12. $x^4 \cdot x^3 =$ ____.

13. $(x^4)^3 =$ ____.

14. $(x^2y^3)^2 =$ ____.

15. $\left(\dfrac{x}{y}\right)^4 =$ ____.

16. $x^0 =$ ____.

17. $(3x)^0 =$ ____.

18. $x^{-2} =$ ____.

19. $5x^{-2} =$ ____.

20. $\dfrac{x^8}{x^6} =$ ____.

21. $\dfrac{x^6}{x^8} =$ ____.

In Exercises 22–25, classify each polynomial as a monomial, a binomial, or a trinomial.

22. $3x^2 + 2x - 1$

23. $4x$

24. $3x^2 + 5x$

25. $5x^2 + 3x - 4$

In Exercises 26–29, state the degree of each monomial.

26. $2x^2y^3z$

27. $-5xy$

28. 3^2x^3y

29. $4^3x^2y^5z$

In Exercises 30–35, state the degree of each polynomial.

30. $-3x + 9$

31. $5x - 8$

32. $-5x^2 - 5x + 3$

33. $2x^2 + 4x - 1$

34. $4x^2y^3 - 5xy + 2$

35. $3x^2y - 2xy + 4$

In Exercises 36–39, simplify each expression by combining like terms.

36. $3x^2 + 3x - 5x^2 - 2 + 4x$

37. $4y^2 + 2y + 6y - 2y^2$

38. $-3ab + 2ab - 5b + b$

39. $-2xy - 5xy + 6x - 4y$

40. The STOre feature of the graphing calculator was used to evaluate an expression. The screen display is pictured.

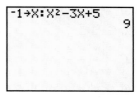

    a. What is the value of the polynomial?

    b. What was the polynomial that was being evaluated?

    c. What was the value used for the variable?

41. The STOre feature of the graphing calculator was used to evaluate an expression. The screen display is pictured.

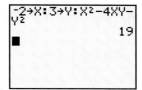

    a. What is the value of the polynomial?

    b. What was the polynomial that was being evaluated?

    c. What were the values used for the variables?

In Exercises 42–47, evaluate each polynomial for the given value(s) of the variable(s). Check your results using the STOre feature of your graphing calculator.

42. $3x^2 - 4x + 2$; $x = 2$

43. $5x^2 - 8x + 1$; $x = 4$

44. $-x^2 + 3x - 4$; $x = -1$

45. $-x^2 - 5x + 2$; $x = -2$

46. $3x^2 - 2xy - y^2$; $x = 3$; $y = -1$

47. $2x^2 - 3xy - y^2$; $x = -2$; $y = 1$

In Exercises 48–53, simplify each expression by performing the indicated division.

48. $\dfrac{3x}{6}$

49. $\dfrac{5x}{15}$

50. $\dfrac{8x^2}{2}$

51. $\dfrac{10y^2}{2}$

52. $\dfrac{3x + 6y}{3}$

53. $\dfrac{6x^2 + 9x + 3}{3}$

PRACTICE *In Exercises 54–75, use the laws of exponents to simplify each expression. All results should be expressed with positive exponents only.*

54. $\left(\dfrac{a^4}{b^6}\right)^2$

55. $\left(\dfrac{x^3}{y^2}\right)^4$

56. $\dfrac{5y^3}{y^2}$

57. $\dfrac{2x^4}{x^3}$

58. $\dfrac{a^5b^4}{a^3b}$

59. $\dfrac{x^4y^3}{x^2y}$

60. $(y^2z^3)(xy^2z)$

61. $(a^3b^5c)(a^2b)$

62. $(3x^2)^3$

63. $(4y^2)^4$

64. $\dfrac{(m^3n^2)^2}{(mn)^4}$

65. $\dfrac{(x^2y^3)^3}{(xy)^5}$

66. $2x^2y^3(xy)^2$

67. $-5xy^4(x^2y)^3$

68. $\dfrac{2x^2y(x^5)^0}{4xy(y^6)^0}$

69. $\dfrac{5x^3y^2(x^5)^0}{3xy(y^4)^0}$

70. $(ab^2)^{-3}$

71. $(xy^3)^{-2}$

72. $(2x^{-3}y^2)^{-2}$

73. $(3a^{-2}b^3)^{-3}$

74. $(3x^{-4}y^2)(2x^3y)$

75. $(2x^{-2}y)(-3x^2y^{-3})$

In Exercises 76–97, simplify each expression by removing all grouping symbols and combining like terms.

76. $(2x + 4) + (6x - 1)$

77. $(5x - 2) + (8x + 4)$

78. $6(3x - 1) - 4(2x + 3)$

79. $2(4y + 3) - 2(6y - 1)$

80. $(3a^2 + 2b - 4) - (4a^2 - 3b)$

81. $(6b^2 - 2a + 6) - (3b^2 - 5a)$

82. $(x - 5)(x + 2)$

83. $(x + 6)(x - 1)$

84. $(2x + 5)(3x + 4)$

85. $(3x + 2)(5x + 1)$

86. $(2x - 1)(2x + 1)$

87. $(3x - 4)(3x^2 + 2x - 1)$

88. $(3r + 2)(3r^2 - 2r + 5)$

89. $(2b - 5)(2b + 5)$

90. $(x + 4)^2$

91. $(y - 2)^2$

92. $(y - 5)^2$

93. $(x + 3)^2$

94. $(3x + 1)(2x - 4) - 3(x + 5)$

95. $(6x - 1)(x + 5) - 4(x + 2)$

96. $(8y - 3)(y - 1) + (2y + 3)(y - 5)$

97. $(3y - 2)(y + 4) + (3y - 1)(y + 2)$

In Exercises 98–103, divide.

98. $\dfrac{6a^2b + 3ab}{3a}$

99. $\dfrac{5x^2y^2 - 10xy}{5x}$

100. $\dfrac{8x - 16y}{2xy}$

101. $\dfrac{9a + 18b}{3ab}$

102. $\dfrac{12x^3y^2 - 8x^2y - 4y}{4xy}$

103. $\dfrac{12a^2b^2 - 4a^2b - 8b}{4ab}$

7.2 REVIEW OF FACTORING

- REVIEW OF FACTORING TECHNIQUES

Points to remember

1. To factor means to write as a product. The expression $4(x + 5) + y(x + 5)$ is **not** expressed as a product; it is the sum of two terms, $4(x + 5)$ and $y(x + 5)$.
2. Factoring out a negative 1 simply changes the sign of each term of the original polynomial: $-a + b = -1(a - b)$.
3. The factors $(a + b)$ and $(b + a)$ are the same because the operation of addition is commutative.
4. The factors $(a - b)$ and $(b - a)$ are additive inverses (opposites). Thus $a - b = -(b - a)$.
5. The sum of squares, $x^2 + y^2$, does **not** factor over the set of integers. It is a prime polynomial.

REVIEW OF FACTORING TECHNIQUES

To completely factor a polynomial means to write the given polynomial as a *product* of prime polynomial factors. Factoring can always be checked by multiplying factors and ensuring that the product of factors is equivalent to the original polynomial. Remember to check that each polynomial factor is prime before stopping. Also, recall that there should be no *ungrouped* addition or subtraction signs in your final result. Below is a review of techniques that will help you in your factoring.

Factoring Polynomials

1. Always check for and factor out the greatest common factor (GCF) first. If the leading coefficient is negative, factor out a -1 with/as the GCF. Refer to section 5.1 if you need more practice in factoring out a GCF.
2. Count the number of terms in the polynomial, and then follow the given procedures for that specific type of polynomial.

Binomial

If it is the difference of two squares: $a^2 - b^2 = (a - b)(a + b)$.
If it is the difference of two cubes: $a^3 - b^3 = (a - b)(a^2 + ab + b^2)$.
If it is the sum of two cubes: $a^3 + b^3 = (a + b)(a^2 - ab + b^2)$.
If it is *not* one of these problem types it is prime. Refer to Section 5.3 if you need more practice in factoring binomials.

Trinomials

Make sure the trinomial is in descending powers of a variable and then attempt to factor it by either grouping or trial and error. The special product formulas below may be used if desired, but are not required:

$$x^2 + 2xy + y^2 = (x + y)^2$$
$$x^2 - 2xy + y^2 = (x - y)^2$$

Refer to section 5.2 if you need more practice in factoring trinomials.

Four-Term Polynomials

Try to factor by grouping. Refer to Section 5.1 if you need more practice in factoring polynomials that contain four terms.

EXAMPLE 1

Factor: $-12y^2 + 20y$.

SOLUTION

Recall the distributive property: $a(b + c) = ab + ac$. It could also be written as $ab + ac = a(b + c)$. This property allows distribution of multiplication over both addition and subtraction. It is used to factor out the **greatest common factor**.

$$-12 \cdot y^2 = -1 \cdot 2^2 \cdot 3y^2$$

Determine the GCF of the terms of the polynomial.

$$20y = 2^2 \cdot 5 \cdot y$$

$$-1 \cdot 2^2 \cdot y = -4y$$

The GCF, $-4y$, is the product of the common factors and the factor of -1.

$$-12y^2 + 20y = -4y \cdot 3y + (-4y)(-5)$$

Rewrite each term as the product of the GCF and another factor.

$$= -4y(3y - 5)$$

Use the distributive property.

Check the result by verifying that $-4y(3y - 5) = -12y^2 + 20y$.

> **STUDY TIP**
>
> Check your factorization by using the distributive property to multiply the factors. This product should be the same as the simplified form of the original polynomial.

EXAMPLE 2

Factor: $(a + b)x + (a + b)y$.

SOLUTION   The polynomial has two terms: $(a + b)x$ and $(a + b)y$. The greatest common factor of these two terms is the *quantity* $(a + b)$. Therefore,

$$(a + b)x + (a + b)y = (a + b)(x + y)$$

We can verify the result by multiplication:

$$(a + b)(x + y) = (a + b)x + (a + b)y$$

EXAMPLE 3

Factor: $ax + ay + cx + cy$.

SOLUTION   Although there is no factor common to all four terms, there is a common factor of a in $ax + ay$ and a common factor of c in $cx + cy$. We can factor out the a and c as follows:

$$ax + ay + cx + cy = (ax + ay) + (cx + cy)$$ Associative property of addition.

$$= a(x + y) + c(x + y)$$ Distributive property (factoring the GCF of a out of the first two terms and the GCF of c out of the last two terms).

$$= (x + y)(a + c)$$ Distributive property (factoring the GCF of the quantity $(x + y)$ out of each term).

Therefore, $ax + ay + cx + cy = (x + y)(a + c)$.

This type of factoring is called **factoring by grouping**.

EXAMPLE 4

Factor: $6x^3 + 3x - 11x^2$.

SOLUTION $6x^3 + 3x - 11x^2 = 6x^3 - 11x^2 + 3x$ Put the polynomial in descending powers of a variable.

$$= x(6x^2 - 11x + 3)$$ Factor out the GCF of x.

The remaining trinomial, $6x^2 - 11x + 3$, will be factored by grouping. To do this, you will need the trinomial written as $ax^2 + bx + c$. If you prefer to factor by the trial-and-error method, you should refer back to Section 5.2.

To factor the trinomial, identify a, b, and c: $a = 6$, $b = -11$, $c = 3$; and find the product $a \cdot c$: $a \cdot c = 6(3) = 18$. Because the product of the two numbers is positive and the sum of the factors of 18 must equal b, which is a negative 11, the desired factors must both be negative.

Two numbers	Product is 18	Sum is −11
−1, −18	$(-1)(-18) = 18$	$-1 + (-18) = -19$
−2, −9	$(-2)(-9) = 18$	$-2 + (-9) = -11$ ✓

The required numbers are -2 and -9.

Replace the middle term of the trinomial, $-11x$, using -2 and -9 as the coefficients. Do not forget the GCF of x when working with the trinomial. Factor by grouping.

$$x[6x^2 - 2x - 9x + 3] = x[2x(3x - 1) - 3(3x - 1)] \quad \text{Factor by grouping.}$$
$$= x(3x - 1)(2x - 3) \quad\quad\quad \text{Distributive property.}$$

EXAMPLE 5

Factor: $4y^4 - 25z^2$.

SOLUTION Because the binomial $4y^4 - 25z^2$ can be written in the form $(2y^2)^2 - (5z)^2$, it represents the difference of the squares of $2y^2$ and $5z$. Thus, it factors into the sum of these two quantities times the difference of these two quantities.

$$4y^4 - 25z^2 = (2y^2)^2 - (5z)^2$$
$$= (2y^2 + 5z)(2y^2 - 5z)$$

Verify this result by multiplication.

WARNING!   The binomial $25a^2 + 9b^2$ is called the **sum of two squares** because it can be written in the form $(5a)^2 + (3b)^2$. Such binomials cannot be factored if we are limited to integer coefficients. Polynomials that do not factor over the integers are called **irreducible** or **prime polynomials**.

EXAMPLE 6

Factor: $a^3 + 8$.

SOLUTION Since $a^3 + 8$ can be written as $a^3 + 2^3$, we have the sum of two cubes, which factors as follows:

$$F^3 + L^3 = (F + L)(F^2 - FL + L^2) \quad \text{Recall that } F = \text{base of first term}$$
$$\text{and } L = \text{base of last term.}$$
$$a^3 + 2^3 = (a + 2)(a^2 - a2 + 2^2)$$
$$= (a + 2)(a^2 - 2a + 4)$$

Thus, $a^3 + 8$ factors as the product of the sum of a and 2 and the trinomial $a^2 - 2a + 4$.

We can check by distributing.

$$(a + 2)(a^2 - 2a + 4) = a(a^2) + a(-2a) + a(4) + 2(a^2) + 2(-2a) + 2(4)$$
$$= a^3 - 2a^2 + 4a + 2a^2 - 4a + 8$$
$$= a^3 - 2a^2 + 2a^2 + 4a - 4a + 8$$
$$= a^3 + 8$$

EXAMPLE 7

Factor: $a^3 - 64b^3$.

SOLUTION Since $a^3 - 64b^3$ can be written as $a^3 - (4b)^3$, we have the difference of two cubes, which factors as follows:

$$F^3 - L^3 = (F - L)(F^2 + FL + L^2)$$
$$a^3 - (4b)^3 = (a - 4b)[a^2 + a(4b) + (4b)^2]$$
$$= (a - 4b)(a^2 + 4ab + 16b^2)$$

Thus, it factors as $(a - 4b)(a^2 + 4ab + 16b^2)$.
We can check by distributing.

$$(a - 4b)(a^2 + 4ab + 16b^2) = a(a^2) + a(4ab) + a(16b^2) - 4b(a^2) - 4b(4ab) - 4b(16b^2)$$
$$= a^3 + 4a^2b + 16ab^2 - 4a^2b - 16ab^2 - 64b^3$$
$$= a^3 + 4a^2b - 4a^2 + 16ab^2 - 16ab^2 - 64b^3$$
$$= a^3 - 64b^3$$

EXERCISE 7.2

VOCABULARY AND NOTATION *In Exercises 1–3, fill in the blanks to make a true statement.*

1. To _____ means to write as a product.
2. The GCF is the _____ (smallest/largest) factor common to all terms.
3. _____ polynomials are polynomials that cannot be factored over the set of integers.

CONCEPTS

4. Multiply each of the following:
   a. $(x - 3)(x - 3)$
   b. $(x - 4)^2$
   c. $(x - 4)(x + 4)$
   d. $(2x + 3)(x - 5)$
5. Find the GCF of each pair of terms.
   a. $3y^3, 3y^2$ b. $63x^3y^2, 81x^2y^4$
6. Factor out a factor of negative 1 from each expression. Verify by multiplication.
   a. $-5x - 6$
   b. $-x^2 + 6x - 9$
   c. $a - b$ d. $b - a$

PRACTICE *In Exercises 7–14, factor each expression. Check your results by using the distributive property to multiply your final factorization.*

7. $2x + 8$ 8. $3y - 9$
9. $2x^2 - 6x$ 10. $3y^3 + 3y^2$
11. $15x^2y - 10x^2y^2$
12. $63x^3y^2 + 81x^2y^4$
13. $27z^3 + 12z^2 + 3z$
14. $25z^6 - 10z^3 + z^2$

In Exercises 15–18, factor out the negative of the greatest common factor. Check your results by using the distributive property to multiply your final factorization.

15. $-3a - 6$ 16. $-6y + 12$
17. $-6x^2 - 3xy$ 18. $-15y^3 + 25y^2$

In Exercises 19–24, factor by grouping. Factor out all common monomials first. Check your results by multiplication.

19. $ax + bx + ay + by$
20. $ar - br + as - bs$

21. $x^2 + yx + 2x + 2y$

22. $2c + 2d - cd - d^2$

23. $3c - cd + 3d - c^2$

24. $x^2 + 4y - xy - 4x$

In Exercises 25–46, factor each trinomial, if possible. If the coefficient of the first term is negative, begin by factoring out negative 1. Check your results by multiplication.

25. $x^2 + 5x + 6$

26. $y^2 + 7y + 6$

27. $3x^2 + 12x - 63$

28. $2y^2 + 4y - 48$

29. $-a^2 + 4a + 32$

30. $-x^2 - 2x + 15$

31. $6x^2 + 2x - 1$

32. $8x^2 + 2x + 3$

33. $6y^2 + 7y + 2$

34. $6x^2 - 11x + 3$

35. $8a^2 + 6a - 9$

36. $6z^2 + 17z + 12$

37. $a^2 - 3ab - 4b^2$

38. $b^2 + 2bc - 80c^2$

39. $-3a^2 + ab + 2b^2$

40. $-2x^2 + 3xy + 5y^2$

41. $3x^3 - 10x^2 + 3x$

42. $6y^3 + 7y^2 + 2y$

43. $x^4 + 8x^2 + 15$

44. $x^4 + 11x^2 + 24$

45. $y^4 - 13y^2 + 30$

46. $y^4 - 13y^2 + 42$

In Exercises 47–72, factor each expression. Be sure that your answer is completely factored and checks by multiplication.

47. $x^2 - 4$

48. $y^2 - 9$

49. $9y^2 - 64$

50. $16x^4 - 81y^2$

51. $81a^4 + 49b^2$

52. $64x^6 + 121y^2$

53. $(x + y)^2 - z^2$

54. $a^2 - (b - c)^2$

55. $x^4 - y^4$

56. $16a^4 - 81b^4$

57. $2x^2 - 288$

58. $8x^2 - 72$

59. $2x^3 - 32x$

60. $3x^3 - 243x$

61. $x^3 - 8$

62. $y^3 - 27$

63. $y^3 - 64$

64. $x^3y^3 - 125$

65. $125 + a^3$

66. $64 + y^3$

67. $x^3 + 8y^3$

68. $27x^3y^3 + 8$

69. $2x^3y - 16y$

70. $a^4b - ab^4$

71. $(a - b)^3 - 8$

72. $27 - (a - b)^3$

In Exercises 73–82, factor each expression. These problems represent a random mixture of the types of problems reviewed in this section. For more mixed practice, return to Section 5.4. Check your results by multiplication.

73. $-3a^2 - 6ab - 3b^2$

74. $4x^2 + 4xy + y^2$

75. $-x^3 - 2x^2 - x$

76. $36a^2 - 1$

77. $x^2 - 4x + 3$

78. $x^2 - x - 6$

79. $y^3 - 216$

80. $4x^2 + 4x$

81. $-21x^7 + 14x^5 - 7x^4$

82. $125 + b^3$

REVIEW   *In Exercises 83 and 84, simplify each expression using the properties of exponents. Leave no negative exponents in your answers.*

83. $4^{-5} \cdot 4^{-3}$

84. $\left(\dfrac{2x^{-3}}{y^{-2}} \right)^{-2}$

85. Simplify: $3(x^2 + 2x - 1) - 4(x^2 + 3x - 5)$

86. Evaluate $-x^2 + 3xy + y^2$ when $x = -1$ and $y = -3$

7.3 REVIEW OF EQUATIONS

- ALGEBRAIC SOLUTIONS OF LINEAR EQUATIONS IN ONE VARIABLE
- IDENTITIES AND CONTRADICTIONS
- LITERAL EQUATIONS AND FORMULAS
- PROBLEM SOLVING
- ABSOLUTE VALUE EQUATIONS
- FACTORABLE POLYNOMIAL EQUATIONS
- EQUATIONS WITH RATIONAL EXPRESSIONS

Points to remember

1. The addition and subtraction properties of equality allow you to add or subtract the same quantity from both sides of an equation.
2. The multiplication and division properties of equality allow you to multiply or divide both sides of an equation by the same nonzero quantity.
3. Equations involving absolute value will have
   a. Two solution parts if the absolute value is equal to a positive constant.
   b. One solution if the absolute value is equal to 0.
   c. No solution if the absolute value is equal to a negative number.
4. When the equation involves a factorable polynomial, the zero product property allows you to set the factored polynomial equal to zero and then solve by setting each factor equal to zero.
5. When solving equations involving a rational expression, multiply through the equation by the least common denominator to clear fractions. Solve the resulting equation.

▌▌▌ ALGEBRAIC SOLUTIONS OF LINEAR EQUATIONS IN ONE VARIABLE

To **solve** an equation means to find the value(s) of the variable(s) that will make the equation a true statement. When solving linear equations in one variable it is best to first simplify both sides of the equation and then begin transforming the equation by using the properties of equality to write a series of equivalent equations. To solve any linear equation, we must isolate the variable on one side. This can be a multistep process that may require combining like terms. As we solve equations, we will use the following steps:

Steps for Solving a Linear Equation in One Variable

1. Use the distributive property to remove grouping symbols.
2. Combine like terms on each **side** of the equation.
3. Clear fractions, if desired, by multiplying each **term** on both sides of the equation by the LCD.
4. Use the addition or subtraction property of equality to collect all terms containing the variable on one side of the equation and all constants on the other side.
5. Use the multiplication or division property of equality to isolate the variable. The equation should be in the form:

   variable = constant

6. Check by ensuring that the proposed solution produces a true statement when the variable is replaced by the constant. Always use the original equation when checking.

EXAMPLE 1

Solve: $3(x + 2) - 2x = 4x$.

SOLUTION

$$3x + 6 - 2x = 4x$$ Use the distributive property to remove grouping symbols.

$$x + 6 = 4x$$ Combine like terms.

$$x - x + 6 = 4x - x$$ Use the subtraction property of equality to get all variable terms together.

$$6 = 3x$$ Simplify each side.

$$\frac{6}{3} = \frac{3x}{3}$$ Use the division property of equality to isolate the variable.

$$2 = x$$ Simplify each side.

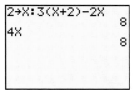

Confirm your solution by storing the value of 2 for the variable x and using the STOre feature of your calculator to evaluate each side of the equation. Because each side of the equation evaluates to the same value for $x = 2$, we conclude that 2 is the solution of the equation.

You may check the solution(s) of your equation in this section either analytically (paper and pencil), by the STOre feature, or by the TABLE feature. If you have completed Chapters 1–6, you may also graphically solve your equations as a means of checking your analytical solution. The review of graphical solutions occurs in the next section of this chapter.

IDENTITIES AND CONTRADICTIONS

The equation in Example 1 is called a **conditional equation** because it is true for some, but not all, real numbers. An equation that is true for all real numbers is called an **identity**. Regardless of the value used to replace the variable, the equation will be true. For example, the equation $x + x = 2x$ is an identity because it is true regardless of the real number substituted in for the variable x. Its solution is **all real numbers**. This can be expressed with the symbol $\mathbb{R}$, which means "the set of all real numbers."

An equation that has no solution is called a **contradiction**. The equation $x = x + 1$ is a contradiction because no number is equal to itself plus 1. Its solution is **no solution**, which can be represented by the symbol $\varnothing$ representing the **null set** or the set that contains no elements.

WARNING! Do not confuse the **type** of equation with its **solution**. The terms *conditional*, *identity*, and *contradiction* refer to the type. Remember the solution or root is all real numbers that make the equation a true statement.

LITERAL EQUATIONS AND FORMULAS

Equations with several variables are called **literal equations**. Often, these equations are **formulas** such as $A = lw$, the formula for finding the area of a rectangle.

EXAMPLE 2

Solve $A = lw$ for l.

SOLUTION To isolate l on the left-hand side, we undo the multiplication of w by dividing both sides of the equation by w.

$$A = lw$$

$$\frac{A}{w} = \frac{\overset{1}{\cancel{l}w}}{\underset{1}{\cancel{w}}} \qquad \text{Divide both sides by } w.$$

$$\frac{A}{w} = l \qquad \text{Simplify.}$$

$$l = \frac{A}{w}$$

EXAMPLE 3

Solve $3x + 2y = 8$ for y.

SOLUTION

$$2y = -3x + 8 \qquad \text{Subtract } 3x \text{ from both sides.}$$

$$\frac{2y}{2} = \frac{-3x + 8}{2} \qquad \text{or} \qquad \frac{2y}{2} = \frac{-3x}{2} + \frac{8}{2} \qquad \text{Divide both sides by 2.}$$

$$y = -\frac{3}{2}x + 4 \qquad \text{Simplify.}$$

▊ PROBLEM SOLVING

Application problems illustrate the power of algebra in solving practical, everyday problems. In order to solve any problem, you first need a game plan.

Problem-Solving Strategies

1. Read the problem several times and organize the facts.

   What is known?
   What are you asked to find?
   It may help you to organize the facts under the two headings, **known** and **unknown**.

2. a. If there is only one unknown quantity, choose a variable to represent that quantity. Clearly define what the variable is and what it represents with a "Let" statement.

    b. When there is more than one unknown, choose a variable to represent one of the quantities and express all other important quantities in terms of the chosen variable.

3. Translate the information expressed in the words of the problem to an algebraic equation.
4. Solve the equation.
5. Answer the question posed in the problem.
6. Check your results in terms of both the equation and the words of the problem. Make sure your solution makes sense within the context of the problem.

EXAMPLE 4

Bonnie is a mathematics teacher who loves to read. She's currently reading *War and Peace*. While placing her bookmark in the book she notices that the sum of the page numbers is 2,065. Between which two pages is the bookmark?

SOLUTION

A table can help us organize the information. In this problem, we begin with the pages 1,000 and 1,001. (This is simply a guess.) The table allows us to generalize from the specific to the general and write an algebraic equation that we can solve to answer the question posed.

Left page no.	1,000	1,002	x
Right page no.	1,001	1,003	$x + 1$
Sum	2,001	2,005	$x + (x + 1)$

Let x = left page number

$x + 1$ = right page number

$$x + (x + 1) = 2{,}065$$
$$2x + 1 = 2{,}065$$
$$2x + 1 - 1 = 2{,}065 - 1$$
$$2x = 2{,}064$$
$$x = 1{,}032 \qquad \text{Thus, } x + 1 = 1{,}033$$

The bookmark is between pages 1,032 and 1,033.

Calculator confirmation: Once the equation is constructed, enter the left side at the y1= prompt. Set the table to start at 1,000 and to increment by 1. Enter 2,065 at the y2= prompt. Scroll down until the value in the Y1 column equals the value in the Y2 column.

Similar triangles are triangles in which corresponding angles are equal and lengths of corresponding sides are in proportion. Because all pairs of corresponding sides are in proportion in similar triangles, we can use proportions to solve application problems that involve similar triangles. This will enable us to measure sides of triangles indirectly, as seen in the following application.

EXAMPLE 5

Measuring a tree   On a sunny day, a large tree casts a shadow of 24 feet at the same time a vertical yardstick casts a shadow of 2 feet. Find the height of the tree.

SOLUTION   Refer to Figure 7-1, which shows the triangles determined by the tree and its shadow and the yardstick and its shadow. Because the triangles have the same shape, they are similar, and the measures of their corresponding sides are in proportion. We can let h represent the height of the tree and find h by setting up and solving the following proportion.

$$\frac{h}{3} = \frac{24}{2}$$

$2h = 3(24)$ In a proportion, the product of the extremes is equal to the product of the means.

$2h = 72$ Simplify.

$h = 36$ Divide both sides by 2.

The tree is 36 feet tall.

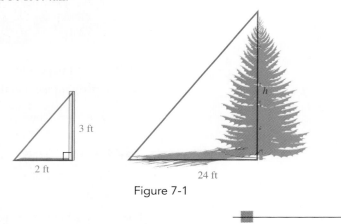

Figure 7-1

ABSOLUTE VALUE EQUATIONS

Consider the absolute value equation $|x| = 5$. The variable x can have a value of either 5 or -5 because both of these numbers are a distance of 5 units from 0 on the number line.

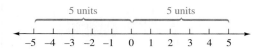

Therefore, the solutions to the equation are $x = 5$ or $x = -5$. This leads to the following generalization:

Absolute Value Equations
If $k > 0$, then $|x| = k$ is equivalent to $x = k$ or $x = -k$

It is important to note two things about the given algorithm:

1. Absolute value is isolated on one side of the equation.
2. k is a positive constant.

EXAMPLE 6

Solve the equation $|x - 3| + 5 = 12$.

SOLUTION   Remember to first isolate the absolute value on the left-hand side.

$$|x - 3| = 7$$

The equation $|x - 3| = 7$ indicates that a point on the number line with a coordinate of $x - 3$ is 7 units from the origin. Thus, $x - 3$ can be either 7 or -7.

$$x - 3 = 7 \qquad \text{or} \qquad x - 3 = -7$$
$$x = 10 \qquad \text{or} \qquad x = -4$$

The solutions of the equation $|x - 3| + 5 = 12$ are 10 and -4. If either of these numbers is substituted in for x in the original equation $|x - 3| + 5 = 12$, the equation is satisfied.

$|x - 3| + 5 = 12$	$|x - 3| + 5 = 12$
$|10 - 3| + 5 \overset{?}{=} 12$	$|-4 - 3| + 5 \overset{?}{=} 12$
$|7| + 5 \overset{?}{=} 12$	$|-7| + 5 \overset{?}{=} 12$
$7 + 5 \overset{?}{=} 12$	$7 + 5 \overset{?}{=} 12$
$12 = 12$	$12 = 12$

Confirm the solution by using the TABLE feature:

Plot1 Plot2 Plot3
\Y1◻abs(X-3)+5
\Y2◻12
\Y3=
\Y4=
\Y5=
\Y6=
\Y7=

TABLE SETUP
TblStart=-4
⊿Tbl=1
Indpnt: **Auto** Ask
Depend: **Auto** Ask

X	Y1	Y2
-4	12	12
-3	11	12
-2	10	12
-1	9	12
0	8	12
1	7	12
2	6	12
X=-4		

X	Y1	Y2
10	12	12
11	13	12
12	14	12
13	15	12
14	16	12
15	17	12
16	18	12
X=10		

EXAMPLE 7

Solve the equation $\left|7x + \dfrac{1}{2}\right| = -4$.

SOLUTION   Careful examination of the algorithm reveals that our constant k must be positive. In this example, $k = -4$, thus the algorithm cannot be applied. If we refer back to the definition of absolute value as it relates to the number line, the problem translates to "what number(s) are a *negative four* units from zero?" This does not make sense because not only is distance always positive but we also know from the definition that absolute value is always equal to 0 or some positive number. The solution is $\varnothing$.

FACTORABLE POLYNOMIAL EQUATIONS

The solution (or root) of an equation is the value (or values) of the variable that make the equation a true statement. To find the solutions of an equation that is quadratic or of higher degree, the **zero factor property** is used.

Zero Factor Property

If $a \cdot b = 0$ and a and b are real numbers, it follows that $a = 0$ or $b = 0$.

This property simply states that if a product of factors is zero, *at least one* of the factors must be equal to zero. To apply this property, the polynomial equations must be set equal to zero. A polynomial equation is said to be in **standard form** when one side of the equation is zero.

Solving Equations by Factoring

1. Simplify both sides of the equation by removing grouping symbols and combining like terms. Fractions may be eliminated by multiplying each term on both sides of the equation by the LCD.
2. Put the equation in standard form (set one side equal to zero).
3. Factor the polynomial.
4. Set each factor containing a variable equal to zero.
5. Solve the resulting equations.
6. Check solutions either analytically or with the graphing calculator.

EXAMPLE 8   Solve: $x^3 - 2x^2 = 63x$.

SOLUTION

$$x^3 - 2x^2 - 63x = 0$$ Put the equation in standard form.

$$x(x^2 - 2x - 63) = 0$$ Begin factoring the polynomial by factoring out the GCF of x.

$$x(x - 9)(x + 7) = 0$$ Factor the trinomial.

$x = 0$ | $x - 9 = 0$ | $x + 7 = 0$ Use the zero factor property to set each factor equal to 0.

$x = 0$ | $x = 9$ | $x = -7$ Solve each linear equation.

Solutions: $0, 9, -7$; Solution set: $\{0, 9, -7\}$
    The verification using the TABLE feature is illustrated.
    The left side of the original equation is entered at the y1 prompt and the right side is entered at the y2 prompt.

The TABLE is set up to begin with an x value of -7. Because the values in the Y1 and Y2 columns are equivalent, -7 is verified as a solution. Scrolling through the TABLE verifies the other roots of 0 and 9.

X	Y1	Y2
-7	-441	-441
-6	-288	-378
-5	-175	-315
-4	-96	-252
-3	-45	-189
-2	-16	-126
-1	-3	-63

X= -7

X	Y1	Y2
0	0	0
1	-1	63
2	0	126
3	9	189
4	32	252
5	75	315
6	144	378

X=0

X	Y1	Y2
3	9	189
4	32	252
5	75	315
6	144	378
7	245	441
8	384	504
9	567	567

X=9

▌▌ EQUATIONS WITH RATIONAL EXPRESSIONS

When solving a linear equation, we use the addition, subtraction, multiplication, and division properties of equality to rewrite the original equation as a series of equivalent equations until the variable is isolated on one side of the equation. For equations of degree two or higher, the zero factor property is applied. The solutions, or roots, of the equation are the values of the variable that make the equation a true statement.

Care must be taken when solving equations containing rational expressions with variables in the denominator. Sometimes, when solving rational equations false solutions result. These are called **extraneous solutions** and are rejected. Any value of the variable that makes a denominator equal to zero is an extraneous solution.

 WARNING!   When algebraically solving equations that contain variables in the denominators, solutions **must** be verified through a checking process. Extraneous solutions **must** be rejected. For this reason, it is suggested that excluded values be determined first.

The following steps can be used:

Solving Rational Equations Algebraically

1. Determine any and all excluded values by setting each denominator containing a variable equal to zero and solving for the variable.
2. Determine the LCD.
3. Multiply each *term* on both sides of the equation by the LCD to clear fractions.
4. Simplify both sides of the equation.
5. Solve the equation. Remember, linear (first-degree) equations are solved by transformation, whereas quadratic (second-degree) and higher-order equations require factoring and the application of the zero factor property after the equation is put in standard form and the polynomial is factored.
6. Reject any extraneous solutions (those solutions that equal the excluded values found in step 1).
7. Check solutions in the original equation to verify solutions.

EXAMPLE 9

Solve the equation $\dfrac{2}{x-1} + 4 = \dfrac{4}{x^2-1} + 3$.

SOLUTION   Determine the excluded values by setting each denominator equal to zero and solving for the variable.

$$x - 1 = 0 \qquad \text{and} \qquad x^2 - 1 = 0$$
$$x = 1 \qquad\qquad\qquad (x-1)(x+1) = 0$$
$$x - 1 = 0 \quad \text{or} \quad x + 1 = 0$$
$$x = 1 \quad \text{or} \quad x = -1$$

Therefore, $x \neq 1, x \neq -1$.

Because $x^2 - 1 = 0$ is a quadratic equation, we factor the polynomial side and use the zero factor property to find the solutions. The excluded values are 1 and -1.

We now begin solving the rational equation. To determine the LCD, we factor denominators that are not prime.

$$\dfrac{2}{x-1} + 4 = \dfrac{4}{x^2-1} + 3$$

$$\dfrac{2}{x-1} + 4 = \dfrac{4}{(x-1)(x+1)} + 3 \qquad \text{The LCD is } (x-1)(x+1).$$

We begin solving by multiplying both sides of the equation by the LCD $(x-1)(x+1)$.

$$(x-1)(x+1)\left(\dfrac{2}{x-1} + 4\right) = (x-1)(x+1)\left(\dfrac{4}{(x-1)(x+1)} + 3\right)$$

Use the distributive property to remove groupings.

$$(x-1)(x+1)\cdot\dfrac{2}{x-1} + (x-1)(x+1)\cdot 4 = (x-1)(x+1)\cdot\dfrac{4}{(x-1)(x+1)} + (x-1)(x+1)\cdot 3$$

$$2(x+1) + 4(x^2 - 1) = 4 + 3(x^2 - 1) \qquad$$ Simplify by dividing out common factors and performing multiplications when no common factors divide out.

$$2x + 2 + 4x^2 - 4 = 4 + 3x^2 - 3 \qquad$$ Simplify by removing grouping.

$$4x^2 + 2x - 2 = 3x^2 + 1 \qquad$$ Combine like terms.

$$x^2 + 2x - 3 = 0 \qquad$$ Because the equation is quadratic, use the subtraction property of equality to put it in standard form.

$$(x+3)(x-1) = 0 \qquad$$ Factor the polynomial side.

$$x + 3 = 0 \quad \text{or} \quad x - 1 = 0 \qquad$$ Use the zero factor property to set each factor equal to 0 and solve for the variable.

$$x = -3 \quad \text{or} \quad x = 1$$

Because the number 1 is an excluded value, it is rejected as a root. The solution is -3.

The solution of -3 is verified below by demonstrating both sides of the equation have a value of 3.5 when x is -3.

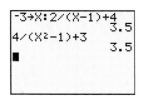

The fact that 1 is an extraneous root is also verified below:

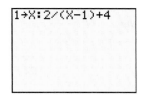   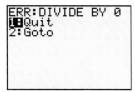

Before ENTER is pressed.   After ENTER is pressed.

The TABLE feature show us -3 is a solution because the expressions entered at the y1 and y2 prompts evaluate to the same constant (3.5) for $x = -3$. It also confirms the fact that 1 cannot be a solution. The ERROR message displayed indicates that you have made an error — in this case, "asking" the calculator to evaluate expressions for a value of the variable that makes those expressions undefined.

EXERCISE 7.3

Read the information in Technology 1.1 and 1.2 for your particular calculator before beginning the exercises.

VOCABULARY AND NOTATION   *In Exercises 1–7, fill in the blanks to make a true statement.*

1. To _____ an equation means to find the value(s) of the variables(s) that will make the equation a true statement.

2. A _____ equation is true for some, but not all, real numbers.

3. An _____ equation is an equation that is true for all acceptable values of the variable.

4. A _____ is an equation that is never true for any value of the variable and thus has no solution.

5. A _____ is an equation involving several variables.

6. The symbol ___ denotes an empty set.

7. The symbol ___ denotes the set of all real numbers.

CONCEPTS   *In Exercises 8–12, determine if the given value(s) are or are not solutions to the equations. Use the STOre or TABLE features of your calculator.*

8. $4(x - 5) = \dfrac{1}{2}(x + 1) - 2$; 41

9. $|2x - 1| = 8; \dfrac{9}{2}, -\dfrac{7}{2}$

10. $-|x + 5| - 6 = 0; -5$

11. $6x^2 - 11x - 10 = 0; \dfrac{5}{2}, -\dfrac{2}{3}$

12. $\dfrac{6}{x} + \dfrac{5}{x} = 5; 2$

PRACTICE   *In Exercises 13–22, solve each linear equation and verify results with either the STOre or the TABLE features of your graphing calculator.*

13. $4(x - 1) = 28$

14. $3(x + 7) = 42$

15. $13(x - 9) - 2 = 7x - 5$

16. $\dfrac{8(x - 5)}{3} = 2(x - 4)$

17. $\dfrac{1}{3}x - 2 = \dfrac{1}{2}x + 2$

18. $\dfrac{7}{4}(x + 3) = \dfrac{3}{8}(x - 3)$

19. $6(x - 2) = 3(4 + x) + 3x$

20. $4(2x + 1) = 7x + 3 + x$

21. $6x - 5 = 2(x - 3) + 1 + 4x$

22. $-(2x + 5) - 3 = 4(1 - x) - 12 + 2x$

In Exercises 23–32, solve each literal equation or formula for the indicated variable.

23. $2x + y = 7; y$

24. $3x + y = 9; y$

25. $4x - y = 8; y$

26. $\dfrac{1}{2}x - y = 2; y$

27. $2x - 3y = 5; y$

28. $3x - 5y = 1; y$

29. $x + 2y = 6; y$

30. $2x + 2y = 8; y$

31. $V = \dfrac{1}{3}\pi r^2 h; h$

32. $A = P + Prt; t$

In Exercises 33–42, solve each absolute value equation and verify your results with the calculator.

33. $|x + 1| = 10$

34. $\left|\dfrac{3}{2}x - 4\right| = 9$

35. $|2x - 1| + 4 = 6$

36. $|3x + 5| - 2 = -1$

37. $|4x - 1| = -2$

38. $|5x + 7| + 5 = 1$

39. $2|x - 1| - 2 = 0$

40. $\dfrac{1}{2}|x + 2| - 1 = 0$

41. $|3x + 2| = 0$

42. $|5x - 4| = 0$

In Exercises 43–52, solve each polynomial equation by the zero product property. Verify your results with the calculator.

43. $(x + 1)(x - 2) = 0$

44. $(2x - 3)(x + 4) = 0$

45. $x^2 - 16 = 0$

46. $x^2 - 25 = 0$

47. $x^2 + 21x + 20 = 0$

48. $x^2 - 11x + 30 = 0$

49. $2x^3 - 10x^2 - 48x = 0$

50. $6x^3 + 5x^2 - 4x = 0$

51. $x(x - 12) = -20$

52. $x(x + 6) = -9$

In Exercises 53–62, solve each rational equation. Verify solutions using either the STOre or TABLE features of the calculator.

53. $\dfrac{11}{x} + \dfrac{13}{x} = 12$

54. $\dfrac{5}{x} - \dfrac{4}{x} = 8 + \dfrac{1}{x}$

55. $\dfrac{3}{p + 6} - 2 = \dfrac{7}{p + 6}$

56. $\dfrac{1}{x - 1} + \dfrac{3}{x - 1} = 1$

57. $\dfrac{z^2}{z + 1} + 2 = \dfrac{1}{z + 1}$

58. $\dfrac{x^2}{x + 2} - \dfrac{4}{x + 2} = x$

59. $\dfrac{5}{x - 2} + 7 = \dfrac{2}{x^2 - 4} + 6$

60. $\dfrac{x}{x - 5} - \dfrac{5}{x - 5} = 3$

61. $\dfrac{2p}{3} - \dfrac{1}{p} = \dfrac{2p - 1}{3}$

62. $\dfrac{3r}{2} - \dfrac{3}{r} = \dfrac{3r}{2} + 3$

APPLICATIONS   *In Exercises 63–72, solve each problem and verify results with your calculator. Remember, this section reviews problem solving. You may wish to return to Chapter 4 for specific examples not demonstrated in this section.*

63. Carpentry   A carpenter wants to cut a 20-foot rafter so that one piece is 3 times as long as the other. Where should he cut the board?

64. Geometry   A rectangle is 4 meters longer than it is wide. If the perimeter of the rectangle is 28 meters, find its area.

65. Rental car   Alamo car rental rents cars for $30 per day plus $0.24 a mile. Budget car rental rents the same car for $44 a day plus $0.16 a mile. How many miles would you have to drive in one day for the rental costs to be equal?

66. Height of a tree   A tree casts a shadow of 26 feet at the same time as a 6-foot man casts a shadow of 4 feet (see Illustration 1). Find the height of the tree.

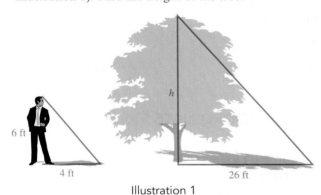

h

6 ft

4 ft   26 ft

Illustration 1

67. Height of a flagpole   A man places a mirror on the ground and sees the reflection of the top of a flagpole, as in Illustration 2. The two triangles in the illustration are similar. Find the height, h, of the flagpole.

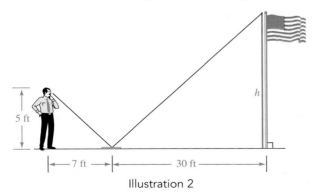

h

5 ft

|← 7 ft →|←   30 ft   →|

Illustration 2

68. Elevation   A sign on a mountain highway near Cloudcroft, New Mexico, indicates a 5% downhill grade. How much decrease in elevation occurs per mile of horizontal distance? (*Hint:* 5280 feet = 1 mile)

Illustration 3

69. Angles of a triangle   The sum of the angles of a triangle is 180°. Find the three angles if the base angles are the same and the third angle is four times one of the base angles.

70. Angles of a triangle   Find the angle measurements of a triangle if the first angle is 2 more than the second angle and the third angle is 3 times the second angle.

71. Investment problem   A retired teacher has a choice of two investment plans: an insured fund that pays 4% interest or a riskier investment that promises a 6% return. The same amount invested at the higher rate would generate an extra $150 per year. How much does the teacher have to invest?

72. Investment problem   A financial adviser recommends investing twice as much in CDs as in a bond fund. A client follows his advice and invests $21,000 in CDs paying 1% more interest than the fund. The CDs generate $420 more interest than the fund. Find the two rates.

REVIEW

73. Simplify each of the following. Leave no negative exponents in your answer.

a. $\dfrac{x^{-2}y^{-4}}{(2x)^{-5}}$ b. $\left(\dfrac{3x^4y^{-2}}{2^{-2}x^{-3}y^2}\right)^{-1}$

74. Simplify: $3(x^2 - 4x + 2) - (x + 2)$

75. Simplify: $(x - 4)(x^2 + 4x + 16)$

76. Factor completely:

a. $9x^3 + 3x^5 - 30x$

b. $x^3 - 343y^3$

c. $y^2 + 2xy - 63x^2$

7.4 GRAPHING AND ITS APPLICATIONS REVISITED

- THE RECTANGULAR COORDINATE SYSTEM
- VIEWING WINDOWS
- GRAPHING EQUATIONS IN TWO VARIABLES
- GRAPHICAL SOLUTIONS OF ONE-VARIABLE EQUATIONS

Points to remember

1. Solutions to one-variable equations can be graphed on a number line. For example, $x = 1$ is the solution to $3x + 2 = 5$. Graphed on a number line this would be

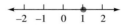

2. Solutions to two-variable equations are graphed on a rectangular coordinate system. For example, the ordered pairs $(3, 2)$, $(2, 3)$, and $(1, 4)$ are only three of an infinite number of solutions to the equation $x + y = 5$. These three ordered pairs are graphed on the coordinate system and connected to form a straight-line graph, as in Figure 7-2. **All** solutions to the equation $x + y = 5$ can be found on this straight line.

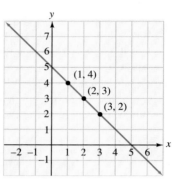

Figure 7-2

THE RECTANGULAR COORDINATE SYSTEM

Ordered pair solutions (x, y), of an equation in two variables can be graphed on a rectangular coordinate system. The system consists of two intersecting number lines called the x- and y-axes. The axes intersect at right angles and divide the coordinate system into four regions called **quadrants**, as shown in Figure 7-3.

Points on the graph are identified by **ordered pairs** of numbers. The order of the numbers, (x, y), is important. The x-**coordinate**, or **abscissa**, is first; and the y-**coordinate**, or **ordinate**, is second.

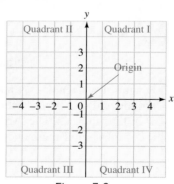

Figure 7-3

EXAMPLE 1

On a rectangular coordinate system, plot the points   **a.** $A(-1, 2)$,   **b.** $B(0, 0)$,   **c.** $C(5, 0)$,   **d.** $D(-\frac{5}{2}, -3)$,   and   **e.** $E(3, -2)$.

SOLUTION   **a.** The point $A(-1, 2)$ has an x-coordinate of -1 and a y-coordinate of 2. To plot point A, we start at the origin and move 1 unit to the left and then 2 units up. (See Figure 7-4.) Point A lies in quadrant II.

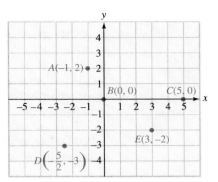

Figure 7-4

b. To plot point $B(0, 0)$, we start at the origin and move 0 units to the right and 0 units up. Because there was no movement, point B is the origin.

c. To plot point $C(5, 0)$, we start at the origin and move 5 units to the right and 0 units up. Point C lies on the x-axis, 5 units to the right of the origin.

d. To plot point $D(-\frac{5}{2}, -3)$, we start at the origin and move $\frac{5}{2}$ units ($\frac{5}{2}$ units $= 2\frac{1}{2}$ units) to the left and then 3 units down. Point D lies in quadrant III.

e. To plot point $E(3, -2)$, we start at the origin and move 3 units to the right and then 2 units down. Point E lies in quadrant IV.

▮▮▮ VIEWING WINDOWS

TECHNOLOGY TIP

See "Technology 3.4: Graphing Basics" for more detail.

When you sketch a graph on your graph paper, the size of your graph is limited by the size of your paper space. Theoretically, each axis extends indefinitely, which we represent by arrows at the end of each axis. However, the number of tic marks you place on each axis limits its visual size. Similar limitations occur when using a graphing calculator to display a rectangular coordinate system. You can limit the visual size of your calculator graph by designating a minimum and maximum value for the x-axis and a minimum and maximum value for the y-axis, as shown in Figure 7-5.

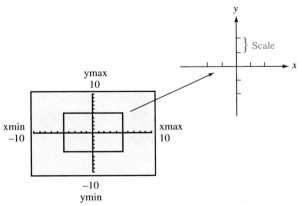

Figure 7-5

To condense this information, we will use the following notation:

$$\begin{array}{cccc} \text{xmin} & \text{xmax} & \text{ymin} & \text{ymax} \\ [-10, 10] & \text{by} & [-10, 10] \end{array}$$

Observe that there are 10 tic marks to the left, right, above, and below the origin on the calculator graph. The distance between the tic marks is called the *scale*. On this graph, the scale for both the *x*- and *y*-axes is 1. We would write this as xscl = 1 and yscl = 1. Unless otherwise noted, it should be assumed xscl = 1 and yscl = 1 on all graphs displayed in this text.

In order to set the limitations on your calculator graph, you should be familiar with the key that accesses this feature. It may be denoted as WINDOW, RANGE, or WIND on your calculator. The viewing window above, $[-10, 10]$ by $[-10, 10]$, is called the **standard viewing window**. It can be set manually by entering the information or automatically using the ZOOM feature.

WARNING!   What you "see" may be deceiving. Each of the calculator graphs displayed below look the same and yet each represents a different-size rectangular coordinate system based on the limitations placed on xmin, xmax, xscl, ymin, ymax, and yscl.

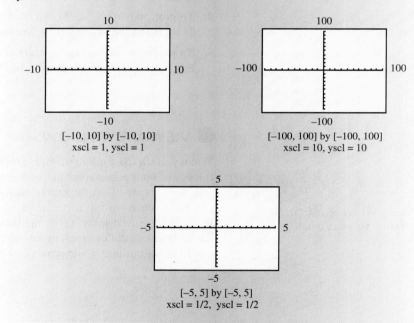

$[-10, 10]$ by $[-10, 10]$
xscl = 1, yscl = 1

$[-100, 100]$ by $[-100, 100]$
xscl = 10, yscl = 10

$[-5, 5]$ by $[-5, 5]$
xscl = 1/2, yscl = 1/2

▮ GRAPHING EQUATIONS IN TWO VARIABLES

Equations may be graphed either with paper and pencil or with a graphing calculator. You should be proficient with both methods. We will review graphing **linear or straight-line equations** in two unknowns.

Standard Form

If A, B, and C are real numbers, where A and B are not both zero, then $Ax + By = C$ is called the **standard form of a linear equation**.

Linear Equation

Any equation that can be written in the form $y = mx + b$, where $y \neq 0$, is called a **linear equation**.

Whether you are graphing an equation of the form $Ax + By = C$ or one of the form $y = mx + b$, you will need at least two ordered pairs to produce a paper-and-pencil graph.

The points where the line intersects the x- and y-axes are called the *intercepts* of the line. Plotting these intercepts and drawing a line through them is called the *intercept method of graphing* and is a convenient method for graphing equations in standard form.

y- and x-Intercepts

The **y-intercept** of a line is the point $(0, b)$ where the line intersects the y-axis. To find b, substitute 0 for x in the equation of the line and solve for y.

The **x-intercept** of a line is the point $(a, 0)$ where the line intersects the x-axis. To find a, substitute 0 for y in the equation of the line and solve for x.

EXAMPLE 2

Graph the equation $3x + 2y = 6$ using the intercept method of graphing.

SOLUTION   To find the y-intercept, we let $x = 0$ and solve for y.

$$3x + 2y = 6$$
$$3(0) + 2y = 6 \qquad \text{Substitute 0 for } x.$$
$$2y = 6 \qquad \text{Simplify.}$$
$$y = 3 \qquad \text{Divide both sides by 2.}$$

The y-intercept is the ordered pair $(0, 3)$.

To find the x-intercept, we let $y = 0$ and solve for x.

$$3x + 2y = 6$$
$$3x + 2(0) = 6 \qquad \text{Substitute 0 for } y.$$
$$3x = 6 \qquad \text{Simplify.}$$
$$x = 2 \qquad \text{Divide both sides by 3.}$$

The x-intercept is the ordered pair $(2, 0)$.

As a check, we plot one more point. If $x = 4$, then

$$3x + 2y = 6$$
$$3(4) + 2y = 6 \qquad \text{Substitute 4 for } x.$$
$$12 + 2y = 6 \qquad \text{Simplify.}$$
$$2y = -6 \qquad \text{Add } -12 \text{ to both sides.}$$
$$y = -3 \qquad \text{Divide both sides by 2.}$$

The point $(4, -3)$ lies on the graph. We plot these three points and draw a line through them. The graph of $3x + 2y = 6$ is shown in Figure 7-6.

$$3x + 2y = 6$$

x	y
0	3
2	0
4	−3

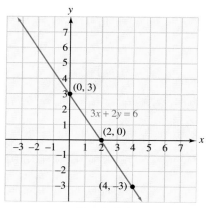

Figure 7-6

EXAMPLE 3

To use the graphing calculator to graph the equation $3x + 2y = 6$ from Example 2, first solve the equation for y in order to enter it on the y= screen. Use the standard viewing window and display the graph of the equation.

SOLUTION   First, solve the equation for y in order to enter it on the y= screen.

$$3x - 3x + 2y = 6 - 3x$$
$$2y = 6 - 3x$$
$$\frac{2y}{2} = \frac{6 - 3x}{2}$$
$$y = 3 - \frac{3}{2}x$$
$$y = -\frac{3}{2}x + 3$$

TECHNOLOGY TIP

The integer viewing window can be automatically set by accessing it from the ZOOM menu. See "Technology 3.4: Graphing Basics" for details.

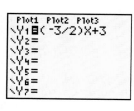

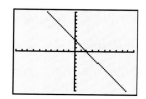

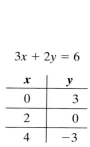

You may now access the TABLE feature of the calculator to compare your table of values, as constructed in Example 2, to those of the calculator.

If you access the TRACE feature you can use the left and right cursor keys to trace along the line. You will notice that the coordinate values are cumbersome decimal values.

For this reason, we will *not* use the standard viewing window when we want to trace. The calculator has a preset window called ZInteger that yields integer values when tracing (see "Technology 3.4"). To set these values manually, enter $[-47, 47]$ by $[-31, 31]$ on your WINDOW screen now (TI-86 users enter

$[-63, 63]$ by $[-31, 31]$), with x- and y-scales equal to 10. Trace and observe the x- and y-coordinates.

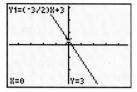 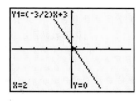 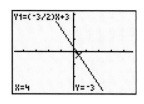

 EXAMPLE 4   Graph the equation $y = 3$.

SOLUTION   We can write the equation $y = 3$ in standard form as $0x + y = 3$. Because the coefficient of x is 0, the numbers assigned to x have no effect on y. The value of y is always 3. For example, if we replace x with -3, we get

$$0x + y = 3$$
$$0(-3) + y = 3$$
$$0 + y = 3$$
$$y = 3$$

The table in Figure 7-7 gives several ordered pairs that satisfy the equation $y = 3$. After plotting these pairs (x, y) and joining them with a line, we see that the graph of $y = 3$ is a horizontal line, parallel to the x-axis and intersecting the y-axis at 3. The y-intercept is $(0, 3)$, and there is no x-intercept.

$y = 3$

x	y
-3	3
0	3
2	3
4	3

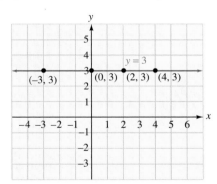

Figure 7-7

The horizontal line $y = 3$ is graphed below in the ZInteger viewing window with two of the four ordered pairs displayed by using the TRACE feature.

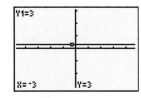 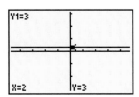

EXAMPLE 5   Graph the equation $x = -2$.

SOLUTION   We can write the equation $x = -2$ in standard form as $x + 0y = -2$. Because the coefficient of y is 0, the values of y have no effect on x. The value of x is always -2. A table of values and the graph are shown in Figure 7-8.

The graph of $x = -2$ is a vertical line, parallel to the y-axis and intersecting the x-axis at -2. The x-intercept is $(-2, 0)$, and there is no y-intercept.

$x = -2$

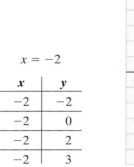

x	y
-2	-2
-2	0
-2	2
-2	3

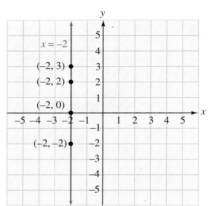

Figure 7-8

From the results of Examples 4 and 5, we have the following facts:

Equations of Lines Parallel to the Coordinate Axes
The equation $y = b$ represents a horizontal line that intersects the y-axis at $(0, b)$. If $b = 0$, the line is the x-axis.

The equation $x = a$ represents a vertical line that intersects the x-axis at $(a, 0)$. If $a = 0$, the line is the y-axis.

Example 5 was not graphed on the calculator because the calculator will graph nonvertical lines but will not graph vertical lines. Vertical lines are not **functions** and the calculator is designed to graph only functions.

Function
A function is a set of ordered pairs in which each x-value in the ordered pair (x, y) is assigned a unique y-value.

The ordered pairs used to graph the line $x = -2$ in Example 5 were $(-2, 3), (-2, 2)$, $(-2, 0), (-2, -2)$. The x-value of -2 does not have a unique y-value; it has many y-values. Thus, vertical lines are not functions.

EXAMPLE 6

Is the formula for converting Fahrenheit temperature to Celsius a function? The formula is

$$y = \frac{5}{9}(x - 32)$$

where y = degrees Celsius and x = degrees Fahrenheit.

SOLUTION

We need to answer the question "does each x produce a unique y?" Select a value for x, say 41°. Then

$$y = \frac{5}{9}(41 - 32)$$

$$y = \frac{5}{9}(9) \qquad \text{Subtraction produces a unique result.}$$

$$y = 5 \qquad \text{Multiplication produces a unique result.}$$

Because the operations of addition, subtraction, multiplication, and division produce unique results, any equation that applies these operations to x will produce unique results for y, and thus a function.

Linear Function
Any linear equation that can be written in the form $y = mx + b$, where $y \neq 0$ and m and b are real numbers, is called a **linear function**.

EXAMPLE 7

Intelix Home Video produces videotapes of weddings, baptisms, etc. For each tape produced there is a basic fixed cost of $100 and a variable cost of $2 per tape. The tapes then sell for $20 each. Write an equation to represent the production costs, y, for x tapes. Then write an equation to represent the revenue, y, for x tapes. Explore these graphs and their meaning in the context of the problem.

SOLUTION

a. The cost equation would be $y = 100 + 2x$. Graph this equation in the viewing window $[-20, 74]$ by $[-5, 181]$ (TI-86: $[-20, 106]$ by $[-5, 181]$) with x- and y-scales equal to 10. TRACE to find the production cost for 23 tapes. At the point (23, 146), the cost for producing 23 tapes is $146.

b. The revenue equation would be $y = 20x$. Graph this equation on the same screen as the cost equation and TRACE on the **revenue** equation to determine the revenues produced for 23 tapes. Use the [▼] and [▲] keys to toggle from one graphical display to the other. The point (23, 460) indicates that the production of 23 tapes produces a revenue of $460.

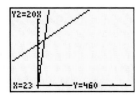

TECHNOLOGY TIP

Use the [▶] and [◀] keys to TRACE along the graph. The calculator will trace to a point not visible in your designated viewing window. The coordinates at the bottom of the screen indicate the cursor position.

c. Does the company sustain a loss or make a profit when 23 tapes are sold? Because the production cost is $146 and the revenues are $460, the company makes a profit of $314.

d. How many tapes must be produced and sold for the company to break even (cost = revenue)? Attempting to trace to the point of intersection is futile because the *x*-value at the point of intersection is not an integer. (The window you were given was designed to produce integer values for *x* when tracing.) The *x*-value of the point of intersection of the two equations represents the solution to the equation cost = revenue (i.e., $100 + 2x = 20x$). Use the INTERSECT feature of your calculator to locate the point of intersection displayed at the right. The company would need to produce approximately 5.6 tapes to break even.

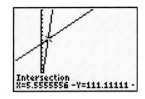

TECHNOLOGY TIP

See "Technology 3.5: INTERSECT Method for Solving Equations" for specific help in using the INTERSECT feature of your calculator.

When discussing functions and their applications, we are often interested in just the *x*-values of the function or just the *y*-values. These classifications are called the **domain** and **range** of the function.

Domain of a Function
The **domain** of a function is the set of all *x*-values of the ordered pairs of the function.

Range of a Function
The **range** of a function is the set of all *y*-values of the ordered pairs of the function.

EXAMPLE 8

In the conversion chart, identify the domain and range.

x = number of cups	$\frac{1}{2}$	1	$1\frac{1}{2}$	2	$2\frac{1}{2}$
y = number of pints	$\frac{1}{4}$	$\frac{1}{2}$	$\frac{3}{4}$	1	$1\frac{1}{4}$

SOLUTION   The domain is $\{\frac{1}{2}, 1, 1\frac{1}{2}, 2, 2\frac{1}{2}\}$. The braces indicate we have a **set**. The range is $\{\frac{1}{4}, \frac{1}{2}, \frac{3}{4}, 1, 1\frac{1}{4}\}$.

▌▌▌ GRAPHICAL SOLUTIONS OF ONE-VARIABLE EQUATIONS

It is important that you understand the difference between graphing linear equations in two variables and using a graphical display to interpret the solution of a linear equation in one variable.

To graphically solve an equation in one variable, we examine the graphical representation of the algebraic expression on each side of the equation. Each side of the equation is entered at the y1 and y2 prompts respectively. The solution occurs when y1 = y2, which is at the point of intersection of the two graphs. It is the **x-value** of the ordered pair that is the solution to the one-variable equation.

EXAMPLE 9

Graphically solve $2(x + 2) - x = 10 - x$.

SOLUTION

Let y1 = $2(x + 2) - x$ and y2 = $10 - x$.

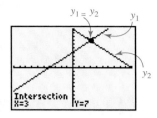

When solving graphically, begin in the standard viewing window and adjust the window as needed to display the point of intersection. Use the INTERSECT feature to display the graph at the right. The display at the bottom of the screen indicates that the ordered pair $(3, 7)$ is common to both graphs. However, we have merely graphed the one-variable expressions of the left and right sides of an equation. It is the value $x = 3$ that is the solution to our equation. Checking analytically verifies this:

$$2(x + 2) - x = 10 - x$$
$$2(\mathbf{3} + 2) - 3 \ ? \ 10 - 3$$
$$2(5) - 3 \ ? \ 7$$
$$10 - 3 \ ? \ 7$$
$$7 = 7\checkmark$$

The y-value displayed on the calculator screen can be interpreted to mean that both sides of the equation will evaluate to 7 when x is replaced with 3.

Solution: 3; Solution set: {3}

EXAMPLE 10

Graphically solve $|3x + 1| = 4$.

SOLUTION

Let y1 = abs $(3x + 1)$
    y2 = 4

Use the INTERSECT feature **twice** to determine both solutions.

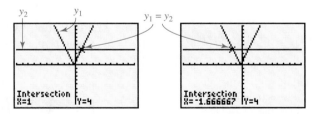

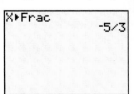

Remember, when the INTERSECT feature is used, the value computed for x is stored in x until you store another value there or until you use the INTERSECT feature again. You can return to the home screen and convert the decimal value displayed into a simplified fraction.

Solutions: $1, -\frac{5}{3}$; Solution set: $\{1, -\frac{5}{3}\}$

EXAMPLE 11

Graphically solve $x^2 - x - 12 = 0$.

SOLUTION

Let $y1 = x^2 - x - 12$

$y2 = 0$

Use the INTERSECT feature **twice** to determine both solutions.

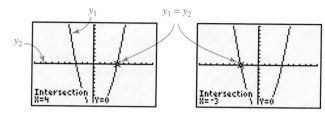

Remember that although only one graph appears to be visible, the equation $y2 = 0$ is actually the x-axis.

Solution: $-3, 4$; Solution set: $\{-3, 4\}$

The two points of intersection in Example 11 occurred at $x = 4$, $y = 0$ and $x = -3$, $y = 0$. These ordered pairs $(4, 0)$ and $(-3, 0)$ represent the x-intercepts for the **two-variable** equation $y = x^2 - x - 12$. The x-intercept(s) represent the root(s) or zero(s) of the equation. The ZERO (or ROOT) feature of the graphing calculator offers not only another approach to graphical solutions of one-variable equations but also an accurate means of determining x-intercepts of any equation in two variables.

EXAMPLE 12

Graphically solve $x^2 - x - 12 = 0$ using the ZERO feature of the calculator.

SOLUTION

In order to use this feature of the calculator, the equation **must** be set equal to zero. Enter the nonzero side as y1.

$y1 = x^2 - x - 12$

Use the ZERO feature twice to determine both of the solutions (i.e., roots or zeros) of the equation as displayed on the two screens below.

TECHNOLOGY TIP

See "Technology 7.4: ROOT/ ZERO Method for Solving" at the end of this section for more details on using the ZERO feature. Some calculators may designate this feature as the ROOT feature.

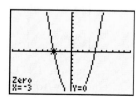 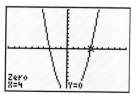

It is now the x-intercepts that are displayed at the bottom of the screen, along with the word *zero*. Remember, it is the x-value that is the solution to your equation in one variable.

Solution: $-3, 4$; Solution set: $\{-3, 4\}$

When solving graphically, it is best to let the format of your equation determine whether you use the INTERSECT or ZERO feature. For example, $x^2 - x - 12 = 0$ is in the required format for the ZERO feature, whereas $|3x + 1| = 4$ is not because neither side of the equation is zero. The less algebraic manipulation that occurs before you solve graphically, the higher your accuracy level.

Having two approaches to obtain graphical solutions provides two advantages. First, you can select the method that requires no algebraic manipulation on your part, thus decreasing the chance of human error when determining the solutions. Second, when the graphical display of one approach is difficult to interpret, you have an alternative approach.

EXAMPLE 13   Graphically solve $x^3 + x^2 = 4x + 4$.

SOLUTION   Since neither side of the equation is zero, we will use the INTERSECT feature.

$$\text{Let } y1 = x^3 + x^2$$
$$y2 = 4x + 4$$

The graphical display does not clearly show all points of intersection. We would need to adjust the viewing window before using the INTERSECT feature. However, if we rewrite the equation to set one side equal to zero, we can produce a much improved graphical display.

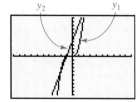

$$x^3 + x^2 = 4x + 4$$
$$x^3 + x^2 - 4x = 4x - 4x + 4$$
$$x^3 + x^2 - 4x = 4$$
$$x^3 + x^2 - 4x - 4 = 4 - 4$$
$$x^3 + x^2 - 4x - 4 = 0$$

Now solve using the ZERO feature.

$$\text{Let } y1 = x^3 + x^2 - 4x - 4$$

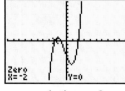   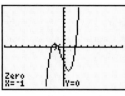   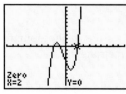

solution: -2 solution: -1 solution: 2

Conclusion: The solutions are $-2, -1, 2$. The solution set is $\{-2, -1, 2\}$.

We can make several observations at this time.

1. Conditional linear equations reviewed in Example 9 will have one real root because linear equations are also first-degree equations.
2. Quadratic equations such as in Example 12 will have at most two real roots because quadratic equations are second-degree equations.

3. Third-degree polynomial equations such as in Example 13 will have at most three real roots.

If all polynomial equations are set equal to zero, then the degree of the polynomial not only predicts the maximum number of real roots but it will also predict the maximum number of x-intercepts the graph of the polynomial will have.

Zeros of a Polynomial Equation
A polynomial equation of degree n will have at most n distinct roots/zeros.

EXERCISE 7.4

Read the information in Technology 3.4, 3.5, and 7.4 before beginning the exercises.

VOCABULARY AND NOTATION   *In Exercises 1–7, fill in the blanks to make a true statement.*

1. Points on a graph are identified by _____ of numbers.

2. Another name for the x-coordinate is _____ and for the y-coordinate is _____ .

3. The _____ window is defined to be $[-10, 10]$ by $[-10, 10]$.

4. A _____ in the form $Ax + By = C$ is said to be in _____ form.

5. The _____ of a line is the point $(a, 0)$ where the line intersects the x-axis.

6. The _____ of a line is the point $(0, b)$ where the line intersects the y-axis.

7. A _____ is a set of ordered pairs in which each x-value in the domain is assigned a unique y-value in the range.

CONCEPTS

8. Is the ordered pair $(4, 2)$ a solution to the equation $y = |x - 4| + 2$?

9. Plot the following ordered pairs on a rectangular co-ordinate system:
   a. $(2, 5)$ b. $(-3, 1)$
   c. $(-4, 2)$ d. $(5, 0)$
   e. $(-5, 3)$ f. $(0, -2)$

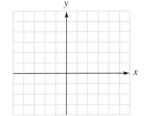

In Exercises 10–13, determine if each ordered pair would be visible if graphed on your calculator in the designated viewing window.

10. $[-15, 12]$ by $[-2, 2]$
    a. $(-12, 6)$ b. $(4, 1)$

11. $[0, 150]$ by $[-10, 50]$
    a. $(4, 45)$ b. $(5, 150)$

12. $[-25, 5]$ by $[25, 125]$
    a. $(-18, 27)$ b. $(-26, 20)$

13. $[-20, 20]$ by $[-10, 5]$
    a. $(35, 10)$ b. $(-15, -5)$

14. Each of the following represent solutions to an equation. Based on the given information, is the equation a function?
    a. $\{(4, 7), (2, 1), (-5, -7), (0, 3)\}$
    b.

X	Y1	
-3	12	
-2	7	
-1	4	
0	3	
1	4	
2	7	
3	12	

X= -3

15. A table of values for a function is displayed at the right.
    a. What is the domain?

    b. What is the range?

X	Y1	
-3	4	
-2	-1	
-1	-4	
0	-5	
1	-4	
2	-1	
3	4	

X= -3

PRACTICE *In Exercises 16–23, use the intercept method to construct a paper-and-pencil graph of the equation.*

16. $4x - 2y = 5$

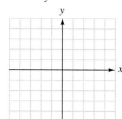

17. $x + 3y = 6$

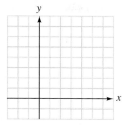

18. $x + 2y = -1$

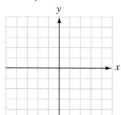

19. $2x - y = -2$

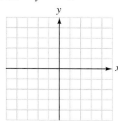

20. $3x + 1 = 4$

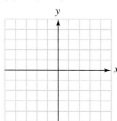

21. $4 - x = 5$

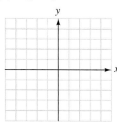

22. $y = -2$

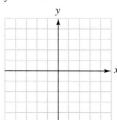

23. $y = 4$

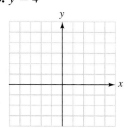

In Exercises 24–31, graph the given equations on your calculator and sketch the screen display. Use the ZInteger viewing window and TRACE to label the x- and y-intercepts that have integer coordinates.

24. $f(x) = 4x + 12$

25. $f(x) = -3x + 6$

26. $3x + y = 18$

27. $-2x + y = 24$

28. $4x + 6y = 36$

29. $5x - 8y = 80$

30. $6x - 7y = 42$

31. $5x + 9y = 45$

In Exercises 32–37, based on the displayed ordered pairs, determine if the graph displayed is a function.

32.

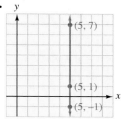

33.

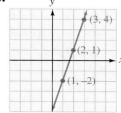

34.

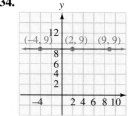

35.

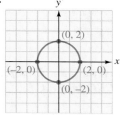

36.

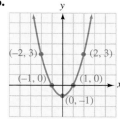

37.

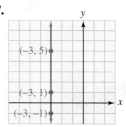

In Exercises 38–41, determine the domain and range of each function.

38. $\{(2, 4), (3, 5), (-1, 2), (4, 2)\}$

39. $\{(-8, 5), (9, 2), (7, -3), (-\frac{1}{2}, \frac{2}{3}), (-4.5, 3)\}$

40.

X	Y₁		

X	Y1
-3	12
-2	7
-1	4
0	3
1	4
2	7
3	12

X= -3

41.

X	Y1
-6	31
-5.5	25.25
-5	20
-4.5	15.25
-4	11
-3.5	7.25
-3	4

X= -6

*In Exercises 42–57, graphically solve each equation using the INTERSECT method. There are problems in this group that you have **not** yet learned how to solve analytically. Therefore, set your viewing window to ensure all existing points of intersection are visible. With the INTERSECT and ROOT features, you have the power to solve many equations.*

42. $\dfrac{x-9}{5} = 4 - x$

43. $2 + \dfrac{x}{3} = \dfrac{5}{6} - x$

44. $4x - 5 = 7x + 4 - 3x$

45. $2(x-8) - 3 = 4x + 10 - 2x$

46. $x^2 - 24 = 2x$

47. $x^2 = -2x + 8$

48. $\dfrac{x+6}{x} = -7$

49. $\dfrac{x-3}{5} = \dfrac{8}{x}$

50. $|6x + 7| = 18$

51. $|3x - 5| = 10$

52. $-|4x + 7| = -11$

53. $-|2x - 8| = -12$

54. $|x + 3| + |x - 5| = 15$

55. $|x + 2| + |x-3| = 12$

56. $\sqrt{4x - 3} = 7$

57. $5 = \sqrt{3x + 4}$

In Exercises 58–73, graphically solve each equation using the ZERO method. Begin in the standard viewing window and adjust your window as needed. (Hint: When solving equations in 66–73, the degree of the polynomial will predict the maximum number of possible roots if you set the equation equal to zero.)

58. $\dfrac{x-9}{5} - 4 + x = 0$

59. $2 + \dfrac{x}{3} - \dfrac{5}{6} + x = 0$

60. $\dfrac{x+6}{x} + 7 = 0$

61. $\dfrac{x-3}{5} - \dfrac{8}{x} = 0$

62. $|6x + 7| - 18 = 0$

63. $|3x - 5| - 10 = 0$

64. $-|4x + 7| + 11 = 0$

65. $-|2x - 8| + 12 = 0$

66. $x^2 - 2x - 24 = 0$

67. $x^2 + 2x - 8 = 0$

68. $15x^2 - 5x - 10 = 0$

69. $-6x^2 - 7x + 20 = 0$

70. $x^3 + x^2 - 30x = 0$

71. $x^3 + 3x^2 - x - 3 = 0$

72. $2x^3 - 5x^2 - 2x + 5 = 0$

73. $2x^3 - 9x^2 - 8x + 36 = 0$

APPLICATIONS   *In Exercises 74 and 75, graph the linear equation that models each problem and use your calculator to answer the questions posed.*

74. Mr. Hodge is part owner in a newly formed communications business. His monthly salary is $400 plus 10% of the business's net revenue for the month.

  a. Write a linear equation to model the problem where y represents Mr. Hodge's total salary for the month and x represents the amount of net revenue for the business.

  b. Graph the linear equation and state your viewing window.

  c. Use the VALUE feature to determine his salary if the net revenue is $9000.

  d. Resketch the graph in the following window: [0, 94000] by [−10, 10000] (x- and y-scales equal 0).

  e. TRACE to determine the business net revenue if Mr. Hodge's earnings were $6,500.

75. A car rental agency charges a flat fee of $45 plus 24 cents per mile.

  a. Write an equation to model the total cost, y, as related to miles driven, x.

  b. Graph on your calculator and indicate your viewing window.

  c. TRACE to determine the approximate cost (to the nearest cent) of traveling 131 miles.

  d. Use VALUE to determine the exact cost of traveling 131 miles.

REVIEW

76. Write an equation in one variable to model the problem and then determine the indicated value. What number must be added to the numerator of $\frac{6}{7}$ and subtracted from the denominator to produce the fraction $\frac{8}{5}$?

77. If a barrel contains 42 gallons of crude oil and a country has crude oil reserves of 2.61×10^{11} barrels, write an expression in scientific notation to indicate the number of gallons of crude oil reserves.

78. Simplify: $\dfrac{(2x^{-4}y^2)^3}{4x^{-2}y^{-1}}$.

79. Use the STOre feature of your calculator to evaluate $\dfrac{(2x^{-4}y^2)^3}{4x^{-2}y^{-1}}$ for $x = \dfrac{1}{2}$ and $y = \dfrac{1}{5}$.

80. Solve algebraically for x:

  a. $|3x + 5| - 6 = -2$ **b.** $2x^2 + x - 3 = 0$

ROOT/ZERO Method for Solving Equations

The ROOT/ZERO feature on the TI-86 and the TI-83 Plus work similarly. If using the TI-86, watch for parenthetical information that specifies the differences for your calculator.

If an equation is set equal to zero, when the nonzero side of the equation is graphed, the real roots or zeros are the x-values at the point(s) where the graph crosses the horizontal axis (x-axis). This method is *perfect* for equations of any type that are already set equal to zero.

1. To solve the equation $x^2 + 2x - 8 = 0$ using the root/zero method, enter the polynomial (in this case, the left side of the equation) at the **y1=** prompt. Nothing will be entered at the **y2=** prompt. Press **[ZOOM] [6:ZStandard]** (TI-86 users press **[2nd] [M3] (ZOOM) [F4] (ZSTD)**) to automatically set the standard viewing window. Your display should match the one given (the TI-86 screen will still have the GRAPH menu displayed at the bottom). Circle the points where the graph crosses the horizontal (x) axis. These are the roots or zeros.

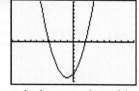

2. To access the ROOT/ZERO option, press **[2nd] [CALC] [2:zero]** (TI-86 users press **[MORE] [F1] (MATH) [F1] (ROOT)**).

    a. **Set left bound:** The screen display asks for a left bound. This is an x-value smaller than the expected root. You may either enter a value for the variable x at the prompt or move the cursor to the left of the left-hand root and press **[ENTER]**. Because the roots are determined on the horizontal axis, a left bound is always determined by locating a value for x to the *left of the desired root*. At the top of the screen a ▶ marker will appear to designate the location of the left bound once ENTER has been pressed.

    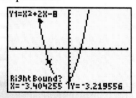

    b. **Set right bound:** Similarly, the right bound is always determined by moving the cursor to the *right of the root* or by entering a value for x that is larger than the desired root and pressing **[ENTER]**. Again, a ◀ marker is at the top of the screen to designate the location of the bound.

    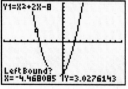

    Note: If the bound markers do not point toward each other, ▶ ◀, then you will get an "ERROR:bounds" message. If this happens, start the ROOT/ZERO calculation over. **Care must be taken to ensure that only one root occurs between the set boundaries.**

    c. **Locate first root:** Move the cursor to the approximate location where the graph crosses the x-axis for your guess. When you press **[ENTER]** the calculator will search for the root within the area marked by ▶ and ◀. The root is $x = -4$.

    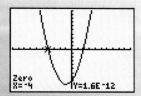

Note: The calculator should display x = −4, y = 0, which means that the expression entered at **y=** has a value of 0 when *x* = −4. **To verify that the *y*-value is actually zero, scroll through the TABLE to x = −4 (or use EVAL) and note that the *y*-value is 0 when *x* = −4.** In this particular problem, round-off error has occurred at the *y*-value.

Round-off error: There will be times when the calculator will be very close to zero but will not display zero exactly. This occurred on the pictured screen. The *y*-value, y = 1.6 E −12, would be Y = −.0000000000016, which is, for all practical purposes, a zero. Therefore, when using the graph screen to solve equations/inequalities, you should be aware that the display coordinate values approximate the actual mathematical coordinates. The accuracy of these display values is determined by the height and width of the pixel space being displayed.

d. **Locate subsequent roots:** Repeat the entire process outlined above to determine the right-hand root. This root is x = 2. The screen displays below illustrate the process. Be aware that the bounds set below are arbitrary; the only requirement is that the bounds be to the left and right, respectively, of the desired root.

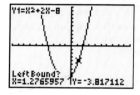
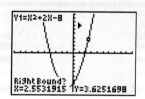
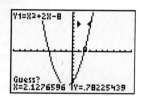
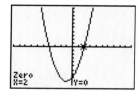

7.5   REVIEW OF INEQUALITIES

- BASIC DEFINITIONS AND SOLUTIONS OF SIMPLE LINEAR INEQUALITIES
- SOLVING COMPOUND INEQUALITIES
- SOLVING ABSOLUTE-VALUE INEQUALITIES

BASIC DEFINITIONS AND SOLUTIONS OF SIMPLE LINEAR INEQUALITIES

Points to Remember

1. An **inequality** is a mathematical expression that states that two quantities are not necessarily equal.
2. Recall the inequality symbols:

Inequality Symbols		
<	means	"is less than"
>	means	"is greater than"
≤	means	"is less than or equal to"
≥	means	"is greater than or equal to"

3. The **solution of an inequality** is any number that makes the inequality a true statement and can be a region of numbers that we will graph on the number line.
4. The **critical point** p on a number line graph is that number that divides the graph into the region of solution values versus nonsolution values. Open circles indicate that the critical points are **not** included as a solution value and closed circles indicate that the critical points **are** included as solution values.

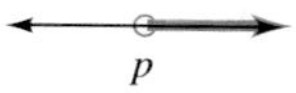

p

Figure 7-9

5. **Interval notation** is used to describe the shaded region on the graph.
6. Recall that in interval notation parentheses are used to indicate solutions that are not included and correspond to open circles on number line graphs (and are always used to represent infinity). Square brackets, [ ], are used to indicate solutions that are included and correspond to closed circles on number line graphs.
7. Recall the general notation information for simple inequalities:

Summary of Unbounded Intervals			
Inequality	**Graph**	**Interval Notation**	**Set Notation**
$x > a$	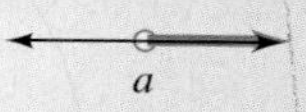	(a, ∞)	$\{x \mid x > a\}$
$x \geq a$	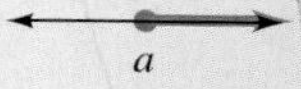	$[a, \infty)$	$\{x \mid x \geq a\}$
$x < a$	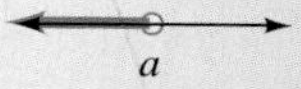	$(-\infty, a)$	$\{x \mid x < a\}$
$x \leq a$	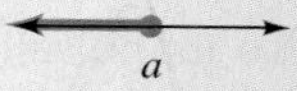	$(-\infty, a]$	$\{x \mid x \leq a\}$

8. A linear inequality is solved exactly like a linear equation **except** if you multiply or divide by a negative value the inequality symbol is reversed.

EXAMPLE 1

Solve $4x + 7 \leq -5$ and express the solution as both a number line graph and in interval notation.

SOLUTION

$$4x + 7 \leq -5$$
$$4x + 7 - 7 \leq -5 - 7 \qquad \text{Subtract 7 from both sides.}$$
$$4x \leq -12 \qquad \text{Combine like terms.}$$
$$x \leq -3 \qquad \text{Divide both sides by 4.}$$

The solution consists of all real numbers less than **and including** -3. A solid circle at -3 indicates that we include -3 and shade to the left of -3. (See Figure 7-10.)

-3

Figure 7-10

A square bracket will be used in the interval notation to indicate that -3 is part of the solution. The interval, from left to right on the graph, is $-\infty$ to -3: $(-\infty, -3]$. In set notation this would be $\{x \mid \leq -3\}$.

EXAMPLE 2

Solve $4 - 6x < -2$ and express the solution both as a number line graph and in interval notation.

SOLUTION

$$4 - 6x < -2$$

$$4 - 6x - 4 < -2 - 4 \qquad \text{Subtract 4 from both sides.}$$

$$-6x < -6 \qquad \text{Combine like terms.}$$

$$\frac{-6x}{-6} > \frac{-6}{-6} \qquad \begin{array}{l}\text{Divide both sides by } -6 \text{ and reverse the direction}\\ \text{of the } < \text{ symbol.}\end{array}$$

$$x > 1$$

In the last step, we divided both sides of the inequality by -6. Any time we divide or multiply both sides by a negative value in an inequality, we must reverse the inequality symbol. The graph of the solution appears in Figure 7-11.

Figure 7-11

The parenthesis at the critical point of 1 corresponds to the open circle, indicating that we do not include 1 in our solution: $(1, \infty)$. In set notation this would be $\{x \mid x > 1\}$.

To graphically solve this same inequality we will graph both the left and right sides of the inequality and determine the point of intersection just as we do when we are solving equations graphically. The critical point in the inequality corresponds to the x-value at the point of intersection. Record the critical point on your number line graph. (See Figure 7-12.)

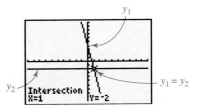

Figure 7-12

Remember, the inequality has been entered on the y-edit screen as $y1 = 4 - 6x$ and $y2 = -2$. Thus, instead of being written as $4 - 6x < -2$, we could write it as $y1 < y2$.

To graphically solve the inequality we want to determine the x-values for which $y1 < y2$. Remember, the INTERSECT feature has already shown us the value for which $y1 = y2$. In general, $y1 < y2$ where the graph of $y1$ is below the graph

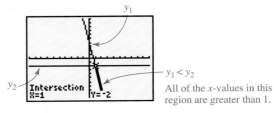

of $y2$. This should be the section that is highlighted on the graph displayed.

The *x*-values greater than 1 satisfy the requirement that $y1 < y2$. Combine this information with the critical point on the number line to produce the final number line graph:

SOLVING COMPOUND INEQUALITIES

Points to Remember

1. **Compound inequalities** are inequalities that are linked by the words **and** or **or**.
2. When solving **and** statements, you will be looking for the **intersection** (common values) of the solutions to simple linear inequalities.
3. When solving **or** statements, you will be looking for the **union** (all represented values) of the solutions to the simple inequalities.
4. **And** statements are more formally called **conjunctions**.
5. Inequalities that involve the use of the word **or** are more formally called **disjunctions**.
6. **Continued inequalities** are **and** statements that can be written in abbreviated form: "$x > 3$ **and** $x < 6$" becomes the **continued** inequality "$3 < x < 6$."
7. Recall the general notation information for compound inequalities:

Summary of Bounded Intervals			
Inequality	**Number Line Graph**	**Interval Notation**	**Set Notation**
$a < x < b$		(a, b) **open interval**	$\{x \mid a < x < b\}$
$a \le x \le b$		$[a, b]$ **closed interval**	$\{x \mid a \le x \le b\}$
$a < x \le b$		$(a, b]$ **half-open interval**	$\{x \mid a < x \le b\}$
$a \le x < b$		$[a, b)$ **half-open interval**	$\{x \mid a \le x < b\}$

EXAMPLE 3

Solve the compound inequality $3(x + 1) < 4$ **and** $3(x + 1) > -4$.

SOLUTION

We can solve each inequality separately as before or we can combine the two inequalities into a continued inequality because they share the common side of $3(x + 1)$.

$$-4 < 3(x + 1) < 4$$ Be sure that the double inequality makes sense. Is $-4 < 4$?

$$-4 < 3x + 3 < 4$$ Remove grouping symbols.

$$-4 - 3 < 3x + 3 - 3 < 4 - 3$$ Subtract 3 from all three parts of the inequality.

$$-7 < 3x < 1$$ Simplify.

$$\frac{-7}{3} < x < \frac{1}{3}$$ Divide all three parts of the inequality by 3.

The solution expressed as a number line graph is

$$\xleftarrow{\quad}\overset{-\frac{7}{3}}{\circ}\!\!-\!\!\overset{\frac{1}{3}}{\circ}\xrightarrow{\quad}$$

Expressed in interval notation, the solution is $\left(-\frac{7}{3}, \frac{1}{3}\right)$. The set notation would be $\{x \mid -\frac{7}{3} < x < \frac{1}{3}\}$.

EXAMPLE 4

Solve the compound inequality $2x - 5 > 3$ **and** $4 - \frac{1}{3}x < 1$.

SOLUTION   Solve each simple inequality.

$$2x - 5 > 3 \quad \textbf{and} \quad 4 - \frac{1}{3}x < 1$$

$$2x > 8 \qquad\qquad -\frac{1}{3}x < -3$$

$$x > 4 \qquad\qquad\quad x > 9$$

The graphs are shown below

$$x > 4 \qquad \xleftarrow{\quad}\overset{}{\underset{4}{\circ}}\!\!-\!\!\!\xrightarrow{\quad}$$

$$x > 9 \qquad \xleftarrow{\quad}\overset{}{\underset{9}{\circ}}\!\!-\!\!\!\xrightarrow{\quad}$$

The solution is the set of all values that satisfy both inequalities: all real numbers greater than 4 and also greater than 9.

$$\xleftarrow{\quad}\overset{}{\underset{9}{\circ}}\!\!-\!\!\!\xrightarrow{\quad}$$

Expressed in interval notation, this is $(9, \infty)$ and in set notation this would be $\{x \mid x > 9\}$.

EXAMPLE 5

Solve the compound inequality $4x - 1 < 3$ **and** $3x + 1 > 10$.

SOLUTION   Solve each simple inequality.

$$4x - 1 < 3 \quad \textbf{and} \quad 3x + 1 > 10$$

$$4x < 4 \qquad\qquad 3x > 9$$

$$x < 1 \qquad\qquad\quad x > 3$$

Express each inequality as a graph.

$$x < 1 \qquad \xleftarrow{\quad}\!\!-\!\!\overset{}{\underset{1}{\circ}}\xrightarrow{\quad}$$

$$x > 3 \qquad \xleftarrow{\quad}\overset{}{\underset{3}{\circ}}\!\!-\!\!\!\xrightarrow{\quad}$$

These two graphs have no common points of intersection. There are no values of x that are both less than one and greater than three. The solution is the empty set, $\varnothing$.

The word *or* in a disjunction indicates that only one of the inequalities needs to be true to make the statement true.

EXAMPLE 6

Solve the compound inequality $4x - 1 < 3$ **or** $3x + 1 \geq 10$.

SOLUTION   Solve each simple inequality.

$$4x - 1 < 3 \quad \textbf{or} \quad 3x + 1 \geq 10$$
$$4x < 4 \qquad\qquad 3x \geq 9$$
$$x < 1 \qquad\qquad x \geq 3$$

Graph each inequality.

$x < 1$

$x \geq 3$

The word **or** indicates that the solution can come from either of the inequalities; thus, we take the union of the two.

The interval notation would be

$$(-\infty, 1) \cup [3, \infty).$$

The set notation would be $\{x \mid x < 1 \text{ or } x \geq 3\}$.

EXAMPLE 7

Solve the compound inequality $\dfrac{1}{2}(6x - 2) + 1 > 3$ **or** $4\left(\dfrac{1}{4}x - 1\right) < 2$.

SOLUTION   Solve each inequality.

$$\frac{1}{2}(6x - 2) + 1 > 3 \quad \textbf{or} \quad 4\left(\frac{1}{4}x - 1\right) < 2$$
$$3x - 1 + 1 > 3 \qquad\qquad x - 4 < 2$$
$$3x > 3 \qquad\qquad x < 6$$
$$x > 1$$

The graph of each inequality would be

$x > 1$

$x < 6$

The union of these two inequalities would be every real number

Expressed in interval notation this would be $(-\infty, \infty)$ and the set would be $\mathbb{R}$.

▌ SOLVING ABSOLUTE-VALUE INEQUALITIES

Points to Remember

1. The fundamental definition of absolute value refers to the distance from 0 on the number line.
2. The technique for solving an absolute value inequality statement that is "less than/less than or equal to":

Solving Absolute Value Inequalities: |x| < k

If k is a positive number then $|x| < k$ is equivalent to

$-k < x < k$

3. The technique for solving an absolute value inequality statement that is "greater than/greater than or equal to":

Solving Absolute Value Inequalities: |x| > k

If k is a positive number then $|x| > k$ is equivalent to

$x > -k$ or $x < -k$

EXAMPLE 8

Solve $|3x - 1| < 8$ both analytically and graphically.

SOLUTION

Analytical Solution:

We write the inequality as a continued inequality: $|3x - 1| < 8$ is equivalent to $-8 < 3x - 1 < 8$ and solve for x.

$-8 < 3x - 1 < 8$

$-7 < 3x < 9$ Add 1 to all 3 parts.

$-\dfrac{7}{3} < x < 3$ Divide all 3 parts by 3.

Graphical Solution:

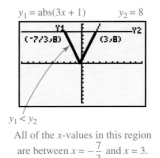

$y_1 = \text{abs}(3x + 1)$ $y_2 = 8$

$(-7/3, 8)$ $(3, 8)$

$y_1 < y_2$

All of the x-values in this region are between $x = -\dfrac{7}{3}$ and $x = 3$.

Any number between $-\frac{7}{3}$ and 3 is a solution. This is the interval $(-\frac{7}{3}, 3)$. The graph is shown in Figure 7-13.

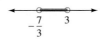

$-\dfrac{7}{3}$ 3

Figure 7-13

The set would be $\{x \mid -\frac{7}{3} < x < 3\}$.

TECHNOLOGY TIP

Because of the two points of intersection, the intersection points have been typed on the screen as ordered pairs rather than displayed at the bottom of the screen after using the INTERSECT feature twice.

EXAMPLE 9

Solve $|2x + 5| + 1 > 4$ both analytically and graphically.

SOLUTION   **Analytical Solution**

We begin by subtracting 1 from both sides of the inequality to isolate the absolute value.

$$|2x + 5| + 1 > 4$$
$$|2x + 5| > 3$$

$|2x + 5| > 3$ is equivalent to

$$2x + 5 > 3 \quad \textbf{or} \quad 2x + 5 < -3$$

Solve each simple inequality for x and determine the union of this **or** statement for the solution.

$$2x + 5 > 3 \qquad \textbf{or} \qquad 2x + 5 < -3$$
$$2x > -2 \qquad\qquad\qquad 2x < -8$$
$$x > -1 \qquad\qquad\qquad x < -4$$

Thus, $x < -4$ *or* $x > -1$, the union of which is

Expressed in interval notation this is $(-\infty, -4) \cup (-1, \infty)$. The set notation is $\{x \mid x < -4 \text{ or } x > -1\}$.

Graphical Solution

$$y1 = |2x + 5| + 1$$
$$y2 = 4$$

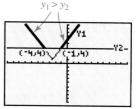

All of the x-values are less than $x = -4$ or greater than $x = -1$.

⚡ **WARNING!**   The expression involving the absolute value $|ax + b|$ must be isolated before applying either of the properties for solving absolute-value inequalities.

EXAMPLE 10

Solve the inequality $|6x - 1| \geq -3$.

SOLUTION   **Analytical Solution**

Because the absolute value of any number is nonnegative and because any nonnegative number is larger than -3, the inequality is true for all x. The solution set, $\mathbb{R}$, is the interval $(-\infty, \infty)$ whose graph appears in Figure 7-14.

Figure 7-14

Graphical Solution

$$y_1 = |6x - 1| \qquad y_2 = -3$$

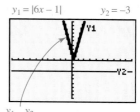

Every portion of the graph of y_1 is above the graph of y_2.

EXERCISE 7.5

VOCABULARY AND NOTATION   *In Exercises 1–10, fill in the blanks to make a true statement.*

1. An _____ is a mathematical statement that indicates that two quantities are not necessarily equal.

2. Notation used to describe the region that has been shaded on the number line graph is called _____ notation.

3. The number line graph is divided into regions by the _____.

4. _____ are compound inequalities that are linked by the word *or*.

5. _____ are compound inequalities that are linked by the word *and*.

6. Absolute-value expressions can never be less than _____.

7. The solution to the inequality $|x| > -1$ is _____.

8. The inequality $x < 4$ is equivalent to the interval notation _____.

9. The interval notation $[-2, 5)$ is equivalent to the inequality _____.

10. The number line graph 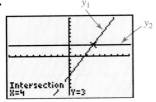 is equivalent to the interval notation _____.

CONCEPTS   *In Exercises 11–14, the solution to y1 = y2 can be determined from each of the following graphs. Based on the displayed information, determine a number line graph (if possible) of the solution to each of the indicated inequalities.*

11.

a. $y1 < y2$

b. $y1 \le y2$

c. $y1 > y2$

d. $y1 \ge y2$

12.

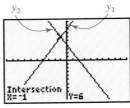

a. $y1 < y2$

b. $y1 \le y2$

c. $y1 > y2$

d. $y1 \ge y2$

13.

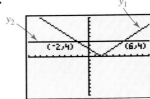

a. $y1 < y2$

b. $y1 \le y2$

c. $y1 > y2$

d. $y1 \ge y2$

14.

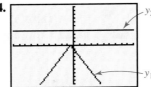

a. $y1 < y2$

b. $y1 \le y2$

c. $y1 > y2$

d. $y1 \ge y2$

PRACTICE   *In Exercises 15–34, solve each inequality and express the solution using interval notation, a number line graph, and set notation. Verify by solving graphically.*

15. $2 - x > 5$

16. $3 - x < 1$

17. $\dfrac{x}{4} \le -3$

18. $\dfrac{1}{4}x > -2$

19. $3x + 1 \le 13$

20. $2x - 3 \ge 7$

21. $\dfrac{x}{2} + 1 > 2$

22. $\dfrac{x}{3} + 2 < 5$

23. $3 - 2x \ge 17$

24. $1 - 6x \le 25$

25. $3x - 3 > 2x + 1$

26. $1 - x > 2x + 7$

27. $\dfrac{x}{2} + 1 < \dfrac{3}{2}$

28. $1 + \dfrac{x}{3} > \dfrac{2}{3}$

29. $1 - x \le 2 + x$

30. $x - 1 \le 2 - x$

31. $2(x - 3) < 3(x + 2)$

32. $3(3 + x) < 5(5 + x)$

33. $\dfrac{1}{2}x \le 2(x - 3)$

34. $x - \dfrac{3}{2} \le \dfrac{1}{2}(x + 2)$

In Exercises 35–56, solve each compound inequality and express the solution using interval notation, a number line graph, and set notation.

35. $-1 < x - 3 < 4$

36. $2 \le x + 5 < 6$

37. $-5 \le 2x + 1 \le 7$

38. $-2 < 3x - 2 \le 10$

39. $-1 < -2x < 5$

40. $-2 < 3 - x < 2$

41. $x + 2 \le -3 \text{ or } x + 2 \ge 3$

42. $x - 5 < -1 \text{ or } x - 5 > 1$

43. $2x - 5 < -3 \text{ or } 2x - 5 > 3$

44. $3x + 1 \le -5 \text{ or } 3x + 1 \ge 5$

45. $3x - 2 \le 6 \text{ or } 2x - 1 < 7$

46. $6 - x > -3 \text{ or } 4x - 3 < 11$

47. $2 - x < 8 \text{ or } 3x - 1 < -1$

48. $3 - x < 0 \text{ or } 2x - 3 \ge 5$

49. $x + 2 > -3 \text{ and } x + 2 < 3$

50. $x - 6 < -3 \text{ and } x + 5 > 1$

51. $4x + 6 \ge 6 \text{ and } 3x - 2 \le 4$

52. $5x - 3 > 10 \text{ and } 5 - 3x < -3$

53. $2x - 3 \le 5 \text{ and } -3x - 5 < 4$

54. $-3x - 7 > -1 \text{ and } 7 - 5x \le 12$

55. $6 - x \ge 2 \text{ and } x - 6 < 2$

56. $1 - 6x > 7 \text{ and } 1 - 2x < -1$

In Exercises 57–74, solve each inequality and express the solution using interval notation, a number line graph, and set notation. Verify by solving graphically.

57. $|x| + 2 < 5$

58. $|x| - 3 \le 1$

59. $|x + 2| < 5$

60. $|x - 3| \le 1$

61. $|3x - 2| < 4$

62. $|4 - 2x| < 5$

63. $|4 - x| - 3 \le -2$

64. $|3 - x| + 5 \le 2$

65. $|2x - 3| - 3 < 2$

66. $|5x + 4| + 1 < 2$

67. $|x + 3| \ge 2$

68. $|x - 5| \ge 1$

69. $|2x - 9| > 5$

70. $|5x - 4| > 6$

71. $|3x - 5| + 4 > 1$

72. $|4 - 5x| - 3 > -2$

73. $|2x + 1| + 2 \ge 3$

74. $|3x + 4| + 1 \ge 5$

REVIEW

75. Graph $y = 2x + 3$ and determine the x- and y-intercepts.

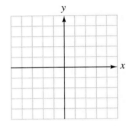

76. Solve $4x - 5y = 3$ for y and write the solution in $y = mx + b$ format to show that this is the equation of a linear function.

77. Is $(2, 3)$ an ordered-pair solution to the equation $4x - 5y = 3$?

78. Which of the following graphs represent the graph of a function?

a.

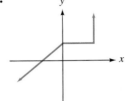

b.

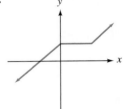

c.

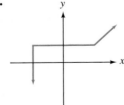

CHAPTER SUMMARY

CONCEPTS

REVIEW EXERCISES

Section 7.1

Points to Remember
1. Natural numbers can be used as exponents to indicate repeated multiplications. The exponent states the number of times the base is to be used as a factor.
2. A negative is not a part of a base unless grouping is used.
3. Any nonzero base raised to the zero power, x^0, is equal to 1.
4. Negative exponents are defined as:

$$x^{-n} = \frac{1}{x^n} \quad \text{and} \quad \frac{1}{x^{-n}} = x^n$$

5. The following rules define operations with expressions containing exponents.

    Product Rule: $x^m \cdot x^n = x^{m+n}$

    Power to a Power Rule:
    $(x^m)^n = x^{mn}$

    Product to a Power and Quotient to a Power:
    $(xy)^n = x^n y^n$
    and
    $\left(\dfrac{x}{y}\right)^n = \dfrac{x^n}{y^n}$

    Quotient Rule: $\dfrac{x^m}{x^n} = x^{m-n}$

6. Do not leave negative exponents in a result.
7. To simplify an expression means to remove grouping symbols and to combine like terms.
8. To evaluate an expression, substitute given values for the specified variables and then simplify the resulting arithmetic expression following the order of operations.

Review of Exponents and Polynomials

Simplify each expression. Use positive exponents in all results.

1. **a.** $(3x^3y^2)^2$
   b. $(-2x^3y^3)^3$

2. **a.** $(3x^2y)(-2x^3y^2)$
   b. $(6xy^3)(-2x^3y^{-1})$

3. **a.** $\left(\dfrac{2x^3}{y^{-3}}\right)^4$
   b. $\left(\dfrac{5x^{-2}}{y^{-3}}\right)^2$

4. **a.** $\dfrac{3^{-1}x^3y^0z^3}{x^{-2}yz^3}$
   b. $\dfrac{12x^5y^2z^{-1}}{4x^3x^5y^{-4}z^{-5}}$

Evaluate each expression below for the specified value of the variable(s). Use the STOre feature of your graphing calculator to confirm your results.

5. $4x^2 - 2x + 4; x = -4$

6. $-3x^2y + 4x - y^2; x = 3, y = -1$

7. Simplify each expression.
    a. $(x^2 + 3x - 1) + (2x^2 - x + 5)$
    b. $(2x^2 - 5x - 4) - (-x^2 + 2x - 1)$
    c. $4(x^2 - 3x + 2) - 5(3x^2 + 2x - 1)$

8. Simplify each expression.
    a. $(2x + 5)(2x - 5)$
    b. $(3x - 4)(3x + 2)$
    c. $(2x - 7)^2$

9. Simplify each expression.
    a. $\dfrac{5x^2y^3 - 10xy}{5x}$
    b. $\dfrac{16a^3b^2c - 4a^4b^3c^2}{2abc}$
    c. $\dfrac{8x^2 - 5x + 2}{2x}$

9. Use the commutative and associative properties and the product rule for exponents to multiply monomials.

10. To multiply a monomial by a polynomial, use the distributive property to multiply each term of the polynomial by the monomial. To multiply two polynomials, each term of one polynomial is multiplied by each term of the other polynomial. Like terms are then combined.

11. To divide polynomials, common factors may be divided out. To divide a polynomial by a monomial, divide each *term* of the polynomial by the monomial.

Section 7.2

Review of Factoring

Points to Remember

1. To factor means to write as a product. The expression $4(x + 5) + y(x + 5)$ is **not** expressed as a product; it is the sum of two terms $4(x + 5)$ and $y(x + 5)$.

2. Factoring out a negative 1 simply changes the sign of each term of the original polynomial.

3. The factors $(a + b)$ and $(b + a)$ are the same because the operation of addition is commutative.

4. The factors $(a - b)$ and $(b - a)$ are additive inverses (opposites). Thus $a - b = -(b - a)$.

5. The sum of squares, $x^2 + y^2$, does **not** factor over the set of integers. It is a prime polynomial.

Factoring Polynomials

1. Always check for and factor out the greatest common factor (GCF) first. If the leading coefficient is negative, factor out a -1 with/as the GCF.

10. Factor each polynomial completely.

a. $xy + 2y + 4x + 8$

b. $ab + bc + 3a + 3c$

c. $z^2 - 16$

d. $2x^4 - 98$

e. $y^2 + 21y + 20$

f. $-y^2 + 5y + 24$

g. $15x^2 - 57x - 12$

h. $x^3 + 343$

i. $x^2 + 121$

j. $6x^2 - 17x - 5$

2. Count the number of terms in the polynomial and then follow the given procedures for that specific type of polynomial as indicated in the review on page 415.

Section 7.3

Points to Remember

1. The addition and subtraction properties of equality allow you to add or subtract the same quantity from both sides of an equation.

2. The multiplication and division properties of equality allow you to multiply or divide both sides of an equation by the same nonzero quantity.

3. Equations involving absolute value will have

   a. two solution parts if the absolute value is equal to a positive constant.

   b. one solution if the absolute value is equal to 0.

   c. no solution if the absolute value is equal to a negative number.

4. When the equation involves a factorable polynomial, the zero product property allows you to set the factored polynomial equal to zero and then solve by setting each factor equal to zero.

5. When solving equations involving a rational expression, find the least common denominator for all rational expressions and multiply both sides of the equation by the LCD to clear the fractions.

Review of Equations

11. Solve each equation below algebraically and confirm results with either the TABLE or STOre feature of the graphing calculator.

   a. $2(x - 1) + 5 = \dfrac{1}{2}(2x - 3)$

   b. $\dfrac{x - 4}{5} + 1 = \dfrac{2x}{3}$

   c. $(x + 2)(x - 5) = (x - 4)(x - 1)$

   d. $2x^2 - x - 1 = 0$

   e. $6a^2 - 6 = 0$

   f. $x^3 + 5x^2 + 6x = 0$

   g. $|3x + 1| = 10$

   h. $\left|\dfrac{3}{2}x - 4\right| = -9$

   i. $\dfrac{4}{x} + \dfrac{5}{x} = 8$

   j. $\dfrac{x}{x + 2} + \dfrac{1}{x - 1} = 1$

12. Solve each equation for the indicated variable.

   a. $p = 2l + 2w;\ w$

   b. $4x - y = 2;\ y$

   c. $x + 2y = 6;\ y$

Section 7.4

Graphing and Its Applications Revisited

Solutions to two variable equations are graphed on a rectangular coordinate system. For example, the ordered pairs $(3, 2)$, $(2, 3)$, and $(1, 4)$ are only three of an infinite number of solutions to the equation $x + y = 5$. These three ordered pairs are graphed on the coordinate system and connected to form a straight-line graph. **All** solutions to the equation $x + y = 5$ can be found on this straight line.

To determine the x-intercept for the equation of a line, substitute 0 for y and solve for x.

To determine the y-intercept for the equation of a line, substitute 0 for x and solve for y.

The standard viewing window for a graphing calculator is $[-10, 10]$ by $[-10, 10]$.

A function is a collection of ordered pairs in which each x-value is assigned a unique y-value.

Both the INTERSECT and ROOT features can be used to graphically solve equations in one unknown.

13. Graph the linear equation using x- and y-intercepts.

a. $2x + 5y = 10$

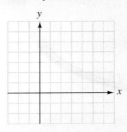

b. $3x - 4y = 12$

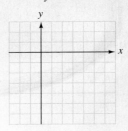

c. $3x - y = 6$

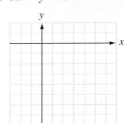

d. $5x + y = 7$

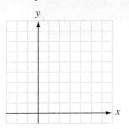

14. Graphically solve each equation using either the INTERSECT or ZERO/ROOT feature of the calculator. Begin in the standard viewing window and adjust the window when necessary.

a. $4(x - 1) = 8$ **b.** $3(x + 7) = -6$ **c.** $|3x + 1| = 10$

d. $\left|\dfrac{3}{2}x - 4\right| = 9$ **e.** $x^2 - 11x + 30 = 0$ **f.** $x^2 - 5x - 24 = 0$

15. Determine if the pictured graph is that of a function.

a.

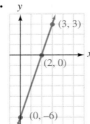

b.

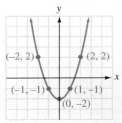

Section 7.5

Review of Inequalities

Points to Remember

1. The solution of an inequality is any number that makes the inequality a true statement.
2. A linear inequality is solved exactly like a linear equation

16. Solve each linear inequality and express the solution using interval notation, a number line graph, and in set notation. Verify results with the graphing calculator.

a. $2x - 7 \geq 7$

b. $4 - 3x > -2$

except, if you multiply or divide by a negative value, the inequality symbol is reversed.

3. The critical point on the number line graph is that number that divides the graph into the region of solution values versus non-solution values. Open circles indicate that the critical points are **not** included as a solution value and closed circles indicate that the critical points **are** included as solution values.

4. Interval notation is used to describe the shaded region on the number line graph.

5. Compound inequalities are inequalities that are linked by the words **and** or **or**.

6. When solving **and** statements (called conjunctions), you are looking for the intersection (common values) of the solutions to the component simple linear inequalities. These can be written in abbreviated form: "$x > 3$ **and** $x < 6$" becomes the **continued** inequality "$3 < x < 6$."

7. When solving **or** statements (called disjunctions), you are looking for the union (all represented values) of the solutions to the simple inequalities.

8. Solving absolute-value inequalities: $|x| < k$
   If k is a positive number, then $|x| < k$ is equivalent to $-k < x < k$

9. Solving absolute-value inequalities: $|x| > k$
   If k is a positive number, then $|x| > k$ is equivalent to $x < -k$ or $x > k$

c. $3(x + 3) \geq 4(x - 1)$

d. $x - \dfrac{3}{2} \leq \dfrac{1}{2}(x + 2)$

17. Solve each compound inequality and express the solution using interval notation, a number line graph, and set notation.
   a. $3 < 2x - 1 < 7$
   b. $x + 3 \geq 5$ and $-x - 2 \geq 7$
   c. $3x - 2 < 4$ or $5x - 1 \geq 14$
   d. $-2x - 1 > 5$ or $3x + 7 < -17$

18. Solve each absolute value inequality and express the solution using interval notation, a number line graph, and set notation.
   a. $|2x - 4| + 3 \geq 8$
   b. $|x + 2| < 5$
   c. $|3x - 4| + 6 \leq 2$
   d. $|2 - 3x| + 8 \geq 5$

CHAPTER TEST

In Problems 1 and 2, simplify, using the laws of exponents. Leave no parentheses or negative exponents in your final results.

1. $\dfrac{-12y^3(xz)^2}{18x^3(y^2z)^2}$ **2.** $\dfrac{x^2y^{-2}}{x^3z^2}$

Factor completely.

3. a. $a^3 - 125$
   b. $2x^2 - 32$
   c. $6x^2 + x - 2$

4. Simplify.
   a. $3(2x - 5) - 2(x - 4) + 6$
   b. $2x(x - 4) + 3x(x - 1) + 4x^2 - 8x$
   c. $(x - 4) - (x + 3) + 3x - 1$

5. Simplify.
   a. $(2x + 3)^2 - (x + 2)(x - 2)$
   b. $(x - 4)^2 - 3(x + 5)^2$
   c. $\dfrac{6x^2 - 9x + 12}{3x}$
   d. $\dfrac{2(2x - 6) - 4(3x + 1)}{8}$

In Problems 6 and 7, solve each equation and verify your results with your calculator.

6. $9(x + 4) + 4 = 4(x - 5)$

7. $\dfrac{y - 1}{5} + 2 = \dfrac{2y - 3}{3}$

8. Solve $A = p + prt$ for t.

9. Solve $3x - 7y = 8$ for y.

In Problems 10 and 11, write an equation to model the problem and solve either algebraically or graphically.

10. A 20-foot pipe is to be cut into three pieces. One piece is to be twice as long as another, and the third piece is to be six times as long as the shortest. Find the length of the longest piece.

11. A rectangle with a perimeter of 26 centimeters is 5 centimeters longer than it is wide. Find its area.

In Problems 12 and 13, solve each equation and verify results with the graphing calculator.

12. $|2x + 3| = 11$ **13.** $\dfrac{3}{5} + \dfrac{7}{x + 2} = 2$

In Problems 14 and 15, graph the given equation.

14. $3x - y = 6$ **15.** $y = 5$

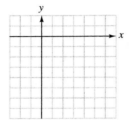   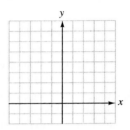

In Problems 16 and 17, use the graphical display to determine the solution to the equation or inequality.

16. a. $y_1 = y_2$
   b. $y_1 < y_2$
   c. $y_1 > y_2$

17. a. $y_1 = y_2$
   b. $y_1 < y_2$
   c. $y_1 > y_2$

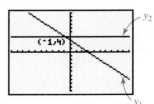   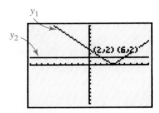

18. Determine if the pictured graph is a function or not:

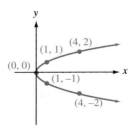

In Problems 19–25, solve each inequality and express the solution in interval notation and as a number line graph.

19. $\dfrac{1}{4}x - 2 \le -2$

20. $-2 \le 4 - x < 5$

21. $5 - 2x > 6$ or $2x + 2 > 1$

22. $4x + 1 < x$ and $2x - 5 < 1$

23. $|2x - 5| < 4$

24. $\left|\dfrac{1}{3}x - 1\right| \ge 5$

25. $\left|\dfrac{1}{2}x + 3\right| + 4 \le 2$

CAREERS & MATHEMATICS

© J. A. Giordano/CORBIS SABA

Statistician

Statisticians devise and carry out experiments and surveys and interpret the numerical results. They then apply their knowledge to subject areas such as economics, human behavior, natural science, and engineering.

Often, they are able to obtain accurate information about a group of people or things by surveying a small portion, called a **sample**, rather than the whole group. This technique requires statisticians to decide where and how to get the data, how to determine the type and size of the sample group, and how to develop the survey questionnaire or reporting form.

Qualifications

A bachelor's degree in statistics or mathematics is the minimum educational requirement for many beginning jobs in statistics. For other entry-level statistics jobs, however, it is preferable to have a bachelor's degree with a major in an applied field such as economics or natural science and a minor in statistics. Because computers are used extensively for statistical applications, a strong background in computer science is recommended.

Job Outlook

Employment opportunities for persons who combine training in statistics with knowledge of a field of application such as manufacturing, engineering, scientific research, or business are expected to be favorable through the year 2010.

Example Application

A researcher wishes to estimate the mean (average) property tax paid by homeowners living in the city. To do so, he decides to select a **random sample** of homeowners and compute the mean tax paid by the homeowners in that sample. How large must the sample be for the researcher to be 95% certain that his computed sample mean will be within $35 of the population mean — that is, within $35 of the average tax paid by *all* homeowners in the city? Assume that the standard deviation, σ, of all tax bills is known to be $120.

Solution

From elementary statistics, the researcher has the formula

$$\frac{3.84\sigma^2}{N} < E^2$$

where σ^2, called the **population variance**, is the square of the standard deviation, E is the maximum acceptable error, and N is the sample size. He substitutes 120 for σ and 35 for E in the previous formula and solves for N.

$$\frac{(3.84)(120^2)}{N} < 35^2$$

$$\frac{55{,}296}{N} < 1{,}225 \qquad \text{Simplify.}$$

$$55{,}296 < 1{,}225N \qquad \text{Multiply both sides by } N.$$

$$\frac{55{,}296}{1{,}225} < N \qquad \text{Divide both sides by 1,225.}$$

$$45.1 < N \qquad \text{Simplify.}$$

To be 95% certain that the sample mean will be within $35 of the population mean, the researcher must sample *more than* 45.1 homeowners: the sample must contain at least 46 homeowners.

EXERCISES

1. What sample size must the researcher choose to keep the error less than $20?

2. What must the sample size be to keep the error less than $10?

3. If the researcher cuts the acceptable error in half, by what factor is the sample size increased?

4. If the researcher decides to double the acceptable error, by what factor is the sample size decreased?

_____

(*Answers:* **1.** 139 **2.** 553 **3.** 4 **4.** $\frac{1}{4}$)

CHAPTER

8

Functions

© Owen Franken/CORBIS

InfoTrac Project

Do a keyword search on "linear relationships" and find the article "Storage characteristics and nutritive value changes in Bermudagrass hay as affected by moisture content and density of rectangular bales." According to the article, when the moisture level was 325 g, the visible mold was 3.73. At a moisture level of 178 g, the visible mold was 1.13. Using this information, develop an equation showing the relationship between moisture level and visible mold. Round the slope to the nearest thousandth. Use your equation to predict the mold level for a moisture level of 248 g. How does your prediction compare with the actual measurement of mold reported in the article for a moisture level of 248 g? According to the article, what effect did bale density have on nutritive value?

Complete this project after studying Section 8.4.

PERSPECTIVE

Graphs in Space

In an *xy*-coordinate system, graphs of equations containing the two variables *x* and *y* are lines or curves. Other equations have more than two variables, and graphing them often requires some ingenuity and perhaps the aid of a computer. Graphs of equations with the three variables *x*, *y*, and *z* are viewed in a three-dimensional coordinate system with three axes. The coordinates of points in a three-dimensional coordinate system are ordered triples (x, y, z). For example, the points $P(2, 3, 4)$ and $Q(-1, 2, 3)$ are plotted in Illustration 1.

Graphs of equations in three variables are not lines or curves, but flat planes or curved surfaces. Only the simplest of these equations can be conveniently graphed by hand; a computer provides the best images of others. The graph in Illustration 2 is called a paraboloid. Illustration 3 models a portion of the vibrating surface of a drum head.

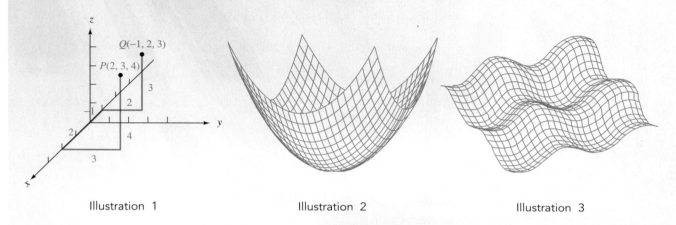

Illustration 1   Illustration 2   Illustration 3

Computer programs for producing three-dimensional graphs are readily available. Some of the more powerful are Maple, Mathematica, MathCad, and Derive. Perhaps your school has such a program available for student use. With a brief introduction on the program's use, you can easily create several interesting graphs.

8.1 RELATIONS AND FUNCTIONS

- RELATIONS
- FUNCTIONS
- VERTICAL LINE TEST

- FUNCTION NOTATION
- LINEAR FUNCTIONS

 ### RELATIONS

In a linear equation such as $y = 3x + 5$ we say that *y* is expressed in terms of *x*. The quantities *x* and *y* are related as an ordered pair (x, y) with *x* denoted as the independent variable and *y* as the dependent variable. A set of ordered pairs that establishes a correspondence between the numbers *x* and the values of *y* is called a **relation**.

Relation

A **relation** is a set of ordered pairs. The set of all values of x for which the relation is defined is called the **domain** of the relation. The set of all numbers, y, for which the relation is defined is called the **range** of the relation.

EXAMPLE 1

The following are examples of relations:

a. $A = \{(-2, 1), (0, 1), (1, 3), (4, 2)\}$

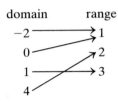

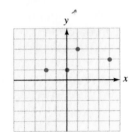

domain: $\{-2, 0, 1, 4\}$; range: $\{1, 2, 3\}$

b. $B = \{(4, 1), (4, -2), (4, 3)\}$

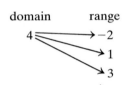

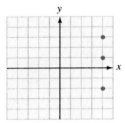

domain: $\{4\}$; range: $\{-2, 1, 3\}$

c. $y = 3x + 5$

domain = {all real numbers} (denoted symbolically as $\mathbb{R}$)

range = $\mathbb{R}$

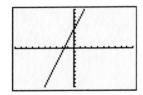

Remember: When finding ordered pair solutions to $y = 3x + 5$ you may select *any* real number for x, substituted into the equation, and produce a real number for y.

▎▎ FUNCTIONS

The relations illustrated in Example 1 differ in one critical way. The x-values in relation b are all the same, while the y-values are all different. However, in both relation a and relation c, each x-value corresponds to a unique value of y. When each x-value is paired with exactly one y-value, we call the relation a **function** and say that y is a function of x.

Function

A **function** is a relation that assigns to each number x exactly one value y.

The **domain** of the function is the set of all numbers x (the independent variable) for which the function is defined. The **range** is the set of all possible y-values (the dependent variable).

WARNING! A function is always a relation because it consists of ordered pairs. However, a relation is not necessarily a function because a relation places no restriction on the relationship between the *x*- and *y*-values.

EXAMPLE 2

Determine which of the following relations is also a function.

a. $\{(-4, 5), (2, 3), (3, 5), (-2, 1)\}$ **b.** $\{(2, 8), (-1, 3), (4, 1), (2, -6)\}$

c. $y = 2x - 1$ **d.** $y^2 = x$

e. You are paid \$5.75 per hour to sack groceries. If *P* represents your pay and *n* the number of hours worked, then $P = 5.75n$.

SOLUTION   **a.** Function; each *x* is assigned *exactly* one *y*-value.

domain   range

-4 → 1
-2 → 3
      → 5
2
3

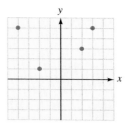

b. Not a function; when $x = 2$, $y = 8$ and $y = -6$.

domain   range

-1 → -6
2 → 1
4 → 3
    → 8

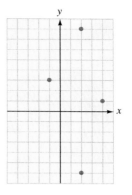

c. Function; each value selected for *x* will produce exactly one *y*-value.

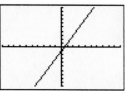

d. Not a function; for example, when $x = 4$, $y = 2$ or $y = -2$; when $x = 25$, $y = 5$ or $y = -5$.

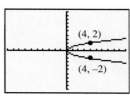

(4, 2)

(4, -2)

e. Function; for each number of hours that you work, *n*, you will receive exactly one amount of pay.

▓ VERTICAL LINE TEST

Examples 2b and 2d were not functions because there existed values of x corresponding to more than one value of y. When this happens, we can observe from the graph of the relation that the ordered pairs containing the same x-value lie in the path of a vertical line. Thus, the **vertical line test** can be used to determine whether the graph of an equation represents a function. If any vertical line intersects a graph in more than one point, the graph cannot represent a function, because a single x is paired with more than one value of y.

The graph in Figure 8-1(a) represents a function, because every vertical line that intersects the graph does so in exactly one point. However, the graph in Figure 8-1(b) does not represent a function, because some vertical lines intersect the graph in more than one point.

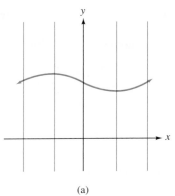

(a)

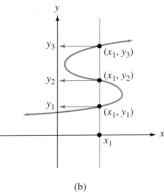

(b)

Figure 8-1

EXAMPLE 3

Use the vertical line test to determine which of the following relations is also a function.

a.

b.

c.

SOLUTION   **a.** Function

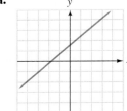

b. Not a function

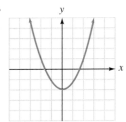

c. Function

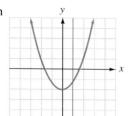

EXAMPLE 4

Determine the domain and range of each relation displayed.

a. 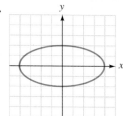   **b.** 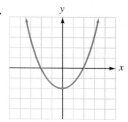   **c.**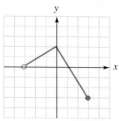

SOLUTION   **a.** 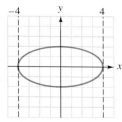   **b.** 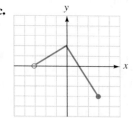   **c.**

Domain: $\{x \mid -4 \leq x \leq 4\}$ or $[-4, 4]$

Domain: $\mathbb{R}$ or $(-\infty, \infty)$

Domain: $(-3, 3]$ or $[x \mid -3 < x \leq 3]$

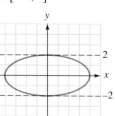   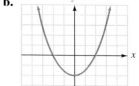   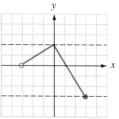

Range: $\{y \mid -2 \leq y \leq 2\}$ $[-2, 2]$

Range: $\{y \mid y \geq -2\}$ $[-2, \infty)$

Range: $\{y \mid -3 \leq y \leq 2\}$ or $[-3, 2]$

▌▌ FUNCTION NOTATION

When we say that y is a function of x, we are emphasizing that each x-value in the domain is assigned to a unique y-value in the range. The notation used to specify that y is a function of x is $y = f(x)$.

Function Notation

The notation $y = f(x)$ denotes that the variable y is a function of x.

The notation $f(x)$ is read as "f of x."

 WARNING! The notation $f(x)$ does not mean "f times x."

We use function notation to simplify English statements that are used to give algebraic directions. For example, "find y when $x = 2$ in the function $y = 2x + 1$" can be expressed as "if $f(x) = 2x + 1$, find $f(2)$." The variable x is replaced by the constant 2 in the algebraic expression $2x + 1$. Therefore, $f(2) = 2(2) + 1$ and $f(2) = 5$. This information could also be expressed as the ordered pair $(2, 5)$. As an ordered pair, it indicates a position in the rectangular coordinate system.

EXAMPLE 5

Let $f(x) = 4x + 3$. Find **a.** $f(3)$, **b.** $f(-1)$, **c.** $f(0)$, and **d.** $f(r)$.

SOLUTION **a.** We replace x with **3**:

$$f(x) = 4x + 3$$
$$f(3) = 4(3) + 3$$
$$= 12 + 3$$
$$= 15$$

The ordered pair is $(3, 15)$.

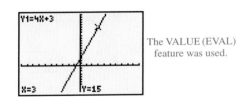

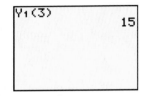

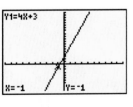

The VALUE (EVAL) feature was used.

b. We replace x with -1:

$$f(x) = 4x + 3$$
$$f(-1) = 4(-1) + 3$$
$$= -4 + 3$$
$$= -1$$

The ordered pair is $(-1, -1)$.

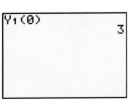

c. We replace x with **0**:

$$f(x) = 4x + 3$$
$$f(0) = 4(0) + 3$$
$$= 0 + 3$$
$$= 3$$

TECHNOLOGY TIP

TI-83 Plus: To use the Y(X) notation, a function must be entered on the y= screen. To evaluate the function for a given x-value, make sure you are on the home screen and press [VARS] [▶] to highlight the Y-VARS submenu. Press [1] to select the FUNCTION option. Select [1] for y1 if your equation was entered at y1 on the y= screen. Finally, enter the value of x in parentheses after the Y1 prompt on the home screen and press [ENTER].

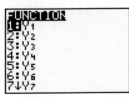

TI-86: To use the Y(X) notation, a function must be entered on the y= screen. To evaluate the function for a given x-value, make sure you are on the home screen and press [2nd] [ALPHA] [Y] [1] [( ] [the value of x] [ )] [ENTER].

The ordered pair is $(0, 3)$.

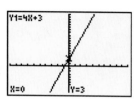

d. We replace x with r:

$$f(x) = 4x + 3$$
$$f(r) = 4(r) + 3$$
$$= 4r + 3$$

The letter f used in the notation $y = f(x)$ represents the word *function*. However, other letters can be used to represent functions. The notations $y = g(x)$ and $y = h(x)$ also denote functions involving the variable x.

EXAMPLE 6   If $g(x) = x^2 - 2x$, find   **a.** $g\left(\dfrac{2}{5}\right)$ **b.** $g(s)$ **c.** $g(s - 1)$ **d.** $g(-t)$.

SOLUTION   **a.** We replace x with $\dfrac{2}{5}$.

$$g(x) = x^2 - 2x$$
$$g\left(\frac{2}{5}\right) = \left(\frac{2}{5}\right)^2 - 2\left(\frac{2}{5}\right)$$
$$= \frac{4}{25} - \frac{4}{5}$$
$$= -\frac{16}{25}$$

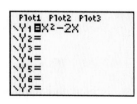

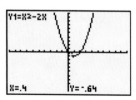

$g\left(\frac{2}{5}\right) = -\frac{16}{25}$ means the ordered pair $\left(\frac{2}{5}, -\frac{16}{25}\right)$ is a point on the graph of $g(x) = x^2 - 2x$.

b. We replace x with s:

$$g(x) = x^2 - 2x$$
$$g(s) = s^2 - 2s$$

c. We replace x with $s - 1$:

$$g(x) = x^2 - 2x$$
$$g(s - 1) = (s - 1)^2 - 2(s - 1)$$
$$= (s^2 - 2s + 1) - 2s + 2$$
$$= s^2 - 4s + 3$$

d. We replace x with $-t$:

$$g(x) = x^2 - 2x$$
$$g(-t) = (-t)^2 - 2(-t)$$
$$= t^2 + 2t$$

EXAMPLE 7

Let $f(x) = 4x - 1$. Find **a.** $f(3) + f(2)$ and **b.** $f(a) - f(b)$.

SOLUTION **a.** We find $f(3)$ and $f(2)$ separately.

$$f(x) = 4x - 1 \qquad\qquad f(x) = 4x - 1$$
$$f(3) = 4(3) - 1 \qquad\quad f(2) = 4(2) - 1$$
$$= 12 - 1 \qquad\qquad\quad = 8 - 1$$
$$= 11 \qquad\qquad\qquad\quad = 7$$

We then add the results to obtain

$$f(3) + f(2) = 11 + 7$$
$$= 18$$

b. We find $f(a)$ and $f(b)$ separately.

$$f(x) = 4x - 1 \qquad\qquad f(x) = 4x - 1$$
$$f(a) = 4a - 1 \qquad\qquad f(b) = 4b - 1$$

We then subtract the results to obtain

$$f(a) - f(b) = (4a - 1) - (4b - 1)$$
$$= 4a - 1 - 4b + 1 \qquad \text{Remove parentheses.}$$
$$= 4a - 4b \qquad\qquad\quad \text{Combine like terms.}$$

▍▍ LINEAR FUNCTIONS

In Chapters 3 and 7 we determined ordered pair solutions for linear equations in x and y. These solutions were then graphed, and the graph of the equation was determined to be a straight line. All straight-line graphs, except for vertical lines, are linear functions.

Linear Functions
A **linear function** is a function defined by an equation that can be written in the form

$$y = mx + b \qquad \text{or} \qquad f(x) = mx + b$$

EXAMPLE 8

Solve the equation $3x + 2y = 10$ for y to show that it defines a linear function. Then graph it and state its domain and range. (See Figure 8-2.)

SOLUTION We solve the equation for y as follows:

$$3x + 2y = 10$$
$$2y = -3x + 10 \qquad \text{Subtract } 3x \text{ from both sides.}$$
$$y = -\frac{3}{2}x + 5 \qquad \text{Divide both sides by 2.}$$

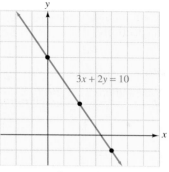

$3x + 2y = 10$

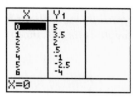

Figure 8-2

Domain: The set of all real numbers, $\mathbb{R}$.

Justification: If any x-value is selected on the x-axis, a corresponding y-value can be determined. Likewise, any x-value substituted into the function will yield exactly one y-value because the only two operations applied to the x-value are multiplication and addition.

Range: The set of all real numbers, $\mathbb{R}$.

Justification: If any y-value is selected on the y-axis, a corresponding x-value can be determined.

EXERCISE 8.1

VOCABULARY AND NOTATION *In Exercises 1–7, fill in the blanks to make the given statements true.*

1. A set of ordered pairs that establishes a correspondence between numbers x and y is called a _____.

2. The _____ is the set of all values of x for which a relation is defined.

3. The _____ is the set of all values of y for which a relation is defined.

4. A relation in which each x-value is paired with exactly one y-value is called a _____.

5. In a linear equation, solutions are expressed as _____ of numbers (x, y).

6. The notation $f(x)$ is read as "_____" and does not mean "f times x."

7. In the function defined as $f(x) = 3x - 5$, "$f(x)$" corresponds with the ___-coordinate.

CONCEPTS

In Exercises 8–11, use the given description to write a set of ordered pairs. State the domain and the range of the relation.

8. The first coordinate is a natural number between 2 and 7. The second coordinate is double the first coordinate.

9. The first coordinate is an integer between -4 and 2. The second coordinate is 3 less than the first coordinate.

10. The first coordinate is the number of hours worked in 4 consecutive weeks: 15, 25, 30, and 25. The second coordinate is weekly pay at $8.50 per hour, respectively.

11. The first coordinate is the rate of speed driven for 3 hours: 50 mph, 60 mph, and 70 mph. The second coordinate is the distance traveled, respectively.

12.

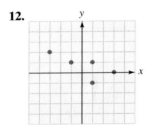

a. List the ordered pairs displayed.

b. State the domain of the relation.

c. State the range of the relation.

d. Is the pictured relation a function? Why or why not?

In Exercises 13–16, complete the given tables.

13.

x	−3	−2	0	1	4
$f(x) = 2x - 1$					
(x, y)					

14.

x	−3	−2	−1	0	1
$f(x) = x^2 - 1$					
(x, y)					

15.

x	−8	−6	−4	2	6
$f(x) = \lvert x + 1 \rvert$					
(x, y)					

16.

x	−8	−6	−4	2	6
$f(x) = \lvert x \rvert + 1$					
(x, y)					

17.

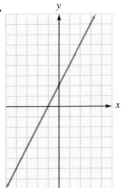

Use the graph to find each of the following:

a. $f(-4)$ **b.** $f(-2)$

c. $f(0)$ **d.** $f(3)$

18. The graph of a function is pictured, with coordinates of a point displayed at the bottom of the graph screen. Find the value of each function specified below.

a.

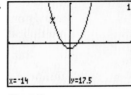

$f(-14) =$ _____

b.

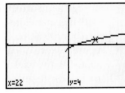

$f(22) =$ ___

c.

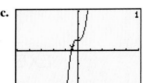

$f(-5) =$ _____

d.

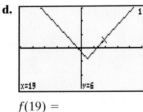

$f(19) =$ ___

PRACTICE   *In Exercises 19–22, determine which relations represent functions. Determine the domain and range.*

19. $\{(2, 5), (3, -1), (-4, 7)\}$

20. $\{(5, -1), (7, -1), (3, -1)\}$

21. $\{(-1, 2), (-2, 3), (-1, -2)\}$

22. $\{(5, 7), (5, 8), (5, 9)\}$

In Exercises 23–30, tell whether each equation determines y to be a function of x.

23. $y = 2x + 3$

24. $y = 4x - 1$

25. $y = 3x^2$

26. $y = 3 + 7x^2$

27. $y^2 = x + 1$

28. $y^2 = 3 - 2x$

29. $y = |x|$

30. $x = |y|$

In Exercises 31–34, display the graph of each function in the Zinteger viewing window. Find the domain and the range.

31. $y = -x^2 + 5$

32. $y = (x - 10)^2$

33. $y = |x - 15| - 10$

34. $y = -|x + 20|$

In Exercises 35–42, determine if the relation displayed is also a function, and determine the domain and range.

35.

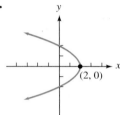

36.

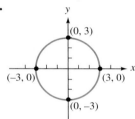

37.

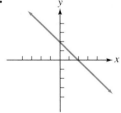

38.

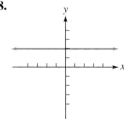

39.

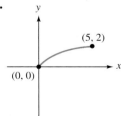

40.

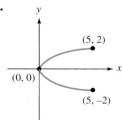

41.

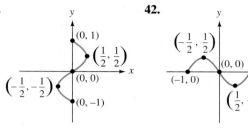

42.

In Exercises 43–46, determine if the relation indicated by the table of values is also a function.

43.

x	y
4	1
3	2
−1	1
4	−1

44.

x	y
−1	5
2	−4
−2	−4
3	7

45.

x	y
−2	3
4	3
−1	3
0	3

46.

x	y
3	−2
3	4
3	−1
3	0

In Exercises 47–52, find   a. $f(2)$ b. $f(-3)$ c. $f(a)$ d. $f(a + 2)$.

47. $f(x) = 3x + 2$

48. $f(x) = -4x + 1$

49. $f(x) = x^2 - 2x$

50. $f(x) = x^2 + x - 3$

51. $f(x) = \dfrac{3}{x - 4}$

52. $f(x) = \dfrac{x}{x^2 + 2}$

In Exercises 53–62, the function $y = f(x)$ is displayed in either graphic or tabular form. Determine $f(x)$.

53. $f(-4) = ?$

54. $f(0) = ?$

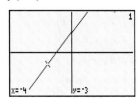

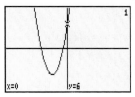

55. $f(-3) = ?$ **56.** $f(-2) = ?$

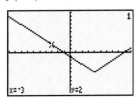   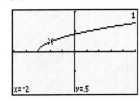

57. $f(3) = ?$

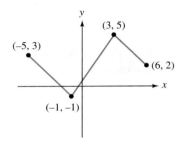

58. $f(8) = ?$

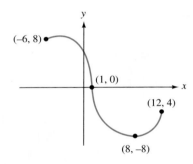

59. $f(-4) = ?$ **60.** $f(5) = ?$

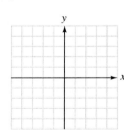

61. $f(-3) = ?$ **62.** $f(-7) = ?$

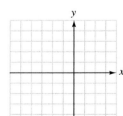

In Exercises 63–70, $f(x) = 2x + 1$. Find each of the following.

63. $f(3) + f(2)$ **64.** $f(1) - f(-1)$
65. $f(b) - f(a)$ **66.** $f(b) + f(a)$
67. $f(b) - 1$ **68.** $f(b) - f(1)$
69. $f(0) - f(-\frac{1}{2})$ **70.** $f(a) + f(2a)$

In Exercises 71–74, use your calculator to evaluate each of the given functions.

71. $g(x) = 3x - 1$ and $t(x) = 3x^2 - 2x + 5$
   a. $g(-3.6)$ **b.** $g(\frac{1}{2})$
   c. $t(8.1)$ **d.** $t(\frac{2}{3})$

72. $h(x) = \frac{2}{3}x - 5$ and $p(x) = \frac{1}{2}x^2 - 4x + 3$
   a. $h(12)$ **b.** $h(6)$
   c. $p(3)$ **d.** $p(100)$

73. $h(x) = \sqrt{2x - 4}$ and $q(x) = \frac{x^2 + x}{2}$
   a. $h(10)$ **b.** $h(20)$
   c. $q(26)$ **d.** $q(20)$

74. $f(x) = 2x^4 - 3x^3 - 2x^2 + 5x - 1$
and
$g(x) = 5x^4 - 3x^3 - 8$
   a. $f(-9)$ **b.** $f(1)$
   c. $g(-8)$ **d.** $g(-1)$

In Exercises 75–80, determine which equation defines a linear function. If the equation is a linear function, write it in $y = mx + b$ form and graph the equation.

75. $y = 3x^2 + 2$ **76.** $y = \dfrac{x - 3}{2}$

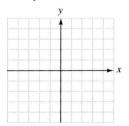

77. $x = 3y - 4$ **78.** $x = \dfrac{8}{y}$

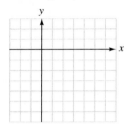

79. $4x - 2y = 8$

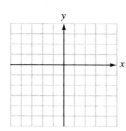

80. $6x - y = 2$

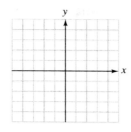

REVIEW

REVIEW

81. Is 5 part of the solution of the inequality $2x + 4 \geq 9$? Justify your response.

82. Simplify the expression $5(3x + 4y)$.

83. Evaluate the expression $3x + 2y$ when $x = -2$ and $y = 5$.

84. Is the ordered pair $(-2, 5)$ a solution of the equation $3x + 2y = 4$? Justify your response.

85. Is -7 a solution of the equation $x^2 - 49 = 0$? Justify your response.

8.2 FUNCTIONS AND GRAPHS

- GRAPH ANALYSIS
- APPLICATIONS

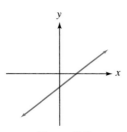

Figure 8-3

▎▎▎ GRAPH ANALYSIS

To analyze a graph means to look at the relation or function and how it behaves. The person analyzing the graph should:

1. obtain a clear picture of the graph by ensuring that all x- and y-intercepts are displayed, as well as any high or low points,
2. determine the domain, and
3. determine the range.

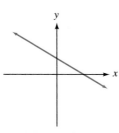

Figure 8-4

When we describe a graph, we must agree to examine the graph from left to right. For example, the graph of the linear function (Figure 8-3) is said to be an **increasing function** because as the x-value increases, the y-value also increases. Observe that the graph rises from left to right. Conversely, a **decreasing function** (Figure 8-4) is one whose graph falls from left to right, and as the x-values increase, the y-values decrease. All functions are described in terms of increasing values of x because we are reading the graph from left to right. Notice that whether a function is increasing or decreasing depends on the y-values (the values of the function).

EXAMPLE 1

Graph $y = 3\sqrt{x + 5}$ in the ZInteger viewing window. Is the function increasing or decreasing?

SOLUTION TRACE along the function from left to right. As x increases, so does y. We say the function is increasing on the interval $[-5, \infty)$.

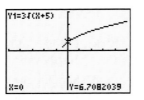
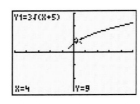
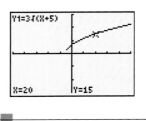

As we analyze graphs, the following definitions will be helpful.

Basic Definitions

x-intercept: a point where the graph intersects the x-axis; $y = 0$, therefore the coordinates are of the form $(x, 0)$

y-intercept: a point where the graph intersects the y-axis; $x = 0$, therefore the coordinates are of the form $(0, y)$

Absolute maximum: the y-coordinate of the highest point of the function

Relative maximum: the y-coordinate of the highest point of the function within an interval (neighborhood) of x-values

Absolute minimum: the y-coordinate of the lowest point of the function

Relative minimum: the y-coordinate of the lowest point of the function within an interval (neighborhood) of x-values

Turning point: a point where the graph changes from increasing to decreasing or decreasing to increasing

EXAMPLE 2

Determine the intervals where the function is increasing and where it is decreasing in Figure 8-5.

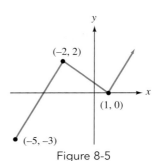

Figure 8-5

SOLUTION   Remember to read the graph from left to right. The function is increasing on the interval $[-5, -2]$ because as the x-value increases from -5 to -2, the y-value increases from -3 to $+2$. The graph also increases on the interval $[1, \infty)$ because as the x-values increase, so do the y-values. The function is decreasing on the interval $[-2, 1]$ because as the x-value increases from -2 to $+1$, the y-value decreases from $+2$ to 0.

The points $(-2, 2)$ and $(1, 0)$ are the turning points of the graph because it changes from increasing to decreasing at the point $(-2, 2)$ and from decreasing to increasing at the point $(1, 0)$. The relative maximum is 2 and the relative minimum is 0. There is no absolute maximum and the absolute minimum is -3.

EXAMPLE 3

Find the turning point (absolute minimum) for the graph of the function $f(x) = x^2 - 2x + 4$.

SOLUTION

Graph $y1 = x^2 - 2x + 4$. Under your CALC menu, access the "minimum" feature and follow the prompts. Keep in mind that the prompts on your calculator are referencing x-values of the function. Thus, "lower bound" or "left bound" indicates an x-value that is lower, or to the left of, the desired x-coordinate.

TECHNOLOGY TIP

TI-83 Plus: To find the coordinates of a minimum, at the graphical display press [2nd] [CALC] [3:minimum]. At the LEFT BOUND prompt, enter an x-value smaller than the x-value of the displayed minimum and press [ENTER]. At the RIGHT BOUND prompt, enter an x-value larger than the x-value of the displayed minimum and press [ENTER]. At the GUESS prompt, you may guess by moving the cursor to the point on the graph that appears to be the desired minimum; press [ENTER]. The calculator will then display the coordinates of the lowest point of the graph. (Watch out for round-off error. See Technology 7.4, page 447.) To find a maximum, choose option 4 from the CALC menu and proceed as above.

TI-86: To find the coordinates of a minimum, at the graphical display (with one line of menu displayed) press [MORE] [F1] (MATH) and then [F4] (FMIN). At the LEFT BOUND prompt, enter an x-value smaller than the x-value of the displayed minimum and press [ENTER]. At the RIGHT BOUND prompt, enter an x-value larger than the x-value of the displayed minimum and press [ENTER]. At the GUESS prompt, you may guess by moving the cursor to the point on the graph that appears to be the desired minimum; press [ENTER]. The calculator will then display the coordinates of the lowest point of the graph. (Watch out for round-off error. See Technology 7.4, page 447.) To find a maximum, choose the FMAX option at the F5 position and proceed as above.

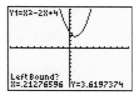

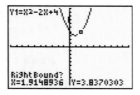

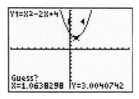

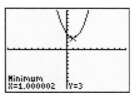

We believe the minimum occurs when $x = 1$. We suspect that round-off error has occurred. To confirm the minimum of 1, use the VALUE (EVAL) feature to confirm that when $x = 1$, $y = 3$. (Other options that could have been used would include the TABLE, the VARS feature, or the STOre feature.) Therefore, the minimum is 3.

EXAMPLE 4

Graph the function $y = 3\sqrt{x + 4}$ and

a. State the coordinates of the x- and y-intercepts (if any).

b. State the domain of the function.

c. State the range of the function.

d. Is the function increasing or decreasing?

e. What is the absolute maximum?

f. What is the absolute minimum?

SOLUTION   Graph the function in the ZInteger viewing window.

a. To find the y-intercept, we press TRACE and its coordinates will be displayed most of the time (without having to move the cursor). If the y-intercept is not displayed on your calculator, trace to it. (Remember, $x = 0$.) The coordinates are $(0, 6)$. Now trace to the x-intercept. (Remember, the y-coordinate will equal 0.) The coordinates are $(-4, 0)$.

b. We determine by tracing that the domain of the function appears to be $[-4, \infty)$.

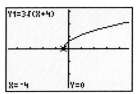   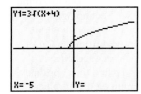

Notice that when $x = -5$ there
is no y-value displayed, nor does
the cursor appear on the graph.

Substituting any value between -4 and -5 (such as -4.5) for x gives the square root of a negative number, which is not defined in the real number system.

c. The range of the function appears to be $[0, \infty)$. Examination of the function, $y = 3\sqrt{x + 4}$, reveals that y cannot be negative.

d. Tracing from left to right, as well as observation of the graphical display, reveals that as x-values increase, the y-values also increase. Therefore, the function is increasing.

e. The graph has no absolute maximum. (As x increases, y increases without bound.)

f. The graph has an absolute minimum. The x-intercept of $(-4, 0)$ appears to be the lowest point. We say that 0 is the minimum.

▌ APPLICATIONS

EXAMPLE 5

Suppose that the ball shown in Figure 8.6 is thrown straight up with a velocity of 128 feet per second. The function $f(x) = 128x - 16x^2$ gives the relation between the height, $f(x)$, of the ball and the time, x, as measured in seconds.

a. Graph this function in an appropriate viewing window. Using the TRACE feature to assist you, describe how the time and height are related.

b. What are the coordinates of the turning point of the graph?

c. What is the absolute maximum?

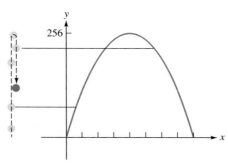

Figure 8-6

SOLUTION

a. Tracing was used to determine an appropriate viewing window of $[-5, 10]$ by $[-10, 275]$. (Remember, you must be able to see x- and y-intercepts and critical features such as turning points.) Tracing from $x = 0$ to $x = 8$, we learn that as the seconds increase, so does the height until we reach the turning point. Then the function becomes decreasing because as the seconds *increase* the height *decreases*.

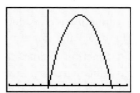

TECHNOLOGY TIP

You may want to review Section 3.5 for more information on setting appropriate viewing windows.

b. The coordinate of the turning point is (4, 256). The maximum feature under the CALC menu was used.

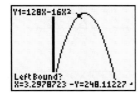

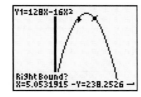

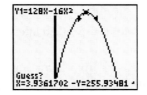

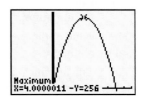

c. The absolute maximum is 256.

Again, verify through the TABLE or other feature of the calculator that the actual maximum is (4, 256). Interpreted within the context of the problem, this means that the ball will reach its maximum height of 256 feet in 4 seconds.

EXAMPLE 6

If x represents time and y represents location, the graph in Figure 8-7 can be used to illustrate a person walking through the mall. Describe what is happening from point A to B, B to C, C to D, and D to E.

SOLUTION

A to B: The person is walking briskly because location has changed dramatically in a short amount of time.

B to C: Time is passing, but the location has not changed. The person could be talking with someone, looking in a window, or resting on a bench.

C to D: Location has changed. Moreover, the person has retraced previous steps to go back for something.

D to E: The person is again moving forward, but much more slowly.

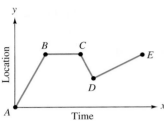

Figure 8-7

EXERCISE 8.2

VOCABULARY AND NOTATION *In Exercises 1–8, fill in the blanks to make the given statements true.*

1. A function is considered to be _____ if an increase in the x-value corresponds to an increase in the y-value.

2. A function is considered to be _____ if, as the x-value increases, the y-value decreases.

3. When looking at graphs, you should scan them from _____ to _____.

4. The y-value of the highest point of a graph is called its _____ maximum.

5. The y-value of the highest point of a graph within a specified interval of x-values is a _____ maximum.

6. The y-value of the lowest point on a graph is called an _____ minimum.

7. The y-value of the lowest point within a specified interval of x-values is a _____ minimum.

8. The point at which a graph changes from increasing to decreasing or decreasing to increasing is called a _____ point.

CONCEPTS

In Exercises 9–16, determine if the function is increasing or decreasing and over what interval it is increasing or decreasing. The viewing window is the standard viewing window.

9.
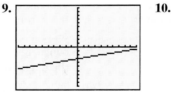

10.
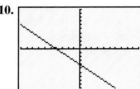

11.
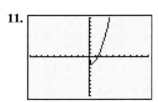

12.
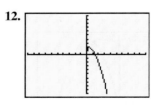

13.
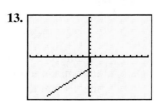

14.
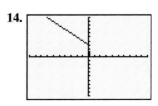

15.

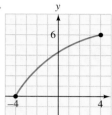

16.

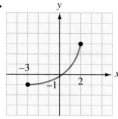

25. $f(x) = x^3 - 3x + 2$

26. $f(x) = x^3 - 3x - 4$

In Exercises 27–32, use Illustration 1 to respond to the questions.

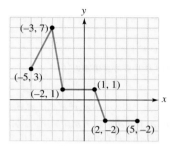

Illustration 1

PRACTICE   *In Exercises 17 and 18, determine the interval(s) where the function is (a) increasing, (b) decreasing, and (c) neither (constant).*

17.

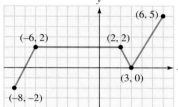

18.

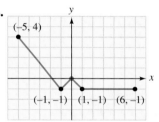

27. For what interval(s) of x is the graph
   a. increasing?
   b. decreasing?
   c. constant?

28. Is the graph a function? Why or why not?

29. a. What is the domain?   **b.** What is the range?

30. On the interval $[-5, -2]$,
   a. What are the coordinates of the turning point?

   b. Is it an absolute maximum or a relative maximum?

31. Why can you *not* determine $f(-8)$?

32. Determine $f(5)$.

In Exercises 19–22, graph each function and determine the turning point using the max or min feature of your calculator. Verify with the TABLE to avoid round-off errors.

19. $y = x^2 - 2x + 5$

20. $y = 5x^2 - 2x - 4$

21. $y = 4 - |x + 2|$

22. $y = 6 - |x - 5|$

In Exercises 23–26, use your calculator to determine the relative maximum and relative minimum of each function. As applicable, round your result to the nearest tenth.

23. $f(x) = x^3 - 3x$

24. $f(x) = x^3 - 5x$

In Exercises 33–38, use Illustration 2 to respond to the questions.

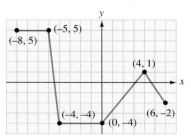

Illustration 2

33. Is the graph a function: Why or why not?

34. For what intervals of x is the graph

   a. increasing?

   b. decreasing?

   c. constant?

35. On the interval $[0, 6]$, what are the coordinates of the turning point? Is it a relative maximum or minimum?

36. What is the domain? What is the range?

37. Determine $f(-5)$.

38. Why can you *not* determine $f(7)$?

*In Exercises 39–44, the graph of a function is pictured. Use the graph to find **a.** the domain and range; **b.** the intervals on which it is increasing or decreasing; **c.** relative and absolute maximum(s) and relative and absolute minimum(s), if any.*

39.

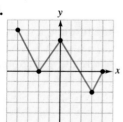

40.

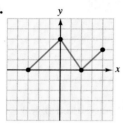

41.

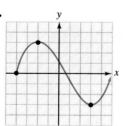

42.

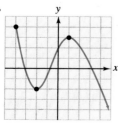

43.

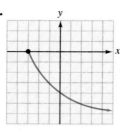

44.

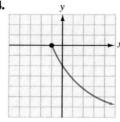

In Exercises 45–50, state the coordinates of the x- and y-intercepts (if any) for each specified function.

45. $f(x) = 3x + 2$

46. $f(x) = 2x - 1$

47. $f(x) = (x - 2)^2 + 1$

48. $f(x) = (x + 2)^2 + 1$

49. $f(x) = \sqrt{x + 4}$

50. $f(x) = \sqrt{x + 1}$

51. The function displayed in Illustration 3 describes the water level, y, in an aquarium over a period of time, x. In complete sentences, write a scenario that could explain the changes in water level.

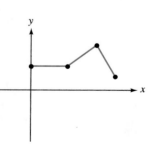

Illustration 3

52. The function in Illustration 4 describes the height, y, over a period of time, x, of a ball that a child has thrown into the air. In complete sentences, write a scenario that could explain what has happened to the ball.

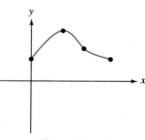

Illustration 4

APPLICATIONS *In Exercises 53–55, use your graphing calculator to respond to the questions.*

53. A traveling circus has a human cannonball act as its grand finale. The equation $y = -.01x^2 + .64x + 9.76$, where y = height in feet and x = horizontal distance traveled in feet, represents the flight path of the human cannonball. Display a graphical representation of this equation and use the TRACE feature to answer the questions. (ZInteger is the preferred viewing window.)

a. What do the x-values represent?

b. What do the y-values represent?

c. The curve represents the flight path of the human cannonball. TRACE along the path to the right. How far, *approximately*, has the human cannonball traveled horizontally when he hits the ground?

d. The human cannonball is traveling at speeds up to 65 mph. To land on the ground would mean certain death. If he uses a net for his landing, how far will he have traveled horizontally if the net is 11 feet above the ground?

e. What is the maximum height that he reaches during the course of his flight?

f. How far has he traveled horizontally when he reaches this maximum height?

g. The human cannonball is shot out of the cannon head first, so all of the distances are measured from his head. TRACE along the curve to $x = 0$ and $y = 9.76$. Explain the meaning of these two values.

54. Bridges are often supported by arches in the shape of a parabola. The equation $y = \frac{10}{7}x - \frac{2}{49}x^2$, where y = height and x = distance from the base of the arch, provides a model for a specific parabolic arch that supports a bridge.

a. What do the x-intercepts represent? (*Hint:* Your calculator has a ROOT or ZERO feature that should be used to determine the x-intercept.)

b. What is the highest point under the arch?

c. If an average vehicle is no more than 6 feet high, can it drive under the arch? Explain.

d. If a two-lane road is 20 feet wide and has 2 shoulders of width 5 feet each, will it fit between the bases of the arch? Explain.

e. Can an average-height vehicle drive in either lane and not scrape the paint off the roof of the car? That is to say, if this 20-foot-wide road is centered under the arch, is the arch at least 6 feet above the ground at all points in its width?

55. The local community theater is considering increasing the price of its tickets to cover increases in costuming and stage effects. They must be careful because a ticket price that is too low will mean that expenses are not covered, and yet a ticket price that is too high will discourage people from attending. They estimate the total profit, y, by the formula $y = -x^2 + 35x - 150$, where x is the cost of the ticket.

a. What is the amount that should be charged for a ticket to maximize the profit?

b. What is the maximum profit?

c. If $13 is charged for each ticket, what will the profit be?

d. If they predict a profit of $150.00 on a play, how much was charged per ticket? Scroll through the y-values in the TABLE to answer the question.)

REVIEW

56. Solve the inequality $3(x + 5) - 8 \le 5x$ and specify your solution

a. as a number line graph.

b. using interval notation.

57. Describe how you would find the x- and y-intercepts of an equation in two variables both algebraically and graphically.

58. Solve the equation $3x^2 - 10x - 8 = 0$ both algebraically and graphically.

59. Graph the function $y = |x + 2| - 3$ by plotting its x-intercepts and turning point. (*Suggestion:* Use your calculator's features to help you determine these points.)

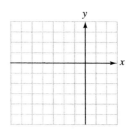

60. Solve the equation $|x + 2| - 3 = 5$ both algebraically and graphically.

8.3 SLOPE OF A NONVERTICAL LINE

- INTRODUCTION/OVERVIEW
- SLOPES OF HORIZONTAL AND
  VERTICAL LINES

- SLOPES OF PARALLEL AND
  PERPENDICULAR LINES
- INTERPRETATION OF SLOPE

▌▌ INTRODUCTION/OVERVIEW

Throughout this text you have been presented with equations of various forms and asked to interpret data that related to the equation. In order to successfully apply what you have learned to real-life situations, you must also be able to take specific data and construct the equation. This section and the next will help you develop the tools necessary for writing the most fundamental equation: the equation of a line.

Consider the following chart that relates the number of pounds of apples to the purchase price:

Pounds	1	2	3	4	5
Cost	$0.40	$0.80	$1.20	$1.60	$2.00

We can generalize this to: x pounds cost $0.40x$. Thus the equation modeling this information is $y = 0.40x$, where y is the cost and x is the number of pounds of apples.

Graph this equation in the ZInteger viewing window and TRACE along the line. As x increases, so does y.

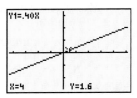 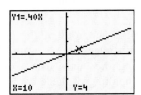 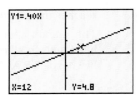

Thus the function is increasing. Note also that for each change of 1 unit in x (i.e., 1 pound of apples), there is a 0.4 change in y (an increase of $0.40 per pound). This comparison ratio of

$$\frac{\text{change in } y}{\text{change in } x}$$

determines the *rate* of increase in an increasing linear function and also the rate of decrease in a decreasing linear function. In linear functions, this rate is called the **slope** of the line. The "change in y-coordinates" is often called Δy (delta y) and the "change in x-coordinates" is often called Δx.

EXAMPLE 1

Graph $y = \dfrac{3}{4}x$ in the ZInteger viewing window. By tracing, determine Δy (change in y-coordinates) and Δx (change in x-coordinates) and compute the slope of the line.

SOLUTION   As we trace from $(0, 0)$ to $(4, 3)$ to $(8, 6)$ to $(12, 9)$, we observe the y is changing 3 units for every 4 units of change in x. (See Figure 8-8.)

The ratio of $\dfrac{\Delta y}{\Delta x}$ is $\dfrac{3}{4}$.

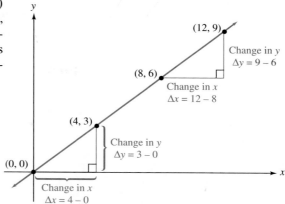

Figure 8-8

The slope, m, can also be defined as $m = \dfrac{\text{rise}}{\text{run}}$. As we move from one point to another on a linear function, the ratio $\dfrac{\Delta y}{\Delta x}$ remains constant. Thus the slope of a nonvertical line can be determined by any two points on the line. Furthermore, if points $P(x_1, y_1)$ and $Q(x_2, y_2)$ are points on any nonvertical line we can say that

$$m = \frac{\Delta y}{\Delta x} = \frac{y_2 - y_1}{x_2 - x_1}$$

(See Figure 8-9.) Δy is called the **rise** of the line between P and Q and Δx is called the **run**.

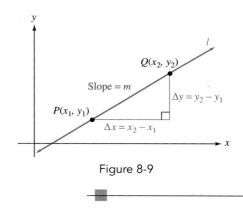

Figure 8-9

EXAMPLE 2

Use the graph in Figure 8-10(a) to determine the slope of the line passing through P and Q.

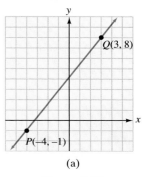

(a)

Figure 8-10

SOLUTION Use the graph in Figure 8-10(b) to determine the slope $\dfrac{\Delta y}{\Delta x}$.

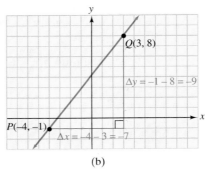

(b)

Figure 8-10

$$m = \frac{\Delta y}{\Delta x} = \frac{y_2 - y_1}{x_2 - x_1} = \frac{-1 - 8}{-4 - 3} = \frac{-9}{-7} = \frac{9}{7}$$

The slope is $\frac{9}{7}$.

EXAMPLE 3

If a line passes through the point $(0, 5)$ and has a slope of 4, determine one other point on the line and graph the line.

SOLUTION If the line has a slope of 4, then $\dfrac{\Delta y}{\Delta x} = \dfrac{4}{1}$. Plot the point $(0, 5)$ and count off $\dfrac{\Delta y}{\Delta x}\left(\text{i.e., } m = \dfrac{\text{rise}}{\text{run}}\right)$. (See Figure 8-11.)

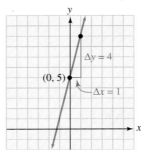

A second point would be $(1, 9)$.

Figure 8-11

EXAMPLE 4

In Figure 8-12, point P has coordinates $(-2, 5)$ and point Q has coordinates of $(6, -7)$. Find the slope of the line PQ.

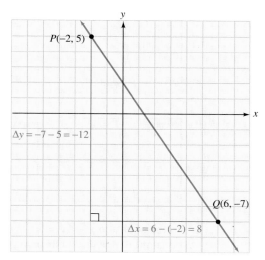

Figure 8-12

SOLUTION We can let $P(x_1, y_1) = P(-2, 5)$ and $Q(x_2, y_2) = Q(6, -7)$. Then

$$m = \frac{\Delta y}{\Delta x}$$

$$= \frac{y_2 - y_1}{x_2 - x_1}$$

$$= \frac{-7 - 5}{6 - (-2)} \qquad \text{Substitute } -7 \text{ for } y_2, 5 \text{ for } y_1, 6 \text{ for } x_2, \text{ and } -2 \text{ for } x_1.$$

$$= \frac{-12}{8}$$

$$= -\frac{3}{2}$$

The slope of the line is $-\frac{3}{2}$. We would obtain the same result if we had let $P(x_1, y_1) = P(6, -7)$ and $Q(x_2, y_2) = Q(-2, 5)$.

WARNING! When finding the slope of a line, always subtract the y-values and the x-values in the same order, or your answer will have the wrong sign.

EXPLORING THE CONCEPT Graph the linear function $y = 4x + 5$ in the ZInteger viewing window.

a. TRACE to determine the y-intercept.

b. Using two other coordinates from the line, determine the slope.

c. Draw conclusions about the slope and y-intercept as they relate to the equation $y = 4x + 5$.

CONCLUSION   **a.**

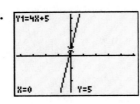

y-intercept: $(0, 5)$

b.

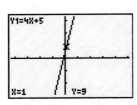   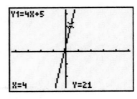

$$m = \frac{21 - 9}{4 - 1} = \frac{12}{3} = 4$$

c.   In the equation $y = 4x + 5$, the coefficient of x is the slope of the line and the constant of 5 is the y-intercept.

A line is determined by a point and its slope.

Slope-Intercept Form

A linear equation in the form $y = mx + b$ is in **slope-intercept form**. The coefficient of x, denoted as m, represents the slope of the line, and the y-intercept is b.

EXAMPLE 5   Find the slope and the y-intercept of the line determined by the equation $3x - 4y = 12$.

SOLUTION   Solve $3x - 4y = 12$ for y to write the equation in slope-intercept form.

$$3x - 4y = 12$$
$$-4y = -3x + 12$$
$$y = \frac{3}{4}x - 3$$

Therefore, the slope is $\frac{3}{4}$ and the coordinates of the y-intercept are $(0, -3)$.

An alternative approach would be to determine the y-intercept by letting $x = 0$ and solving for y.

$$3(0) - 4y = 12$$
$$-4y = 12$$
$$y = -3$$

The coordinates of the y-intercept are $(0, -3)$.

Compute the slope by determining one other ordered pair and using the slope formula.

If $y = 0$, then $x = 4$, and the point $(4, 0)$ is on the line (see Figure 8-13). We can then let $P(x_1, y_1) = P(0, -3)$, $Q(x_2, y_2) = Q(4, 0)$ and use the formula for slope to obtain

$$m = \frac{\Delta y}{\Delta x}$$

$$= \frac{y_2 - y_1}{x_2 - x_1}$$

$$= \frac{0 - (-3)}{4 - 0} \qquad \text{Substitute 0 for } y_2, -3 \text{ for } y_1, 4 \text{ for } x_2, \text{ and 0 for } x_1.$$

$$= \frac{3}{4}$$

The slope of the line is $\frac{3}{4}$.

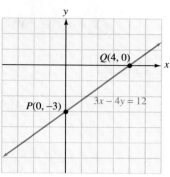

Figure 8-13

3x − 4y = 12, P(0, −3), Q(4, 0)

EXPLORING THE CONCEPT

Graph each of the following groups of lines in the ZInteger viewing window.

i. $y = \frac{1}{2}x, y = x, y = 3x$

ii. $y = -\frac{1}{2}x, y = -x, y = -3x$

a. As the positive coefficient of x (slope) gets larger, what happens to the graph of the line?

b. As the negative coefficient of x gets smaller, what happens to the graph of the line?

c. What conclusions can you draw about the slope of a linear function and its classification as an increasing or decreasing function?

CONCLUSIONS

a. As the coefficient of x (slope) gets larger, the line becomes steeper. The line is also rising when viewed from left to right.

b. When the coefficient is negative, the line falls when viewed from left to right. The smaller the negative coefficient the faster the line falls.

c. Linear functions with positive slope are increasing functions, and linear functions with negative slopes are decreasing functions.

SLOPES OF HORIZONTAL AND VERTICAL LINES

If $P(x_1, y_1)$ and $Q(x_2, y_2)$ are points on the horizontal line shown in Figure 8-14(a), then $y_1 = y_2$, and the numerator of the fraction

$$\frac{y_2 - y_1}{x_2 - x_1} \qquad \text{On a horizontal line, } x_2 \neq x_1.$$

is 0. Thus, the value of the fraction is 0, and the slope of the horizontal line is 0.

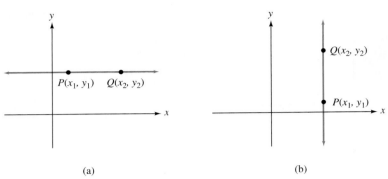

(a)   (b)

Figure 8-14

If $P(x_1, y_1)$ and $Q(x_2, y_2)$ are two points on the vertical line shown in Figure 8-14(b), then $x_1 = x_2$, and the denominator of the fraction

$$\frac{y_2 - y_1}{x_2 - x_1} \qquad \text{On a vertical line, } y_2 \neq y_1.$$

is 0. Since the denominator of a fraction cannot be 0, a vertical line has no defined slope.

Slopes of Horizontal and Vertical Lines

Horizontal lines (lines with equations of the form $y = b$) have a slope of 0.

Vertical lines (lines with equations of the form $x = a$) have no defined slope.

If a line rises as we follow it from left to right, as in Figure 8-15(a), its slope is positive. If a line drops as we follow it from left to right, as in Figure 8-15(b), its slope is negative. If a line is horizontal, as in Figure 8-15(c), its slope is 0. If a line is vertical, as in Figure 8-15(d), it has no defined slope.

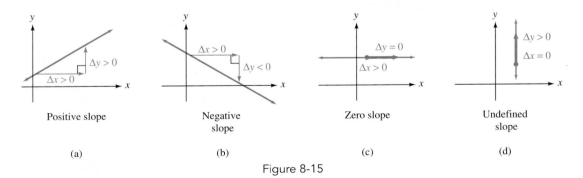

Positive slope   Negative   Zero slope   Undefined
                          slope   slope

(a)   (b)   (c)   (d)

Figure 8-15

▮ SLOPES OF PARALLEL AND PERPENDICULAR LINES

We have observed that it only takes a point and a slope to determine a given line. We now want to see what occurs when two lines share the same slope value but pass through different points.

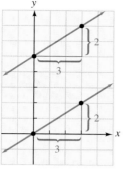

Figure 8-16

Graph the line l_1 that has a slope of $\frac{2}{3}$ and passes through $(0, 0)$ and the line l_2 that also has a slope of $\frac{2}{3}$ but passes through $(0, 5)$. (See Figure 8-16.)

Lines that have the same slope but pass through different points are parallel. It is also true that when two lines are parallel they have the same slope.

Slopes of Parallel Lines

Nonvertical parallel lines have the same slope, and lines having the same slope are parallel.

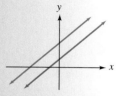

Since vertical lines are parallel, lines with no defined slope are parallel.

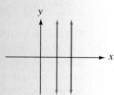

EXAMPLE 6

If the line passing through $P(3, -2)$ and $Q(-3, 4)$ is parallel to the line passing through $R(-2, 5)$ and $S(3, y)$, find y.

SOLUTION   Since the lines PQ and RS are parallel, they have equal slopes (see Figure 8-17). To find y, we find the slope of each line, set them equal, and solve the resulting equation.

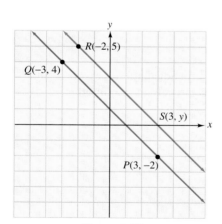

Figure 8-17

$$\text{Slope of PQ} \qquad \text{Slope of RS}$$

$$\frac{4 - (-2)}{-3 - 3} = \frac{y - 5}{3 - (-2)}$$

$$\frac{6}{-6} = \frac{y - 5}{5}$$

$$-1 = \frac{y - 5}{5}$$

$$-5 = y - 5$$

$$0 = y$$

Thus, $y = 0$. The line passing through $P(3, -2)$ and $Q(-3, 4)$ is parallel to the line passing through $R(-2, 5)$ and $S(3, 0)$.

EXPLORING THE CONCEPT

Graph each of the following pairs of lines in the ZInteger viewing window. Decide if the lines appear parallel, perpendicular, or neither. Then compare the slopes as indicated in the equation. What conclusions can you draw about the lines and the slopes of those lines that appear to be parallel or perpendicular?

a. $y = \dfrac{3}{7}x + 10$ and $y = \dfrac{3}{7}x - 8$

b. $y = \dfrac{3}{7}x + 10$ and $y = -\dfrac{3}{7}x - 8$

c. $y = \dfrac{3}{7}x + 10$ and $y = \dfrac{7}{3}x - 8$

d. $y = \dfrac{3}{7}x + 10$ and $y = -\dfrac{7}{3}x - 8$

CONCLUSIONS **a.**

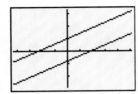

Lines appear parallel; both slopes are $\frac{3}{7}$; lines with the same slope are parallel.

b.

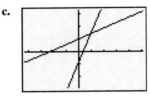

Lines are neither parallel nor perpendicular.

c.

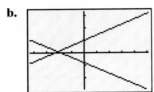

Lines are neither parallel nor perpendicular.

d.

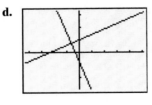

Lines appear perpendicular; slopes are $\frac{3}{7}$ and $-\frac{7}{3}$, negative reciprocals of one another.

Slopes of Perpendicular Lines

If two nonvertical lines are perpendicular, their slopes are negative reciprocals.

If the slopes of two lines are negative reciprocals the lines are perpendicular.

Because a horizontal line is perpendicular to a vertical line, a line with a slope of 0 is perpendicular to a line with an undefined slope.

▮▮▮ INTERPRETATION OF SLOPE

Many applications of mathematics involve equations of lines and their slopes. We will begin by looking at applications/interpretations of slope in this section and continue with applications of lines in the next section.

EXAMPLE 7

Cost of Carpet   The cost to purchase 10 square yards of carpet is $400 and the cost to purchase 20 square yards of carpet is $800. Assume that the relationship between the number of yards, x, and the cost, y, is linear and

a. find the slope of the line,

b. write the equation of the line using the slope-intercept form and graph it, and

c. determine the meaning of the slope in the context of the problem.

SOLUTION   **a.** We begin by organizing the given information into ordered pairs:

$x_1 = 10$ square yards and $y_1 = \$400$, which yields the ordered pair (10,400)

$x_2 = 20$ square yards and $y_2 = \$800$, which yields the ordered pair (20,800)

We can find the slope of the line:

$$m = \frac{\Delta y}{\Delta x}$$

$$= \frac{y_2 - y_1}{x_2 - x_1}$$

$$= \frac{800 - 400}{20 - 10} \qquad \text{Substitute 800 for } y_2, \text{ 400 for } y_1, \text{ 20 for } x_2, \text{ and 10 for } x_1.$$

$$= \frac{400}{10} \qquad \text{Simplify.}$$

$$= 40 \qquad \text{Divide.}$$

The slope of the line is 40.

b. The slope-intercept form of the equation is $y = mx + b$. In order to determine the y-intercept, we need to know y (the cost of the carpet) if $x = 0$. If 0 yards are purchased, the cost y will be 0. Thus, $b = 0$.

$$m = 40 \qquad b = 0$$

$$y = 40x + 0$$

The equation of the line in slope-intercept form is $y = 40x$.
      Figure 8-18 shows a table of ordered pairs and the graph.

c. To help us determine the meaning of the slope, we will examine the equation $y = 40x$. If 1 square yard of carpet is purchased then

$$y = 40(1) = \$40$$

Thus, we can see that the slope represents the cost per square yard of carpet.

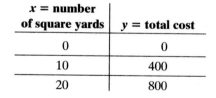

x = number of square yards	y = total cost
0	0
10	400
20	800

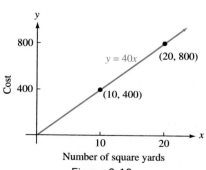

Figure 8-18

Slope always represents a **rate of change**. In Example 7, the slope gives the rate of change of the cost of the carpet in dollars per square yard.

If temperature changes with time (measured in hours), the concept of slope gives the rate of change of temperature in degrees per hour. If the amount of money in an account is increasing over the years, slope gives the rate of growth in dollars per year.

EXERCISE 8.3

VOCABULARY AND NOTATION   *In Exercises 1–10, fill in the blanks to make the given statements true.*

1. The comparison ratio of change in ___ over change in ___ determines the slope of a line.

2. Slope is sometimes defined as _____ over run.

3. The formula for slope $\dfrac{\Delta y}{\Delta x} =$ _____ .

4. If, when viewed from left to right, a linear function has increasing *x*-values, its slope is _____ .

5. If, when viewed from left to right, a linear function has decreasing *y*-values, its slope is _____ .

6. A linear equation in the form $y = mx + b$ is in _____ form. The coefficient of *x*, *m*, represents the _____ of the line, whereas *b* represents the

   _____ .

7. The slope of a horizontal line is _____ .

8. The slope of a vertical line is _____ .

9. If two lines are parallel, their slopes are _____ .

10. If two lines are perpendicular, their slopes are _____ of one another.

CONCEPTS   *In Exercises 11–14, find the slope of the line whose displayed TABLE is pictured.*

11.
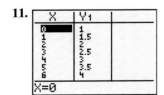

12.

X	Y₁
1	-4.333
2	-3.667
3	
4	-2.333
5	-1.667
6	-1
7	-.3333

X=1

13.
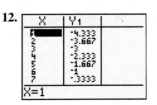

14.
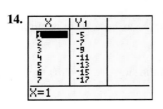

In Exercises 15–19, classify each of the lines as having positive, negative, 0, or no defined value for its slope.

15.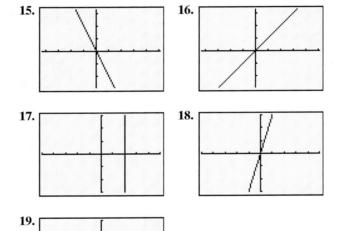

16.

17.

18.

19.

In Exercises 20–23, determine whether the graph of y1 or y2 has the greater absolute value for slope.

20.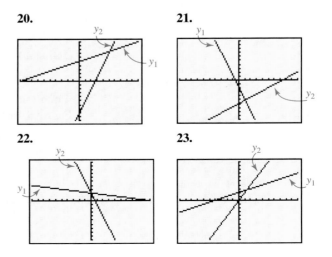

21.

22.

23.

PRACTICE *In Exercises 24–37, when possible, find the slope of the line that passes through the given points.*

24. $(-1, 8), (6, 1)$ **25.** $(-5, -8), (3, 8)$

26. $(3, -1), (-6, 2)$ **27.** $(0, -8), (-5, 0)$

28. $(7, 5), (-9, 5)$ **29.** $(-7, -5), (-7, -2)$

30. $(2, 5), (4, 8)$ **31.** $(6, 9), (9, 5)$

32. $(-2, -6), (3, -8)$ **33.** $(-5, -2), (4, 6)$

34. $(3, -4), (5, -4)$ **35.** $(5, 6), (5, -2)$

36. $\left(\dfrac{2}{3}, \dfrac{3}{4}\right), \left(\dfrac{1}{2}, \dfrac{1}{3}\right)$ **37.** $\left(\dfrac{1}{4}, \dfrac{2}{3}\right), \left(\dfrac{1}{2}, \dfrac{3}{4}\right)$

In Exercises 38–45, determine two solutions of the given equation and use these points to determine the slope of the line through them.

38. $2x + 3y = 6$ **39.** $5x - 3y = 15$

40. $3x + y = 8$ **41.** $4x + y = 9$

42. $3x + 2 = 6$ **43.** $4x + 8 = 9$

44. $3y + 2 = 6$ **45.** $4y - 8 = 2$

In Exercises 46–51, use the pictured graph to determine the slope of each line.

46.
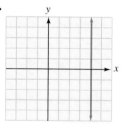

47.
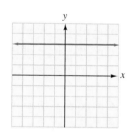

48.
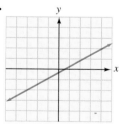

49.
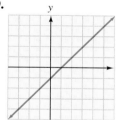

50.
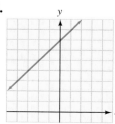

51.
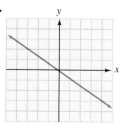

In Exercises 52–59, determine the unknown coordinate so that the line passing through the given points has the specified slope.

52. $(6, 3)$ and $(x, -2)$; slope is 2

53. $(3, 2)$ and $(x, -5)$; slope is 4

54. $(6, 3)$ and $(-4, y)$; slope is $\frac{4}{5}$

55. $(2, 4)$ and $(-5, y)$; slope is $\frac{5}{7}$

56. $(3, 2)$ and $(x, 5)$; slope is not defined

57. $(-5, 6)$ and $(x, -2)$; slope is not defined

58. $(-2, -1)$ and $(5, y)$; slope is 0

59. $(3, 2)$ and $(6, y)$; slope is 0

In Exercises 60–67, find the slope and y-intercept of the line determined by the given equation.

60. $3x + 2y = 12$

61. $2x + 3y = 12$

62. $2x - y = 6$

63. $3x - y = 9$

64. $3x = 4y - 2$

65. $4x = 3y - 8$

66. $y = \dfrac{x - 4}{2}$

67. $x = \dfrac{3 - y}{4}$

In Exercises 68 and 69, find the slope of a line parallel to the given line.

68. $y = 3x - 5$ **69.** $2x - y = 6$

In Exercises 70–73, find the slope of a line perpendicular to the given line.

70. $4x - y = 2$ **71.** $y = \dfrac{1}{3}x + 1$

72. $y = \dfrac{4}{3}x + 2$ **73.** $8x - y = 6$

In Exercises 74–79, determine if the pair of lines determined by the given equation are parallel, perpendicular, or neither.

74. $y = 3x - 4; y = -\dfrac{1}{3}x + 2$

75. $y = 5x + 2; y = -5x + 2$

76. $2x - y = 7; 2x + y = 8$

77. $x - 3y = 5; 3x + y = 6$

78. $4x - y = 1; 4x - y = -6$

79. $5x + 2y = 1; 10x + 4y = 12$

80. Find the equation of the *x*-axis and its slope.

81. Explain why the graphing calculator will graph a horizontal line but not a vertical line.

82. Find the equation of the *y*-axis and its slope.

In Exercises 83–86, use the slope-intercept form of an equation (y = mx + b) to write the equation of a line with the given description.

83. slope = 4; *y*-intercept = (0, −2)

84. slope = $\frac{1}{5}$; *y*-intercept = (0, 4)

85. Perpendicular to a line whose slope is $\frac{2}{3}$ and has a *y*-intercept of (0, 5).

86. Parallel to a line whose slope is 5 and has a *y*-intercept of (0, −2).

APPLICATIONS

87. When a college started an aviation program, the administration agreed to predict enrollments using the graph of a straight line. If the enrollment during the first year was 12, and the enrollment during the fifth year was 26, find the rate of growth per year (the slope of the line). (See Illustration 1.)

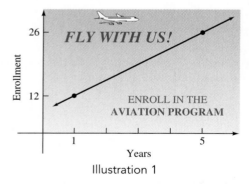

Illustration 1

88. Grade of a road   Find the slope of the road shown in Illustration 2. (*Hint:* 1 mi = 5,280 ft.)

Illustration 2

89. Pitch of a roof   Find the pitch of the roof shown in Illustration 3.

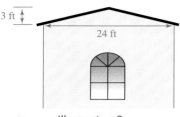

Illustration 3

90. Physical fitness   Find the slope of the treadmill shown in Illustration 4 for each setting listed in the table.

Illustration 4

91. Wheelchair ramps   Illustration 5 shows two designs for a ramp to make a platform wheelchair accessible.

a. Find the slope of the ramp shown in design 1.

b. Find the slope of each part of the ramp shown in design 2.

c. Give one advantage and one disadvantage of each design.

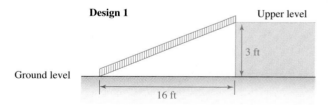

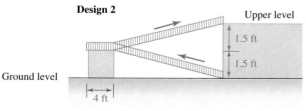

Illustration 5

92. **Slope of a ladder**   A ladder reaches 18 feet up the side of a building with its base 5 feet from the building. Find the slope of the ladder.

93. The first day of the fair, admission is $5 and it costs $0.75 to ride each ride. If y represents the total cost and x the number of rides, we can say that $y = 5 + 0.75x$ models the cost.

    a. What is the slope of this line and how does it relate to the context of the problem?

    b. If the initial cost was $7 and the cost per ride was $0.50, what would the equation be?

    c. Would it be possible, in reality, for the equation that models the cost of attending the fair to ever have a negative value for the slope? Why or why not?

94. A local printer charges $200 plus $1.25 per invitation to print custom-designed wedding invitations.

    a. Write an equation to model this application, where y represents the cost and x the number of invitations.

    b. What does the slope of the linear equation represent?

    c. Speculate what it would mean if your bill was only $200.

95. Assume that an antique purchased for $3,000 increases in value in a straight-line fashion at a rate of $250 per year.

    a. Write an equation to model this increase, where y represents the cost and x the subsequent years.

    b. What does the slope of your equation represent?

    c. In terms of the antique, what would it mean if the slope were negative instead of positive?

96. In June, the latest desktop computer technology could be purchased for $3,000. Throughout the summer and into the fall, the cost of processor components began to decrease at a rate of $100 per month.

    a. Assuming the decrease was a straight-line model, write an equation to model that cost, y, over x months.

    b. What does the slope represent in your equation?

    c. How would your equation change if prices were rising instead of falling?

8.4 EQUATIONS OF LINES

- REVIEW OF EQUATIONS IN SLOPE-INTERCEPT FORM
- POINT-SLOPE FORM OF THE EQUATION OF A LINE
- THE MIDPOINT FORMULA
- STRAIGHT-LINE DEPRECIATION

▓ REVIEW OF EQUATIONS IN SLOPE-INTERCEPT FORM

Previously, we looked at linear applications whose basic information allowed us to write a model equation whose form was $y = mx + b$, slope-intercept form. This meant that we had to be given the y-intercept (the constant of the equation) and the rate at which the x variable was changing (the slope). From this basic information, we are able to determine other points that satisfy the model equation and subsequently graph the model.

EXAMPLE 1

The first day admission to the fair is $5 plus $0.75 per ride. Write the equation to model this situation and graph the line.

SOLUTION   The constant cost is the $5 admission fee and the variable cost is $0.75 per ride.

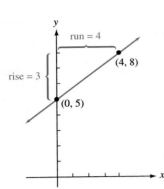

Let y = total cost, x = number of rides

$$y = .75x + 5 \qquad \text{or} \qquad y = \frac{3}{4}x + 5$$

$$\text{Slope} = \frac{3}{4} = \frac{\text{rise}}{\text{run}}$$

y-intercept: $(0, 5)$

POINT-SLOPE FORM OF THE EQUATION OF A LINE

In applications, we must be presented with enough information to construct an equation that models the situation. However, the information rarely comes in the straightforward form of slope and y-intercept. We know that it only requires two points to determine the graph of a line. It also only requires two points to determine the equation of a line.

The slope of a line is a constant. This means that when the coordinates of *any* two points are substituted into the slope formula, the result will be the same constant regardless of the coordinates used. Recall the slope formula,

$$m = \frac{y_2 - y_1}{x_2 - x_1}$$

If we have a line with slope m and the point (x_1, y_1) is on the line, then for the point (x, y) to be on the line it must satisfy the equation

$$m = \frac{y - y_1}{x - x_1}$$

If we clear the fraction by multiplying both sides of the equation by the quantity $x - x_1$, assuming $x - x_1 \neq 0$, we have the equation

$$m(x - x_1) = y - y_1$$

or

$$y - y_1 = m(x - x_1) \qquad \text{Symmetric property of equality.}$$

We call $y - y_1 = m(x - x_1)$ the **point-slope form** of an equation of a line. It will be used to write equations of lines where the given information produces a point and a slope.

Point-Slope Form of the Equation of a Line
The equation of the line passing through $P(x_1, y_1)$ and with slope m is

$$y - y_1 = m(x - x_1)$$

We can see that if our line has slope m and a y-intercept of $(0, b)$, the equation above will generate the slope-intercept form:

$$y - y_1 = m(x - x_1)$$
$$y - b = m(x - 0)$$
$$y - b = mx$$
$$y = mx + b \qquad \text{Slope-intercept form}$$

EXAMPLE 2

Use the point-slope form to write the equation of a line with a slope of $-\dfrac{2}{3}$ and passing through $P(-4, 5)$.

SOLUTION We substitute $-\dfrac{2}{3}$ for m, -4 for x_1, and 5 for y_1 in the point-slope form and simplify.

$$y - y_1 = m(x - x_1)$$

$$y - 5 = -\frac{2}{3}[x - (-4)] \qquad \text{Substitute } -\tfrac{2}{3} \text{ for } m, -4 \text{ for } x_1, \text{ and } 5 \text{ for } y_1.$$

$$y - 5 = -\frac{2}{3}(x + 4) \qquad -(-4) = 4$$

$$y - 5 = -\frac{2}{3}x - \frac{8}{3} \qquad \text{Use the distributive property to remove parentheses.}$$

$$y = -\frac{2}{3}x + \frac{7}{3} \qquad \text{Add 5 to both sides.}$$

The equation of the line is $y = -\dfrac{2}{3}x + \dfrac{7}{3}$.

Verify your equation with your calculator.

You may use your calculator to determine if the given point $(-4, 5)$ is on the graph of $y = -\dfrac{2}{3}x + \dfrac{7}{3}$. You can either locate the coordinates $(-4, 5)$ in the TABLE or use the VALUE feature of your calculator.

TECHNOLOGY TIP

Remember that when using VALUE or EVAL, your x-value must be between the values defined for x-minimum and x-maximum on the window screen.

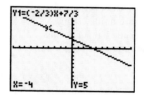

EXAMPLE 3

Use the point-slope form of the equation of a line to write the equation of the line passing through $P(-5, 4)$ and $Q(8, -5)$.

SOLUTION The two basic pieces of information you need to write the equation of a line are the slope and a point. Thus, we must first find the slope of the line:

STUDY TIP

Place a piece of paper over the solution to an example in your text. Read the example and try to work the problem on your paper. If you get stuck, slide the paper down and jump-start yourself.

$$m = \frac{y_2 - y_1}{x_2 - x_1}$$

$$= \frac{-5 - 4}{8 - (-5)} \qquad \text{Substitute } -5 \text{ for } y_2, 4 \text{ for } y_1, 8 \text{ for } x_2, \text{ and } -5 \text{ for } x_1.$$

$$= -\frac{9}{13}$$

Because the line passes through both P and Q, we can choose either point and substitute its coordinates into the point-slope form. If we choose $P(-5, 4)$, we substitute -5 for x_1, 4 for y_1, and $-\frac{9}{13}$ for m and simplify.

$$y - y_1 = m(x - x_1)$$

$$y - 4 = -\frac{9}{13}[x - (-5)] \qquad \text{Substitution.}$$

$$y - 4 = -\frac{9}{13}(x + 5) \qquad -(-5) = 5$$

$$y - 4 = -\frac{9}{13}x - \frac{45}{13} \qquad \text{Use the distributive property to remove parentheses.}$$

$$y = -\frac{9}{13}x + \frac{7}{13} \qquad \text{Add 4 to both sides.}$$

The equation of the line is $y = -\frac{9}{13}x + \frac{7}{13}$.

Now that we have the equation of the line, we can verify, using the TABLE feature, that the coordinates of the given points satisfy the equation of the line.

Because we have changed our setting for the independent variable from automatic to ASK, we must enter the x-values for which we desire corresponding y-values. We enter the x-value of -5 and press ENTER to see the corresponding y-value. We then enter the x-value of 8, press ENTER, and see the corresponding value for y.

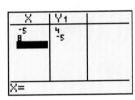

The TABLE display confirms the fact that the points $(-5, 4)$ and $(8, -5)$, given initially, satisfy our linear equation.

EXAMPLE 4

Write the equation of a line passing through the point $P(-2, 5)$ and parallel to the line $y = 8x - 2$.

SOLUTION

Since our line will be parallel to $y = 8x - 2$, we know that the slope of our line will be 8 ($m = 8$), the same as the slope of the given line. We now proceed as before, using $m = 8$ and the coordinates of the point $P(-2, 5)$; substitute 8 in for m, -2 for x_1, and 5 for y_1.

$$y - y_1 = m(x - x_1)$$
$$y - 5 = 8[x - (-2)] \qquad \text{Substitution.}$$
$$y - 5 = 8(x + 2) \qquad -(-2) = 2$$
$$y - 5 = 8x + 16 \qquad \text{Use the distributive property to remove parentheses.}$$
$$y = 8x + 21 \qquad \text{Add 5 to both sides.}$$

Verify by graphing $y = 8x + 21$ and locating $(-2, 5)$ in the TABLE (choose AUTO from the TBLSET screen and use -2 as the initial, starting value) or use the VALUE feature from the graph screen to evaluate the function when $x = -2$.

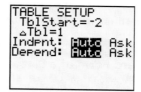

EXAMPLE 5

Write the equation of the line passing through the point $P(-2, 5)$ and perpendicular to $y = 8x - 2$.

SOLUTION   The slope of the given line is 8. Thus, the slope of the desired line must be $-\frac{1}{8}$, which is the negative reciprocal of 8.

We substitute -2 for x_1, 5 for y_1, and $-\frac{1}{8}$ for m into the point-slope form and simplify.

$$y - y_1 = m(x - x_1)$$

$$y - 5 = -\frac{1}{8}[x - (-2)] \qquad \text{Substitution.}$$

$$y - 5 = -\frac{1}{8}(x + 2)$$

$$y - 5 = -\frac{1}{8}x - \frac{1}{4}$$

$$y = -\frac{1}{8}x + \frac{19}{4}$$

The displays below, illustrating the use of the TABLE feature and the VALUE feature, confirm that the point $(-2, 5)$ is on the line with equation $y = -\frac{1}{8}x + \frac{19}{4}$.

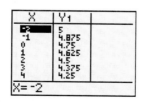
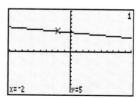

Graphing the lines with equations $y = -\frac{1}{8}x + \frac{19}{4}$ and $y = 8x - 2$ on a square screen shows that they appear to be perpendicular.

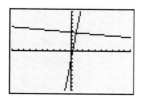

The final form for the equation of nonvertical lines can be slope-intercept ($y = mx + b$) or standard form ($Ax + By = C$). We consistently use slope-intercept form because it is the most convenient when using a graphing calculator.

We summarize the various forms for the equation of a line as follows:

Standard Form of a Linear Equation
$Ax + By = C$
A and B cannot both be 0; A, B, and C are integers; $A > 0$.

Slope-Intercept Form of a Linear Equation
$y = mx + b$
The slope is m and the y-intercept is b.
Remember, the *coordinates* of the y-intercept are $(0, b)$.

Point-Slope Form of a Linear Equation
$y - y_1 = m(x - x_1)$
The slope is m and the line passes through the point with coordinates (x_1, y_1).

A Horizontal Line
$y = b$
The slope is 0, and the y-intercept is b, $(0, b)$

A Vertical Line
$x = a$
There is no defined slope, and the x-intercept is a, $(a, 0)$

▮ THE MIDPOINT FORMULA

If point M in Figure 8-19 lies midway between points $P(x_1, y_1)$ and $Q(x_2, y_2)$, point M is called the **midpoint** of segment PQ. To find the coordinates of M, we average the x-coordinates and average the y-coordinates of P and Q.

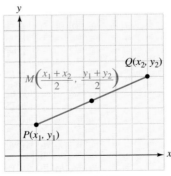

Figure 8-19

The Midpoint Formula
The midpoint of the line segment $P(x_1, y_1)$ and $Q(x_2, y_2)$ is the point M with coordinates of $\left(\dfrac{x_1 + x_2}{2}, \dfrac{y_1 + y_2}{2} \right)$.

EXAMPLE 6   Find the midpoint of the segment joining $P(-2, 3)$ and $Q(3, -5)$.

SOLUTION   To find the midpoint, we average the x-coordinates and the y-coordinates to get

$$\frac{x_1 + x_2}{2} = \frac{-2 + 3}{2} \quad and \quad \frac{y_1 + y_2}{2} = \frac{3 + (-5)}{2}$$

$$= \frac{1}{2} \qquad\qquad = -1$$

The midpoint of segment PQ is the point $M(\frac{1}{2}, -1)$.

▍ STRAIGHT-LINE DEPRECIATION

For tax purposes, many businesses use the equation of a line to find the declining value of aging equipment. This method is called **straight-line depreciation** and assumes a constant decrease in value each year.

EXAMPLE 7   **Value of a lathe**   A machine shop buys a lathe for $1,970 and expects it to last for 10 years. The lathe then can be sold as scrap for an estimated **salvage value** of $270. If y represents the value of the lathe after x years of use and y and x are related by the equation of a line,

  a. Find the equation of the line.

  b. Find the value of the lathe after $2\frac{1}{2}$ years.

  c. Find the economic meaning of the y-intercept of the line.

  d. Find the economic meaning of the slope of the line.

SOLUTION   **a.** To find the equation of the line, we first find two points on the line and then find the line's slope. We can then use either the point-slope form of a linear equation OR the slope-intercept form to find the equation of the line.

When the lathe is new, its age x is 0 and its value y is $1,970. When the lathe is 10 years old, $x = 10$ and its value is $y = 270. The line passes through the points $(0, 1970)$ and $(10, 270)$, as shown in Figure 8-20, so the slope of the line is

$$m = \frac{y_2 - y_1}{x_2 - x_1}$$

$$= \frac{270 - 1970}{10 - 0}$$

$$= \frac{-1700}{10}$$

$$= -170$$

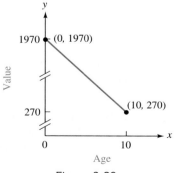

Figure 8-20

In this particular problem, we have two options to use when writing the equation of the line:

Option 1: We know $m = -170$ and can use the point-slope form to find the equation of the line. We can use either given point for (x_1, y_1), and choose $(0, 1{,}970)$. Substituting, we have

$$y - y_1 = m(x - x_1)$$
$$y - \mathbf{1970} = -170(x - \mathbf{0})$$
$$y = -170x + 1970$$

Option 2: We know $m = -170$ and that the y-intercept is $(0, 1{,}970)$. Thus, $b = 1{,}970$. We substitute these values into the slope-intercept form of a linear equation:

$$y = mx + b$$
$$y = -170x + \mathbf{1970}$$

b. To find the age of the lathe after $2\frac{1}{2}$ years, we substitute **2.5** for x into the equation of the line, $y = -170x + 1{,}970$:

$$y = -170(\mathbf{2.5}) + 1{,}970$$
$$y = -425 + 1{,}970$$
$$y = 1{,}545$$

Therefore, the lathe will be worth \$1,545 when it is $2\frac{1}{2}$ years old.

We could have used the graphing calculator to answer the question. Enter the equation at the y1 prompt and use either the VALUE feature or the TABLE to find the value of y when x has a value of 2.5. This is illustrated in the screens displayed below:

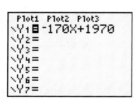 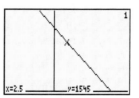 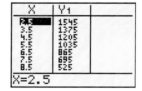

Window values of
[–10, 15] by [–10, 2500]

c. The y-intercept of the graph is $(0, b)$, where b is the value of y when $x = 0$. Thus, b is the value of a 0-year-old lathe, which is the lathe's original cost.

d. Each year, the value of the lathe decreases by \$170, because the slope of the line is -170. The slope of the depreciation line is the **annual depreciation rate**.

EXERCISE 8.4

VOCABULARY AND NOTATION   *In Exercises 1–6, fill in the blanks to make each statement true.*

1. When a linear equation is written in the form $y = mx + b$, it is in _____ form. The coefficient of x is the _____ of the line, and b is the _____.

2. When a linear equation is written in the form $y - y_1 = m(x - x_1)$, it is in _____ form, where m is the _____ of the line and _____ are the coordinates of a point on the line.

3. When a linear equation is written in the form $Ax + By = C$, we say it is written in _____ form.

4. A horizontal line has the equation _____. Its slope is always _____.

5. A vertical line has the equation _____. Its slope is always _____.

6. The method of using an equation of a line to find the declining value of aging equipment is called

_____.

CONCEPTS

7. A TABLE of values is given. Graph the values. Then use the graph and/or the TABLE to answer the questions below:

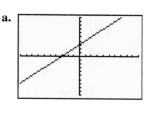

a. Does the graph indicate a linear relationship?

b. What is the slope?

c. What is the *y*-intercept?

d. The equation of the line is *y* = ___ *x* + ___ .

8. A house was purchased for $120,000. Its value in 5 years in predicted to be $150,000. Illustration 1 is a graph with points marked.

a. State the coordinates of the two marked points.

b. What is the average rate of appreciation per year? (*Hint:* Find the slope.)

c. The *y*-intercept is _____.

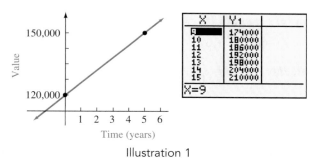

Illustration 1

d. The equation of the line is *y* = _____ *x* + _____ .

e. The equation of the line was entered at the y1 prompt on the y= screen. Part of the TABLE is displayed. What will the value of the house be in 10 years? After how many years will the house have a value of $204,000?

9. Match each equation (stated in slope-intercept form) with its graphical display. (*Hint:* The sign of *m*, slope, and *b*, the *y*-intercept, will be of great help.)

$y = x + 3$ $y = x - 3$ $y = -x + 5$ $y = x + 5$
$y = 5x - 4$ $y = 5x + 4$

a. 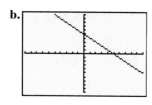   **b.**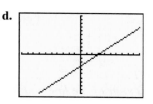

c. 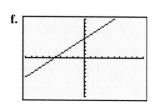   **d.**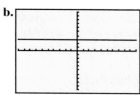

e. 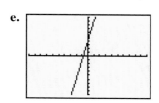   **f.**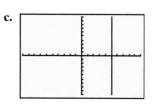

10. Match each equation with its graphical display.
$y = 3$ $y = -4$ $y = 8$ $x = 5$ $x = -6$ $x = -2$

a. 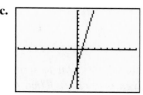   **b.**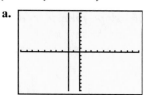

c. 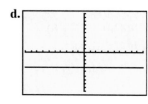   **d.**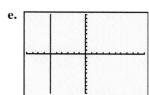

e. 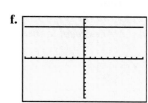   **f.**

PRACTICE

In Exercises 11–18, find the coordinates of the midpoint of segment PQ.

11. $P(6, 8)$, $Q(12, 16)$

12. $P(10, 4)$, $Q(2, -2)$

13. $P(2, 4), Q(5, 8)$

14. $P(5, 9), Q(8, 13)$

15. $P(a, b), Q(4a, 3b)$

16. $P(a + b, b), Q(-b, -a)$

17. $P(a - b, b), Q(a + b, 3b)$

18. $P(3a, a + b), Q(a + 2b, a - b)$

In Exercises 19–24, use the point-slope form to write the equation of the line with the given properties. Then transform the equation into standard form.

19. $m = 5$, passing through $P(0, 7)$

20. $m = -8$, passing through $P(0, -2)$

21. $m = -3$, passing through $P(2, 0)$

22. $m = 4$, passing through $P(-5, 0)$

23. $m = \frac{3}{2}$, passing through $P(2, 5)$

24. $m = -\frac{2}{3}$, passing through $P(-3, 2)$

In Exercises 25–30, use the point-slope form to write the equation of the line passing through the two given points. Then transform the equation into slope-intercept form.

25. $P(0, 0), Q(4, 4)$

26. $P(-5, -5), Q(0, 0)$

27. $P(3, 4), Q(0, -3)$

28. $P(4, 0), Q(6, -8)$

29. $P(-2, 4), Q(3, -5)$

30. $P(3, -5), Q(-1, 12)$

In Exercises 31–36, write each equation in slope-intercept form to find the slope and the y-intercept. Then use the slope and the y-intercept to draw the line.

31. $y + 1 = x$

32. $x + y = 2$

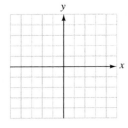

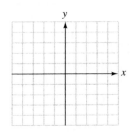

33. $y = \frac{2}{3}x + \frac{9}{2}$

34. $5x - 4y = 10$

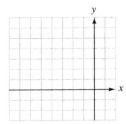

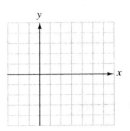

35. $3(y - 4) = -2(x - 3)$

36. $-4(2x + 3) = 3(3y + 8)$

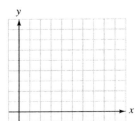

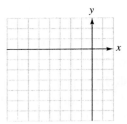

In Exercises 37–42, write the equation of the line that passes through the given point and is parallel to the given line. Your equation should be written in slope-intercept form.

37. $P(0, 0), y = 4x - 7$

38. $P(0, 0), x = -3y - 12$

39. $P(2, 5), 4x - y = 7$

40. $P(-6, 3), y + 3x = -12$

41. $P(4, -2), x = \frac{5}{4}y - 2$

42. $P(1, -5), x = -\frac{3}{4}y + 5$

In Exercises 43–48, write the equation of the line that passes through the given point and is perpendicular to the given line. Your equation should be written in slope-intercept form.

43. $P(0, 0), y = 4x - 7$

44. $P(0, 0), x = -3y - 12$

45. $P(2, 5), 4x - y = 7$

46. $P(-6, 3), y + 3x = -12$

47. $P(4, -2), x = \frac{5}{4}y - 2$

48. $P(1, -5), x = -\frac{3}{4}y + 5$

49. Find the equation of the line perpendicular to the line $y = 3$ and passing through the midpoint of the segment joining $(2, 4)$ and $(-6, 10)$.

50. Find the equation of the line parallel to the line $y = -8$ and passing through the midpoint of the segment joining $(2, -4)$ and $(8, 12)$

51. Find the equation of the line parallel to the line $x = 3$ and passing through the midpoint of the segment joining $(2, -4)$ and $(8, 12)$.

52. Find the equation of the line perpendicular to the line $x = 3$ and passing through the midpoint of the segment joining $(-2, 2)$ and $(4, -8)$.

53. Solve $Ax + By = C$ for y and thereby show that the slope of its graph is $-\frac{A}{B}$ and its y-intercept is $(0, \frac{C}{B})$.

54. Show that the x-intercept of the graph of $Ax + By = C$ is $(\frac{C}{A}, 0)$.

APPLICATIONS   *In Exercises 55–65, assume straight-line depreciation or straight-line appreciation.*

55. Finding a depreciation equation   A taxicab was purchased for $24,300. Its salvage value at the end of its 7-year useful life is expected to be $1,900 (see Illustration 2). Find the depreciation equation.

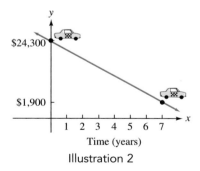

Illustration 2

56. Finding a depreciation equation   A computer system was purchased for $7,900 and depreciated over its 4-year useful life. Its expected salvage value is $300. Find its depreciation equation.

57. Finding an appreciation equation   An apartment building was purchased for $475,000. The owners ex-

pect the property to double in value in 10 years (see Illustration 3). Find the appreciation equation.

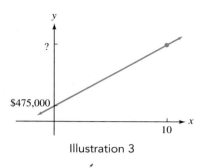

Illustration 3

58. Finding an appreciation equation   A house purchased for $112,000 is expected to double in value in 12 years. Find its appreciation equation.

59. Finding a depreciation equation   Find the depreciation equation for the TV in the want ad in Illustration 4.

For Sale: 3-year-old 54-inch TV, $3,900 new. Asking $1,890. Call 875-5555. Ask for Mike.

Illustration 4

60. Depreciating a word processor   A word processor costs $555 when new and is expected to be worth $80 after 5 years. What will it be worth after 3 years?

61. Finding salvage value   A copier cost $1050 when new and will be depreciated at the rate of $120 per year. If the useful life of the copier is 8 years, find its salvage value.

62. Finding annual rate of depreciation   A truck that cost $27,600 when new will have no salvage value after 12 years. Find its annual rate of depreciation.

63. Finding the value of antiques   An antique table is expected to appreciate $40 each year. If the table will be worth $450 in 2 years, what will it be worth in 13 years?

64. Finding the value of antiques   An antique clock is expected to be worth $350 after 2 years and $530 after 5 years. What will the clock be worth after 7 years?

65. Finding the purchase price of real estate   A cottage that was purchased 3 years ago is now appraised at $47,700. If the property has been appreciating $3500 per year, find its original purchase price.

66. Charges for computer repair A computer repair company charges a fixed amount, plus an hourly rate, for a service call. Use the information in Illustration 5 to find the hourly rate.

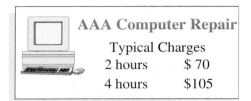

AAA Computer Repair

Typical Charges

2 hours	$ 70
4 hours	$105

Illustration 5

67. Charges for automobile repair An auto repair shop charges an hourly rate, plus the cost of parts. If the cost of labor for a $1\frac{1}{2}$-hour radiator repair is $69, find the cost of labor for a 5-hour transmission overhaul.

68. Finding printer charges A printer charges a fixed setup cost, plus $1 for every 100 copies. If 700 copies cost $52, how much will it cost to print 1000 copies?

69. Estimating the cost of a rain gutter A neighbor says that an installer of rain gutters charges $60, plus a dollar amount per foot. If the neighbor paid $435 for the installation of 250 feet of gutter, how much will it cost you to have 300 feet installed?

70. Predicting fires A local fire department recognizes that city growth and the number of reported fires are related by a linear equation. City records show that 300 fires were reported in a year when the local population was 57,000 persons, and 325 fires were reported

in a year when the population was 59,000 persons. How many fires can be expected when the population reaches 100,000 persons?

REVIEW *In Exercises 71–74, solve each equation in two ways: (a) algebraically, showing all steps; and (b) graphically, copying the screen display and all displayed information from the ZERO/ROOT screen.*

71. $2x^2 + x - 3 = 0$ **72.** $6x^2 - x - 2 = 0$
73. $4x^2 - 9 = 0$ **74.** $2x^2 - x = 0$

For each problem below, write an equation that models the words of the problem and solve it either algebraically (showing all steps in the solving process) or graphically (sketching the graph display, specifying window values, and copying all displayed information from the use of the INTERSECT or ZERO/ROOT feature). Regardless of the method of solution, write the answer to the question posed in a complete sentence.

75. Mixing alloys In 60 ounces of alloy for watch cases, there are 20 ounces of gold. How much copper must be added to the alloy so that a watch case weighing 4 ounces, made from the new alloy, will contain exactly 1 ounce of gold?

76. Mixing coffee To make a mixture of 80 pounds of coffee worth $272, a grocer mixes coffee worth $3.25 a pound with coffee worth $3.85 a pound. How many pounds of the cheaper coffee should the grocer use?

 8.5   VARIATION

- DIRECT VARIATION
- INVERSE VARIATION
- JOINT VARIATION
- COMBINED VARIATION

DIRECT VARIATION

To introduce direct variation, we consider the formula for finding the cirumference of a circle when given the diameter, d: $C = \pi d$. To produce a table of values using the graphing calculator, we translate the formula to $y = \pi x$, where y is the circumference and x is the diameter.

In this formula, $C = \pi d$, we say that the variables C and d **vary directly**, or that they are **directly proportional**. This is because, as one variable gets larger, so does the other, in a predictable way. In this example, the constant π is called the **constant of variation** or the **constant of proportionality**.

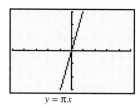

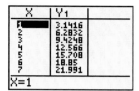

$y = \pi x$

Looking at the TABLE, we see that when the diameter of the circle is 3 ($x = 3$), the circumference is approximately 9.4 ($y \approx 9.4$). As the diameter increases, the circumference also increases. When $x = 6$, $y \approx 18.8$, therefore, when the diameter doubles, so does the circumference. Thus, the function is increasing because the y-values are increasing as the x-values are increasing. Algebraically speaking, if we double the diameter we will create another circle with a circumference that is doubled.

Direct Variation

The words *y varies directly with x* or *y is directly proportional to x* mean that for some nonzero constant, k, $y = kx$. The constant, k, is called the **constant of variation** or the **constant of proportionality**.

Another example of direct variation is Hooke's law from physics. Hooke's law states that the distance a spring will stretch varies directly with the force that is applied to it. If d represents a distance and f represents a force, Hooke's law is expressed mathematically as

$$d = kf$$

where k is the constant of variation. If the spring stretches 10 inches when a weight of 6 pounds is attached, we can find k as follows:

$$d = kf$$
$$10 = k(6) \qquad \text{Substitute 10 for } d \text{ and 6 for } f.$$
$$\frac{5}{3} = k \qquad \text{Divide both sides by 6 and simplify.}$$

Graphical: Our equation now becomes $d = \dfrac{5}{3}f$ or $y = \dfrac{5}{3}x$. Graphed in ZInteger, we can TRACE to $y = 35$ to find the force (x) required to stretch a spring 35 inches. From the graph, we see that it would require a force of 21 pounds.

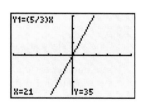

Analytically: To solve the problem analytically, we can solve the equation $d = kf$ for f, with $d = 35$ and $k = \frac{5}{3}$.

$$d = kf$$
$$35 = \frac{5}{3}f \qquad \text{Substitute 35 for } d \text{ and } \tfrac{5}{3} \text{ for } k.$$
$$105 = 5f \qquad \text{Multiply both sides by 3.}$$
$$21 = f \qquad \text{Divide both sides by 5.}$$

Thus, the force required to stretch the spring a distance of 35 inches is 21 pounds.

EXAMPLE 1

Travel distance   The distance traveled in a given time is directly proportional to speed. If a car travels 70 miles at 30 miles per hour, how far will it travel in the same time at 45 miles per hour?

SOLUTION   The words *distance is directly proportional to speed* can be expressed by the equation

$$d = ks$$

where d is distance, k is the constant of variation, and s is the speed. To find k, we substitute 70 for d, 30 for s, and solve for k.

$$d = ks$$
$$70 = k(30)$$
$$\frac{7}{3} = k \qquad \text{Divide both sides by 30 and simplify.}$$

Graphical: Our variation equation now becomes $d = \frac{7}{3}s$ or $y = \frac{7}{3}x$. This equation is graphed as shown below in the ZInteger viewing window. From the graph, we can determine the distance, y, for any integer speed, x, by tracing to that x-value.

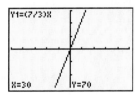

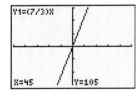

For a speed of 30 miles per hour the distance traveled would be 70 miles.

For a speed of 45 miles per hour, the distance is 105 miles.

TECHNOLOGY TIP

Remember to adjust your window to include the desired x-value before using the VALUE (EVAL) feature.

For speeds that are not integers, use the VALUE (EVAL) option or the TABLE feature.

Analytical: To find the distance traveled at 45 miles per hour, we substitute $\frac{7}{3}$ for k and 45 for s in the equation and simplify.

$$d = ks$$
$$d = \frac{7}{3}(45)$$
$$d = 105$$

In the time it took to go 70 miles at 30 miles per hour, the car could travel 105 miles at 45 miles per hour.

Not all direct variation equations are linear. The next example illustrates a variation equation that is quadratic.

EXAMPLE 2

Area of a circle   The area of a circle provides a familiar nonlinear application of direct variation involving a power x: $A = \pi r^2$. In the formula, our constant of variation is π. Translated to a graphical format we have $y = \pi x^2$.

a. Explore the graph of "y varies directly as the square of x" and determine which values of x are in the domain of the formula for the area of a circle.

b. Determine the area when the radius is 6.1 centimeters. Use both the graphical display and the TABLE feature.

SOLUTION

a. Begin by graphing in a window that is twice the size of the ZDecimal window. Values of x less than zero are not applicable to the area formula because this would mean that the radius is a negative measurement. Thus, our domain is restricted to the interval $[0, \infty)$. For this domain, our function is increasing.

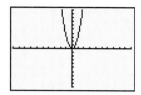

TECHNOLOGY TIP

Set the ZDecimal window and then multiply all entry values by 2 except for x-scl and y-scl.

b. In this particular window you cannot trace to $x = 6.1$. You may use the VALUE (EVAL) feature. The area is approximately 116.89866 square centimeters when the radius is 6.1 centimeters.

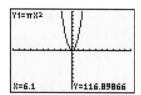

From the TABLE we can see that the accuracy level of our response has decreased because the TABLE rounds the area to 116.9. However, if we cursor over to the displayed y-value and look at the bottom of the screen, we see a more accurate approximation.

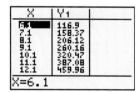

▊ INVERSE VARIATION

In the formula $w = \frac{12}{l}$, w gets smaller as l gets larger, and w gets larger as l gets smaller. Because these variables vary in opposite directions in a predictable way, we say that the variables **vary inversely**, or that they are **inversely proportional**. The constant 12 is the **constant of variation**.

Inverse variation is an example of an application of rational functions. Variables will occur in the denominators of fractions, so you must be careful to watch for excluded values. Because of this, graphs should be viewed in both connected and in dot mode.

Inverse Variation

The words *y varies inversely with x* or *y is inversely proportional to x* mean that $y = \frac{k}{x}$ for some constant k. The constant k is called the **constant of variation**.

EXAMPLE 3

Illumination   The intensity I of light received from a light source varies inversely with the square of the distance from the light source. If the intensity from a light source 4 feet from an object is 8 candelas, find the intensity at a distance of 2 feet.

SOLUTION   The words *intensity varies inversely with the square of the distance d* can be expressed by the equation

$$I = \frac{k}{d^2}$$

To find k, we substitute 8 for I and 4 for d, and solve for k.

$$I = \frac{k}{d^2}$$

$$8 = \frac{k}{4^2}$$

$$128 = k \qquad \text{Multiply both sides by } 4^2.$$

Graphical: Our variation equation becomes $I = \dfrac{128}{d^2}$

or $y = \dfrac{128}{x^2}$, $(x \neq 0)$ for graphical purposes. The prob-

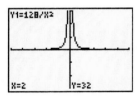

lem asks for the intensity at a distance of 2 feet, so we begin by graphing in the ZInteger viewing window. By tracing, we can determine that the intensity will be 32 candelas.

Note that for the application, the domain must be restricted to $[0, \infty)$; it would not be valid for our distance to be represented by a negative value of x. Thus, the portion of the graph in the second quadrant is not applicable to this particular application. For the domain $[0, \infty)$, the function is decreasing.

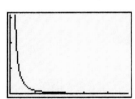

Analytical: To find the intensity when the object is 2 feet from the light source, we substitute 2 for d and 128 for k and simplify.

$$I = \frac{k}{d^2}$$

$$I = \frac{128}{2^2}$$

$$I = 32$$

The intensity at 2 feet is 32 candelas.

JOINT VARIATION

There are times when one variable varies with the product of several variables. For example, the area of a triangle varies directly with the product of its base and height:

$$A = \frac{1}{2}bh$$

Such variation is called **joint variation**. Because more than two variables are involved, we will not examine these graphically.

Joint Variation

If one variable varies directly with the product of two or more variables, the relationship is called **joint variation**. If y varies jointly with x and z, then $y = kxz$. The constant is called the **constant of variation**.

EXAMPLE 4

The volume, V, of a cone varies jointly with its height, h, and the area of its base, B. If $v = 6$ cm^3 when $h = 3$ cm and $B = 6$ cm^2, find V when $h = 2$ cm and $B = 8$ cm^2.

SOLUTION   The words *V varies jointly with h and B* can be expressed by the equation

$$V = khB$$

The relationship can also be read as "V is directly proportional to the product of h and B."

We can find k by substituting 6 for V, 3 for h, and 6 for B.

$$V = khB$$
$$6 = k(3)(6)$$
$$6 = k(18)$$
$$\frac{1}{3} = k$$

To find V when $h = 2$ and $B = 8$, we substitute these values into the formula $V = \frac{1}{3}hB$.

$$V = \frac{1}{3}hB$$
$$V = \frac{1}{3}(2)(8)$$
$$= \frac{16}{3} \text{ cm}^3$$

When $h = 2$ and $B = 8$, the volume is $5\frac{1}{3}$ cm^3.

▌▌ COMBINED VARIATION

Many problems involve a combination of direct and inverse variation. Such variation is called **combined variation**.

EXAMPLE 5

Highway construction   The time it takes to build a highway varies directly with the length of the road but inversely with the number of workers. If it takes 100 workers 4 weeks to build 2 miles of highway, how long will it take 80 workers to build 10 miles of highway?

SOLUTION

We can let t represent the time in weeks, l represent the length in miles, and w represent the number of workers. The relationship among these variables can be expressed by the equation

$$t = \frac{kl}{w}$$

STUDY TIP

An effective way to see if you understand a problem is to restate it in your own words.

We substitute 4 for t, 100 for w, and 2 for l to find k:

$$4 = \frac{k(2)}{100}$$

$$400 = 2k \qquad \text{Multiply both sides by 100.}$$

$$200 = k \qquad \text{Divide both sides by 2.}$$

We now substitute 80 for w, 10 for l, and 200 for k in the equation $t = \frac{kl}{w}$, and simplify:

$$t = \frac{kl}{w}$$

$$t = \frac{200(10)}{80}$$

$$t = 25$$

It will take 25 weeks for 80 workers to build 10 miles of highway.

Keep in mind that problems involving variation often consist of three parts:

1. Writing the variation equation from the words of the problem — this serves as a formula, or model.
2. Finding the value of k, the constant of variation (if it is not given).
3. Answering the question posed in the problem.

EXERCISE 8.5

VOCABULARY AND NOTATION   *In Exercises 1–6, fill in the blanks to make the given statements true.*

1. The equation $y = kx$ is an example of _____ variation.
2. This equation $y = kx$ is also a _____ function.
3. The equation $y = \frac{k}{x}$ is an example of _____ variation.

4. The equation $y = \frac{k}{x}$ is also an example of a _____ function.
5. The equation $y = kxz$ is an example of _____ variation, because one variable varies directly with the value of two variables.

6. The equation $y = \frac{kx}{z}$ is an example of _____ variation. In this example, y varies _____ with x and _____ with z.

CONCEPTS *In Exercises 7–12, tell whether each equation is an example of direct variation, inverse variation, joint variation, or combined variation and identify the constant of variation.*

7. $y = \dfrac{8}{x}$

8. $y = \dfrac{1}{2}xz$

9. $y = \dfrac{3x}{z}$

10. $y = \dfrac{1}{2}x$

11. $y = 3xz$

12. $y = \dfrac{3x}{z^2}$

PRACTICE *In Exercises 13–22, express each sentence as a formula.*

13. A varies directly with the square of p.

14. z varies inversely with the cube of t.

15. v varies inversely with the cube of r.

16. r varies directly with the square of s.

17. B varies jointly with m and n.

18. C varies jointly with x, y, and z.

19. P varies directly with the square of a, and inversely with the cube of j.

20. M varies inversely with the cube of n, and jointly with x and the square of z.

21. The force of attraction between two masses, m_1 and m_2, varies directly with the product of m_1 and m_2 and inversely with the square of the distance between them.

22. The force of wind on a vertical surface varies jointly with the area of the surface and the square of the velocity of the wind.

In Exercises 23–30, express each formula in words. In each formula, k is the constant of variation.

23. $L = kmn$

24. $P = \dfrac{km}{n}$

25. $E = kab^2$

26. $U = krs^2 t$

27. $X = \dfrac{kx^2}{y^2}$

28. $Z = \dfrac{kw}{xy}$

29. $R = \dfrac{kL}{d^2}$

30. $e = \dfrac{kPL}{A}$

APPLICATIONS

31. Area of a circle The area of a circle varies directly with the square of its radius, and the constant of variation is π. Find the area of the circle with a radius of 6 inches.

32. Electronics The power (in watts) lost in a resistor in the form of heat is directly proportional to the square of the current (in amperes) passing through it. The constant of proportionality is the resistance (in ohms). What power is lost in a 5-ohm resistor carrying a 3-ampere current?

33. Electronics The voltage (in volts) measured across a resistor is directly proportional to the current (in amperes) flowing through the resistor. The constant of variation is the resistance (in ohms). If 6 volts is measured across a resistor carrying a current of 2 amperes, find the resistance.

34. Temperature and pressure The temperature of a gas varies directly with its pressure. A temperature of 260° K produces a pressure of 65 pounds per square inch. What is the constant of proportionality?

35. Falling objects An object in free fall travels a distance s that is directly proportional to the square of the time t. If an object falls 1,024 feet in 8 seconds, how far will it fall in 10 seconds?

36. Finding distance The distance that a car can travel is directly proportional to the number of gallons of gasoline it consumes. If a car can go 288 miles on 12 gallons of gasoline, how far can it go on a full tank of 18 gallons?

37. Geometry For a fixed area, the length of a rectangle is inversely proportional to its width. A rectangle has a width of 18 feet and a length of 12 feet. If the length is increased to 16 feet and the area remains the same, find the width.

38. Gas pressure Under a constant temperature, the volume occupied by a gas is inversely proportional to the pressure applied. If the gas occupies a volume of 20 cubic inches under a pressure of 6 pounds per square inch, find the volume when the gas is subjected to a pressure of 10 pounds per square inch.

39. Value of a car   The value of a car usually varies inversely with its age. If a car is worth $7,000 when it is 3 years old, how much will it be worth when it is 7 years old?

40. Organ pipes   The frequency of vibration of air in an organ pipe is inversely proportional to the length of the pipe. (See Illustration 1.) If a pipe 2 feet long vibrates 256 times per second, how many times per second will a 6-foot pipe vibrate?

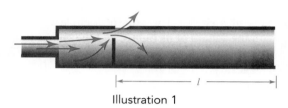

Illustration 1

41. Building construction   The deflection of a beam is inversely proportional to its width and the cube of its depth. If the deflection of a 4-inch-by-4-inch beam is 1.1 inches, find the deflection of a 2-inch-by-8-inch beam positioned as in Illustration 3.

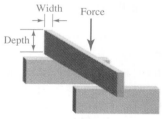

Illustration 2

42. Building construction   Find the deflection of the beam in Exercise 41 when the beam is positioned as in Illustration 3.

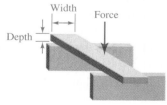

Illustration 3

43. Geometry   The area of a rectangle varies jointly with its length and width. If both the length and width are tripled, by what factor is the area multiplied?

44. Geometry   The volume of a rectangular solid varies jointly with its length, width, and height. If the length is doubled, the width is tripled, and the height is doubled, by what factor is the volume multiplied?

45. Storing oil   The number of gallons of oil that can be stored in a cylindrical tank varies jointly with the height of the tank and the square of the radius of its base. The constant of proportionality is 23.5. Find the number of gallons that can be stored in the cylindrical tank in Illustration 4.

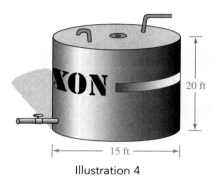

Illustration 4

46. Area of a triangle   The area of a triangle varies jointly as its base and the height drawn to that base. The constant of proportionality is $\frac{1}{2}$. If a triangle with a base of 10 ft and a height of 4 ft has an area of 20 ft^2, Find the area of a triangle with a base of 3 ft and a height of 8 ft.

47. Costs of a trucking company   The costs incurred by a trucking company vary jointly with the number of trucks in service and the number of hours they are used. When 4 trucks are used for 6 hours each, the costs are $1,800. Find the costs of using 10 trucks, each for 12 hours.

48. Moving company   The costs incurred by a moving company vary jointly with the number of employees used to pack the items prior to moving and the number of hours each employee works. When 5 people work 6 hours each, the cost is $300. What would the cost be for 6 people?

49. Gas pressure   The pressure of a certain amount of gas is directly proportional to the temperature (measured in degrees Kelvin) and inversely proportional to the volume. A sample of gas at a pressure of 1 atmosphere occupies a volume of 1 cubic meter at a temperature of 273° Kelvin. When heated, the gas expands to twice its volume but the pressure remains constant. To what temperature is it heated?

50. Tension A stone twirled at the end of a string is kept in its circular path by the tension of the string. The tension, T, is directly proportional to the square of the speed, s, and inversely proportional to the radius, r, of the circle. In Illustration 5, the tension is 32 pounds when the speed is 8 feet/second and the radius is 6 feet. Find the tension when the speed is 4 feet/second and the radius is 3 feet.

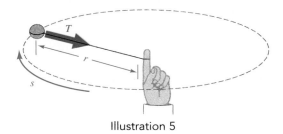

Illustration 5

51. Centripetal force When an object moves in a circular path, the centripetal force, F, varies directly as the square of the velocity, v, and inversely as the radius, r, of the circle. If $F = 64$ when $v = 2$ and $r = 5$, find F when $v = 3$ and $r = 4$.

52. Car skidding The force needed to keep a car from skidding on a curve varies inversely as the radius of the curve and jointly as the weight of the car and the square of the speed. It takes 3,000 pounds of force to keep a 2,000 pound car from skidding on a curve of radius 500 ft at 30 mph. What force is needed to keep the same car from skidding on a curve of radius 800 feet at 60 mph?

53. Discuss the similarities and differences between direct variation, inverse variation, joint variation, and combined variation.

54. As temperature increases on the Fahrenheit scale, it also increases on the Celsius scale. Is this direct variation? Explain.

55. As the cost of a purchase (less than \$5) increases, the amount of change received from a five-dollar bill decreases. Is this inverse variation? Explain.

REVIEW *Specify the domain and the range of each function specified below (you may use either set notation or interval notation). Each function is represented in three ways — with a definition, a graphical display, and a TABLE. Use one or all three representations to aid you in determining domain and range.*

56. $f(x) = x + 3$

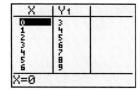

57. $f(x) = x^2 + 3$

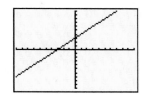

58. $f(x) = \sqrt{x + 3}$

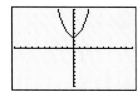

59. $f(x) = |x + 3|$

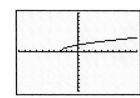

CHAPTER SUMMARY

CONCEPTS

REVIEW EXERCISES

Section 8.1

Relations and Functions

A *relation* is a set of ordered pairs that establishes a correspondence between the numbers x and the values of y.

When each x-value is paired with a unique y-value we call the relation a *function*.

The *domain* is the set of all numbers x for which the function is defined.

The *range* is the set of all possible y-values.

When given the graph of an equation, the vertical line test can be used to determine if the graph is a function. If any vertical line intersects the graph at more than one point, the graph is *not* a function.

The notation $y = f(x)$ denotes that the variable y is a function of x.

1. State whether each equation defines y to be a function of x. Find the domain and range.
    a. $y = 4x - 1$
    b. $x = 3y - 10$
    c. $y = 3x^2 + 1$
    d. $y = \dfrac{4}{2 - x}$
    e. $x = \dfrac{y + 3}{2}$
    f. $y^2 = 4x$

2. Use the vertical line test to determine whether each graph represents a function.

    a.
    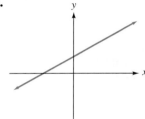

    b.
    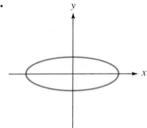

    c.
    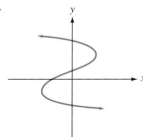

    d.
    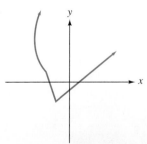

3. Assume that $f(x) = 3x + 2$ and $g(x) = x^2 - 4$. Find each value.
    a. $f(-3)$ b. $g(8)$ c. $g(-2)$ d. $f(5)$

4. Determine the value of $f(x)$ from the displayed function $y = f(x)$.

a.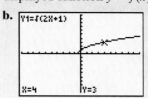

b.

All straight-line graphs, except for vertical lines, are functions. A *linear function* is defined by an equation that can be written in the form $y = mx + b$ or $f(x) = mx + b$.

5. Tell which equations define linear functions.

a. $y = 3x + 2$

b. $y = \dfrac{x + 5}{4}$

c. $4x - 3y = 12$

d. $y = x^2 - 25$

Functions and Graphs

Graphs are always "read" from left to right. If, as the x-value increases the y-value also increases, the function is said to be *increasing*. The graph rises (when viewed from left to right).

If, as the x-values increase, the y-values decrease, the function is said to be decreasing. The graph falls (when viewed from left to right).

6. Determine the interval(s) over which the function is increasing, decreasing, and/or constant.

a.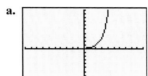

b.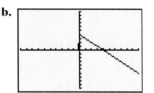

c.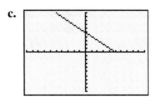

d.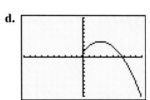

e.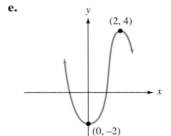

f.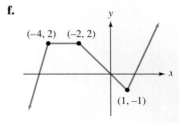

The *turning points* of a graph are the points where the graph changes from increasing to decreasing or from decreasing to increasing. It is also called a *relative maximum* or a *relative minimum* and can be determined with the graphing calculator.

The *absolute maximum* and the *absolute minimum* are the *y*-coordinates of the highest and lowest points of the function, respectively.

7. Graph each function using the *x*-intercepts and the turning points (relative maximum/relative minimum).

a. $y = -x^2 + 36$

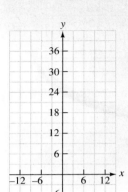

b. $y = |x + 3| - 5$

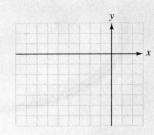

c. $y = -2x^3 + 8x^2 - 6x$

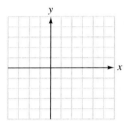

d. $y = (x - 3)(2x + 5)(x - 1)$

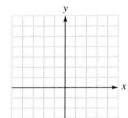

8. Suppose that the ball shown in Illustration 1 is thrown straight up with a velocity of 128 feet per second. The function $y = 128x - 16x^2$ models the relationship between the number of feet above the ground, *y*, and the time, *x*, measured in seconds.

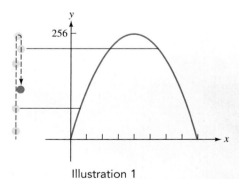

Illustration 1

a. Graph in an appropriate viewing window (one in which all interesting features as well as *x*- and *y*-intercepts are visible).

b. Determine the relative maximum.

c. At the relative maximum, interpret the meanings of the *x*- and *y*-values of the coordinates within the context of the problem.

d. Explain why the two sets of coordinates $(3.2, 245.76)$ and $(4.8, 245.76)$ both appear on the graph.

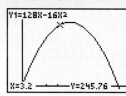
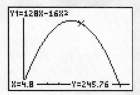

e. Explain the meaning of the x-intercept.

9. A man wants to build the rectangular pen shown in Illustration 2 to house his dog. To save fencing, he intends to use one side of his garage and 80 feet of fencing.

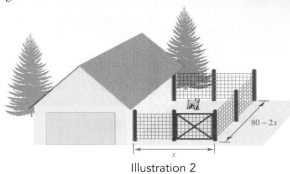

Illustration 2

a. Write an equation where y is the area and x is the width to model this problem.

b. Sketch the graph of the equation using the x-intercepts and the turning point.

c. Find the maximum area he can enclose.

d. What is the width when area is maximized?

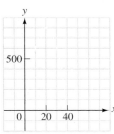

Section 8.3

Slope of a Nonvertical Line

The slope of a line is its rate of increase or decrease.

$$m = \frac{\Delta y}{\Delta x} = \frac{y_2 - y_1}{x_2 - x_1} = \frac{\text{rise}}{\text{run}}$$

A line is determined by a point and its slope. A linear equation in the form $y = mx + b$ is in *slope-intercept form*. The coefficient of x, m, represents the slope of the line, and b denotes the y-intercept.

10. Find the slope of the line passing through points P and Q.

a. $P(2, 5)$ and $Q(5, 8)$.   **b.** $P(-3, -2)$ and $Q(6, 12)$.

c. $P(-3, 4)$ and $Q(-5, -6)$.   **d.** $P(5, -4)$ and $Q(-6, -9)$.

e. $P(3, 2)$ and $Q(5, 1)$.

11. Find the slope of the graph of each equation, if one exists.

a. $2x - 3y = 18$ **b.** $2x + y = 8$

c. $-2(x - 3) = 10$ **d.** $3x + 1 = 7$

e. $2(y - 1) = 8$

Lines with positive slope are increasing functions, and the lines rise when viewed from left to right. Lines with negative slope are decreasing functions, and the lines fall when viewed from left to right.

Horizontal lines have equations of the form $y = b$ and have a slope of 0.

Vertical lines have equations of the form $x = a$ and have no defined slope.

If two lines are parallel, their slopes are equal. If two nonvertical lines are perpendicular, their slopes are negative reciprocals.

12. Write the equation of the line with the given properties.

    a. Through $(0, -5)$ and with slope $\frac{2}{3}$.

    b. Through $(0, 2)$ and with slope $-\frac{1}{2}$.

13. Determine if the displayed graph has positive, negative, 0, or no defined slope.

    a. 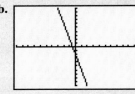   **b.**

    c. 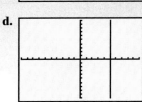   **d.**

14. Tell whether the lines with the given slopes are parallel, perpendicular, or neither.

    a. $m_1 = 4, m_2 = -\frac{1}{4}$ **b.** $m_1 = 0.5, m_2 = \frac{1}{2}$

    c. $m_1 = 0.5, m_2 = -\frac{1}{2}$ **d.** $m_1 = 5, m_2 = -0.2$

Equations of Lines

The equation of the line passing through $P(x_1, y_1)$ and with slope m is $y - y_1 = m(x - x_1)$. This is called the *point-slope form* of a linear equation. The *standard form* of a linear equation is $Ax + By = C$ (A and B cannot both be zero).

The equation of a line can also be used to find the declining value of an item (such as a car or aging equipment in a business). The method is called *straight-line depreciation*. The rate of depreciation (or appreciation, increase in value) is given by the slope of the line.

The midpoint of the line segment $P(x_1, y_1)$ and $Q(x_2, y_2)$ is the point, M, with coordinates of

$$\left(\frac{x_1 + x_2}{2}, \frac{y_1 + y_2}{2}\right).$$

15. Write the equation of the line with the given properties. Write each equation in standard form.

    a. Slope of 3; passing through $P(-8, 5)$.

    b. Passing through $(-2, 4)$ and $(6, -9)$.

    c. Passing through $(-3, -5)$; parallel to the graph of $3x - 2y = 7$.

    d. Passing through $(-3, -5)$; perpendicular to the graph of $3x - 2y = 7$.

16. Write the equation of the pictured lines. Write each equation in standard form.

    a.   **b.**

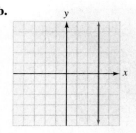

17. Find the midpoint of the line segment joining $P(-3, 5)$ and $Q(6, 11)$.

18. A computer purchased in 1998 for $2,500 has a value of $850 3 years later.

    a. What is the rate of depreciation?

    b. Write a linear equation to model this depreciation.

    c. Determine a good viewing window to display the graph of this equation.

19. In 1940, a cameo pin retailed for $15. The same cameo has increased in value in a linear pattern, and 58 years later is valued at $200. Find the rate of appreciation.

Section 8.5	Variation

The words *y varies directly as x* or *y is directly proportional to x* mean that for some nonzero constant k, $y = kx$.

The letter k is called the *constant of variation* or the constant of proportionality.

The words *y varies inversely with x* or *y is inversely proportional to x* mean that $y = \frac{k}{x}$ for some constant of variation, k. Problems involving direct and inverse variation can be examined graphically.

Joint variation occurs when more than two variables are involved; one variable varies directly with the product of several variables.

Problems that involve a combination of direct and inverse variation are called *combined variation* problems.

20. Assume that x varies directly as y. If $x = 12$ when $y = 2$, find the value of x when $y = 12$.

21. Assume that x varies inversely with y. If $x = 24$ when $y = 3$, find the value of y when $x = 12$.

22. Assume that x varies jointly with y and z. Find the constant of variation if $x = 24$ when $y = 3$ and $z = 4$.

23. Assume that x varies directly with t and inversely with y. Find the constant of variation if $x = 2$ when $t = 8$ and $y = 64$.

24. Physicians use the body-mass index (BMI), the accepted measure of a person's body composition, as a weight/height comparison. The BMI varies directly as an individual's weight in pounds and inversely as the square of their height in inches. If a person weighing 182 pounds at 69″ tall has a BMI of 26.95, find the BMI (to the nearest whole number) of a person weighing 132 pounds at 55″ height.

25. The volume of a cylinder varies jointly with its height and the area of its base. If the height is tripled and the area of the base is doubled, by what factor is the volume multiplied?

CHAPTER TEST

1. Does the equation $y = |x + 1|$ determine y to be a function of x?

2. Determine the domain and range of the displayed function.

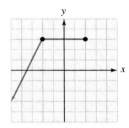

3. If $f(x) = x^2 + 4x$, find $f(-3)$.

4. Given the function at the right, determine the intervals over which the function is
  a. increasing
  b. decreasing
  c. constant

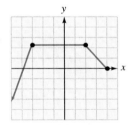

5. A child, standing on the deck of his house, throws a ball straight up with an initial velocity of 32 feet per second.

The equation $y = -16x^2 + 32x + 10$ gives the height, y, of the ball x seconds after it is thrown.

a. Graph in an appropriate viewing window.

b. Find the maximum height reached by the ball.

c. Find the time it takes the ball to hit the ground.

d. What is the turning point of this function?

6. Find the slope of the graph of $x = \dfrac{3y - 8}{2}$.

7. Find the slope of the line passing through $P(-5, -6)$ and $Q(5, -2)$.

8. Write the equation of a line whose slope is $\frac{4}{5}$ and passes through $(0, 3)$.

9. Are the lines $y = \dfrac{2}{3}x + 4$ and $y = \dfrac{3}{2}x - 4$ parallel, perpendicular, or neither?

10. Write the equation of a line perpendicular to the y-axis passing through $(3, -1)$.

11. Write the equation of a line passing through $(4, 1)$ and $(-7, 6)$.

12. Does $y(x + 3) + 4 = x(y - 2)$ define a linear function? Explain your reasoning.

13. Find the midpoint of the line segment joining $P(4, -2)$ and $Q(-3, 3)$.

14. Express as a formula: y varies directly with w and inversely with the square of z.

15. Express as a formula: The volume, V, of a cone varies jointly as the area of its base and its height. The constant of proportionality is $\frac{1}{3}$.

16. The power (in watts) lost in a resistor in the form of heat is directly proportional to the square of the current (in amperes) passing through it. If k is the constant of variation, express this relationship with an equation.

17. In the previous equation, if the constant of proportionality is 5, answer the following questions.

a. On the interval $[0, \infty)$ is the variation function increasing or decreasing?

b. Locate the point $(3, 45)$ on the graph. To what do the values 3 and 45 correspond within the context of the problem?

18. x varies directly with y. If $x = 30$ when $y = 4$, find x when $y = 9$.

19. v varies inversely with t. If $v = 55$ when $t = 20$, find t when $v = 75$.

CUMULATIVE REVIEW

CHAPTERS 7 AND 8
Vocabulary / Concepts

1. To _____ a polynomial means to rewrite the indicated sum or difference of terms as a product of factors.

2. Simplify. Results should contain only positive exponents.

    a. $\dfrac{1}{x^{-3}}$ **b.** $\dfrac{x^8}{x^{-2}}$ **c.** $x^{-5} \cdot x^0$ **d.** $\dfrac{x^{-2}}{x^5}$

3. To _____ a polynomial means to remove grouping symbols and to combine like terms.

4. To _____ a polynomial means to find the value when given values for the variable(s).

5. The _____ of an equation is the value(s) of the variable(s) that make the equation a true statement.

6. A _____ is a set of ordered pairs in which each x-value in the domain is assigned a unique y-value in the range.

7. Each of the following represent solutions to an equation. Based on the given information, is the equation a function?

    a. $\{(1, 1), (1, -1), (0, 0), (4, 2), (4, -2)\}$

    b.

X	Y1
-3	9
-2	4
-1	1
0	0
1	1
2	4
3	9

X= -3

8. *AND* statements represent the _____ of the set of values that satisfy two simple inequalities.

9. *OR* statements represent the _____ of the set of values that satisfy two simple inequalities.

10. Write $|x + 2| \leq 4$ as a compound inequality.

11. State which of the following graphs are functions.

a.

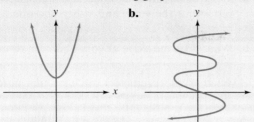

b.

c.

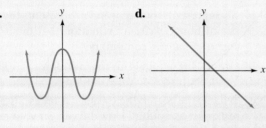

d.

e.

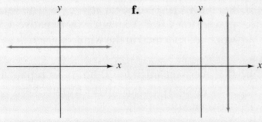

f.

12. If $f(x) = x^2 - 3x + 2$, find each of the following:

  a. $f(-1)$ **b.** $f(3)$ **c.** $f(-3)$ **d.** $f(8)$

13. Write the equation $x = 3y - 4$ in $y = mx + b$ form.

14. The slope of a line is defined as the change in ___ over the change in ___.

15. A horizontal line is written in the form _____.

16. A vertical line is written in the form _____.

17. The slope of a horizontal line is ___.

18. The slope of a vertical line is _____.

19. When a linear equation is written in the form $y = mx + b$, it is in _____ form. The coefficient of x is the _____ of the line, and b is the _____.

20. When a linear equation is written in the form $y - y_1 = m(x - x_1)$ it is in _____ form, where m is the _____ of the line and _____ are the coordinates of a point on the line.

Practice

21. a. Simplify $\dfrac{(2x^{-4}y^2)^3}{4x^{-2}y^{-1}}$.

  b. Evaluate both the original monomial and the simplified monomial when $x = \frac{1}{2}$ and $y = \frac{1}{5}$ using the STOre feature of your calculator.

In Exercises 22–26, solve each equation algebraically. Confirm your solutions with either the STOre or TABLE feature of the graphing calculator.

22. $\dfrac{8(x - 5)}{3} = 2(x - 4)$

23. $2x^2 - 7x = 4$

24. $2|x + 1| - 2 = 0$

25. $3 + |2x - 1| = 2$

26. $\dfrac{x}{x - 2} - \dfrac{5}{x - 2} = 3$

27. Solve the compound inequality $-2 \leq \dfrac{5 - 3x}{2} \leq 2$ and express the solution as a number line graph and in interval notation. Confirm your solution graphically.

28. Solve the inequality $3|2x + 5| \geq 9$. Express your solution in interval notation and as a number line graph. Confirm your solution graphically.

29. If $f(x) = 3x - 1$, find $f(2) - f(3)$.

30. For what intervals of x is the accompanying graph

  a. increasing?

  b. decreasing?

  c. constant?

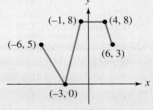

31. Graph the equation $2x + 3y = 6$ using x- and y-intercepts.

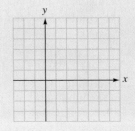

32. Determine the domain and range of the function specified below:

$$\{(1, 5), (2, 5), (-1, -1), (-3, -5)\}$$

33. Find the slope of the line whose displayed TABLE is pictured.

X	Y1
0	-4
1	-3.5
2	-3
3	-2.5
4	-2
5	-1.5
6	-1

X=0

34. Find the slope and the y-intercept of the line with the equation $x = \dfrac{3}{4} - \dfrac{1}{4}y$.

35. Determine if the pair of lines determined by the given equations are parallel, perpendicular, or neither.

$$x - 2y = -18; \; 2x + y = -6$$

In Exercises 36–40, write the equation of the line with the given properties. Then transform the equation into standard form.

36. The line with slope $\frac{2}{3}$ and passing through $(0, 5)$.

37. The line with slope 3 and passing through $(5, 1)$.

38. The line with slope 0, perpendicular to the line with equation $x = 5$ and passing through the point $(6, 2)$.

39. The line passing through the points with coordinates $(3, 2)$ and $(-1, 4)$.

40. The line through $(-6, 3)$ and perpendicular to the line with equation $y + 3x = -12$.

41. Solve the equation $-3x + 15 = \sqrt{x - 5}$ using the INTERSECT feature of your calculator. Check your solution in the TABLE.

42. Solve the equation $\sqrt{2x + 5} - 4 = 0$ using the ROOT/ZERO feature of your calculator. Check your solution in the TABLE.

43. Solve the equation $x^3 - x^2 - 12x = 0$ graphically using the INTERSECT or ROOT/ZERO feature of the graphing calculator. Check your solutions in the TABLE.

44. Graph the function $y = 3(x - 1)^2 + 6$ and determine the coordinates of the minimum.

45. Assume that an antique purchased for $300 increases at a rate of $15 per year.

   a. Write an equation to model this increase, where y represents the value and x the subsequent years.

   b. What does the slope of your equation represent?

   c. What is the predicted value of the antique in 20 years?

46. A computer service charges a fixed amount of $45 and $20 per hour for repair.

   a. If y represents total cost and x represents the number of hours, write an equation that models the cost of repairing computers.

   b. Graph the model in the window $[0, 30]$ by $[-10, 300]$.

   c. What is the cost to the owner if it takes 5 hours to repair his computer? How did you get your result?

   d. Set the TABLE to begin at $x = -2$. The value displayed for y is 5. Explain why negative x-values were not specified in the window given in part (b).

   e. If it took technicians 20 hours to repair a computer, what is the cost to the owner? How did you arrive at your result?

Applications

47. Car rental   It costs $45 per day to rent a car plus $0.24 per mile. If you had $100 to spend for one day of rental, how many miles could be driven?

48. Calculating grades   A student has test scores of 78%, 80%, and 85%. What must she make on the fourth test to maintain an average of 70% or higher?

49. Area of a circle   The area of a circle varies directly with the square of its radius, and the constant of variation is π. Find the area of a circle with a radius of 9 mm.

50. Value of a car   The value of a car varies inversely with age. If a car is worth $8,000 when it is 5 years old, how much will it be worth after 10 years?

CAREERS & MATHEMATICS

© Robert Essel NYC/CORBIS

Economist   Economists study the way a society uses resources such as land, labor, raw materials, and machinery to provide goods and services. They analyze the results of their research to determine the costs and benefits of making, distributing, and using resources in a particular way. Some economists are theoreticians who use mathematical models to explain the causes of recession and inflation. Most economists, however, are concerned with practical applications of economic policy in a particular area.

Qualifications   Economists must thoroughly understand economic theory, mathematical methods of economic analysis, and basic statistical procedures. Training in computer science is highly recommended.

Job Outlook   Employment of economists is expected to grow faster than the average for all occupations during the next decade. Opportunities should be best for economists in business and industry, research organizations, and consulting firms.

Example Application   An electronics firm manufactures tape recorders, receiving $120 for each unit it makes. If x represents the number of recorders produced, then the income received is determined by the *revenue function*, given by the linear equation

$$R(x) = 120x$$

The manufacturer has determined that the *fixed costs* for advertising, insurance, utilities, and so on, are $12,000 per month, and the *variable cost* for materials is $57.50 for each machine produced. Thus, the *cost function* is given by the linear equation

$$C(x) = variable\ cost\ +\ fixed\ costs$$
$$= 57.50x + 12,000$$

The company's profit is the amount by which revenue exceeds costs. It is determined by the *profit function*, given by the equation

$$Profit = revenue - costs$$
$$P(x) = R(x) - C(x)$$
$$= 120x - (57.50x + 12,000)$$
$$= 62.50x - 12,000$$

If $P(x) > 0$, the company is making money. If $P(x) < 0$, it is operating at a loss. How many recorders must the company manufacture to break even?

Solution   Graph the profit function $y = P(x) = 62.50x - 12,000$. The break-even point is that value of x that gives a profit of zero. It is the x-intercept of the graph of the profit function. (See Illustration 1.) To find it, set $P(x)$ equal to 0 and solve for x.

$$P(x) = 0$$
$$62.50x - 12,000 = 0$$
$$62.50x = 12,000 \qquad \text{Add 12,000 to both sides.}$$
$$x = 192 \qquad \text{Divide both sides by 62.50}$$

The company must manufacture and sell 192 tape recorders each month to break even.

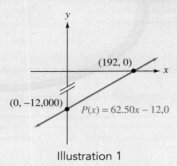

Illustration 1

EXERCISES

1. Find the revenue and the cost of manufacturing 192 units, and verify that the revenue and the cost are equal.

2. Determine the company's profit if it manufactures 150 units each month.

3. Determine the company's profit if it manufactures 400 units each month.

4. How many units must be manufactured each month to produce a total profit of $47,375?

(*Answers:* **1.** $23,040 revenue and cost   **2.** $2625 loss   **3.** $13,000 profit   **4.** 90 units)

Rational Exponents and Radicals

© Royalty-Free/CORBIS

 InfoTrac

Do a subject guide search on "alternating current." Click on "View periodical references," and find the article "AC fundamentals" from *EC&M Electrical Construction & Maintenance*, Jan 1, 2003, v102. In an alternating current circuit, voltage, current, and impedance are represented by complex numbers. Using the formula found in the article, find the voltage (in volts) of a circuit with a current of $5 - 6.4i$ amperes and an impedance of $3.2 + 4.25i$ ohms. Now find the current of a circuit with voltage of $6.25 - 4.00i$ volts and an impedance of $5.3 + 6.75i$ ohms.

Complete this project after studying Section 9.7.

PERSPECTIVE

Don't Buck the System

Pythagoras was a teacher. Although it was unusual for schools at that time, his classes were coeducational. According to some legends, Pythagoras married one of his students. He and his followers formed a secret society with two rules: membership was for life, and members could not reveal the secrets they knew.

Much of their teaching was good mathematics, but some ideas were strange. To them, numbers were sacred. Because beans were used as counters to represent numbers, Pythagoreans refused to eat beans. They also believed that the only numbers were the whole numbers. To them, fractions were not numbers; $\frac{2}{3}$ was just a way of comparing the whole numbers 2 and 3. They believed that whole numbers were the building blocks of the universe, just as we believe atoms are. The basic Pythagorean doctrine was "all things are number," and they meant *whole* number.

The Pythagorean theorem was an important discovery of the Pythagorean school, yet it caused some division in the ranks. The right triangle in Illustration 1 has two legs of length 1. By the Pythagorean theorem, the length of the hypotenuse is $\sqrt{2}$. One of their own group, Hippasus of Metapontum, discovered that $\sqrt{2}$ is an irrational number: There are *no* whole numbers a and b that make the fraction $\frac{a}{b}$ exactly equal to $\sqrt{2}$. This discovery was not appreciated by the other Pythagoreans. How could everything in the universe be described with whole numbers, when the side of this simple triangle couldn't? The Pythagoreans had a choice. Either revise and expand their beliefs, or cling to the old. According to legend, the group was at sea at the time of the discovery. Rather than upset the system, they threw Hippasus overboard.

Illustration 1

The ability to work with roots and radicals enables us to solve many more real-world problems. This chapter will examine the algebra of radicals and radical functions. At the conclusion of the chapter, we will extend our real-number system to include the complex number system.

9.1   INTRODUCTION TO ROOTS AND RADICALS

- SQUARE ROOTS AND RADICALS
- CUBE ROOTS
- *n*TH ROOTS
- SQUARE ROOTS AND *n*TH ROOTS OF EXPRESSIONS WITH VARIABLES

▮ SQUARE ROOTS AND RADICALS

The opposite of squaring a number is finding the **square root** of a number. For example, the square of 3, 3^2, is equivalent to the number 9. Conversely, the positive square root of 9 is 3. Some other examples of square roots are:

4 is a square root of 16 because $4^2 = 16$.

5 is a square root of 25 because $5^2 = 25$.

8 is a square root of 64 because $8^2 = 64$.

a^3 is the square root of a^6 because $(a^3)^2 = a^6$.

$7xy$ is a square root of $49x^2y^2$ because $(7xy)^2 = 49x^2y^2$.

All positive real numbers have two square roots, one positive and one negative. We symbolize the nonnegative or **principal** square root with a **radical sign**, $\sqrt{}$. An expression containing a radical sign is a **radical expression**. The expression under the radical sign is called the **radicand**.

$$\text{radical sign} \rightarrow \sqrt{a} \leftarrow \text{radicand}$$

Square Roots

If $x > 0$, the *principal square root* of x is the positive square root of x and is denoted as $\sqrt{x}$. The negative or opposite of the principal square root is denoted as $-\sqrt{x}$.

The principal square root of 0 is 0: $\sqrt{0} = 0$.

 EXAMPLE 1

Simplify:

a. $\sqrt{81}$ **b.** $-\sqrt{81}$ **c.** $\sqrt{225}$ **d.** $\sqrt{\dfrac{1}{4}}$

e. $\sqrt{\dfrac{16}{121}}$ **f.** $\sqrt{0.04}$ **g.** $-\sqrt{0.0009}$ **h.** $\sqrt{-4}$

SOLUTION

a. $\sqrt{81} = 9$ because $9^2 = 81$.

b. $-\sqrt{81} = -9$. The negative in front of the radical indicates the negative, or opposite, of the square root of 81. The square root of 81 is 9, and its opposite is -9.

c. $\sqrt{225} = 15$ because $15^2 = 225$.   **d.** $\sqrt{\dfrac{1}{4}} = \dfrac{1}{2}$ because $\left(\dfrac{1}{2}\right)^2 = \dfrac{1}{4}$.

e. $\sqrt{\dfrac{16}{121}} = \dfrac{4}{11}$ because $\left(\dfrac{4}{11}\right)^2 = \dfrac{16}{121}$.

f. $\sqrt{0.04} = 0.2$ because $(0.2)^2 = 0.04$.

g. $-\sqrt{0.0009} = -0.03$ because $(0.03)^2 = 0.0009$ and its opposite is -0.03.

h. $\sqrt{-4}$ is undefined in the real-number system because there is no real number that can be multiplied *by itself* to obtain a result that is negative (a positive number times a positive number is positive and a negative number times a negative number is also positive).

TECHNOLOGY TIP

The $\sqrt{}$ key can be used to find square roots. As long as the calculator is in **real** mode, an error message results when $\sqrt{-4}$ is entered.

```
ERR:NONREAL ANS
1▮Quit
2:Goto
```

This confirms our definition that negative radicands are undefined in the real-number system.

TI-86   Your display for $\sqrt{-4}$ would appear as $(0, 2)$. This display would be interpreted in the Complex Number System. This is addressed in Section 9.7.

The numbers 1, 4, 9, 16, and 25 are examples of **perfect squares**, so called because their square roots are rational numbers. Numbers like 3 are *not* perfect squares because $\sqrt{3}$ is not equivalent to a rational number. If using the calculator to evaluate $\sqrt{3}$, the number displayed is an *approximation*. It is an irrational number and is a nonrepeating, nonterminating decimal.

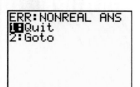

```
√(3)
          1.732050808
```

The decimal terminates *on the screen* because that is all the display that programming allows. It is imperative that when approximating roots you *clearly* indicate when a result is an approximation. This can be done by using a symbol for approximation, either $\approx$ or $\doteq$. Therefore, $\sqrt{3} \doteq 1.732$.

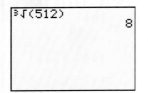
■ CUBE ROOTS

Finding roots can be extended from square roots to cube roots. Again, finding the cube root of a number is the opposite of cubing the number.

$\quad$ 4 is a cube root of 64 because $4^3 = 64$.

$\quad$ $3x^2y$ is a cube root of $27x^6y^3$ because $(3x^2y)^3 = 27x^6y^3$.

$\quad$ -2 is a cube root of -8 because $(-2)^3 = -8$.

To denote cube roots, we indicate the **index** or **order**, as shown below:

$$\text{index} \rightarrow \sqrt[3]{a} \leftarrow \text{radicand}$$

Cube Roots

The **cube root** of a is denoted as $\sqrt[3]{a}$ and is defined by $\sqrt[3]{a} = b$ if $b^3 = a$.

All real numbers have exactly one cube root. That root is either a rational number (a perfect cube) or an irrational number.

EXAMPLE 2

Simplify. If a root is irrational, approximate it to two decimal places.

a. $\sqrt[3]{125}$ **b.** $\sqrt[3]{\dfrac{1}{8}}$ **c.** $\sqrt[3]{-\dfrac{1}{8}}$

d. $\sqrt[3]{-27x^3}$ **e.** $\sqrt[3]{0.216x^3y^6}$ **f.** $\sqrt[3]{13}$

SOLUTION   **a.** $\sqrt[3]{125} = 5$ because $5^3 = 5 \cdot 5 \cdot 5 = 125$.

b. $\sqrt[3]{\dfrac{1}{8}} = \dfrac{1}{2}$ because $\left(\dfrac{1}{2}\right)^3 = \dfrac{1}{2} \cdot \dfrac{1}{2} \cdot \dfrac{1}{2} = \dfrac{1}{8}$.

c. $\sqrt[3]{-\dfrac{1}{8}} = -\dfrac{1}{2}$ because $\left(-\dfrac{1}{2}\right)^3 = \left(-\dfrac{1}{2}\right)\left(-\dfrac{1}{2}\right)\left(-\dfrac{1}{2}\right) = -\dfrac{1}{8}$.

d. $\sqrt[3]{-27x^3} = -3x$ because $(-3x)^3 = (-3x)(-3x)(-3x) = -27x^3$.

e. $\sqrt[3]{0.216x^3y^6} = 0.6xy^2$ because
$(0.6xy^2)^3 = (0.6xy^2)(0.6xy^2)(0.6xy^2) = 0.216x^3y^6$.

f. $\sqrt[3]{13} = 2.351334688\ldots$ The equality sign was used because the three dots at the end of the approximation indicate that the expression goes on indefinitely without repeating. Because the directions instructed us to state the irrational result correct to two decimal places, we may write $\sqrt[3]{13} \approx 2.35$. Notice that a symbol for approximation is used so that it is clear that 2.35 is an approximate value rather than an exact value.

■ *n*TH ROOTS

Just as there are square roots and cube roots, there are fourth, fifth, sixth roots, and so on. The notation $\sqrt[n]{a}$ indicates the **nth root** of a number, where n is a natural number greater than 1 ($n > 1$). The letter n is the index and specifies the order of the root. The index of 2 is omitted when writing square roots.

When the index is odd, the radical $\sqrt[n]{a}$ ($n > 1$) is often called an **odd root**. Every real number has exactly one real, nth root when n is odd. Moreover, when the radicand is positive, the root is positive. When the radicand is negative, the root is negative.

When the index is even, the radical $\sqrt[n]{a}$ ($n > 1$) is called an **even root**. The notation $\sqrt[n]{a}$ represents the positive, or principal root. When finding even roots we often use absolute value symbols to guarantee that the nth root is positive:

$$\sqrt[4]{(-3)^4} = |-3| = 3.$$

When the index is *even* the radicand must be positive for the radical to be defined within the real-number system. The basic definitions for the nth roots of real numbers are summarized as follows:

Definition of $\sqrt[n]{a^n}$

If n is a real number and $n > 1$, then

if n is an odd natural number, $\sqrt[n]{a^n} = a$.

if n is an even natural number, $\sqrt[n]{a^n} = |a|$.

EXAMPLE 3

Simplify each of the following roots. Verify results with a calculator.

a. $\sqrt[4]{625}$ **b.** $\sqrt[4]{-81}$ **c.** $\sqrt[3]{-27}$ **d.** $-\sqrt[6]{64}$ **e.** $\sqrt[5]{27}$

SOLUTION

a. $\sqrt[4]{625} = 5$ because $5^4 = 625$;

$\sqrt[4]{625} = \sqrt[4]{5^4} = |5| = 5.$

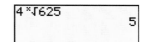

b. $\sqrt[4]{-81}$ is undefined for real numbers.

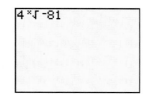
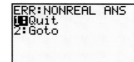

c. $\sqrt[3]{-27} = -3$ because $(-3)^3 = -27$;

$\sqrt[3]{-27} = \sqrt[3]{(-3)^3} = -3.$

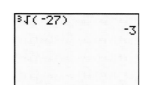

d. $-\sqrt[6]{64} = -2$ because $2^6 = 64$. The negative in front of the radical indicates we want the opposite of the indicated root.

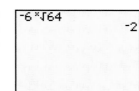

e. $\sqrt[4]{27} \approx 2.2795$

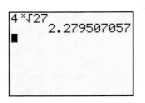

SQUARE ROOTS AND *n*TH ROOTS OF EXPRESSIONS WITH VARIABLES

If $x \neq 0$, the positive (principal) square root of x^2 is denoted as $\sqrt{x^2}$. *Incorrectly*, students often assume that $\sqrt{x^2} = x$. This assumption is a result of the idea that squaring a number and extracting a square root are inverse operations, with one operation "undoing" the other. The fallacy of this argument is demonstrated when taking the square root of a variable because we do not know if the variable represents a positive or a negative number. For example, if $x = 2$, then $\sqrt{2^2} = 2$, or $\sqrt{x^2} = x$. However, if $x = -2$, then $\sqrt{(-2)^2} \neq -2$; rather, $\sqrt{(-2)^2} = 2$. The use of absolute value guarantees extracting the principal root. Care must be taken when the index of a radical is even and variables are contained in the radicand.

EXAMPLE 4

Assuming x is a real number, simplify the following:

a. $\sqrt{16x^2}$ b. $\sqrt{(x + 4)^2}$ c. $\sqrt{x^2 + 2x + 1}$

d. $\sqrt{x^4}$ e. $\sqrt[5]{x^5}$ f. $\sqrt[4]{16x^4}$ g. $\sqrt{x^6}$

SOLUTION

a. $16x^2 = \sqrt{(4x)^2} = |4x| = 4|x|$. Absolute value symbolism is necessary because x could represent a negative value. The use of absolute value ensures the principal, positive root.

b. $\sqrt{(x + 4)^2} = |x + 4|$. Absolute value ensures the principal root.

c. $\sqrt{x^2 + 2x + 1} = \sqrt{(x + 1)^2} = |x + 1|$. Absolute value ensures the principal root.

d. $\sqrt{x^4} = \sqrt{(x^2)^2} = x^2$. Because x^2 is *always* positive (regardless of the value of x) absolute value symbols are unnecessary.

e. $\sqrt[5]{x^5} = \sqrt[5]{(x)^5} = x$. Because the index is odd, absolute value symbolism is unnecessary.

f. $\sqrt[4]{16x^4} = \sqrt[4]{(2x)^4} = |2x| = 2|x|$. Absolute value ensures the principal root.

g. $\sqrt{x^6} = \sqrt{(x^3)^2} = |x^3|$. Absolute value ensures the principal root.

The definitions concerning $\sqrt[n]{a}$ are summarized as follows:

Summary of Definitions of $\sqrt[n]{a}$

If n is a real number and $n > 1$, then

if $a > 0$, then $\sqrt[n]{a}$ is a positive number such that $(\sqrt[n]{a})^n = a$.
if $a = 0$, then $\sqrt[n]{a} = 0$.
if $a < 0$ and n is odd, then $\sqrt[n]{a}$ is the real number such that $(\sqrt[n]{a})^n = a$.
if $a < 0$ and n is even, then $\sqrt[n]{a}$ is not a real number.

 WARNING! Care must be taken when extracting even roots of radicands that contain variables. If the variable(s) is not restricted to positive numbers, absolute value symbolism *may* be needed.

EXERCISE 9.1

VOCABULARY AND NOTATION *In Exercises 1–6, fill in the blanks to make a true statement.*

1. When finding the square root of a number, we are always looking for the nonnegative or _____ square root.

2. The expression under a radical sign is called the _____.

3. If an indicated root is perfect, the result is an _____ result. If an indicated root is *not* perfect, the calculator can be used to obtain an _____ result, which is an _____ number.

4. The order of a root is specified by the _____ of the radical.

5. In the expression $\sqrt[5]{x}$, 5 is the _____ and x is the _____.

6. If $x > 0$ and $n > 1$, then $(\sqrt[n]{x})^n =$ ___.

CONCEPTS

7. Which of the indicated roots below are undefined for real numbers?
   a. $\sqrt[3]{-8}$ b. $\sqrt{-8}$
   c. $\sqrt[5]{-36}$ d. $\sqrt[4]{-16}$

8. Classify each of the following as rational or irrational.
   a. $\sqrt{8}$ b. $\sqrt[3]{8}$
   c. $\sqrt[5]{32}$ d. $\sqrt[4]{256}$

9. a. How many real cube roots does a positive number have?

   b. How many real cube roots does a negative number have?

   c. How many real square roots does a negative number have?

10. Assuming x is a real number, simplify the following:
    a. $\sqrt{x^2}$ b. $\sqrt[4]{y^4}$
    c. $\sqrt[3]{a^3}$ d. $\sqrt{(x+3)^2}$
    e. $\sqrt{(x+5)^2}$

PRACTICE *In Exercises 11–52, simplify each radical, if possible. If the calculator is used on the real number parts, copy the screen display exactly as it appears. Assume that all variables represent positive real numbers.*

11. $\sqrt{121}$ 12. $\sqrt{144}$

13. $\sqrt[3]{1}$ 14. $\sqrt[3]{-8}$

15. $\sqrt[3]{8a^3}$ 16. $\sqrt[3]{-27x^6}$

17. $\sqrt{\dfrac{1}{9}}$ 18. $-\sqrt{\dfrac{4}{25}}$

19. $-\sqrt[5]{243}$ 20. $-\sqrt[4]{625}$

21. $\sqrt{0.16}$ 22. $\sqrt{-49}$

23. $-\sqrt[5]{-\dfrac{1}{32}}$ 24. $\sqrt[6]{-729}$

25. $-\sqrt{64}$ 26. $-\sqrt{1}$

27. $\sqrt[3]{-125}$ 28. $\sqrt[3]{512}$

29. $\sqrt[3]{125}$ 30. $-\sqrt{-4}$

31. $\sqrt[3]{-1000p^3q^3}$

32. $\sqrt[3]{343a^6b^3}$

33. $\sqrt[4]{\dfrac{16}{625}}$

34. $\sqrt[5]{-\dfrac{243}{32}}$

35. $\sqrt{0.25}$

36. $\sqrt{-25}$

37. $\sqrt[4]{81}$

38. $\sqrt[6]{64}$

39. $\sqrt{(-4)^2}$

40. $\sqrt{(-9)^2}$

41. $\sqrt[3]{-\dfrac{8}{27}}$

42. $\sqrt[3]{\dfrac{125}{216}}$

43. $\sqrt[3]{-\dfrac{1}{8}m^6n^3}$

44. $\sqrt[3]{\dfrac{27}{1000}a^6b^6}$

45. $\sqrt[5]{-32}$

46. $\sqrt[6]{729}$

47. $-\sqrt{\dfrac{25}{49}}$

48. $\sqrt{\dfrac{49}{81}}$

49. $\sqrt[3]{0.064}$

50. $\sqrt[3]{0.001}$

51. $\sqrt[4]{-256}$

52. $-\sqrt[4]{\dfrac{81}{256}}$

In Exercises 53–68, simplify each radical. Assume that all variables are unrestricted and use absolute value symbols as necessary.

53. $\sqrt{4x^2}$

54. $\sqrt[4]{16y^4}$

55. $\sqrt[3]{8a^3}$

56. $\sqrt[6]{64x^6}$

57. $\sqrt[4]{x^{12}}$

58. $\sqrt[8]{x^{24}}$

59. $\sqrt{a^2 + 6a + 9}$

60. $\sqrt{x^2 + 10x + 25}$

61. $\sqrt[4]{\dfrac{1}{16}x^4}$

62. $\sqrt[4]{\dfrac{1}{81}x^8}$

63. $\sqrt[5]{-x^5}$

64. $\sqrt[3]{-x^6}$

65. $\sqrt[3]{-27a^6}$

66. $\sqrt[5]{-32x^5}$

67. $\sqrt{(-5b)^2}$

68. $\sqrt{(-8c)^2}$

In Exercises 69–76, a student's calculator screen is displayed.
(a) Write the problem that was being simplified as it would appear in a mathematics text.
*(b) Identify the displayed result as a **rational** or an **irrational** number.*

69.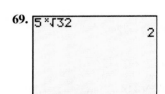

70.

71.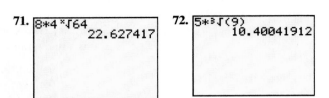

72.

73.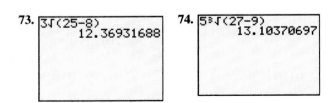

74.

75.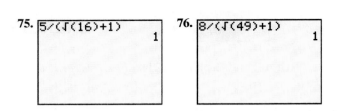

76.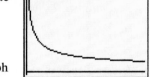

In Exercises 77–82, insert =, >, or < to describe the given relationship. Use your calculator as appropriate.

77. $\sqrt{25 - 16}$ $\sqrt{25} - \sqrt{16}$

78. $\sqrt{25 - 9}$ $\sqrt{25} - \sqrt{9}$

79. $\sqrt{25} + \sqrt{9}$ $\sqrt{34}$

80. $\sqrt{36} + \sqrt{16}$ $\sqrt{52}$

81. $\sqrt{25} - \sqrt{16}$ $\sqrt{64} - \sqrt{15}$

82. $\sqrt{100}$ $4\sqrt{25}$

APPLICATIONS

83. Medicine The approximate pulse rate, p (in beats per minute), of an adult who is t inches tall is given by the formula $p = \dfrac{590}{\sqrt{t}}$. The graph is displayed in the window $[0, 84]$ by $[-20, 500]$.

a. From the general shape of the graph, as a person gets taller, what happens to the pulse rate?

b. Interpret the screen display within the context of the problem:

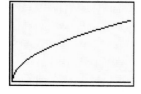

84. Law enforcement   The police sometimes use the formula $s = k\sqrt{L}$ to estimate the speed, s, in miles per hour of a car involved in an accident. In this formula, L is the length of the skid in feet, and k is a constant related to the condition of the pavement. For wet pavement, $k \approx 3.24$. The formula can be written as $s = 3.24\sqrt{L}$. The graph is displayed in the window $[0, 600]$ by $[0, 100]$.

a. From the general shape of the graph, what can you conclude about the relationship between the length of skid marks on the pavement and the speed of the vehicle?

b. Interpret the accompanying graphical display within the context of the problem.

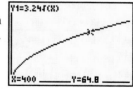

REVIEW   *In Exercises 85–90, simplify each expression.*

85. $\left(\dfrac{3x}{y}\right)^2$

86. $(-5x^2y)^3$

87. $(3x^3y^2)^2$

88. $-(2a^2b^3c)^4$

89. $\left(\dfrac{2}{3}a\right)^3$

90. $(x + 2y)^2$

9.2   RATIONAL EXPONENTS AND SIMPLIFYING RADICAL EXPRESSIONS

- REVIEW OF EXPONENTS
- RATIONAL EXPONENTS
- PRODUCT AND QUOTIENT RULES FOR RADICALS
- SIMPLIFYING RADICAL EXPRESSIONS

▮ REVIEW OF EXPONENTS

Previously, we have defined integer exponents. Positive integer exponents indicate the number of times that a base is to be used as a factor in a product. For example, x^5 means that the base, x, is to be used as a factor five times. Therefore, $x^5 = x \cdot x \cdot x \cdot x \cdot x$. A nonzero base raised to the zero power equals 1. For example, $8^0 = 1$ and $x^0 = 1$. Finally, a base raised to a negative exponent is defined as $x^{-n} = \dfrac{1}{x^n}$. The properties of exponents are summarized as follows:

Exponents

Definitions

If n is a natural number, then

$$x^n = \overbrace{x \cdot x \cdot x \cdot \ldots \cdot x}^{n \text{ factors of } x}$$

$$x^0 = 1 \quad (x \neq 0)$$

0^0 is undefined

$$x^{-n} = \dfrac{1}{x^n}$$

$$\dfrac{1}{x^{-n}} = x^n$$

Properties

If m and n are natural numbers and there are no divisions by 0:

Product Rule: $x^m \cdot x^n = x^{m+n}$

Power to a Power Rule: $(x^m)^n = x^{mn}$

Product to a Power and Quotient to a Power:

$$(xy)^n = x^n y^n \qquad \left(\dfrac{x}{y}\right)^n = \dfrac{x^n}{y^n}$$

Quotient Rule: $\dfrac{x^m}{x^n} = x^{m-n}$

We now turn our attention to a definition for rational exponents.

RATIONAL EXPONENTS

EXPLORING THE CONCEPT

Simplify each expression using the graphing calculator. Include all rational exponents in parentheses to ensure correct results.

a. $3^{1/2} \cdot 3^{1/2}$

b. $\sqrt{3} \cdot \sqrt{3}$

c. $4^{1/3} \cdot 4^{1/3} \cdot 4^{1/3}$

d. $\sqrt[3]{4} \cdot \sqrt[3]{4} \cdot \sqrt[3]{4}$

CONCLUSION

a.
```
3^(1/2)*3^(1/2)
              3
```

b.
```
√(3)√(3)
              3
```

c.
```
4^(1/3)*4^(1/3)*
4^(1/3)
              4
```

d.
```
³√(4)³√(4)³√(4)
              4
```

TECHNOLOGY TIP

When confirming analytical work with rational exponents with the calculator, enclose exponents in parentheses. The graphing calculator should only be used initially to confirm analytical processes.

This leads to the following definition:

Definition of $a^{1/n}$

If n is a positive integer greater than 1 and $\sqrt[n]{a}$ is a real number, then $\sqrt[n]{a} = a^{1/n}$.

We notice that the denominator of the rational number used as an exponent becomes the index of the radical.

EXAMPLE 1

Rewrite each expression as a radical and simplify, if possible.

a. $9^{1/2}$ b. $-\left(\dfrac{16}{9}\right)^{1/2}$ c. $(-64)^{1/3}$ d. $16^{1/4}$

e. $\left(\dfrac{1}{32}\right)^{1/5}$ f. $(-32x^5)^{1/5}$ g. $(3x^2y)^{1/4}$

SOLUTION

a. $9^{1/2} = \sqrt{9} = 3$ b. $-\left(\dfrac{16}{9}\right)^{1/2} = -\sqrt{\dfrac{16}{9}} = -\dfrac{4}{3}$

c. $(-64)^{1/3} = \sqrt[3]{-64} = -4$ d. $16^{1/4} = \sqrt[4]{16} = 2$

e. $\left(\dfrac{1}{32}\right)^{1/5} = \sqrt[5]{\dfrac{1}{32}} = \dfrac{1}{2}$

f. $(-32x^5)^{1/5} = \sqrt[5]{-32x^5} = \sqrt[5]{(-2x)^5} = -2x$

g. $(3x^2y)^{1/4} = \sqrt[4]{3x^2y}$

The definition also allows us to move from radical form to exponential form.

EXAMPLE 2   Rewrite each expression with a rational exponent.

a. $\sqrt{6}$ b. $\sqrt[4]{5x}$ c. $\sqrt[4]{5}\,x$

d. $\sqrt[3]{18}$ e. $\sqrt[5]{\dfrac{1}{3}}$

SOLUTION   a. $\sqrt{6} = 6^{1/2}$ b. $\sqrt[4]{5x} = (5x)^{1/4}$ c. $\sqrt[4]{5}\,x = 5^{1/4}x$

d. $\sqrt[3]{18} = 18^{1/3}$ e. $\sqrt[5]{\dfrac{1}{3}} = \left(\dfrac{1}{3}\right)^{1/5}$

All of the rational exponents considered thus far have had numerators of 1. We can extend our definition of $a^{1/n}$ to include fractional exponents with numerators other than 1. This can be done by combining the definition of a rational exponent with previously discussed properties of exponents.

Consider $4^{3/2}$. It can be rewritten as follows:

$$4^{3/2} = (4^{1/2})^3 = (\sqrt{4})^3 = 2^3 = 8 \quad \text{or} \quad 4^{3/2} = (4^3)^{1/2} = \sqrt{4^3} = \sqrt{64} = 8$$

This leads to the following definition:

Definition of $a^{m/n}$

If m and n are positive integers greater than 1, and m/n is in simplest form,

$$a^{m/n} = \sqrt[n]{a^m} = (\sqrt[n]{a})^m$$

This definition allows us to interpret $a^{m/n}$ in two ways:

1. $a^{m/n}$ means the **nth root** of the **mth power** of a.
2. $a^{m/n}$ means the **mth power** of the **nth root** of a.

The denominator of the rational exponent becomes the index of the radical. The numerator, m, becomes the power of either the radicand **or** the entire radical. Therefore, $4^{3/2} = (\sqrt{4})^3 = 2^3 = 8$ or $4^{3/2} = \sqrt{4^3} = \sqrt{64} = 8$. Either method leads to a correct result.

EXAMPLE 3   Simplify each of the following by first writing in radical form. Verify results with the calculator.

a. $27^{2/3}$ b. $\left(\dfrac{1}{16}\right)^{3/4}$ c. $(-8x^3)^{4/3}$

SOLUTION   a. $27^{2/3} = (\sqrt[3]{27})^2$ or $27^{2/3} = \sqrt[3]{27^2}$

$= 3^2$ $= \sqrt[3]{729}$

$= 9$ $= 9$

$$\left(\frac{1}{16}\right)^{3/4} = \left(\sqrt[4]{\frac{1}{16}}\right)^3 \quad \text{or} \quad \left(\frac{1}{16}\right)^{3/4} = \sqrt[4]{\left(\frac{1}{16}\right)^3}$$

b.

$$= \left(\frac{1}{2}\right)^3 \qquad\qquad = \sqrt[4]{\frac{1}{4096}}$$

$$= \frac{1}{8} \qquad\qquad\qquad = \frac{1}{8}$$

c.

$$(-8x^3)^{4/3} = \left(\sqrt[3]{-8x^3}\right)^4 \quad \text{or} \quad (-8x^3)^{4/3} = \sqrt[3]{(-8x^3)^4}$$

$$= (\sqrt[3]{(-2x)^3})^4 \qquad\qquad = \sqrt[3]{4096x^{12}}$$

$$= (-2x)^4 \qquad\qquad\qquad = \sqrt[3]{(16x^4)^3}$$

$$= 16x^4 \qquad\qquad\qquad = 16x^4$$

To be consistent with the definition of negative integer exponents, we define $a^{-m/n}$ as follows:

Definition of $a^{-m/n}$

If m and n are positive integers, m/n is in simplest form, and $a^{1/n}$ is a real number, then

$$a^{-m/n} = \frac{1}{a^{m/n}} \quad \text{and} \quad \frac{1}{a^{-m/n}} = a^{m/n}, \quad a \neq 0$$

EXAMPLE 4

Simplify each expression and verify your results with the graphing calculator.

a. $64^{-1/2}$ **b.** $16^{-3/2}$ **c.** $(-27)^{-2/3}$ **d.** $(-16)^{-3/4}$

SOLUTION **a.** $64^{-1/2} = \dfrac{1}{64^{1/2}} = \dfrac{1}{\sqrt{64}} = \dfrac{1}{8}$

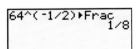

b. $16^{-3/2} = \dfrac{1}{16^{3/2}} = \dfrac{1}{(\sqrt{16})^3} = \dfrac{1}{4^3} = \dfrac{1}{64}$

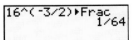

c. $(-27)^{-2/3} = \dfrac{1}{(-27)^{2/3}} = \dfrac{1}{(\sqrt[3]{-27})^2} = \dfrac{1}{(-3)^2} = \dfrac{1}{9}$

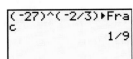

d. $(-16)^{-3/4} = \dfrac{1}{(-16)^{3/4}} = \dfrac{1}{(\sqrt[4]{-16})^3}$

Because the *index* is even and the *radicand* is negative, the expression is undefined in the real-number system. The TI-83 Plus calculator screen confirms our conclusion:

▐▌ PRODUCT AND QUOTIENT RULES FOR RADICALS

The radical expressions studied thus far have been either perfect roots, which evaluate to **rational numbers**, or those that are **irrational numbers**, for which decimal approximations may be given.

We now turn our attention to expressing radicals in simplest radical form. To write radicals in simplest radical form, we introduce both a product rule and a quotient rule for radicals. These are developed next.

Because $a^{1/n} \cdot b^{1/n} = (ab)^{1/n}$, we may convert from exponential form to radical form:

$$\sqrt[n]{a} \cdot \sqrt[n]{b} = \sqrt[n]{ab}$$

For example,

$$\sqrt{5} \cdot \sqrt{5} = \sqrt{5^2} = 5 \quad \text{and} \quad \sqrt[3]{7x} \cdot \sqrt[3]{49x^2} = \sqrt[3]{7x \cdot 7^2 x^2} = \sqrt[3]{7^3 x^3} = 7x$$

This leads to the following rule:

Product Rule for Radicals

If $\sqrt[n]{a}$ and $\sqrt[n]{b}$ are real numbers, then $\sqrt[n]{a} \cdot \sqrt[n]{b} = \sqrt[n]{ab}$. Because equality is symmetric, this could also be written $\sqrt[n]{ab} = \sqrt[n]{a} \cdot \sqrt[n]{b}$.

This rule states that as long as all radicals represent real numbers, *the nth root of the product of two numbers is equal to the product of the nth roots of the numbers.*

WARNING! The product rule applies to the *n*th root of the *product* of two numbers. **There is no such property for sums or differences.** For example,

$$\sqrt{9 + 4} \neq \sqrt{9} + \sqrt{4} \quad \text{and} \quad \sqrt{9 - 4} \neq \sqrt{9} - \sqrt{4}$$
$$\sqrt{13} \neq 3 + 2 \qquad\qquad \sqrt{5} \neq 3 - 2$$
$$\sqrt{13} \neq 5 \qquad\qquad\qquad \sqrt{5} \neq 1$$

Thus, $\sqrt{a + b} \neq \sqrt{a} + \sqrt{b}$ and $\sqrt{a - b} \neq \sqrt{a} - \sqrt{b}$.

Another property of radicals involves quotients. Because $\dfrac{a^{1/n}}{b^{1/n}} = \left(\dfrac{a}{b}\right)^{1/n}$ it

follows that $\dfrac{\sqrt[n]{a}}{\sqrt[n]{b}} = \sqrt[n]{\dfrac{a}{b}}, (b \neq 0)$. For example, $\dfrac{\sqrt{8x^3}}{\sqrt{2x}} = \sqrt{\dfrac{8x^3}{2x}} = \sqrt{4x^2} = 2x$,

$(x > 0)$, and $\dfrac{\sqrt[3]{54x^5}}{\sqrt[3]{2x^2}} = \sqrt[3]{\dfrac{54x^5}{2x^2}} = \sqrt[3]{27x^3} = 3x$. This leads to the following rule:

Quotient Rule for Radicals

If $\sqrt[n]{a}$ and $\sqrt[n]{b}$ are real numbers, then $\sqrt[n]{\dfrac{a}{b}} = \dfrac{\sqrt[n]{a}}{\sqrt[n]{b}}, b \neq 0$. Because equality is symmetric, this could also be written $\dfrac{\sqrt[n]{a}}{\sqrt[n]{b}} = \sqrt[n]{\dfrac{a}{b}}, b \neq 0$.

As long as radicals represent real numbers, *the nth root of the quotient of two numbers is equal to the quotient of their nth roots.*

▍ SIMPLIFYING RADICAL EXPRESSIONS

A radical is said to be in simplest radical form when each of the following statements is true:

Simplified Form of a Radical Expression
A radical expression is in simplest form when

1. no radical appears in the denominator of a fraction.
2. the radicand contains no fractions or negative numbers.
3. each factor in the radicand appears to a power that is less than the index of the radical.

The third statement in the preceding box implies that no factors of the radicand can contain perfect powers. When a constant is written in exponential form, perfect roots often become more apparent. Factoring constants into primes enables us to identify perfect roots more easily and frees us from memorizing long lists of perfect squares, cubes, fourths, etc. For example, to simplify $\sqrt[3]{343x^3y^6}$, we factor the constant, 343, into primes and rewrite the radicand as follows: $\sqrt[3]{343x^3y^6} = \sqrt[3]{7^3x^3y^6}$. Applying our definition of rational exponents allows us to continue thusly: $\sqrt[3]{7^3x^3y^6} = 7^{3/3}x^{3/3}y^{6/3} = 7xy^2$.

To write radical expressions in simplest radical form, use the following steps:

Simplifying a Radical Expression

1. Factor the constant factors of the radicand into primes. Use exponential notation rather than writing repeated factors.
2. Use the product rule and laws of exponents to rewrite the radical as a product of two radicals with radicands as specified below:
    a. The first radicand is comprised of factors that are perfect roots. The exponents of the factors are evenly divisible by the index of the radical.
    b. The second radicand is comprised of factors that have powers *smaller* than the index of the radical.
3. Simplify the radical containing perfect roots by rewriting the radicand with rational exponents and reducing those exponents.
4. The final result is the product of a rational number and a radical with a radicand that contains no factors that are perfect roots.

EXAMPLE 5

Simplify each of the following. Verify results with your calculator.

a. $\sqrt{12}$ **b.** $\sqrt{98}$ **c.** $\sqrt[3]{54}$

SOLUTION

a. $\sqrt{12} = \sqrt{2^2 \cdot 3^1}$ Factor the radicand into primes.

$\quad\quad\quad = \sqrt{2^2} \cdot \sqrt{3^1}$ Use the product rule to write the radical as a product of two radicals — one of which contains perfect powers.

$\quad\quad\quad = 2^{2/2} \cdot \sqrt{3}$ Use the definition of rational exponents to express the radical containing perfect powers in exponential form.

$\quad\quad\quad = 2\sqrt{3}$ Simplify the exponent, thus expressing the radical as the product of a rational number and a radical that contains no factors that are perfect roots.

This result is read as "2 square roots of 3" and is the product of 2 and $\sqrt{3}$. Verification that the original radical and our result are equal is shown by the calculator.

Both are equivalent to the same irrational number.

```
√(12)
          3.464101615
2√(3)
          3.464101615
```

b. $\sqrt{98} = \sqrt{7^2 \cdot 2}$

$\quad\quad\quad = \sqrt{7^2} \cdot \sqrt{2}$

$\quad\quad\quad = 7^{2/2}\sqrt{2}$

$\quad\quad\quad = 7\sqrt{2}$

Verification:

```
√(98)
          9.899494937
7√(2)
          9.899494937
```

c. $\sqrt[3]{54} = \sqrt[3]{3^3 \cdot 2}$

$\quad\quad\quad = \sqrt[3]{3^3} \cdot \sqrt[3]{2}$

$\quad\quad\quad = 3^{3/3} \cdot \sqrt[3]{2}$

$\quad\quad\quad = 3\sqrt[3]{2}$

Verification:

```
³√(54)
          3.77976315
3³√(2)
          3.77976315
```

Many students eliminate the step of rewriting the perfect root in exponential form. However, be careful about taking too many shortcuts.

EXAMPLE 6

Simplify: $\sqrt{\dfrac{15}{49x^2}}$, $(x > 0)$.

SOLUTION

The quotient rule allows us to rewrite the radical as a quotient of square roots.

$$\sqrt{\frac{15}{49x^2}} = \frac{\sqrt{15}}{\sqrt{49x^2}}$$

$$= \frac{\sqrt{5^1 \cdot 3^1}}{\sqrt{7^2 \cdot x^2}}$$ Factor the constant in the radicand into primes.

$$= \frac{\sqrt{15}}{7^{2/2}x^{2/2}}$$ The exponents of 1 on the factors of 5 and 3 assure us that 15 has no factors that are perfect powers. Therefore, we simply write the radicand as 15.

$$= \frac{\sqrt{15}}{7x}$$

Note: Since $x > 0$, no absolute value symbol is necessary.

EXAMPLE 7

Simplify each expression. Assume that all variables represent positive numbers, thus no absolute value symbol is necessary.

a. $\sqrt{128a^5}$ **b.** $\sqrt[3]{24x^5}$ **c.** $\dfrac{\sqrt{45xy^2}}{\sqrt{5x}}$ **d.** $\dfrac{\sqrt[3]{-432x^5}}{\sqrt[3]{8x}}$

SOLUTION

a.
$$\sqrt{128a^5} = \sqrt{2^7 \cdot a^5}$$ Factor the constant in the radicand into primes.

$$= \sqrt{2^6 \cdot a^4}\sqrt{2^1 \cdot a^1}$$ Use the product rule and laws of exponents to rewrite the radical as the product of two radicals, one of which contains perfect powers and the other containing factors with exponents smaller than the index of the radical.

$$= 2^{6/2}a^{4/2}\sqrt{2a}$$ Rewrite the radical containing perfect powers with rational exponents, thus extracting the root.

$$= 2^3a^2\sqrt{2a}$$ Simplify exponents.

$$= 8a^2\sqrt{2a}$$ Raise constant bases to specified powers.

b.
$$\sqrt[3]{24x^5} = \sqrt[3]{2^3 \cdot 3x^5}$$ Factor the constant in the radicand into primes.

$$= \sqrt[3]{2^3x^3}\sqrt[3]{3x^2}$$ Apply the product rule and laws of exponents to rewrite the radical as the product of two radicals.

$$= 2^{3/3}x^{3/3}\sqrt[3]{3x^2}$$ Rewrite the radical containing perfect powers with rational exponents.

$$= 2x\sqrt[3]{3x^2}$$ Simplify powers.

c.
$$\dfrac{\sqrt{45xy^2}}{\sqrt{5x}} = \sqrt{\dfrac{45xy^2}{5x}}$$ Use the quotient rule to rewrite the quotient of square roots as the square root of a quotient.

$$= \sqrt{9y^2}$$ Simplify the fraction.

$$= \sqrt{3^2y^2}$$ Factor the constant in the radicand into primes.

$$= 3^{2/2}y^{2/2}$$ Rewrite the radical using rational exponents.

$$= 3y$$ Simplify exponents.

Note: As you become more comfortable with finding roots, $\sqrt{9y^2}$ may be replaced with the equivalent expression, $3y$, without the intermediate steps.

d.
$$\dfrac{\sqrt[3]{-432x^5}}{\sqrt[3]{8x}} = \sqrt[3]{\dfrac{-432x^5}{8x}}$$ Use the quotient rule to rewrite the quotient of cube roots as the cube root of a quotient.

$$= \sqrt[3]{-54x^4}$$ Simplify the fraction in the radicand.

$$= \sqrt[3]{(-3)^3 \cdot 2x^4}$$ Factor the constant in the radicand into primes.

$$= \sqrt[3]{(-3)^3x^3}\sqrt[3]{2^1x^1}$$ Use the product rule to rewrite the radical as the product of two radicals.

$$= (-3)^{3/3}x^{3/3}\sqrt[3]{2x}$$ Rewrite the radical containing perfect powers in exponential form.

$$= -3x\sqrt[3]{2x}$$ Simplify exponents.

EXERCISE 9.2

VOCABULARY AND NOTATION *In Exercises 1–4, fill in the blanks to make the statement true.*

1. Positive integer exponents indicate the number of times a _____ is to be used as a _____.
2. Any nonzero base raised to the zero power equals _____.
3. Radical expressions that simplify to perfect roots are _____ numbers.
4. If a root is not perfect, it is a(n) _____ number, and its root may be approximated.

CONCEPTS *In Exercises 5–10, express each exponential expression as a radical.*

5. $7^{1/3}$ (verify with your calculator that your two expressions are equal).
6. $26^{1/2}$ (verify with your calculator that your two expressions are equal).
7. $3x^{1/4}$
8. $4ab^{1/6}$
9. $(3x)^{1/4}$
10. $(4ab)^{1/6}$

In Exercises 11–16, express each radical as an exponential expression.

11. $\sqrt[4]{3a}$
12. $3\sqrt[4]{a}$
13. $\sqrt[6]{\frac{1}{7}abc}$
14. $\sqrt[7]{\frac{3}{8}p^2q}$
15. $\sqrt[3]{a^2 - b^2}$
16. $\sqrt{x^2 + y^2}$

PRACTICE *In Exercises 17–36, simplify each expression, if possible. Verify results using the calculator, copying the screen display exactly as it appears. Negative exponents should not appear in results.*

17. $4^{1/2}$
18. $25^{1/2}$
19. $8^{1/3}$
20. $125^{1/3}$
21. $\left(\frac{1}{8}\right)^{1/3}$
22. $\left(\frac{1}{16}\right)^{1/4}$
23. $-16^{1/4}$
24. $(-16)^{1/4}$

25. $0^{1/3}$
26. $(a^{1/3} + b^{1/3})^0$
27. $81^{3/4}$
28. $100^{3/2}$
29. $\left(\frac{1}{8}\right)^{2/3}$
30. $\left(\frac{4}{9}\right)^{3/2}$
31. $4^{-1/2}$
32. $8^{-1/3}$
33. $4^{-3/2}$
34. $25^{-5/2}$
35. $\left(\frac{27}{8}\right)^{-4/3}$
36. $\left(\frac{25}{49}\right)^{-3/2}$

In Exercises 37–48, simplify each expression, if possible. Assume that all variables represent positive numbers.

37. $(25x^4)^{3/2}$
38. $(27a^3b^3)^{2/3}$
39. $\left(\frac{8x^3}{27}\right)^{2/3}$
40. $\left(\frac{27}{64y^6}\right)^{2/3}$
41. $(16x^2)^{-3/2}$
42. $(81c^4)^{-3/2}$
43. $(-27y^3)^{-2/3}$
44. $(-8z^9)^{-2/3}$
45. $(-32p^5)^{-2/5}$
46. $(16q^6)^{-5/2}$
47. $\left(-\frac{8x^3}{27}\right)^{-1/3}$
48. $\left(\frac{16}{81y^4}\right)^{-3/4}$

In Exercises 49–78, express each radical in simplest radical form. Assume that all variables represent positive numbers.

49. $\sqrt{6} \cdot \sqrt{6}$
50. $\sqrt{11} \cdot \sqrt{11}$
51. $\sqrt[3]{5x^2} \cdot \sqrt[3]{25x}$
52. $\sqrt[4]{25a} \sqrt[4]{25a^3}$
53. $\frac{\sqrt{180ab^4}}{\sqrt{5ab^2}}$
54. $\frac{\sqrt{112ab^3}}{\sqrt{7ab}}$
55. $\frac{\sqrt[3]{189a^4}}{\sqrt[3]{7a}}$
56. $\frac{\sqrt[3]{243x^7}}{\sqrt[3]{9x}}$
57. $\sqrt{20}$
58. $\sqrt{8}$
59. $-\sqrt{200}$
60. $-\sqrt{250}$
61. $\sqrt[3]{80}$
62. $\sqrt[3]{270}$
63. $\sqrt[3]{-81}$
64. $\sqrt[3]{-72}$
65. $\sqrt[5]{96}$
66. $\sqrt[7]{256}$
67. $\sqrt{\frac{7}{9}}$
68. $\sqrt{\frac{3}{4}}$

69. $\sqrt[5]{\dfrac{3}{32}}$

70. $\sqrt[6]{\dfrac{5}{64}}$

71. $\sqrt{175a^2b^3}$

72. $\sqrt{128a^3b^5}$

73. $\sqrt[4]{32x^{12}y^4}$

74. $\sqrt[5]{64x^{10}y^5}$

75. $\sqrt{\dfrac{z^2}{16x^2}}$

76. $\sqrt{\dfrac{b^4}{64a^8}}$

77. $\sqrt[4]{\dfrac{5x}{16z^4}}$

78. $\sqrt[3]{\dfrac{11a^2}{125b^6}}$

REVIEW Simplify

79. $(3x^2y)^{-2}$

80. $\dfrac{3a^2b^{-4}c^3}{6a^{-2}b^{-3}c^3}$

81. $-\left(\dfrac{2xy}{4x^2y^3}\right)^{-3}$

Evaluate each of the following (a) algebraically, and (b) by using the STOre feature of the graphing calculator.

82. $6x^2y^{-3}$; $x = -2$, $y = -3$

83. $\left(\dfrac{3x}{6y}\right)^{-2}$; $x = -1$, $y = 4$

9.3   ADDING AND SUBTRACTING RADICAL EXPRESSIONS

- ADDING AND SUBTRACTING RADICAL EXPRESSIONS

ADDING AND SUBTRACTING RADICAL EXPRESSIONS

Radical expressions with the same *index* and *radicand* can be added and subtracted using the distributive property. To simplify the expression $3\sqrt{2} + 5\sqrt{2}$, we use the distributive property to factor out the common factor of $\sqrt{2}$ and then simplify:

$$3\sqrt{2} + 5\sqrt{2} = (3 + 5)\sqrt{2}$$
$$= 8\sqrt{2}$$

Oftentimes, radicals with the same index but different radicands can be combined once each radical is written in simplest radical form. For example, to simplify the expression $\sqrt{27} - \sqrt{12}$, we simplify each radical and then combine, if possible:

$$\sqrt{27} - \sqrt{12} = \sqrt{3^2 \cdot 3} - \sqrt{2^2 \cdot 3}$$
$$= 3\sqrt{3} - 2\sqrt{3}$$
$$= (3 - 2)\sqrt{3}$$
$$= \sqrt{3}$$

These examples suggest the following rules for adding and subtracting radicals.

Adding and Subtracting Radical Expressions

1. Simplify each radical.
2. Use the distributive property to add or subtract radicals that have the same index and radicand.

EXAMPLE 1   Simplify: $2\sqrt{12} - 3\sqrt{48} + 3\sqrt{3}$.

SOLUTION   We simplify each radical, and then combine those that have radicands and indices that are the same.

$$2\sqrt{12} - 3\sqrt{48} + 3\sqrt{3} = 2\sqrt{2^2 \cdot 3} - 3\sqrt{2^4 \cdot 3} + 3\sqrt{3}$$
$$= 2(2)\sqrt{3} - 3(4)\sqrt{3} + 3\sqrt{3}$$
$$= 4\sqrt{3} - 12\sqrt{3} + 3\sqrt{3}$$
$$= (4 - 12 + 3)\sqrt{3}$$
$$= -5\sqrt{3}$$

Therefore, $2\sqrt{12} - 3\sqrt{48} + 3\sqrt{3} = -5\sqrt{3}$. Our result can be confirmed with the calculator:

```
2√(12)-3√(48)+3√
(3)
          -8.660254038
-5√(3)
          -8.660254038
```

EXAMPLE 2

Simplify: $\sqrt[3]{16} - \sqrt[3]{54} + \sqrt[3]{24}$.

SOLUTION   We simplify each radical separately and then combine those that have radicands and indices that are the same.

$$\sqrt[3]{16} - \sqrt[3]{54} + \sqrt[3]{24} = \sqrt[3]{2^3 \cdot 2} - \sqrt[3]{3^3 \cdot 2} + \sqrt[3]{2^3 \cdot 3}$$
$$= 2\sqrt[3]{2} - 3\sqrt[3]{2} + 2\sqrt[3]{3}$$
$$= (2 - 3)\sqrt[3]{2} + 2\sqrt[3]{3}$$
$$= -\sqrt[3]{2} + 2\sqrt[3]{3}$$

Therefore, $\sqrt[3]{16} - \sqrt[3]{54} + \sqrt[3]{24} = -\sqrt[3]{2} + 2\sqrt[3]{3}$. Our result can be confirmed with the calculator:

```
3√(16)-3√(54)+3√
(24)
           1.624578091
-3√(2)+23√(3)
           1.624578091
```

EXAMPLE 3

Simplify: $\sqrt[3]{16x^4} + \sqrt[3]{54x^4} - \sqrt[3]{-128x^4}$.

SOLUTION   We simplify each radical separately and then combine those that have radicands and indices that are the same.

$$\sqrt[3]{16x^4} + \sqrt[3]{54x^4} - \sqrt[3]{-128x^4} = \sqrt[3]{2^4 x^4} + \sqrt[3]{3^3 \cdot 2x^4} - \sqrt[3]{(-4)^3 \cdot 2x^4}$$
$$= \sqrt[3]{2^3 x^3 \cdot 2x} + \sqrt[3]{3^3 x^3 \cdot 2x} - \sqrt[3]{(-4)^3 x^3 \cdot 2x}$$
$$= 2x\sqrt[3]{2x} + 3x\sqrt[3]{2x} + 4x\sqrt[3]{2x}$$
$$= (2x + 3x + 4x)\sqrt[3]{2x}$$
$$= 9x\sqrt[3]{2x}$$

EXAMPLE 4

$3\sqrt[3]{24} - 8\sqrt[3]{81} - 2\sqrt[3]{\dfrac{3}{27}}.$

SOLUTION   We simplify each radical expression separately and combine like radicals.

$$3\sqrt[3]{24} - 8\sqrt[3]{81} - 2\sqrt[3]{\frac{3}{27}} = 3\sqrt[3]{2^3 \cdot 3} - 8\sqrt[3]{3^3 \cdot 3} - 2\frac{\sqrt[3]{3}}{\sqrt[3]{3^3}}$$

$$= 3\sqrt[3]{2^3}\sqrt[3]{3} - 8\sqrt[3]{3^3}\sqrt[3]{3} - 2\frac{\sqrt[3]{3}}{3}$$

$$= 3(2)\sqrt[3]{3} - 8(3)\sqrt[3]{3} - \frac{2}{3}\sqrt[3]{3}$$

$$= 6\sqrt[3]{3} - 24\sqrt[3]{3} - \frac{2}{3}\sqrt[3]{3}$$

$$= \frac{18}{3}\sqrt[3]{3} - \frac{72}{3}\sqrt[3]{3} - \frac{2}{3}\sqrt[3]{3}$$

$$= \left(\frac{18}{3} - \frac{72}{3} - \frac{2}{3}\right)\sqrt[3]{3}$$

$$= -\frac{56}{3}\sqrt[3]{3}$$

Therefore, $3\sqrt[3]{24} - 8\sqrt[3]{81} - 2\sqrt[3]{\dfrac{3}{27}} = -\dfrac{56}{3}\sqrt[3]{3}$. Our result can be confirmed with the calculator:

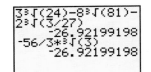

Throughout the rest of this chapter, we will assume all variables represent positive numbers.

EXERCISE 9.3

VOCABULARY AND NOTATION *Fill in the blanks to make the statements true.*

1. Radical expressions such as $\sqrt[5]{2x}$ and $3\sqrt[5]{2x}$ are called _____ radical expressions because they have the same radicands and same index.

2. Radical expressions with the same _____ and radicands can be added (or subtracted) using the _____ property.

CONCEPTS *In Exercises 3–6, classify each of the following statements as true or false.*

3. $\sqrt{5} + \sqrt{5} = 2\sqrt{5}$

4. $5\sqrt[3]{6} + 4\sqrt[3]{6} = 9\sqrt[3]{12}$

5. $\sqrt{3} - 3\sqrt{5} = -2\sqrt{5}$

6. $\sqrt[4]{2} - 1\sqrt[4]{2} = -\sqrt[4]{2}$

In Exercises 7–8, fill in the blanks to complete the problem.

7. $\sqrt[3]{54} = \sqrt[3]{\blacksquare^3 \cdot \blacksquare}$

$= \sqrt[3]{\blacksquare^3} \cdot \sqrt[3]{\blacksquare}$

$= 3\sqrt[3]{\blacksquare}$

8. $\sqrt[5]{32x^6} = \sqrt[5]{\blacksquare^5 \cdot x^{\blacksquare} \cdot x^{\blacksquare}}$

$= \sqrt[5]{\blacksquare^5} \cdot \sqrt[5]{x^{\blacksquare}} \cdot \sqrt[5]{x^{\blacksquare}}$

$= 2x\sqrt[5]{x^{\blacksquare}}$

PRACTICE *In Exercises 9–54, add or subtract as the signs indicate. When possible, check results with the calculator, recording screen displays exactly as they appear.*

9. $4\sqrt{2} + 6\sqrt{2}$

10. $6\sqrt{5} + 3\sqrt{5}$

11. $8\sqrt[5]{7} - 7\sqrt[5]{7}$

12. $10\sqrt[6]{12} - \sqrt[6]{12}$

13. $8\sqrt{x} + 6\sqrt{x}$

14. $10\sqrt{xy} - 2\sqrt{xy}$

15. $\sqrt{3} + \sqrt{27}$

16. $\sqrt{8} + \sqrt{32}$

17. $\sqrt{2} - \sqrt{8}$

18. $\sqrt{20} - \sqrt{125}$

19. $\sqrt{98} - \sqrt{50}$

20. $\sqrt{72} - \sqrt{200}$

21. $3\sqrt{24} + \sqrt{54}$

22. $\sqrt{18} + 2\sqrt{50}$

23. $\sqrt[3]{24} + \sqrt[3]{3}$

24. $\sqrt[3]{16} + \sqrt[3]{128}$

25. $\sqrt[3]{32} - \sqrt[3]{108}$

26. $\sqrt[3]{80} - \sqrt[3]{10,000}$

27. $2\sqrt[3]{125} - 5\sqrt[3]{64}$

28. $3\sqrt[3]{27} + 12\sqrt[3]{216}$

29. $3\sqrt[3]{-54} + 8\sqrt[3]{-128}$

30. $5\sqrt[3]{-81} - 7\sqrt[3]{-375}$

31. $14\sqrt[4]{32} - 15\sqrt[4]{162}$

32. $23\sqrt[4]{768} + \sqrt[4]{48}$

33. $3\sqrt[4]{512} + 2\sqrt[4]{32}$

34. $4\sqrt[4]{243} - \sqrt[4]{48}$

35. $\sqrt{98} - \sqrt{50} - \sqrt{72}$

36. $\sqrt{20} + \sqrt{125} - \sqrt{80}$

37. $\sqrt{18} + \sqrt{300} - \sqrt{243}$

38. $\sqrt{80} - \sqrt{128} + \sqrt{288}$

39. $2\sqrt[3]{16} - \sqrt[3]{54} - 3\sqrt[3]{128}$

40. $\sqrt[4]{48} - \sqrt[4]{243} - \sqrt[4]{768}$

41. $\sqrt{25y^2z} - \sqrt{16y^2z}$

42. $\sqrt{25yz^2} + \sqrt{9yz^2}$

43. $\sqrt{36xy^2} + \sqrt{49xy^2}$

44. $2\sqrt{2x} - \sqrt{8x}$

45. $2\sqrt[3]{64a} + 2\sqrt[3]{8a}$

46. $3\sqrt[4]{x^4y} - 2\sqrt[4]{x^4y}$

47. $\sqrt{y^5} - \sqrt{9y^5} - \sqrt{25y^5}$

48. $\sqrt{8y^7} + \sqrt{32y^7} - \sqrt{2y^7}$

49. $\sqrt[5]{x^6y^2} + \sqrt[5]{32x^6y^2} + \sqrt[5]{x^6y^2}$

50. $\sqrt[3]{xy^4} + \sqrt[3]{8xy^4} - \sqrt[3]{27xy^4}$

51. $\sqrt{x^2 + 2x + 1} + \sqrt{x^2 + 2x + 1}$

52. $\sqrt{4x^2 + 12x + 9} + \sqrt{9x^2 + 6x + 1}$

53. $\sqrt{3x^2 + 6x + 3} + \sqrt{3x^2}$

54. $\sqrt{5x^2 + 10x + 5} - \sqrt{5x^2 + 20x + 20}$

REVIEW *In Exercises 55–57, solve each equation (a) algebraically and (b) graphically (copy the screen display and specify WINDOW values that are different from ZStandard).*

55. $3(x + 4) - 2x = 5x - 6$

56. $x^2 + 5x = 6$

57. $|x + 4| = 6$

In Exercises 58–60, solve each inequality. Specify the solution (a) as a number line graph and (b) using interval notation.

58. $2x - 1 > 9$

59. $-8 \le -4x \le 20$

60. $x + 1 < 0$ or $2x - 6 > 0$

9.4 **MULTIPLYING AND DIVIDING RADICAL EXPRESSIONS**

- MULTIPLYING A MONOMIAL BY A MONOMIAL
- MULTIPLYING A POLYNOMIAL BY A MONOMIAL
- MULTIPLYING A POLYNOMIAL BY A POLYNOMIAL
- RATIONALIZING DENOMINATORS

▐▐▐ MULTIPLYING A MONOMIAL BY A MONOMIAL

The product rule ensures that radical expressions with the same index can be multiplied. Because radical expressions represent real numbers, properties of real numbers are applicable to radical expressions. The commutative and associative properties along with the product rule are used when multiplying radical expressions. It is suggested that *multiplication* be performed *before* simplification.

Multiplication of Radical Expressions

1. The product rule allows the multiplication of radical expressions that have the same index.
2. Simplify radical expressions *after* the operation of multiplication has been performed.

EXAMPLE 1   Multiply: $3\sqrt{6} \cdot 2\sqrt{3}$.

SOLUTION   We use the commutative and associative properties of multiplication coupled with the product rule to multiply.

$$3\sqrt{6} \cdot 2\sqrt{3} = (3 \cdot 2)(\sqrt{6} \cdot \sqrt{3}) \qquad \text{Commutative and associative properties.}$$

$$= 6\sqrt{18} \qquad \text{Simplify and apply the product rule.}$$

$$= 6\sqrt{3^2 \cdot 2} \qquad \text{Simplify the radical.}$$

$$= 6 \cdot 3\sqrt{2} \qquad \text{Extract the perfect root of 3.}$$

$$= 18\sqrt{2} \qquad \text{Multiply the integers 6 and 3.}$$

Therefore, $3\sqrt{6} \cdot 2\sqrt{3} = 18\sqrt{2}$. Our result can be confirmed with the calculator:

```
3√(6)*2√(3)
        25.45584412
18√(2)
        25.45584412
```

▐▐▐ MULTIPLYING A POLYNOMIAL BY A MONOMIAL

To multiply a polynomial by a monomial, we use the distributive property to remove parentheses and then simplify each term.

EXAMPLE 2   Multiply: $3\sqrt{3}(4\sqrt{8} - 5\sqrt{10})$.

SOLUTION $3\sqrt{3}(4\sqrt{8} - 5\sqrt{10})$

$$= 3\sqrt{3} \cdot 4\sqrt{8} - 3\sqrt{3} \cdot 5\sqrt{10} \qquad \text{Use the distributive property.}$$

$$= 12\sqrt{24} - 15\sqrt{30} \qquad \text{Use the commutative and associative properties and the product rule to perform the indicated multiplication.}$$

$$= 12\sqrt{2^2 \cdot 2 \cdot 3} - 15\sqrt{2 \cdot 5 \cdot 3} \qquad \text{Factor radicands to simplify radicals.}$$

$$= 12 \cdot 2\sqrt{6} - 15\sqrt{30}$$

Notice that the factorization is used to identify perfect roots. Once they are extracted and we are assured the radicands have no other factors that are perfect squares, we multiply the factors in the radicand.

$$= 24\sqrt{6} - 15\sqrt{30}$$

Therefore, $3\sqrt{3}(4\sqrt{8} - 5\sqrt{10}) = 24\sqrt{6} - 15\sqrt{30}$.
Our result can be confirmed with the calculator:

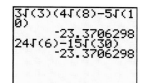

▐ MULTIPLYING A POLYNOMIAL BY A POLYNOMIAL

To multiply a polynomial by a polynomial we extend the use of the distributive property.

EXAMPLE 3

Multiply: $(\sqrt{7} + \sqrt{2})(\sqrt{7} - 3\sqrt{2})$.

SOLUTION

Distribute the multiplication (the mnemonic FOIL may be used). We then compute products, simplify radicals, and add and subtract as applicable.

$$(\sqrt{7} + \sqrt{2})(\sqrt{7} - 3\sqrt{2}) = \sqrt{7}\sqrt{7} - \sqrt{7} \cdot 3\sqrt{2} + \sqrt{2} \cdot \sqrt{7} - \sqrt{2} \cdot 3\sqrt{2}$$
$$= \sqrt{7^2} - 3\sqrt{14} + \sqrt{14} - 3\sqrt{2^2}$$
$$= 7 - 2\sqrt{14} - 3 \cdot 2$$
$$= 1 - 2\sqrt{14}$$

Therefore, $(\sqrt{7} + \sqrt{2})(\sqrt{7} - 3\sqrt{2}) = 1 - 2\sqrt{14}$.
Our result can be confirmed with the calculator:

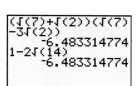

EXAMPLE 4

Multiply $(\sqrt{3}x - \sqrt{5})(\sqrt{2}x + \sqrt{10})$.

SOLUTION

$$(\sqrt{3}x - \sqrt{5})(\sqrt{2}x + \sqrt{10}) = \sqrt{3}\sqrt{2}x^2 + \sqrt{3}\sqrt{10}x - \sqrt{5}\sqrt{2}x - \sqrt{5}\sqrt{10}$$
$$= \sqrt{6}x^2 + \sqrt{30}x - \sqrt{10}x - \sqrt{50}$$
$$= \sqrt{6}x^2 + \sqrt{30}x - \sqrt{10}x - \sqrt{5^2}\sqrt{2}$$
$$= \sqrt{6}x^2 + \sqrt{30}x - \sqrt{10}x - 5\sqrt{2} \quad \text{or}$$
$$x^2\sqrt{6} + x\sqrt{30} - x\sqrt{10} - 5\sqrt{2}$$

WARNING! It is important to draw radical signs carefully so that they completely cover the radicands, but no more than the radicands. To avoid confusion, we often write an expression such as $\sqrt{3}x$ in the form $x\sqrt{3}$.

▍ RATIONALIZING DENOMINATORS

For a radical expression to be in simplified form, there cannot be a fraction in the radicand or a radical in the denominator of a fraction. The process of removing a radical from the denominator of a fraction is called **rationalizing the denominator**.

To rationalize a denominator, multiply both the numerator and the denominator by a radical that will make the denominator a perfect root. Factoring radicands in the denominator into primes is often helpful in identifying the required factors.

EXAMPLE 5   Rationalize each denominator.

a. $\sqrt{\dfrac{20}{7}}$ **b.** $\dfrac{4}{\sqrt[3]{2}}$ **c.** $\dfrac{\sqrt[3]{5x}}{\sqrt[3]{18x^2}}$

SOLUTION   **a.** $\sqrt{\dfrac{20}{7}} = \dfrac{\sqrt{20}}{\sqrt{7}}$ Quotient rule.

Because 7 is prime and we are looking for a square root, multiply the numerator and the denominator by $\sqrt{7}$.

$$= \frac{\sqrt{20}}{\sqrt{7}} \cdot \frac{\sqrt{7}}{\sqrt{7}}$$

$$= \frac{\sqrt{140}}{\sqrt{7^2}}$$ Product rule.

$$= \frac{\sqrt{140}}{7}$$

The denominator is rationalized. We now simplify the numerator.

$$\frac{\sqrt{140}}{7} = \frac{\sqrt{2^2 \cdot 5 \cdot 7}}{7}$$

$$= \frac{2\sqrt{5 \cdot 7}}{7}$$

$$= \frac{2\sqrt{35}}{7}$$

```
√(20/7)
        1.690308509
(2√(35))/7
        1.690308509
```

Therefore, $\sqrt{\dfrac{20}{7}} = \dfrac{2\sqrt{35}}{7}$. Our result can be verified with the calculator.

b. $\dfrac{4}{\sqrt[3]{2}}$

Since the denominator is a cube root and is the prime number 2, we will multiply the numerator and the denominator by powers of 2 so that a perfect cube will be under the radical sign.

$$\frac{4}{\sqrt[3]{2}} \cdot \frac{\sqrt[3]{2^2}}{\sqrt[3]{2^2}} = \frac{4\sqrt[3]{2^2}}{\sqrt[3]{2^3}}$$ Product rule.

$$= \frac{4\sqrt[3]{4}}{2}$$ Extract the perfect root of 2 from the denominator.

$$= 2\sqrt[3]{4}$$ Reduce by dividing out the common factor of 2.

Therefore, $\dfrac{4}{\sqrt[3]{2}} = 2\sqrt[3]{4}$. Our result can be verified by the calculator.

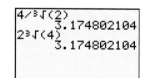

c. $$\frac{\sqrt[3]{5x}}{\sqrt[3]{18x^2}} = \frac{\sqrt[3]{5x}}{\sqrt[3]{3^2 \cdot 2x^2}}$$ Factor the radicand in the denominator to help identify the factors necessary to make the denominator a perfect cube.

From the factorization, we see that we need one more factor of 3, two more factors of 2, and one more factor of x. The necessary radical is $\sqrt[3]{3 \cdot 2^2 x}$. We now multiply the numerator and the denominator by this radical.

$$\frac{\sqrt[3]{5x}}{\sqrt[3]{3^2 \cdot 2x^2}} \cdot \frac{\sqrt[3]{3 \cdot 2^2 x}}{\sqrt[3]{3 \cdot 2^2 x}} = \frac{\sqrt[3]{5 \cdot 3 \cdot 2^2 x^2}}{\sqrt[3]{3^3 \cdot 2^3 x^3}}$$ Product rule.

$$= \frac{\sqrt[3]{60x^2}}{3 \cdot 2x}$$ The prime factors in the numerator assure us there are no factors of 60 that are perfect cubes. We therefore multiply the factors to get a result of 60. We also extract the perfect cubes from the denominator.

$$= \frac{\sqrt[3]{60x^2}}{6x}$$ Multiply $3 \cdot 2$ to get 6.

To rationalize binomial denominators containing square roots, we multiply the numerator and denominator by the **conjugate** of the binomial in the denominator. Conjugate binomials are binomials with opposite signs between the two terms.

Conjugate Binomials
The conjugate of the binomial $a + b$ is $a - b$, and the conjugate of $a - b$ is $a + b$.

The use of conjugate binomials in rationalizing denominators is illustrated in the next two examples.

EXAMPLE 6   Rationalize the denominator of $\dfrac{1}{\sqrt{2} + 1}$.

SOLUTION   We multiply the numerator and denominator of the fraction by $\sqrt{2} - 1$, which is the conjugate of the denominator.

$$\frac{1}{\sqrt{2}+1} = \frac{1(\sqrt{2}-1)}{(\sqrt{2}+1)(\sqrt{2}-1)}$$

Multiply the numerator and the denominator by the conjugate of the denominator.

$$= \frac{\sqrt{2}-1}{(\sqrt{2})^2 - 1}$$

$(\sqrt{2}+1)(\sqrt{2}-1) =$
$(\sqrt{2})^2 - \sqrt{2} + \sqrt{2} - 1 = (\sqrt{2})^2 - 1$

$$= \frac{\sqrt{2}-1}{2-1}$$

$(\sqrt{2})^2 = 2$

$$= \sqrt{2} - 1$$

$\frac{\sqrt{2}-1}{2-1} = \frac{\sqrt{2}-1}{1} = \sqrt{2} - 1$

Our result can be verified with the calculator:

```
1/(√(2)+1)
        .4142135624
√(2)-1
        .4142135624
```

EXAMPLE 7

Rationalize the denominator of $\dfrac{\sqrt{x}+\sqrt{2}}{\sqrt{x}-\sqrt{2}}$.

SOLUTION   We multiply the numerator and denominator by $\sqrt{x} + \sqrt{2}$, which is the conjugate of $\sqrt{x} - \sqrt{2}$, and simplify.

$$\frac{\sqrt{x}+\sqrt{2}}{\sqrt{x}-\sqrt{2}} = \frac{(\sqrt{x}+\sqrt{2})(\sqrt{x}+\sqrt{2})}{(\sqrt{x}-\sqrt{2})(\sqrt{x}+\sqrt{2})}$$

$$= \frac{x + \sqrt{2x} + \sqrt{2x} + 2}{x - 2}$$

Use the distributive property.

$$= \frac{x + 2\sqrt{2x} + 2}{x - 2}$$

EXERCISE 9.4

VOCABULARY AND NOTATION   *Fill in the blanks to make the statements true.*

1. The process of removing a radical from the denominator of a fraction is called _____ the denominator.

2. Binomials with opposite signs between the terms are called _____ binomials.

CONCEPTS   *In Exercises 3–4, fill in the blanks to make the statements true.*

3. To rationalize the denominator of $\dfrac{2}{\sqrt[3]{5}}$, multiply the numerator and denominator of the fraction by _____.

4. To rationalize the denominator of $\dfrac{3}{1+\sqrt{3}}$, multiply the numerator and denominator of the fraction by _____.

In Exercises 5–8, simplify the multiplication of the conjugates.

5. $(1 + \sqrt{2})(1 - \sqrt{2})$

6. $(\sqrt{7} - \sqrt{5})(\sqrt{7} + \sqrt{5})$

7. $(\sqrt{x} - 1)(\sqrt{x} + 1)$

8. $(\sqrt{x} - \sqrt{y})(\sqrt{x} + \sqrt{y})$

PRACTICE *In Exercises 9–32, multiply and simplify. All variables represent positive numbers. When possible, check results with the calculator, recording screen displays exactly as they appear.*

9. $\sqrt{2}\sqrt{8}$ **10.** $\sqrt{3}\sqrt{27}$

11. $\sqrt{5}\sqrt{10}$ **12.** $\sqrt{7}\sqrt{35}$

13. $\sqrt{3}\sqrt{6}$ **14.** $\sqrt{11}\sqrt{33}$

15. $\sqrt[3]{5}\sqrt[3]{25}$ **16.** $\sqrt[3]{7}\sqrt[3]{49}$

17. $\sqrt[3]{9}\sqrt[3]{3}$ **18.** $\sqrt[3]{16}\sqrt[3]{4}$

19. $\sqrt[3]{2}\sqrt[3]{12}$ **20.** $\sqrt[3]{3}\sqrt[3]{18}$

21. $\sqrt{ab^3}\sqrt{ab}$ **22.** $\sqrt{8x}\sqrt{2x^3y}$

23. $\sqrt{5ab}\sqrt{5a}$ **24.** $\sqrt{15rs^2}\sqrt{10r}$

25. $\sqrt[3]{5r^2s}\sqrt[3]{2r}$ **26.** $\sqrt[3]{3xy^2}\sqrt[3]{9x^3}$

27. $\sqrt[3]{a^2b}\sqrt[3]{16ab^5}$ **28.** $\sqrt[3]{3x^4y}\sqrt[3]{18x}$

29. $\sqrt{x(x + 3)}\sqrt{x^3(x + 3)}$

30. $\sqrt{y^2(x + y)}\sqrt{(x + y)^3}$

31. $\sqrt[3]{6x^2(y + z)^2}\sqrt[3]{18x(y + z)}$

32. $\sqrt[3]{9x^2y(z + 1)^2}\sqrt[3]{6xy^2(z + 1)}$

In Exercises 33–52, perform each multiplication and simplify. All variables represent positive numbers.

33. $3\sqrt{5}(4 - \sqrt{5})$

34. $2\sqrt{7}(3\sqrt{7} - 1)$

35. $3\sqrt{2}(4\sqrt{3} + 2\sqrt{7})$

36. $-\sqrt{3}(\sqrt{7} - \sqrt{5})$

37. $-2\sqrt{5x}(4\sqrt{2x} - 3\sqrt{3})$

38. $3\sqrt{7t}(2\sqrt{7t} + 3\sqrt{3t^2})$

39. $(\sqrt{2} + 1)(\sqrt{2} - 3)$

40. $(2\sqrt{3} + 1)(\sqrt{3} - 1)$

41. $(4\sqrt{3x} + 3)(2\sqrt{3x} - 5)$

42. $(7\sqrt{2y} + 2)(3\sqrt{2y} - 5)$

43. $(\sqrt{5z} + \sqrt{3})(\sqrt{5z} + \sqrt{3})$

44. $(\sqrt{3p} - \sqrt{2})(\sqrt{3p} + \sqrt{2})$

45. $(\sqrt{3x} - \sqrt{2y})(\sqrt{3x} + \sqrt{2y})$

46. $(\sqrt{3m} + \sqrt{2n})(\sqrt{3m} + \sqrt{2n})$

47. $(2\sqrt{3a} - \sqrt{5b})(\sqrt{3a} + 3\sqrt{5b})$

48. $(5\sqrt{2p} - \sqrt{3q})(2\sqrt{2p} + 2\sqrt{3q})$

49. $(3\sqrt{2r} - 2)^2$

50. $(2\sqrt{3t} + 5)^2$

51. $-2\sqrt{3x}(\sqrt{3x} + \sqrt{3})^2$

52. $\sqrt{2}(2\sqrt{5x} + 3\sqrt{3})^2$

In Exercises 53–76, rationalize each denominator. All variables represent positive numbers.

53. $\sqrt{\dfrac{1}{7}}$ **54.** $\sqrt{\dfrac{5}{3}}$ **55.** $\sqrt{\dfrac{2}{3}}$

56. $\sqrt{\dfrac{3}{2}}$ **57.** $\dfrac{\sqrt{5}}{\sqrt{8}}$ **58.** $\dfrac{\sqrt{3}}{\sqrt{50}}$

59. $\dfrac{\sqrt{8}}{\sqrt{2}}$ **60.** $\dfrac{\sqrt{27}}{\sqrt{3}}$ **61.** $\dfrac{1}{\sqrt[3]{2}}$

62. $\dfrac{2}{\sqrt[3]{6}}$ **63.** $\dfrac{3}{\sqrt[3]{9}}$ **64.** $\dfrac{2}{\sqrt[3]{a}}$

65. $\dfrac{\sqrt[3]{2}}{\sqrt[3]{9}}$ **66.** $\dfrac{\sqrt[3]{9}}{\sqrt[3]{54}}$

67. $\dfrac{\sqrt{8x^2y}}{\sqrt{xy}}$ **68.** $\dfrac{\sqrt{9xy}}{\sqrt{3x^2y}}$

69. $\dfrac{\sqrt{10xy^2}}{\sqrt{2xy^3}}$ **70.** $\dfrac{\sqrt{5ab^2c}}{\sqrt{10abc}}$

71. $\dfrac{\sqrt[3]{4a^2}}{\sqrt[3]{2a^2b}}$ **72.** $\dfrac{\sqrt[3]{9x}}{\sqrt[3]{3x^2y}}$

73. $\dfrac{1}{\sqrt[3]{3}}$ **74.** $\dfrac{1}{\sqrt[5]{2}}$

75. $\dfrac{4}{\sqrt[5]{16}}$ **76.** $\dfrac{4}{\sqrt[4]{32}}$

In Exercises 77–104, perform each division by rationalizing the denominator and simplifying. All variables represent positive numbers.

77. $\dfrac{1}{\sqrt{2} - 1}$ **78.** $\dfrac{2}{\sqrt{3} - 1}$

79. $\dfrac{-6}{\sqrt{5} + 4}$ **80.** $\dfrac{-10}{\sqrt{5} - 1}$

81. $\dfrac{2}{\sqrt{3}+1}$

82. $\dfrac{2}{\sqrt{5}+1}$

83. $\dfrac{25}{\sqrt{6}+1}$

84. $\dfrac{50}{\sqrt{7}+1}$

85. $\dfrac{\sqrt{2}}{\sqrt{5}+3}$

86. $\dfrac{\sqrt{3}}{\sqrt{3}-2}$

87. $\dfrac{\sqrt{7}}{2-\sqrt{5}}$

88. $\dfrac{\sqrt{11}}{3+\sqrt{7}}$

89. $\dfrac{2}{\sqrt{7}-\sqrt{5}}$

90. $\dfrac{5}{\sqrt{7}-\sqrt{2}}$

91. $\dfrac{20}{\sqrt{3}+1}$

92. $\dfrac{36}{\sqrt{5}+2}$

93. $\dfrac{\sqrt{3}+1}{\sqrt{3}-1}$

94. $\dfrac{\sqrt{2}-1}{\sqrt{2}+1}$

95. $\dfrac{\sqrt{7}-\sqrt{2}}{\sqrt{2}+\sqrt{7}}$

96. $\dfrac{\sqrt{3}+\sqrt{2}}{\sqrt{3}-\sqrt{2}}$

97. $\dfrac{2}{\sqrt{x}+1}$

98. $\dfrac{3}{\sqrt{x}-2}$

99. $\dfrac{x}{\sqrt{x}-4}$

100. $\dfrac{2x}{\sqrt{x}+1}$

101. $\dfrac{2z-1}{\sqrt{2z}-1}$

102. $\dfrac{3t-1}{\sqrt{3t}+1}$

103. $\dfrac{\sqrt{x}-\sqrt{y}}{\sqrt{x}+\sqrt{y}}$

104. $\dfrac{\sqrt{x}+\sqrt{y}}{\sqrt{x}-\sqrt{y}}$

REVIEW   *In Exercises 105–108, solve each equation algebraically and verify by solving graphically.*

105. $\dfrac{2}{3-x}=1$

106. $5(x-4)=-5(x-4)$

107. $\dfrac{8}{y-2}+\dfrac{3}{2-y}=-\dfrac{1}{y}$

108. $\dfrac{2}{x-2}+\dfrac{1}{x+1}=\dfrac{1}{x^2-x-2}$

9.5   RADICAL EQUATIONS AND THEIR APPLICATIONS

- THE POWER RULE
- EQUATIONS CONTAINING RADICALS
- THE PYTHAGOREAN THEOREM
- THE DISTANCE FORMULA

THE POWER RULE

To solve equations that contain radicals we will use the **power rule**.

Power Rule

If x, y, and n are real numbers, and $x=y$, then $x^n=y^n$.

If we raise both *sides* of an equation to the same power, the resulting equation is true. However, the resulting equation *may* or *may not* be equivalent to the original equation. This is illustrated below.

1. Consider the equation $x=3$, which has a solution set of $\{3\}$. If we square both sides of the equation, we obtain
2. $x^2=9$, which has a solution set of $\{3,-3\}$.

The equations in 1 and 2 above are not equivalent because they have different solution sets. All we are guaranteed when raising both sides of an equation to a power is that all solutions of the original equation are solutions of the new equation. However, the new equation may have extra, extraneous roots. For this reason, roots *must* be checked in the original equation.

▐ EQUATIONS CONTAINING RADICALS

A **radical equation** is an equation containing a radical with a variable in the radicand. Example 1 illustrates the solution of a radical equation.

EXAMPLE 1

Solve the equation $\sqrt{x + 3} = 4$. Check the analytic solution by solving graphically.

SOLUTION

To eliminate the radical, apply the power rule by squaring both *sides* of the equation.

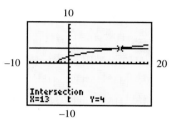

$$(\sqrt{x + 3})^2 = 4^2 \quad \text{Square both sides of the equation (apply the power rule).}$$

$$x + 3 = 16 \quad \text{Subtract 3 from both sides.}$$

$$x = 13$$

Check the analytic solution by using the INTERSECT feature of the calculator.

TECHNOLOGY TIP

Recall that to use the INTERSECT feature for solving equations, enter the left side of the equation at the y1 prompt and the right side at the y2 prompt. Access the INTERSECT feature and respond to the prompts.

To algebraically solve a radical equation, follow these steps:

Solving Radical Equations

1. Isolate one radical on one side of the equation.
2. Raise both *sides* of the equation to the same power as the index of the radical.
3. Simplify both sides of the equation.
4. a. If there are no radicals left in the equation, solve.
   b. If the equation still contains radicals, repeat steps 1–3.
5. Check all proposed solutions and reject any extraneous roots.

EXAMPLE 2

Solve the equation $\sqrt{3x + 1} + 1 = x$.

SOLUTION

$$\sqrt{3x + 1} + 1 = x$$

$$\sqrt{3x + 1} = x - 1 \quad \text{Subtract 1 from both sides of the equation to isolate the radical.}$$

$$(\sqrt{3x + 1})^2 = (x - 1)^2 \quad \text{Raise both sides of the equation to the same power as the index of the radical to eliminate the radical.}$$

$$3x + 1 = x^2 - 2x + 1 \quad \text{Recall that } (x - 1)^2 = (x - 1)(x - 1) \text{ and the distributive property must be applied.}$$

$$0 = x^2 - 5x \quad \text{Subtract } 3x \text{ and 1 from both sides because the equation is quadratic and must be solved by factoring.}$$

$$0 = x(x - 5) \quad \text{Factor the polynomial } x^2 - 5x.$$

$$x = 0 \quad \text{or} \quad x - 5 = 0 \quad \text{Set each factor equal to zero and solve for the variable.}$$

$$x = 5$$

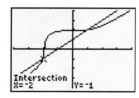

The equation *appears* to have the two solutions of 0 and 5. A graphical check will confirm one, both, or neither of the solutions.

The graphical display confirms the root of 5 and also shows that 0 is *not* a root. Therefore, the solution is simply 5, {5}. Zero is considered an extraneous root.

EXAMPLE 3

Solve the equation $\sqrt[3]{x^3 + 7} = x + 1$. Check by solving the equation graphically.

SOLUTION   To eliminate the radical, we cube both *sides* of the equation.

$$\sqrt[3]{x^3 + 7} = x + 1$$

$(\sqrt[3]{x^3 + 7})^3 = (x + 1)^3$ Cube both sides to eliminate the radical.

$x^3 + 7 = x^3 + 3x^2 + 3x + 1$ Remember, $(x + 1)^3 = (x + 1)(x + 1)(x + 1)$.

$0 = 3x^2 + 3x - 6$ Subtract x^3 and 7 from both sides of the equation.

$0 = x^2 + x - 2$ Divide both sides by 3 to make factoring easier.

$0 = (x + 2)(x - 1)$ Factor the trinomial.

$x + 2 = 0$ or $x - 1 = 0$ Set each factor equal to zero and solve.

$x = -2$ | $x = 1$

The solutions appear to be -2 and 1. A graphical solution will confirm one, both, or neither of the solutions.

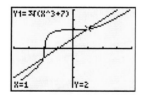

The INTERSECT feature confirms both solutions of -2 and 1: $\{-2, 1\}$. Remember, move the cursor close to the desired point of intersection. The graphs displayed are in the window $[-4, 4]$ by $[-4, 4]$.

When equations contain two or more radicals, the algebra involved can become rather lengthy. Be extra careful when solving these equations algebraically.

EXAMPLE 4

Solve the equation $\sqrt{x} + \sqrt{x + 2} = 2$.

SOLUTION   To remove the radicals, we must square both sides of the equation. This is easier to do if one radical is on each side of the equation. We begin by subtracting $\sqrt{x}$ from both sides to isolate a radical.

$$\sqrt{x} + \sqrt{x+2} = 2$$

$$\sqrt{x+2} = 2 - \sqrt{x}$$ Subtract $\sqrt{x}$ from both sides.

$$(\sqrt{x+2})^2 = (2 - \sqrt{x})^2$$ Square both sides to eliminate one square root.

$$x + 2 = 4 - 4\sqrt{x} + x$$ $\quad (2 - \sqrt{x})(2 - \sqrt{x}) = 4 - 2\sqrt{x} - 2\sqrt{x} + x$
$$= 4 - 4\sqrt{x} + x$$

$$2 = 4 - 4\sqrt{x}$$ Subtract x from both sides.

$$-2 = -4\sqrt{x}$$ Subtract 4 from both sides.

$$\frac{1}{2} = \sqrt{x}$$ Divide both sides by -4.

$$\frac{1}{4} = x$$ Square both sides.

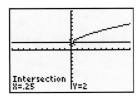

The graphical display and use of the INTERSECT feature of the calculator confirm the analytic solution of $\frac{1}{4}$: $\{\frac{1}{4}\}$.

EXAMPLE 5

Solve the equation $\sqrt{x+2} + \sqrt{2x} = \sqrt{18-x}$.

SOLUTION Because the right side of the equation contains a single radical, begin by squaring both sides. We then proceed as follows:

$$\sqrt{x+2} + \sqrt{2x} = \sqrt{18-x}$$

$$(\sqrt{x+2} + \sqrt{2x})^2 = (\sqrt{18-x})^2$$ Square both sides to eliminate one square root.

$$x + 2 + 2\sqrt{x+2}\sqrt{2x} + 2x = 18 - x$$

$$3x + 2 + 2\sqrt{x+2}\sqrt{2x} = 18 - x$$ Combine like terms.

$$2\sqrt{x+2}\sqrt{2x} = 16 - 4x$$ Subtract $3x$ and 2 from both sides.

$$\sqrt{x+2}\sqrt{2x} = 8 - 2x$$ Divide both sides by 2.

$$(\sqrt{x+2}\sqrt{2x})^2 = (8 - 2x)^2$$ Square both sides to eliminate the other square roots.

$$(x + 2)2x = 64 - 32x + 4x^2$$

$$2x^2 + 4x = 64 - 32x + 4x^2$$ Remove grouping symbols.

$$0 = 2x^2 - 36x + 64$$ Write the equation in standard quadratic form.

$$0 = x^2 - 18x + 32$$ Divide both sides by 2.

$$0 = (x - 16)(x - 2)$$ Factor the trinomial.

$$x - 16 = 0 \quad \text{or} \quad x - 2 = 0$$ Set each factor equal to 0.

$$x = 16 \quad | \quad x = 2$$

$$\{2\}$$

The INTERSECT feature of the calculator confirms the solution of 2. Sixteen is an extraneous root.

TECHNOLOGY TIP

The STOre feature, VARS capability, and TABLE features of the calculator can all be used to confirm solutions to equations. Experiment with your calculator and decide which method(s) you prefer.

▮ THE PYTHAGOREAN THEOREM

A **right triangle** is a triangle that contains a right (90°) angle. The two sides that form the right angle are called **legs** of the triangle. The side opposite the right angle is called the **hypotenuse** of the triangle and is the longest side.

Pythagorean Theorem

In any right triangle, the square of the length of the hypotenuse is equal to the sum of the squares of the lengths of the legs.

Symbolically, if a and b represent the legs of a right triangle and c is the hypotenuse, $a^2 + b^2 = c^2$.

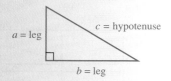

 WARNING! Many students simply remember "a² + b² = c²," disregarding the critical fact that a and b represent the lengths of the **legs** of the triangle and c is the length of the hypotenuse. Remember the *words* of the Pythagorean theorem: $(\text{leg})^2 + (\text{leg})^2 = (\text{hypotenuse})^2$.

EXAMPLE 6

Fighting fires   To fight a forest fire, the forestry department plans to clear a rectangular fire break around the fire, as shown in Figure 9-1. Crews are equipped with mobile communications with a 3,000-yard range. Can crews at points A and B remain in radio contact?

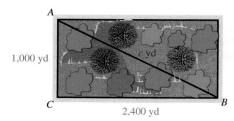

Figure 9-1

SOLUTION   Points A, B, and C form a right triangle. The distance from A to C is 1,000 yards and is a leg of the right triangle. The distance from C to B is 2,400 yards and forms the other leg of the right triangle. We are looking for the distance from A to B, the hypotenuse.

$$(\textbf{leg})^2 + (\textbf{leg})^2 = (\text{hypotenuse})^2$$
$$\textbf{2,400}^2 + \textbf{1,000}^2 = h^2$$
$$5,760,000 + 1,000,000 = h^2$$
$$6,760,000 = h^2$$
$$\sqrt{6,760,000} = \sqrt{h^2}$$
$$2,600 = h$$

Remember, the opposite of squaring a number is finding the square root. (Section 9.1)

Two fire crews are 2,600 yards apart. Because this distance is less than the range of the radios, the crews can communicate.

▉ THE DISTANCE FORMULA

With the *distance formula*, we can find the distance between any two points that are graphed on a rectangular coordinate system.

To find the distance d between points $P(x_1, y_1)$ and $Q(x_2, y_2)$ shown in Figure 9-2, we construct the right triangle PRQ. The distance between P and R is $|x_2 - x_1|$, and the distance between R and Q is $|y_2 - y_1|$. We apply the Pythagorean theorem to the right triangle PRQ to get

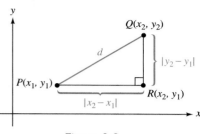

Figure 9-2

$$[d(PQ)]^2 = |x_2 - x_1|^2 + |y_2 - y_1|^2$$

Read $d(PQ)$ as "the distance between P and Q."

$$= (x_2 - x_1)^2 + (y_2 - y_1)^2$$

Because $|x_2 - x_1|^2 = (x_2 - x_1)^2$ and $|y_2 - y_1|^2 = (y_2 - y_1)^2$.

or

$$d(PQ) = \sqrt{(x_2 - x_1)^2 + (y_2 - y_1)^2}$$

This equation is the *distance formula*. Because it is one of the most important formulas in mathematics, take the time to memorize it.

Distance Formula

The distance between two points $P(x_1, y_1)$ and $Q(x_2, y_2)$ is given by the formula

$$d(PQ) = \sqrt{(x_2 - x_1)^2 + (y_2 - y_1)^2}$$

EXAMPLE 7

Find the distance between points $P(-2, 3)$ and $Q(4, -5)$.

SOLUTION

To find the distance, we can use the distance formula by substituting 4 for x_2, -2 for x_1, -5 for y_2, and 3 for y_1.

$$\begin{aligned}
d(PQ) &= \sqrt{(x_2 - x_1)^2 + (y_2 - y_1)^2} \\
&= \sqrt{[4 - (-2)]^2 + (-5 - 3)^2} \\
&= \sqrt{(4 + 2)^2 + (-5 - 3)^2} \\
&= \sqrt{6^2 + (-8)^2} \\
&= \sqrt{36 + 64} \\
&= \sqrt{100} \\
&= 10
\end{aligned}$$

The distance between points P and Q is 10 units.

EXAMPLE 8

Building a freeway   In a city, streets run north and south and avenues run east and west. Streets and avenues are 750 feet apart. The city plans to construct a straight freeway from the intersection of 21st Street and 4th Avenue to the intersection of 111th Street and 60th Avenue. How long will the freeway be?

SOLUTION   We can represent the roads of the city by the coordinate system in Figure 9-3, where the units on each axis represent 750 feet. We represent the end of the freeway at 21st Street and 4th Avenue by the point $(x_1, y_1) = (21, 4)$. The other end is $(x_2, y_2) = (111, 60)$.

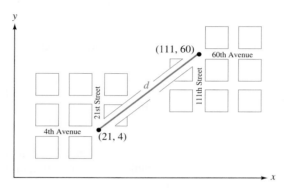

Figure 9-3

We can now use the distance formula to find the length of the freeway.

$$d = \sqrt{(x_2 - x_1)^2 + (y_2 - y_1)^2}$$
$$d = \sqrt{(111 - 21)^2 + (60 - 4)^2}$$
$$= \sqrt{8100 + 3136}$$
$$= \sqrt{11{,}236}$$
$$= 106 \qquad\qquad \text{Use a calculator to find the square root.}$$

Because each unit represents 750 feet, the length of the freeway is $106 \cdot 750 = 79{,}500$ feet. Since 5,280 feet = 1 mile, we can divide 79,500 by 5,280 to convert 79,500 feet to 15.056818 miles. The freeway will be about 15 miles long.

EXERCISE 9.5

VOCABULARY AND NOTATION   *Fill in each blank to make a true statement.*

1. A radical equation is an equation that contains a _____ with a variable in the _____.

2. A _____ triangle is one that contains a _____ angle.

3. The legs of a right triangle are the sides that form the _____ angle.

4. The _____ is the longest side of a right triangle and is opposite the _____ angle.

5. The Pythagorean theorem states that, in a right triangle, the sum of the squares of the lengths of the _____ is equal to the square of the length of the _____.

6. The distance formula can be used to find the distance between two points in the _____ coordinate system.

CONCEPTS

7. To isolate the variable x, match each equation with the necessary process.

a. $x^2 = 4$ **i.** Divide each side of the equation by 5.

b. $\sqrt{x} = 4$ **ii.** Add 5 to both sides of the equation.

c. $x + 5 = 4$ **iii.** Extract the square root of both sides of the equation.

d. $x - 5 = 4$ **iv.** Square both sides of the equation.

e. $5x = 4$ **v.** Subtract 5 from both sides of the equation.

f. $\dfrac{x}{5} = 4$ **vi.** Multiply both sides of the equation by 5.

8. When solving radical equations, we raise both sides of the equation to the same power as the index of the radical. Consider the equation $\sqrt{2x - 3} = x - 9$. Which of the following is the square of the right *side* of the equation?

a. $x^2 - 18$ **b.** $x^2 - 81$

c. $x^2 - 9x + 81$ **d.** $x^2 - 18x + 81$

9. Consider the two equations $\sqrt{2x} = 5$ and $\sqrt[3]{2x} = 5$. Explain the similarities and differences in the solving processes for these two equations.

10. The first step in solving $\sqrt{x} + 5 = 8$ is to

a. square each term.

b. square both sides.

c. subtract 5 from both sides.

In Exercises 11–18, state the solution of each radical equation from the pictured graphs.

11.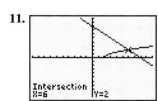

12.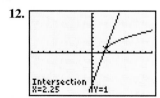

13.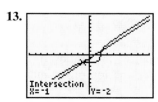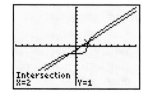

14.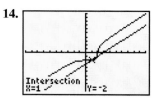

15.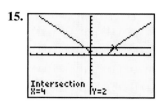

16.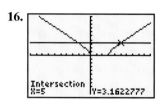

17.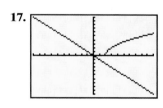

18.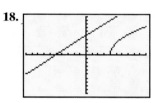

PRACTICE   *In Exercises 19–62, solve each equation algebraically. Check results by solving graphically.*

19. $\sqrt{5x - 6} = 2$ **20.** $\sqrt{7x - 10} = 12$

21. $\sqrt[3]{7n - 1} = 3$ **22.** $\sqrt[3]{12m + 4} = 4$

23. $\sqrt[4]{10p + 1} = \sqrt[4]{11p - 7}$

24. $\sqrt[4]{10x + 2} = 2\sqrt[4]{2}$ **25.** $\sqrt{6x + 1} + 2 = 7$

26. $\sqrt{6x + 13} - 2 = 5$ **27.** $2\sqrt{4x + 1} = \sqrt{x + 4}$

28. $\sqrt{3x + 12} = \sqrt{5x - 2}$

29. $2\sqrt{x} = \sqrt{5x - 16}$ **30.** $3\sqrt{x} = \sqrt{3x + 12}$

31. $\sqrt{y + 2} = 4 - y$

32. $\sqrt{22y + 86} = y + 9$

33. $r - 9 = \sqrt{2r - 3}$

34. $-x - 3 = 2\sqrt{5 - x}$

35. $\sqrt{-5x + 24} = 6 - x$

36. $\sqrt{-x + 2} = x - 2$

37. $\sqrt[4]{x^4 + 4x^2 - 4} = x$

38. $\sqrt[4]{8x - 8} + 2 = 0$

39. $\sqrt{v} + \sqrt{3} = \sqrt{v + 3}$ **40.** $\sqrt{x} + 2 = \sqrt{x + 4}$

41. $2 + \sqrt{u} = \sqrt{2u + 7}$

42. $5r + 4 = \sqrt{5r + 20} + 4r$

43. $r - 9 = (2r - 3)^{1/2}$

44. $-x - 3 = 2(5 - x)^{1/2}$

45. $(-5x + 24)^{1/2} = 6 - x$

46. $(-x + 2)^{1/2} = x - 2$

47. $\sqrt{x}\sqrt{x + 16} = 15$

48. $\sqrt{x}\sqrt{x + 6} = 4$

49. $(x^3 - 7)^{1/3} = x - 1$

50. $(x^3 + 56)^{1/3} - 2 = x$

51. $\sqrt{2y + 1} = 1 - 2\sqrt{y}$

52. $\sqrt{u} + 3 = \sqrt{u - 3}$

53. $\sqrt{y + 7} + 3 = \sqrt{y + 4}$

54. $1 + \sqrt{z} = \sqrt{z + 3}$

55. $\sqrt{6x + 1} - 3\sqrt{x} = -1$

56. $\sqrt{4x + 1} - \sqrt{6x} = -1$

57. $\sqrt{x - 1} + \sqrt{x + 2} = 3$

58. $\sqrt{16x + 1} + \sqrt{8x + 1} = 12$

59. $\sqrt{\sqrt{x} + \sqrt{x + 8}} = 2$

60. $\sqrt{\sqrt{2y} - \sqrt{y - 1}} = 1$

61. $\dfrac{6}{\sqrt{x + 5}} = \sqrt{x}$

62. $\dfrac{\sqrt{2x}}{\sqrt{x + 2}} = \sqrt{x - 1}$

Solve the equations in Exercises 63–66, graphically. Copy your graphical display to justify your result.

63. $\sqrt{x} + \sqrt{3x + 2} + 4 = \sqrt{2x - 5} - \sqrt{5x - 1}$

64. $\sqrt{x + 4} + \sqrt{x + 2} + 4 = \sqrt{2x} - \sqrt{5x - 1} + 2$

65. $\sqrt{x} + \sqrt{4x - 4} = \sqrt{6x - 8} - \sqrt{2x - 4}$

66. $\sqrt{2x - 9} + 3 - \sqrt{x} = \sqrt{5x + 4} - \sqrt{3x - 11}$

In Exercises 67–70, the lengths of two sides of the right triangle ABC shown in Illustration 1 are given. Find the length of the missing side.

67. $a = 6$ ft and $b = 8$ ft

68. $a = 10$ cm and $c = 26$ cm

69. $b = 18$ m and $c = 82$ m

70. $a = 14$ in. and $c = 50$ in.

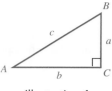

Illustration 1

In Exercises 71–80, find the distance between P and Q.

71. $P(-2, -8), Q(3, 4)$ **72.** $P(-5, -2), \ Q(7, \ 3)$

73. $P(6, 8), Q(12, 16)$ **74.** $P(10, 4), Q(2, -2)$

75. $Q(-3, 5), P(-5, -5)$

76. $Q(2, -3), P(4, -8)$

77. $P(5, 8), Q(1, 2)$

78. $P(-4, -3), Q(-1, 5)$

79. $P(-5, -3), Q(-1, 1)$

80. $P(8, 10), Q(6, 4)$

81. Geometry Show that a triangle with vertices at $(-2, 4), (2, 8),$ and $(6, 4)$ is isosceles.

82. Geometry The square in Illustration 2 has an area of 18 square units, and its diagonals lie on the x- and y-axes. Find the coordinates of each corner of the square.

83. Geometry Show that a triangle with vertices at $(2, 3),$ $(-3, 4),$ and $(1, -2)$ is a right triangle. (*Hint:* If the Pythagorean relation holds, then the triangle is a right triangle.)

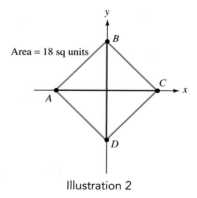

Illustration 2

APPLICATIONS

84. Sailing Refer to the sailboat in Illustration 3. How long must a rope be to fasten the top of the mast to the bow?

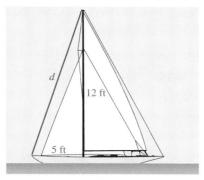

Illustration 3

85. Carpentry   The gable end of the roof shown in Illustration 4 is divided in half by a vertical brace. Find the distance from eaves to peak.

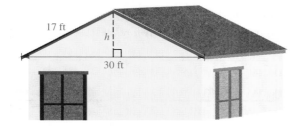

Illustration 4

In Exercises 86–89, the baseball diamond shown in Illustration 5 is a square, 90 feet on a side.

86. Baseball   How far must a catcher throw the ball to throw out a runner stealing second base?

87. Baseball   In baseball, the pitcher's mound is 60 feet, 6 inches from home plate. How far from the mound is second base?

88. Baseball   If the third baseman fields a ground ball 10 feet directly behind third base, how far must he throw the ball to throw the batter out at first base?

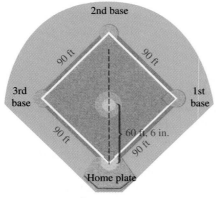

Illustration 5

89. Baseball   A shortstop fields a grounder at a point one-third of the way from second base to third base, How far will he have to throw the ball to make an out at first base?

90. Packing a tennis racquet   The diagonal d of a rectangular box with dimensions $a \times b \times c$ is given by $d = \sqrt{a^2 + b^2 + c^2}$. Will the tennis racquet pictured in Illustration 6 fit into the box?

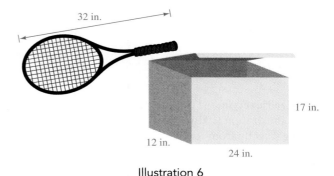

Illustration 6

91. Shipping packages   A delivery service won't accept a package for shipping if any dimension exceeds 21 inches. An archeologist wants to ship a 36-inch femur. Will it fit in a 3-inch-tall box that has a 21-inch-square base?

92. Shipping packages   Can the archeologist in Exercise 91 ship the femur in a cubical box 21 inches on an edge?

93. Reach of a ladder   The base of the 37-foot ladder in Illustration 7 is 9 feet from the wall. Will the top reach a window ledge that is 35 feet above the ground?

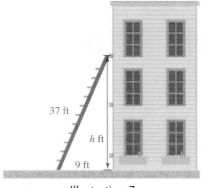

Illustration 7

94. Telephone service   The telephone cable in Illustration 8 runs from A to B to C to D. How much cable is required to run from A to D directly?

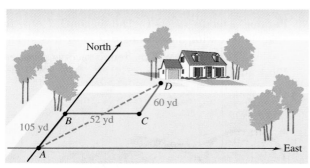

Illustration 8

95. Electric service   The power company routes its lines as in Illustration 9. How much wire could be saved by going directly from A to E?

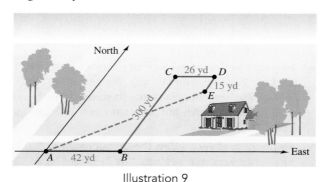

Illustration 9

96. Supporting a weight   A weight placed on the tight wire in Illustration 10 pulls the center down 1 foot. By how much is the wire stretched? Round the answer to the nearest hundredth of a foot.

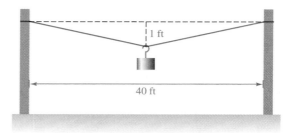

Illustration 10

97. Geometry   The side, s, of a square with area A square feet is given by the formula $s = \sqrt{A}$. Find the perimeter of a square with an area of 49 square feet.

REVIEW   *In Exercises 98–101, determine if each of the graphs is a function or strictly a relation.*

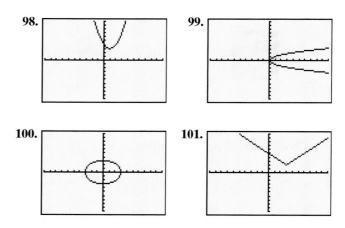

9.6 RADICAL FUNCTIONS

- FUNCTION NOTATION
- DOMAIN AND RANGE

- INTERPRETING GRAPHS OF RADICAL FUNCTIONS

FUNCTION NOTATION

Recall that a **relation** is a correspondence that assigns to each number x one or more values, y. A **function** is a relation that assigns to each value of x **exactly** one value, y.

When we state that "y is a function of x," we are asserting that each x-value in the domain is assigned to a **unique** y-value in the range. The notation used to specify that "y is a function of x" is $f(x)$. This notation is useful for evaluation. For example, if $f(x) = \sqrt{x + 2}$, then $f(2) = \sqrt{2 + 2} = \sqrt{4} = 2$.

EXAMPLE 1

If $f(x) = \sqrt{x + 4}$, find each of the following: **a.** $f(0)$ **b.** $f(5)$ **c.** $f(7)$

SOLUTION   **a.** Because $f(x) = \sqrt{x + 4}$, to find the value of the function when $x = 0, f(0)$, we substitute 0 in for x and simplify.

$$f(x) = \sqrt{x + 4}$$
$$f(0) = \sqrt{0 + 4}$$
$$f(0) = \sqrt{4}$$
$$f(0) = 2$$

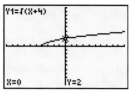

This can be confirmed graphically using the VALUE (EVAL) feature of the calculator.

b. Because $f(x) = \sqrt{x + 4}$, to find the value of the function when $x = 5, f(5)$, we substitute 5 in for x and simplify.

$$f(x) = \sqrt{x + 4}$$
$$f(5) = \sqrt{5 + 4}$$
$$f(5) = \sqrt{9}$$
$$f(5) = 3$$

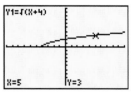

This can be confirmed graphically using the VALUE (EVAL) feature of the calculator.

c. Because $f(x) = \sqrt{x + 4}$, to find the value of the function when $x = 7, f(7)$, we substitute 7 in for x and simplify.

$$f(x) = \sqrt{x + 4}$$
$$f(7) = \sqrt{7 + 4}$$
$$f(7) = \sqrt{11}$$

The radical expression $\sqrt{11}$ is an exact result. However, its value can be approximated. This can be done from the home screen, as illustrated below.

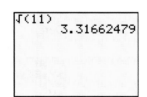

Use of the VALUE (EVAL) option and the graphical display also allows us an approximation.

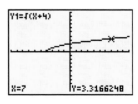

The notation $y = f(x)$ gives two different notations for the same quantity. Therefore, $y = \sqrt[3]{x + 3} - 2$ is the same statement as $f(x) = \sqrt[3]{x + 3} - 2$. Either notation specifies a pairing of two values using a definition, or rule.

EXAMPLE 2

If $f(x) = \sqrt[3]{x+3} - 2$, find **a.** $f(-4)$, **b.** $f(-3)$, **c.** $f(-2)$, **d.** $f(5)$, and **e.** $f(7)$.

SOLUTION   Graph the function in the window ZDecimal $\times$ 2, $[-9.4, 9.4]$ by $[-6.2, 6.2]$ for the TI-83 Plus, and $[-12.2, 12.2] \times [-6.2, 6.2]$ for the TI-86. Tracing, we note the following:

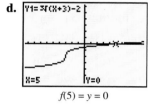
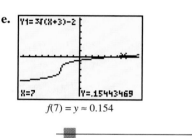

a. $f(-4) = y = -3$

b. $f(-3) = y = -2$

c. $f(-2) = y = -1$

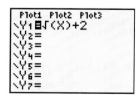
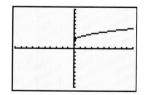
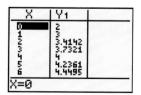

d. $f(5) = y = 0$

e. $f(7) = y \approx 0.154$

■ DOMAIN AND RANGE

The domain of a function is the set of all possible values for the independent variable (x). The range is the set of all possible values for the dependent variable (y). Three tools that help determine the domain and range are the definition of the function, its graphical display, and the table of values. Experiment with all three to ensure success in determining the domain and the range of a function.

EXAMPLE 3

Determine the domain and the range of the function $y = \sqrt{x} + 2$.

SOLUTION

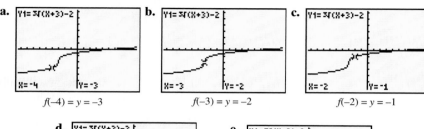

The graph appears in the first quadrant. From the display, we first make a guess (an *educated* guess) that the domain is $x \geq 0$ and $y \geq 0$ because both x and y are positive in the first quadrant.

The TABLE display supports this guess, in a limited way; however, it should be noted that both the graph and the TABLE information are limited because of the display capabilities of the graphing calculator.

Examining the function, $f(x) = \sqrt{x} + 2$, we note that x *must* be 0 or positive to be defined in the real-number system. Therefore, our domain, $x \geq 0$, appears to be correct.

Scrolling down the TABLE confirms the fact that each x-value in the domain is paired with a single y-value in the range. These y-values increase as x increases. However, scrolling *up* the TABLE (looking at x-values smaller than 0), we see that an ERROR message is consistently displayed for these negative x-values. Again, this serves as confirmation of our domain. However, scrolling in this direction also reveals that the *smallest* y-value displayed is 2. We therefore modify our original conjecture of a range from $y \geq 0$ to $y \geq 2$.

We can confirm this algebraically since the smallest possible value for x is 0:

$$y = \sqrt{x} + 2$$
$$y = \sqrt{0} + 2$$
$$y = 0 + 2$$
$$y = 2$$

Therefore, our domain is $[0, \infty)$ and our range is $[2, \infty)$.

Suggestions for Using the Graphing Calculator to Determine Domain and Range

1. Examine the graphical display in a window that shows the x-intercept(s), y-intercept, and any peaks or valleys of the graph. A good starting place is to note the quadrants that contain graphical display.
2. Examine the values displayed in the TABLE for the function. It is suggested that values be examined for $x = 0$, for positive x-values, and for x-values that are negative.
3. Examine the definition (or rule) of the function. **Think** about any values for x for which the function would not be defined. Think about any maximum/minimum values for either x or y (often the peaks and valleys of the graphical display are helpful).

EXAMPLE 4

State the domain and the range of the function $y = \sqrt{x + 2}$.

SOLUTION   We begin, as before, by examining the graphical display, the TABLE, and the function itself.

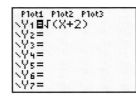
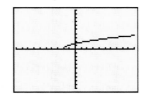
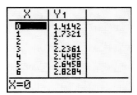

The graphical display reveals the graph in quadrants I and II, which suggests that the function is defined for some negative x-values.

Close examination of the definition of the function, $y = \sqrt{x + 2}$, reveals that the radicand of $x + 2$ must be 0 or positive. Therefore, $x + 2 \geq 0$ if and only if $x \geq -2$. We conclude that our domain must be $[-2, \infty)$. This is confirmed when we scroll up the TABLE to negative values for x.

The range appears to be all positive numbers and 0. Again, examination of the graphical display, function definition, and the TABLE confirm our conjecture.

Therefore, we conclude that the domain is $[-2, \infty)$ and the range is $[0, \infty)$.

▌ INTERPRETING GRAPHS OF RADICAL FUNCTIONS

EXPLORING THE CONCEPT

The distance x (in feet) that an object will fall in y seconds is given by the formula $y = \sqrt{\frac{x}{16}}$. Graph the equation in the window $[-75.2, 75.2]$ by $[-6.2, 6.2]$ (TI-86: $[-100.8, 100.8]$ by $[-6.2, 6.2]$). The scale on the x-axis is 10, whereas the scale on the y-axis is 1.

Examining the general shape of the graph and being aware that x represents height and y represents time, we see that as x increases, so does y. Interpreted within the context of our problem, this means that as the object falls a greater distance, the time increases.

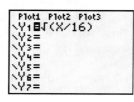
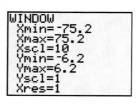
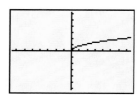

TRACE along the graph to answer the questions that appear below **and** answer each question using analytic methods.

a. If a stone is dropped from a height of 16 feet, how long will it take to hit the ground?

b. Approximately how long will it take a stone to reach the ground if it is dropped from a height of 32 feet?

c. After 2 seconds have passed, how far has the stone fallen?

CONCLUSION   **a.** Graphical interpretation:

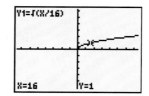

Interpreting the graphical display within the context of the problem, we TRACE until the value of the variable x is equal to 16 (remember, x represents the height). The corresponding y-value is 1. Because y represents time, we can conclude that it would take 1 second for the stone to hit the ground.

Analytical solution: Because our equation is a function, we can rewrite $y = \sqrt{\frac{x}{16}}$ as $f(x) = \sqrt{\frac{x}{16}}$, using function notation. Because we know that the

TECHNOLOGY TIP

The VALUE (EVAL) feature of the calculator could also be used to answer the question, as could the TABLE.

stone was dropped from a height of 16 feet, we are actually being asked to evaluate the function when $x = 16$, or

$$f(16) = \sqrt{\frac{16}{16}} = \sqrt{1} = 1$$

Again, interpreting the *equation* within the context of the problem, we conclude that it would take 1 second for the stone to fall 16 feet.

b. Graphical interpretation:

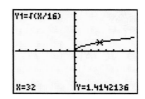

Again, you must be aware that x represents the distance and y represents time. Tracing to a value of $x = 32$ gives a corresponding y-value of approximately 1.4. Therefore, it takes approximately 1.4 seconds for the stone to travel 32 feet.

Analytical solution: Using function notation, and recalling what the variable x and y represent within the context of the problem, we are looking for the values of the function when x has a value of 32.

$$f(32) = \sqrt{\frac{32}{16}} = \sqrt{2} \approx 1.414213562$$

Therefore, it takes the stone approximately 1.4 seconds to fall 32 feet.

c. Graphical interpretation:

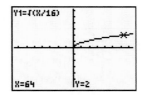

TECHNOLOGY TIP

We could also have entered the given y-value of 2 at the y2 prompt and used the INTER-SECT feature to answer the question posed.

Because we are given time (which is represented by the y-coordinate), we TRACE until $y = 2$. The corresponding x-value is 64. Therefore, the stone will have fallen 64 feet in 2 seconds.

Analytical solution: Recall the function: $f(x) = \sqrt{\frac{x}{16}}$. In this particular example we are given the *value* of the function (2) and are asked to find x. Therefore, we substitute 2 in for $f(x)$ and solve the resulting equation for x.

$$f(x) = \sqrt{\frac{x}{16}}$$

$$2 = \sqrt{\frac{x}{16}} \qquad \text{Substitution}$$

$$4 = \frac{x}{16} \qquad \text{Square both sides to remove the radical.}$$

$$64 = x \qquad \text{Multiply both sides by 16 to isolate the variable.}$$

A quick check reveals that 64 is **not** an extraneous root. It takes 2 seconds to fall 64 feet.

The preceding exploration illustrates that if a good window is used, tracing along a graph and interpreting the display allows you to answer specific questions. However, tracing is not the only way to interpret a graphical display nor is it the only way to obtain a desired result. Make sure you read the technology notes, which provide other solving options using the graphing calculator.

EXERCISE 9.6

VOCABULARY AND NOTATION *Fill in the blanks to make a true statement.*

1. A _____ is a relation that assigns to each value of x exactly one y-value.

2. The mathematical statement $f(x) = \sqrt{x + 3}$ is the same as the mathematical statement ___ $= \sqrt{x + 3}$.

3. The variable ___ is often referred to as the *independent variable*. The _____ is the set of all possible values for this variable.

4. The variable ___ is often referred to as the *dependent variable*. The _____ is the set of all possible values for this variable.

CONCEPTS *In Exercises 5–7, use the given display to fill in the blanks.*

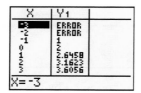

5. $f(-1) =$ ___

6. The ERROR message displayed for an x-value of -2 implies that -2 is not a part of the _____ of the function.

7. An approximation (to the nearest thousandth) of $f(3)$ is _____.

In Exercises 8–11, match each graphical display with its equation.

8. $y = \sqrt{x + 4}$ 9. $y = \sqrt{x} + 4$

10. $y = \sqrt{x - 3}$ 11. $y = \sqrt{x} - 3$

a.

b.

c.

d.

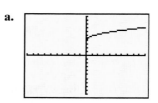
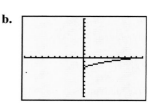

In Exercises 12–19, match each graphical display with its equation.

12. $y = 3\sqrt{x}$ 13. $y = -3\sqrt{x}$

14. $y = -3\sqrt{x} + 4$ 15. $y = 3\sqrt{x + 4}$

a.

b.

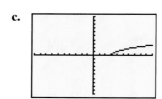
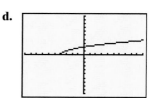

c.

d.

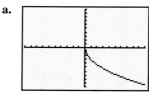
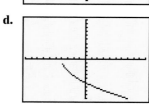

16. $f(x) = \sqrt{1 - x}$ 17. $f(x) = -\sqrt{1 - x} + 3$

18. $f(x) = \sqrt{1 - x} - 3$ 19. $f(x) = -\sqrt{1 - x}$

a.

b.

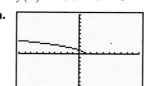
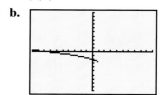
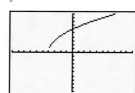

c. 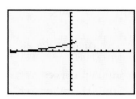 **d.**

In Exercises 20–23, use the graph of $y = \sqrt{x}$ as your reference graph.

20. Graph $y = \sqrt{x} + 2$, $y = \sqrt{x} + 5$, and $y = \sqrt{x} - 4$ on your graphing calculator and sketch the graphs on your paper.

   a. What effect does adding or subtracting a positive constant have on the reference graph?

   b. Does this affect the domain or the range of the function?

   c. Write an equation of a square root function with domain $[0, \infty)$ and range $[-6, \infty)$.

21. Graph $y = \sqrt{x + 2}$, $y = \sqrt{x + 5}$, and $y = \sqrt{x - 4}$ on your graphing calculator and sketch the graphs on your paper.

   a. What effect does adding or subtracting a positive constant to the variable under the radical sign have on the reference graph?

   b. Does this affect the domain or the range of the function?

   c. Write an equation of a square root function with domain $[-6, \infty)$ and range $[0, \infty)$.

22. Graph $y = \sqrt{1 - x}$, $y = \sqrt{3 - x}$, $y = \sqrt{5 - x}$, and $y = \sqrt{-5 - x}$ on your graphing calculator and sketch the graphs on your paper.

   a. What effect does a negative variable with a positive constant added or subtracted to it under the radical have on the reference graph?

   b. Does this affect the domain or the range of the function?

   c. Write an equation of a square root function with domain $(-\infty, -3]$ and range $[0, \infty)$.

23. Graph $y = 2\sqrt{x}$, $y = 4\sqrt{x}$, and $y = 5\sqrt{x}$ on your graphing calculator and sketch the graphs on your paper.

   a. What effect does multiplying the radical by a positive constant have on the reference graph?

   b. Does this affect the domain or the range of the function?

   c. Graph $y = -2\sqrt{x}$, $y = -4\sqrt{x}$, and $y = -5\sqrt{x}$ on your graphing calculator and sketch the graphs on your paper. What effect does multiplying the radical by a negative constant have on the reference graph?

   d. Does this affect the domain or the range?

In Exercises 24–29, write a square root equation that meets the specified criteria. Use your graphing calculator to confirm your equation.

24. Write the equation of a square root function with domain $[-3, \infty)$ and range $[3, \infty)$.

25. Write the equation of a square root function with domain $[2, \infty)$ and range $[2, \infty)$.

26. Write the equation of a square root function with domain $[-6, \infty)$ and range $(-\infty, 2]$.

27. Write the equation of a square root function with domain $[1, \infty)$ and range $(-\infty, -4)$.

28. Write the equation of a square root function with domain $(-\infty, 6]$ and range $[0, \infty)$.

29. Write the equation of a square root function with domain $(-\infty, -2]$ and range $[3, \infty)$.

PRACTICE *In Exercises 30–41, $f(x) = \sqrt{2x - 3}$ and $g(x) = \sqrt{3x - 3} + 2$. Find each of the following (a) algebraically, (express results in simplest radical form) and (b) by using an appropriate feature of your calculator.*

30. $f(1)$

31. $f(2)$

32. $g(1)$

33. $g(10)$

34. $g(73)$

35. $g(19)$

36. $g(-8)$

37. $g(-71)$

38. $f(-2)$

39. $f(-1)$

40. $f(10)$

41. $f(14)$

In Exercises 42–61, specify the domain and the range of each of the functions specified below.

42. $f(x) = \sqrt{x} + 2$

43. $f(x) = \sqrt{x + 2}$

44. $f(x) = \sqrt{x} - 5$

45. $f(x) = \sqrt{x - 5}$

46. $f(x) = 2\sqrt{x} + 3$

47. $f(x) = 2\sqrt{x + 3}$

48. $f(x) = \sqrt{3x - 2}$

49. $f(x) = \sqrt{5x - 3}$

50. $f(x) = \sqrt{2x - 5} + 4$

51. $f(x) = \sqrt{3x - 8} - 2$

52. $f(x) = \sqrt{7x + 2} + 5$

53. $f(x) = \sqrt{6x - 5} - 4$

54. $f(x) = 3\sqrt{2x - 1} + 3$

55. $f(x) = 5\sqrt{3x - 6} + 2$

56. $f(x) = -3\sqrt{2x - 1} + 3$

57. $f(x) = -5\sqrt{3x - 6} + 2$

58. $f(x) = \sqrt[3]{x + 2} - 1$

59. $f(x) = \sqrt[3]{x - 3} + 4$

60. $f(x) = -5\sqrt[3]{x + 2} - 1$

61. $f(x) = -2\sqrt[3]{x + 2} - 1$

62. Discuss the uses of the graphical display, the value feature, and the TABLE option of the calculator for evaluating functions.

63. a. Graph $y = \sqrt{x} \cdot \sqrt{x}$ in a standard viewing window.

    b. Graph $y = x$ in a standard viewing window.

    c. Explain why the graphical displays differ.

APPLICATIONS *In Exercises 64–69, use your graphing calculator or algebraic methods to answer the question posed in each problem.*

64. Banked curves The highway curve shown in Illustration 1 is banked at 8° and will accommodate traffic traveling s mph if the radius of the curve is r feet, according to the formula $s = 1.45\sqrt{r}$. If highway engineers expect 65-mph traffic, what radius should they specify?

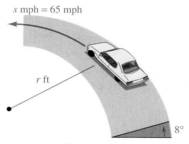

s mph = 65 mph

r ft

8°

Illustration 1

65. Horizon distance The higher a lookout tower is built, the farther an observer can see. That distance d (called the *horizon distance*, measured in miles) is related to the height h of the observer (measured in feet) by the formula $d = 1.4\sqrt{h}$. How tall must the lookout tower in Illustration 2 be to see the edge of the forest, 25 miles away?

Illustration 2

66. Producing power The power generated by a certain windmill is related to the velocity of the wind by the formula $v = \sqrt[3]{\dfrac{P}{0.02}}$, where P is the power (in watts) and v is the velocity of the wind (in mph). Find the speed of the wind when the windmill is generating 500 watts of power.

67. Carpentry During construction, carpenters often brace walls, as shown in Illustration 3, where the length of the brace is given by the formula $l = \sqrt{f^2 + h^2}$. If a carpenter nails a 10-foot brace to the wall 6 feet above the floor, how far from the base of the wall should he nail the brace to the floor?

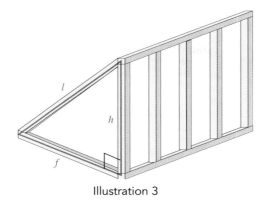

Illustration 3

68. Marketing   The number of wrenches that will be produced at a given price can be predicted by the formula $s = \sqrt{5x}$, where s is the supply (in thousands) and x is the price (in dollars). If the demand, d, for wrenches can be predicted by the formula $d = \sqrt{100 - 3x^2}$, find the equilibrium price.

69. Marketing   The number of footballs that will be produced at a given price can be predicted by the formula $s = \sqrt{23x}$, where s is the supply (in thousands) and x is the price (in dollars). If the demand, d, for footballs can be predicted by the formula $d = \sqrt{312 - 2x^2}$, find the equilibrium price.

REVIEW   *Graph each of the following using the x- and y-intercepts and max/minimums as applicable.*

70. $2x + 3y = 6$

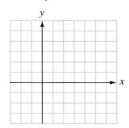

71. $-x^2 - 2x + 8 = y$

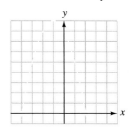

72. $y = |x + 2| - 4$

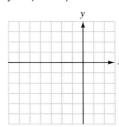

73. $y = \sqrt{x + 4}$

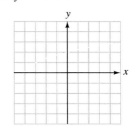

74. If $f(x) = 3x^2 - 4x + 2$, find each of the following:
   a. $f(0)$ **b.** $f(-3)$
   c. $f(2)$ **d.** $f(\frac{1}{2})$

9.7 COMPLEX NUMBERS

- BASIC DEFINITIONS
- POWERS OF i
- COMPLEX NUMBERS
- ARITHMETIC OF COMPLEX NUMBERS
- RATIONALIZING DENOMINATORS

▚ BASIC DEFINITIONS

Thus far, all of our work with simplifying and evaluating expressions as well as solving equations has involved only **real numbers**. Recall from Section 9.1 expressions like $\sqrt{-4}$ are undefined in the real number system.

For years, it was believed that numbers like $\sqrt{-1}$, $\sqrt{-3}$, $\sqrt{-4}$, and $\sqrt{-9}$ were nonsense. The great mathematician Sir Isaac Newton (1642–1727) called them impossible. In the 17th century, René Descartes (1596–1650) called them **imaginary numbers**. Today, we no longer think of imaginary numbers as being fictitious. In fact, imaginary numbers have many important uses, such as describing the behavior of alternating current in electronics.

Imaginary Unit
The imaginary number $\sqrt{-1}$ is often denoted by the letter i. It is defined as the square root of -1. Therefore, $i = \sqrt{-1}$ and $i^2 = -1$.

This definition allows us to express numbers like $\sqrt{-4}$ as multiples of i.

EXAMPLE 1

Simplify each of the following and verify results with the calculator (make sure the MODE setting has been changed on the TI-83 Plus): **a.** $\sqrt{-25}$ **b.** $\sqrt{-3}$.

SOLUTION   **a.** $\sqrt{-25} = \sqrt{-1} \cdot \sqrt{25} = i \cdot 5 = 5i$

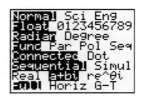
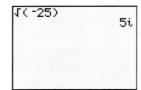

b. $\sqrt{-3} = \sqrt{-1} \cdot \sqrt{3} = i\sqrt{3} = \sqrt{3}i$

To avoid confusion between $\sqrt{3}i$ and $\sqrt{3i}$, $i\sqrt{3}$ is an acceptable way to write the result. When the calculator is used, the following is displayed:

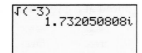

The result displayed is an approximation, whereas $i\sqrt{3}$ is an exact result.

WARNING!   If a and b are both negative, then $\sqrt{a} \cdot \sqrt{b} \neq \sqrt{ab}$. For example, $\sqrt{-16} \cdot \sqrt{-4} \neq \sqrt{64}$, rather $\sqrt{-16} \cdot \sqrt{-4} = 4i \cdot 2i = 8i^2 = -8$.

▮ POWERS OF *i*

The powers of i produce an interesting pattern.

EXPLORING THE CONCEPT

a. Use the definition of i^2 and the laws of exponents to evaluate $i^1, i^2, i^3, \ldots i^8$.
b. State the pattern revealed from part (a).
c. Suggest a method for simplifying powers of i.

CONCLUSION   **a.** $i^1 = i$

$i^2 = -1$

$i^3 = i^2 \cdot i = -1 \cdot i = -i$

$i^4 = i^2 \cdot i^2 = -1 \cdot -1 = 1$

$$i^5 = i^1 \cdot i^4 = i \cdot 1 = i$$
$$i^6 = i^2 \cdot i^4 = -1 \cdot 1 = -1$$
$$i^7 = i^3 \cdot i^4 = -i \cdot 1 = -i$$
$$i^8 = i^4 \cdot i^4 = 1 \cdot 1 = 1$$

b. The powers of i repeat in a pattern of four results: $i, -1, -i, 1$.

c. **Powers of i**

If n is a natural number that leaves a remainder of r when divided by 4, then $i^n = i^r$. When n is divisible by 4, the remainder, r, is 0 and $i^0 = 1$.

EXAMPLE 2

Simplify each of the powers of i: **a.** i^9 **b.** i^{24} **c.** i^{55}

SOLUTION   **a.** Dividing 9 by 4 leaves a remainder of 1. Therefore, $i^9 = i^1 = i$.

b. Dividing 24 by 4 leaves no remainder. Therefore, $i^{24} = i^0 = 1$. Remember, any base (except 0) raised to the zero power is 1.

c. Dividing 55 by 4 leaves a remainder of 3. Therefore, $i^{55} = i^3 = -i$.

▍▍▍ COMPLEX NUMBERS

Defining $\sqrt{-1}$ equal to i allows the expansion of the real-number system to the **complex-number** system.

Complex Numbers

A complex number is *any* number that can be written in the form $a + bi$, where a and b are real numbers and $i = \sqrt{-1}$. If $b = 0$, then $a + bi$ is a real number. If $b \neq 0$ and $a = 0$, the complex number $a + bi = 0 + bi = bi$ and is an imaginary number.

In the complex number $a + bi$, a is the *real part* and b is the *imaginary part*.

The relationship between the real numbers, the imaginary numbers, and the complex numbers is shown in Figure 9-4 below.

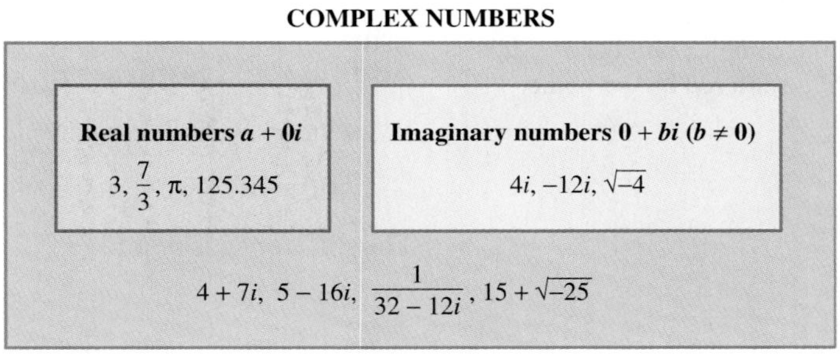

COMPLEX NUMBERS

Real numbers $a + 0i$	Imaginary numbers $0 + bi$ $(b \neq 0)$
$3, \dfrac{7}{3}, \pi, 125.345$	$4i, -12i, \sqrt{-4}$

$$4 + 7i, \quad 5 - 16i, \quad \frac{1}{32 - 12i}, \quad 15 + \sqrt{-25}$$

Figure 9-4

Two complex numbers are equal if and only if their real parts are equal and their imaginary parts are equal.

Equality of Complex Numbers

The complex numbers $a + bi$ and $c + di$ are equal if and only if $a = c$ and $b = d$.

EXAMPLE 3

Classify each statement as true or false. Verify your response both analytically and with the graphing calculator.

a. $2 + 3i = \sqrt{4} + \sqrt{-9}$ **b.** $6 + 4i = \dfrac{12}{2} + \sqrt{-8}$

SOLUTION   **a.** $2 + 3i = \sqrt{4} + \sqrt{-9}$

$$= 2 + \sqrt{-1} \cdot \sqrt{9}$$
$$= 2 + i \cdot 3$$
$$= 2 + 3i$$

The original statement is true.

```
2+3i=√(4)+√(-9)
                 1
```

b. $6 + 4i = \dfrac{12}{2} + \sqrt{-8}$

$$= 6 + \sqrt{-1} \cdot \sqrt{2^2 \cdot 2}$$
$$= 6 + i \cdot 2\sqrt{2}$$
$$= 6 + 2i\sqrt{2}$$

The original statement is false.

```
6+4i=12/2+√(-8)
                 0
```

TECHNOLOGY TIP

The equal sign is located under the TEST menu. After entering the equation and pressing ENTER, either a 1 or a zero will appear. A 1 indicates the statement is true, whereas a 0 indicates a false statement.

TECHNOLOGY TIP

TI-83 Plus: Once the calculator is set from real to $a + bi$ on the MODE screen, complex numbers may be entered on the home screen. The unit, i, is located above the decimal point in yellow. To access it, press [2nd] [•].

TI-86: To enter a complex number, put the real part, a, a comma, and then the imaginary part, b, in parentheses. Therefore, the complex number $8 + 2i$ would be entered on the home screen as (8, 2).

ARITHMETIC OF COMPLEX NUMBERS

Complex numbers can be added or subtracted by adding or subtracting their real parts and imaginary parts.

Addition and Subtraction of Complex Numbers

If $a + bi$ and $c + di$ are complex numbers, their sum is

$$(a + bi) + (c + di) = (a + c) + (b + d)i$$

The difference is defined as

$$(a + bi) - (c + di) = a + bi - c - di = (a - c) + (b - d)i$$

EXAMPLE 4

Add or subtract the given complex numbers as the signs indicate. Verify results with your calculator.

a. $(8 + 4i) + (12 + 8i)$ b. $(7 - 4i) + (9 + 2i)$

c. $(-6 + i) - (3 - 4i)$ d. $(2 - 4i) - (-4 + 3i)$

SOLUTION

a. $(8 + 4i) + (12 + 8i) = 8 + 4i + 12 + 8i$
$$= (8 + 12) + (4i + 8i)$$
$$= (8 + 12) + (4 + 8)i$$
$$= 20 + 12i$$

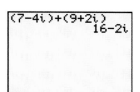

b. $(7 - 4i) + (9 + 2i) = 7 - 4i + 9 + 2i$
$$= (7 + 9) + (-4i + 2i)$$
$$= (7 + 9) + (-4 + 2)i$$
$$= 16 + (-2)i$$
$$= 16 - 2i$$

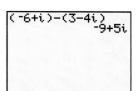

c. $(-6 + i) - (3 - 4i) = -6 + i - 3 + 4i$
$$= (-6 - 3) + (i + 4i)$$
$$= (-6 - 3) + (1 + 4)i$$
$$= -9 + 5i$$

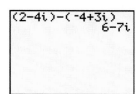

d. $(2 - 4i) - (-4 + 3i) = 2 - 4i + 4 - 3i$
$$= (2 + 4) + (-4i - 3i)$$
$$= (2 + 4) + (-4 - 3)i$$
$$= 6 + (-7)i$$
$$= 6 - 7i$$

To multiply complex numbers (say, by another complex number, an imaginary number, or a real number), use the distributive property. Remember, $i^2 = -1$ and powers of i are *not* retained in results.

Multiplying Complex Numbers

Complex numbers are multiplied as if they were binomials, with $i^2 = -1$.

$$(a + bi)(c + di) = ac + a \cdot di + bi \cdot c + bi \cdot di$$
$$= ac + adi + bci + bdi^2$$
$$= ac + (ad + bc)i + bd(-1)$$
$$= (ac - bd) + (ad + bc)i$$

EXAMPLE 5

Multiply and verify results with your calculator.

a. $6(3 + 4i)$

b. $6i(5 - 2i)$

c. $(2 + 3i)(3 - 2i)$

d. $(2 - 3i)^2$

e. $(-4 + 2i)(2 + i)$

SOLUTION   **a.** $6(3 + 4i) = 6 \cdot 3 + 6 \cdot 4i$

$= 18 + 24i$

```
6(3+4i)
              18+24i
```

b. $6i(5 - 2i) = 6i \cdot 5 + 6i \cdot -2i$

$= 30i - 12i^2$

$= 30i - 12(-1)$

$= 30i + 12$

$= 12 + 30i$

```
6i(5-2i)
              12+30i
```

c. $(2 + 3i)(3 - 2i) = 2 \cdot 3 + 2 \cdot -2i + 3i \cdot 3 + 3i \cdot -2i$

$= 6 - 4i + 9i - 6i^2$

$= 6 + 5i - 6(-1)$

$= 6 + 5i + 6$

$= 12 + 5i$

```
(2+3i)(3-2i)
              12+5i
```

d. $(2 - 3i)^2 = (2 - 3i)(2 - 3i)$

$= 2 \cdot 2 + 2 \cdot -3i - 3i \cdot 2 - 3i \cdot -3i$

$= 4 - 6i - 6i + 9i^2$

$= 4 - 12i + 9(-1)$

$= -5 - 12i$

```
(2-3i)²
              -5-12i
```

e. $(-4 + 2i)(2 + i) = -4 \cdot 2 - 4 \cdot i + 2i \cdot 2 + 2i \cdot i$

$= -8 - 4i + 4i + 2i^2$

$= -8 + 2(-1)$

$= -8 - 2$

$= -10$

```
(-4+2i)(2+i)
              -10
```

Notice that the product of two complex numbers can be simply a real number.

To divide a complex number by a nonzero real number, both the real part and the imaginary part are divided by that real number.

Dividing a Complex Number by a Real Number

If $a + bi$ is a complex number and c is a nonzero real number, then

$$\frac{a + bi}{c} = \frac{a}{c} + \frac{b}{c}i$$

EXAMPLE 6

Divide and verify results with your calculator:

a. $\dfrac{6 + 9i}{3}$ **b.** $\dfrac{-8 - 3i}{-2}$

SOLUTION   **a.** $\dfrac{6 + 9i}{3} = \dfrac{6}{3} + \dfrac{9}{3}i$

$= 2 + 3i$

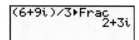

b. $\dfrac{-8 - 3i}{-2} = \dfrac{-8}{-2} - \dfrac{3}{-2}i$

$= 4 + \dfrac{3}{2}i$

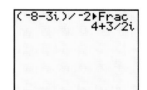

▮ RATIONALIZING DENOMINATORS

Because the definition of i is that $i = \sqrt{-1}$, i is not left in the denominator of a fractional result. Multiplying the numerator and the denominator by i, when the denominator is a monomial containing an *odd* power of i, removes i from the denominator. Recall that even powers of i may be replaced by either 1 or -1, as appropriate. If the denominator is a binomial, the numerator and the denominator must be multiplied by its **complex conjugate** to remove i.

Complex Conjugates

The complex numbers $a + bi$ and $a - bi$ are called **complex conjugates** of each other. Their product is always a real number, as illustrated below ($i^2 = -1$):

$$(a + bi)(a - bi) = a \cdot a + a \cdot -bi + bi \cdot a + bi \cdot -bi$$
$$= a^2 - abi + abi - b^2i^2$$
$$= a^2 - b^2(-1)$$
$$= a^2 + b^2$$

EXAMPLE 7

Rationalize the denominators and express results in $a + bi$ form. Verify results with your calculator.

a. $\dfrac{6i}{i^5}$ **b.** $\dfrac{2}{3 + i}$

c. $\dfrac{3 - i}{2 + i}$ **d.** $\dfrac{4 + \sqrt{-16}}{2 + \sqrt{-4}}$

SOLUTION

a. $\dfrac{6i}{i^5} = \dfrac{6i}{i^5} \cdot \dfrac{i}{i}$ Multiply the numerator and the denominator by i so that the denominator will have an even power of i.

$$= \dfrac{6i^2}{i^6}$$

$$= \dfrac{6(-1)}{-1}$$ Substitution: $i^6 = i^2 = -1$.

$$= 6$$

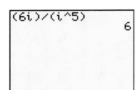

An alternative method is as follows:

$$\dfrac{6i}{i^5} = \dfrac{6}{i^4} = \dfrac{6}{1} = 6$$

Either method yields a correct result, as verified above.

b. $\dfrac{2}{3 + i} = \dfrac{2}{3 + i} \cdot \dfrac{3 - i}{3 - i}$

$$= \dfrac{2(3 - i)}{(3 + i)(3 - i)}$$

$$= \dfrac{6 - 2i}{9 - 3i + 3i - i^2}$$

$$= \dfrac{6 - 2i}{9 - (-1)}$$

$$= \dfrac{6 - 2i}{10}$$

$$= \dfrac{6}{10} - \dfrac{2}{10}i$$

$$= \dfrac{3}{5} - \dfrac{1}{5}i$$

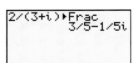

c.
$$\frac{3 - i}{2 + i} = \frac{3 - i}{2 + i} \cdot \frac{2 - i}{2 - i}$$

$$= \frac{(3 - i)(2 - i)}{(2 + i)(2 - i)}$$

$$= \frac{6 - 3i - 2i + i^2}{4 - 2i + 2i - i^2}$$

$$= \frac{6 - 5i - 1}{4 - (-1)}$$

$$= \frac{5 - 5i}{5}$$

$$= \frac{5}{5} - \frac{5}{5}i$$

$$= 1 - i$$

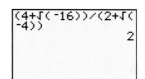

d.
$$\frac{4 + \sqrt{-16}}{2 + \sqrt{-4}} = \frac{4 + 4i}{2 + 2i}$$

$$= \frac{4 + 4i}{2 + 2i} \cdot \frac{2 - 2i}{2 - 2i}$$

$$= \frac{(4 + 4i)(2 - 2i)}{(2 + 2i)(2 - 2i)}$$

$$= \frac{8 - 8i + 8i - 8i^2}{4 - 4i + 4i - 4i^2}$$

$$= \frac{8 - 8(-1)}{4 - 4(-1)}$$

$$= \frac{8 + 8}{4 + 4}$$

$$= \frac{16}{8}$$

$$= 2$$

STUDY TIP

Come up with a test-taking strategy and practice several strategies prior to the actual test.

a. Make out a sample test (or use the test in the text).
b. Try working each problem slowly and carefully "straight through."
c. Now try going through the test and doing the problems that are easiest for you first. Then go back and work the ones that are more difficult.

 WARNING!   To avoid mistakes, always put complex numbers in $a + bi$ form before performing any complex arithmetic.

EXERCISE 9.7

VOCABULARY AND NOTATION   *Fill in the blanks to make a true statement.*

1. The imaginary number $\sqrt{-1}$ is defined as the imaginary unit ___.

2. Defining $\sqrt{-1}$ allows us to expand the real number system to the _____ number system.

3. In the complex number, $a + bi$, a is the _____ part and b is the _____ part.

4. Two complex numbers are equal if and only if their _____ parts are equal and their _____ parts are equal.

5. When rationalizing a denominator that contains a complex number, multiply the numerator and the denominator by the _____ to remove i.

CONCEPTS

6. Powers of i repeat in a cycle of 4 values:

$i^1 =$ ___ $i^2 =$ ___ $i^3 =$ ___ $i^4 =$ ___

7. Simplify each power of i:

a. $i^5 =$ ___   b. $i^8 =$ ___

c. $i^{14} =$ ___   d. $i^{11} =$ ___

In Exercises 8–19, simplify each expression.

8. $\sqrt{-25}$

9. $\sqrt{-36}$

10. $-\sqrt{-75}$

11. $-\sqrt{-200}$

12. $\sqrt{-98}$

13. $\sqrt{-18}$

14. $\sqrt{-27}$

15. $-\sqrt{-32}$

16. i^{21}

17. i^{19}

18. i^{27}

19. i^{22}

*In Exercises 20–25, classify each equation as true or false. Justify your results analytically **and** by using the TEST option of your calculator.*

20. $3 + 7i = \sqrt{9} + (5 + 2)i$

21. $\sqrt{4} + \sqrt{25}i = 2 - (-5)i$

22. $8 + 5i = 2^3 + \sqrt{25}i^3$

23. $4 - 7i = -4i^2 + 7i^3$

24. $\sqrt{4} + \sqrt{-4} = 2 - 2i$

25. $\sqrt{-9} - i = 4i$

PRACTICE   *In Exercises 26–73, perform the indicated operations. All results should be written in a + bi form. Verify results with your calculator.*

26. $(3 + 4i) + (5 - 6i)$

27. $(5 + 3i) - (6 - 9i)$

28. $(7 - 3i) - (4 + 2i)$

29. $(8 + 3i) + (-7 - 2i)$

30. $(1 + i) - 2i + (5 - 7i)$

31. $(-9 + i) - 5i + (2 + 7i)$

32. $(5 + 3i) - (3 - 5i) + \sqrt{-1}$

33. $(8 + 7i) - (-7 - \sqrt{-64}) + (3 - i)$

34. $(-8 - \sqrt{3}i) - (7 - 3\sqrt{3}i)$

35. $(2 + 2\sqrt{2}i) + (-3 - \sqrt{2}i)$

36. $3i(2 - i)$

37. $-4i(3 + 4i)$

38. $(2 - 4i)(3 + 2i)$

39. $(3 - 2i)(4 - 3i)$

40. $(2 + \sqrt{2}i)(3 - \sqrt{2}i)$

41. $(5 + \sqrt{3}i)(2 - \sqrt{3}i)$

42. $(8 - \sqrt{-1})(-2 - \sqrt{-16})$

43. $(-1 + \sqrt{-4})(2 + \sqrt{-9})$

44. $(2 + i)^2$

45. $(3 - 2i)^2$

46. $i(5 + i)(3 - 2i)$

47. $i(-3 - 2i)(1 - 2i)$

48. $(2 + i)(2 - i)(1 + i)$

49. $(3 + 2i)(3 - 2i)(i + 1)$

50. $\dfrac{4}{5i^3}$

51. $\dfrac{3}{2i}$

52. $\dfrac{3i}{8\sqrt{-9}}$

53. $\dfrac{5i^3}{2\sqrt{-4}}$

54. $\dfrac{-3}{5i^5}$

55. $\dfrac{-4}{6i^7}$

56. $\dfrac{3 + 6i}{3}$

57. $\dfrac{5 + 10i}{5}$

58. $\dfrac{8 - 4i}{2}$

59. $\dfrac{9 - 6i}{3}$

60. $\dfrac{-8 + 9i}{3}$

61. $\dfrac{-6 - 5i}{5}$

62. $\dfrac{5}{2 - i}$

63. $\dfrac{26}{3 - 2i}$

64. $\dfrac{13i}{5 + i}$

65. $\dfrac{2i}{5 + 3i}$

66. $\dfrac{-12}{7 - \sqrt{-1}}$

67. $\dfrac{4}{3 + \sqrt{-1}}$

68. $\dfrac{5i}{6 + 2i}$

69. $\dfrac{-4i}{2 - 6i}$

70. $\dfrac{3 - 2i}{3 + 2i}$

71. $\dfrac{2 + 3i}{2 - 3i}$

72. $\dfrac{\sqrt{5} - \sqrt{3}i}{\sqrt{5} + \sqrt{3}i}$

73. $\dfrac{\sqrt{3} + \sqrt{2}i}{\sqrt{3} - \sqrt{2}i}$

74. a. Show, using analytical methods, that $1 - 5i$ and its conjugate are solutions of $x^2 - 2x = -26$.

  b. Use the STOre feature of your calculator to verify your analytical results.

75. a. Show, using analytical methods, that $3 - 2i$ and its conjugate are solutions of $x^2 = 6x - 13$.

  b. Use the STOre feature of your calculator to verify your analytical results.

76. a. Show, using analytical methods, that i and its conjugate are solutions of $x^4 = 3x^2 + 4$.

  b. Use the STOre feature of your calculator to verify your analytical results.

77. a. Show, using analytical methods, that $2 + i$ and its conjugate are *not* solutions of $x^2 + x + 1 = 0$.

  b. Use the STOre feature of your calculator to verify your analytical results.

REVIEW *In Exercises 78–83, simplify each of the polynomials.*

78. $(3x + 2) + (5x - 8)$

79. $(3x^2 + 5x - 2) - (8x^2 - 2x + 4)$

80. $3x^2(x - 4)$

81. $(3x + 2)(5x - 8)$

82. $\dfrac{3x^2 - 6x + 9}{3}$

83. $\dfrac{x^2 - 1}{x - 1}$

CHAPTER SUMMARY

CONCEPTS

REVIEW EXERCISES

Section 9.1

The expression $\sqrt[n]{a}$ specifies the nth root of a number, when $n > 1$. The letter n is the *index*, or order of the root.

The expression under the radical sign is the *radicand*. The index of 2 is omitted when writing square roots.

$$\sqrt[n]{a^n} = a \text{ if } n \text{ is odd}$$

$$\sqrt[n]{a^n} = |a| \text{ if } n \text{ is even}$$

When n is odd, the radicand may be positive, negative, or 0. When n is even, the radicand must be 0 or positive to be defined in the real-number system.

Introduction to Roots and Radicals

1. Simplify each radical.

   a. $\sqrt{49}$
   b. $-\sqrt{121}$
   c. $-\sqrt{36}$
   d. $\sqrt{225}$
   e. $\sqrt[3]{-27}$
   f. $-\sqrt[3]{216}$
   g. $\sqrt[4]{625}$
   h. $\sqrt[5]{-32}$

2. Simplify each radical. Assume that all variables represent positive numbers.

   a. $\sqrt{25x^2}$
   b. $\sqrt{x^2 + 4x + 4}$
   c. $\sqrt[3]{27a^6b^2}$
   d. $\sqrt[4]{256a^8y^4}$

Section 9.2

The link between rational exponents and radicals is stated below:

$$a^{m/n} = \sqrt[n]{a^m} = (\sqrt[n]{a})^m$$

A radical expression is simplified when no radical appears in the denominator of a fraction, the radicand contains no fractions or negative numbers, and each factor in the radicand appears to a power that is less than the index of the radical.

The product rule states

$$\sqrt[n]{ab} = \sqrt[n]{a} \cdot \sqrt[n]{b}$$

whereas the quotient rule specifies

$$\sqrt[n]{\frac{a}{b}} = \frac{\sqrt[n]{a}}{\sqrt[n]{b}}$$

Rational Exponents and Simplifying Radical Expressions

3. Simplify each expression. Assume that all variables represent positive numbers.

   a. $(-8)^{1/3}$
   b. $-8^{2/3}$
   c. $\left(\frac{4}{9}\right)^{3/2}$
   d. $\frac{1}{25^{5/2}}$
   e. $\left(\frac{4}{9}\right)^{-3/2}$
   f. $(27x^3y)^{1/3}$
   g. $(81x^4y^2)^{1/4}$
   h. $(25x^3y^4)^{3/2}$
   i. $(8u^2v^3)^{-2/3}$

4. Simplify each expression. Assume that all variables represent positive numbers.

   a. $\sqrt{240}$
   b. $\sqrt[3]{54}$
   c. $\sqrt[4]{32}$
   d. $\sqrt[5]{96}$
   e. $\sqrt[3]{16x^5y^4}$
   f. $\sqrt[3]{54x^7y^3}$
   g. $\frac{\sqrt{32x^3}}{\sqrt{2x}}$
   h. $\frac{\sqrt[3]{16x^5}}{\sqrt[3]{2x^2}}$
   i. $\sqrt[3]{\frac{2a^2b}{27x^3}}$
   j. $\sqrt{\frac{17xy}{64a^4}}$

Section 9.3 — Adding and Subtracting Radical Expressions

To add or subtract radical expressions, simplify each radical and use the distributive property to add or subtract radicals that have the same index and radicand.

5. Simplify and combine radicals as indicated. Assume that all variables represent positive numbers.

a. $2\sqrt[3]{3} - \sqrt[3]{24}$ **b.** $\sqrt[4]{32} + 2\sqrt[4]{162}$

c. $2x\sqrt{8} + 2\sqrt{200x^2} + \sqrt{50x^2}$

d. $\sqrt[3]{54} - 3\sqrt[3]{16} + 4\sqrt[3]{128}$

Section 9.4 — Multiplying and Dividing Radical Expressions

The product rule ensures that radical expressions with the same index can be multiplied. It is suggested that multiplication be done before simplification.

For a radical expression to be in simplified form, there cannot be a fraction under the radical or a radical in the denominator of a fraction.

To rationalize the denominator, multiply both the numerator and the denominator by a radical that will make the denominator a perfect root.

To rationalize binomial denominators, multiply the numerator and the denominator by the conjugate of the binomial in the denominator.

6. Simplify each expression. Assume that all variables represent positive numbers.

a. $\sqrt[3]{3} \cdot \sqrt[3]{9}$ **b.** $-\sqrt[3]{2x^2} \cdot \sqrt[3]{4x}$

c. $\sqrt{2}(\sqrt{8} - 3)$ **d.** $\sqrt{5}(\sqrt{2} - 1)$

e. $(\sqrt{2} + 1)(\sqrt{2} - 1)$ **f.** $(2\sqrt{u} + 3)(3\sqrt{u} - 4)$

7. Rationalize each denominator.

a. $\dfrac{1}{\sqrt{3}}$ **b.** $\dfrac{\sqrt[3]{uv}}{\sqrt[3]{u^5v^7}}$

c. $\dfrac{2}{\sqrt{2} - 1}$ **d.** $\dfrac{\sqrt{2}}{\sqrt{3} - 1}$

Section 9.5 — Radical Equations and Their Applications

A radical equation is an equation that contains one or more radicals, with a variable in the radicand. The power rule allows algebraic solutions of radical equations: If x, y, and n are real numbers and $x = y$, then $x^n = y^n$.

Raising both sides of an equation to the same power can introduce extraneous roots. Therefore, all supposed solutions *must* be checked.

8. Solve each equation both algebraically and graphically.

a. $\sqrt{x + 3} = \sqrt{2x - 19}$

b. $x = \sqrt{12x - 27}$

c. $\sqrt{x + 1} + \sqrt{x} = 2$

d. $\sqrt[3]{x^3 + 8} = x + 2$

e. $\sqrt{x^2 + 3x} = -2$

The *Pythagorean theorem* states that the sum of the squares of the legs of a right triangle is equal to the square of the hypotenuse.

$$(\text{leg})^2 + (\text{leg})^2 = (\text{hypotenuse})^2$$

The distance formula allows us to find the distance between any two points in the rectangular coordinate system:

$$d = \sqrt{(x_2 - x_1)^2 + (y_2 - y_1)^2},$$

where (x_1, y_1) and (x_2, y_2) are the coordinates of two points.

9. Sailing A technique called *tacking* allows a sailboat to make progress into the wind. A sailboat follows the course in Illustration 1. Find d, the distance the boat advances into the wind.

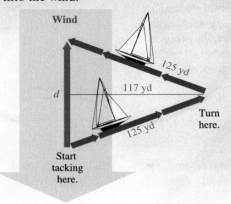

Illustration 1

10. Find the distance between the points

    a. $P(-2, -7)$ and $Q(4, 1)$.   **b.** $P(-3, -5)$ and $Q(6, 8)$.

Radical Functions

A function is a relation that assigns to each value of x exactly one value, y.

The domain of a function is the set of all possible values for the independent variable, x, whereas the range is the set of all possible values for the dependent variable, y.

The definition of the function, the TABLE, and the graph of the function are often useful in determining domains and ranges.

11. If $f(x) = \sqrt{2x + 4} - 1$ and $g(x) = \sqrt[3]{2x}$, find each of the following. Round irrational results to the nearest hundredth.

    a. $f(6)$

    b. $g(4)$

    c. $g(32)$

    d. $f(3)$

    e. $g(8)$

12. State the domain and the range of each of the following.

    a. $f(x) = \sqrt{x - 5} - 4$

    b. $f(x) = \sqrt{2x + 4}$

    c. $f(x) = \sqrt[3]{x + 2}$

    d. $f(x) = \sqrt[3]{x + 4} - 3$

The horizontal distance d (measured in miles) is related to the height h (measured in feet) of the observer by the formula $d = 1.4\sqrt{h}$.

13. View from a submarine A submarine periscope extends 4.7 feet above the surface. How far is the horizon?

14. View from a submarine How far out of the water must a submarine periscope extend to provide a 4-mile horizon?

Section 9.7

Complex Numbers

The letter i is used to denote the imaginary number defined below:

$$\sqrt{-1} = i \quad \text{and} \quad i^2 = -1$$

The powers of i repeat in the pattern, i, -1, $-i$, 1. This allows us to simplify powers of i. If n is a natural number that leaves a remainder of r when divided by 4, then $i^n = i^r$.

Defining i allows the expansion of the real-number system to the complex-number system. A complex number is any number that can be written in the form $a + bi$.

Two complex numbers $a + bi$ and $c + di$ are equal if and only if $a = c$ and $b = d$.

Complex numbers can be added or subtracted by adding or subtracting their real parts and their imaginary parts.

To multiply complex numbers, use the distributive property.

The imaginary unit i is never left in the denominator of a fraction. To rationalize the denominator, multiply the numerator and the denominator by i when the denominator is a monomial containing an odd power of i. If the denominator is a binomial, the numerator and the denominator must be multiplied by its complex conjugate to remove i.

15. Simplify and express results in $a + bi$ form.

  a. $(5 + 4i) + (7 - 12i)$

  b. $(-6 - 40i) - (-8 + 28i)$

  c. $(-32 + \sqrt{-144}) - (64 + \sqrt{-81})$

  d. $(-8 + \sqrt{-8}) + (6 - \sqrt{-32})$

16. Simplify and express results in $a + bi$ form.

  a. $(2 - 7i)(-3 + 4i)$

  b. $(-5 + 6i)(2 + i)$

  c. $(5 - \sqrt{-27})(-6 + \sqrt{-12})$

  d. $(2 + \sqrt{-128})(3 - \sqrt{-98})$

17. Simplify to $a + bi$ form.

  a. $\dfrac{3}{4i}$

  b. $\dfrac{-2}{5i^3}$

  c. $\dfrac{42 + 7i}{7}$

  d. $\dfrac{13 + 4i}{2}$

  e. $\dfrac{6}{2 + i}$

  f. $\dfrac{7}{3 - i}$

  g. $\dfrac{3}{5 + \sqrt{-4}}$

  h. $\dfrac{2}{3 - \sqrt{-9}}$

CHAPTER TEST

In Problems 1–5, simplify each expression. Assume that all variables represent positive numbers.

1. $\sqrt{250x^3y^5}$

2. $\dfrac{\sqrt[3]{24x^{15}y^4}}{\sqrt[3]{y}}$

3. $\sqrt{\dfrac{3a^5}{48a^7}}$

4. $\sqrt[3]{27x^3}$

5. $\sqrt{18x^4y^9}$

6. Simplify each expression. Write results without using negative exponents.

  a. $16^{1/4}$ **b.** $36^{-3/2}$ **c.** $\left(-\dfrac{8}{27}\right)^{-2/3}$

In Problems 7–10, simplify each expression containing radicals. Assume that all variables represent positive numbers.

7. $\sqrt{12} - \sqrt{27}$

8. $2\sqrt[3]{40} - \sqrt[3]{5000} + 4\sqrt[3]{625}$

9. $2\sqrt{48y^5} - 3y\sqrt{12y^3}$

10. $2\sqrt{xy}(3\sqrt{x} + \sqrt{xy^3})$

11. Rationalize each denominator.

  a. $\dfrac{1}{\sqrt{5}}$ **b.** $\dfrac{6}{\sqrt[3]{9}}$

  c. $\dfrac{-4\sqrt{2}}{\sqrt{5}+3}$ **d.** $\dfrac{3x-1}{\sqrt{3x}-1}$

In Problems 12 and 13, solve each equation both algebraically and graphically.

12. $\sqrt[3]{6x+4} - 4 = 0$

13. $1 - \sqrt{x} = \sqrt{x-3}$

14. Find the distance between P and Q if $P(-2, 5)$ and $Q(22, 12)$.

15. Shipping crates   The diagonal brace on the shipping crate shown in Illustration 1 below is 53 inches. Find the height, h, of the crate.

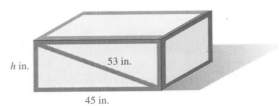

45 in.

Illustration 1

16. If $f(x) = 2\sqrt{x} + 1$, find each of the following. Round any irrational results to the nearest thousandth.

  a. $f(3)$ **b.** $f(9)$

  c. $f(-20)$ **d.** $f(13)$

17. Find the domain and the range of each function specified below.

  a. $f(x) = \sqrt{x-6} + 2$

  b. $f(x) = -\sqrt{x}$

  c. $f(x) = \sqrt[3]{x}$

18. Add or subtract the complex numbers as the signs indicate.

  a. $(2 + 4i) + (-3 + 7i)$

  b. $(3 - \sqrt{-9}) - (-1 + \sqrt{-16})$

19. Multiply the complex numbers below.

  a. $2i(3 - 4i)$

  b. $(3 + 2i)(-4 - i)$

20. Rationalize the denominators of each of the expressions below.

  a. $\dfrac{1}{i\sqrt{2}}$ **b.** $\dfrac{2+i}{3-i}$

CAREERS & MATHEMATICS

© Image 100/Royalty-Free/CORBIS

Electrical/Electronic Engineer

Electrical engineers design, develop, test, and supervise the manufacture of electrical and electronic equipment. Electrical engineers who work with electronic equipment are often called **electronic engineers**.

Electrical engineers generally specialize in a major area, such as power-distributing equipment, integrated circuits, computers, electrical equipment manufacturing, or communications. Besides manufacturing and research, development, and design, many are employed in administration and management, technical sales, or teaching.

Qualifications

A bachelor's degree in engineering is generally acceptable for beginning engineering jobs. College graduates with a degree in natural science or mathematics also may qualify for some jobs.

Engineers should be able to work as part of a team and should have creativity, an analytical mind, and a capacity for detail. In addition, engineers should be able to express themselves well — both orally and in writing.

Job Outlook

Employment of electrical engineers is expected to increase about as fast as the average for all occupations through the year 2010. Although increased demand for computers, communications equipment, and electronics is expected to be the major contributor to this growth, demand for electrical and electronic consumer goods, along with increased research and development in new types of power generation, should create additional jobs.

Example Application

In a radio, an inductor and a capacitor are used in a resonant circuit to select a desired station at frequency f and reject all others. The inductance L and the capacitance C determine the inductive reactance X_L and the capacitive reactance X_C of that circuit, where

$$X_L = 2\pi f L \qquad \text{and} \qquad X_C = \frac{1}{2\pi f C}$$

The radio station selected will be at the frequency f for which the inductive reactance and capacitive reactance are equal. To find that frequency, solve for f in terms of L and C.

Solution

Since X_L and X_C are to be equal, we can solve the system of equations

$$\begin{cases} X_L = 2\pi f L \\ X_C = \dfrac{1}{2\pi f C} \\ X_L = X_C \end{cases}$$

by substitution to express f in terms of L and C.

$$X_L = X_C$$

$$2\pi f L = \frac{1}{2\pi f C}$$

$$(2\pi f L)(2\pi f C) = 1 \qquad \text{Multiply both sides by } 2\pi f C.$$

$$4\pi^2 f^2 L C = 1 \qquad \text{Simplify.}$$

$$f^2 = \frac{1}{4\pi^2 LC} \qquad \text{Divide both sides by } 4\pi^2 LC.$$

$$f = \sqrt{\frac{1}{4\pi^2 LC}} \qquad \text{Take the positive square root of both sides.}$$

$$f = \frac{1}{2\pi\sqrt{LC}} \qquad \text{Simplify.}$$

EXERCISES

1. At what frequency will a 0.0001 farad capacitor and a 0.005 henry inductor resonate? (*Hint:* $C = 0.0001$ and $L = 0.005$.)

2. If the inductor and the capacitor in Exercise 1 were both doubled, what would the resonant frequency become?

3. At what frequency will a 0.0008 farad capacitor and a 0.002 henry inductor resonate?

4. If the inductor of Exercise 3 were doubled and the capacitance reduced by one-half, what would the resonant frequency be?

(*Answers:*   **1.** 225 hertz   **2.** 113 hertz   **3.** 126 hertz   **4.** 126 hertz)

CHAPTER

10

Quadratics

© Royalty-Free/CORBIS

 InfoTrac Projects

Do a subject guide search on "parabola." Choose "parabola" and "view periodical references." Find the article "Galileo's trajectory." (Italian astronomer Galileo Galilei's theory on trajectory and the concept of parabola.) According to the article, what is the best trajectory for firing a missile from a cannon? The author seems to think that instead of following a parabolic path, the missile follows what other conic section path?

Assume that a missile fired at 450 feet per second from a height of 20 feet follows the path of a parabola with the equation $s = -16t^2 + 450t + 20$, where s is in feet. Find the vertex to find the maximum height achieved by the missile. If t is in seconds, how long will it take to reach the maximum height?

Complete this project after studying Sections 10.3 and 10.4.

PERSPECTIVE

From Ridicule to Respect

The Pythagoreans (ca. 500 B.C.) understood the universe as a harmony of whole numbers. They did not classify fractions as numbers and were upset that $\sqrt{2}$ was not the ratio of whole numbers. For 2000 years, little progress was made in the understanding of the various kinds of numbers.

The father of algebra, François Vieta (1540–1603), understood the whole numbers, fractions, and certain irrational numbers. But he was unable to accept negative numbers and certainly not imaginary numbers.

Girolamo Cardano (1501–1576) was more daring, even though negative and imaginary numbers puzzled him. He attacked the problem *Divide 10 into two parts whose product is 40.* "Of course, that is impossible," said Cardano. "Nevertheless, we will operate." (Cardano had some medical training!) He found solutions: the two numbers $5 + i\sqrt{15}$ and $5 - i\sqrt{15}$. (Can you show that their sum is 10 and their product is 40?) Cardano was impressed. "These numbers," he said, "are truly sophisticated." Unfortunately, he decided that this kind of arithmetic would be useless. Cardano was ahead of his time, but he didn't appreciate the value of his discovery.

René Descartes (1596–1650) thought these numbers to be nothing more than figments of his imagination, so he called them *imaginary numbers.* Leonhard Euler (1707–1783) used the letter i for $\sqrt{-1}$; Augustin Cauchy (1789–1857) used the term *conjugate*; and Karl Gauss (1777–1855) first used the world *complex.*

Today, we accept complex numbers without question, and use them freely in science, economics, medicine, and industry. But it took many centuries and the work of many mathematicians to make them respectable.

This chapter studies methods appropriate for finding solutions of quadratic (second-degree) equations that are not factorable. Graphs and tables are used to model problem solutions of quadratic equations. The discussion then extends to families of graphs and further to quadratic and other nonlinear inequalities.

10.1 SQUARE ROOT PROPERTY AND COMPLETING THE SQUARE

- REVIEW OF SOLUTIONS OF QUADRATICS BY FACTORING
- SQUARE ROOT PROPERTY
- COMPLETING THE SQUARE

REVIEW OF SOLUTIONS OF QUADRATICS BY FACTORING

Earlier in this text quadratic equations were solved by factoring. Recall that a **quadratic equation (second-degree equation)** is one that *can be* written in the form $ax^2 + bx + c = 0$, where a, b, and c are real numbers and $a \neq 0$. This is called **standard form**.

To solve a quadratic equation by factoring, the equation was first written in standard form, the polynomial was factored, and the zero factor property was applied. Consider the solution of the quadratic equation $x^2 = 16$.

$$x^2 - 16 = 0$$

Subtract 16 from both sides so that the equation is in standard form.

$$(x - 4)(x + 4) = 0$$

Factor the polynomial side.

$$x - 4 = 0 \quad \text{or} \quad x + 4 = 0$$

Apply the zero factor property and set each factor equal to zero.

$$x = 4 \quad | \quad x = -4$$

Solve.

The solution set is $\{4, -4\}$. Use of the STOre feature confirms both solutions.

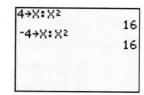

Many quadratic equations do not factor easily. For example, it would be difficult to factor $x^2 - 3 = 0$ because it cannot be factored using only integers. Therefore, we will develop methods for solving all quadratic equations.

▮ SQUARE ROOT PROPERTY

TECHNOLOGY TIP

Change the MODE setting of your calculator from real to $a + bi$. Because the set of real numbers is a subset of the set of complex numbers, the calculator will display all roots (both real and complex).

Reconsider the quadratic equation $x^2 = 16$. We can take the square root of both sides to solve; however, we must consider both the positive **and** the negative square roots of 16. This is necessary to find *all* values of the variable that make the statement $x^2 = 16$ true. Therefore,

$$x^2 = 16$$
$$x = \pm\sqrt{16}$$
$$x = \pm 4$$

The solution set is $\{4, -4\}$. This illustrates the **square root property**,

Square Root Property

If c is a real number, the equation $x^2 = c$ has two solutions, $x = \sqrt{c}$ and $x = -\sqrt{c}$. This is often expressed as $x = \pm\sqrt{c}$.

EXAMPLE 1

Solve the equation $x^2 - 12 = 0$.

SOLUTION   We can write the equation as $x^2 = 12$ and use the square root property.

$$x^2 - 12 = 0$$
$$x^2 = 12$$

Add 12 to both sides.

$$x = \pm\sqrt{12}$$

Apply the square root property.

$$x = \pm 2\sqrt{3}$$

Simplify the radical.

The solutions of the equation are $2\sqrt{3}$ and $-2\sqrt{3}$. The solution set is $\{2\sqrt{3}, -2\sqrt{3}\}$.

STUDY TIP

Try to schedule a free hour immediately after your algebra class to go over your notes and begin the homework assignment. It is important to reinforce the concepts learned in class as soon as possible after class.

The STOre feature of the calculator can be used to check the solutions:

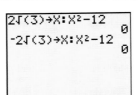

TECHNOLOGY TIP

The ROOT/ZERO method and the Intersect method can both be used to find the solutions to the displayed equation. Remember, the only restriction on the use of the ROOT/ZERO feature is that your equation is set equal to zero.

The approximate solutions to the equation $x^2 - 12 = 0$ could also have been found using the ROOT/ZERO feature of the graphing calculator.

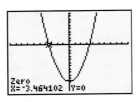 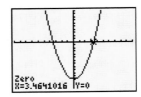 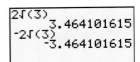

EXAMPLE 2

Solve the equation $(x - 3)^2 - 1 = 15$.

SOLUTION   We begin by adding 1 to both sides so that the equation will be in the form $x^2 = c$ (quantity squared = constant).

$(x - 3)^2 = 16$	Add 1 to both sides.
$x - 3 = \pm\sqrt{16}$	Apply the square root property.
$x - 3 = \pm 4$	Simplify the radical.
$x = 3 \pm 4$	Solve for x (add 3 to both sides).
$x = 3 + 4$ or $x = 3 - 4$	Consider both cases indicated by the $\pm$ sign.
$x = 7$ | $x = -1$	

The solution set is $\{7, -1\}$.

The STOre feature can be used to check solutions:

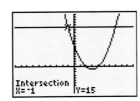

The solutions could also be found using the INTERSECT feature of the calculator:

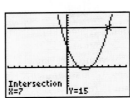

EXAMPLE 3

Solve $(3x + 2)^2 = -16$.

SOLUTION

$$(3x + 2)^2 = -16$$

$3x + 2 = \pm\sqrt{-16}$ Square root property.

$3x + 2 = \pm 4i$ Simplify the radical.

$3x = -2 \pm 4i$ Subtract 2 from both sides.

$x = \dfrac{-2 \pm 4i}{3}$ Divide both sides by 3.

Stated in $a + bi$ form, the solution set is $\left\{ -\frac{2}{3} + \frac{4i}{3}, -\frac{2}{3} - \frac{4i}{3} \right\}$.

The STOre feature can be used to check the solutions.

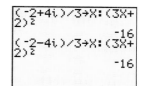

When we try to solve the equation graphically, we obtain the following display in the window $[-10, 10]$ by $[-20, 10]$:

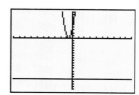

The graphs of the two sides of the equation *do not* intersect. Previously, this meant that the equation was a contradiction and had no solution. However, the check above contradicts that fact. Examination of the roots shows that they are complex. The rectangular coordinate system includes real numbers *only*. It will only show real-number solutions. The graphs shown above can be interpreted to mean that the roots (solutions) of the equation are not real numbers.

▌ COMPLETING THE SQUARE

The square root property can be used to solve any quadratic equation that is in the form $x^2 = c$ (quantity squared = constant). The process of **completing the square** allows us to solve any quadratic equation and is based on the special products that form perfect square trinomials:

$$x^2 + 2ax + a^2 = (x + a)^2 \qquad \text{and} \qquad x^2 - 2ax + a^2 = (x - a)^2$$

EXPLORING
THE CONCEPT

a. Rewrite each perfect square trinomial as a binomial quantity squared.

$x^2 + 10x + 25$

$x^2 - 6x + 9$

$x^2 + 8x + 16$

$x^2 - 12x + 36$

b. Compare the *terms* of the binomial quantity with the terms of the original trinomials.

c. Compare the middle term of the trinomial with the binomial terms.

CONCLUSIONS   **a.** $(x + 5)^2$

$(x - 3)^2$

$(x + 4)^2$

$(x - 6)^2$

b. The terms of the binomial quantities are the square roots of the first and third terms of the trinomial.

$x^2 + 10x + 25$	$x^2 - 6x + 9$
$(x)^2 + 10x + (\mathbf{5})^2$	$(x)^2 - 6x + (\mathbf{3})^2$
$(x + \mathbf{5})^2$	$(x - \mathbf{3})^2$
$x^2 + 8x + 16$	$x^2 - 12x + 36$
$(x)^2 + 8x + (\mathbf{4})^2$	$(x)^2 - 12x + (\mathbf{6})^2$
$(x + \mathbf{4})^2$	$(x - \mathbf{6})^2$

c. The middle term of the *trinomial* is twice the product of the terms in the *binomial*.

$$x^2 + 10x + 25 = (x + 5)^2 \qquad\qquad x^2 - 6x + 9 = (x - 3)^2$$
$$10x = 2(5 \cdot x) \qquad\qquad -6x = 2(-3 \cdot x)$$
$$x^2 + 8x + 16 = (x + 4)^2 \qquad\qquad x^2 - 12x + 36 = (x - 6)^2$$
$$8x = 2(4 \cdot x) \qquad\qquad -12x = 2(-6 \cdot x)$$

In the process of completing the square, we will algebraically manipulate a trinomial to that we can write it in the form of a perfect square trinomial.

EXAMPLE 4   Use the process of completing the square to solve $x^2 + 2x = 5$.

SOLUTION   To make $x^2 + 2x$ a perfect square trinomial, we will add the square of $\frac{1}{2}$ the coefficient of x to both sides of the equation.

$$\frac{1}{2}(2) = 1 \qquad\qquad \text{Find } \tfrac{1}{2} \text{ the coefficient of } x.$$

$$x^2 + 2x + (1)^2 = 5 + (1)^2 \qquad \text{Add the square of } \tfrac{1}{2} \text{ the coefficient of } x \text{ to both sides of the equation.}$$

$$(x + 1)^2 = 6 \qquad\qquad \text{Factor the perfect square trinomial on the left and simplify the right side of the equation.}$$

$$x + 1 = \pm\sqrt{6} \qquad\qquad \text{Apply the square root property.}$$

$$x = -1 \pm \sqrt{6} \qquad\qquad \text{Solve for } x.$$

Solution set: $\{-1 + \sqrt{6}, -1 - \sqrt{6}\}$.

Check solutions using the STOre feature:

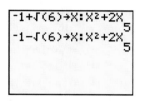

The solutions can also be approximated graphically:

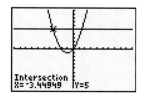

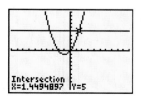

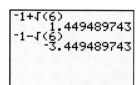

EXAMPLE 5

Solve $2x^2 + 4x + 1 = 0$ by completing the square.

SOLUTION   Notice that the trinomials in the Exploring the Concepts exercise and in Example 4 had leading coefficients of 1. To make the coefficient of x^2 equal to 1, we begin by dividing through the equation by 2.

$$2x^2 + 4x + 1 = 0$$

$$\frac{2x^2}{2} + \frac{4x}{2} + \frac{1}{2} = \frac{0}{2}$$ Divide through the equation by 2.

$$x^2 + 2x + \frac{1}{2} = 0$$ Simplify.

$$x^2 + 2x = -\frac{1}{2}$$ Subtract $\frac{1}{2}$ from both sides.

$$x^2 + 2x + (1)^2 = -\frac{1}{2} + (1)^2$$ Get $\frac{1}{2}$ the coefficient of x and add its square to both sides of the equation.

$$(x + 1)^2 = \frac{1}{2}$$ Factor the polynomial side of the equation and simplify constants on the right.

$$x + 1 = \pm\sqrt{\frac{1}{2}}$$ Square root property.

$$x + 1 = \pm\frac{\sqrt{2}}{2}$$ Simplify the radical (make sure the denominator is rationalized).

$$x = -1 \pm \frac{\sqrt{2}}{2}$$ Solve for x.

The solution set is $\left\{-1 + \dfrac{\sqrt{2}}{2}, -1 - \dfrac{\sqrt{2}}{2}\right\}$.

The STOre feature can be used to check the solutions:

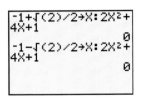

The solutions can also be approximated using the ROOT/ZERO feature of the calculator:

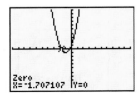 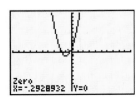 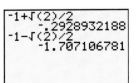

The methods used in Example 5 suggest steps that could be used when completing the square to solve quadratic equations of the form $ax^2 + bx + c = 0$.

Completing the Square

1. Make sure the coefficient of x^2 is 1. If it is not, divide both *sides* of the equation by the coefficient of x^2 to make it 1.
2. The constant term should be on the right side of the equal sign (terms containing variables will be located on the left).
3. Complete the square by
   a. Identifying the coefficient of the linear term (x).
   b. Find one-half of this coefficient.
   c. Add the square of $\frac{1}{2}$ of this coefficient to *both* sides of the equation.
4. Factor the polynomial (on the left) as the square of a binomial.
5. Use the square root property to solve the resulting equation.

EXAMPLE 6

Solve $3x^2 + 4x + 2 = 0$ by completing the square.

SOLUTION

$$\frac{3x^2}{3} + \frac{4x}{3} + \frac{2}{3} = \frac{0}{3}$$

Divide through the equation by the coefficient of x^2.

$$x^2 + \frac{4}{3}x + \frac{2}{3} = 0$$

Simplify.

$$x^2 + \frac{4}{3}x = -\frac{2}{3}$$

The constant term should be on the right side of the equation.

$$x^2 + \frac{4}{3}x + \left(\frac{2}{3}\right)^2 = -\frac{2}{3} + \left(\frac{2}{3}\right)^2$$

Take one-half of the coefficient of x and add its square to both sides of the equation.

$$\left(x + \frac{2}{3}\right)^2 = -\frac{2}{9}$$

Write the polynomial side as a binomial quantity squared and simplify the right side.

$$x + \frac{2}{3} = \pm\sqrt{-\frac{2}{9}} \qquad \text{Apply the square root property.}$$

$$x + \frac{2}{3} = \pm\frac{\sqrt{2}}{3}i \qquad \text{Simplify the radical.}$$

$$x = -\frac{2}{3} \pm \frac{\sqrt{2}}{3}i \qquad \text{Solve for } x.$$

The solution set is $\left\{ -\frac{2}{3} + \frac{\sqrt{2}}{3}i, -\frac{2}{3} - \frac{\sqrt{2}}{3}i \right\}$.

The STOre feature can be used to check the solutions:

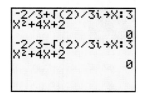

The fact that the solutions are complex and not real is confirmed by the graphical display:

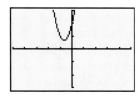

EXERCISE 10.1

VOCABULARY AND NOTATION   *Fill in the blanks to make the given statements true.*

1. A quadratic equation is one that can be written in the form _____. The constant, represented by ___, cannot equal 0. Otherwise, the equation would be linear rather than quadratic.

2. The _____ property allows us to find the roots of equations that can be put into the form $x^2 = c$.

3. The _____ of an equation is/are the value(s) of the variable that make the equation a true statement.

4. The process of _____ allows us to solve any quadratic equation.

5. The _____ feature of the graphing calculator can be used to check both real and complex roots of quadratic equations.

6. Quadratic equations that have _____ roots cannot be confirmed using the graphical display of the graphing calculator.

CONCEPTS

7. The graphs of several quadratic functions are displayed below. Specify the roots as *real* or *imaginary*.

a.
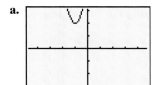

b.
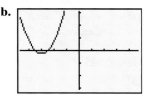

c.
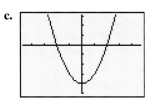

d.
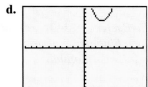

8. a. Graph the quadratic equation $y = x^2 - 9$ by completing the table of values, graphing the points whose coordinates are represented in the table, and connecting the points with a smooth curve.

x	y
0	
2	
−2	
3	
−3	
4	
−4	

b. Graphically, the solutions of the equation can be determined by looking at the x-intercepts. What are the x-intercepts, the solutions, of the equation $x^2 - 9 = 0$?

In Exercises 9 and 10, consider the specified equations:

a. $x^2 = 49$ **b.** $x^2 = -16$

c. $x^2 = 6$ **d.** $x^2 = 8$

e. $x^2 = -5$

9. Which of the given equations have solutions that are real numbers? Explain how you arrived at your conclusion.

10. Which of the given equations have solutions that are imaginary numbers? Explain how you arrived at your conclusion.

In Exercises 11–16, use the square root property to solve each equation. Check roots using the STOre feature of the graphing calculator.

11. $x^2 = 25$ **12.** $x^2 = 49$

13. $x^2 = 20$ **14.** $x^2 = 27$

15. $x^2 = -4$ **16.** $x^2 = -36$

In Exercises 17–22, (a) add the proper constant to make each binomial a perfect square trinomial and (b) factor the resulting trinomial.

17. $x^2 + 10x$

18. $x^2 + 8x$

19. $x^2 - 6x$

20. $x^2 - 12x$

21. $x^2 - 5x$

22. $x^2 + 3x$

PRACTICE *In Exercises 23–34, use the square root property to solve each quadratic equation. Check solutions using the STOre feature of the graphing calculator.*

23. $3x^2 - 16 = 0$

24. $5x^2 - 49 = 0$

25. $(x + 1)^2 = 27$

26. $(x + 1)^2 = 8$

27. $(x + 3)^2 = -18$

28. $(x + 4)^2 = -25$

29. $(2x + 3)^2 = 40$

30. $(3x + 5)^2 = 28$

31. $(3x + 1)^2 = -13$

32. $(2x - 4)^2 = -16$

33. $(x - 1)^2 - 4 = 0$

34. $(x - 7)^2 - 9 = 0$

In Exercises 35–48, solve each equation by completing the square. Check solutions by using the STOre feature or graphically.

35. $x^2 + 2x - 8 = 0$ **36.** $x^2 + 6x + 5 = 0$

37. $x^2 - 6x + 8 = 0$ **38.** $x^2 + 8x + 15 = 0$

39. $x^2 - 2x + 5 = 0$

40. $x^2 - 2x + 10 = 0$

41. $2x^2 - x = -1$ **42.** $2x^2 - 5x = 2$

43. $3x^2 - 11x = -10$ **44.** $2x^2 = x - 6$

45. $5x^2 = -3x - 2$

46. $x^2 = 2x - 6$

47. $5x^2 + 2x = 6$ **48.** $3x^2 + 4x = 2$

49. When solving equations graphically, either the INTERSECT or ROOT/ZERO feature of the calculator may be used.

     a. Discuss which feature you prefer and **why** you prefer it.

     b. Discuss the advantages and disadvantages of each feature.

50. Solve the equation $x^2 - 12 = 0$ both analytically (by hand) and graphically. Which process gives an exact result? Explain. (*Hint:* Use the STOre feature to confirm the exact roots and to show that the irrational approximations are truly just approximations.)

51. Solve the equation $x^2 - x - 12 = 0$

    a. by factoring

    b. by completing the square

    c. graphically

52. Find a quadratic equation with roots of 3 and 5.

53. Find a quadratic equation with solutions of -4 and 6.

54. Find a third-degree equation with solutions of 2, 3, and -4.

55. Find a fourth-degree equation with solutions of 3, -3, 4, and -4.

APPLICATIONS

56. Integer problem The product of two consecutive, even, positive integers is 288. Find the integers. (*Hint:* If one integer is x, the next consecutive even integer is $x + 2$.)

57. Integer problem The product of two consecutive, odd, negative integers is 143. Find the integers. (*Hint:* If one integer is x the next consecutive, odd integer is $x + 2$.)

58. Integer problem The sum of the squares of two consecutive, positive integers is 85. Find the integers. (*Hint:* If one integer is x, the next consecutive, positive integer is $x + 1$.)

59. Integer problem The sum of the squares of three consecutive, positive integers is 77. Find the integers. (*Hint:* If one integer is x, the next consecutive, positive integer is $x + 1$, and the third is $x + 2$.)

60. Building fences Mr. Maier is constructing a dog run for his miniature dachshund (who *hates* the outdoors). He'd like the length to be 3 times the width, and the area is to be 147 square feet.

    a. What are the dimensions of the dog run?

    b. How many feet of fencing must he buy?

61. Building sheds Mr. MacLean is building a shed for his yard tools. He wants it to be in the shape of a rectangle, with the length being 4 feet more than the width. The area needs to be 96 square feet. What are the dimensions of the shed?

62. Pricing concert tickets Tickets to a community fundraiser cost $4, and the projected attendance is 300 people. It is further projected that for every $0.10 increase in ticket price, the average attendance will decrease by 5.

    a. The chart below models the data from the problem. Fill in the last row of the chart, letting x represent the number of price increases.

Ticket price	Attendance	Revenue
$4	300	$4 \cdot 300$
$4 + 1(.10)$	$300 - 1(5)$	$[4 + 1(.10)][300 - 1(5)]$
$4 + 2(.10)$	$300 - 2(5)$	$[4 + 2(.10)][300 - 2(5)]$
$4 + 3(.10)$	$300 - 3(5)$	$[4 + 3(.10)][300 - 3(5)]$
$\vdots$	$\vdots$	$\vdots$

    b. How many price increases will yield nightly receipts of $1248?

    c. What would the price of a ticket be?

63. Setting bus fares A bus company has 3,000 passengers daily, paying a $0.25 fare. For each nickel increase in fare, the company estimates it will lose 80 passengers.

    a. The chart below models the data from the problem. Fill in the last row of the chart, letting x represent the number of price increases.

Fare	Number of passengers	Daily revenue
.25	3,000	$.25(3000)$
$.25 + 1(.05)$	$3,000 - 1(80)$	$[.25 + 1(.05)][3,000 - 1(80)]$
$.25 + 2(.05)$	$3,000 - 2(80)$	$[.25 + 2(.05)][3,000 - 2(80)]$
$.25 + 3(.05)$	$3,000 - 3(80)$	$[.25 + 3(.05)][3,000 - 3(80)]$
$\vdots$	$\vdots$	$\vdots$

    b. How many price increases will yield a daily revenue of at least $994?

    c. What will the fare price be?

64. Computing revenue The *Gazette's* profit is $20 per year for each of its 3,000 subscribers. Management estimates that the profit per subscriber will increase $0.01 for each additional subscriber over the current 3,000. How many subscribers will bring a total profit of $120,000? (*Hint:* Use the methods developed in Exercises 62 and 63 above to work this problem.)

65. Finding interest rates A woman invests $1,000 in a mutual fund for which interest is compounded annually at a rate r. After 1 year she deposits an additional $2,000. After 2 years the balance in the account is $1,000(1 + r)^2 + 2,000(1 + r)$. If this amount is $3,368.10, find r.

66. Framing a picture   The frame surrounding the picture in Illustration 1 has a constant width, and the area of the frame equals the area of the picture. How wide is the frame?

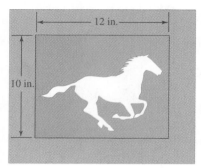

Illustration 1

REVIEW

67. The pictured table is that of a linear function.

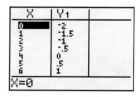

a. State the *x*-intercept of the function.

b. State the *y*-intercept of the function.

c. Write an equation of the function in slope-intercept form.

In Exercises 68–70, classify each pictured display as that of a linear function, an absolute value function, or a radical function.

68.

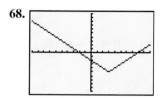

69.

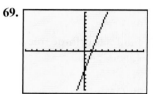

70.

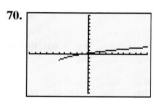

10.2   THE QUADRATIC FORMULA AND THE DISCRIMINANT

- THE QUADRATIC FORMULA
- THE DISCRIMINANT

▐▐▐ THE QUADRATIC FORMULA

Although all quadratic equations can be solved by completing the square, the method is sometimes tedious. The **quadratic formula** will give the solutions of a quadratic equation with much less effort. To develop this formula, we will complete the square on the standard form of a quadratic equation.

A quadratic equation is an equation that can be written in the form $ax^2 + bx + c = 0$, with $a \neq 0$. If we solve this standard form for x by completing the square, we are solving every possible quadratic equation at one time. The solution of the standard quadratic equation results in the quadratic formula.

To solve the equation $ax^2 + bx + c = 0$, with $a \neq 0$, we proceed as follows:

$$ax^2 + bx + c = 0$$

$$\frac{ax^2}{a} + \frac{bx}{a} + \frac{c}{a} = \frac{0}{a}$$ Since $a \neq 0$, we divide both sides by a.

$$x^2 + \frac{bx}{a} = -\frac{c}{a}$$ Simplify and subtract $\frac{c}{a}$ from both sides.

$$x^2 + \frac{b}{a}x + \left(\frac{b}{2a}\right)^2 = \left(\frac{b}{2a}\right)^2 - \frac{c}{a}$$ Complete the square on x and add $\left(\frac{b}{2a}\right)^2$ to both sides.

$$x^2 + \frac{b}{a}x + \frac{b^2}{4a^2} = \frac{b^2}{4a^2} - \frac{4ac}{4aa}$$ Remove parentheses and get a common denominator of $4a^2$ on the right-hand side.

$$\left(x + \frac{b}{2a}\right)^2 = \frac{b^2 - 4ac}{4a^2}$$ Factor the left-hand side and add the fractions on the right-hand side.

We can solve this equation by using the square root property.

$$x + \frac{b}{2a} = \sqrt{\frac{b^2 - 4ac}{4a^2}} \qquad \text{or} \qquad x + \frac{b}{2a} = -\sqrt{\frac{b^2 - 4ac}{4a^2}}$$

$$x + \frac{b}{2a} = \frac{\sqrt{b^2 - 4ac}}{2a} \qquad\qquad\qquad x + \frac{b}{2a} = -\frac{\sqrt{b^2 - 4ac}}{2a}$$

$$x = -\frac{b}{2a} + \frac{\sqrt{b^2 - 4ac}}{2a} \qquad\qquad x = -\frac{b}{2a} - \frac{\sqrt{b^2 - 4ac}}{2a}$$

$$= \frac{-b + \sqrt{b^2 - 4ac}}{2a} \qquad\qquad\qquad = \frac{-b - \sqrt{b^2 - 4ac}}{2a}$$

These two values of x are the solutions to the equation $ax^2 + bx + c = 0$, with $a \neq 0$. They are usually written as a single expression called the **quadratic formula**. Read the symbol $\pm$ as "plus or minus."

Quadratic Formula

If $a \neq 0$, the solutions of $ax^2 + bx + c = 0$ are given by the formula

$$x = \frac{-b \pm \sqrt{b^2 - 4ac}}{2a}.$$

WARNING!   Be sure to draw the fraction bar under *both parts* of the numerator and be sure to draw the radical sign *exactly* over $b^2 - 4ac$.

EXAMPLE 1

Solve each of the given equations by using the quadratic formula.

a. $x^2 - 4x - 3 = 0$

b. $\frac{1}{2}x^2 = 4x - \frac{7}{2}$

c. $3x^2 + 2x + 2 = 0$

d. $9x^2 + 12x = -4$

SOLUTION **a.** Identify a (the coefficient of the square term), b (the coefficient of the linear term), and c (the constant term).

$$x^2 - 4x - 3 = 0$$

$$a = 1, b = -4, c = -3$$

$$x = \frac{-(-4) \pm \sqrt{(-4)^2 - 4(1)(-3)}}{2(1)}$$ Substitute values into the quadratic formula.

$$x = \frac{4 \pm \sqrt{16 + 12}}{2}$$ Simplify.

$$x = \frac{4 \pm \sqrt{28}}{2}$$ Simplify.

$$x = \frac{4 \pm 2\sqrt{7}}{2}$$ Simplify the radical.

$$x = \frac{2(2 \pm \sqrt{7})}{2}$$ Simplify the fraction (notice you *must* first factor the numerator so that only common *factors* are divided out).

$$x = 2 \pm \sqrt{7}$$

The solutions are $x = 2 + \sqrt{7}$ and $x = 2 - \sqrt{7}$.
Solution set: $\{2 + \sqrt{7}, 2 - \sqrt{7}\}$.

Check: The STOre feature of the calculator ensures a correct solution:

```
2+√(7)→X:X²-4X-3
                 0
2-√(7)→X:X²-4X-3
                 0
```

b. Begin by putting the quadratic equation in standard form:

$$\frac{1}{2}x^2 - 4x + \frac{7}{2} = 0$$

We could use the values $a = \frac{1}{2}$, $b = -4$, and $c = \frac{7}{2}$ in the quadratic formula. However, coefficients that are integers leave most students less open to error.

$$2\left(\frac{1}{2}x^2 - 4x + \frac{7}{2}\right) = 2(0)$$ Multiply through the equation by the LCD of 2.

$$x^2 - 8x + 7 = 0$$ Simplify.

$$a = 1, b = -8, c = 7$$ Identify the values of a, b, and c.

$$x = \frac{-(-8) \pm \sqrt{(-8)^2 - 4(1)(7)}}{2(1)}$$ Substitute.

$$x = \frac{8 \pm \sqrt{64 - 28}}{2}$$ Simplify.

$$x = \frac{8 \pm \sqrt{36}}{2}$$ Simplify.

$$x = \frac{8 \pm 6}{2}$$ Simplify the radical.

Do not leave solutions in this form; separate and consider *both* solutions as indicated by the $\pm$ sign. This is illustrated below.

$$x = \frac{8 + 6}{2} = \frac{14}{2} = 7 \quad \text{and} \quad x = \frac{8 - 6}{2} = \frac{2}{2} = 1$$

The solution set is {1, 7}.

Check: Using the STOre feature:

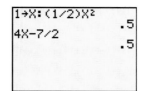

Using the graphical capabilities:

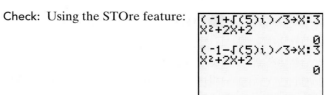

$[-10, 10]$ by $[-10, 30]$

c. $3x^2 + 2x + 2 = 0$

$a = 3, b = 2, c = 2$	Identify the values of a, b, and c.
$x = \dfrac{-2 \pm \sqrt{(2)^2 - 4(3)(2)}}{2(3)}$	Substitute.
$x = \dfrac{-2 \pm \sqrt{4 - 24}}{6}$	Simplify.
$x = \dfrac{-2 \pm \sqrt{-20}}{6}$	Simplify.
$x = \dfrac{-2 \pm 2i\sqrt{5}}{6}$	Simplify the radical.
$x = \dfrac{2(-1 \pm i\sqrt{5})}{6}$	Factor the numerator so that the fraction can be simplified by dividing out common factors.
$x = \dfrac{-1 \pm i\sqrt{5}}{3}$	

The solutions are $x = \dfrac{-1 + i\sqrt{5}}{3}$ and $x = \dfrac{-1 - i\sqrt{5}}{3}$. These could also be expressed in $a + bi$ form as $-\dfrac{1}{3} + \dfrac{\sqrt{5}}{3}i$ and $-\dfrac{1}{3} - \dfrac{\sqrt{5}}{3}i$.

Solution set: $\{-\frac{1}{3} + \frac{\sqrt{5}}{3}i, -\frac{1}{3} - \frac{\sqrt{5}}{3}i\}$.

Check: Using the STOre feature:

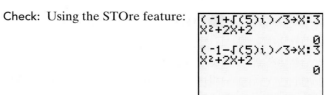

d. Begin by putting the equation $9x^2 + 12x = -4$ in standard form and identifying the values of a, b, and c.

$$9x^2 + 12x + 4 = 0$$

$$a = 9, b = 12, c = 4$$

$$x = \frac{-12 \pm \sqrt{(12)^2 - 4(9)(4)}}{2(9)} \qquad \text{Substitute.}$$

$$x = \frac{-12 \pm \sqrt{144 - 144}}{18} \qquad \text{Simplify.}$$

$$x = \frac{-12 \pm \sqrt{0}}{18} \qquad \text{Simplify.}$$

$$x = \frac{-12 \pm 0}{18} \qquad \text{Simplify the radical.}$$

Because either adding or subtracting the number 0 to -12 does not change the value of the numerator, this equation has only one root, $x = \dfrac{-12}{18} = -\dfrac{2}{3}$, which is often called a *double root*.

The solution set is $\left\{-\frac{2}{3}\right\}$.

Check: Using the STOre feature: Using the graphical capabilities:

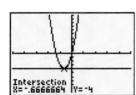

Notice that round-off error has occurred in the graphical solution: $-.6666664$ *is not equal* to $-\frac{2}{3}$.

Remember:

1. The quadratic equation must be in standard form before a, b, and c are identified.
2. To ensure that the quantity represented by b is squared, enclose b in parentheses *before* squaring.
3. The numerator of the quadratic formula is the *entire* expression $-b \pm \sqrt{b^2 - 4ac}$, and the radicand is the *entire* expression $b^2 - 4ac$.
4. When simplifying results, factor the numerator and divide out common factors.

▮ THE DISCRIMINANT

Sometimes it is helpful to know the type of solutions a quadratic equation will have, whether the roots are real (rational or irrational), imaginary, or equal. Consider the solutions of the quadratic equations in Example 1.

Equation	Roots	Number/Type of roots
$x^2 - 4x - 3 = 0$	$2 + \sqrt{7}, 2 - \sqrt{7}$	two distinct real, irrational roots
$x^2 - 8x + 7 = 0$	$7, 1$	two distinct real, rational roots
$3x^2 + 2x + 2 = 0$	$\dfrac{-1 + i\sqrt{5}}{3}, \dfrac{-1 - i\sqrt{5}}{3}$	two distinct imaginary roots
$9x^2 + 12x + 4 = 0$	$-\dfrac{2}{3}$	one real, rational root, often called a *double root*

What caused the differences in the types of roots? Looking back at the solutions, it is apparent that the discriminator was the quantity underneath the radical sign. From the quadratic formula, we see that this expression is $b^2 - 4ac$. This quantity is called the **discriminant**.

Discriminant

If $b^2 - 4ac > 0$ **and** is a perfect square,
   then the quadratic equation will have two distinct real roots (Example 1b). Moreover, if a, b, and c are rational, the roots will be rational.
If $b^2 - 4ac > 0$ **and** is *not* a perfect square,
   then the quadratic equation will have two distinct real, irrational roots (Example 1a).
If $b^2 - 4ac = 0$,
   then the quadratic equation will have one real root, often called a *double root* (Example 1d). Again, if a, b, and c are rational, the roots will be rational.
If $b^2 - 4ac < 0$,
   then the quadratic equation will have two distinct imaginary roots. These roots are complex conjugates (Example 1c).

EXAMPLE 2

Use the discriminant to determine the number and type of roots of each quadratic equation.

a. $15x^2 = 14x + 8$ **b.** $25x^2 + 10x = -1$

c. $x^2 - 2x + 5 = 0$ **d.** $x^2 - 2x = 5$

SOLUTION   **a.** Put the equation is standard form and identify a, b, and c.

$$15x^2 - 14x - 8 = 0$$
$$a = 15, b = -14, c = -8$$

Evaluate the discriminant, $b^2 - 4ac$, for the given values of a, b, and c.

$$b^2 - 4ac = (-14)^2 - 4(15)(-8)$$
$$= 196 + 480$$
$$= 676$$

The discriminant is positive, which means the equation will have two distinct, real roots. Next, we must determine if the roots are rational or irrational. This is done by checking to see if 676 is a perfect square. If it is, the roots are rational because the equation's coefficients are rational; if it is not, the roots are irrational.

$$\sqrt{676} = 26$$

therefore, the roots are rational. This equation has two distinct real, rational roots.

b. Put the equation in standard form and identify a, b, and c.

$$25x^2 + 10x + 1 = 0$$
$$a = 25, b = 10, c = 1$$

Evaluate the discriminant, $b^2 - 4ac$, for the given values of a, b, and c.

$$b^2 - 4ac = (10)^2 - 4(25)(1)$$
$$= 100 - 100$$
$$= 0$$

The value of the discriminant is 0, which means the equation will have one real, rational root. This root is often called a *double root*. This root is rational because the coefficients are rational.

c. Because the equation

$$x^2 - 2x + 5 = 0$$

is in standard form, we simply identify a, b, and c.

$$a = 1, b = -2, c = 5$$

Evaluate the discriminant, $b^2 - 4ac$, for the given values of a, b, and c.

$$b^2 - 4ac = (-2)^2 - 4(1)(5)$$
$$= 4 - 20$$
$$= -16$$

The discriminant is negative, which means the equation will have two distinct, imaginary roots. These roots are complex conjugates.

d. Put the equation in standard form and identify a, b, and c.

$$x^2 - 2x - 5 = 0$$
$$a = 1, b = -2, c = -5$$

Evaluate the discriminant, $b^2 - 4ac$, for the given values of a, b, and c.

$$b^2 - 4ac = (-2)^2 - 4(1)(-5)$$
$$= 4 + 20$$
$$= 24$$

The discriminant is positive, which means the equation will have two distinct, real roots. Next, we must determine if the roots are rational or irrational. This is done by checking to see if 24 is a perfect square. If it is, the roots are rational; if it is not, the roots are irrational.

$$\sqrt{24} \approx 4.898979486$$

therefore, the roots are irrational. This equation has two distinct real, irrational roots.

EXAMPLE 3

Find the value of k for which the quadratic equation $kx^2 + (2k + 4)x + 9 = 0$ has a double root.

SOLUTION

If the quadratic equation has a double root, then the solutions must be equal and hence the value of the discriminant must be zero: $b^2 - 4ac = 0$.

Identify the values of a, b, and c: $a = k$, $b = 2k + 4$, and $c = 9$. Then solve the equation $b^2 - 4ac = 0$ using the given values of a, b, and c.

$(2k + 4)^2 - 4(k)(9) = 0$	Substitution.
$4k^2 + 16k + 16 - 36k = 0$	Remove grouping symbols.
$4k^2 - 20k + 16 = 0$	Combine like terms.
$k^2 - 5k + 4 = 0$	Divide both sides of the equation by 4.
$(k - 4)(k - 1) = 0$	Factor the quadratic to solve by the zero factor property.
$k - 4 = 0 \quad$ and $\quad k - 1 = 0$	Set each factor equal to zero and solve.
$k = 4 \qquad | \qquad k = 1$	

To produce quadratic equations with double roots, k could have a value of 4 ($4x^2 + 12x + 9 = 0$) or a value of 1 ($x^2 + 6x + 9 = 0$).

TECHNOLOGY TIP

It is often convenient to store values for a, b, and c; evaluate the discriminant; and then recall the store command to change values for a, b, and c.

Solving Quadratic Equations

1. Using the quadratic formula or completing the square will provide solutions to any quadratic equation.
2. If $b = 0$ (there is no x, or linear term), use the square root property.
3. If the polynomial $ax^2 + bx + c$ is factorable, factoring may be used.
4. If required solutions are to be exact, algebraic methods are preferred.
5. If approximations are acceptable, graphical methods may be more expedient.
6. *All* exact solutions may be checked, or confirmed, using the STOre feature of the calculator.

EXERCISE 10.2

VOCABULARY AND NOTATION *Fill in the blanks to make a true statement.*

1. When a quadratic equation is in standard form, $ax^2 + bx + c = 0$, the quadratic formula states.

   $x = $ _____.

2. When solving a quadratic equation, the roots are $1 + \sqrt{5}$ and $1 - \sqrt{5}$. This is an _____ result. (exact/approximate)

3. When solving the same quadratic equation as in Exercise 2, a student used the ROOT/ZERO feature of the graphing calculator and stated that the roots are 3.236068 and -1.236068. These results are _____. (exact/approximate)

4. The formula for the discriminant is _____.

5. If the value of the discriminant is positive and a perfect square and the coefficients are rational, the equation will have two distinct real, _____ roots.

6. If the value of the discriminant is positive and is not a perfect square, the equation has two distinct real, _____ roots.

7. If the value of the discriminant is 0 and the coefficients are rational, the equation has one real _____ root, often called a _____ root.

8. If the value of the discriminant is negative, the equation has two distinct _____ roots. These are _____.

CONCEPTS   *Consider the equation $y = x^2 - x - 12$ whose graph is pictured below. A table of values is also provided.*

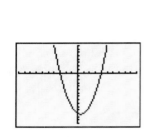

x	y
-4	8
-3	0
-2	-6
-1	-10
0	-12
1	-12
2	-10
3	-6
4	0
5	8

9. State the x-coordinates of the x-intercepts.

10. Complete the solution of $x^2 - x - 12 = 0$, where $a = 1, b = -1, c = -12$.

$$x = \frac{-(-1) \pm \sqrt{\blacksquare^2 - 4\,\blacksquare\,\blacksquare}}{2\,\blacksquare}$$

$$x = \frac{\blacksquare \pm \sqrt{1 + 48}}{\blacksquare}$$

$$x = \frac{1 \pm \sqrt{49}}{2}$$

$$x = \frac{1 \pm \blacksquare}{2}$$

$$x = \blacksquare \quad \text{or} \quad x = \blacksquare$$

11. The solutions, or roots, of the equation are the same as _____.

12. Explain why the displayed equation $y1 = y2$ has no real number solutions.

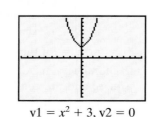

$$y1 = x^2 + 3, \ y2 = 0$$

PRACTICE   *In Exercises 13–36, use the quadratic formula to solve each quadratic equation. Check your solutions, or roots, by using the STOre feature of the calculator.*

13. $x^2 + 3x + 2 = 0$

14. $x^2 - 3x + 2 = 0$ **15.** $x^2 + 9 = 0$

16. $x^2 + 16 = 0$ **17.** $x^2 + 12x = -36$

18. $y^2 - 18y = -81$ **19.** $3x^2 = -16$

20. $2x^2 = -25$ **21.** $x^2 + 2x + 2 = 0$

22. $x^2 + 3x + 3 = 0$

23. $5x^2 + 5x + 1 = 0$

24. $4w^2 + 6w + 1 = 0$

25. $2x^2 + x + 1 = 0$

26. $3x^2 + 2x + 1 = 0$

27. $16y^2 + 8y - 3 = 0$

28. $16x^2 + 16x + 3 = 0$

29. $3x^2 - 4x = -2$

30. $2x^2 + 3x = -3$

31. $\dfrac{x^2}{2} + \dfrac{5}{2}x = -1$

32. $-3x = \dfrac{x^2}{2} + 2$

33. $3x^2 - 2x = -3$

34. $5x^2 = 2x - 1$

35. $8u = -4u^2 - 3$

36. $4t + 3 = 4t^2$

In Exercises 37–44, use the discriminant to determine what type of solutions exist for each quadratic equation. Do not solve the equation.

37. $4x^2 - 4x + 1 = 0$

38. $5x^2 + x + 2 = 0$

39. $2x^2 = 4x - 1$

40. $x(2x - 3) = 20$

41. $6x^2 - 5x - 6 = 0$

42. $3x^2 + 10x - 2 = 0$

43. $9x^2 = 12x - 4$

44. $x(x - 3) = -10$

In Exercises 45–52, find the value(s) of k that will make the solutions of each given quadratic equation equal.

45. $x^2 + kx + 9 = 0$

46. $9x^2 + 4 = -kx$

47. $(k - 1)x^2 + (k - 1)x + 1 = 0$

48. $(k + 4)x^2 + 2kx + 9 = 0$

49. $kx^2 - 12x + 4 = 0$

50. $9x^2 - kx + 25 = 0$

51. $(k + 3)x^2 + 2kx + 4 = 0$

52. $(k + 15)x^2 + (k - 30)x + 4 = 0$

53. Use the discriminant to determine whether the solutions of $1492x^2 + 1776x - 1984 = 0$ are real numbers.

54. Use the discriminant to determine whether the solutions of $1776x^2 - 1492x + 1984 = 0$ are real numbers.

55. Determine k such that the solutions of $3x^2 + 4x = k$ are imaginary numbers.

56. Determine k such that the solutions of $kx^2 - 4x = 7$ are imaginary numbers.

In Exercises 57–64, solve each equation graphically. Round to the nearest hundredth where necessary. Specify window values used as well as calculator feature (INTERSECT or ROOT/ZERO).

57. $2.3x^2 + 16.2x = 21.5$

58. $3.4x^2 + 10.2x = 16.5$

59. $\dfrac{2x}{x + 1} - \dfrac{3x}{x + 2} = -3$

60. $\dfrac{3x}{x - 1} - \dfrac{x}{x + 2} = 5$

61. $(x + 5)(2x + 3) = 2x + 5$

62. $(x + 1)(x + 3) = 2x^2 + 3x$

63. $0.2x + \dfrac{5}{x} = 8$

64. $0.4x + \dfrac{6}{x} = 6$

65. Discuss ways to use the graphing calculator to check solutions of quadratic equations. Your discussion should address the use of the VARS, STOre, and TABLE features as well as the use of the graphical display. Be sure to address roots that are rational, irrational, and imaginary.

66. Does the quadratic formula hold for equations with complex coefficients? Solve $x^2 + 2ix - 5 = 0$ using the quadratic formula. Verify the solutions.

67. Does the quadratic formula hold for equations with complex coefficients? Solve $2x^2 - 4ix + 6 = 0$ using the quadratic formula. Verify the solutions.

In Exercises 68–71, solve each quadratic equation and verify that the sum of the solutions is $-\frac{b}{a}$ and that the product of the solutions is $\frac{c}{a}$.

68. $12x^2 - 5x - 2 = 0$

69. $8x^2 - 2x - 3 = 0$

70. $2x^2 + 5x + 1 = 0$

71. $3x^2 + 9x + 1 = 0$

APPLICATIONS

72. Discriminant

  a. Solve the quadratic equation $x^2 + 2\sqrt{3}x + 2 = 0$ by completing the square.

  b. Identify a, b, and c and find the discriminant.

  c. What type of root(s) would you expect, given the discriminant value in part (b)?

  d. Explain why your answers in parts (a) and (c) are not a contradiction.

73. Discriminant

  a. Solve the quadratic equation $x^2 - 2\sqrt{2}x + 2 = 0$ by completing the square.

  b. Identify a, b, and c and find the discriminant.

  c. What type of root(s) would you expect, given the discriminant value in **b**?

  d. Explain why your answers in parts (a) and (c) are not a contradiction.

REVIEW

74. a. Graph the function $f(x) = x^2 + 5x - 6$.

  b. Find the x-intercepts of the function.

  c. Solve the equation $x^2 + 5x - 6 = 0$ by factoring.

  d. Compare the x-intercepts of the function (found in part (b)) and the solutions of the equation (found in part (c)).

75. a. Graph the function $f(x) = 3x + 6$.

b. Find the x-intercept of the function.

c. Solve the equation $3x + 6 = 0$.

 d. Compare the x-intercept of the function and the solution of the equation.

76. a. Graph the function $f(x) = |3x + 6| - 2$.

b. Find the x-intercepts of the function.

c. Solve the equation $|3x + 6| - 2 = 0$.

 d. Compare the x-intercepts of the function and the solutions of the equation.

77. a. Graph the function $f(x) = \sqrt{5x - 10} - 4$.

b. Find the x-intercept of the function.

c. Solve the equation $\sqrt{5x - 10} - 4 = 0$.

 d. Compare the x-intercept of the function and the solution of the equation.

10.3 APPLICATIONS OF QUADRATIC FUNCTIONS AND SOLUTIONS OF EQUATIONS USING QUADRATIC METHODS

- APPLICATIONS OF QUADRATIC FUNCTIONS

- SOLUTIONS OF EQUATIONS USING QUADRATIC METHODS

▌ APPLICATIONS OF QUADRATIC FUNCTIONS

In the previous sections we examined solutions of quadratic equations. We found that the roots of the equations are the x-intercepts of the graph of the function associated with the equation. Examination of quadratic functions reveals information about the general behavior of quadratic models. We will relate the algebraic processes previously developed to the applications presented in this section.

The graph of a quadratic function is a curve called a **parabola**. All graphs of quadratic functions are parabolic in shape, opening either up or down, as shown in the following graphs:

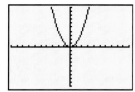   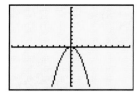

The turning point of the curve is also the highest or the lowest point (depending on the opening of the parabola). This point is called the **vertex** of the parabola.

EXAMPLE 1

Suppose the ball shown in Figure 10-1 is thrown straight up with a velocity of 128 feet per second. The quadratic function $y = 128x - 16x^2$ gives the relationship between x and y, where x represents the time measured in seconds, and y represents the number of feet the ball is above the ground.

a. Graph the function in an appropriate window. Make sure the *x*-intercepts, the *y*-intercept, and the vertex are clearly visible.

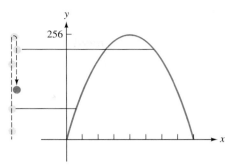

Figure 10-1

b. How long does it take for the ball to hit the ground?

c. What is the maximum height of the ball?

d. How high is the ball after 2 seconds?

e. At what other time does the ball reach the same height?

SOLUTION   **a.** Graphing the quadratic equation in the standard viewing window reveals little information. The methods previously developed (tracing to find maximum or minimum points and *x*-intercepts) are laborious and time-consuming. To obtain a good graph, think about what *x* and *y* represent. The variable *x* is time, measured in seconds, whereas *y* (the value of the function at *x*, or $f(x)$) is the height. Set the TABLE to begin at 0 and to be incremented by 1. Examining the table, we see that 256 is a maximum (for the given increment).

Using this information, reset the window to an *x*-minimum of 0 and an x-maximum of 10 (we are using our table values to formulate an educated guess). Our y-minimum is 0, and we will use 300 as a maximum. Both scales have been set equal to 0 to eliminate distortion of the axes.

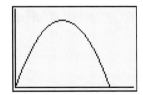

b. Recall that our function is $y = 128x - 16x^2$. The ball will hit the ground when $y = 0$. We are looking for the *x*-value corresponding to the value of 0 for *y*. These points are the *x*-intercepts of the graph. The ROOT/ZERO feature enables us to find the *x*-intercepts. Interpreting the display within the context of the problem, we can conclude that the ball will hit the ground after 8 seconds.

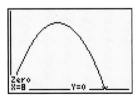

The equation $0 = 128x - 16x^2$ could have been solved algebraically by any of the methods presented in Sections 10.1 and 10.2. Of course, whether solving algebraically or graphically, we only find the *x*-intercept that "makes sense" within the context of the problem and the graphical display. The other *x*-intercept is 0, which is the initial placement of the ball; it has traveled 0 feet in 0 seconds.

TECHNOLOGY TIP

TI-83 Plus: The maximum option is located under the CALC menu. TI-86: To access the maximum option, press [GRAPH] [MORE] [F1] [MATH] [F5] [FMAX].

c. To find the maximum height of the ball, the maximum option of the calculator can be used. Bounds are set to the left and the right of the maximum point.

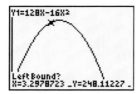 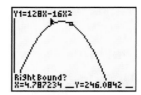 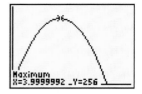

There may be round-off error (as in the third screen), so be careful. Because the variable y represents the height of the ball, the maximum height is 256 feet.

Currently, we have no algebraic process in place to find a maximum.

d. To find the height after 2 seconds, we are finding the value of the function $y = 128x - 16x^2$ when $x = 2$. The value feature is the perfect option to use. Interpretation of the display reveals that the height of the ball is 192 feet after 2 seconds.

Algebraically, we would simply substitute the value of 2 in for x in the definition of the function and solve for y.

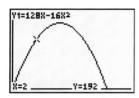

e. To find the other x-value when $y = 192$, we are asked to find the other solution of the equation $192 = 128x - 16x^2$. Entering the value of 192 at the y2 prompt shows *both* points of intersection with the curve. Using the INTERSECT feature and interpreting the screen display, we see that the ball reaches a height of 192 feet again after 6 seconds have elapsed.

Algebraically, we would solve the equation $192 = 128x - 16x^2$ using methods introduced previously.

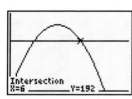

 EXAMPLE 2

A man wants to build the rectangular pen shown in Figure 10-2 to house his dog. To save fencing, he intends to use one side of his garage. Find the maximum area that he can enclose with 80 feet of fencing.

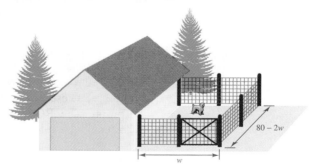

Figure 10-2

SOLUTION   We can let the width of the desired area be represented by w. Then, the length is represented by $80 - 2w$. The area of a rectangular region is the length times the width. Therefore, $A = w(80 - 2w)$.

We enter the function $y = x(80 - 2x)$ at the y= screen, remembering that the variable x represents the width of the desired region and y represents the area.

To find a good window, we again scroll through TABLE values while observing x- and y-values. We set the window to $[0, 40]$ by $[0, 1000]$ to obtain the pictured graph.

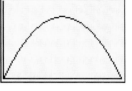

Use of the maximum feature of the calculator and interpretation of the graphical display allows us to conclude that the pen should be 20 feet wide and 40 feet long for a maximum area of 800 square feet.

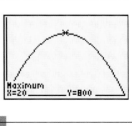

In the previous examples, the maximum feature of the graphing calculator was used to find the vertex of graphs of parabolas that opened downward. No algebraic methods have been developed, thus far, to find these vertices. The following Exploring the Concept feature will help you develop a method for finding the vertices of a parabola.

EXPLORING THE CONCEPT

a. Graph $f(x) = -x^2 - 2x + 24$ in the window $[-10, 10]$ by $[-10, 30]$ and use the maximum feature of the graphing calculator to find the vertex.

b. Find the x-intercepts using either algebraic methods or the graphing calculator.

c. Find the average of the x-intercepts.

d. Find the value of the function when x equals the average of the x-intercepts.

e. Compare the results of part (d) to part (a).

CONCLUSIONS

a.

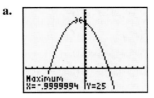

The vertex has coordinates $(-1, 25)$. (Round-off error has occurred in the calculator graph. Verify the ordered pair using the STOre feature.)

b.

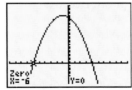 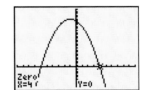

The x-intercepts are -6 and 4.

Algebraically:

$$0 = -x^2 - 2x + 24$$

$$x^2 + 2x - 24 = 0$$

$$(x + 6)(x - 4) = 0$$

$$x + 6 = 0 \quad \text{and} \quad x - 4 = 0$$

$$x = -6 \quad | \quad x = 4$$

c. The average of the x-intercepts is $\dfrac{-6 + 4}{2} = \dfrac{-2}{2} = -1.$

d. Since $f(x) = -x^2 - 2x + 24$, then

$$f(-1) = -(-1)^2 - 2(-1) + 24$$

$$= -1 + 2 + 24$$

$$= 25$$

Therefore, $f(-1) = 25$.

e. The vertex of the parabola can be found by averaging the x-intercepts and then finding the value of the function when x equals the average of the x-intercepts.

   The average of the x-intercepts is the x-coordinate of the vertex, whereas the value of the function at this average is the y-coordinate of the vertex.

SOLUTIONS OF EQUATIONS USING QUADRATIC METHODS

Many equations can be put into quadratic form and then solved with the techniques used for solving quadratic equations. For example, to solve $x^4 - 5x^2 + 4 = 0$, we can proceed as follows:

$$x^4 - 5x^2 + 4 = 0$$

$$(x^2)^2 - 5(x^2) + 4 = 0$$

$$u^2 - 5u + 4 = 0 \qquad \text{Let } u = x^2$$

$$(u - 4)(u - 1) = 0$$

$$u - 4 = 0 \quad \text{or} \quad u - 1 = 0$$

$$u = 4 \quad | \quad u = 1$$

Since $u = x^2$, it follows that $x^2 = 4$ or $x^2 = 1$. Thus, we can use the square root property to find the roots. Because

$$x^2 = 4 \quad \text{or} \quad x^2 = 1$$

it follows that

$$x = 2 \quad \text{or} \quad x = -2 \qquad x = 1 \quad \text{or} \quad x = -1.$$

Solution set: $\{2, -2, 1, -1\}$.

The equation could also be solved graphically. The ROOT/ZERO feature would simply have to be used four times to find all of the roots. It should be remembered that the x-intercepts of a function are the real solutions of the equation.

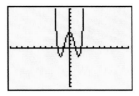

EXAMPLE 3

Solve the equation $x - 7x^{1/2} + 12 = 0$.

SOLUTION

The equation is not quadratic. However, if u^2 is substituted for x and u is substituted for $x^{1/2}$, the equation $x - 7x^{1/2} + 12 = 0$ becomes a quadratic equation that can be solved by factoring:

$$u^2 - 7u + 12 = 0 \qquad \text{Substitute } u^2 \text{ for } x \text{ and } u \text{ for } x^{1/2}.$$

$$(u - 3)(u - 4) = 0 \qquad \text{Factor.}$$

$$u - 3 = 0 \quad \text{or} \quad u - 4 = 0 \qquad \text{Set each factor equal to zero.}$$

$$u = 3 \qquad\qquad u = 4 \qquad \text{Solve.}$$

Because $x = u^2$, it follows that

$$x = 3^2 \quad \text{or} \quad x = 4^2$$

$$x = 9 \qquad\qquad x = 16$$

The equation has two solutions: $\{9, 16\}$.

The equation could also be solved graphically. The ROOT/ZERO feature would simply have to be used twice to find both of the roots. It should be remembered that the x-intercepts of a function are the real solutions of the equation.

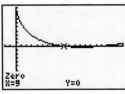 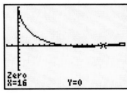

$[-2, 20]$ by $[-10, 10]$

EXAMPLE 4

Solve the equation $x^{2/3} - 5x^{1/3} + 6 = 0$.

SOLUTION

Because $x^{2/3} = (x^{1/3})^2$, we will let $t = x^{1/3}$, which means that $t^2 = x^{2/3}$. Substituting these values in makes our equation $t^2 - 5t + 6 = 0$, which can be solved by factoring:

$$(t - 3)(t - 2) = 0$$

$$t - 3 = 0 \quad \text{or} \quad t - 2 = 0$$

$$t = 3 \qquad\qquad t = 2$$

Because $t = x^{1/3}$, we have

$$x^{1/3} = 3 \quad \text{or} \quad x^{1/3} = 2$$

$$(x^{1/3})^3 = (3)^3 \qquad (x^{1/3})^3 = (2)^3$$

$$x = 27 \qquad\qquad x = 8$$

The solution set is $\{27, 8\}$.

The equation could also be solved graphically. The ROOT/ZERO feature would simply have to be used twice to find both of the roots. It should be remembered that the *x*-intercepts of a function are the real solutions of the equation.

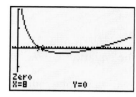 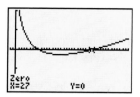

(The window values used in the calculator graphs shown are $[-2, 40]$ by $[-2, 2]$.)

EXAMPLE 5

Solve $\dfrac{24}{x} + \dfrac{12}{x + 1} = 11$.

SOLUTION   Setting each denominator equal to zero gives us excluded values of 0 and -1.

$$\frac{24}{x} + \frac{12}{x + 1} = 11$$

$$x(x + 1)\left(\frac{24}{x} + \frac{12}{x + 1}\right) = x(x + 1)(11)$$ Multiply each term on both sides by the LCD of $x(x + 1)$.

$$24(x + 1) + 12x = (x^2 + x)11$$ Simplify.

$$24x + 24 + 12x = 11x^2 + 11x$$ Use the distributive property to remove parentheses.

$$36x + 24 = 11x^2 + 11x$$ Combine like terms.

$$0 = 11x^2 - 25x - 24$$ Subtract 36x and 24 from both sides.

$$0 = (11x + 8)(x - 3)$$ Factor.

$$11x + 8 = 0 \quad \text{or} \quad x - 3 = 0$$ Set each factor equal to 0.

$$x = -\frac{8}{11} \qquad\qquad x = 3$$

Solution set: $\left\{-\dfrac{8}{11}, 3\right\}$.

EXERCISE 10.3

VOCABULARY AND NOTATION   *Fill in the blanks to make a true statement.*

1. The roots of a quadratic equation are the _____ of the graph of the associated quadratic function.

2. The graph of a quadratic function is a curve called a _____. This curve will open either ____ or _____.

3. The turning point of the graph of a quadratic function is called the _____. It is the _____ or _____ point, depending on how the curve opens.

CONCEPTS A ball is dropped from the Sears Tower in Chicago, a height of approximately 1,454 feet. The pull of gravity is approximately 32 ft/sec². If we disregard air resistance, the approximate height of the ball from the ground is $h = -16t^2 + 1,454$, where h is height and t is time in seconds. A graph modeling the relationship between the height of the ball and elapsed time is shown in Illustration 1.

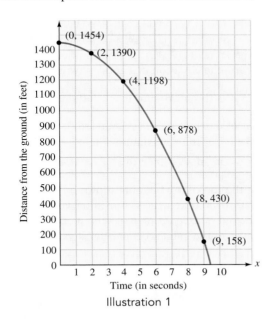

Illustration 1

Match each phrase in Exercises 4–7 with one of these positions on the parabola: x-intercept, y-intercept, point with coordinates (4, 1198).

4. The position of the ball when it first reaches the ground.

5. The position of the ball 4 seconds after release.

6. The position of the ball before it is released.

7. The ball is 1198 feet from the ground after 4 seconds.

In Exercises 8 and 9, use Illustration 1 to answer the questions.

8. After how many seconds is the ball 430 feet from the ground?

9. After 6 seconds, how far is the ball from the ground?

PRACTICE *In Exercises 10 and 11, use the given TABLES to answer the questions.*

10. The function $f(x) = x^2 - 2x - 15$ was entered at the y1= prompt.

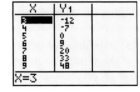

a. Graph the equation on graph paper.

b. State the coordinates of the x-intercepts.

c. State the coordinates of the y-intercept.

d. State the coordinates of the vertex (turning point).

e. $f(2) = ?$

f. If $f(x) = -12$, then $x = ?$ or $x = ?$

11. The function $f(x) = -x^2 + 8x - 16$ was entered at the y1= prompt.

a. Graph the equation on graph paper.

b. State the coordinates of the x-intercept.

c. State the coordinates of the y-intercept.

d. State the coordinates of the vertex (turning point).

e. $f(2) = ?$

f. If $f(x) = -1$, then $x = ?$ or $x = ?$

In Exercises 12–15, use the given graph of the area of a circle ($A = \pi r^2$), where y represents the area and x represents the radius.

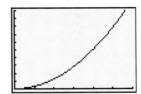

12. As the radius of the circle increases, what happens to the area?

13. If the area is approximately 78.5 square units, what is the radius of the circle? Express this information using an ordered pair and using function notation.

14. If the radius of the circle is 7 units, what is the area? Express this information using an ordered pair and using function notation.

15. Explain why the authors chose to display only part of the parabolic curve.

In Exercises 16–20, use the given graph relating the height, y, (in feet) of a model rocket to its time, x, to answer the given questions. The equation $y = -16x^2 + 750x$ is graphed in a window $[0, 60]$ by $[0, 10000]$.

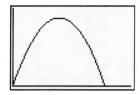

16. Two x-intercepts are apparent, one with coordinates (0, 0) and the other with coordinates (46.875, 0). Interpret each of these values within the context of the problem.

17. The pictured display gives approximations of the coordinates of the vertex of the parabola. Interpret these values (rounded to the nearest unit) within the context of the problem.

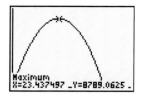

18. Interpret the given graphical displays within the context of the problem.

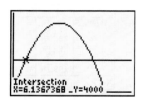 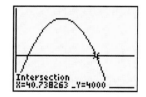

19. After 20 seconds, how high is the rocket? Explain in detail how you obtained your result.

20. After how many seconds (approximately) is the rocket 8,000 feet high? Explain in detail how you obtained your result.

APPLICATIONS *Solve each of the following either algebraically or graphically. Make sure that anyone reading your solution could logically follow your reasoning.*

21. If a ball is thrown straight up with an initial velocity of 48 feet per second, its height, y, after x seconds is given by the equation $y = 48x - 16x^2$.

  a. Graph the equation in a good viewing window.

  b. What is the maximum height attained by the ball?

  c. To the nearest tenth of a second, how long does it take to reach its maximum height?

  d. How long does it take the ball to reach the ground?

22. From the top of the building shown in Illustration 2, a ball is thrown straight up with an initial velocity of 32 feet per second. The equation $s = -16t^2 + 32t + 48$ gives the height, s, of the ball t seconds after it was thrown.

Illustration 2

  a. Find the maximum height of the ball.

  b. Find the time it takes for the ball to hit the ground.

23. A farmer wants to fence in three sides of a rectangular field (as shown in Illustration 3) with 1,000 feet of fencing. The other sides of the rectangle will be a river. If the enclosed area is to be a maximum, determine the size of the field.

Illustration 3

PRACTICE *In Exercises 24–51, solve each equation either graphically or by using algebraic methods.*

24. $x^4 - 17x^2 + 16 = 0$

25. $x^4 - 10x^2 + 9 = 0$

26. $x^4 - 3x^2 = -2$

27. $x^4 - 29x^2 = -100$

28. $x^4 = 6x^2 - 5$

29. $x^4 = 8x^2 - 7$

30. $x^4 = 6x^2 - 5$

31. $2x^4 + 24 = 26x^2$

32. $2x + x^{1/2} - 3 = 0$

33. $2x - x^{1/2} - 1 = 0$

34. $3x + 5x^{1/2} + 2 = 0$

35. $3x - 4x^{1/2} + 1 = 0$

36. $x^{2/3} + 5x^{1/3} + 6 = 0$

37. $x^{2/3} - 7x^{1/3} + 12 = 0$

38. $x^{2/3} - 2x^{1/3} - 3 = 0$

39. $x^{2/3} + 4x^{1/3} - 5 = 0$

40. $x + 5 + \dfrac{4}{x} = 0$

41. $x - 4 + \dfrac{3}{x} = 0$

42. $x + 1 = \dfrac{20}{x}$

43. $x + \dfrac{15}{x} = 8$

44. $\dfrac{1}{x-1} + \dfrac{3}{x+1} = 2$

45. $\dfrac{6}{x-2} - \dfrac{12}{x-1} = -1$

46. $\dfrac{1}{x+2} + \dfrac{24}{x+3} = 13$

47. $\dfrac{3}{x} + \dfrac{4}{x+1} = 2$

48. $x^{-4} - 2x^{-2} + 1 = 0$

49. $4x^{-4} + 1 = 5x^{-2}$

50. $x + \dfrac{2}{x-2} = 0$

51. $x + \dfrac{x+5}{x-3} = 0$

REVIEW *In Exercises 52–59, graph each of the following functions on graph paper. You may use your graphing calculator to help you obtain points; however, make sure your hand-drawn graph is complete.*

52. Graph the function
$y = \dfrac{1}{2}x + 4.$

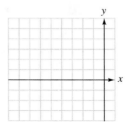

53. Graph the function
$y = \dfrac{1}{2}x - 3.$

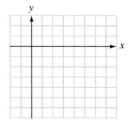

54. Graph the function
$y = (x - 3)^2 + 4.$

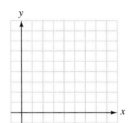

55. Graph the function
$y = (x - 3)^2 - 3.$

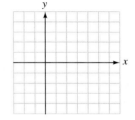

56. Graph the function
$y = \sqrt{x} + 4.$

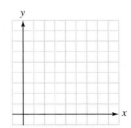

57. Graph the function
$y = \sqrt{x} - 3.$

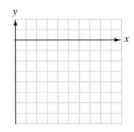

58. Graph the function
$y = |x| + 4.$

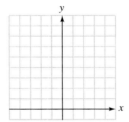

59. Graph the function
$y = |x| - 3.$

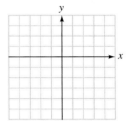

10.4 GRAPHS OF QUADRATIC FUNCTIONS

- GRAPHS OF $f(x) = ax^2$
- GRAPHS OF $f(x) = ax^2 + k$
- GRAPHS OF $f(x) = a(x - h)^2$

- SUMMARY OF GRAPHS OF $f(x) = a(x - h)^2 + k$
- GRAPHS OF $f(x) = ax^2 + bx + c$

▊ GRAPHS OF $f(x) = ax^2$

A **quadratic function** is a second-degree polynomial function of the form $y = f(x) = ax^2 + bx + c$ ($a \neq 0$), where a, b, and c are real numbers. The graphs of quadratic functions are called **parabolas**.

EXPLORING THE CONCEPT

a. Examine the graphs of the quadratic functions

$$f(x) = x^2 \qquad g(x) = 3x^2 \qquad h(x) = \frac{1}{3}x^2 \qquad t(x) = -3x^2$$

b. Complete a table of values for each of the functions for x-values of -2, -1, 0, 1, 2.

c. State the vertex of each of the parabolas.

d. Does the graph of each parabola open up or down?

e. What effect does changing the value of a (the coefficient of the squared term) have on the graphs?

CONCLUSIONS

a.

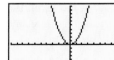

$f(x) = x^2$

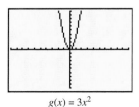

$g(x) = 3x^2$

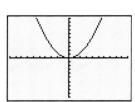

$h(x) = \frac{1}{3}x^2$

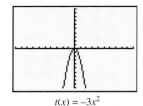

$t(x) = -3x^2$

b.

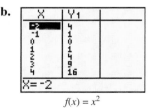

$f(x) = x^2$

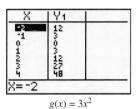

$g(x) = 3x^2$

X	Y₁
-2	1.3333
-1	.33333
0	0
1	.33333
2	1.3333
3	3
4	5.3333

X= -2

$h(x) = \dfrac{1}{3}x^2$

X	Y₁
-2	-12
-1	-3
0	0
1	-3
2	-12
3	-27
4	-48

X= -2

$t(x) = -3x^2$

c. The vertex of $f(x) = (0, 0)$.
The vertex of $g(x) = (0, 0)$.
The vertex of $h(x) = (0, 0)$.
The vertex of $t(x) = (0, 0)$.

d. $f(x)$, $g(x)$, and $h(x)$ open up; $t(x)$ opens down.

e. When $a > 0$ (positive), the parabola opens up; when $a < 0$ (negative), the parabola opens down. The graph of $h(x) = \frac{1}{3}x^2$ is wider than the graph of $f(x) = x^2$. The graph of $g(x) = 3x^2$ is narrower than the graph of $f(x) = x^2$. The graph of $t(x) = -3x^2$ is the same width as $g(x) = 3x^2$; however, the vertex is a maximum rather than a minimum.

The results are summarized below:

Horizontal Stretches of Graphs of Quadratic Functions
If f is a function and a a real number, then

- the graph of $f(x) = ax^2$ opens up when a is positive and down when a is negative.
- if $|a| > 1$, then the graph of $f(x) = ax^2$ is narrower than the graph of $f(x) = x^2$; it is shrunk by a factor of $|a|$.
- if $0 < |a| < 1$, then the graph of $f(x) = ax^2$ is wider than the graph of $f(x) = x^2$; it is increased by a factor of $|a|$.

GRAPHS OF $f(x) = ax^2 + k$

The next Exploring the Concept section examines the effect of adding or subtracting a constant to the squared term of a quadratic function.

EXPLORING THE CONCEPT

a. Examine the graphs of the quadratic functions

$$f(x) = x^2 \qquad g(x) = x^2 + 4 \qquad h(x) = x^2 - 5$$

b. Complete a table of values for each of the functions for x-values of $-2, -1, 0, 1, 2$.

c. State the vertex of each of the parabolas.

d. What effect does adding or subtracting a positive constant have on the function?

CONCLUSIONS   **a.**

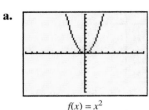

$f(x) = x^2$

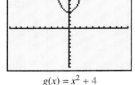

$g(x) = x^2 + 4$

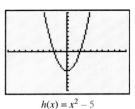

$h(x) = x^2 - 5$

b.

X	Y₁
-2	4
-1	1
0	0
1	1
2	4
3	9
4	16
X= -2	

$f(x) = x^2$

X	Y₁
-2	8
-1	5
0	4
1	5
2	8
3	13
4	20
X= -2	

$g(x) = x^2 + 4$

X	Y₁
-2	-1
-1	-4
0	-5
1	-4
2	-1
3	4
4	11
X= -2	

$h(x) = x^2 - 5$

c. The vertex of $f(x) = (0, \mathbf{0})$.
The vertex of $g(x) = (0, \mathbf{4})$.
The vertex of $h(x) = (0, \mathbf{-5})$.

d. The constant simply translates (moves) the parabola up or down the y-axis. The translation is up when a positive constant is added; down when a positive constant is subtracted.

The results are summarized below:

Vertical Translations of Graphs of Quadratic Functions
If f is a function and k is a positive number, then

- the graph of $f(x) + k$ is identical to the graph or $y = f(x)$ except that it is translated k units upward.
- the graph of $f(x) - k$ is identical to the graph of $y = f(x)$, except that it is translated k units downward.

EXAMPLE 1

Write the equation of each quadratic function described:

a. a quadratic function that is shifted 5 units vertically and is stretched so that it is 5 times wider than the graph of $f(x) = x^2$

b. a quadratic function that is shrunk so as to be $\frac{1}{8}$ as wide as $f(x) = x^2$ and is translated 4 units down

SOLUTION   **a.** To shift the graph of $f(x) = x^2$ five units vertically, we add 5. The function is now $f(x) = x^2 + 5$. To be stretched so that the graph is five times wider, $a = \frac{1}{5}$. Therefore, the function is $f(x) = \frac{1}{5}x^2 + 5$.

The graphing calculator confirms our translations and stretches are correct:

$[-10, 10]$ by $[-5, 15]$

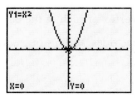

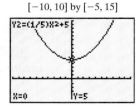

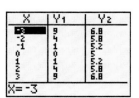

Notice the vertex of $f(x) = x^2$ is $(0, 0)$, whereas the vertex of $f(x) = \frac{1}{5}x^2 + 5$ is $(0, 5)$.

b. To shift the graph of $f(x) = x^2$ down 4 units, subtract 4. The function is now $f(x) = x^2 - 4$. To be $\frac{1}{8}$ as wide, multiply x^2 by 8. Therefore, the function is $f(x) = 8x^2 - 4$.

    The graphing calculator confirms our conjectures.

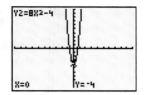

Notice the vertex of $f(x) = x^2$ is $(0, 0)$, whereas the vertex of $f(x) = 8x^2 - 4$ is $(0, -4)$.

▮ GRAPHS OF $f(x) = a(x - h)^2$

What happens to the graph of $f(x) = x^2$ when a positive constant is added to or subtracted from x *before* squaring?

EXPLORING THE CONCEPT

a. Examine the graphs of the quadratic functions

$$f(x) = x^2 \qquad g(x) = (x + 3)^2 \qquad h(x) = (x - 4)^2$$

b. Complete a table of values for each of the functions for x-values of $-2, -1, 0, 1, 2$.

c. State the vertex of each of the parabolas.

d. What effect does adding or subtracting a constant to x before squaring have on the function?

CONCLUSIONS

a.

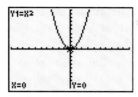
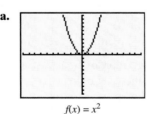
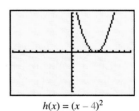

$f(x) = x^2$ $g(x) = (x + 3)^2$ $h(x) = (x - 4)^2$

b.

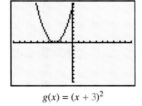

$f(x) = x^2$ $g(x) = (x + 3)^2$ $h(x) = (x - 4)^2$

c. The vertex of $f(x) = (\mathbf{0}, 0)$.
    The vertex of $g(x) = (\mathbf{-3}, 0)$.
    The vertex of $h(x) = (\mathbf{4}, 0)$.

d. The constant translates (moves) the graph of $f(x) = x^2$ to the left when a positive constant is added to or a negative constant is subtracted from x before squaring, and to the right when a positive constant is subtracted or a negative constant is added to x.

The results are summarized below.

Horizontal Translations of Graphs of Quadratic Functions

If f is a function, and h is a constant,

- the graph of $f(x - h)$ is identical to the graph of $y = f(x)$ except that it is translated h units to the right when h is positive.
- the graph of $f(x - h)$ is identical to the graph of $y = f(x)$, except that it is translated h units to the left when h is negative.

EXAMPLE 2

Write the equation of each quadratic function described:

a. a quadratic function that is translated 3 units to the right and 6 units down from the graph of $f(x) = x^2$

b. a quadratic function that is translated 8 units left and 5 units up from the graph of $f(x) = x^2$

SOLUTION

a. To shift the graph of $y = x^2$ three units horizontally to the right, we subtract 3 from x before squaring. The function is now $f(x) = (x - 3)^2$. To shift the function down six units, we subtract 6. The desired function is $f(x) = (x - 3)^2 - 6$.

The graphing calculator confirms our translations are correct.

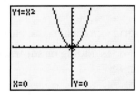 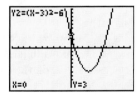

The vertex of $f(x) = (x - 3)^2 - 6$ is $(3, -6)$. The coordinates specified at the bottom of the pictured screen are those of the y-intercept.

b. To shift the graph of $f(x) = x^2$ eight units horizontally to the left, we add 8 to the value of x before squaring. The function is now $f(x) = (x + 8)^2$. To shift the graph up five units, we add 5. The desired function is $f(x) = (x + 8)^2 + 5$.

The graphing calculator confirms our translations are correct.

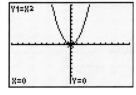 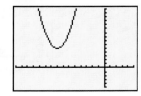

The vertex of $f(x) = (x + 8)^2 + 5$ is $(-8, 5)$.

SUMMARY OF GRAPHS OF $f(x) = a(x - h)^2 + k$

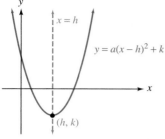

Figure 10-3

When a quadratic function is written in the form $f(x) = a(x - h)^2 + k$, its vertex is easily identified as (h, k). Graphs of quadratic functions are symmetric about an imaginary vertical line through the vertex. This line is called the **axis of symmetry** and has *equation* $x = h$ (see Figure 10-3). It is *not* a part of the graph of the quadratic function, but is often helpful when graphing.

Vertex and Axis of Symmetry of a Parabola

The graph of the equation $y = a(x - h)^2 + k$, where $a \neq 0$, is a parabola with vertex at (h, k).

The parabola opens upward when $a > 0$ and downward when $a < 0$. The axis of symmetry is the line $x = h$.

EXAMPLE 3

Graph the quadratic function $f(x) = 3(x + 5)^2 - 2$. State the coordinates of the vertex and the equation of the axis of symmetry.

SOLUTION

The vertex is $(-5, -2)$, and the axis of symmetry is $x = -5$. The parabola has been shrunk by a factor of $\frac{1}{3}$ and is narrower than the graph of $f(x) = x^2$. However, specific coordinates need to be found, graphed, and connected with a smooth curve.

It is suggested that two x-values larger than -5 and two x-values smaller than -5 be used (see Figure 10-4).

x	y
-7	10
-6	1
-5	-2
-4	1
-3	10

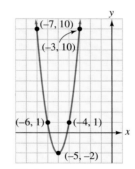

Figure 10-4

Remember, the vertex must be on the graph of the parabola and coordinates of enough points must be found on either side of the vertex to form a smooth curve.

GRAPHS OF $f(x) = ax^2 + bx + c$

When the equation of a quadratic function is in the form $f(x) = y = a(x - h)^2 + k$, the vertex and axis of symmetry are easily identifiable. For this reason, when graphing quadratic functions, we often put them in this form by completing the square on x.

EXAMPLE 4

Graph: $f(x) = 2x^2 - 4x - 1$.

SOLUTION   We complete the square on x to write the function in the form $f(x) = a(x - h)^2 + k$.

$f(x) = 2x^2 - 4x - 1$

$f(x) = 2(x^2 - 2x) - 1$ Factor 2 from $2x^2 - 4x$ so that the coefficient of x^2 is 1.

$f(x) = 2(x^2 - 2x + \mathbf{1}) - 1 - \mathbf{2}$ Complete the square on x. Since this adds 2 to the right-hand side, we also subtract 2 from that side to maintain equality.

$f(x) = 2(x - 1)^2 - 3$ Factor $x^2 - 2x + 1$ and combine terms.

To graph, we identify the vertex $(1, -3)$, and complete a table of values by choosing two x-values smaller than the x-value of 1, and two x-values larger than 1. We then plot the points and connect them with a smooth curve (see Figure 10-5).

x	$f(x)$
-1	5
0	-1
1	-3
2	-1
3	-5

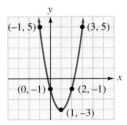

Figure 10-5

The graphing calculator, coupled with our knowledge of completing the square, can be used to write the function in the desired form, $f(x) = a(x - h)^2 + k$.

Graph the function with the calculator. Make sure your display shows a complete graph. Because the parabola opens up, use the minimum feature to find the vertex.

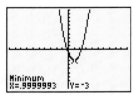

Be aware that there can be round-off error. Set the TABLE to begin at $x = 1$ (what we believe the x-coordinate of the vertex to be) and increment by 1. Scroll up until the vertex is centered on the screen.

X	Y₁
1	-3
2	-1
3	5
4	15
5	29
6	47
7	69

X=1

X	Y₁
-2	15
-1	5
0	-1
1	-3
2	-1
3	5
4	15

X=-2

Notice the symmetry of y-values about $y = -3$.

Therefore, to write the equation $y = 2x^2 - 4x - 1$ in the form $f(x) = a(x - h)^2 + k$, we replace h and k with the x- and y-values of the vertex:

$$f(x) = a(x - 1)^2 - 3$$

The only value we need is *a*. Recall that when completing the square we began by factoring out the coefficient of x^2. This is *always* the value of *a*. Therefore, the desired equation is $f(x) = 2(x - 1)^2 - 3$.

Experiment with the different forms of quadratic functions, and how you can algebraically and graphically move from one form to the other.

EXERCISE 10.4

VOCABULARY AND NOTATION   *Fill in each blank to make a true statement.*

1. A quadratic function that is a second-degree polynomial function can be written in the form _____, where ___ ≠ 0.

2. The turning point, the highest or lowest point of a parabola, is called the _____.

3. The axis of symmetry is an imaginary line passing through the vertex (h, k) and has equation _____.

4. When $a > 0$, the graph of $y = f(x) = a(x - h)^2 + k$ opens ____.

5. When $a < 0$, the graph of $y = f(x) = a(x - h)^2 + k$ opens _____.

6. In the graph of $f(x) = y = a(x - h)^2 + k$, the vertex is located at the coordinates _____.

CONCEPTS   *In Exercises 7–12, match each graph below with the phrase that best describes it.*

a.

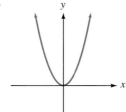

b.

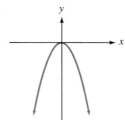

c.

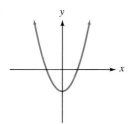

d.

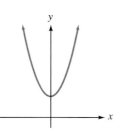

e.

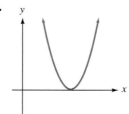

f.

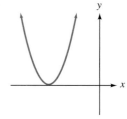

7. The graph of the function $f(x) = x^2$.

8. The graph of a function of the form $f(x) = -x^2$.

9. The graph of a function of the form $f(x) = (x - h)^2$.

10. The graph of a function of the form $f(x) = (x + h)^2$.

11. The graph of a function of the form $f(x) = x^2 + k$.

12. The graph of a function of the form $f(x) = x^2 - k$.

PRACTICE   *In Exercises 13–28, graph each function on graph paper. Be sure to provide a table of values.*

13. $y = x^2$

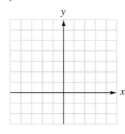

14. $y = -x^2$

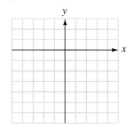

15. $y = -2x^2$

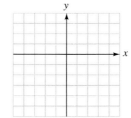

16. $y = 2x^2$

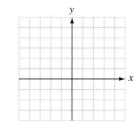

17. $y = -\frac{1}{2}x^2$

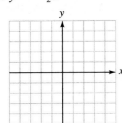

18. $y = \frac{1}{2}x^2$

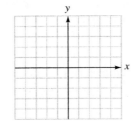

27. $y = \frac{1}{2}(x - 1)^2 + 2$

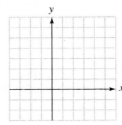

28. $y = \frac{1}{2}(x + 1)^2 + 3$

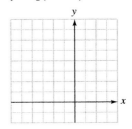

19. $y = x^2 + 2$

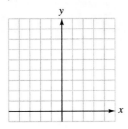

20. $y = x^2 - 3$

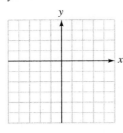

In Exercises 29–34, tell if the parabola opens up or down, state the coordinates of the vertex, and the axis of symmetry of the graph of the equation. If necessary, complete the square on x to write the equation in the form $y = a(x - h)^2 + k$. Do not graph the equation.

29. $y = (x - 1)^2 + 2$

30. $y = 2(x - 2)^2 - 1$

31. $y = 2(x + 3)^2 - 4$

32. $y = -3(x + 1)^2 + 3$

33. $y = -3x^2$

34. $y = 3x^2 - 3$

21. $y = (x + 2)^2$

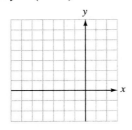

22. $y = (x - 2)^2$

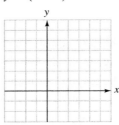

23. $y = -(x - 2)^2$

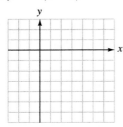

24. $y = -(x + 2)^2$

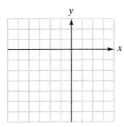

In Exercises 35–42, tell if the parabola opens up or down, state the coordinates of the vertex, the axis of symmetry, and write the equation in the form $y = a(x - h)^2 + k$. Do not graph the equation.

35. $y = x^2 - 2x + 3$

36. $y = 2x^2 - 8x + 7$

37. $y = 2x^2 + 12x + 14$

38. $y = -3x^2 - 6x$

39. $y = 2x^2 - 4x$

40. $y = 3x^2 + 6x$

41. $y = -4x^2 + 16x + 5$

42. $y = 5x^2 + 20x + 25$

25. $y = (x + 3)^2 - 4$

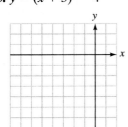

26. $y = (x - 3)^2 - 2$

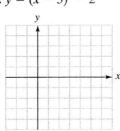

In Exercises 43–46, use your calculator to solve each equation. Solutions should be approximations to the nearest hundredth.

43. $2x^2 + x - 4 = 0$

44. $2x^2 - 5x - 3 = 0$

45. $0.5x^2 - 0.7x - 3 = 0$

46. $2x^2 - 0.5x - 2 = 0$

47. If $a > 0$, do changes in h and k in the equation $f(x) = a(x - h)^2 + k$ affect the domain, the range, or both? In what way? Be specific in your discussion.

48. If $a < 0$, do changes in h and k in the equation $f(x) = a(x - h)^2 + k$ affect the domain, the range, or both? In what way? Be specific in your discussion.

49. The graph of $x = y^2 - 2y$ is a parabola.

   a. Graph the equation.

   b. Explain why the graph is not a function.

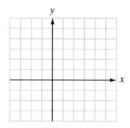

50. Consider the linear function $y = ax + b$, with $y = x$ as the parent function.

   a. Using different values for a (both positive and negative), describe the effect these differing values have on the function when it is compared to $y = x$.

   b. Using different values for b (both positive and negative), describe the effect these differing values have on the function when compared to $y = x$.

51. Consider the absolute value function $y = a|x - h| + k$, with $y = |x|$ as the parent function.

   a. Using different values for a (both positive and negative), describe the effect these differing values have on the function when it is compared to $y = |x|$.

   b. Using different values of k (both positive and negative), describe the effect these differing values have on the function when compared to $y = |x|$.

   c. Using different values for h (both positive and negative), describe the effect these differing values have on the function when compared to $y = |x|$.

52. Consider the square root function $y = a\sqrt{x - h} + k$, with $y = \sqrt{x}$ as the parent function.

   a. Using different values for a (both positive and negative), describe the effect these differing values have on the function when it is compared to $y = \sqrt{x}$.

   b. Using different values for k (both positive and negative), describe the effect these differing values have on the function when compared to $y = \sqrt{x}$.

   c. Using different values for h (both positive and negative), describe the effect these differing values have on the function when compared to $y = \sqrt{x}$.

APPLICATIONS

53. Maximizing revenue   The revenue received for selling x radios is given by the formula

$$R = -\frac{x^2}{728} + 9x$$

   a. Graph the function on the calculator and determine a good window.

   b. Find the maximum of the function.

   c. How many radios must be sold to obtain the maximum revenue?

   d. What is the maximum revenue?

54. Maximizing revenue   The revenue received for selling x stereos is given by the formula

$$R = -\frac{x^2}{5} + 80x - 1000$$

   a. Graph the function on the calculator and determine a good window.

   b. Find the maximum of the function.

   c. How many stereos must be sold to obtain the maximum revenue?

   d. What is the maximum revenue?

REVIEW   *Solve each inequality below. Specify solutions as number line graphs **and** by using interval notation.*

55. $-3x - 1 \leq 8$.

56. $3(x - 2) > 2(x + 7)$

57. $-2 - x + 4 < 5$

58. $3x + 2 < 8$ or $2x - 3 > 11$

59. $-10 \leq 3x + 2 \leq 8$

10.5   QUADRATIC AND OTHER INEQUALITIES

- SOLVING QUADRATIC INEQUALITIES
- SOLVING OTHER NONLINEAR INEQUALITIES
- GRAPHS OF INEQUALITIES IN TWO VARIABLES

SOLVING QUADRATIC INEQUALITIES

To solve a quadratic inequality such as $x^2 + x - 6 < 0$, we want to find all the values of x that will make the inequality a true statement.

If the inequality were an equality (an equation), its solutions could be found by factoring. We will find the values of x that make the *equation $x^2 + x - 6 = 0$* true.

$$x^2 + x - 6 = 0$$
$$(x + 3)(x - 2) = 0$$
$$x + 3 = 0 \quad \text{or} \quad x - 2 = 0$$
$$x = -3 \quad | \quad x = 2$$

Placing the solutions to the equation associated with the inequality serves to partition the number line. We will call these points (solutions of the associated equations) **critical points**. Because they are not a part of the solution to the *inequality $x^2 + x - 6 < 0$*, we will enclose them with open circles.

The number line is now partitioned into three regions that we will call A, B, and C,

If the value of the polynomial $x^2 + x - 6$ is less than zero (or negative) for *one* point in a region, it will be negative for *all* points in the region. If the value of $x^2 + x - 6$ is positive for one value in the region, it will be positive for all values in the region.

We will now test a point from each region in the factored form of the polynomial (because that is easiest). Because we are looking for the polynomial to be negative when it is evaluated, we need to remember that if the product contains an *even* number of negative signs, the result is positive. If the product contains an *odd* number of negative signs, the result is negative.

Region	Test point	$(x + 3)(x - 2) < 0$	True/False
region A	test −5	$(-5 + 3)(-5 - 2) < 0?$ $(-2)(-7) < 0$ ✗	false
region B	test 0	$(0 + 3)(0 - 2) < 0?$ $(3)(-2) < 0$ ✔	true
region C	test 6	$(6 + 3)(6 - 2) < 0?$ $(9)(4) < 0$ ✗	false

Because our point tested true in region B, we shade that region on the number line graph.

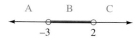

The solution can be expressed in interval notation as $(-3, 2)$.

Interpreting the graphical representation of $x^2 + x - 6$ can also provide the solution to the inequality. The x-intercepts are the solutions of the *equation* $x^2 + x - 6 = 0$, $y_1 = y_2$. Use of the ROOT/ZERO feature provides their values, −3 and 2.

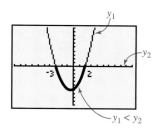

The x-axis has equation $y_2 = 0$. Therefore, the part of the graph that depicts values of the function that are positive is located **above** the x-axis in quadrants I and II, where y has positive values. The part of the graph **below** the x-axis (the part of the parabola containing the vertex, which is heavily shaded) depicts values of the function that are negative (because they are located **below** the x-axis in quadrants III and IV where y has negative values). The **x-values** that produce this part of the graph are those located *between* the critical points of −3 and 2. Therefore the solution of the inequality $x^2 + x - 6 < 0$ is $\{x \mid -3 < x < 2\}$ or

Expressed in interval notation, this is $(-3, 2)$.

The TABLE also provides a way to actually see how the critical points partition the number line. To examine values of the function $y_1 = x^2 + x - 6$ for x-values *smaller* than −3, we set the TABLE to begin at −7. All x-values less than −3 yield y-values (functional values) that are positive.

To examine values of the function between the two critical points, scroll down the TABLE until $x = -3$ is at the top. We can now see that when x has a value between −3 and 2, the y-values are negative. Therefore, the x-values between −3 and 2 are solutions of the inequality $x^2 + x - 6 < 0$.

Continuing to scroll down the TABLE, we see that for x-values larger than 2 (the critical point on the right), y-values (the value of the function) are positive. Therefore, these x-values are *not* part of the solution.

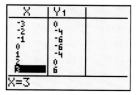

Methods for solving quadratic inequalities algebraically are summarized as follows:

Solving Quadratic Inequalities Algebraically

1. Ensure one side of the inequality is set equal to 0.
2. Change the inequality sign to an equal sign. Solve this equation.
3. The solutions of the equation associated with the inequality are critical points and should be located on a number line.
4. If the original inequality is $>$ or $<$, enclose the critical points with open circles. If the original inequality is $\geq$ or $\leq$, enclose the critical points with closed circles.
5. The critical points have partitioned the number line into regions. Test a point from each region in the original inequality.
6. Shade regions for which the test points result in a true statement. These are solutions of the inequality.

EXAMPLE 1

Solve the inequality $x^2 + 2x \geq 3$.

SOLUTION   Rewrite the inequality so that one side is equal to 0:

$$x^2 + 2x - 3 \geq 0$$
$$x^2 + 2x - 3 = 0 \qquad \text{Write the equation associated with the inequality and solve it.}$$
$$(x + 3)(x - 1) = 0$$

$$x + 3 = 0 \qquad \text{or} \qquad x - 1 = 0$$
$$x = -3 \qquad | \qquad x = 1$$

The critical points are -3 and 1. They are located on a number line and enclosed with closed circles because the inequality includes equality ($\geq$).

We will now test a point from each region:

Region	Test point	$(x + 3)(x - 1) \geq 0$	True/False
region A	test -4	$(-4 + 3)(-4 - 1) \geq 0?$ $(-1)(-5) \geq 0✔$	true
region B	test 0	$(0 + 3)(0 - 1) \geq 0?$ $(3)(-1) \geq 0✘$	false
region C	test 8	$(8 + 3)(8 - 1) \geq 0?$ $(11)(7) \geq 0✔$	true

Shade the regions that tested true. Express the solution using interval notation $(-\infty, -3] \cup [1, \infty)$.

We can confirm our solutions graphically. After finding the x-intercepts, it can be seen that x-values *smaller* than -3 or *larger* than 1 yield the heavily shaded portions of the graphical display. Therefore, the solution found algebraically is correct.

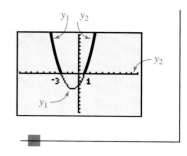

▊ SOLVING OTHER NONLINEAR INEQUALITIES

EXAMPLE 2

Use a graphical approach to solve $(x + 3)(x - 2)(x - 6) < 0$.

SOLUTION

Let $y_1 = (x + 3)(x - 2)(x - 6)$ and $y_2 = 0$. The x-intercepts are $-3, 2,$ and 6 and can be found by using the ROOT/ZERO feature or INTERSECT feature of the calculator or by solving the associated equation $(x + 3)(x - 2)(x - 6) = 0$. (The window used is $[-10, 10]$ by $[-30, 50]$.) The values ($-3, 2,$ and 6) are the critical points

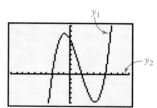

of the graph; they are located on a number line and enclosed with open circles because the original example is strictly an inequality.

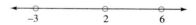

Looking at the graphical display, we see that the graph of $(x + 3)(x - 2)(x - 6)$ is below the x-axis (less than 0) to the left of the critical point -3 and between the critical points 2 and 6. These parts of the graph are heavily shaded. Therefore, the solution of $(x + 3)(x - 2)(x - 6) < 0$ is

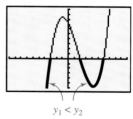

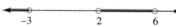

This would be written in interval notation as $(-\infty, -3) \cup (2, 6)$. We suggest that you check your solution by looking at values for y1 in the TABLE, as was done in the introductory problem of this section.

When working with inequalities that contain rational expressions, care must be taken to *not* include any values in the solution that would make a fraction undefined. Therefore, we suggest that you **first** find these excluded values and located them on the number line. They, too, serve as critical points and **are always enclosed with an open circle because they are *never* part of the solution**. Otherwise, these inequalities are solved using the same methods described previously for solving quadratic inequalities.

EXAMPLE 3

Solve: $\dfrac{x^2 - 3x + 2}{x - 3} \geq 0$.

SOLUTION

When the denominator, $x - 3$, equals 0, the rational expression is undefined. Therefore, x cannot have a value of 3; 3 is an excluded value and is a critical point on our number line graph. Critical points that are excluded values are **always** enclosed by open circles.

Write the associated equation and solve for x.

$$\dfrac{x^2 - 3x + 2}{x - 3} = 0$$ Multiply both sides by the quantity $(x - 3)$ to clear the fraction. The *only* way the rational expression can equal 0 is for the numerator to be 0.

$$x^2 - 3x + 2 = 0$$ Solve the equation.

$$(x - 2)(x - 1) = 0$$

$$x - 2 = 0 \quad \text{or} \quad x - 1 = 0$$
$$x = 2 \qquad\qquad x = 1$$

The x-values of 2 and 1 are solutions of the equation. They are also critical points, and they are located on the number line and are enclosed with closed circles because the inequality is $\geq$.

Test a point from each region in the original inequality.

Region	Test point	$\dfrac{(x - 2)(x - 1)}{x - 3} \geq 0$	True/False
region A	test 0	$\dfrac{(-2)(-1)}{-3} \geq 0$ ✗	false
region B	test 1.5	$\dfrac{(-0.5)(0.5)}{-1.5} \geq 0$ ✔	true
region C	test 2.5	$\dfrac{(0.5)(1.5)}{-0.5} \geq 0$ ✗	false
region D	test 5	$\dfrac{(3)(4)}{2} \geq 0$ ✔	true

Shade the regions that tested true.

Write the solution using interval notation: $[1, 2] \cup (3, \infty)$.

Interpreting the graphical displays of rational functions can be difficult and takes a lot of practice. Consider the graphical display of $\dfrac{x^2 - 3x + 2}{x - 3} \geq 0$.

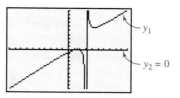

The vertical "line" that appears at $x = 3$ is simply where the calculator connected two pieces of the graph over the excluded value of $x = 3$. It is **not** a part of the graphical display. This becomes evident if you try to TRACE along it. A better display results if the calculator is changed from connected to dot mode.

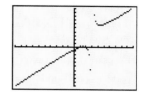

However, you must remember that 3 is an excluded value and is not part of the solution. It does, however, help partition the number line because it is a critical point. Because the x-intercepts are located so close together, care must be taken when setting bounds if using the ROOT/ZERO option.

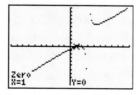 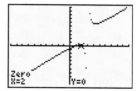

The only parts of the graph above the x-axis are the small piece between the two roots (shown below in the window $[0, 3]$ by $[-1, 1]$) and the piece of graph "hanging" in the first quadrant.

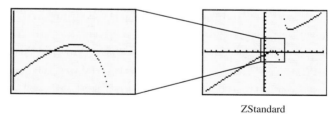

ZStandard

Therefore, the solutions pictured on the number line are confirmed.

To examine values of $y1 = \dfrac{x^2 - 3x + 2}{x - 3}$ in the TABLE, increment by $\frac{1}{2}$ and scroll to values smaller than 1, between 1 and 2, and larger than 3. Examination of these values confirms our number line graph of the solutions.

X	Y1
-2	-2.4
-1.5	-1.944
-1	-1.5
-.5	-1.071
0	-.6667
.5	-.3
1	0
X=-2	

X	Y1
0	-.6667
.5	-.3
1	0
1.5	.16667
2	0
2.5	-1.5
3	ERROR
X=3	

X	Y1
3	ERROR
3.5	7.5
4	6
4.5	5.8333
5	6
5.5	6.3
6	6.6667
X=6	

▮ GRAPHS OF INEQUALITIES IN TWO VARIABLES

The graph of an equation in two variables consists of the points (ordered pairs) that satisfy the equation. The graph of an inequality in two variables consists of an area of the plane bounded by the equation. For example, the inequality $y \geq x - 5$ consists of the area of the plane containing all ordered pairs such that $y > x - 5$ and all the ordered pairs on the line $y = x - 5$.

EXAMPLE 4

Graph the inequality $y \geq x - 5$.

SOLUTION   Since $y \geq x - 5$ means $y = x - 5$ or $y > x - 5$, we begin by graphing the line $y = x - 5$. The graph appears in Figure 10-6.

$$y = x - 5$$

x	y
0	−5
5	0

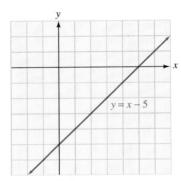

Figure 10-6

Since $y \geq x - 5$ also means y can be greater than $x - 5$, what coordinates satisfy the inequality? The coordinates of the origin produce a true inequality. We can verify this by letting x and y be 0 in the given inequality.

$$y \geq x - 5$$
$$0 \geq 0 - 5 \qquad \text{Substitute 0 for } x \text{ and 0 for } y.$$
$$0 \geq -5$$

Because $0 \geq -5$, the coordinates of the origin satisfy the original inequality. In fact, the coordinates of every point on the same side of the line as the origin satisfy the inequality. The graph of $y \geq x - 5$ is the **half-plane**, the area of the plane bounded by a line, that is shaded in Figure 10-7. Since the boundary line $y = x - 5$ is included, we draw it with a solid line.

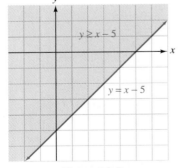

Figure 10-7

If the inequality to be graphed had been a strict inequality, strictly greater than or strictly less than, we would have to indicate on the graph that the equation was *not* part of the solution set. To do this, the boundary line (or boundary curve as in the next example) would be designated as a dashed line to show that the points on it are *not* part of the solution.

EXAMPLE 5   Graph the inequality $y < -x^2 + 4$.

SOLUTION   Using our knowledge of quadratics, we know that the graph of $y = -x^2 + 4$ is a parabola with vertex at $(0, 4)$. Because the coefficient of x^2 is negative one, the parabola opens downward. We construct a table of values and graph the parabola (see Figure 10-8).

x	y
-2	0
-1	3
0	4
1	3
2	0

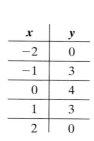

Figure 10-8

Because we are graphing an inequality ($<$), we draw the curve with a dashed line rather than a solid line. This is because points on the curve *do not* satisfy the inequality.

To determine whether to shade inside the parabolic curve, or outside the curve, choose a point to test in the original inequality. Test the origin, $(0, 0)$:

$$0 < -0^2 + 4?$$
$$0 < 4 \text{ ✔}$$

The point tests true; therefore, shade inside the curve (Figure 10-9):

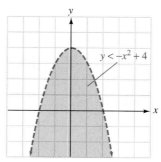

Figure 10-9

The graph style icon on the calculator can be set to shade "below" because the inequality is $<$. The calculator-generated graph confirms our hand-drawn solution.

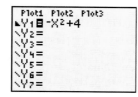

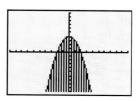

EXAMPLE 6

Graph the inequality $x \le |y|$.

SOLUTION   Using our knowledge of graphs of absolute value equations, we carefully choose values so as to form the "V" shape that is typical of the graphs. We will choose values for y and substitute them into the inequality to find corresponding x-values. We use a solid line because equality is included (see Figure 10-10).

x	y
0	0
1	1
1	−1
2	2
2	−2

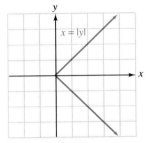

Figure 10-10

To determine whether to shade inside the figure or outside, we choose a test point $(1, 0)$, and substitute into the original inequality:

$$x \le |y|$$
$$1 \le |0|$$
$$1 \le 0 \; ✗$$

We see that the point tests false. Therefore, we *do not* shade the area inside the figure, but, rather shade the area outside the figure (see Figure 10-11).

Figure 10-11

EXERCISE 10.5

VOCABULARY AND NOTATION   *Fill in each blank to make a true statement.*

1. When solving nonlinear inequalities, solve the associated equation to find the _____ points.

2. The solutions of the equation associated with a nonlinear inequality _____ the number line.

3. If the inequality symbol is $>$ or $<$, solutions to the associated equation are enclosed with _____ circles. If the symbol is $\ge$ or $\le$, solutions to the associated equation are enclosed with _____ circles.

4. When a nonlinear inequality contains one or more rational expressions, values that make the denominator equal zero must be _____. They also serve to partition the number line and are always enclosed with an _____ circle.

5. Solutions of a two-variable inequality are graphed on a rectangular _____.

6. A two-variable inequality that uses $<$ or $>$ will have a graph with a _____ line/curve.

CONCEPTS   *In Exercises 7–10, state the excluded values for each rational inequality.*

7. $\dfrac{x - 3}{x} \ge 0$

8. $\dfrac{x^2 - 5x + 6}{x - 5} < 0$

9. $\dfrac{x^2 + x - 12}{(x - 2)(x + 3)} < 0$

10. $\dfrac{x - 5}{x^2 + 2} \ge 0$

In Exercises 11–14, interpret the graphical display and state the solution of the quadratic inequality as either all real numbers (ℝ) or ∅.

11.
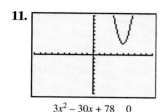
$3x^2 - 30x + 78 \quad 0$

12.
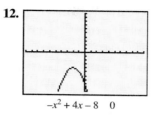
$-x^2 + 4x - 8 \quad 0$

13.
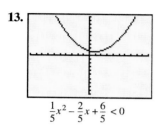
$\dfrac{1}{5}x^2 - \dfrac{2}{5}x + \dfrac{6}{5} < 0$

14.
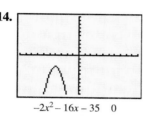
$-2x^2 - 16x - 35 \quad 0$

In Exercises 15–18, match each graphical display with the appropriate inequality.

a.
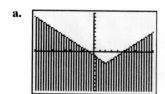

b.
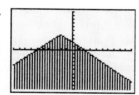

c.
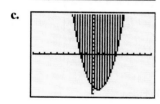

d.
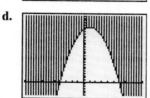

15. $y \geq x^2 - 2x - 8$

16. $y \geq -x^2 + 2x + 24$

17. $y \leq |x - 2| - 3$

18. $y \leq -|x + 2| + 4$

PRACTICE *In Exercises 19–56, solve each inequality. Give each result in interval notation and graph the solution set.*

19. $x^2 - 5x + 4 < 0$

20. $x^2 - 3x - 4 > 0$

21. $x^2 - 8x + 15 > 0$

22. $x^2 + 2x - 8 < 0$

23. $x^2 + x - 12 \leq 0$

24. $x^2 + 7x + 12 \geq 0$

25. $x^2 + 2x \geq 15$

26. $x^2 - 8x \leq -15$

27. $-x^2 + 8x > 18$

28. $x^2 + 6x \geq -9$

29. $x^2 \geq 9$

30. $x^2 \geq 16$

31. $2x^2 - 50 < 0$

32. $3x^2 - 243 < 0$

33. $x^4 - 34x^2 + 225 \leq 0$

34. $x^4 - 52x^2 + 576 \geq 0$

35. $x^4 - 2x^3 - 25x^2 + 26x + 120 > 0$

36. $x^4 + 3x^3 - 28x^2 - 36x + 144 < 0$

37. $x^3 - 5x^2 - 12x + 36 > 0$

38. $x^3 - 3x^2 - 25x + 75 > 0$

39. $x^3 + 6x^2 - 9x - 54 \leq 0$

40. $x^3 + 4x^2 - 17x - 60 \leq 0$

41. $\dfrac{1}{x} < 2$

42. $\dfrac{1}{x} > 3$

43. $\dfrac{4}{x} \geq 2$

44. $-\dfrac{6}{x} < 12$

45. $\dfrac{x^2 - x - 12}{x - 1} < 0$

46. $\dfrac{x^2 + x - 6}{x - 4} \geq 0$

47. $\dfrac{x^2 + x - 20}{x + 2} \geq 0$

48. $\dfrac{x^2 - 10x + 25}{x + 5} < 0$

49. $\dfrac{3}{x - 2} < \dfrac{4}{x}$

50. $\dfrac{-5}{x + 2} \geq \dfrac{4}{2 - x}$

51. $\dfrac{x}{x + 4} \leq \dfrac{1}{x + 1}$

52. $\dfrac{x}{x + 9} \geq \dfrac{1}{x + 1}$

53. $\dfrac{x}{x + 16} > \dfrac{1}{x + 1}$

54. $\dfrac{x}{x + 25} < \dfrac{1}{x + 1}$

55. $(x + 2)^2 > 0$

56. $(x - 3)^2 < 0$

In Exercises 57–76, graph each inequality on graph paper.

57. $2x - y > 2$

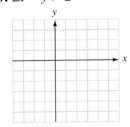

58. $3x - 2y > 6$

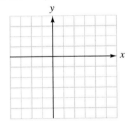

59. $4x + 3y \leq 12$

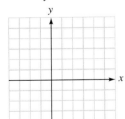

60. $5x + 4y \geq 20$

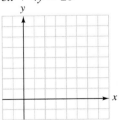

61. $3(x + y) + x < 6$

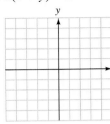

62. $2(x - y) - y \geq 4$

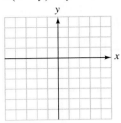

63. $4x - 4(x + 2y) \geq -6y - 8$

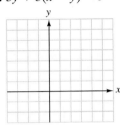

64. $3y + 3(x - y) < 9$

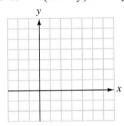

65. $y < x^2 + 1$

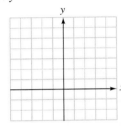

66. $y > x^2 - 3$

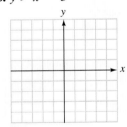

67. $y \leq x^2 + 5x + 6$

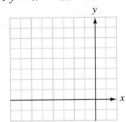

68. $y \geq x^2 + 5x + 4$

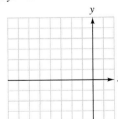

69. $x \geq y^2 - 3$

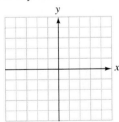

70. $x \le y^2 + 1$

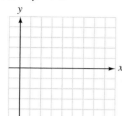

71. $-x^2 - y + 6 > -x$

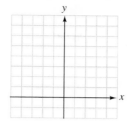

76. $y > |x| - 2$

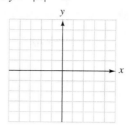

72. $y > (x + 3)(x - 2)$

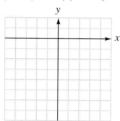

73. $y < |x + 4|$

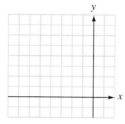

REVIEW *Solve each of the following variation problems.*

77. The area of a circle varies directly with the square of its radius, and the constant of variation is π. Find the area of a circle with a radius of 8 inches.

78. An object in free fall travels a distance, s, that is directly proportional to the square of the time, t. If an object falls 375 feet in 5 seconds, how far does it fall in 7 seconds?

74. $y \ge |x - 3|$

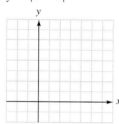

75. $y \le -|x| + 2$

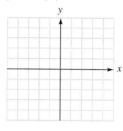

79. The value of a car usually varies inversely with its age. If a car is worth $9,000 when it is 3 years old, how much will it be worth when it is 9 years old?

80. The distance that a car can travel is directly proportional to the number of gallons of gasoline it consumes. If a car can go 360 miles on 15 gallons of gasoline, how far can it go on a full tank of 18 gallons?

CHAPTER SUMMARY

CONCEPTS	REVIEW EXERCISES

Section 10.1

A quadratic equation is one that can be written in the form $ax^2 + bx + c = 0, a \neq 0$.

If factorable, the equations can be solved using the zero factor property.

The square root property allows solutions of equations that may or may not be factorable. It states: If c is a real number, the equation $x^2 = c$ has two solutions, $x = \sqrt{c}$ and $x = -\sqrt{c}$. This is often expressed as $x = \pm\sqrt{c}$.

The process of completing the square may be used to solve *any* quadratic equation. In this process the coefficient of x^2 must be 1 and the constant term is moved to the right side of the equation. The "square" is completed by adding $\frac{1}{2}$ the coefficient of the linear term, squared, to both sides of the equation. The polynomial is then factored as a binomial quantity squared, and the square root property is used to determine the roots.

Square Root Property and Completing the Square

1. Solve each equation by factoring or by using the square root property.

   a. $12x^2 + x - 6 = 0$

   b. $15x^2 + 2x - 8 = 0$

   c. $6x^2 + 17x + 5 = 0$

   d. $(x + 2)^2 = 36$

   e. $(x + 4)^2 = 18$

   f. $(x + 4)^2 = -4$

   g. $(2x + 1)^2 = 12$

   h. $(2x - 1)^2 - 5 = 19$

2. Solve each equation by completing the square.

   a. $x^2 + 6x + 8 = 0$

   b. $2x^2 - 9x + 7 = 0$

   c. $x^2 + 8x = -25$

   d. $-2x^2 + 4 = -7x$

   e. $(x + 2)(x - 3) = 1$

   f. $5x^2 - 3x + 1 = 0$

   g. $5x^2 + 5x + 1 = 0$

   h. $4w^2 + 6w + 1 = 0$

Section 10.2

If $a \neq 0$, the solutions of $ax^2 + bx + c = 0$ are given by the formula:

$$x = \frac{-b \pm \sqrt{b^2 - 4ac}}{2a}$$

The Quadratic Formula and the Discriminant

3. Solve each equation using the quadratic formula.

   a. $x^2 - 8x - 9 = 0$ b. $x^2 - 10x = 0$

   c. $2x^2 + 13x - 7 = 0$ d. $3x^2 - 7 = 20x$

   e. $\frac{3}{4}x = x^2 - 2$ f. $(2x - 1)^2 = 24$

   g. $3x^2 + 3x + 3 = 0$ h. $4x^2 - 4x - 1 = 0$

The value of the discriminant, $b^2 - 4ac$, can be used to determine the number and type of roots of a quadratic equation.

If $b^2 - 4ac > 0$ and is a perfect square, the equation will have two distinct real, rational roots if the coefficients are rational.

If $b^2 - 4ac > 0$ and is not a perfect square, the equation will have two distinct real, irrational roots.

If $b^2 - 4ac = 0$, the equation will have one real rational root, often called a *double root*.

If $b^2 - 4ac < 0$, the equation will have two distinct imaginary roots. They are complex conjugates.

4. Use the discriminant to determine what type of solutions exist for each equation.
   a. $3x^2 + 4x - 3 = 0$
   b. $4x^2 - 5x + 7 = 0$

5. Find the values of k that will make the solutions of $(k - 8)x^2 + (k + 16)x = -49$ equal.

6. Find the values of k such that the solutions of $3x^2 + 4x = k + 1$ are real numbers.

Section 10.3

Applications of Quadratic Functions and Solutions of Equations Using Quadratic Methods

Solutions of quadratic equations are the x-intercept(s) of the graph of the equation.

Many equations can be put into quadratic form and then solved using methods previously presented for solving quadratic equations. A careful substitution of variables can be used to put the equation in quadratic form.

7. **Dimensions of a rectangle**   A rectangle is 2 centimeters longer than it is wide. If both the length and width are doubled, its area is increased by 72 square centimeters. Find the dimensions of the original rectangle.

8. If a rocket is launched straight up into the air with an initial velocity of 112 feet per second, its height after t seconds is given by the formula $h = 112t - 16t^2$, where h represents the height of the rocket in feet.
   a. Graph the function on your graphing calculator in an appropriate viewing window.
   b. After how many seconds does the rocket touch the ground?
   c. After 2 seconds have passed, how high is the rocket?
   d. At what other time is the rocket the same height as it is after 2 seconds?
   e. What is the maximum height of the rocket?

9. Solve each equation.
   a. $x - 13x^{1/2} + 12 = 0$
   b. $a^{2/3} + a^{1/3} - 6 = 0$
   c. $\dfrac{1}{x + 1} - \dfrac{1}{x} = -\dfrac{1}{x + 1}$
   d. $\dfrac{6}{x + 2} - \dfrac{6}{x + 1} = 5$

Section 10.4

Graphs of Quadratic Functions

The graphs of quadratic functions are parabolas. The parabolas will either open up or down. The turning point (the highest or lowest point) is called the *vertex* of the parabola.

If $a \neq 0$, the graph of $y = a(x - h)^2 + k$ is a parabola with vertex at (h, k). It opens upward when $a > 0$ and downward when $a < 0$.

The axis of symmetry is $x = h$.

To graph a parabola, graph the vertex, two points with x-coordinates smaller than that of the vertex, and two points with x-coordinates greater than the x-coordinate of the vertex. Connect all points with a smooth curve.

To put a quadratic equation into the form $y = a(x - h)^2 + k$, either complete the square in x or use the capabilities of the graphing calculator in combination with the algebra.

10. Graph each equation. State the coordinates of the vertex, the axis of symmetry, and write the equation in the form $y = a(x - h)^2 + k$.

a. $y = 2x^2 - 3$ **b.** $y = -2x^2 - 1$

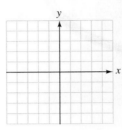 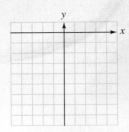

c. $y = -4x^2 + 16x - 15$ **d.** $y = 5x^2 + 10x - 1$

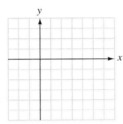 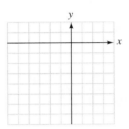

11. a. Find the dimensions of the rectangle of maximum area that can be constructed with 100 feet of fencing.

b. Find the maximum area.

Section 10.5

Quadratic and Other Inequalities

To solve quadratic inequalities, begin by ensuring that one side of the inequality equals 0. Locate solutions to the associated equation on the number line. These are critical points that partition the number line into regions. Test a point from each region in the original inequality and shade regions on the number line that test true. These are the solutions of the inequality.

When solving other nonlinear inequalities, find any excluded values and place them on the number line as critical points.

12. Solve each quadratic inequality below. Express solutions both as a number line graph and using interval notation.

a. $x^2 + 2x - 35 > 0$

b. $x^2 + 7x - 18 < 0$

c. $x^2 + 2 \leq 3x$

d. $3x^2 + 16x \geq -5$

13. Solve each inequality below. Express solutions both as a number line graph and using interval notation.

a. $(x - 2)(x + 1)(x + 4) < 0$

b. $\dfrac{3}{x} \leq 5$

These points are always enclosed with an open circle. The remainder of the solution is completed as when solving quadratic inequalities.

To graph an inequality in two variables, begin by graphing the associated equation. Use a dashed line if the inequality is $>$ or $<$. The graph of the equation separates the plane into two regions. Test a point in one of the regions. If it tests true, shade the region containing the point. If it tests false, shade the other region.

c. $\dfrac{2x^2 - x - 28}{x - 1} > 0$

d. $\dfrac{x}{x + 4} \geq 2x$

14. Graph each inequality.

a. $3x + y \geq 3$

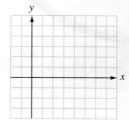

b. $y < 4$

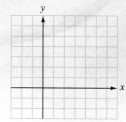

c. $y < \dfrac{1}{2}x^2 - 1$

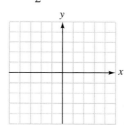

d. $y \geq -|x|$

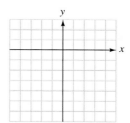

CHAPTER TEST

1. Solve by factoring.

   a. $x^2 + 3x + 2 = 0$

   b. $12x^2 + 5x = 2$

2. Solve by using the square root property.

   a. $(x - 3)^2 = -2$

   b. $3x^2 = 27$

3. Solve by completing the square or using the quadratic formula.

   a. $x^2 + 3x - 18 = 0$

   b. $x(6x + 19) = -15$

   c. $x^2 + 4x + 1 = 0$

   d. $x^2 - 5x - 3 = 0$

4. Determine whether the solutions of $3x^2 + 5x + 17 = 0$ are real or imaginary. ***Do not solve the equation.***

5. For what value(s) of k are the solutions of $4x^2 - 2kx + k - 1 = 0$ equal?

6. One leg of a right triangle is 14 inches longer than the other, and the hypotenuse is 26 inches. Find the length of the shorter leg.

7. Solve the equation $2x - 3x^{1/2} + 1 = 0$.

8. Consider the equation $y = \dfrac{1}{2}x^2 + 5$.

   a. Rewrite the equation in the form $y = a(x - h)^2 + k$.

   b. State the vertex.

   c. State the axis of symmetry.

   d. Graph the figure.

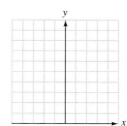

9. Graph the inequality $y \leq -x^2 + 3$.

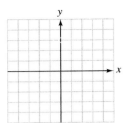

In Problems 10 and 11, solve each inequality and express the solution as both a number line graph and in interval notation.

10. $x^2 - 2x - 8 > 0$

11. $\dfrac{x - 2}{x + 3} \leq 0$

CUMULATIVE REVIEW

CHAPTERS 9 AND 10
Vocabulary/Concepts

1. The expression written under the radical sign is called the _____.

2. If an indicated root of a number is perfect, the result is always a _____ and is an exact result. If an indicated root is not perfect, the calculator can be used to obtain an approximation, which is an _____ number.

3. The Pythagorean theorem states that the sum of the squares of the _____ of a right triangle equals the square of the _____.

4. Powers of i repeat in a cycle of 4 values:

    ___, ___, ___, ___.

5. The _____ of an equation is/are the value(s) of the variable that make the equation a true statement.

6. Graphically, the ___-intercepts are the solutions of an equation when one side is equal to 0.

7. The graph of a quadratic function is a curve called a _____. The turning point of this figure is called its _____.

8. Classify each of the following as rational or irrational.
    a. $\sqrt{64}$ b. $\sqrt{32}$
    c. $\sqrt[5]{32}$ d. $\sqrt[4]{18}$

9. Express each radical as an exponential expression.
    a. $3\sqrt{ab}$ b. $\sqrt{3ab}$
    c. $\sqrt[3]{x^5}$ d. $\sqrt[6]{\dfrac{1}{3}p^5 q}$

10. Classify each of the following as *true* or *false*.
    a. $\sqrt{3} + \sqrt{3} = \sqrt{9} = 3$
    b. $5\sqrt[3]{2} - \sqrt[3]{2} = 4\sqrt[3]{2}$
    c. $\sqrt{25 - 9} = \sqrt{25} - \sqrt{9}$

11. Match each graphical display with its equation.
    a. $y = \sqrt{x + 5}$
    b. $y = \sqrt{x} + 5$
    c. $y = -\sqrt{x + 5}$
    d. $y = \sqrt{x - 5}$

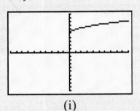
(i)

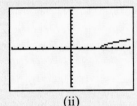
(ii)

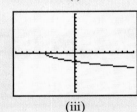
(iii)

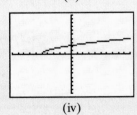
(iv)

12. Use the given display to fill in the blanks below. A radical function was entered at the y1 prompt.

X	Y1
-1	ERROR
0	1
1	2
2	2.4142
3	2.7321
4	3
5	3.2361

X= -1

a. $f(1) =$ ___.

b. An approximation to the nearest thousandth of $f(3) =$ _____.

c. The ERROR message displayed for an x-value of -1 implies that -1 is not a part of the _____ of the function.

13. Simplify each of the following:

a. $\sqrt{-4}$

b. i^{15}

c. $6 + \sqrt{-25}$

d. $(8 + 3i) - 2(5 - 3i)$

14. The graphs of four quadratic functions are displayed below. Specify the roots as real or imaginary.

a.

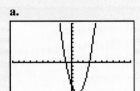

b.

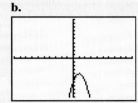

c.

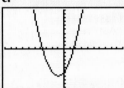

d.

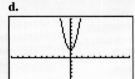

15. State the vertex of each parabola whose equation is given below.

a. $y = 3(x - 2)^2 + 4$

b. $y = -2(x + 4)^2 - 5$

16. Write the equation of the parabola whose graph is shifted 3 units to the right and 4 units up from the graph of $y = x^2$.

17. Write the equation of the parabola that is shifted 5 units left and 6 units down from the graph of $y = x^2$.

18. Write the equation of the parabola that has a vertical stretch by a factor of $\dfrac{1}{6}$ when compared to the parabola whose graph is $y = x^2$.

19. Which of the graphs below is of the inequality $y < (x - 1)^2 + 3$?

a.

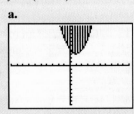

b.

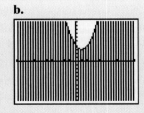

c.

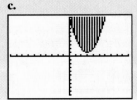

d.

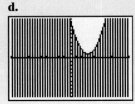

20. Which inequality graphed below is the solution of $\dfrac{x + 2}{x - 1} \le 0$?

a.

b.

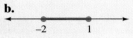

c.

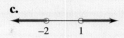

d.

Practice *In Exercises 21–27, simplify each expression (being sure to rationalize denominators as applicable).*

21. $\sqrt{175a^4b^3c^5}$

22. $\sqrt[4]{\dfrac{5x}{16z^5}}$

23. $\sqrt{18} + \sqrt{400} - \sqrt{243}$

24. $\sqrt[3]{6x^2y^4} \cdot \sqrt[3]{9xy^2}$

25. $\dfrac{\sqrt[3]{2}}{\sqrt[3]{3}}$

26. $\dfrac{\sqrt{3} + 1}{\sqrt{3} - 1}$

27. $\dfrac{-4i}{2 - 3i}$

28. If $f(x) = \sqrt{2x + 2}$ and $g(x) = \sqrt{3x + 3} - 1$, find each of the following (express your result in simplest radical form).

a. $f(1)$

b. $g(2)$

c. $f(5)$

29. Solve the equation $\sqrt{x} + 6 = -x + 8$ algebraically. Verify your result(s) using the STOre feature of the calculator. Copy your screen display exactly as it appears to justify your work.

30. Graph $y = \sqrt{x} + 3$ on graph paper and specify the domain and the range.

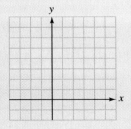

31. Solve the quadratic equation $2x^2 - 5x - 3 = 0$

    a. by factoring

    b. by completing the square

    c. by using the quadratic formula

32. Find a quadratic equation with roots of -3 and $\dfrac{2}{3}$ (express the equation with integral coefficients).

33. The function $f(x) = x^2 - 5x + 6$ was entered at the y1 prompt on the calculator. Its TABLE display is pictured. Use the display to answer the following questions.

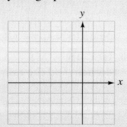

    a. State the coordinates of the x-intercept(s).

    b. State the coordinates of the y-intercept(s).

    c. $f(5) =$ ___.

    d. If $f(x) = 6$, then $x =$ ___ or ___.

34. Solve the equation $x^4 = 6x^2 - 5$.

35. Solve the equation $x^{2/3} - 2x^{1/3} - 3 = 0$.

36. Solve the equation $\dfrac{1}{x + 2} + \dfrac{24}{x + 3} = 13$.

37. Solve the equation $3(x - 4)^2 + 2 = 5$

    a. algebraically   **b.** graphically

38. Graph the function $y = (x + 3)^2 - 4$ on graph paper. Provide a table of values for your graph.

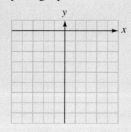

39. Graph the function $y = -2x^2 - 1$ on graph paper. Provide a table of values for your graph.

40. Consider the equation $y = 2x^2 - 8x + 13$.

    a. Does the graph of the parabolic curve open up or down?

    b. State the coordinates of the vertex.

    c. State the equation of the axis of symmetry.

    d. Write the equation in the form $y = a(x - h)^2 + k$.

41. Consider the equation $(x - 8)^2 + 3 = 8$.

    a. Solve the equation graphically, rounding solutions to the nearest hundredth, if necessary.

    b. Solve the equation algebraically, expressing results in simplest radical form.

42. Solve the equation below graphically. Copy your screen display to justify your results. Include the WINDOW settings if they are different from ZStandard.

$$\sqrt{x} - \sqrt{5x + 4} + 6 = \sqrt{18x - 48} + \sqrt{16x}$$

43. Solve the equation below graphically. Copy your screen display to justify your results. Include the WINDOW settings if they are different from ZStandard (ensure the graphical display is clear in the defined window). Confirm your solution(s) with the STOre feature of the calculator.

$$\sqrt[4]{x} = \sqrt{\dfrac{x}{4}}$$

44. Solve the given equation graphically. Round results to the nearest hundredth.

$$\sqrt{1.3x} + 4 = 1.2x + 3.2$$

45. Consider the rounded solution from Exercise 44. If you used the STOre feature to check your solution, what would be the result? Explain fully.

46. Solve the equation $2.3x^2 - 5.2x - 7 = 0$ graphically. Round results to the nearest hundredth.

47. If a ball is thrown straight up in the air with an initial velocity of 50 ft/sec, its height, y, after x seconds is given by the equation $y = 50x - 16x^2$.

    a. Graph the equation in a good viewing window.

    b. What is the maximum height attained by the ball (rounded to the nearest foot)?

    c. To the nearest tenth of a second, how long does it take the ball to reach this height?

    d. To the nearest tenth of a second, how long does it take the ball to reach the ground?

Applications

48. Anchoring flagpoles   A cable is to be used to support a flagpole that is 17 feet tall. The cable is 18 feet long. How far from the base of the pole should the anchor stake be placed? Round to the nearest tenth.

49. Bowling   The velocity, v, of an object after it has fallen d feet is given by the equation $v^2 = 64d$. An inexperienced bowler lifts the ball 4 feet. With what velocity does it strike the alley?

50. Investments   A woman invests $2,000 in a mutual fund for which interest is compounded annually at a rate r. After 1 year she deposits an additional $3,000. After 2 years, the balance in the account is $2,000(1 + r)^2 + 3,000(1 + r)$. If this amount is $5,211.80, find r.

51. Maximizing revenue   The revenue received for selling x stereos is given by the formula

$$R = -\frac{x^2}{1000} + 10x$$

a.   How many stereos must be sold to obtain the maximum revenue?

b.   Find the maximum revenue.

CAREERS & MATHEMATICS

Photographer

Photographers use cameras and film to portray people, objects, places, and events. Some specialize in scientific, medical, or engineering photography and provide illustrations and documentation for publications and research reports. Others specialize in portrait, fashion, or industrial photography and provide the pictures for catalogs and other publications. Photojournalists capture newsworthy events, people, and places, and their work is seen in newspapers and magazines as well as on television.

© Photodisc Collection/Getty Images

Qualifications

Employers want applicants with a broad technical understanding of photography, as well as imagination, creativity, and a good sense of timing. Some knowledge of mathematics, physics, and chemistry is helpful for understanding lenses, films, lighting, and development processes. Industrial or scientific photography and photojournalism generally require a college degree in journalism or photography.

Job Outlook

Job opportunities are expected to increase about as fast as the average through the year 2010. Business and industry will need photographers to provide visual aids for meetings and reports, sales campaigns, and public relations work. Law enforcement agencies and scientific and medical research organizations will also require photographers with appropriate technical skills.

Example Application

In each photographic lens, there is an adjustable circular opening called the **aperture**, which controls the amount of light that passes through the lens. Various lenses — wide-angle, close-up, and telephoto — are distinguished by their **focal length**. The diameter, d, of the aperture and the focal length, f, of the lens determine the ***f*-number** of the lens by the formula

$$f\text{-number} = \frac{f}{d}$$

Thus, a lens with a focal length of 12 centimeters and an aperture with a diameter of 6 centimeters has an f-number of $\frac{12}{6}$, or 2. It would be an $f/2$ lens.

Find the f-number of this lens if the *area* of its aperture is cut in half so as to admit only half as much light.

SOLUTION

First determine the area of a circle with diameter d. Then substitute 6 for d to find the area of the aperture.

$$A = \pi r^2 \qquad \text{The formula for the area of a circle.}$$

$$A = \pi \left(\frac{d}{2}\right)^2 \qquad \text{Substitute } \tfrac{d}{2} \text{ for } r.$$

$$A = \pi \left(\frac{6}{2}\right)^2 \qquad \text{Substitute 6 for } d.$$

$$A = 9\pi$$

To find the diameter of the circle with area equal to one-half of 9π, substitute $\frac{9\pi}{2}$ for A in the formula for the area of a circle and solve for d.

$$A = \pi r^2 \qquad \text{The formula for the area of a circle.}$$

$$\frac{9\pi}{2} = \pi\left(\frac{d}{2}\right)^2 \qquad \text{Substitute } \tfrac{9\pi}{2} \text{ for } A \text{ and } \tfrac{d}{2} \text{ for } r.$$

$$\frac{9\pi}{2} = \frac{\pi d^2}{4}$$

$$18 = d^2 \qquad \text{Multiply both sides by 4.}$$

$$3\sqrt{2} = d \qquad \sqrt{18} = \sqrt{9}\sqrt{2} = 3\sqrt{2}$$

If the diameter of the aperture were reduced from 6 centimeters to $3\sqrt{2}$ centimeters, the area (and the light admitted) would be cut in half. The f-number of the lens is now found by letting $f = 12$ (as before) and $d = 3\sqrt{2}$ in the formula for an f-number.

$$f\text{-number} = \frac{f}{d} = \frac{12}{3\sqrt{2}} = 2\sqrt{2} \approx 2.8$$

An $f/2.8$ lens admits one-half the light admitted by an $f/2$ lens.

If the light were cut in half again, the f-number would be $2.8\sqrt{2}$, or 4. The next f-number, representing another halving of the light admitted, would be $4\sqrt{2}$, or 5.6. These numbers,

$$f/2, \ f/2.8, \ f/4, \ f/5.6, \ f/8, \ f/11, \ f/16$$

are called **f-stops**. They are well known to all professional photographers.

EXERCISES

1. What would be the f-number of a lens with focal length of 20 cm and aperture with diameter of 5 cm?

2. If the focal length of the lens of Exercise 1 were doubled and the aperture held constant, what would be the f-number?

3. What diameter would give a lens with focal length 55 mm an f-number of $f/3.5$?

4. The **speed** of a lens is the square of the reciprocal of the f-number. How many times faster is an $f/2$ than an $f/4.5$ lens?

Functions and Their Inverses

© Photodisc Collection/Getty Images

InfoTrac Project

Do a keyword search on "exponential function." Find the article "Mathskit: The exponential function: Fading into the sunset." Read the article through the section titled "Foam." Suppose the equation for the number of bubbles remaining after t seconds can be represented by $N = N_0 e^{-t}$, where N_0 is the number of bubbles originally in the glass. If there were 100,000 bubbles originally, find the number of bubbles remaining after 2 seconds, and after 4 seconds. If there are 15,000 bubbles left in the glass, how much time has lapsed?

Complete this project after studying Section 11.6.

PERSPECTIVE

Another Graph Paper

Because an exponential function increases rapidly, it is difficult to fit the graph into a reasonable space. For example, between $x = 0$ and $x = 2$, the exponential function $y = 10^x$ changes from a low of 1 to a high of 100.

On the **semilogarithmic graph paper** in Illustration 1, the vertical scale is marked so that the distance between 1 and 2 is the logarithm of 2, the distance between 1 and 3 is log 3, and so on. On this logarithmic scale, the distance between 1 and 2 represents a doubling of y. That same distance, between 2 and 4, also represents a doubling, as does that be-

tween 4 and 8, and between 8 and 16, and so on. On a logarithmic y-axis, equal vertical distances represent increasing y by the same *factor*.

Illustration 2 shows the graphs of three exponential functions in the first quadrant of a Cartesian coordinate system. All of them quickly run off of the grid. Illustration 3 shows the graphs of the same exponential functions on a semilog graph. Note how easily the coordinate system accommodates large values of y. Note also that the graph of an exponential function on a semilog coordinate system is a straight line.

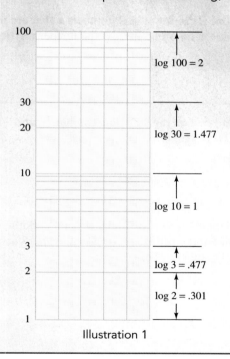

Illustration 1

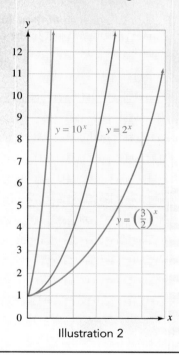

Illustration 2

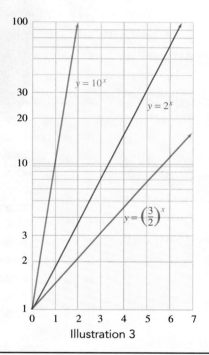

Illustration 3

11.1 ALGEBRA OF FUNCTIONS

- SUM, DIFFERENCE, PRODUCT, AND QUOTIENTS OF FUNCTIONS
- COMPOSITION OF FUNCTIONS
- IDENTITY FUNCTION

SUM, DIFFERENCE, PRODUCT, AND QUOTIENTS OF FUNCTIONS

A function is a rule that can be described by using an algebraic formula. Functions can be added, subtracted, multiplied, and divided to form new functions.

Operations on Functions

If the domains and ranges of functions f and g are included in the set of real numbers, then

the **sum** of f and g, denoted as $f + g$, is defined by

$$(f + g)(x) = f(x) + g(x)$$

the **difference** of f and g, denoted as $f - g$, is defined by

$$(f - g)(x) = f(x) - g(x)$$

the **product** of f and g, denoted as $f \cdot g$, is defined by

$$(f \cdot g)(x) = f(x)g(x)$$

the **quotient** of f and g, denoted as $\frac{f}{g}$, is defined by

$$\left(\frac{f}{g}\right)(x) = \frac{f(x)}{g(x)} \quad (g(x) \neq 0)$$

The domain of each of these functions is the set of real numbers x that are in the domain of both f and g. In the case of the quotient, there is the further restriction that $g(x) \neq 0$.

EXAMPLE 1

Let $f(x) = 2x^2 + 1$ and $g(x) = 5x - 3$. Find each of the following combined functions and its domain: **a.** $f + g$ and   **b.** $f - g$.

SOLUTION   **a.** $(f + g)(x) = f(x) + g(x)$

$$= (2x^2 + 1) + (5x - 3)$$

$$= 2x^2 + 5x - 2$$

The domain of $f + g$ is the set of real numbers that are in the domain of both f and g. Because the domain of both f and g is the set of real numbers, the domain of $f + g$ is also the set of real numbers: $\mathbb{R}$ or $(-\infty, \infty)$.

b. $(f - g)(x) = f(x) - g(x)$

$$= (2x^2 + 1) - (5x - 3)$$

$$= 2x^2 + 1 - 5x + 3 \qquad \text{Remove parentheses.}$$

$$= 2x^2 - 5x + 4 \qquad \text{Combine like terms.}$$

Because the domain of both f and g is the set of real numbers, the domain of $f - g$ is also the set of real numbers: $\mathbb{R}$ or $(-\infty, \infty)$.

EXPLORING THE CONCEPT

Enter $f(x) = \frac{1}{10}x - 15$ and $g(x) = \frac{1}{100}x^2 + 5$ on the y= screen of your calculator.

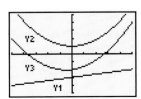

At the Y3= prompt, enter Y1 + Y2 to indicate the addition $f(x) + g(x)$. Display the graphs of the two functions and their sum in the ZInteger viewing window.

a. TRACE to each indicated x-value and record the corresponding y-value. (*Hint:* The left and right cursor keys move you *along* the graph while the up and down cursor keys move you from one graph to another. The number displayed at the top of the screen indicates which function you are tracing on.)

Y1:	(0, _____)		Y1:	(−5, _____)
Y2:	(0, _____)		Y2:	(−5, _____)
Y1 + Y2 = Y3:	(0, _____)		Y1 + Y2 = Y3:	(−5, _____)
Y1:	(8, _____)			
Y2:	(8, _____)			
Y1 + Y2 = Y3:	(8, _____)			

b. What pattern do you observe?

c. The same functions are displayed in the TABLE below. Does the same pattern occur for all values of x displayed in the TABLE?

X	Y1	Y2	Y3	
-5	-15.5	5.25	-10.25	
-4	-15.4	5.16	-10.24	
-3	-15.3	5.09	-10.21	
-2	-15.2	5.04	-10.16	
-1	-15.1	5.01	-10.09	
0	-15	5	-10	
1	-14.9	5.01	-9.89	
X= -5				

d. What do you predict will happen to the pattern when $f(x) - g(x)$ is considered?

e. What do you predict will happen to the pattern when $f(x) \cdot g(x)$ is considered?

f. What do you predict will happen to the pattern when $f(x)/g(x)$ is considered?

CONCLUSIONS

a.

Y1: (0, −15)	Y1: (−5, −15.5)	Y1: (8, −14.2)
Y2: (0, 5)	Y2: (−5, 5.25)	Y2: (8, 5.64)
Y3: (0, −10)	Y3: (−5, −10.25)	Y3: (8, −8.56)

b. The sum of the values of Y1 and Y2 is equal to the value of Y3.

c. The same pattern of sums occurs for all x-values displayed in the TABLE.

d. If the functions are subtracted, $f(x) - g(x)$, then the difference of the values of Y1 and Y2 should be equal to the value of Y3.

e. If the functions are multiplied, $f(x) \cdot g(x)$, then the product of the values of Y1 and Y2 should be equal to the value of Y3.

f. If the functions are divided, $f(x)/g(x)$, then the quotient of the values of Y1 and Y2 should be equal to the value of Y3.

EXAMPLE 2   Let $f(x) = 2x^2 + 1$ and $g(x) = 5x - 3$. Find each combined function and its domain: **a.** $f \cdot g$ and   **b.** $\frac{f}{g}$.

SOLUTION   **a.** $(f \cdot g)(x) = f(x)g(x)$

$$= (2x^2 + 1)(5x - 3)$$

$$= 10x^3 - 6x^2 + 5x - 3$$

The domain of $f \cdot g$ is the set of real numbers that are in the domain of both f and g. Because the domain of both f and g is the set of real numbers, the domain of $f \cdot g$ is also the set of real numbers: $\mathbb{R}$ or $(-\infty, \infty)$.

b. $\left(\dfrac{f}{g}\right)(x) = \dfrac{f(x)}{g(x)}$

$$= \frac{2x^2 + 1}{5x - 3}$$

Because the denominator of the fraction cannot be 0, $x \neq \frac{3}{5}$. Thus, the domain of $\frac{f}{g}$ is the set of real numbers, except $\frac{3}{5}$: $(-\infty, \frac{3}{5}) \cup (\frac{3}{5}, \infty)$.

We can abbreviate this information by using set notation to write $\{x \mid x \neq \frac{3}{5}\}$, which is read as "the set of all real numbers, x, such that x does not equal $\frac{3}{5}$."

The fraction $\frac{3}{5}$ is said to be the **excluded value** of the quotient and is determined by setting the denominator equal to zero and solving for x.

$$5x - 3 = 0$$

$$5x = 3$$

$$x = \frac{3}{5}$$

The value of $x = \frac{3}{5}$ will thus make the denominator equal to zero.

▌▌▌ COMPOSITION OF FUNCTIONS

Composition of functions is the "building" of a new function from two previously defined functions. Consider the following two tables that can be used to convert yards to feet and feet to inches.

Domain of f	x = number of yards	1	2	3	x
Range of f	$f(x)$ = number of feet	3	6	9	$3x$

The function rule for this conversion is $f(x) = 3x$.

Domain of g	f = number of feet	3	6	9	f
Range of g	$g(f)$ = number of inches	36	72	108	$12f$

The function rule for this conversion is $g(f) = 12f$.

Study the two charts. Mathematically speaking, the range of the function f is the domain of the function g. Remember that in the function notation $f(x)$, x represents the values in the domain and $f(x)$ the values in the range. Thus, in symbolic notation we have

$$g(f(x))$$

for our composition function.

If we create a chart that combines the two functions and converts yards directly into inches we have:

Domain of f	x = number of yards	1	2	3	x
Range of g	$g(f(x))$ = number of inches	36	72	108	$12(3x)$

The function rule for this conversion is $g(f(x)) = g(3x) = 12(3x) = 36x$. The notation $g(f(x))$ can also be written as $(g \circ f)(x)$ and expressed as "g composition f."

Composite Function
The **composition function** $f \circ g$ is defined by $(f \circ g)(x) = f(g(x))$. The composite function $g \circ f$ is defined by $(g \circ f)(x) = g(f(x))$.

EXAMPLE 3

Let $f(x) = 2x^2 + 1$ and $g(x) = x - 3$. Find each of the following composition functions and its domain: **a.** $(f \circ g)(x)$ and   **b.** $(g \circ f)(x)$.

SOLUTION   **a.** $(f \circ g)(x) = f(g(x))$
$$= f(x - 3)$$
$$= 2(x - 3)^2 + 1$$
$$= 2(x^2 - 6x + 9) + 1$$
$$= 2x^2 - 12x + 18 + 1$$
$$= 2x^2 - 12x + 19$$

Note that the graph of $(f \circ g)(x)$ is the function $y = 2x^2 + 1$ shifted/translated 3 units to the right.

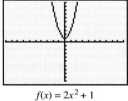
$f(x) = 2x^2 + 1$
Domain: $\mathbb{R}$

$g(x) = x - 3$
Domain: $\mathbb{R}$

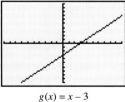
$f(g(x)) = Y1(Y2)$
Domain: $\mathbb{R}$

b. $(g \circ f)(x) = g(f(x))$
$$= g(2x^2 + 1)$$
$$= (2x^2 + 1) - 3$$
$$= 2x^2 - 2$$

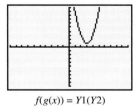
$g(f(x)) = Y2(Y1)$
Domain: $\mathbb{R}$

From parts (a) and (b) we can draw the conclusion that, unlike addition and multiplication of functions, in general, composition of functions is not a commutative operation.

EXAMPLE 4

A boy throws a stone into a pond causing a circular ripple to move out from the point of impact. If the radius of the circle is increasing at a rate of 3 inches per second, write a function that defines the area of the circle in terms of time and determine the area of the circle 5 seconds after impact.

SOLUTION   To find the area of the circle we use the formula $A = A(r) = \pi r^2$. The notation $A(r)$ indicates the area is a function of the radius. Remember, we need the area to be a function of time, and it is the operation of composition that will allow us to do this.

Because the radius is increasing at a rate of 3 inches per second, the radius is a function of time. Thus, $r = r(t) = 3t$. The area of the circle in terms of time (not radius) can be defined as the following composition function:

$$(A \circ r)(t) = A(r(t)) = A(3t)$$
$$= \pi(3t)^2$$
$$= \pi \cdot 9t^2$$
$$= 9\pi t^2$$

Thus, the area of the circle in terms of time is: $A(t) = 9\pi t^2$.

To find the area after 5 seconds we have

$$A(r(5)) = 9\pi(5)^2 = 225\pi \qquad \text{225π is approximately 707 square inches}$$

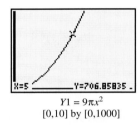 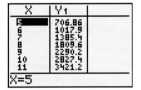

$Y1 = 9\pi x^2$

$[0,10]$ by $[0,1000]$

▍IDENTITY FUNCTION

A special function, called the **identity function**, is defined by the equation $I(x) = x$. Under this function, the value that corresponds to any real number x is x itself. If f is any function, the composition of f with the identity function is just the function f:

$$(f \circ I)(x) = (I \circ f)(x) = f(x)$$

EXAMPLE 5

Let f be any function and I be the identity function, $I(x) = x$. Show that
a. $(f \circ I)(x) = f(x)$ and   **b.** $(I \circ f)(x) = f(x)$.

SOLUTION   **a.** $(f \circ I)(x)$ means $f(I(x))$. Because $I(x) = x$, we have

$$(f \circ I)(x) = f(I(x)) = f(x)$$

b. $(I \circ f)(x)$ means $I(f(x))$. Because I passes any number unchanged, we have

$$I(f(x)) = f(x) \qquad \text{and} \qquad (I \circ f)(x) = I(f(x)) = f(x)$$

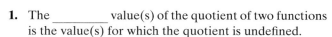

EXERCISE 11.1

VOCABULARY AND NOTATION *Fill in the blanks to make a true statement.*

1. The _____ value(s) of the quotient of two functions is the value(s) for which the quotient is undefined.

2. _____ of functions builds a new function from two previously defined functions.

3. The _____ function is defined by the equation $I(x) = x$. For any function f, $f \circ I = f$.

CONCEPTS *In Exercises 4 and 5, use your results from Exploring the Concept to determine $f + g$, $f - g$, $f \cdot g$, and $\frac{f}{g}$.*

4. $f = \{(4, 5), (8, -2), (-3, 1), (-6, -2)\}$

    $g = \{(4, 8), (8, 3), (-3, 5), (-6, 4)\}$

5. $f = \{(-2, 1), (5, 3), (7, -1), (-5, -2), (4, -3)\}$

    $g = \{(-5, 6), (4, 1), (-2, 8), (7, 6), (5, 3)\}$

In Exercises 6–12, $f(x) = 4x^2 - 8x$ and $g(x) = 4x$.

6. What are the domains and ranges of each of the functions $f(x)$ and $g(x)$?

7. Graph the indicated function in the standard viewing window of your calculator. How do the domains and ranges compare to the domains and ranges of $f(x)$ and $g(x)$?
    a. $f(x) + g(x)$
    b. $f(x) - g(x)$
    c. $f(x) \cdot g(x)$

8. State the domain of $f(x)/g(x)$ and sketch the graph in the standard viewing window of your calculator as well as on graph paper using the x- and y-intercepts.

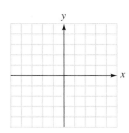

9. A table of values for $f(x)/g(x)$ is displayed. Why does $x = 0$ yield an error message?

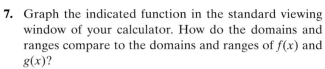

10. What should be done to your hand-drawn graph to reflect the fact that $x = 0$ is not on the graph?

11. A function that can be traced without lifting your pencil is called a **continuous function**. The point at which a break in the graph occurs is the **point of discontinuity**. What is the point of discontinuity for $f(x)/g(x)$?

12. How is the point of discontinuity related to the information given when you stated the domain of the function $f(x)/g(x)$?

TECHNOLOGY TIP

TI-83 Plus: Set your calculator viewing window to ZInteger. From the FORMAT screen turn off the axes, then view the graph and its point of discontinuity. TI-86: [GRAPH] [MORE] [F3] (FORMT).

PRACTICE *In Exercises 13–20, determine the domain of each function.*

13. $f(x) = 3x - 1$

14. $f(x) = 6x + 3$

15. $f(x) = 2(x - 4)^2 + 2$

16. $f(x) = 3(x - 6)^2 - 4$

17. $f(x) = \dfrac{x^2 - 4}{x - 2}$

18. $f(x) = \dfrac{x^2 - 9}{x + 3}$

19. $f(x) = \sqrt{2x - 1}$

20. $f(x) = \sqrt{3x - 1}$

In Exercises 21–28, determine each function (in simplified form) and specify the domain of the new function: **a.** $f + g$
b. $f - g$ **c.** fg **d.** $\dfrac{f}{g}$

21. $f(x) = 3x - 6$ and $g(x) = x - 2$

22. $f(x) = 4x + 8$ and $g(x) = x + 2$

23. $f(x) = x^2 + 1$ and $g(x) = x^2 - 1$

24. $f(x) = x^2 - 4$ and $g(x) = x^2 + 4$

25. $f(x) = \dfrac{1}{x + 3}$ and $g(x) = \dfrac{1}{x^2 - 9}$

26. $f(x) = \dfrac{1}{x + 2}$ and $g(x) = \dfrac{1}{x^2 - 4}$

27. $f(x) = \sqrt{x - 2}$ and $g(x) = \sqrt{5 - x}$ (*Hint:* To find the domain of each combined function use a graphical display and your algebraic knowledge.)

28. $f(x) = \sqrt{x + 3}$ and $g(x) = \sqrt{2 - x}$ (*Hint:* To find the domain of each combined function use a graphical display and your algebraic knowledge.)

In Exercises 29–40, $f(x) = 2x + 1$ and $g(x) = x^2 - 1$. Find each value.

29. $(f \circ g)(2)$

30. $(g \circ f)(2)$

31. $(g \circ f)(-3)$

32. $(f \circ g)(-3)$

33. $(f \circ g)(0)$

34. $(g \circ f)(0)$

35. $(g \circ f)\left(\dfrac{1}{2}\right)$

36. $(f \circ g)\left(\dfrac{1}{2}\right)$

37. $(f \circ g)(x)$

38. $(g \circ f)(x)$

39. $(g \circ f)(2x)$

40. $(f \circ g)(2x)$

41. Is composition of functions associative? Choose functions f, g, and h and determine if
$[f \circ (g \circ h)](x) = [(f \circ g) \circ h](x)$.

42. If $f(x) = x + 1$ and $g(x) = 2x - 5$, show that
$(f \circ g)(x) \neq (g \circ f)(x)$.

In Exercises 43–46, show that $(f \circ g)(x) = x$ and that $(g \circ f)(x) = x$.

43. $f(x) = 2x + 1$ and $g(x) = \frac{1}{2}x - \frac{1}{2}$

44. $f(x) = 3x - 2$ and $g(x) = \frac{1}{3}x + \frac{2}{3}$

45. $f(x) = \frac{1}{5}x - \frac{3}{5}$ and $g(x) = 5x + 3$

46. $f(x) = \frac{1}{2}x + \frac{1}{2}$ and $g(x) = 2x - 1$

47. If $f(x) = x^2 + 2x - 3$, find $f(a)$, $f(h)$, and $f(a + h)$. Then show that $f(a + h) \neq f(a) + f(h)$.

48. If $f(x) = x^2 + 2$, find $\dfrac{f(x + h) - f(x)}{h}$.

49. If $f(x) = x^3 - 1$, find $\dfrac{f(x + h) - f(x)}{h}$.

APPLICATIONS

50. Volume   Sand is dropped off a conveyor belt to form a conical pile whose height is always twice its base radius. The volume of the conical pile, expressed as a function of the radius, r, is $V(r) = \frac{2}{3}\pi r^3$ cubic feet. If the base radius of the pile is increasing at a rate of 1.5 feet per minute, express the volume as a function of time. What is the volume after 4 minutes?

51. Volume   If the volume of a right circular cone is $V = \frac{1}{3}Bh$ (where B is the area of the base and h is the height), explain how to obtain the formula $V(r) = \frac{2}{3}\pi r^3$ used in Exercise 50.

52. Volume   A party store uses helium to fill balloons. If a spherical balloon is filled so that the radius of the sphere is increasing at the rate of 1.25 inches per second, express the volume of the sphere as a function of time. What is the volume of the balloon to the nearest hundredth after 5 seconds? (*Hint:* $V(r) = \frac{4}{3}\pi r^3$)

53. **Area** A radiator is leaking antifreeze onto the garage floor, creating a circular pool. Suppose that the radius is growing at a constant rate of 2 inches per minute. Express the area of the circular pool as a function of time, *t*, where *t* is the number of minutes elapsed since the leak began. Find the area of the pool after 7 minutes.

54. **World record** On April 18, 1999, students at Clarksville High School in Clarksville, Tennessee, broke the world record for the longest paper-clip chain created within a 24-hour period. The students added $1\frac{1}{4}$-inch-long paper clips at the rate of 38,480 per hour; therefore, the length of the chain can be expressed as a function of time, *t* (in hours). The previous world record was 16.88 miles in a 24-hour period. How long was the chain made by the Clarksville students? (*Hint:* Length is a function of the number of paper clips and rate is a function of hours.)

Courtesy of Leaf-Chronicle, Clarksville, TN

Record-breaking paper-clip chain

REVIEW

55. Determine the slope of a line passing through the points (3, 5) and (−2, 1).

56. Write the equation of the line determined by the two points in Exercise 55.

57. Determine the domain and range of each of the following functions:

   a. $f(x) = 4x + 2$

   b. $f(x) = \sqrt{x - 5} + 2$

   c. $f(x) = (x - 5)^2 + 2$

 58. What is the difference between finding the roots of the equation $-2(x + 3)^2 + 2 = 0$ and finding the *x*-intercepts of the function $f(x) = -2(x + 3)^2 + 2$?

11.2 INVERSE FUNCTIONS

- INVERSE RELATIONS
- ONE-TO-ONE FUNCTIONS
- INVERSE FUNCTIONS

▍ INVERSE RELATIONS

In the previous section, we constructed a composite function by comparing tables of measurement conversions from yards to feet and feet to inches. The composite function was formed by combining the two tables to create a new function that converted yards to inches. The composition of two functions *always* produces another function.

Consider what happens when we take the table of measurements used to convert yards to feet.

Domain	a = number of yards	1	2	3	a
Range	b = number of feet	3	6	9	$3a$

If we want to convert feet to yards, the table becomes

Domain	b = number of feet	3	6	9	$3a$
Range	a = number of yards	1	2	3	a

By reversing the a and b elements we have created an inverse relationship where inverse basically means "undoing." Although these two relations that are inverses of one another are also functions, that will not always be the case. In general, if a relation contains the point (a, b), then the **inverse relation** will contain the point (b, a).

All relations have inverse relations because the definition of a relation is that it is a set of ordered pairs.

Inverse Relations

If R is any relation and R^{-1} is the relation obtained from R by interchanging the components of each ordered pair of R, then R^{-1} is called the **inverse relation of R**. The domain of R^{-1} is the range of R, and the range of R^{-1} is the domain of R.

EXAMPLE 1

If the relation defined by converting x minutes to y seconds produces the ordered pairs $\{(\frac{1}{2}, 30), (1, 60), (\frac{3}{2}, 90), (2, 120)\}$, find the inverse relation that is defined by the conversion of seconds to minutes.

SOLUTION   Inverse relations are determined by reversing the domain and range elements. Thus, the inverse relation would be

$$\{(30, \tfrac{1}{2}), (60, 1), (90, \tfrac{3}{2}), (120, 2)\}$$

This means that 30 seconds is equal to $\frac{1}{2}$ a minute, 60 seconds is equal to 1 minute, and so on.

EXAMPLE 2

Show that $y = |x|$ and $x = |y|$ are inverse relations and then graph each relation.

SOLUTION   Look at a table of values for $y = |x|$.

x	-3	-2	-1	0	1	2	3
y	3	2	1	0	1	2	3

Now reverse the domain and range values:

x	3	2	1	0	1	2	3
y	-3	-2	-1	0	1	2	3

Each of these domain and range pairs satisfies $x = |y|$. Thus, if (a, b) is an ordered pair in the relation $y = |x|$, we have $b = |a|$, and (b, a) is an ordered pair in the relation $x = |y|$.

The graph of $y = |x|$ in the standard viewing window is displayed. To *draw* (not *graph*) the inverse, $x = |y|$, on your calculator, use the DrawInv Feature.

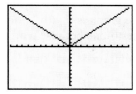 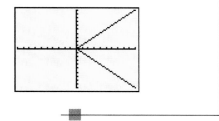

ONE-TO-ONE FUNCTIONS

In order for a function, $f(x)$, to have an inverse that is also a function, the function $f(x)$ must be a one-to-one function. Recall that a function is a relation (set of ordered pairs) in which each value x in the domain is paired with exactly one value y in the range. In order for a function to be **one-to-one**, we must extend the definition so that each y-value is paired with exactly one x-value.

One-to-One Function

A function is called **one-to-one** if and only if each value of y in the range corresponds to only one number x in the domain.

We use a **vertical line test** to determine if each x in the domain has a unique y in the range. A similar test, called the **horizontal line test**, can be used to determine if each y in the range has a unique x in the domain.

Horizontal Line Text

If any horizontal line that intersects the graph of a function does so only once, the function is one-to-one.

EXAMPLE 3

Determine if the graphs of the following relations are functions and then determine if they are one-to-one functions.

a.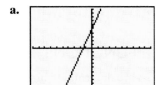

b.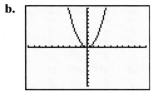

c.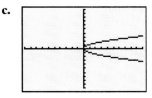

d.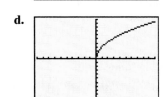

e.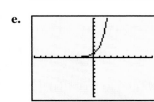

SOLUTION   **a.** The graph is that of a function because it passes the vertical line test. It is one-to-one because it also passes the horizontal line test.

b. The graph is that of a function because it passes the vertical line test. It is *not* one-to-one because it does not pass the horizontal line test.

c. The graph is simply a relation.

d. The graph is that of a function because it passes the vertical line test. It is one-to-one because it also passes the horizontal line test.

e. The graph is that of a function because it passes the vertical line test. It is one-to-one because it also passes the horizontal line test.

EXPLORING THE CONCEPT

Use your calculator to *draw* the inverse of each of the following functions:
a. $y = 3x + 5$ **b.** $y = x^2$ **d.** $y = 3\sqrt{x}$ **e.** $y = 2^x$. (The functions were labeled in this manner because they correspond to the graphs in Example 3.)
Compare your inverses to the graphs of a, b, d, and e in Example 3.

Which of the inverses are function?
Which of the graphs are one-to-one functions?
What conclusion can you draw?

CONCLUSION   **a.**

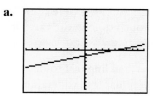

b.

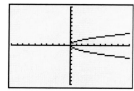

d.

e.

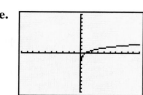

The graphs in Example 3 that are one-to-one functions are a, d, and e. Similarly, the inverses that are functions are also a, d, and e.

Conclusion: Functions that are one-to-one have inverses that are also functions.

▓ INVERSE FUNCTIONS

Inverse functions are determined in the same way that inverse relations are determined — by switching the domain and range values. We will also use the notation $f^{-1}(x)$, read *f inverse* of x, to denote the inverse of the function f.

Inverse Function

The inverse of the one-to-one function $f(x)$ is denoted as $f^{-1}(x)$. It consists of all ordered pairs (b, a) where (a, b) belongs to $f(x)$.

EXAMPLE 4

Find the inverse function for $f(x) = 4x + 2$.

SOLUTION

We know that $f^{-1}(x)$ exists because the graph of $y = 4x + 2$ passes both the vertical and horizontal line tests. To find $f^{-1}(x)$, we interchange the x and y variables to obtain

$$x = 4y + 2 \qquad \text{This equation is the inverse of } y = 4x + 2.$$

Solve this for y so that the $f^{-1}(x)$ notation may be used:

$$x = 4y + 2$$
$$x - 2 = 4y \qquad \text{Subtract 2 from both sides of the equation.}$$
$$\frac{x - 2}{4} = y \qquad \text{Divide both sides by 4.}$$
$$\frac{1}{4}x - \frac{1}{2} = y$$
$$y = \frac{1}{4}x - \frac{1}{2}$$

Thus, the function $f(x) = 4x + 2$ has $f^{-1}(x) = \frac{1}{4}x - \frac{1}{2}$ as its inverse.

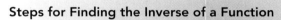

Steps for Finding the Inverse of a Function

1. If the function is given as a set of ordered pairs, interchange the x- and y-values to obtain the inverse function.
2. If the function is given as an equation:
   a. Interchange the variables x and y.
   b. Solve the resulting equation for y, if possible.
   c. This equation is $y = f^{-1}(x)$, which defines the inverse function.

To reinforce the concept stated in the definition that if the ordered pair (a, b) belongs to $f(x)$ then the ordered pair (b, a) belongs to $f^{-1}(x)$, do the following, using the function given in Example 4, $f(x) = 4x + 2$:
   Find $y = f(3)$.

$$y = 4(3) + 2$$
$$y = 14$$

Thus, $(3, 14)$ belongs to $f(x)$.
   Now find $y = f^{-1}(14)$.

$$y = \frac{1}{4}(14) - \frac{1}{2}$$
$$y = \frac{7}{2} - \frac{1}{2}$$
$$y = 3$$

Thus, $(14, 3)$ belongs to $f^{-1}(x)$.

EXPLORING
THE CONCEPT

We know that inverse functions "undo" each other. We now want to explore what occurs when we take the composition of a function and its inverse.

$$\text{Let } f(x) = 4x + 2 \qquad \text{and} \qquad f^{-1}(x) = \frac{1}{4}x - \frac{1}{2}$$

a. Find $(f \circ f^{-1})(3)$ **b.** Find $(f^{-1} \circ f)(3)$

c. Find $(f \circ f^{-1})(x)$ **d.** Find $(f^{-1} \circ f)(x)$

e. What conclusions can you draw about $f \circ f^{-1}$ and $f^{-1} \circ f$?

f. Given $g(x)$ and $g^{-1}(x)$, what would be the value of $g[g^{-1}(4)]$?

CONCLUSIONS

a. $(f \circ f^{-1})(3) = f[f^{-1}(3)]$

$$= f\left[\frac{1}{4} \cdot 3 - \frac{1}{2}\right]$$

$$= f\left[\frac{1}{4}\right]$$

$$= 4\left(\frac{1}{4}\right) + 2$$

$$= 1 + 2$$

$$= 3$$

$$(f \circ f^{-1})(3) = 3$$

b. $(f^{-1} \circ f)(3) = f^{-1}[f(3)]$

$$= f^{-1}[4 \cdot 3 + 2]$$

$$= f^{-1}[14]$$

$$= \frac{1}{4}(14) - \frac{1}{2}$$

$$= 3$$

$$(f^{-1} \circ f)(3) = 3$$

c. $(f \circ f^{-1})(x) = f[f^{-1}(x)]$

$$= f\left[\frac{1}{4} \cdot x - \frac{1}{2}\right]$$

$$= 4\left(\frac{1}{4}x - \frac{1}{2}\right) + 2$$

$$= x - 2 + 2$$

$$= x$$

$$(f \circ f^{-1})(x) = x$$

d. $(f^{-1} \circ f)(x) = f^{-1}[f(x)]$

$$= f^{-1}[4x + 2]$$

$$= \frac{1}{4}(4x + 2) - \frac{1}{2}$$

$$= x + \frac{1}{2} - \frac{1}{2}$$

$$= x$$

$$(f^{-1} \circ f)(x) = x$$

e. Conclusions: Composition of a function and its inverse will always yield x, the identity function. Thus, composition of a function and its inverse is commutative. (This is the **only** condition under which composition is commutative.) To show that two functions are inverses of one another, you must show that $f[f^{-1}(x)] = f^{-1}[f(x)] = x$.

f. $g[g^{-1}(4)] = 4$ since $g[g^{-1}(x)] = x$.

EXAMPLE 5

If $f(x) = \dfrac{1}{3}x + 8$, find $f^{-1}(x)$ and graph both functions and $y = x$ in the ZInteger viewing window.

SOLUTION

$$f(x) = \frac{1}{3}x + 8$$

$$y = \frac{1}{3}x + 8$$

$$x = \frac{1}{3}y + 8 \qquad \text{Interchange } x \text{ and } y.$$

$$x - 8 = \frac{1}{3}y \qquad \text{Subtract 8 from both sides.}$$

$$3x - 24 = y \qquad \text{Multiply both sides by 3.}$$

$$f^{-1}(x) = 3x - 24$$

The graph of any function, $f(x)$, and its inverse, $f^{-1}(x)$, will be symmetric about the line $y = x$. This is always true because we defined inverse functions as one-to-one functions such that if (a, b) belongs to the function then (b, a) belongs to the inverse, and the points (a, b) and (b, a) are symmetric with respect to the line $y = x$.

TECHNOLOGY TIP

Any "square" viewing window is acceptable. The pictured graph is in ZInteger.

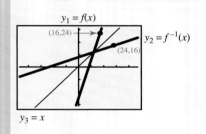

$y_1 = f(x)$
(16,24)
$y_2 = f^{-1}(x)$
(24,16)
$y_3 = x$

EXERCISE 11.2

VOCABULARY AND NOTATION *Fill in each blank to make a true statement.*

1. If a relation contains a point (a, b), then the _____ will contain the point (b, a).

2. A function that is _____ is a function in which each y-value in the range is paired with a unique x-value in the domain.

3. The _____ test is used to determine if each y-value in the range is paired with a unique x-value in the domain.

4. The notation _____ is used to denote the inverse of the function $f(x)$.

5. The domain of R^{-1} is the _____ of R, and the range of R^{-1} is the _____ of R.

6. The graphs of a function and its inverse are symmetrical about the line _____.

CONCEPTS

7. True or false: Every relation has an inverse. Justify your response.

8. True or false: Every function has an inverse function. Justify your response.

In Exercises 9–12, determine the inverse relation of each set of ordered pairs (x, y).

9. $\{(4, 5), (2, 8), (-3, 7), (6, -9)\}$

10. $\{(1, 2), (2, 3), (1, 3), (1, 5)\}$

11. $\{(-4, 2), (-3, 2), (0, 2), (1, 2)\}$

12. $\{(1, 4), (1, 5), (1, 6), (1, 7)\}$

13. The following relation defines the conversion of kilograms to pounds. Find the inverse relation that defines the conversion of pounds to kilograms.

Kilograms	1	2	3	3.5	4
Pounds	2.2	4.4	6.6	7.7	8.8

14. The following relation defines the conversion of degrees in Fahrenheit to degrees in Celsius. Find the inverse relation that defines the conversion of degrees in Celsius to degrees in Fahrenheit.

Fahrenheit	23	32	41	50	59	68	77
Celsius	-5	0	5	10	15	20	25

PRACTICE   *In Exercises 15–26, determine which graphs represent one-to-one functions.*

15.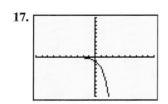

16.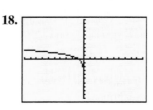

17.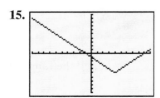

18.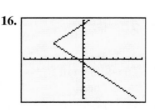

19.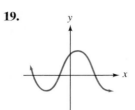

20.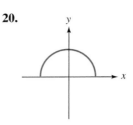

21.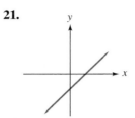

22.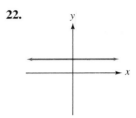

23.

x	4	5	6	1
y	2	-1	3	-1

24.

x	2	3	4	5
y	4	4	4	4

25.

x	1	3	5	7
y	2	4	6	8

26.

x	-2	1	3	-4
y	1	-2	-4	5

In Exercises 27–32, graph each of the given functions on your graphing calculator. From the graph, determine if the function is one-to-one.

27. $f(x) = 3x + 2$ **28.** $f(x) = 2x^2 + 3x - 5$

29. $f(x) = 3|x + 2| - 4$ **30.** $f(x) = x^3 - 4$

31. $f(x) = 3\sqrt{x - 2}$ **32.** $f(x) = 2^x$

In Exercises 33–40, find the inverse of the relation determined by the given equation and tell whether that inverse relation is a function. If the inverse relation is a function, express it in the form $y = f^{-1}(x)$.

33. $y = 3x + 1$

34. $y + 1 = 5x$

35. $x + 4 = 5y$

36. $x = 3y + 1$

37. $y = \dfrac{x - 4}{5}$

38. $y = \dfrac{2x + 6}{3}$

39. $4x - 5y = 20$

40. $3x + 5y = 15$

In Exercises 41–48, find the inverse of each linear function. Graph both the function and its inverse on graph paper on the same set of axes, as well as the axis of symmetry, $y = x$.

41. $y = 4x + 3$ **42.** $x = 3y - 1$

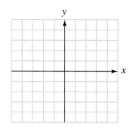   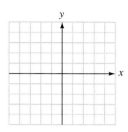

43. $x = \dfrac{y - 2}{3}$ **44.** $y = \dfrac{x + 3}{4}$

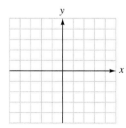   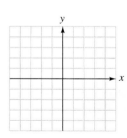

45. $3x - y = 5$ **46.** $2x + 3y = 9$

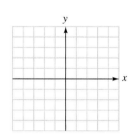   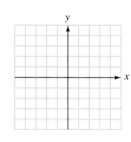

47. $3(x + y) = 2x + 4$

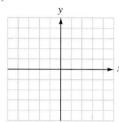

48. $-4(y - 1) + x = 2$

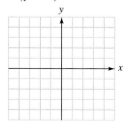

In Exercises 49–56, graph each function, $f(x)$ and its inverse $f^{-1}(x)$ in the ZInteger viewing window. Then graph $f(f^{-1}(x))$. (Hint: Graph the composition $y1(y2)$ where $y1 = f(x)$ and $y2 = f^{-1}(x)$). Compare your calculator display to your hand-drawn graphs in Exercises 41–48.

49. $y = 4x + 3$ **50.** $x = 3y - 1$

51. $x = \dfrac{y - 2}{3}$ **52.** $y = \dfrac{x + 3}{4}$

53. $3x - y = 5$ **54.** $2x + 3y = 9$

55. $3(x + y) = 2x + 4$ **56.** $-4(y - 1) + x = 2$

In Exercises 57–60, show that each of the pairs of equations are inverse functions. For each pair, $f(f^{-1}(x)) = x$.

57. $y = 2x + 6$
$\quad y = \dfrac{1}{2}x - 3$

58. $y = \dfrac{2}{3}x + 1$
$\quad y = \dfrac{3x - 3}{2}$

59. $y = \dfrac{x + 5}{3}$
$\quad y = 3x - 5$

60. $y = x - 7$
$\quad y = x + 7$

In Exercises 61–70, find the inverse of the relation determined by each equation. Tell whether the inverse relation is a function. Verify your inverse by both graphing the inverse on graph paper and drawing the inverse (via the DrawInv feature) on your calculator.

61. $y = x^2 + 4$

62. $x = y^2 - 2$

63. $x = y^2 - 4$

64. $y = x^2 + 5$

65. $y = x^3$

66. $xy = 4$

67. $y = \pm\sqrt{x}$

68. $y = \sqrt[3]{x}$

69. $y = \sqrt{x}$

70. $4y^2 = x - 3$

REVIEW

71. Evaluate:

   a. 4^3 **b.** -4^4 **c.** 4^{-4}

   d. $4^{1/2}$ **e.** $-4^{1/2}$ **f.** $(-4)^{-1/2}$

72. Express in simplest radical form:

   a. $a^{2/5}$ **b.** $4^{-2/5}$

   c. $8a^{1/3}$ **d.** $(8a)^{1/3}$

73. Graph each of the following and determine their domains and ranges.

   a. $f(x) = x^2$ **b.** $f(x) = x^3$

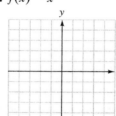 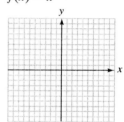

   c. $f(x) = \sqrt{x}$

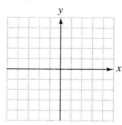

74. Write the equation of a function that has the same size/shape as $f(x) = x^2$ but is reflected in the x-axis, translated 3 units left and 4 units up.

11.3 EXPONENTIAL FUNCTIONS

- BASIC CONCEPTS OF EXPONENTS
- GRAPHS OF EXPONENTIAL FUNCTIONS
- SOLVING EXPONENTIAL EQUATIONS
- APPLICATIONS OF EXPONENTIAL FUNCTIONS
- BASE-e EXPONENTIAL FUNCTIONS

▮ BASIC CONCEPTS OF EXPONENTS

Natural-number exponents were defined in Section 1.2 and rational exponents were defined in Section 9.2. From these definitions we note that

$$x^5 = x \cdot x \cdot x \cdot x \cdot x$$

and

$$x^{2/5} = \sqrt[5]{x^2}$$

We now want to extend our definition of exponents to include all real-number values. It should be noted that all exponent rules previously addressed for natural- and rational-number exponents are also valid for all real-number exponents.

Consider $3^{\sqrt{2}}$. Because $\sqrt{2}$ is approximately 1.414213562, we can approximate $3^{\sqrt{2}}$ by expressing it as $3^{1.4}$, $3^{1.41}$, $3^{1.414}$, etc., depending on the desired degree of accuracy.

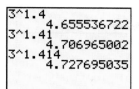   vs.

This can also be supported using a graph and letting $y = 3^x$:

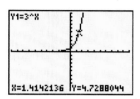

TECHNOLOGY TIP

The VALUE feature under the CALC menu was used. Let $x = \sqrt{2}$ at the prompt. This graph is displayed in the standard viewing window.

EXAMPLE 1

Evaluate the following powers in one of two ways: using the laws of exponents when possible or using the exponentiation power of the calculator. Then, verify your results graphically.

a. $4^{1/2}$ **b.** $8^{2/3}$ **c.** $5^{\sqrt{2}}$ **d.** $0.5^{-\sqrt{2}}$

SOLUTION   **a.** $4^{1/2} = \sqrt{4} = 2$

Definition of $a^{1/n} = \sqrt[n]{a}$ (Section 9.2)

When $y = 4^x$ and $x = \frac{1}{2}$, then $y = 2$.

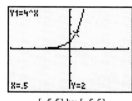

[–5,5] by [–5,5]

b. $8^{2/3} = (\sqrt[3]{8})^2 = (2)^2 = 4$

Definition of $a^{m/n} = (\sqrt[n]{a})^m$ (Section 9.2)

When $y = 8^x$ and $x = \frac{2}{3}$, then $y = 4$.

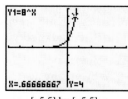

[–5,5] by [–5,5]

c.

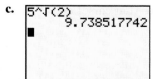

When $y = 5^x$ and $x = \sqrt{2}$, then $y \approx 9.738517$.

[–5,5] by [–5,10]

d.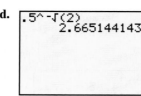
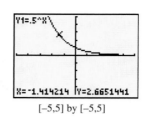

[–5,5] by [–5,5]

When $y = 0.5^x$ and $x = -\sqrt{2}$, then $y \approx 2.665144$.

█ GRAPHS OF EXPONENTIAL FUNCTIONS

Functions with variables as exponents are called **exponential functions**.

Exponential Function

A function of the form $f(x) = b^x$, where $b > 0$, $b \neq 1$, and x is a real number is called an **exponential function** with base b.

You should note that the base, b, of an exponential function cannot be 1 because $f(x) = 1^x = 1$ for all values of x. Thus, if $b = 1$, we have a constant function, not an exponential function. Be careful not to confuse functions with real number powers with exponential functions.

Exponential Function	*Not an Exponential Function*
$f(x) = 2^x$	$f(x) = x^2$
$f(x) = 3^x$	$f(x) = x^3$
$f(x) = \left(\dfrac{1}{2}\right)^x$	$f(x) = x^{1/2}$

EXPLORING THE CONCEPT

Set the viewing window of your calculator to $[-4.7, 4.7]$ by $[-10, 10]$. TI-86: $[-6.3, 6.3]$ by $[-10, 10]$. Examine each of the exponential functions where the base b is greater than 1.

$$f(x) = 2^x \qquad f(x) = 3^x \qquad f(x) = 8^x$$

a. Begin with your cursor at $x = -3$ and TRACE along the graph of each function. As x increases, what happens to y? Verify your conclusion by scrolling through the TABLE and observing the behavior of the y-values.

b. TRACE (or scroll through the TABLE) to determine if the y-value can ever be equal to 0. Why or why not?

c. What is the domain and the range of each function?

d. Compare the three graphs. How are they similar? How are they different? (*Hint:* Place your cursor at $(0, 0)$ and Zoom In once.)

CONCLUSIONS

a. As x increases, so does y; however, y increases very rapidly as compared to x.

b. The y-value will never equal zero because $y = b^x (b \neq 0)$, and a nonzero base raised to any power is a nonzero number.

c. Domain = $(-\infty, \infty)$ and range = $(0, \infty)$.

d. All graphs represent increasing functions. All graphs have the same basic shape and they all pass through $(0, 1)$. All graphs $y = b^x$ contain the point $(1, b)$. For $x > 0$, the graph rises faster for larger bases.

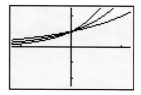

Because the graphs approach the x-axis but never cross it (i.e., $y \neq 0$), we say that the x-axis is the horizontal asymptote of each function. Thus, the equation of the asymptote is $y = 0$.

We can also observe from these graphs that the y-intercept is $(0, 1)$. This is true for all exponential functions $f(x) = b^x$, since

$f(x) = b^0$ Let $x = 0$ and find the y-intercept.

$f(x) = 1$ $b^0 = 1$ since $b \neq 0$.

Exponential functions of the form $f(x) = b^x$ where $b > 1$ are called **exponential growth** functions because y increases rapidly as x increases.

EXAMPLE 2

Graph $y = 4^x$ on graph paper and verify with your calculator.

SOLUTION   Construct a table of values based on the y-intercept of $(0, 1)$.

x	y
-1	$\frac{1}{4}$
$-\frac{1}{2}$	$\frac{1}{2}$
0	1
$\frac{1}{2}$	2
1	4

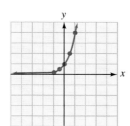

Note: Values of $\pm\frac{1}{2}$ were selected because we can compute exact values of $4^{1/2}$ and $4^{-1/2}$. Integer values could have been selected instead, but the y-axis would need to be increased vertically.

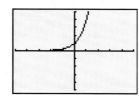

EXPLORING THE CONCEPT

Set your viewing window to $[-4.7, 4.7]$ by $[-10, 10]$. TI-86: $[-6.3, 6.3]$ by $[-10, 10]$. Examine each of the exponential functions where the base b is between 0 and 1.

$$f(x) = \left(\frac{1}{10}\right)^x \qquad f(x) = \left(\frac{1}{2}\right)^x \qquad f(x) = \left(\frac{4}{5}\right)^x$$

a. Begin with your cursor at the leftmost visible point on the graph and TRACE along each function. As x increases, what happens to y? Verify your conclusions by scrolling through the TABLE and observing the behavior of the y-values.

b. TRACE (or scroll through the TABLE) to determine if the y-value will ever be 0. Is the line $y = 0$ the horizontal asymptote?

c. What is the domain and range of each function?

d. Compare the three graphs. How are they similar? How are they different? (*Hint:* Place your cursor at $(0, 0)$ and Zoom In once).

CONCLUSIONS

a. As x increases, y decreases.

b. Because y never equals 0 (a nonzero base raised to a nonzero power will never be zero), the line $y = 0$ is the horizontal asymptote.

c. Domain $= (-\infty, \infty)$ and range $= (0, \infty)$.

d. All graphs represent decreasing functions. They all have the same basic shape and pass through $(0, 1)$. All graphs $y = b^x$ contain $(1, b)$. The closer the b value is to zero, the faster the graph decreases.

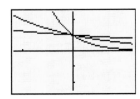

What we've discovered about exponential functions and their graphs is summarized below.

Exponential Functions: $f(x) = b^x$, $b \neq 1$

1. There is no x-intercept and the y-intercept has coordinates $(0, 1)$.
2. The domain is the set of all real numbers, denoted $(-\infty, \infty)$ or $\mathbb{R}$.
3. The range is $(0, \infty)$.
4. The graph approaches the x-axis, but does not touch it nor does it cross it.
5. The function is one-to-one.
6. For $f(x) = b^x$, $b > 1$, the function is increasing.
7. For $f(x) = b^x$, $0 < b < 1$, the function is decreasing.

EXAMPLE 3

Graph $y = \left(\dfrac{1}{4}\right)^x$ on graph paper and verify with your calculator.

SOLUTION   Construct a table of values based on a y-intercept of $(0, 1)$.

x	y
-1	4
$-\frac{1}{2}$	2
0	1
$\frac{1}{2}$	$\frac{1}{2}$
1	$\frac{1}{4}$
2	$\frac{1}{16}$

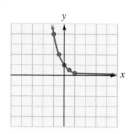

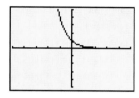 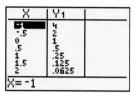

Exponential functions of the form $f(x) = b^x$, where $0 < b < 1$, are called **exponential decay** functions because y decreases as x increases.

▍ SOLVING EXPONENTIAL EQUATIONS

To solve exponential equations, we use the exponential property of equality.

Exponential Property of Equality

If $b > 0$ and $b \neq 1$, then $b^r = b^s$ is equivalent to $r = s$.
If $r \neq s$, then $b^r \neq b^s$.

EXAMPLE 4

Solve each of the following:

a. $2^x = 4$ **b.** $3^x = 3^{4x+5}$ **c.** $\left(\dfrac{1}{16}\right)^x = 4^{3x-1}$

SOLUTION   **a.** $2^x = 4$

$2^x = 2^2$ Rewrite the equation so that both sides have the same base.

$x = 2$ Based on the property, if $b^r = b^s$, then $r = s$.

This can be verified most efficiently by substitution.

b. $3^x = 3^{4x+5}$

$x = 4x + 5$ If $b^r = b^s$, then $r = s$.

$x = -\dfrac{5}{3}$

Verify graphically:

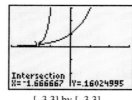

$y_1 = 3^x$

$y_2 = 3^{(4x + 5)}$

[–3,3] by [–3,3]

c. $\left(\dfrac{1}{16}\right)^x = 4^{3x-1}$

$\left(\dfrac{1}{4^2}\right)^x = 4^{3x-1}$ Write each base as a power of four.

$(4^{-2})^x = 4^{3x-1}$ $\dfrac{1}{a^m} = a^{-m}$ for $m > 0$.

$4^{-2x} = 4^{3x-1}$ $(a^m)^n = a^{mn}$

$-2x = 3x - 1$ If $b^r = b^s$, then $r = s$.

$-5x = -1$

$x = \dfrac{1}{5}$

Verify graphically:

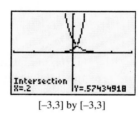

$y_1 = (1/16)^x$

$y_2 = 4^{(3x - 1)}$

[–3,3] by [–3,3]

APPLICATIONS OF EXPONENTIAL FUNCTIONS

A mathematical description of an observed event is called a **model** of that event. Many events that change with time are modeled by functions defined by equations of the form

$$y = f(t) = ab^{kt}$$ Remember that ab^{kt} means $a(b^{kt})$.

where a, b, and k are constants and t represents time. If the function is an increasing function, then y is said to **grow exponentially**. If the function is a decreasing function, then y is said to **decay exponentially**.

One example of exponential growth is **compound interest**. When interest is allowed to accumulate in a savings account, the balance will grow exponentially according to the following model.

Formula for Compound Interest

If the present value P is deposited in an account, and interest is paid k times a year at an annual interest rate r, the amount of money, A, in the account after t years is given by $A = P\left(1 + \dfrac{r}{k}\right)^{kt}$.

EXAMPLE 5

Saving for college   To save for college, parents invest $12,000 for their newborn child in a mutual fund that should average 4% annual interest. If the quarterly dividends are reinvested, how much will be available in 18 years?

SOLUTION   We substitute 12,000 for P, 0.04 for r, and 18 and t into the formula for compound interest and find A. Since interest is paid quarterly, $k = 4$.

$$A = P\left(1 + \frac{r}{k}\right)^{kt}$$

$$A = 12{,}000\left(1 + \frac{0.04}{4}\right)^{4(18)}$$

$$= 12{,}000(1 + 0.1)^{72}$$

$$= 12{,}000(1.1)^{72}$$

$$= 24{,}565.19$$

```
12000(1+0.04/4)^
(4*18)
        24565.19175
```

In 18 years, the account will contain more than $24,500.

▮ BASE-e EXPONENTIAL FUNCTIONS

When banks compound interest quarterly, the account balance increases four times a year. In other situations, quantities grow continuously. For example, it is possible for a bank to compound money continuously, and the population growth of a town increases steadily (not in abrupt jumps at every quarter).

We use the irrational number e, named after the Swiss mathematician Leonhard Euler, to predict the future value of a quantity that increases or decreases continuously. The value of e can be best approximated by the expression $\left(1 + \frac{1}{x}\right)^{x}$. As x gets increasingly larger, $\left(1 + \frac{1}{x}\right)^{x}$ gets closer and closer to the value of e, which is approximately 2.71828182846.

If we compound interest continuously using the formula $A = P\left(1 + \frac{r}{k}\right)^{kt}$, it can be shown that $\left(1 + \frac{r}{k}\right)^{kt}$ approaches the value of e^{rt} (mathematical justification of this will not be done until a later course in algebra). Thus, $A = P\left(1 + \frac{r}{k}\right)^{kt}$, which represents a finite number of compounding periods, can be rewritten as

$$A = Pe^{rt}$$

where compounding occurs continuously.

Formula for Continuous Compounding

$A = Pe^{rt}$, where P is the present value of the quantity, A is the future value, r is the rate of interest, and t is the time in years.

Previously in Example 5, we determined that $24,565.19 would be in a college fund if $12,000 were compounded quarterly for 18 years at a 4% interest rate. We will now see how the future value of the $12,000 is affected if we compound continuously.

EXAMPLE 6

Saving for college   Compute the future value of $12,000 invested for 18 years at 4% compounded continuously.

SOLUTION

$$A = Pe^{rt}$$
$$A = 12{,}000\,e^{.04(18)}$$
$$A = \$24{,}653.20$$

```
12000→P:.04→R:18
→T:Pe^(RT)
          24653.19853
```

After 18 years, the account will contain $24,653.20. This is $88.01 more than the result in Example 5, which was compounded quarterly.

Note: TI-86: You must include the operation of multiplication between the R and T in the expression (RT).

In the **Malthusian model for population growth**, future population is related to the present population by the formula for exponential growth. This is the same formula as was used for continuous compounding. However, P represents the current population and A is the future population.

EXAMPLE 7

City planning   The population of a city is currently 15,000, but changing economic conditions are causing the population to decrease 2% each year. If this trend continues by the Malthusian model, find the population in 30 years.

SOLUTION

Since the population is decreasing 2% each year, the annual growth rate is -2%, or -0.02. We can substitute -0.02 for r, 30 for t, and 15,000 for P in the formula for exponential growth and find A.

$$A = Pe^{rt}$$
$$A = 15{,}000\,e^{(-.02)(30)}$$
$$A = 15{,}000\,e^{(-0.6)}$$
$$A = 8232.17$$

```
15000→P:-.02→R:3
0→T:Pe^(RT)
          8232.174541
```

In 30 years, city planners expect a population of approximately 8232 persons.

EXERCISE 11.3

VOCABULARY AND NOTATION *Fill in each blank to make a true statement.*

1. Functions with variables as exponents are called _____ functions.

2. An exponential _____ function is in the form $f(x) = b^x$, where $b > 1$.

3. An exponential _____ function is in the form $f(x) = b^x$, where $0 < b < 1$.

4. The x-axis forms the horizontal _____ for all exponential functions because exponential functions never have a y-value of 0.

CONCEPTS

5. Evaluate each of the following with the laws of exponents when possible or by the calculator.

   a. $25^{1/2}$

   b. $16^{3/4}$

   c. $25^{-\sqrt{2}}$

   d. $\left(\dfrac{1}{3}\right)^{\sqrt{3}}$

6. Several exponential expressions have been simplified using the calculator and then verified graphically. Match the correct graphical display to its corresponding simplification.

   a.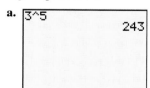

   i.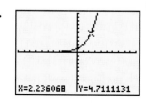

   b.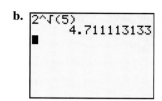

   ii.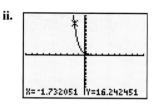

   c.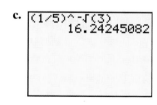

   iii.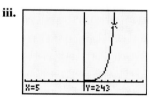

7. Classify each function as "exponential function" or "not an exponential function."

   a. $f(x) = x^4$

   b. $f(x) = 4^x$

   c. $f(x) = \left(\dfrac{1}{2}\right)^x$

   d. $f(x) = \dfrac{1}{2}x^2$

8. Label each graph as an increasing or decreasing function and as exponential growth or exponential decay.

   a.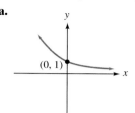

   b.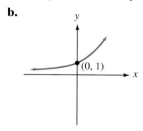

9. Under what conditions will an exponential function represent exponential decay?

10. Under what conditions will an exponential function represent exponential growth?

11. Explain why all exponential functions will have a y-intercept of $(0, 1)$.

12. The irrational number e has an approximate value of 2.71828182846, it can be best approximated by the expression $\left(1 + \dfrac{1}{x}\right)^x$. Fill in the table below and watch the expression $\left(1 + \dfrac{1}{x}\right)^x$ grow increasingly closer to the value of e as x grows larger.

x	$\left(1 + \dfrac{1}{x}\right)^x$
1	
10	
100	
1,000	
10,000	
100,000	
1,000,000	

13. Solve each of the following equations by rewriting each side of the equation as an exponential expression with a common base.

a. $2^x = 16$ **b.** $3^x = 243$

14. Solve each of the following equations by rewriting each side of the equation as an exponential expression with a common base.

a. $4^x = 1024$ **b.** $27^x = 81$

PRACTICE   *In Exercises 15–22, graph each exponential function on graph paper. Verify your graphs with your graphing calculator.*

15. $f(x) = 3^x$ **16.** $f(x) = 4^x$

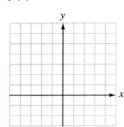   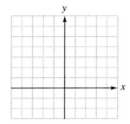

17. $f(x) = 5^x$ **18.** $f(x) = 6^x$

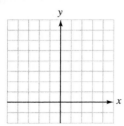   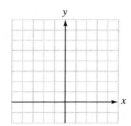

19. $f(x) = \left(\dfrac{1}{3}\right)^x$ **20.** $f(x) = \left(\dfrac{1}{4}\right)^x$

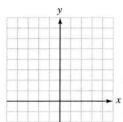   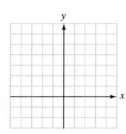

21. $f(x) = \left(\dfrac{1}{5}\right)^x$ **22.** $f(x) = \left(\dfrac{1}{6}\right)^x$

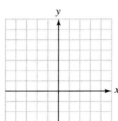   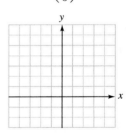

In Exercises 23–26, refer to the following graphs of the functions and graph each of the given functions on graph paper. The pictured functions are graphed in the window $[-5, 10]$ by $[-2, 10]$.

$f(x) = 2^x$ $f(x) = 3^x$

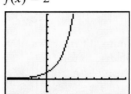   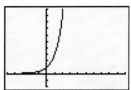

23. Graph each function below on graph paper and verify with your calculator.

a. $f(x) = 2^x + 2$ **b.** $f(x) = 3^x - 2$

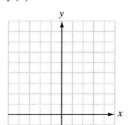   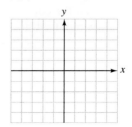

c. In each part of **a.** and **b.**, a constant was either added to or subtracted from the basic exponential function of $f(x) = b^x$. What effect did the constant have on the graph of the function?

24. Graph each function below on graph paper and verify with your calculator.

a. $f(x) = 3(2^x)$ **b.** $f(x) = 2(3^x)$

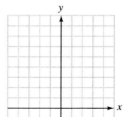   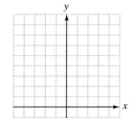

 **c.** In parts **a.** and **b.**, the basic function was multiplied by an integer value. What effect did this have on the graph of the function?

25. Graph each function on graph paper and verify with your calculator.

a. $f(x) = 2^{-x}$

b. $f(x) = 3^{-x}$

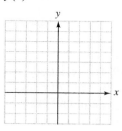

 **c.** What effect does the negative exponent have on the graph of the function? Also explain why the function changes from one of exponential growth, in Exercise 24, to exponential decay in Exercise 25.

26. Graph each of the functions and verify with your calculator.

a. $f(x) = 2^{x+5}$

b. $y = 3^{x-4}$

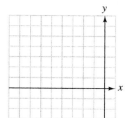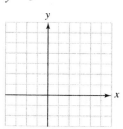

 **c.** How is the graph affected by adding/subtracting a constant to the exponent?

In Exercises 27–34, let $f(x) = 2^{2x+1}$ and $g(x) = \left(\frac{1}{2}\right)^{3x}$. Evaluate each of the following:

27. $f(4)$

28. $f(5)$

29. $g(-2)$

30. $g(-3)$

31. $f(-1)$

32. $f(-2)$

33. $g(1)$

34. $g(-1)$

In Exercises 35–54, solve each equation algebraically and verify with the graphing calculator.

35. $2^3 = 2^{5x+7}$

36. $3^5 = 3^{4-2x}$

37. $9 = 3^{1/2x+5}$

38. $16 = 4^{3x-2}$

39. $32 = 4^{6x-5}$

40. $27 = 9^{2x+4}$

41. $4^{5x-1} = 8^{3x+2}$

42. $125^{x-3} = 25^{2x+7}$

43. $125^{6-2x} = 625^{3x+1}$

44. $81^{4x-1} = 27^{6x-2}$

45. $5^{-x} = 125$

46. $4^{-x} = 64$

47. $e^{2x} = e^{x+1}$

48. $e^{x-2} = e^{3x}$

49. $\dfrac{3^{x+1}}{3^{1-x}} = \dfrac{1}{9}$

50. $\dfrac{4^{x+1}}{4^{1-x}} = \dfrac{1}{16}$

51. $\dfrac{3^{2x+2}}{27^x} = \dfrac{1}{9^x}$

52. $\dfrac{2^{3x-2}}{16^x} = \dfrac{1}{4^x}$

53. $3^{x-1} \cdot 9^{x-4} = 27$

54. $2^{x-2} \cdot 4^{x-1} = 8$

APPLICATIONS In Exercises 55–60, assume there are no deposits or withdrawals.

55. Compound interest An initial deposit of $10,000 earns 2% interest, compounded quarterly. How much will be in the account after 10 years?

56. Compound interest An initial deposit of $10,000 earns 2% interest, compounded monthly. How much will be in the account after 10 years?

57. Comparing savings plans How much more interest could $1,000 earn in 5 years, compounded quarterly, if the annual interest rate were $3\frac{1}{2}\%$ instead of 3%?

58. Comparing savings plans Which institution in Illustration 1 provides the better investment?

Fidelity Savings & Loan

earn 2.25%
compounded monthly

Union Trust

Money Market account
paying 2.35%
compounded annually

Illustration 1

59. Compound interest If $1 had been invested on July 4, 1776 at 5% interest, compounded annually, what would it be worth on July 4, 2076?

60. Frequency of compounding $10,000 is invested in each of two accounts, both paying 3% annual interest. In the first account, interest compounds quarterly; and in the second account, interest compounds daily. Find the difference between the accounts after 20 years.

61. Fish population A population of fish is growing according to the Malthusian model. How many fish will there be in 10 years if the annual growth rate is 3% and the initial population is 2,700 fish?

62. Town population The population of a town is 1,350. The town is expected to grow according to the Malthusian model, with an annual growth rate of 6%. Find the population of the town in 20 years.

63. World population The population of the world is approximately 5 billion. If the population is growing according to the Malthusian model, with an annual growth rate of 1.8%, what will be the population of the world in 30 years?

64. World population The population of the world is approximately 5 billion. If the population is growing according to the Malthusian model, with an annual growth rate of 1.8%, what will be the population of the world in 60 years?

65. Bacteria cultures A colony of 6 million bacteria is growing in a culture medium (see Illustration 2). The population P after t hours is given by the formula $P = (6 \times 10^6)(2.3)^t$. Find the population after 4 hours.

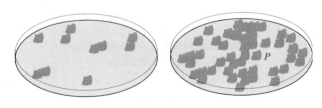

12:00 PM	4:00 PM

Illustration 2

66. Discharging a battery The charge remaining in a battery decreases as the battery discharges. The charge C (coulombs) after t days is given by the formula $C = (3 \times 10^{-4})(0.7)^t$. Find the charge after 5 days.

67. Radioactive decay A radioactive material decays according to the formula $A = A_0 \left(\dfrac{2}{3}\right)^t$, where A_0 is the initial amount present and t is measured in years. Find the amount present in 5 years.

68. Town population The population of North Rivers is decreasing exponentially according to the formula $P = 3,745(0.93)^t$, where t is measured in years from the present date. Find the population in 6 years, 9 months.

69. Depreciation A camping trailer originally purchased for $4,570 is continuously losing value at the rate of 6% per year. Find its value when it is $6\frac{1}{2}$ years old.

70. Depreciation A boat purchased for $7,500 has been continuously decreasing in value at the rate of 2% each year. It is now 8 years, 3 months old. Find its value.

REVIEW

71. Evaluate:
   a. 7^0 **b.** 25^2 **c.** 25^{-2}
   d. $25^{1/2}$ **e.** $25^{-1/2}$

72. Simplify:
   a. $\sqrt{240x^5}$ **b.** $(\sqrt{3x + 1})^2$

73. Solve: $\sqrt{3x + 1} + 1 = x$.

74. Solve: $\dfrac{24}{x} + \dfrac{12}{x + 1} = 11$.

75. Graphically determine the maximum height of a ball thrown straight up from the top of a building with an initial velocity of 32 feet per second if the equation $h = -16t^2 + 32t + 48$ models the situation ($t = $ time, $h = $ height). How long does it take the ball to reach the maximum height?

11.4 LOGARITHMIC FUNCTIONS

- BASIC CONCEPTS OF EXPONENTS AND LOGARITHMS
- CHANGE-OF-BASE FORMULA
- GRAPHS OF LOGARITHMIC FUNCTIONS
- APPLICATIONS OF LOGARITHMS

BASIC CONCEPTS OF EXPONENTS AND LOGARITHMS

When we studied the laws of exponents, we referred to expressions such as 4^5 as exponential expressions. Five is the exponent to which the base of 4 must be raised to get 1024:

$$4^5 = 1024$$

Five is also called the **logarithm** to the base 4 of 1024 and is written as

$$\log_4 1024 = 5$$

Logarithm
If $b > 0$ and $b \neq 1$, then $b^y = x$ is equivalent to $\log_b x = y$.

Remember, a logarithm is just an exponent. The following comparisons can be made between logarithms and exponents:

Logarithmic Form	*Exponential Form*
$\log_7 1 = 0$	$7^0 = 1$
$\log_5 25 = 2$	$5^2 = 25$
$\log_5 \frac{1}{25} = -2$	$5^{-2} = \frac{1}{25}$
$\log_{16} 4 = \frac{1}{2}$	$16^{1/2} = 4$
$\log_{10} 1000 = 3$	$10^3 = 1000$
$\log_{10} \frac{1}{100} = -2$	$10^{-2} = \frac{1}{100}$
$\log_e 148.4131591 = 5$	$e^5 = 148.4131591$

WARNING!   It is impossible to find the logarithm of a negative number or zero. Note that because there is no real number x such that $2^x = -8$, there is no meaning to the expression $\log_2(-8) = x$.

Logarithms of base 10, such as $\log_{10} 1000 = 3$, are called **common logs**. These logarithms often appear in mathematics and science. The base 10 is understood and generally not written. Thus, $\log 1000 = \log_{10} 1000 = 3$.

Because the number e appears in mathematical models of natural events, base-e logarithms are called **natural logarithms**. Natural logarithms are usually denoted by the symbol $\ln x$ instead of $\log_e x$. Thus, $\log_e 148.4131591 = 5$ becomes $\ln 148.4131591 = 5$.

EXAMPLE 1

Find y in each equation.

a. $\log_3 81 = y$

b. $\log_5 \dfrac{1}{5} = y$

c. $\log 10{,}000 = y$

d. $\ln 5 = y$

SOLUTION **a.** Change the equation $\log_3 81 = y$ into the equivalent equation $3^y = 81$ and solve for y.

$$3^y = 81$$
$$3^y = 3^4$$
$$y = 4$$

b. $\log_5 \dfrac{1}{5} = y$ is equivalent to $5^y = \dfrac{1}{5}$.

$$5^y = \frac{1}{5}$$
$$5^y = 5^{-1}$$
$$y = -1$$

`log(10000)`
`                    4`

c. $\log 10{,}000 = y$ is a common logarithm of base 10 and is equivalent to $10^y = 10000$.

$$10^y = 10000$$
$$10^y = 10^4$$
$$y = 4$$

`ln(5)`
`         1.609437912`

d. Use your calculator to compute an approximation for ln 5. Be sure to use the natural log key, LN, and not the common log key, LOG.

EXAMPLE 2

Find the value of x in each equation.

a. $\log_x 81 = 4$ **b.** $\log_4 x = 3$ **c.** $\log x = -2$ **d.** $\log_{1/3} \frac{1}{27} = x$

SOLUTION **a.** $\log_x 81 = 4$ is equivalent to $x^4 = 81$.

$$x^4 = 81$$
$$x^4 = 3^4$$
$$x = 3$$

b. $\log_4 x = 3$ is equivalent to $4^3 = x$. Thus, $x = 64$.

c. $\log x = -2$ is a base 10 common logarithm and is equivalent to $10^{-2} = x$. Thus, $x = \frac{1}{100}$.

d. $\log_{1/3} \frac{1}{27} = x$ is equivalent to $\left(\frac{1}{3}\right)^x = \frac{1}{27}$.

$$\left(\frac{1}{3}\right)^x = \frac{1}{27}$$
$$(3^{-1})^x = 3^{-3}$$
$$3^{-x} = 3^{-3}$$
$$-x = -3$$
$$x = 3$$

▉ CHANGE-OF-BASE FORMULA

Both the LOG and LN keys on your calculator allow for easy calculator approximation of both common and natural logarithms. When the logarithm does not have a base of 10 (or e), we will need the **change-of-base formula** that allows the conversion of a logarithm of any base to a base 10 logarithm for calculator computation.

Change-of-Base Formula

If a, b, and x are positive real numbers, then $\log_b x = \dfrac{\log_a x}{\log_a b}$ $(a \neq 1, b \neq 1)$.

EXAMPLE 3

Use the change-of-base formula to perform the following computations on your calculator. When possible, verify analytically.

a. $\log_7 1 = y$ **b.** $\log_5 25 = y$ **c.** $\log_{16} 4 = y$ **d.** $\log_3 5 = y$

SOLUTION   **a.** $\log_7 1 = \dfrac{\log_{10} 1}{\log_{10} 7} = \dfrac{\log 1}{\log 7}$

```
log(1)/log(7)
              0
```

Thus, $\log_7 1 = 0$.

Analytic verification: $7^0 = 1$.

b. $\log_5 25 = \dfrac{\log 25}{\log 5}$

```
log(25)/log(5)
              2
```

Thus, $\log_5 25 = 2$.

Analytic verification: $5^2 = 25$.

c. $\log_{16} 4 = \dfrac{\log 4}{\log 16}$

```
log(4)/log(16)
             .5
```

Thus, $\log_{16} 4 = 0.5$.

Analytic verification: $16^{1/2} = 4$.

d. $\log_3 5 = \dfrac{\log 5}{\log 3}$

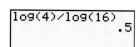

```
log(5)/log(3)
    1.464973521
```

Thus, $\log_3 5 \approx 1.464973521$.

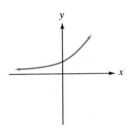

GRAPHS OF LOGARITHMIC FUNCTIONS

All of the exponential functions defined by $y = b^x$ in the previous section had the same basic shape for their graphs, as illustrated at the left. Because the graphs of exponential functions also pass the horizontal line test, we can conclude that exponential functions are one-to-one and thus have an inverse function. The inverse function is $x = b^y$, which was defined at the beginning of this section as being equivalent to $\log_b x = y$.

Before writing this inverse in $y = f^{-1}(x)$ form, we will construct a graph of $y = 2^x$ and its inverse, $x = 2^y$ (see Figure 11-1).

$y = 2^x$

x	y
-2	$\frac{1}{4}$
-1	$\frac{1}{2}$
0	1
1	2
2	4

Domain: $\mathbb{R}$
Range: $\{y \mid y > 0\}$, $(0, \infty)$

$x = 2^y$

x	y
$\frac{1}{4}$	-2
$\frac{1}{2}$	-1
1	0
2	1
4	2

Domain: $\{x \mid x > 0\}$,
$(0, \infty)$ Range: $\mathbb{R}$

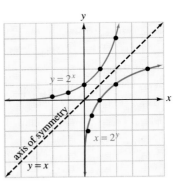

Figure 11-1

Based on the definition of logarithm, $2^y = x$ is equivalent to $\log_2 x = y$. Thus, the inverse of $f(x) = 2^x$ is $f^{-1}(x) = \log_2 x$.

To graph $y = \log_b x$, we find several points (x, y) that satisfy the equation $x = b^y$ and join them with a smooth curve.

EXAMPLE 4

Graph the functions   **a.** $y = \log_3 x$ and   **b.** $y = \log_{1/3} x$.

SOLUTION   **a.**   The equation $y = \log_3 x$ is equivalent to $3^y = x$. Select points (x, y) that satisfy this equation and plot a smooth curve. (Remember to use the change-of-base formula to verify with the calculator.)

x	y
$\frac{1}{9}$	-2
$\frac{1}{3}$	-1
1	0
3	1
9	2

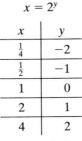

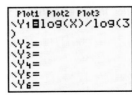

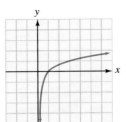

[−1,10] by [−3,3]

b. The equation $y = \log_{1/3} x$ is equivalent to $(\frac{1}{3})^y = x$. Select points (x, y) that satisfy this equation and plot a smooth curve. Verify with your calculator.

x	y
9	−2
3	−1
1	0
$\frac{1}{3}$	1
$\frac{1}{9}$	2

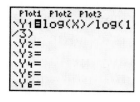

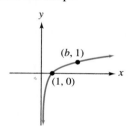

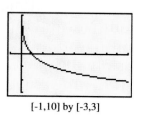

[-1,10] by [-3,3]

We can draw the following conclusions about graphs of logarithmic functions based on these two graphs.

1. If $b > 1$, then $f(x) = \log_b x$ has the shape

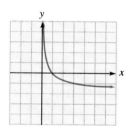

and passes through the points $(0, 1)$ and $(b, 1)$.

2. If $0 < b < 1$ then $f(x) = \log_b x$ has the shape

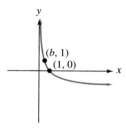

and passes through the points $(1, 0)$ and $(b, 1)$.

▎ APPLICATIONS OF LOGARITHMS

Chemistry: In chemistry, common logarithms are used to express the acidity of solutions. The more acidic a solution, the greater the concentration of hydrogen ions.

This concentration is indicated indirectly by the pH scale, or hydrogen ion index. The pH of a solution is defined by

$$pH = -\log [H^+],$$

where $[H^+]$ is the hydrogen ion concentration in gram-ions per liter.

EXAMPLE 5

Use your calculator to find the pH of pure water, which has a hydrogen ion concentration of 10^{-7} grams-ions per liter.

SOLUTION The hydrogen ion concentration of pure water is 10^{-7} grams-ions per liter, so its pH is

$$pH = -\log (10^{-7})$$

Thus, the pH is 7.

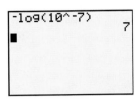

EXAMPLE 6

Find the hydrogen ion concentration of seawater if its pH is 8.5. Note that substances with a pH value greater than 7 are alkaline, and those less than 7 are acidic.

SOLUTION To find the hydrogen ion concentration, solve for $[H^+]$.

$$8.5 = -\log [H^+]$$
$$-8.5 = \log [H^+] \qquad \text{Multiply both sides by } -1.$$
$$[H^+] = 10^{-8.5} \qquad \text{Rewrite the equation in the equivalent exponential form.}$$
$$[H^+] \approx 3.16227766 \times 10^{-9} \qquad \text{Evaluate } 10^{-8.5} \text{ with the calculator.}$$

If a quantity grows exponentially at a certain annual rate, the time required for the quantity to double is called the **doubling time** and is computed using the formula

$$t = \frac{\ln 2}{r}$$

where r is the annual rate and t is the time required for the quantity to double.

EXAMPLE 7

The population of the earth is growing at the rate of approximately 2% per year. If this rate continues, how long will it take the population to double?

SOLUTION Because the population is growing at the rate of 2% per year, we substitute 0.02 for r into the formula for doubling time and simplify.

$$t = \frac{\ln 2}{r}$$

$$t = \frac{\ln 2}{0.02}$$

$$\approx 34.65735903$$

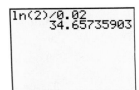

At the current rate of growth, the population of the earth will double in about 35 years.

EXERCISE 11.4

RY AND NOTATION   *Fill in each blank to* *tement true.*

1. A _____ is an exponent.

2. If $b > 0$ and $b \neq 1$, then $\log_b x = y$ is equivalent to _____ .

3. A logarithm of base ____ is called a common logarithm.

4. A _____ logarithm is a log of base ___ and uses the notation _____ instead of $\log_e x$.

5. The _____ formula allows conversion of a logarithm of any base to a base 10 logarithm for calculator computation.

CONCEPTS

6. Change from logarithmic form to exponential form.
   a. $\log_4 256 = 4$
   b. $\log_{121} 11 = \frac{1}{2}$
   c. $\log 10 = 1$
   d. $\log_{1/4} \frac{1}{16} = 2$

7. Write each exponential expression in logarithmic form.
   a. $8^2 = 64$
   b. $4^{-2} = \frac{1}{16}$
   c. $\left(\frac{1}{2}\right)^{-5} = 32$
   d. $10^{-4} = \frac{1}{10,000}$

8. Use the change-of-base formula to correctly match each logarithmic expression to its calculator evaluation.

   i. $\log_7 3$
   ii. $\log 8$
   iii. $\log_3 243$
   iv. $\log_4 64$

   a.
   ```
   log(8)
             .903089987
   ```
   b.
   ```
   log(64)/log(4)
                     3
   ```
   c.
   ```
   log(3)/log(7)
             .5645750341
   ```
   d.
   ```
   log(243)/log(3)
                     5
   ```

PRACTICE   *In Exercises 9–42, find each value of x.*

9. $\log_2 8 = x$

10. $\log_3 9 = x$

11. $\log_4 64 = x$

12. $\log_6 216 = x$

13. $\log_{1/2} \frac{1}{8} = x$

14. $\log_{1/3} \frac{1}{81} = x$

15. $\log_9 3 = x$

16. $\log_{125} 5 = x$

17. $\log_{1/2} 8 = x$

18. $\log_{1/2} 16 = x$

19. $\log_8 x = 2$

20. $\log_7 x = 0$

21. $\log_7 x = 1$

22. $\log_2 x = 3$

23. $\log_{25} x = \frac{1}{2}$

24. $\log_4 x = \frac{1}{2}$

25. $\log_5 x = -2$

26. $\log_3 x = -4$

27. $\log_{36} x = -\dfrac{1}{2}$

28. $\log_{27} x = -\dfrac{1}{3}$

29. $\log_{100} \dfrac{1}{1000} = x$

30. $\log_{5/2} \dfrac{4}{25} = x$

31. $\log_{27} 9 = x$

32. $\log_{12} x = 0$

33. $\log_x 5^3 = 3$

34. $\log_x 5 = 1$

35. $\log_x \dfrac{9}{4} = 2$

36. $\log_x \dfrac{4}{5} = \dfrac{1}{2}$

37. $\log_x \dfrac{1}{64} = -3$

38. $\log_x \dfrac{1}{100} = -2$

39. $\log_{2\sqrt{2}} x = 2$

40. $\log_4 8 = x$

41. $\log 10^3 = x$

42. $\log 10^{-2} = x$

In Exercises 43–54, use a calculator to find each value, if possible. Give answers to four decimal places.

43. $\log 3.25$

44. $\log 0.57$

45. $\log 0.00467$

46. $\log 375.876$

47. $\ln 0.93$

48. $\ln 7.39$

49. $\ln 37.896$

50. $\ln 0.00465$

51. $\log (\ln 1.7)$

52. $\ln (\log 9.8)$

53. $\ln (\log 0.1)$

54. $\log (\ln 0.01)$

In Exercises 55–62, use a calculator to find each value of y, if possible. Give answers to four decimal places.

55. $\log y = 1.4023$

56. $\ln y = 2.6490$

57. $\ln y = 4.24$

58. $\log y = 0.926$

59. $\log y = -3.71$

60. $\ln y = -0.28$

61. $\log y = \ln 8$

62. $\ln y = \log 7$

In Exercises 63–66, graph the function defined by each equation on graph paper. Check your graph with your graphing calculator.

63. $y = \log_3 x$

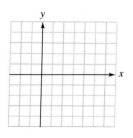

64. $y = \log_{1/3} x$

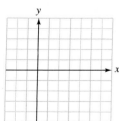

65. $y = \log_{1/2} x$

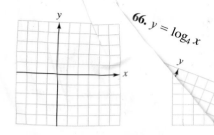

66. $y = \log_4 x$

In Exercises 67–70, graph each pair of invers on graph paper. Check your graph with your calculator.

67. $y = 2^x$, $y = \log_2 x$

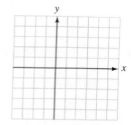

68. $y = \left(\dfrac{1}{2}\right)^x$, $y = \log_{1/2} x$

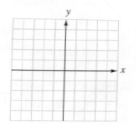

69. $y = \left(\dfrac{1}{4}\right)^x$, $y = \log_{1/4} x$

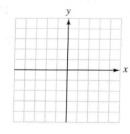

70. $y = 4^x$, $y = \log_4 x$

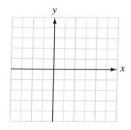

In Exercises 71–74, use a graphing calculator to do each experiment. Describe what you find.

 71. Graph $y = k + \log x$ for many values of k.

 72. Graph $y = k \log x$ for many values of k.

 73. Graph $y = \log (x + k)$ for many values of k.

74. Graph $y = \log kx$ for many values of k.

In Exercises 75–82, use the change-of-base formula to find each logarithm to four decimal places.

75. $\log_3 7$ **76.** $\log_7 3$

77. $\log_{1/3} 3$ **78.** $\log_{1/2} 6$

79. $\log_3 8$ **80.** $\log_5 10$

81. $\log_{\sqrt{2}} \sqrt{5}$ **82.** $\log_\pi e$

APPLICATIONS   *In Exercises 83–90, solve each problem.*

83. pH of a solution   Find the pH of a solution with a hydrogen ion concentration of 1.7×10^{-5} gram-ions.

84. Hydrogen ion concentration of calcium hydroxide   Find the hydrogen ion concentration of a saturated solution of calcium hydroxide whose pH is 13.2.

85. Aquariums   To test for safe pH levels in a fresh-water aquarium, a test strip is compared with the scale shown in Illustration 1. Find the corresponding safe range in the hydrogen ion concentration.

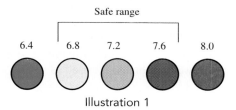

AquaTest pH Kit

Safe range

6.4   6.8   7.2   7.6   8.0

Illustration 1

86. pH of sour pickles   The hydrogen concentration of sour pickles is 6.31×10^{-4}. Find the pH.

87. Population growth   How long will it take the population of River City (Illustration 2) to double?

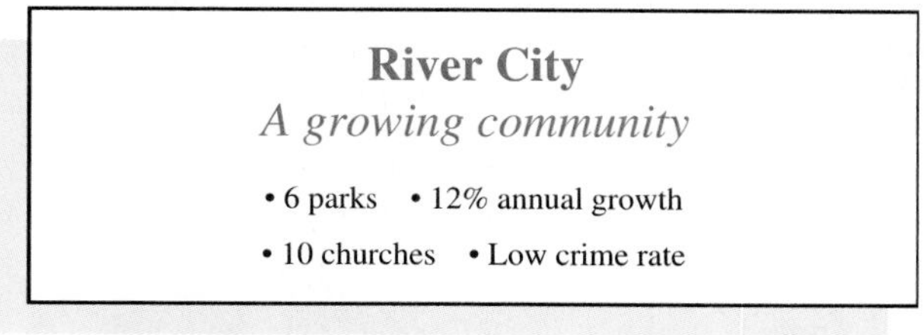

River City

A growing community

- 6 parks   • 12% annual growth
- 10 churches   • Low crime rate

Illustration 2

88. Doubling money   How long will it take $1000 to double if it is invested at an annual rate of 4% compounded continuously?

89. Population growth   A population growing at an annual rate r will triple in a time t given by the formula $t = \dfrac{\ln 3}{r}$. How long will it take the population of a town growing at the rate of 12% per year to triple?

90. Tripling money   Find the length of time for $25,000 to triple if invested at 3% annual interest, compounded continuously.

REVIEW

91. Given the function $y = \sqrt{2x + 5}$
   a. Is the function increasing or decreasing?
   b. What is the domain of the function?

92. Given the radical equation $5 + \sqrt{2x + 5} = x$
   a. Solve analytically.
   b. Justify your analytical solution with a graphical solution.

93. Why is $x = 2$ not a solution to the equation $5 + \sqrt{2x + 5} = x$ but *is* a solution to $x^2 - 12x + 20 = 0$?

94. Solve $x^2 - 12x + 20 = 0$ by
   a. completing the square
   b. quadratic formula
   c. zero-factor property
   d. graphing

11.5   PROPERTIES OF LOGARITHMS

- BASIC PROPERTIES OF LOGARITHMS
- ADDITIONAL PROPERTIES OF LOGARITHMS
- APPLICATIONS

BASIC PROPERTIES OF LOGARITHMS

In Section 11.4 we examined some basic properties of logarithms, which we will now summarize. Because logarithms are exponents, it is not surprising that the properties of exponents have counterparts in logarithmic notation.

Properties of Logarithms

If b is a positive number and $b \neq 1$, then

1. $\log_b 1 = 0$
2. $\log_b b = 1$
3. $\log_b b^x = x$
4. $b^{\log_b x} = x \quad (x > 0)$

These properties follow directly from the definition of logarithms:

1. $\log_b 1 = 0$ because $b^0 = 1$. Thus, $\log_5 1 = 0$ because $5^0 = 1$.
2. $\log_b b = 1$ because $b^1 = b$. Thus, $\log_3 3 = 1$ because $3^1 = 3$.
3. $\log_b b^x = x$ because $b^x = b^x$. Thus, $\log_7 7^3 = 3$ because $7^3 = 7^3$.
4. $b^{\log_b x} = x$ because $\log_b x$ is the exponent to which b is raised to get x. Thus, $b^{\log_b 7} = 7$ because $\log_b 7$ is the power to which b is raised to get 7.

ADDITIONAL PROPERTIES OF LOGARITHMS

Before the widespread use of calculators, logarithms provided the only reasonable way of performing certain long and involved calculations. The advent of hand-held calculators has virtually made these computations obsolete. However, the properties of logarithms that are addressed in this section will provide a foundation for future mathematics courses.

The first property deals with the logarithm of a product and is called the **product property of logarithms**.

Product Property of Logarithms

If M, N, and b are positive numbers and $b \neq 1$, then

$$\log_b MN = \log_b M + \log_b N$$

To prove this, we let $x = \log_b M$ and $y = \log_b N$. Because of the definition of logarithms, these equations can be written in the form

$$M = b^x \quad \text{and} \quad N = b^y$$

Then $MN = b^x b^y$ and the properties of exponents give

$$MN = b^{x+y}$$

Using the definition of logarithms gives

$$\log_b MN = x + y$$

Substituting the values of x and y completes the proof:

$$\log_b MN = \log_b M + \log_b N$$

WARNING! The product property of logarithms asserts that the logarithm of the product of two numbers is equal to the sum of their logarithms. The logarithm of a sum or a difference usually does not simplify. In general,

$$\log_b (M + N) \neq \log_b M + \log_b N \quad \text{and} \quad \log_b (M - N) \neq \log_b M - \log_b N$$

EXAMPLE 1

Use the product property to write each logarithmic expression as a sum of logarithms. Assume that variables represent positive numbers.

  a. $\log_b xyz$ **b.** $\log 6$

SOLUTION   **a.** $\log_b xyz = \log_b (xy)z$ Associative property of multiplication.

$$= \log_b xy + \log_b z \quad \text{Product property of logarithms.}$$

$$= \log_b x + \log_b y + \log_b z \quad \text{Product property of logarithms.}$$

  b. $\log 6 = \log (2 \cdot 3)$

$$= \log 2 + \log 3$$

Calculator verification:

```
log(6)
        .7781512504
log(2)+log(3)
        .7781512504
```

EXAMPLE 2

Use the product property of logarithms to write each sum as a single logarithm.

  a. $\log_b x^3 + \log_b y^{1/2}$ **b.** $\log_b (x + 3) + \log_b (x - 3)$

SOLUTION   **a.** $\log_b x^3 + \log_b y^{1/2} = \log_b (x^3 y^{1/2})$ Product property of logarithms.

$$= \log_b (x^3 \sqrt{y}) \quad \text{Product property of logarithms.}$$

  b. $\log_b (x + 3) + \log_b (x - 3) = \log_b [(x + 3)(x - 3)]$

$$= \log_b (x^2 - 9)$$

Quotient Property of Logarithms

If M, N, and b are positive numbers and $b \neq 1$, then

$$\log_b \frac{M}{N} = \log_b M - \log_b N$$

To prove this, we again let $x = \log_b M$ and $y = \log_b N$. These equations can be written as

$$M = b^x \qquad \text{and} \qquad N = b^y$$

Then $\dfrac{M}{N} = \dfrac{b^x}{b^y}$ and the properties of exponents give

$$\frac{M}{N} = b^{x-y}$$

Using the definition of logarithms gives

$$\log_b \frac{M}{N} = x - y$$

Substituting the values of x and y completes the proof:

$$\log_b \frac{M}{N} = \log_b M - \log_b N$$

WARNING!   The quotient property of logarithms asserts that the logarithm of the quotient of two numbers is equal to the difference of their logarithms. The logarithm of a quotient is not the quotient of the logarithms:

$$\log_b \frac{M}{N} \neq \frac{\log_b M}{\log_b N}$$

EXAMPLE 3

Use the quotient property of logarithms to write each difference as a single logarithm. Assume all variables represent positive numbers. Use the calculator to verify if possible.

a. $\log_b (yx) - \log_b z$ **b.** $\log 20 - \log 2$ **c.** $\log_2 8 + \log_2 5 - \log_2 6$

SOLUTION   **a.** $\log_b (yx) - \log_b z = \log_b \dfrac{yx}{z}$ Quotient property of logarithms.

b. $\log 20 - \log 2 = \log \dfrac{20}{2}$ Quotient property of logarithms.

$\qquad\qquad\qquad = \log 10$ Simplify.

$\qquad\qquad\qquad = 1$ $\log_{10} 10 = y$ implies $10^y = 10$.

Calculator verification:

```
log(20)-log(2)
                 1
log(20/2)
                 1
```

c. $\log_2 8 + \log_2 5 - \log_2 6 = \log_2 (8 \cdot 5) - \log_2 6$ Product property of logarithms.

$$= \log_2 \frac{40}{6}$$ Quotient property of logarithms.

$$= \log_2 \frac{20}{3}$$ Simplify.

Calculator verification:

```
log(8)/log(2)+lo
g(5)/log(2)-log(
6)/log(2)
         2.736965594
log(20/3)/log(2)
         2.736965594
```

Power Property of Logarithms

If M, N, and b are positive numbers and $b \neq 1$, then

$$\log_b M^p = p \log_b M$$

To prove this, we let $x = \log_b M$, write the expression in exponential form, and raise both sides to the pth power:

$$M = b^x$$
$$(M)^p = (b^x)^p$$
$$M^p = b^{px}$$

Using the definition of logarithms gives

$$\log_b M^p = px$$

Substituting the value for x completes the proof:

$$\log_b M^p = p \log_b M$$

EXAMPLE 4

Use the power property to write each difference as a single logarithm. Assume all variables represent positive numbers.

a. $\log_b (x^2 \sqrt[3]{y})$ **b.** $\log_b \dfrac{z^3 \sqrt{x - 2}}{y}$

SOLUTION   **a.** $\log_b (x^2 \sqrt[3]{y}) = \log_b x^2 + \log_b \sqrt[3]{y}$ Product property.

$$= 2 \log_b x + \log_b y^{1/3}$$ Power property.

$$= 2 \log_b x + \frac{1}{3} \log_b y$$ Power property.

b. $\log_b \dfrac{z^3\sqrt{x-2}}{y} = \log_b\left(\dfrac{z^3}{y} \cdot \sqrt{x-2}\right)$ Arithmetic of fractions.

$= \log_b \dfrac{z^3}{y} + \log_b \sqrt{x-2}$ Product property.

$= \log_b z^3 - \log_b y + \log_b \sqrt{x-2}$ Quotient property.

$= \log_b z^3 - \log_b y + \log_b (x-2)^{1/2}$ $\sqrt[m]{x^n} = x^{n/m}$

$= 3\log_b z - \log_b y + \dfrac{1}{2}\log_b (x-2)$ Power property.

EXAMPLE 5

Rewrite each logarithmic expression using the product, quotient, and power properties. Justify the equality of your expressions with the calculator.

a. $\log 5 + \log 3$ **b.** $2\log 3$

c. $\log 10 - \log 2$ **d.** $\frac{1}{2}\log 5$

SOLUTION   **a.** $\log 5 + \log 3 = \log(5\cdot 3)$ Product property.

$= \log 15$

```
log(5)+log(3)
           1.176091259
log(15)
           1.176091259
```

b. $2\log 3 = \log 3^2$ Power property.

$= \log 9$

```
2log(3)
           .9542425094
log(9)
           .9542425094
```

c. $\log 10 - \log 2 = \log\dfrac{10}{2}$ Quotient property.

$= \log 5$

```
log(10)-log(2)
           .6989700043
log(5)
           .6989700043
```

d. $\dfrac{1}{2}\log 5 = \log 5^{1/2}$ Power rule.

$= \log \sqrt{5}$

```
1/2log(5)
           .3494850022
log(√(5))
           .3494850022
```

▌ APPLICATIONS

Electrical engineering   Common logarithms are used in electrical engineering to express the voltage gain (or loss) of an electronic device such as an amplifier. The unit of gain (or loss), called the **decibel**, is defined by the following logarithmic relation.

Decibel Voltage Gain

If E_O is the output voltage of a device and E_I is the input voltage, the decibel voltage gain is given by

$$\text{db gain} = 20 \log \frac{E_O}{E_I}$$

EXAMPLE 6   If the input to an amplifier is 0.5 volt and the output is 40 volts, find the approximate decibel voltage gain of the amplifier.

SOLUTION   We can find the decibel voltage gain by substituting **0.5** for E_I and **40** for E_O into the formula for db gain:

$$\text{db gain} = 20 \log \frac{E_O}{E_I}$$

$$\text{db gain} = 20 \log \frac{40}{0.5}$$

$$= 20 \log 80$$

$$\approx 38 \qquad \text{Use a calculator.}$$

The amplifier provides approximately a 38-decibel voltage gain.

Geology   In seismology, common logarithms are used to measure the intensity of earthquakes on the **Richter scale**. The intensity of an earthquake is given by the following logarithmic function.

Richter Scale

If A is the amplitude of an earthquake (the earth movement, measured in micrometers), and P is the period (the time of one oscillation of the earth's surface measured in seconds), then the intensity R on the Richter scale is

$$R = \log \frac{A}{P}$$

EXAMPLE 7   Find the measure on the Richter scale of an earthquake with an amplitude of 10,000 micrometers (1 centimeter) and a period of 0.1 second.

SOLUTION   We substitute **10,000** for A and **0.1** for P in the Richter scale formula and simplify:

$$R = \log \frac{A}{P}$$

$$R = \log \frac{10,000}{0.1}$$

$$= \log 100,000$$

$$= \log 10^5$$

$$= 5 \log 10 \qquad \text{Use the power property of logarithms.}$$

$$= 5 \qquad \text{Use property 2 of logarithms.}$$

The earthquake measures 5 on the Richter scale.

Physiology   In physiology, experiments suggest that the relationship between the loudness and the intensity of sound is a logarithmic one, known as the **Weber-Fechner law**.

Weber-Fechner Law

If L is the apparent loudness of a sound, I is the actual intensity, and k is a constant, then

$$L = k \ln I$$

EXAMPLE 8

Find the increase in actual intensity that will cause the apparent loudness of a sound to double.

SOLUTION   If the original loudness L_0 is caused by an actual intensity I_0, then

$$L_0 = k \ln I_0$$

To double the apparent loudness, we multiply both sides of the equation by 2.

$$2L_0 = 2k \ln I_0$$

$$= k \ln (I_0)^2 \qquad \text{Power property of logarithms.}$$

Thus, to double the apparent loudness of a sound, the intensity must be squared.

EXERCISE 11.5

VOCABULARY AND NOTATION   *Fill in the blanks to make a true statement.*

1. The _____ property guarantees that $\log_b MN = \log_b M + \log_b N$.

2. If $\log_b \frac{M}{N} = \log_b M - \log_b N$, then the _____ property has been applied.

3. The power property of logarithms ensures that $\log_b M^p =$ _____.

CONCEPTS *In Exercises 4–23, classify each of the following statements as true or false. If false, correct the statement to a true statement. All variables are positive and $b \neq 1$.*

4. $\log_b 0 = 1$

5. $\log_b (x + y) \neq \log_b x + \log_b y$

6. $\log_b xy = (\log_b x)(\log_b y)$

7. $\log_b ab = \log_b a + 1$ **8.** $\log_7 7^7 = 7$

9. $\log^{\log_7 7} = 7$

10. $\dfrac{\log_b A}{\log_b B} = \log_b A - \log_b B$

11. $\log_b (A - B) = \dfrac{\log_b A}{\log_b B}$

12. $3 \log_b \sqrt[3]{a} = \log_b a$

13. $\dfrac{1}{3} \log_b a^3 = \log_b a$ **14.** $\log_b \dfrac{1}{a} = -\log_b a$

15. $\log_b 2 = \log_2 b$

16. If $\log_a b = c$, then $\log_b a = c$

17. If $\log_a b = c$, then $\log_b a = \dfrac{1}{c}$

18. $\log_b (-x) = -\log_b x$

19. If $\log_b a = c$, then $\log_b a^p = pc$

20. $\log_b \dfrac{1}{5} = -\log_b 5$

21. $\log_{4/3} y = -\log_{3/4} y$

22. $\log_b y + \log_{1/b} y = 0$

23. $\log_{10} 10^3 = 3(10^{\log_{10} 3})$

 24. Explain why $\ln (\log 0.9)$ is undefined.

 25. Explain why $\log_b (\ln 1)$ is undefined.

PRACTICE *In Exercises 26–31, use your calculator to verify that the left side of the equation is equal to the right side of the equation. Identify the property that was applied.*

26. $\log [(2.5)(3.7)] = \log 2.5 + \log 3.7$

27. $\ln \dfrac{11.3}{6.1} = \ln 11.3 - \ln 6.1$

28. $\ln (2.25)^4 = 4 \ln 2.25$

29. $\log 45.37 = \dfrac{\ln 45.37}{\ln 10}$

30. $\log \sqrt{24.3} = \dfrac{1}{2} \log 24.3$

31. $\ln 8.75 = \dfrac{\log 8.75}{\log e}$

In Exercises 32–43, assume that x, y, and z are positive numbers. Use the properties of logarithms to write each expression in terms of the logarithms of x, y, and z.

32. $\log_b xyz$

33. $\log_b xz$

34. $\log_b \dfrac{2x}{y}$

35. $\log_b \dfrac{x}{yz}$

36. $\log_b x^3 y^2$

37. $\log_b xy^2 z^3$

38. $\log_b (xy)^{1/2}$

39. $\log_b x^3 y^{1/2}$

40. $\log_b x\sqrt{z}$

41. $\log_b \sqrt{xy}$

42. $\log_b \dfrac{\sqrt[3]{x}}{\sqrt[4]{yz}}$

43. $\log_b \sqrt[4]{\dfrac{x^3 y^2}{z^4}}$

In Exercises 44–51, assume that x, y, and z are positive numbers. Use the properties of logarithms to write each expression as a log of a single quantity.

44. $\log_b (x + 1) - \log_b x$

45. $\log_b x + \log_b (x + 2) - \log_b 8$

46. $2 \log_b x + \dfrac{1}{2} \log_b y$

47. $-2 \log_b x - 3 \log_b y + \log_b z$

48. $-3 \log_b x - 2 \log_b y + \dfrac{1}{2} \log_b z$

49. $3 \log_b (x + 1) - 2 \log_b (x + 2) + \log_b x$

50. $\log_b\left(\dfrac{x}{z} + x\right) - \log_b\left(\dfrac{y}{z} + y\right)$

51. $\log_b (xy + y^2) - \log_b (xz + yz) + \log_b z$

APPLICATIONS

52. Finding output voltage   The db gain of an amplifier is 35. Find the output voltage when the input voltage is 0.05 volt.

53. db gain of an amplifier   Find the db gain of the amplifier shown in Illustration 1.

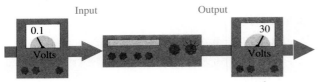

Illustration 1

54. db gain of an amplifier   An amplifier produces an output of 80 volts when driven by an input of 0.12 volts. Find the amplifier's db gain.

55. Earthquakes   An earthquake has amplitude of 5000 micrometers and a period of 0.2 seconds. Find its measure on the Richter scale.

56. Earthquakes   Find the period of an earthquake with amplitude of 80,000 micrometers that measures 6 on the Richter scale.

57. Earthquakes   An earthquake with a period of $\frac{1}{4}$ second measures 4 on the Richter scale. Find its amplitude.

58. Earthquakes   By what factor must the amplitude of an earthquake change to increase its severity by 1 point on the Richter scale? Assume that the period remains constant.

59. Change in loudness   If the intensity of a sound is doubled, find the apparent change in loudness.

60. Change in intensity   If the intensity of a sound is tripled, find the apparent change in loudness.

61. Change in loudness   What change in intensity of sound will cause an apparent tripling of the loudness?

62. Change in intensity   What increase in the intensity of a sound will cause the apparent loudness to be multiplied by 4?

63. Charging a battery   The time (in hours) required to charge a certain battery to a level C (expressed as a fraction of the battery's full charge) is $t = -3.7 \ln (1 - C)$.

How long would it take to bring the battery to 80% of its full charge?

64. Power output   If P_O is the output of an amplifier (in watts) and P_1 is the power input, then the decibel voltage gain is db voltage gain $= 10 \log \dfrac{P_O}{P_1}$. One brand of amplifier produces a 70-watt output with a 0.035-watt input. Find the db voltage gain.

65. Depreciation   In business, equipment is often depreciated using the double declining-balance method. In this method, a piece of equipment with a life expectancy of N years, costing \$$C$, will depreciate to a value of \$$V$ in n years, where n is given by the formula

$$n = \frac{\log V - \log C}{\log\left(1 - \dfrac{2}{N}\right)}$$

A computer that cost \$37,000 has a life expectancy of 5 years. If it has depreciated to a value of \$8000, how old is it?

66. Depreciation   A fax machine worth \$470 when new had a life expectancy of 5 years. If it is now worth \$189, how old is it?

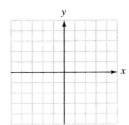 **67.** Explain why an earthquake measuring 7 on the Richter scale is much worse than an earthquake measuring 6.

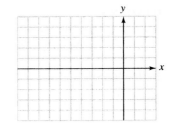 **68.** Graph the logistic function $y = \dfrac{1}{1 + e^{-2x}}$ and discuss its graph.

REVIEW

69. Graphically solve each inequality.

   a. $\dfrac{3}{4}(x + 5) - 3 \le 0$

   b. $\dfrac{3}{4}|x + 5| - 3 \le 0$

   c. $\dfrac{3}{4}(x + 5)^2 - 3 \le 0$

70. Graph each inequality in two variables on graph paper.

   a. $\dfrac{3}{4}(x + 5) - 3 \le y$ **b.** $\dfrac{3}{4}|x + 5| - 3 \le y$

c. $\frac{3}{4}(x + 5)^2 - 3 \le y$

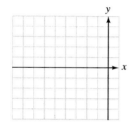

71. Explain the difference in graphically solving an inequality in one variable and graphing an inequality in two variables.

72. In July of 2003, a leading computer firm sold a 2.4 GHz computer for $1,650. From July through October of 2003, the computer decreased in value 13% each month. The value, $V(x)$, of the computer can be represented by the model $V(x) = 1,650(.87^x)$, where $x = 0$ is the value in July and $x = 4$ is the value in October.

  a. Graph the model in an appropriate viewing window.

  b. Determine the value of the computer in each of the 4 months.

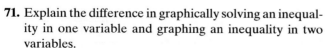

11.6 EXPONENTIAL AND LOGARITHMIC EQUATIONS

- EXPONENTIAL EQUATIONS
- LOGARITHMIC EQUATIONS

- APPLICATIONS OF EXPONENTIAL AND LOGARITHMIC EQUATIONS

 **EXPONENTIAL EQUATIONS**

In Section 11.3 we solved exponential equations $b^r = b^s$ using the exponential property of equality: "for $b > 0$ and $b \ne 1$, $b^r = b^s$ is equivalent to $r = s$." Thus, problems such as $3^x = 81$ can be rewritten as

$$3^x = 3^4$$

and it follows

$$x = 4$$

If, however, we have an equation in which one expression cannot easily be written as a power of the other expression, such as $3^x = 7$, we will use the following property of equality:

Logarithmic Property of Equality

If $M > 0$ and $N > 0$, $\log_b M = \log_b N$, if and only if $M = N$

Both the exponential and logarithmic properties of equality follow from the fact that exponential and logarithmic functions are one-to-one.

EXAMPLE 1   Solve the exponential equation $3^x = 5$.

SOLUTION   Logarithms of equal numbers are equal, so we can take the common logarithm of each side of the equation. The power property of logarithms then provides a way of moving the variable x from its position as an exponent to a position as a coefficient.

Analytic solution:

$$3^x = 5$$

$$\log 3^x = \log 5 \qquad \text{Take the common logarithm of each side.}$$

$$x \log 3 = \log 5 \qquad \text{Power property of logarithms.}$$

$$x = \frac{\log 5}{\log 3} \qquad \text{Divide both sides by log 3.}$$

$$\approx 1.464973521 \qquad \text{Use a calculator.}$$

To four decimal places, $x = 1.4650$.

Graphical solution: Enter the left side of the equation at the y1 prompt and the right side at the y2 prompt. Use the INTERSECT feature to find the approximate value of x.

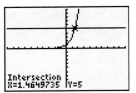

The ROOT/ZERO feature could also be used. However, first rewrite the equation as $3^x - 5 = 0$ and enter the left side of the equation at the y1 prompt.

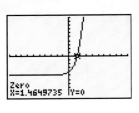

WARNING! A careless reading of the equation $x = \frac{\log 5}{\log 3}$ leads to a common error. The right-hand side of the equation calls for a division, *not a subtraction*. $\frac{\log 5}{\log 3}$ means (log 5) ÷ (log 3). It is the expression $\log \frac{5}{3}$ that means $\log 5 - \log 3$.

EXAMPLE 2

Solve the exponential equation $6^{x-3} = 2^x$.

SOLUTION   Analytic solution:

$$6^{x-3} = 2^x$$

$$\log 6^{x-3} = \log 2^x \qquad \text{Take the common logarithm of each side.}$$

$$(x-3) \log 6 = x \log 2 \qquad \text{Power property of logarithms.}$$

$$x \log 6 - 3 \log 6 = x \log 2 \qquad \text{Use the distributive property.}$$

$$x \log 6 - x \log 2 = 3 \log 6 \qquad \text{Add 3 log 6 and subtract } x \log 2 \text{ from both sides.}$$

$$x(\log 6 - \log 2) = 3 \log 6 \qquad \text{Factor out } x \text{ on the left-hand side.}$$

$$x = \frac{3 \log 6}{\log 6 - \log 2} \qquad \text{Divide both sides by log 6 } - \text{ log 2.}$$

$$x \approx 4.892789261 \qquad \text{Use a calculator.}$$

Graphical solution: Enter the left side of the equation at the y1 prompt and the right side at the y2 prompt. Use the INTERSECT feature to find the approximate value of x. Suggested window: $[-10, 10]$ by $[-10, 60]$ with a scale of 4 on the y-axis.

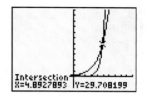

The ROOT/ZERO feature could also be used. However, first rewrite the equation as $6^{x-3} - 2^x = 0$ and enter the left side of the equation at the y1 prompt.

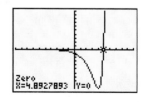

EXAMPLE 3

Solve the exponential equation $2^{x^2 + 2x} = \dfrac{1}{2}$.

SOLUTION

Analytic solution: Because $\frac{1}{2} = 2^{-1}$, we can write the equation in the form $2^{x^2 + 2x} = 2^{-1}$. Equal quantities with equal bases have equal exponents, so we have

$$x^2 + 2x = -1$$
$$x^2 + 2x + 1 = 0 \qquad \text{Add 1 to both sides.}$$
$$(x + 1)(x + 1) = 0 \qquad \text{Factor the trinomial.}$$

$$x + 1 = 0 \quad \text{or} \quad x + 1 = 0 \qquad \text{Set each factor equal to 0.}$$
$$x = -1 \quad | \quad x = -1$$

Therefore, the solution is -1.

Graphical solution: Enter the left side of the equation at the y1 prompt and the right side at the y2 prompt. Use the INTERSECT feature to find the approximate value of x.

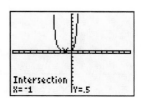

The ROOT/ZERO feature could also be used. However, first rewrite the equation as $2^{x^2 + 2x} - \frac{1}{2} = 0$ and enter the left side of the equation at the y1 prompt.

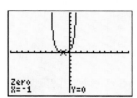

▌▌▌ LOGARITHMIC EQUATIONS

A wide variety of logarithmic equations can be solved by applying the logarithmic properties. It will be important to observe the domain of the variable in the original equation so that extraneous values are not included in the solution.

EXAMPLE 4

Solve the logarithmic equation $\log x + \log (x - 3) = 1$.

SOLUTION   Analytic solution:

$$\log x + \log (x - 3) = 1$$

$$\log x (x - 3) = 1 \qquad \text{Product property of logarithms.}$$

$$x(x - 3) = 10^1 \qquad \text{Use the definition of logarithms to change the equation to exponential form.}$$

$$x^2 - 3x - 10 = 0 \qquad \text{Remove parentheses and subtract 10 from both sides.}$$

$$(x + 2)(x - 5) = 0 \qquad \text{Factor the trinomial.}$$

$$x + 2 = 0 \quad \text{or} \quad x - 5 = 0 \qquad \text{Set each factor equal to 0.}$$

$$x = -2 \quad | \quad x = 5$$

Check: Use the STOre feature to check. The value -2 is extraneous.

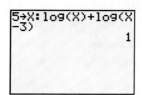
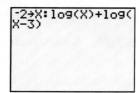
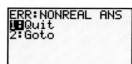

We can conclude that 5 is the only solution.

Graphical solution: Enter the left side of the equation at the y1 prompt and the right side at the y2 prompt. Use the INTERSECT feature to find the approximate value of x. Note that the graphs do not intersect at $x = -2$, verifying that -2 is an extraneous solution. Window values: $[-5, 10]$ by $[-5, 5]$.

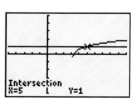

EXAMPLE 5

Solve the logarithmic equation $\log_b (3x + 2) - \log_b (2x - 3) = 0$.

SOLUTION   Analytic solution:

$$\log_b (3x + 2) - \log_b (2x - 3) = 0$$

$$\log_b (3x + 2) = \log_b (2x - 3) \qquad \text{Add } \log_b (2x - 3) \text{ to both sides.}$$

$$3x + 2 = 2x - 3 \qquad \text{If } \log_b r = \log_b s, \text{ then } r = s.$$

$$x = -5 \qquad \text{Subtract } 2x \text{ and } 2 \text{ from both sides.}$$

Check: We could check using the STOre feature of the calculator, or analytically, as follows:

$$\log_b (3x + 2) - \log_b (2x - 3) = 0$$

$$\log_b [3(-5) + 2] - \log_b [2(-5) - 3] \stackrel{?}{=} 0$$

$$\log_b (-13) - \log_b (-13) = 0$$

Because the logarithm of a negative number does not exist, -5 is an extraneous solution and must be discarded. This equation has no roots: $\varnothing$.

EXAMPLE 6   Solve the logarithmic equation $\dfrac{\log(5x - 6)}{\log x} = 2.$

SOLUTION   We can multiply both sides of the equation by $\log x$ to get $\log(5x - 6) = 2 \log x$ and apply the power property of logarithms to get $\log(5x - 6) = \log x^2$. Because $(5x - 6)$ and x^2 have equal logarithms, we can set these expressions equal and solve for x:

$$5x - 6 = x^2$$
$$0 = x^2 - 5x + 6$$
$$0 = (x - 3)(x - 2)$$
$$x - 3 = 0 \quad \text{or} \quad x - 2 = 0$$
$$x = 3 \quad | \quad \quad x = 2$$

Check: We use the STOre feature of the calculator to check these supposed roots. This verifies the fact that both 3 and 2 are actual roots. The roots could also be verified graphically.

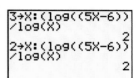

APPLICATIONS OF EXPONENTIAL AND LOGARITHMIC EQUATIONS

Carbon-14 dating   Experiments have determined the time it takes for half of a sample of a given radioactive material to decompose. This time, called the material's **half-life**, is a constant.

When a living organism dies, the oxygen/carbon dioxide cycle common to all living things ceases, and carbon-14, a radioactive isotope with a half-life of 5,700 years, is no longer absorbed. By measuring the amount of carbon-14 present in an ancient object, archeologists can estimate the object's age by using the radioactive decay formula.

Radioactive Decay Formula

If A is the amount of radioactive material present at time t, A_0 was the amount present at $t = 0$, and h is the material's half-life, then

$$A = A_0 2^{-t/h}$$

EXAMPLE 7

How old is a wooden statue that contains only one-third of its original carbon-14 content?

SOLUTION To find the time t when $A = \frac{1}{3}A_0$, we substitute $\dfrac{A_0}{3}$ for A and **5,700** for h into the radioactive decay formula and solve for t:

$$A = A_0 2^{-t/h}$$

$$\frac{A_0}{3} = A_0 2^{-t/5,700}$$

$A_0 = 3A_0 2^{-t/5,700}$ Multiply both sides by 3.

$1 = 3(2^{-t/5,700})$ Divide both sides by A_0.

$\log 1 = \log 3(2^{-t/5,700})$ Take the common logarithm of each side.

$0 = \log 3 + \log 2^{-t/5,700}$ $\log_b 1 = 0$ and product property of logarithms.

$-\log 3 = -\dfrac{t}{5,700}\log 2$ Subtract log 3 from both sides and use the power property of logarithms.

$5,700\left(\dfrac{\log 3}{\log 2}\right) = t$ Multiply both sides by $-\frac{5,700}{\log 2}$.

$t \approx 9,034.286254$ Use a calculator.

The wooden statue is approximately 9,000 years old.

Population growth When there is sufficient food and space, populations of living organisms tend to increase exponentially according to the Malthusian growth model.

Malthusian Growth Model

If P is the population at some time t, P_0 is the initial population at $t = 0$, and k depends on the rate of growth, then

$$P = P_0 e^{kt}$$

EXAMPLE 8

The bacteria in a laboratory culture increased from an initial population of 500 to 1,500 in 3 hours. How long will it take for the population to reach 10,000?

SOLUTION We substitute **500** for P_0, **1,500** for P, and **3** for t and simplify to find k:

$$P = P_0 e^{kt}$$

$1,500 = 500(e^{k3})$ Substitute 1,500 for P, 500 for P_0, and 3 for t.

$3 = e^{3k}$ Divide both sides by 500.

$\ln 3 = \ln e^{3k}$ Take the natural logarithm of both sides.

$\ln 3 = 3k$ $\ln e^{3k} = 3k \ln e = 3k$.

$k = \dfrac{\ln 3}{3}$ Divide both sides by 3.

To find when the population will reach 10,000, we substitute **10,000** for P, **500** for P_0, and $\frac{\ln 3}{3}$ for k in the equation $P = P_0 e^{kt}$ and solve for t:

$$P = P_0 e^{kt}$$

$$10{,}000 = 500 e^{[(\ln 3)/3]t}$$

$$20 = e^{[(\ln 3)/3]t} \qquad \text{Divide both sides by 500.}$$

$$\left(\frac{\ln 3}{3}\right)t = \ln 20 \qquad \text{Change the equation to logarithmic form.}$$

$$t = \frac{3\ln 20}{\ln 3} \qquad \text{Multiply both sides by } \tfrac{3}{\ln 3}.$$

$$\approx 8.180499084 \qquad \text{Use a calculator.}$$

The culture will reach 10,000 bacteria in a little more than 8 hours.

EXERCISE 11.6

VOCABULARY AND NOTATION

1. An equation with a variable in its exponent is called a(n) _____ equation.

2. An equation with a logarithmic expression that contains a variable is a(n) _____ equation.

CONCEPTS

3. Explain why $x = -4$ could not be a solution to $\log(2x + 1) - \log(3x - 2) = 0$.

4. Substitute in several values for b in $\log x = b$ and then explain why $x \neq 0$.

In Exercises 5–8, an exponential or logarithmic function, f, has been graphed at the y1 prompt. Use the graphical display to find the solution $y1 = 0$. Round solutions to the nearest thousandth.

5.

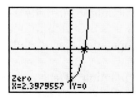

6.

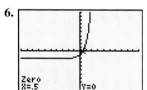

7.

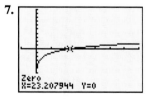

8.

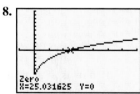

PRACTICE *In Exercises 9–24, solve each exponential equation. Give all results that are not exact. Specify results accurate to 4 decimal places. Check all solutions.*

9. $4^x = 5$

10. $7^x = 12$

11. $13^{x-1} = 2$

12. $5^{x+1} = 3$

13. $2^{x+1} = 3^x$

14. $5^{x-3} = 3^{2x}$

15. $2^x = 3^x$

16. $3^{2x} = 4^x$

17. $7^{x^2} = 10$

18. $8^{x^2} = 11$

19. $8^{x^2} = 9^x$

20. $5^{x^2} = 2^{5x}$

21. $2^{x^2-2x} = 8$

22. $3^{x^2-3x} = 81$

23. $3^{x^2-4x} = \dfrac{1}{81}$

24. $7^{x^2+3x} = \dfrac{1}{49}$

In Exercises 25–42, solve each logarithmic equation. Check each solution with the calculator.

25. $\log 2x = \log 4$

26. $\log 3x = \log 9$

27. $\log (3x + 1) = \log (x + 7)$

28. $\log (x^2 + 4x) = \log (x^2 + 16)$

29. $\log (2x - 3) - \log (x + 4) = 0$

30. $\log (3x + 5) - \log (2x + 6) = 0$

31. $\log \dfrac{4x + 1}{2x + 9} = 0$

32. $\log \dfrac{5x + 2}{2(x + 7)} = 0$

33. $\log x + \log (x - 48) = 2$

34. $\log x + \log (x + 9) = 1$

35. $\log (x + 90) = 3 - \log x$

36. $\log (x - 90) = 3 - \log x$

37. $\log x^2 = (\log x)^2$

38. $\log (\log x) = 1$

39. $\dfrac{\log (3x - 4)}{\log x} = 2$

40. $\dfrac{\log (8x - 7)}{\log x} = 2$

41. $\dfrac{\log (5x + 6)}{2} = \log x$

42. $\dfrac{1}{2} \log (4x + 5) = \log x$

APPLICATIONS   *In Exercises 43–60, solve each problem.*

43. Tritium decay   The half-life of tritium is 12.4 years. How long will it take for 25% of a sample of tritium to decompose?

44. Radioactive decay   In two years, 20% of a radioactive element decays. Find its half-life.

45. Thorium decay   An isotope of thorium, ^{227}Th, has a half-life of 18.4 days. How long will it take 80% of the sample to decompose?

46. Lead decay   An isotope of lead, ^{201}Pb, has a half-life of 8.4 hours. How many hours ago was there 30% more of the original amount of the substance?

47. Carbon-14 dating   The bone fragment shown in the photograph contains 60% of the carbon-14 that it is assumed to have had initially. How old is the bone?

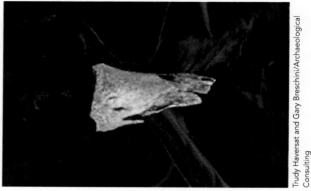

48. Carbon-14 dating   Only 10% of the carbon-14 in a small wooden bowl remains. How old is the bowl?

49. Compound interest   If $500 is deposited in an account paying 4.5% annual interest compounded semiannually, how long will it take for the account to increase to $800?

50. Continuous compound interest   In Exercise 49, how long will it take if the interest is compounded continuously?

51. Compound interest   If $1,300 is deposited in a savings account paying 3% interest compounded quarterly, how long will it take the account to increase to $2,100?

52. Compound interest   A sum of $5,000 deposited in an account grows to $7,000 in 5 years. Assuming continuous compounding, what interest rate is being paid?

53. Comparing investments   Which of these investment opportunities is best: 4.9% compounded annually, 4.8% compounded quarterly, or 4.7% compounded monthly?

54. Bacterial growth   A bacteria culture grows according to the formula $P = P_0 a^t$. If it takes 5 days for the culture to triple in size, how long will it take to double in size?

55. Rodent control   The rodent population in a city is currently estimated at 30,000. If it is expected to double every 5 years, when will the population reach 1 million?

56. Population growth   The population of a city is expected to triple every 15 years. When can the city planners expect the population to double?

57. Bacteria culture   A bacteria culture doubles in size every 24 hours. By how much will it have increased in 36 hours?

58. Oceanography   The intensity I of a light a distance x meters beneath the surface of a lake decreases exponentially. From Illustration 1, find the depth at which the intensity will be 20%.

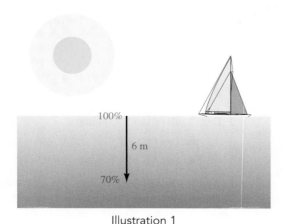

100%

6 m

70%

Illustration 1

59. Newton's law of cooling   Water initially at $100°$ C is left to cool in a room of temperature $60°$ C. After 3 minutes, the water temperature is $90°$. If the water temperature, T, is a function of time t given by $T = 60 + 40e^{kt}$, find k.

60. Newton's law of cooling   Refer to Exercise 59 and find the time for the water temperature to reach $70°$ C.

In Exercises 61–64, use a graphing calculator to solve each equation.

61. $\log x + \log (x - 15) = 2$

62. $\log x + \log (x + 3) = 1$

63. $2 \log (x^2 + 4x) = 1$

64. $\ln (2x + 5) - \ln 3 = \ln (x - 1)$

 65. Explain how to solve the equation $2^x = 7$.

 **66.** Explain how to solve the equation $x^2 = 7$.

REVIEW

67. A ball is thrown into the air with an initial velocity of 80 feet per second. The equation that models the relation between height h after t seconds is $h = -16t^2 + 80t + 3$.

   a. Graph the model in an appropriate viewing window.

   b. Based on the graph, how high is the ball 3 seconds after it is released?

   c. At what other point in time is the ball the same height as it is after 3 seconds?

   d. How long does it take the ball to hit the ground?

68. What is the slope of the line that passes through the y-intercept and the vertex of the parabola in Exercise 67?

69. What is the equation of the line defined in Exercise 68?

CHAPTER SUMMARY

CONCEPTS

REVIEW EXERCISES

Section 11.1

$(f + g)(x) = f(x) + g(x)$
$(f - g)(x) = f(x) - g(x)$
$(f \cdot g)(x) = f(x) \cdot g(x)$
$\left(\dfrac{f}{g}\right)(x) = \dfrac{f(x)}{g(x)},$

provided $g(x) \neq 0$.

The function operations of addition and multiplication are commutative.

The composition of two functions f and g, denoted $(f \circ g)(x)$ or $f(g(x))$, takes the range of the function g and makes it the domain of the function f.

Composition is *not* a commutative operation.

Algebra of Functions

1. Let $f(x) = 2x$ and $g(x) = x^2 - 1$. Find each of the following:
   a. $f + g$ b. $f - g$
   c. $f \cdot g$ d. $\dfrac{f}{g}$

2. Let $f(x) = 2x$ and $g(x) = x^2 + 1$. Find each of the following:
   a. $(f \circ g)(x)$ b. $(g \circ f)(x)$
   c. $(f \circ g)(2)$ d. $(g \circ f)(-1)$

3. Suppose an oil rig is leaking oil into the Gulf of Mexico with the leak forming a circular layer of oil on the surface of the water. If the radius of the circle is increasing at a rate of 5 meters per minute, then the function $f(t) = 5t$ defines the radius as a function of time. Express the area of the circle as a function of time and determine the approximate area (to the nearest foot) 40 minutes after the leak begins.

Section 11.2

If a relation contains the point (a, b), then the inverse will contain the point (b, a).

Every relation has an inverse relation.

In order for a function to be one-to-one, each y-value in the range must be paired with a unique x-value in the domain. The horizontal line test will verify this relationship on a graph.

Inverse Functions

4. If the table defines the relation that converts feet to miles, construct the inverse relation that converts miles to feet.

Feet	1320	2640	3960	5280	6600
Miles	$\frac{1}{4}$	$\frac{1}{2}$	$\frac{3}{4}$	1	$1\frac{1}{4}$

5. Graph each function and determine if it is one-to-one.
   a. $y = 2(x - 3)$ b. $y = x^2 - 4$

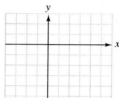

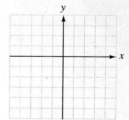

6. Determine if the relation specified is a one-to-one function.
    a. $\{(4, 1), (2, -3), (3, 2), (1, 4)\}$
    b. $\{(2, 3), (3, 2), (4, 2), (1, 5)\}$

In order for a function, f, to have an inverse function, f^{-1}, f must be one-to-one. To algebraically determine the inverse function, reverse the x and y variables in the original function.

7. Algebraically, determine $f^{-1}(x)$ and then graph the function $f(x)$, its inverse $f^{-1}(x)$, and the line of symmetry $f(x) = x$. Verify with your calculator.

    a. $y = 6x - 3$

    b. $y = \frac{1}{2}x + 3$

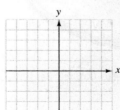
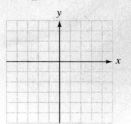

Exponential Functions

An exponential function is a function of the form $f(x) = b^x$, where $b > 0$ and $b \neq 1$. The domain is the set of reals and the range is the set of positive numbers.

8. Graph each exponential function and verify with the calculator.

    a. $f(x) = \left(\frac{6}{5}\right)^x$

    b. $f(x) = \left(\frac{1}{5}\right)^x$

    c. $f(x) = e^x$

Exponential property of equality: If $b > 0$ and $b \neq 1$, then $b^r = b^s$ is equivalent to $r = s$. If $r \neq s$, then $b^r \neq b^s$.

9. Solve for x. Verify solutions graphically.

    a. $2^{x+5} = 2^{x^2-4x}$

    b. $3^{x^2} = 9^{2x}$

    c. $2^{x^2+4x} = \frac{1}{8}$

    d. $\left(\frac{1}{5}\right)^x = 5^{x-3}$

10. Solve graphically. Approximate the solution to the nearest thousandth:
    $3x = -x^x + 5$

Logarithmic Functions

A logarithm is an exponent.

11. **Compound interest**   An initial deposit of $1,000 earns 6% interest, compounded twice a year. How much will be in the account in 1 year?

12. **Interest compounded continuously**   An initial deposit of $1,000 earns 6% interest, compounded continuously. How much will be in the account in one year?

If $b > 0$ and $b \neq 1$, then $b^y = x$ is equivalent to $\log_b x = y$.

The change-of-base formula is necessary for computation or graphing involving logarithms other than base 10 or base e. It states: If a, b, and x are positive real numbers, then

$$\log_b x = \frac{\log_a x}{\log_a b}, \quad (a \neq 1, b \neq 1)$$

The exponential functions $y = 2^x$ and $x = 2^y$ are inverses of one another. Because $x = 2^y$ can be expressed as $\log_2 x = y$ using the definition of logarithms, then $y = 2^x$ and $\log_2 x = y$ are inverses of one another.

13. Find each value.

   a. $\log_3 9$ **b.** $\log_9 \dfrac{1}{3}$ **c.** $\log_5 .04$ **d.** $\ln e^9$

14. Find x.

   a. $\log_2 x = 3$ **b.** $\log_3 x = -2$

   c. $\log_x 9 = 2$ **d.** $\log_x .125 = -3$

   e. $\log_3 \sqrt{3} = x$ **f.** $\log_2 \sqrt{2} = x$

15. Evaluate each of the following on your calculator using the change-of-base formula:

   a. $\log_3 9$

   b. $\log_5 7$ (to the nearest hundredth)

16. Graph each function and its inverse. Verify with your calculator.

   a. $f(x) = 3^x$ and $f^{-1}(x) = \log_3 x$ **b.** $f(x) = \left(\dfrac{1}{2}\right)^x$ and $f^{-1}(x) = \log_{1/2} x$

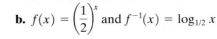

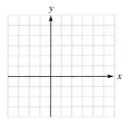 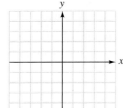

   c. $f(x) = \left(\dfrac{2}{5}\right)^x$ and $f^{-1}(x) = \log_{2/5} x$

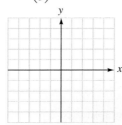

17. pH of grapefruit   The pH of grapefruit juice is approximately 3.1. Find its hydrogen ion concentration.

18. Formula for pH   Some chemistry textbooks define the pH of a solution with the formula

$$\text{pH} = \log_{10} \frac{1}{[\text{H}^+]}$$

Show that this definition is equivalent to the one given in this text.

Section 11.5 — Properties of Logarithms

Properties of logarithms:
If $b > 0$ and $b \neq 1$, then

1. $\log_b 1 = 0$

2. $\log_b b = 1$

3. $\log_b b^x = x$

4. $b^{\log_b x} = x, \quad x > 0$

M, N are both positive:

5. $\log_b MN = \log_b M + \log_b N$

6. $\log_b \dfrac{M}{N} = \log_b M - \log_b N$

7. $\log_b M^p = p \log_b M$

19. Write each expression in terms of logarithms of x, y, and z.

a. $\log_b \dfrac{x^2 y^3}{z^4}$

b. $\log_b \sqrt{\dfrac{x}{yz^2}}$

20. Write each expression as the logarithm of a single quantity.

a. $3 \log_b x - 5 \log_b y + 7 \log_b x$

b. $\dfrac{1}{2} \log_b x + 3 \log_b y - 7 \log_b z$

Section 11.6 — Exponential and Logarithmic Equations

Logarithmic property of equality: If $M > 0$ and $N > 0$, $\log_b M = \log_b N$ if and only if $M = N$.

21. Solve for x and verify solutions graphically.

a. $3^x = 7$

b. $5^{x+2} = 625$

c. $2^x = 3^{x-1}$

d. $\log x + \log(29 - x) = 2$

e. $\dfrac{\log(7x - 12)}{\log x} = 2$

f. $\log 3 - \log(x - 1) = -1$

g. $\ln x = \ln(x - 1)$

h. $\ln x = \ln(x - 1) + 1$

i. $\ln x = \log_{10} x$ (*Hint:* Remember to use the change-of-base formula.)

22. Carbon-14 dating   A wooden statue excavated in Egypt has a carbon-14 content that is two-thirds of that found in living wood. If the half-life of carbon-14 is 5,700 years, how old is the statue?

23. Radioactive decay   One-third of a radioactive material decays in 20 years. Find its half-life.

CHAPTER TEST

In Problems 1–4, $f(x) = x^2 + 5x + 6$ and $g(x) = x + 3$.

1. Find the simplified combined function $f + g$ and state the domain.

2. Find the simplified combined function $f - g$ and state the domain.

3. Find the simplified combined function fg and state the domain.

4. Find the simplified combined function $\dfrac{f}{g}$ and state the domain.

In Problems 5 and 6, $f(x) = 4x$ and $g(x) = x - 1$. Find each function.

5. a. $(f \circ g)(2)$ **b.** $(f \circ g)(x)$

6. a. $(g \circ f)(-3)$ **b.** $(g \circ f)(x)$

In Problems 7 and 8, find the inverse of each function.

7. $3x + 2y = 12$

8. $y = x^2 + 4$, $(x \le 0)$

9. Graph the function $y = \frac{1}{4}x^2 - 3$ on graph paper and use the horizontal line test to decide whether the function is one-to-one.

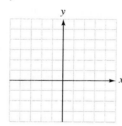

In Problems 10 and 11, graph each function on graph paper.

10. $y = 2^x + 1$ **11.** $y = 2^{-x}$

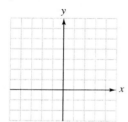 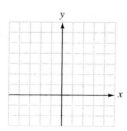

In Problems 12 and 13, solve each equation.

12. A radioactive material decays according to the formula $A = A_0(2)^{-t}$. How much of a 3-gram sample will be left in 6 years?

13. An account contains $2,000 and has been earning 4% interest, compounded continuously. How much will be in the account in 10 years?

14. Graph the function $y = e^x$ on graph paper.

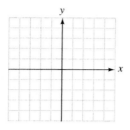

In Problems 15–17, find x.

15. $\log_4 16 = x$ **16.** $\log_x 81 = 4$

17. $\log_3 x = -3$

In Problems 18 and 19, graph each function on graph paper.

18. $y = -\log_3 x$ **19.** $y = \ln x$

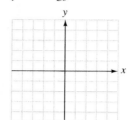 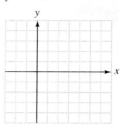

In Problems 20 and 21, write each expression in terms of the logarithms of a, b, and c.

20. $\log a^2bc^3$

21. $\ln \sqrt{\dfrac{a}{b^2 c}}$

In Problems 22 and 23, write each expression as a logarithm of a single quantity.

22. $\dfrac{1}{2} \log (a + 2) + \log b - 2 \log c$

23. $\dfrac{1}{3}(\log a - 2 \log b) - \log c$

In Problems 24 and 25, use the change-of-base formula to find each logarithm. Do not attempt to simplify the result.

24. $\log_5 24$ **25.** $\log_\pi e$

In Problems 26–29, tell whether each statement is true.

26. $\log_a ab = 1 + \log_a b$ **27.** $\dfrac{\log a}{\log b} = \log a - \log b$

28. $\log a^{-3} = \dfrac{1}{3 \log a}$ **29.** $\ln (-x) = -\ln x$

30. Find the pH of a solution with a hydrogen ion concentration of 3.7×10^{-7}. (*Hint:* pH $= -\log [H^+]$.)

31. Find the db gain of an amplifier when $E_0 = 60$ volts and $E_I = 0.3$ volt. (*Hint:* db gain $= 20 \log E_0/E_I$.)

In Problems 32 and 33, solve each equation. Do not simplify the logarithms.

32. $5^x = 3$ **33.** $3^{x-1} = 100^x$

In Problems 34 and 35, solve each equation.

34. $\log (5x + 2) = \log (2x + 5)$

35. $\log x + \log (x - 9) = 1$

CAREERS & MATHEMATICS

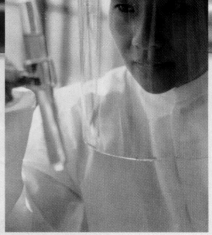

© Photodisc Collection/Getty Images

Chemist Chemists search for new knowledge about substances and put it to practical use. Over half of all chemists work in research and development. In basic research, chemists investigate the properties, composition, and structure of matter and the laws that govern the combination of elements and the reactions of substances. Their research has resulted in the development of a tremendous variety of synthetic materials, of ingredients that have improved other substances, and of processes that help save energy and reduce pollution. In applied research and development, they create new products or improve existing ones, often using knowledge gained from basic research.

Qualifications A bachelor's degree in chemistry or a related discipline is sufficient for many beginning jobs as a chemist. However, graduate training is required for most research jobs, and most college teaching jobs require a Ph.D. degree.

Job Outlook The employment of chemists is expected to grow about as fast as the average for all occupations through the year 2010.

The majority of job openings are expected to be in pharmaceutical and biotechnology firms, primarily in the development of new products. In addition, industrial companies and government agencies will need chemists to help solve problems related to energy shortages, pollution control, and health care.

Example Application A certain weak acid (0.1 M concentration) will break down into free cations (the hydrogen ion, H^+) and anions (A^-). When this acid dissociates, the following equilibrium equation is established.

$$\frac{[H^+][A^-]}{[HA]} = 4 \times 10^{-4}$$

where $[H^+]$, the hydrogen ion concentration, is equal to $[A^-]$, the anion concentration. $[HA]$ is the concentration of the undissociated acid itself. Find $[H^+]$ at equilibrium.

Solution Let x be the concentration of H^+. Then x is also $[A^-]$. From chemistry, it turns out that the concentration $[HA]$ of the undissociated acid is $0.1 - x$. Substituting these concentrations into the equation gives the equation

$$\frac{x^2}{0.1 - x} = 4 \times 10^{-4}$$

which can be solved as follows:

$$\frac{x^2}{0.1 - x} = 0.0004 \qquad \text{Because } 4 \times 10^{-4} = 0.0004.$$

$$x^2 = 0.0004(0.1 - x) \qquad \text{Multiply both sides by } 0.1 - x.$$

$$x^2 = 0.00004 - 0.0004x \qquad \text{Remove parentheses.}$$

$$x^2 + 0.0004x - 0.00004 = 0 \qquad \text{Add } 0.0004x - 0.00004 \text{ to both sides.}$$

Substitute into the quadratic formula with $a = 1$, $b = 0.0004$, and $c = -0.00004$. We only need to consider the positive root.

$$x = \frac{-b + \sqrt{b^2 - 4ac}}{2a}$$

$$x = \frac{-0.0004 + \sqrt{(0.0004)^2 - 4(1)(-0.00004)}}{2(1)}$$

$$x = \frac{-0.0004 + 0.01266}{2}$$

$$x = 0.00613$$

At equilibrium, $[H^+]$ is approximately 6.13×10^{-3} M.

EXERCISES   *Use your calculator to solve these problems.*

1. A saturated solution of hydrogen sulfide (concentration 0.1 M) dissociates into cation H^+ and anion HS^-. When this solution dissociates, the following equilibrium equation is established.

$$\frac{[H^+][HS^-]}{[HHS]} = 1.0 \times 10^{-7}$$

Find $[H^+]$.

2. An HS^- anion dissociates into cation H^+ and anion S^{--} with equilibrium equation

$$\frac{[H^+][S^{--}]}{[HS^-]} = 1.3 \times 10^{-13}$$

Assume the concentration of HS^- to be 1×10^{-4} M. Find $[H^+]$.

3. Suppose the concentration of the acid of the example were 0.2 M. Find the hydrogen ion concentration, $[H^+]$, at equilibrium.

4. Show that the equation of the example has only one meaningful solution by showing that its other solution is negative.

(Answers: **1.** 9.995×10^{-5} **2.** 3.605×10^{-9} **3.** $8.75 \times 10^{-3})$

Systems of Equations and Inequalities

© Arvind Garg/CORBIS

 InfoTrac Project

Do a keyword search on "real estate prices." Write a summary of the article "Cantor Pecorella reports strong demand for Greenwich Village condo tower."

A prospective buyer is trying to decide between buying or renting the cheapest house listed in the article, knowing that, if she buys, upkeep on the condo will cost about 10% of the purchase price each year. Write an equation for the total cost, C for n years, if she buys the condo. If she rents, the rent will be $12,000 a year, with all upkeep expenses paid by the owner. Write an equation for the total cost, C, of renting for n years. Using a system of equations, find the number of years it will take for the cost of renting to equal the cost of purchasing the condo. Find the total cost that will have been put into the condo at this point. What other factors that may help the prospective buyer in her decision has this problem failed to take into consideration?

Complete this project after studying Section 12.2.

PERSPECTIVE

Stay in Bed!

As a child, René Descartes, the inventor of analytic geometry, was frail and often sick. To improve his health, eight-year-old René was sent to a Jesuit school. The headmaster encouraged him to sleep in the morning as long as he wished, often past noon. As a young man, Descartes spent several years as a soldier and world traveler, but his interests included mathematics and philosophy, as well as science, literature, and writing, and taking it easy: The habit of sleeping late continued throughout his life. Descartes claimed that his times of most productive thinking were spent lying in bed, long past breakfast. According to one story, Descartes first thought of analytic geometry as he watched the path of a fly walking on his bedroom ceiling.

Descartes might have lived longer if he had stayed in bed. In 1649, Sweden's Queen Christina decided that she needed a tutor in philosophy, and she requested the service of Descartes. Tutoring would not have been difficult, except that the stubborn queen scheduled her lessons before dawn in her library, with the windows open. The cold Stockholm weather and the early hour was too much for a man who was used to sleeping until noon. Within a few months, Descartes developed a fever and died, probably of pneumonia.

This chapter will consider **systems** of both equations and inequalities and methods for finding solutions to these systems. A system is a group of equations or inequalities. Solutions of systems are those values of the variable(s) that make all the equations (or inequalities) in the system true.

12.1 SOLVING SYSTEMS OF TWO EQUATIONS IN TWO VARIABLES

- SOLVING SYSTEMS GRAPHICALLY
- THE SUBSTITUTION METHOD

- THE ADDITION (ELIMINATION) METHOD

Consider the two linear equations in two variables $x + y = 3$ and $3x - y = 1$. Although there are an infinite number of ordered pairs that satisfy each of these equations, only one ordered pair, $(1, 2)$, satisfies both equations.

The pair of equations,

$$\begin{cases} x + y = 3 \\ 3x - y = 1 \end{cases}$$

is called a **system of equations**. Because the ordered pair $(1, 2)$ satisfies both equations, it is the **solution of the system**. Notice that because the original equations each contained the variables x and y the solution is an ordered pair, (x, y).

 ### SOLVING SYSTEMS GRAPHICALLY

One way to solve a system is by graphing each equation in the same rectangular coordinate system. The coordinates of the point (or points) where the graphs intersect are the solutions of the system. We suggest that you always check solutions in the original equations.

EXAMPLE 1

Solve the system $\begin{cases} x + 2y = 4 \\ 2x - y = 3 \end{cases}$ graphically.

SOLUTION   We graph both equations on a single set of coordinates axes, as shown in Figure 12-1. Although an infinite number of ordered pairs (x, y), satisfy the equation $x + 2y = 4$, and an infinite number of ordered pairs, (x, y) satisfy the equation $2x - y = 3$, only one ordered pair, $(2, 1)$, satisfies both equations in the system.

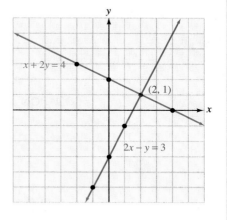

$x + 2y = 4$		
x	y	(x, y)
4	0	$(4, 0)$
0	2	$(0, 2)$
−2	3	$(−2, 3)$

$2x - y = 3$		
x	y	(x, y)
1	−1	$(1, −1)$
0	−3	$(0, −3)$
−1	−5	$(−1, −5)$

Figure 12-1

We can confirm our solution by substituting 2 for x and 1 for y in each of the equations. The resulting statements are true.

Check:

$$
\begin{array}{ll}
x + 2y = 4 & 2x - y = 3 \\
2 + 2(\mathbf{1}) = 4 & 2(\mathbf{2}) - \mathbf{1} = 3 \qquad \text{Substitute 2 for } x \text{ and 1 for } y \\
2 + 2 = 4 & 4 - 1 = 3 \\
4 = 4\ \checkmark & 3 = 3\ \checkmark
\end{array}
$$

We can use the INTERSECT feature of the calculator to solve the system. However, we must first solve each equation in the system for the variable y.

$$x + 2y = 4$$
$$2y = -x + 4 \qquad \text{Subtract } x \text{ from both sides.}$$
$$y = -\frac{1}{2}x + 2 \qquad \text{Divide through the equation by 2.}$$

$$2x - y = 3$$
$$-y = -2x + 3 \qquad \text{Subtract } 2x \text{ from both sides.}$$
$$y = 2x - 3 \qquad \text{Multiply through the equation by } -1 \text{ to make } y \text{ positive.}$$

Enter $y = -\frac{1}{2}x + 2$ at the y1 prompt and $y = 2x - 3$ at the y2 prompt. Use the INTERSECT feature to find the *coordinates* of the ordered pair solution of the system, $(2, 1)$, for this example.

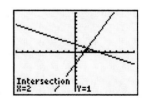

Our solution can be confirmed with the TABLE. Set the TABLE to begin at the x-value of 2 (the increment is immaterial). The expressions entered at the y1 and y2 prompts both evaluate to 1 when x has a value of 2.

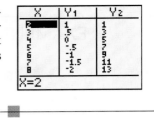

When a system of equations (as in Example 1) has a solution, the system is a **consistent system of equations**. A system with no solution is an **inconsistent system** and is illustrated in the next example.

EXAMPLE 2   Solve the system $\begin{cases} 2x + 3y = 6 \\ 4x + 6y = 24 \end{cases}$.

SOLUTION   We graph both equations on the same set of coordinate axes, as shown in Figure 12-2. The lines do not intersect, so the system does not have a solution. It is an inconsistent system.

$2x + 3y = 6$

x	y	(x, y)
3	0	$(3, 0)$
0	2	$(0, 2)$
−3	4	$(−3, 4)$

$4x + 6y = 24$

x	y	(x, y)
6	0	$(6, 0)$
0	4	$(0, 4)$
−3	6	$(−3, 6)$

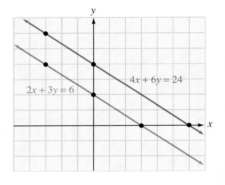

Figure 12-2

To look at the graphical representations on the calculator, begin by solving each equation for the variable y.

$$2x + 3y = 6$$
$$3y = -2x + 6 \qquad \text{Subtract } 2x \text{ from both sides.}$$
$$y = -\frac{2}{3}x + 2 \qquad \text{Divide through the equation by 3.}$$

$$4x + 6y = 24$$
$$6y = -4x + 24 \qquad \text{Subtract } 4x \text{ from both sides.}$$
$$y = -\frac{2}{3}x + 4 \qquad \text{Divide through the equation by 6 and reduce fractions.}$$

Because the slopes are the same, the lines are parallel and have no points of intersection. *Do not conclude that lines are parallel simply from the graphical display. Support the decision of parallelism analytically and confirm graphically.*

Solution set: ∅.

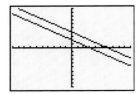

When the equations of a system have different graphs (as in Examples 1 and 2), the equations are **independent equations**. Two equations with the same graph are **dependent equations**.

EXAMPLE 3   Solve the system $\begin{cases} 2y - x = 4 \\ 2x + 8 = 4y \end{cases}$.

SOLUTION   We graph each equation on the same set of coordinate axes, as shown in Figure 12-3.

$2y - x = 4$

x	y	(x, y)
-4	0	$(-4, 0)$
0	2	$(0, 2)$

$2x + 8 = 4y$

x	y	(x, y)
-4	0	$(-4, 0)$
0	2	$(0, 2)$

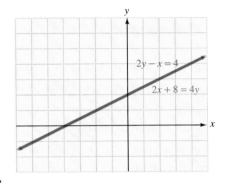

Figure 12-3

Because the graphs coincide, the system has infinitely many solutions. Any ordered pair, (x, y) that satisfies one equation satisfies the other equation as well. Because the two equations have the same graph, they are dependent equations. The solution set could be expressed as $\{(x, y) \mid 2y - x = 4\}$ or $\{(x, y) \mid 2x + 8 = 4y\}$.

When two linear equations in two variables are graphed, the following possibilities can occur:

If the lines are different and intersect, the equations are independent and the system is consistent. *One solution exists.*

If the lines are different and parallel, the equations are independent and the system is inconsistent. *No solution exists.*

If the lines coincide, the equations are dependent and the system is consistent. *Infinitely many simultaneous solutions exist.*

▮ THE SUBSTITUTION METHOD

A system of two equations in two variables can be solved algebraically by either the substitution or the elimination (often called *addition*) method. We will first consider the substitution method.

Substitution Method for Solving Systems of Equations

1. If necessary, solve one equation for one of its variables.
2. Substitute the resulting expression for the variable obtained in step 1 into the *other* equation and solve that equation.
3. Find the value of the other variable by substituting the value of the variable found in step 2 in any equation containing both variables.
4. Check the solution in both of the original equations.

EXAMPLE 4   Use the substitution method to solve $\begin{cases} 3x + y = 4 \\ 2x + 3y = 6 \end{cases}$.

SOLUTION   We begin by selecting an equation and solving for either x or y.

1. $\begin{cases} 3x + y = 4 \\ 2x + 3y = 6 \end{cases}$
2.

Since the y variable in Equation 1 has a coefficient of one, it is the best variable selection because no fractions will occur when we solve for y.

$$3x + y = 4$$
3. $$y = -3x + 4 \qquad \text{Subtract } 3x \text{ from both sides.}$$

We can then substitute $-3x + 4$ for y in Equation 2 and solve for x.

$$2x + 3y = 6$$
$$2x + 3(-3x + 4) = 6 \qquad \text{Substitute } -3x + 4 \text{ for } y.$$
$$2x - 9x + 12 = 6 \qquad \text{Use the distributive property to remove parentheses.}$$
$$-7x + 12 = 6 \qquad \text{Combine like terms.}$$
$$-7x = -6 \qquad \text{Subtract 12 from both sides.}$$
$$x = \frac{6}{7} \qquad \text{Divide both sides by } -7 \text{ and simplify.}$$

We can find y by substituting $\frac{6}{7}$ for x in Equation 3 and simplifying:

$$y = -3x + 4$$
$$y = -3\left(\frac{6}{7}\right) + 4 \qquad \text{Substitute } \tfrac{6}{7} \text{ for } x.$$
$$y = -\frac{18}{7} + \frac{28}{7}$$
$$y = \frac{10}{7}$$

The solution is the ordered pair $(\frac{6}{7}, \frac{10}{7})$. Verify that this solution satisfies each equation in the original system. We can verify by solving graphically. However, the decimals displayed would then have to be converted to fractions. It is quicker to use the STOre feature of the calculator.

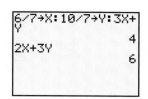

EXAMPLE 5   Use the substitution method to solve the system $\begin{cases} \dfrac{4}{3}x + \dfrac{1}{2}y = -\dfrac{2}{3} \\ \dfrac{1}{2}x + \dfrac{2}{3}y = \dfrac{5}{3} \end{cases}$.

SOLUTION   We first find an equivalent system without fractions by multiplying each side of both equations by 6 to obtain the system

1. $\begin{cases} 8x + 3y = -4 \\ 3x + 4y = 10 \end{cases}$
2.

Because no variable in either equation has a coefficient of 1, it is impossible to avoid fractions when solving for a variable. If we choose Equation 2 and solve it for x, we get

$$3x + 4y = 10$$
$$3x = -4y + 10 \qquad \text{Subtract } 4y \text{ from both sides.}$$
3. $$x = -\frac{4}{3}y + \frac{10}{3} \qquad \text{Divide both sides by 3.}$$

We can then substitute $-\frac{4}{3}y + \frac{10}{3}$ for x in Equation 1 and solve for y:

$$8x + 3y = -4$$
$$8\left(-\frac{4}{3}y + \frac{10}{3}\right) + 3y = -4 \qquad \text{Substitute } -\frac{4}{3}y + \frac{10}{3} \text{ for } x.$$
$$-\frac{32}{3}y + \frac{80}{3} + 3y = -4 \qquad \begin{array}{l}\text{Use the distributive property to remove}\\ \text{parentheses.}\end{array}$$
$$-32y + 80 + 9y = -12 \qquad \text{Multiply both sides by 3.}$$
$$-23y = -92 \qquad \begin{array}{l}\text{Combine like terms and subtract 80 from}\\ \text{both sides.}\end{array}$$
$$y = 4 \qquad \text{Divide both sides by } -23.$$

We can find x by substituting 4 for y in Equation 3 and simplifying:

$$x = -\frac{4}{3}y + \frac{10}{3}$$
$$= -\frac{4}{3}(4) + \frac{10}{3} \qquad \text{Substitute 4 for } y.$$
$$= -\frac{6}{3} \qquad -\frac{16}{3} + \frac{10}{3} = -\frac{6}{3}.$$
$$= -2$$

The solution is the ordered pair $(-2, 4)$. Verify that this solution satisfies each equation in the original system. We can verify the solutions in both equations by using the STOre feature of the graphing calculator.

▐▐▐ THE ADDITION (ELIMINATION) METHOD

The addition (elimination) method is another way to algebraically solve a system of equations. In this method, we combine the equations of the system in a way that will eliminate terms involving one of the variables.

Addition (Elimination) Method for Solving Systems of Equations

1. Write both equations of the system in general form.
2. Multiply the terms of one or both of the equations by constants chosen to make the coefficients of x (or y) differ only in sign.
3. Add the equations and solve the equation that results, if possible.
4. Substitute the value obtained in step 3 into either of the original equations and solve for the remaining variable.
5. The results obtained in steps 3 and 4 are the solution of the system.
6. Check the solution in both of the original equations.

EXAMPLE 6

Use the addition method to solve the system $\begin{cases} 3x + y = 4 \\ 2x + 3y = 6 \end{cases}$.

SOLUTION   This is the system discussed in Example 4. To solve it by addition, we find an equivalent system with coefficients that will force either the x terms or y terms to drop out when we add the equations. Because the y term in the first equation has a coefficient of one, it provides the quickest route to eliminating the y terms. Multiply both sides of the first equation by -3 to get

$$\begin{cases} -9x - 3y = -12 \\ 2x + 3y = 6 \end{cases}$$

When these equations are added, the y terms are eliminated, and we get

$$-7x = -6$$

$$x = \frac{6}{7} \qquad \text{Divide both sides by } -7 \text{ and simplify.}$$

We can find y by substituting $\frac{6}{7}$ for x in the second equation and simplifying:

$$y = -3x + 4$$

$$y = -3\left(\frac{6}{7}\right) + 4 \qquad \text{Substitute } \tfrac{6}{7} \text{ for } x.$$

$$y = -\frac{18}{7} + \frac{28}{7}$$

$$y = \frac{10}{7}$$

The solution is the ordered pair $\left(\frac{6}{7}, \frac{10}{7}\right)$. Verify that this solution satisfies each equation in the original system. We can verify the solutions in both equations by using the STOre feature of the calculator.

EXAMPLE 7

Use the addition method to solve the system $\begin{cases} \dfrac{4}{3}x + \dfrac{1}{2}y = -\dfrac{2}{3} \\ \dfrac{1}{2}x + \dfrac{2}{3}y = \dfrac{5}{3} \end{cases}$.

SOLUTION

This is the system discussed in Example 5. To solve it by addition, we find an equivalent system with no fractions by multiplying both sides of each equation by 6 to obtain

4. $\begin{cases} 8x + 3y = -4 \\ 3x + 4y = 10 \end{cases}$
5.

To eliminate the y terms when we add the equations, we multiply both sides of Equation 4 by 4 and both sides of Equation 5 by -3 to get

$$\begin{cases} 32x + 12y = -16 \\ -9x - 12y = -30 \end{cases}$$

When these equations are added, the y terms drop out, and we get

$$23x = -46$$
$$x = -2 \qquad \text{Divide both sides by 23.}$$

To find y, we substitute -2 for x in either Equation 4 or Equation 5. If we substitute -2 for x in Equation 5, we get

$$3x + 4y = 10$$
$$3(-2) + 4y = 10 \qquad \text{Substitute } -2 \text{ for } x.$$
$$-6 + 4y = 10 \qquad \text{Simplify.}$$
$$4y = 16 \qquad \text{Add 6 to both sides.}$$
$$y = 4 \qquad \text{Divide both sides by 4.}$$

The solution is $(-2, 4)$.

EXAMPLE 8

Solve the system $\begin{cases} y = 2x + 4 \\ 8x - 4y = 7 \end{cases}$ using any method.

SOLUTION

Because the first equation is already solved for y, we use the substitution method.

$$8x - 4y = 7$$
$$8x - 4(2x + 4) = 7 \qquad \text{Substitute } 2x + 4 \text{ for } y.$$

We then solve this equation for x:

$$8x - 8x - 16 = 7 \qquad \text{Use the distributive property to remove parentheses.}$$

$$-16 \neq 7 \qquad \text{Combine like terms.}$$

This impossible result shows that the equations in the system are independent and that the system is inconsistent. Because the system has no solution, the graphs of the equations in the system would be parallel. Solution set: $\varnothing$.

EXAMPLE 9

Solve the system $\begin{cases} 4x + 6y = 12 \\ -2x - 3y = -6 \end{cases}$ by any method.

SOLUTION   Because the equations are written in general form, we use the addition method and multiply both sides of the second equation by 2 to get

$$\begin{cases} 4x + 6y = 12 \\ -4x - 6y = -12 \end{cases}$$

After adding the left-hand sides and the right-hand sides, we get

$$0x + 0y = 0$$
$$0 = 0$$

Here, both the x and y terms drop out. The true statement $0 = 0$ shows that the equations in this system are dependent and that the system is consistent. Note that the equations of the system are equivalent because, when the second equation is multiplied by -2, it becomes the first equation. The line graphs of these equations would coincide. Any ordered pair that satisfies one of the equations also satisfies the other. In set notation this would be $\{(x, y) \mid -2x - 3y = -6\}$.

EXERCISE 12.1

VOCABULARY AND NOTATION   *Fill in the blanks to make a true statement.*

1. If two or more equations are considered at the same time, they are called a _____ of equations.

2. When a system of equations has one or more solutions, it is called a _____ system.

3. If a system has no solutions, it is called an _____ system.

4. If two equations have different graphs, they are called _____ equations.

5. Two equations with the same graph are called _____ equations.

6. If the two equations in a system are equivalent, the systems are called _____ systems.

CONCEPTS   *In Exercises 7–12, state whether the ordered pair is a solution of the system. Justify your response either analytically or by using the STOre feature of the calculator.*

7. $(55, 39)$; $\begin{cases} 2x - 3y = -7 \\ 4x - 5y = 25 \end{cases}$

8. $(-2, -3)$; $\begin{cases} 3x - 2y = 0 \\ 5x - 3y = -1 \end{cases}$

9. $\left(\dfrac{1}{2}, 3\right)$; $\begin{cases} 2x + y = 4 \\ 4x - 3y = 11 \end{cases}$

10. $\left(2, \dfrac{1}{3}\right); \begin{cases} x - 3y = 1 \\ -2x + 6y = -6 \end{cases}$

11. $\left(-\dfrac{2}{5}, \dfrac{1}{4}\right); \begin{cases} 5x - 4y = -6 \\ 8y = 10x + 12 \end{cases}$

12. $\left(-\dfrac{1}{3}, \dfrac{3}{4}\right); \begin{cases} 3x + 4y = 2 \\ 12y = 3(2 - 3x) \end{cases}$

13. Equations were entered at the y1 and y2 prompts of the calculator. State the solution of the system by interpreting the given screen display.

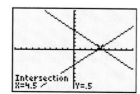

14. Equations were entered at the y1 and y2 prompts of the calculator. State the solution of the system by interpreting the given TABLE display.

15. Equations were entered at the y1 and y2 prompts of the calculator.

    a. State the solution of the system by interpreting the TABLE results.

    b. Are the equations *independent* or *dependent*?

    c. If you were looking at the graphical representation of the equations, would the lines be *parallel* or *coincident*?

16. Equations were entered at the y1 and y2 prompts of the calculator.

    a. State the solution of the system by interpreting the TABLE results.

    b. Are the equations *independent* or *dependent*?

    c. If you were looking at the graphical representation of the equations, would the lines be *parallel* or *coincident*?

PRACTICE   *In Exercises 17–28, solve each system by graphing, if possible. Use hand-drawn graphs. Check your graphs with your calculator.*

17. $\begin{cases} x - y = 4 \\ 2x + y = 5 \end{cases}$ **18.** $\begin{cases} 2x + y = 1 \\ x - 2y = -7 \end{cases}$

19. $\begin{cases} x = 13 - 4y \\ 3x = 4 + 2y \end{cases}$ **20.** $\begin{cases} 3x = 7 - 2y \\ 2x = 2 + 4y \end{cases}$

21. $\begin{cases} x = 3 - 2y \\ 2x + 4y = 6 \end{cases}$

22. $\begin{cases} 3x = 5 - 2y \\ 3x + 2y = 7 \end{cases}$ **23.** $\begin{cases} y = 3 \\ x = 2 \end{cases}$

24. $\begin{cases} 2x + 3y = -15 \\ 2x + y = -9 \end{cases}$

25. $\begin{cases} x = \dfrac{11 - 2y}{3} \\ y = \dfrac{11 - 6x}{4} \end{cases}$ **26.** $\begin{cases} x = \dfrac{1 - 3y}{4} \\ y = \dfrac{12 + 3x}{2} \end{cases}$

27. $\begin{cases} \dfrac{5}{2}x + y = \dfrac{1}{2} \\ 2x - \dfrac{3}{2}y = 5 \end{cases}$ **28.** $\begin{cases} \dfrac{5}{2}x + 3y = 6 \\ y = \dfrac{24 - 10x}{12} \end{cases}$

In Exercises 29–32, use a graphing calculator to solve each system. Give answers to the nearest hundredth.

29. $\begin{cases} y = 3.2x - 1.5 \\ y = -2.7x - 3.7 \end{cases}$

30. $\begin{cases} y = -0.45x + 5 \\ y = 5.55x - 13.7 \end{cases}$

31. $\begin{cases} 1.7x + 2.3y = 3.2 \\ y = 0.25x + 8.95 \end{cases}$

32. $\begin{cases} 2.75x = 12.9y - 3.79 \\ 7.1x - y = 35.76 \end{cases}$

In Exercises 33–44, solve each system by substitution, if possible. Check results using the STOre feature of the calculator.

33. $\begin{cases} y = x \\ x + y = 4 \end{cases}$

34. $\begin{cases} y = x + 2 \\ x + 2y = 16 \end{cases}$

35. $\begin{cases} x - y = 2 \\ 2x + y = 13 \end{cases}$

36. $\begin{cases} x - y = -4 \\ 3x - 2y = -5 \end{cases}$

37. $\begin{cases} x + 2y = 6 \\ 3x - y = -10 \end{cases}$

38. $\begin{cases} 2x - y = -21 \\ 4x + 5y = 7 \end{cases}$

39. $\begin{cases} 3x = 2y - 4 \\ 6x - 4y = -4 \end{cases}$

40. $\begin{cases} 8x = 4y + 10 \\ 4x - 2y = 5 \end{cases}$

41. $\begin{cases} 3x - 4y = 9 \\ x + 2y = 8 \end{cases}$

42. $\begin{cases} 3x - 2y = -10 \\ 6x + 5y = 25 \end{cases}$

43. $\begin{cases} 2x + 2y = -1 \\ 3x + 4y = 0 \end{cases}$

44. $\begin{cases} 5x + 3y = -7 \\ 3x - 3y = 7 \end{cases}$

In Exercises 45–56, solve each system by addition (elimination), if possible. Check results using the STOre feature of the calculator.

45. $\begin{cases} x - y = 3 \\ x + y = 7 \end{cases}$

46. $\begin{cases} x + y = 1 \\ x - y = 7 \end{cases}$

47. $\begin{cases} 2x + y = -10 \\ 2x - y = -6 \end{cases}$

48. $\begin{cases} x + 2y = -9 \\ x - 2y = -1 \end{cases}$

49. $\begin{cases} 8x - 4y = 16 \\ 2x - 4 = y \end{cases}$

50. $\begin{cases} 2y - 3x = -13 \\ 3x - 17 = 4y \end{cases}$

51. $\begin{cases} x = \dfrac{3}{2}y + 5 \\ 2x - 3y = 8 \end{cases}$

52. $\begin{cases} x = \dfrac{2}{3}y \\ y = 4x + 5 \end{cases}$

53. $\begin{cases} \dfrac{x}{2} + \dfrac{y}{2} = 6 \\ \dfrac{x}{2} - \dfrac{y}{2} = -2 \end{cases}$

54. $\begin{cases} \dfrac{x}{2} - \dfrac{y}{3} = -4 \\ \dfrac{x}{2} + \dfrac{y}{9} = 0 \end{cases}$

55. $\begin{cases} \dfrac{3}{4}x + \dfrac{2}{3}y = 7 \\ \dfrac{3}{5}x - \dfrac{1}{2}y = 18 \end{cases}$

56. $\begin{cases} \dfrac{2}{3}x - \dfrac{1}{4}y = -8 \\ \dfrac{1}{2}x - \dfrac{3}{8}y = -9 \end{cases}$

57. Tell which method you would use to solve the following system. Why?

$$\begin{cases} y = 3x + 1 \\ 3x + 2y = 12 \end{cases}$$

58. Use a graphing calculator to solve the system

$$\begin{cases} 11x - 20y = 21 \\ -4x + 7y = 21 \end{cases}$$

59. Two students are solving the equation $2(x + 5) - 6 = 8$ graphically. They both use the INTERSECT feature of the graphing calculator to determine the solution. Their screen is pictured at the right. Blaine states the solution is $x = 2$, $y = 8$, or $(2, 8)$. David states the solution is 2. Who is correct and *why* is he correct? Explain fully.

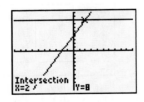

60. Solve the system $\begin{cases} y = 3(x - 1)^2 + 2 \\ y = 2x + 1 \end{cases}$

   a. algebraically   **b.** graphically

REVIEW

61. Solve the equation $2(4x + 5) - 3 = 23$

   a. algebraically

   b. using the INTERSECT feature of the graphing calculator. Specify the viewing window used if it is different from ZStandard.

62. Solve the equation $x^2 + 6x + 8 = 0$

   a. algebraically

   b. using the ROOT/ZERO or INTERSECT feature of the calculator. Specify the viewing window used if it is different from ZStandard.

63. Solve the equation $3\sqrt{x} + 4 = 22$

   a. algebraically

   b. using the INTERSECT feature of the calculator. Specify the viewing window used if it is different from ZStandard.

12.2 APPLICATIONS OF SYSTEMS OF TWO EQUATIONS IN TWO VARIABLES

The use of a system of equations often makes application problems easier to set up. However, it must be remembered that if two variables are used, then two equations must also be specified. The following steps are helpful when setting up applications that involve systems:

Solving Application Problems Using a System of Equations

1. Read the problem several times and analyze the facts. Occasionally, a sketch, chart, or diagram will help in visualizing the problem.
2. Pick different variables to represent the unknown quantities and clearly state what each variable represents.
3. Write two equations involving each of the two variables. This will produce a system of two equations in two variables.
4. Solve the system using the most convenient method.
5. Verify solutions in the words of the problem.

TECHNOLOGY TIP

If the calculator is to be used to find solutions, we suggest that the variables x and y be used to avoid confusion. Common sense is the biggest help in setting a good window. **Think** about what each variable represents and set the viewing window accordingly. Remember, if finding the point of intersection with the calculator becomes too frustrating, algebraic methods may be used, with the calculator providing a check.

EXAMPLE 1

Farming A farmer raises wheat and soybeans on 215 acres. If he wants to plant 31 more acres in wheat than in soybeans, how many acres of each should he plant?

SOLUTION If w represents the number of acres of wheat and s represents the number of acres of soybeans to be planted, we have

The number of acres planted in wheat	+	the number of acres planted in soybeans	=	215 acres
w	+	s	=	215

Since the farmer wants to plant 31 more acres in wheat than in soybeans, we have

The number of acres planted in wheat	−	the number of acres planted in soybeans	=	31 acres
w	−	s	=	31

We can now solve the system

1. $\begin{cases} w + s = 215 \\ w - s = 31 \end{cases}$
2.

by using the addition method.

$$w + s = 215$$
$$\underline{w - s = 31}$$
$$2w = 246$$
$$w = 123 \qquad \text{Divide both sides by 2.}$$

To find s, we substitute 123 for w in Equation 1.

$$w + s = 215$$
$$\mathbf{123} + s = 215 \qquad \text{Substitute 123 for } w.$$
$$s = 92 \qquad \text{Add } -123 \text{ to both sides.}$$

The farmer should plant 123 acres of wheat and 92 acres of soybeans.

Check: The total acreage planted is $123 + 92$, or 215 acres. The area planted in wheat is 31 acres greater than that planted in soybeans, because $123 - 92 = 31$.

EXAMPLE 2

Lawn care An installer of underground irrigation systems wants to cut a 20-foot length of plastic tubing into two pieces. The longer piece is to be 2 feet longer than twice the shorter piece. Find the length of each piece.

SOLUTION We can let s represent the length of the shorter piece and l represent the length of the longer piece (see Figure 12-4).

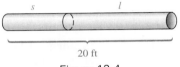

20 ft

Figure 12-4

The length of tubing is 20 feet, so we have

The length of the shorter piece	+	the length of the longer piece	=	20 feet
s	$+$	l	$=$	20

Because the longer piece is 2 feet longer than twice the shorter piece, we have

The length of the longer piece	=	2	$\cdot$	the length of the shorter piece	+	2 feet
l	$=$	2	$\cdot$	s	$+$	2

We can use the substitution method to solve the system.

$$
\begin{aligned}
\textbf{1.} & \quad \begin{cases} s + l = 20 \\ l = 2s + 2 \end{cases} \\
\textbf{2.} &
\end{aligned}
$$

$$s + 2s + 2 = 20 \qquad \text{Substitute } 2s + 2 \text{ for } l.$$
$$3s + 2 = 20 \qquad \text{Combine like terms.}$$
$$3s = 18 \qquad \text{Subtract 2 from both sides.}$$
$$s = 6 \qquad \text{Divide both sides by 3.}$$

The shorter piece should be 6 feet long.

To find the length of the longer piece, we substitute 6 for s in Equation 1 and solve for l.

$$s + l = 20$$
$$6 + l = 20$$
$$l = 14 \qquad \text{Subtract 6 from both sides.}$$

The longer piece should be 14 feet long.

Check: The sum of 6 and 14 is 20. 14 is 2 more than twice 6.

EXAMPLE 3

Gardening   Tom has 150 feet of fencing to enclose a rectangular garden. If the length is to be 5 feet less than 3 times the width, find the area of the garden.

SOLUTION   We can let l represent the length of the garden and w represent the width (see Figure 12-5). The perimeter of a rectangle is two lengths plus two widths, so we have

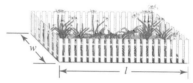

Figure 12-5

| 2 | · | the length of the garden | + | 2 | · | the width of the garden | = | 150 feet |

$$2 \quad \cdot \quad l \quad + \quad 2 \quad \cdot \quad w \quad = \quad 150$$

Because the length is 5 feet less than 3 times the width,

| The length of the garden | = | 3 | · | the width of the garden | − | 5 feet |

$$l \quad = \quad 3 \quad \cdot \quad w \quad - \quad 5$$

We can use the substitution method to solve this system.

1. $\begin{cases} 2l + 2w = 150 \\ l = 3w - 5 \end{cases}$
2.

$$2(3w - 5) + 2w = 150 \qquad \text{Substitute } 3w - 5 \text{ for } l \text{ in Equation 1.}$$
$$6w - 10 + 2w = 150 \qquad \text{Remove parentheses.}$$
$$8w - 10 = 150 \qquad \text{Combine like terms.}$$
$$8w = 160 \qquad \text{Add 10 to both sides.}$$
$$w = 20 \qquad \text{Divide both sides by 8.}$$

The width is 20 feet.

To find the length, we substitute 20 for w in Equation 2 and simplify.

$$l = 3w - 5$$
$$= 3(20) - 5$$
$$= 60 - 5$$
$$= 55$$

The dimensions of the rectangle are 55 feet by 20 feet, and the area of a rectangle is given by the formula

$$A = l \cdot w \qquad \text{Area} = \text{length times width.}$$

so we have

$$A = 55 \cdot 20$$
$$= 1100$$

The garden covers an area of 1100 square feet.

Check: Because the dimensions of the garden are 55 feet by 20 feet, the perimeter is

$$P = 2l + 2w$$
$$= 2(55) + 2(20)$$
$$= 110 + 40$$
$$= 150 \text{ feet}$$

It is also true that 55 feet is 5 feet less than 3 times 20 feet.

EXAMPLE 4

Manufacturing   The setup cost of a machine that mills brass plates is $750. After setup, it costs $0.25 to mill each plate. Management is considering the purchase of a larger machine that can produce the same plate at a cost of $0.20 per plate. If the setup cost of the larger machine is $1,200, how many plates would the company have to produce to make the purchase worthwhile?

SOLUTION   We begin by finding the number of plates (called the **break point**) that will cost equal amounts to produce on either machine. We can let c represent the cost of milling p plates. Then we have

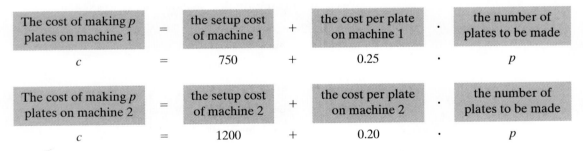

The cost of making p plates on machine 1	=	the setup cost of machine 1	+	the cost per plate on machine 1	·	the number of plates to be made
c	=	750	+	0.25	·	p

The cost of making p plates on machine 2	=	the setup cost of machine 2	+	the cost per plate on machine 2	·	the number of plates to be made
c	=	1200	+	0.20	·	p

Because we want to know when the costs are equal, we can use the substitution method to solve the system

$$\begin{cases} c = 750 + 0.25p \\ c = 1200 + 0.20p \end{cases}$$

$750 + 0.25p = 1200 + 0.20p$	Substitute $750 + 0.25p$ for c in the second equation.
$0.25p = 450 + 0.20p$	Subtract 750 from both sides.
$0.05p = 450$	Subtract $0.20p$ from both sides.
$p = 9,000$	Divide both sides by 0.05.

Current machine
$c = 750 + 0.25p$

New, larger machine
$c = 1{,}200 + 0.20p$

p	c
0	750
1,000	1,000
5,000	2,000

p	c
0	1,200
4,000	2,000
12,000	3,600

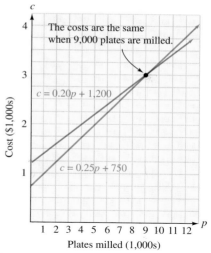

Figure 12-6

If 9,000 plates are milled, the cost will be the same on either machine. If more than 9,000 plates are milled, the cost will be cheaper on the newer machine, and it should be purchased.

EXAMPLE 5

Investments   Terri and Juan earned $550 from a 1-year investment of $15,000. If Terri invested some of the money at 4% interest and Juan invested the rest at 3%, how much did each invest?

SOLUTION   We can let x represent the amount of money invested by Terri and y represent the amount of money invested by Juan. Because the total investment is $15,000, we have

The amount invested by Terri	+	the amount invested by Juan	=	$15,000
x	+	y	=	15,000

The income on x dollars invested at 4% is $0.04x$, the income on y dollars invested at 3% is $0.03y$, and the combined income is $550, so we have

The income on the 4% investment	+	the income on the 3% investment	=	$550
$0.04x$	+	$0.03y$	=	550

Thus, we have the system

1.
2.
$$\begin{cases} x + y = 15{,}000 \\ 0.04x + 0.03y = 550 \end{cases}$$

To solve the system, we use the addition method.

$$-4x - 4y = -60,000 \qquad \text{Multiply both sides of Equation 1 by } -4.$$
$$\underline{4x + 3y = 55,000} \qquad \text{Multiply both sides of Equation 2 by 100.}$$
$$-y = -5,000$$
$$y = 5,000 \qquad \text{Multiply both sides by } -1.$$

To find x, we substitute 5000 for y in Equation 1 and simplify.

$$x + y = 15,000$$
$$x + \mathbf{5000} = 15,000$$
$$x = 10,000 \qquad \text{Subtract 5,000 from both sides.}$$

Terri invested \$10,000 and Juan invested \$5,000.

Check: \$10,000 + \$5,000 = \$15,000

$$0.04(\$10,000) = \$400$$
$$0.03(\$5000) = \$150$$

The combined interest is \$550.

EXAMPLE 6

Boating   A boat traveled 30 kilometers downstream in 3 hours and made the return trip in 5 hours. Find the speed of the boat in still water.

SOLUTION   We can let s represent the speed of the boat in still water and let c represent the speed of the current. Then the rate of speed of the boat while going downstream is $s + c$. The rate of the boat while going upstream is $s - c$. We can organize the information of the problem as in Table 12-1.

TABLE 12-1

	Distance	= Rate	· Time
Downstream	30	$s + c$	3
Upstream	30	$s - c$	5

Because $d = r \cdot t$, the information in the table gives two equations in two variables.

$$\begin{cases} 30 = 3(s + c) \\ 30 = 5(s - c) \end{cases}$$

After removing parentheses and applying the symmetric property, we have

1. $\begin{cases} 3s + 3c = 30 \\ 5s - 5c = 30 \end{cases}$
2.

To solve this system by addition, we multiply Equation 1 by 5 and Equation 2 by 3, add the equations, and solve for s.

$$15s + 15c = 150$$
$$\underline{15s - 15c = 90}$$
$$30s = 240$$
$$s = 8 \qquad \text{Divide both sides by 30.}$$

The speed of the boat in still water is 8 kilometers per hour. Check the result.

EXAMPLE 7

Medical technology   A laboratory technician has one batch of antiseptic that is 40% alcohol and a second batch that is 60% alcohol. She would like to make 8 liters of solution that is 55% alcohol. How many liters of each batch should she use?

SOLUTION

We can let x represent the number of liters to be used from batch 1 and let y represent the number of liters to be used from batch 2.

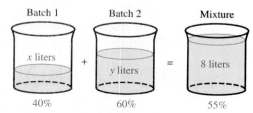

Figure 12-7

1. $x + y = 8$ The number of liters of batch 1 plus the number of liters of batch 2 equals the total number of liters in the mixture.

2. $0.40x + 0.60y = 0.55(8)$ The amount of alcohol in batch 1 plus the amount of alcohol in batch 2 equals the amount of alcohol in the mixture.

We can use addition to solve this system.

$$-40x - 40y = -320 \quad \text{Multiply both sides of Equation 1 by } -40.$$
$$\underline{40x + 60y = 440} \quad \text{Multiply both sides of Equation 2 by 100.}$$
$$20y = 120$$
$$y = 6 \quad \text{Divide both sides by 20.}$$

To find x, we substitute 6 for y in Equation 1 and simplify.

$$x + y = 8$$
$$x + 6 = 8$$
$$x = 2 \quad \text{Subtract 6 from both sides.}$$

The technician should use 2 liters of the 40% solution and 6 liters of the 60% solution. Check the result.

EXERCISE 12.2

Use two equations in two variables to solve each problem.

1. **Integer problem**   One integer is twice another, and their sum is 96. Find the integers.

2. **Integer problem**   The sum of two integers is 38, and their difference is 12. Find the integers.

3. **Integer problem**   Three times one integer plus another integer is 29. If the first integer plus twice the second is 18, find the integers.

4. **Integer problem**   Twice one integer plus another integer is 21. If the first integer plus 3 times the second is 33, find the integers.

5. **Raising livestock**   A rancher raises five times as many cows as horses. If he has 168 animals, how many head of cattle does he have?

6. **Grass seed mixture**   A landscaper used 100 pounds of grass seed containing twice as much bluegrass as rye. She adds 15 more pounds of bluegrass to the mixture

before seeding a lawn. How many pounds of bluegrass did she use?

7. **Buying painting supplies**   Two partial receipts for paint supplies appear in Illustration 1. How much does each gallon of paint and each brush cost?

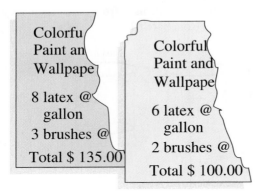

Illustration 1

8. **Buying baseball equipment**   One catcher's mitt and 10 outfielder's gloves cost $239.50. How much does each cost if one catcher's mitt and 5 outfielder's gloves cost $134.50?

9. **Buying contact lens cleaner**   Two bottles of contact lens cleaner and three bottles of soaking solution cost $29.40, and three bottles of cleaner and two bottles of soaking solution cost $28.60. Find the cost of each.

10. **Buying clothing**   Two pairs of shoes and four pairs of socks cost $109, and three pairs of shoes and five pairs of socks cost $160. Find the cost of a pair of socks.

11. **Cutting pipe**   A plumber wants to cut the pipe shown in Illustration 2 into two pieces so that one piece is 5 feet longer than the other. How long should each piece be?

Illustration 2

12. **Cutting lumber**   A carpenter wants to cut a 20-foot board into two pieces so that one piece is four times as long as the other. How long should each piece be?

13. **Splitting the lottery**   Maria and Susan pool their resources to buy several lottery tickets. They agree that Susan should get $50,000 more than Maria because she gave most of the money. If they win $250,000, how much will Susan get?

14. **Figuring inheritances**   According to a will, a man left his older son $10,000 more than twice as much as he left his younger son. If the estate is worth $497,500, how much will the younger son get?

15. **Geometry**   The perimeter of the rectangle shown in Illustration 3 is 110 feet. Find its dimensions.

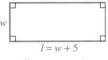

$$l = w + 5$$

Illustration 3

16. **Geometry**   A rectangle is three times as long as it is wide, and its perimeter is 80 centimeters. Find its dimensions.

17. **Geometry**   A rectangle has a length that is 2 feet more than twice its width. If its perimeter is 34 feet, find its area.

18. **Geometry**   A 50-meter path surrounds the rectangular garden shown in Illustration 4. The width of the garden is two-thirds its length. Find its area.

Illustration 4

19. **Choosing a furnace**   A high-efficiency 90+ furnace costs $2,250 and costs an average of $412 per year to operate in Rockford, IL. An 80+ furnace costs only $1,715 but costs $466 per year to operate. Find the break point.

20. **Making tires**   A company has two molds to form tires. One mold has a setup cost of $600 and the other a setup cost of $1,100. The cost to make each tire on the first machine is $15, and the cost to make each tire on the second machine is $13. Find the break point.

21. **Choosing a furnace**   If you intended to live in a house for 7 years, which furnace would you choose in Exercise 19?

22. **Making tires**   If you planned a production run of 500 tires, which mold would you use in Exercise 20?

23. **Investing money**   Bill invested some money at 2% annual interest, and Janette invested some at 3%. If their combined annual interest was $190 on a total investment of $8,000, how much did Bill invest?

24. **Investing money**   Peter invested some money at 3% annual interest, and Martha invested some at 4%. If

their combined investment was $6,000 and their combined annual interest was $210, how much money did Martha invest?

25. Buying tickets   Students can buy tickets to a basketball game for $1. However, the admission for nonstudents is $2. If 350 tickets are sold and the total receipts are $450, how many student tickets were sold?

26. Buying tickets   If receipts for the movie advertised in Illustration 5 were $720 for an audience of 190 people, how many senior citizens attended?

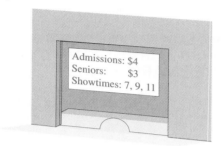

Admissions: $4
Seniors: $3
Showtimes: 7, 9, 11

Illustration 5

27. Boating   A boat can travel 24 miles downstream in 2 hours and can make the return trip in 3 hours. Find the speed of the boat in still water.

28. Aviation   With the wind, a plane can fly 3000 miles in 5 hours. Against the same wind, the trip takes 6 hours. Find the airspeed of the plane (the speed in still air).

29. Aviation   An airplane can fly downwind a distance of 600 miles in 2 hours. However, the return trip against the same wind takes 3 hours. Find the speed of the wind.

30. Finding the speed of a current   It takes a motorboat 4 hours to travel 56 miles down a river, and takes 3 hours longer to make the return trip. Find the speed of the current.

31. Mixing chemicals   A chemist has one solution that is 40% alcohol and another that is 55% alcohol. How much of each must she use to make 15 liters of a solution that is 50% alcohol?

32. Mixing pharmaceuticals   A nurse has a solution that is 25% alcohol and another that is 50% alcohol. How much of each must he use to make 20 liters of a solution that is 40% alcohol?

33. Mixing nuts   A merchant wants to mix the peanuts with the cashews shown in Illustration 6 to get 48 pounds of mixed nuts to sell at $4 per pound. How many pounds of each should the merchant use?

Illustration 6

34. Mixing peanuts and candy   A merchant wants to mix peanuts worth $3 per pound with jelly beans worth $1.50 per pound to make 30 pounds of a mixture worth $2.10 per pound. How many pounds of each should he use?

35. Markdown   A set of golf clubs has been marked down 40% to a sale price of $384. Let r represent the retail price and d represent the discount. Then use the following equations to find the original retail price.

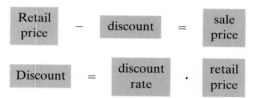

36. Markup   A stereo system retailing at $565.50 has been marked up 45% from wholesale. Let w represent the wholesale cost and m represent the markup. Then use the following equations to find the wholesale cost.

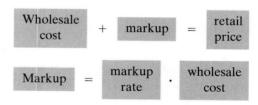

37. Selling radios   An electronics store put two types of car radios on sale. One model sold for $87, and the other sold for $119. During the sale, the receipts for the 25 radios sold were $2,495. How many cheap radios were sold?

38. Selling ice cream   At a store, ice cream cones cost $0.90 and sundaes cost $1.65. One day the receipts for a total of 148 cones and sundaes were $180.45. How many cones were sold?

39. Investing money   An investment of $950 at one rate of interest and $1200 at a higher rate together generate an annual income of $76.50. If the investment rates differ by 1%, find the lower rate. (*Hint:* Treat 1% as 0.01).

40. Motion problem   A man drives for a while at 45 miles per hour. Realizing that he is running late, he increases his speed to 60 miles per hour and completes his 405-mile trip in 8 hours. How long did he drive at 45 miles per hour?

 **41.** Which of the problems in Exercises 1–40 could you have done more easily using a single equation with one variable? What makes a solution method "easier"?

REVIEW   *In Exercises 42–45, solve each inequality and confirm graphically.*

42. $|3x + 6| < 5$

43. $|3x + 6| \geq 5$

44. $|2x + 5| - 2 > -6$

45. $|2x + 5| - 2 < -6$

12.3   SOLVING SYSTEMS OF THREE EQUATIONS IN THREE VARIABLES

- SOLVING SYSTEMS OF THREE EQUATIONS IN THREE VARIABLES

- PROBLEM SOLVING

SOLVING SYSTEMS OF THREE EQUATIONS IN THREE VARIABLES

In this section, we extend the definition of a linear equation to include any equation of the form $ax + by + cz = d$. The solution of a system of three linear equations with three variables is an ordered triple of numbers. For example, the solution of the system

$$\begin{cases} 2x + 3y + 4z = 20 \\ 3x + 4y + 2z = 17 \\ 3x + 2y + 3z = 16 \end{cases}$$

is the ordered triple $(1, 2, 3)$. Verify that the numbers in the triple satisfy each of the three equations in the system.

The graph of an equation of the form $ax + by + cz = d$ is a flat surface called a **plane**. A system of three linear equations in three variables is consistent or inconsistent, depending on how the three planes corresponding to the three equations intersect. Figure 12-8 illustrates some of the possibilities.

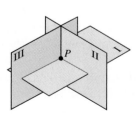

The three planes intersect at a single point P: One solution

(a)

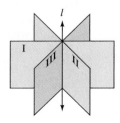

The three planes have a line l in common: An infinite number of solutions

(b)

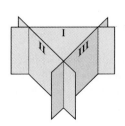

The three planes have no point in common: No solutions

(c)

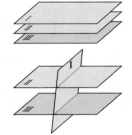

Figure 12-8

To solve a system of three linear equations in three variables, we follow these steps.

Steps to Follow When Solving Three Equations in Three Variables

1. Pick any two equations and eliminate a variable.
2. Pick a different pair of equations and eliminate the same variable.
3. Solve the resulting pair of two equations in two variables.
4. To find the value of the third variable, substitute the values of the two variables found in step 3 into any equation containing all three variables and solve the equation.
5. Check the solution in all three of the original equations.

Example 1 discusses a consistent system of three equations in three variables. Example 2 discusses an inconsistent system.

EXAMPLE 1   Solve the system $\begin{cases} 2x + y + 4z = 12 \\ x + 2y + 2z = 9 \\ 3x - 3y - 2z = 1 \end{cases}$.

SOLUTION   We are given the system

1. $\quad \begin{cases} 2x + y + 4z = 12 \\ x + 2y + 2z = 9 \\ 3x - 3y - 2z = 1 \end{cases}$
2.
3.

If we pick Equations 2 and 3 and add them, the variable z is eliminated:

2. $\quad x + 2y + 2z = 9$
3. $\quad \underline{3x - 3y - 2z = 1}$
4. $\quad 4x - y \quad\quad = 10$

We now pick a different pair of equations (Equations 1 and 3) and eliminate z again. If each side of Equation 3 is multiplied by 2 and the resulting equation is added to Equation 1, z is again eliminated:

1. $\quad 2x + y + 4z = 12$
$\quad\quad \underline{6x - 6y - 4z = 2}$
5. $\quad 8x - 5y \quad\quad = 14$

Equations 4 and 5 form a system of two equations in two variables:

4. $\quad \begin{cases} 4x - y = 10 \\ 8x - 5y = 14 \end{cases}$
5.

To solve this system, we multiply Equation 4 by -5 and add the resulting equation to Equation 5 to eliminate y:

$\quad\quad -20x + 5y = -50$
5. $\quad \underline{\quad 8x - 5y = \quad 14}$
$\quad\quad -12x \quad\quad = -36$

$\quad\quad\quad x = 3 \qquad$ Divide both sides by -12.

To find y, we substitute 3 for x in any equation containing x and y (such as Equation 5) and solve for y:

5. $8x - 5y = 14$

$8(3) - 5y = 14$ Substitute 3 for x.

$24 - 5y = 14$ Simplify.

$-5y = -10$ Subtract 24 from both sides.

$y = 2$ Divide both sides by -5.

To find z, we substitute 3 for x and 2 for y in an equation containing x, y, and z (such as Equation 1) and solve for z:

1. $2x + y + 4z = 12$

$2(3) + 2 + 4z = 12$ Substitute 3 for x and 2 for y.

$8 + 4z = 12$ Simplify.

$4z = 4$ Subtract 8 from both sides.

$z = 1$ Divide both sides by 4.

The solution of the system is $(x, y, z) = (3, 2, 1)$. Verify that these values satisfy each equation in the original system.

EXAMPLE 2   Solve the system $\begin{cases} 2x + y - 3z = -3 \\ 3x - 2y + 4z = 2. \\ 4x + 2y - 6z = -7 \end{cases}$

SOLUTION   We are given the system of equations

1. $\begin{cases} 2x + y - 3z = -3 \\ 3x - 2y + 4z = 2 \\ 4x + 2y - 6z = -7 \end{cases}$
2.
3.

We can multiply Equation 1 by 2 and add the resulting equation to Equation 2 to eliminate y:

$\quad\quad 4x + 2y - 6z = -6$

2. $\quad \underline{3x - 2y + 4z = \quad 2}$

4. $\quad 7x \quad\quad\quad - 2z = -4$

We now add Equations 2 and 3 to eliminate y again:

2. $\quad 3x - 2y + 4z = \quad 2$

3. $\quad \underline{4x + 2y - 6z = -7}$

5. $\quad 7x \quad\quad\quad - 2z = -5$

Equations 4 and 5 form the system

4. $\begin{cases} 7x - 2z = -4 \\ 7x - 2z = -5 \end{cases}$
5.

Because $7x - 2z$ cannot equal both -4 and -5, this system is inconsistent. Thus, it has no solution: solution set, $\varnothing$.

▎▎ PROBLEM SOLVING

If three variables are used to solve an application problem, a system of three equations will be needed.

 EXAMPLE 3

Integer problem   The sum of three integers is 2. The third integer is 2 greater than the second and 17 greater than the first. Find the integers.

SOLUTION   Let a, b, and c represent the integers. Because their sum is 2, we know that

$$a + b + c = 2$$

Because the third integer is 2 greater than the second and 17 greater than the first, we know that

$$c - b = 2$$
$$c - a = 17$$

Put these three equations together to form a system of three equations in three variables:

1. $\quad\begin{cases} a + b + c = 2 \\ - b + c = 2 \\ -a + c = 17 \end{cases}$
2.
3.

We add Equations 1 and 2 to get Equation 4:

4. $\quad a + 2c = 4$

Because Equations 3 and 4 have only the variables a and c, Equations 3 and 4 form a system of two equations in two variables.

3. $\quad\begin{cases} -a + c = 17 \\ a + 2c = 4 \end{cases}$
4.

We add Equations 3 and 4 to get the equation

$$3c = 21$$
$$c = 7$$

We substitute 7 for c in Equation 4 to find a:

4. $\quad a + 2\mathbf{c} = 4$

$$a + 2(\mathbf{7}) = 4$$
$$a + 14 = 4 \qquad \text{Simplify.}$$
$$a = -10 \qquad \text{Add } -14 \text{ to both sides.}$$

We substitute 7 for c in Equation 2 to find b:

2. $\quad -b + \mathbf{c} = 2$

$$-b + \mathbf{7} = 2$$
$$-b = -5 \qquad \text{Subtract 7 from both sides.}$$
$$b = 5 \qquad \text{Divide both sides by } -1.$$

Thus, the integers are $-10, 5$, and 7. These integers have a sum of 2; 7 is 2 greater than 5; and 7 is 17 greater than -10.

EXAMPLE 4

Manufacturing   A company manufactures three types of hammers — good, better, and best. The cost of manufacturing each type is $4, $6, and $7, respectively, and the hammers sell for $6, $9, and $12. Each day, the cost for manufacturing 100 hammers is $520, and the daily revenue from their sales is $810. How many of each type are manufactured?

SOLUTION   We can let x represent the number of good hammers manufactured, y represent the number of better hammers manufactured, and z represent the number of best hammers manufactured. Then

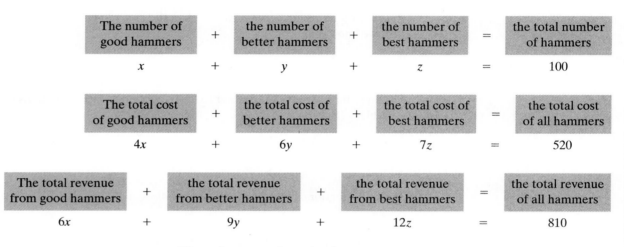

These three equations give the system

1. $\begin{cases} x + y + z = 100 \\ 4x + 6y + 7z = 520 \\ 6x + 9y + 12z = 810 \end{cases}$
2.
3.

If we multiply Equation 1 by -7 and add the result to Equation 2, we get

$$-7x - 7y - 7z = -700$$
$$\underline{4x + 6y + 7z = 520}$$
4. $-3x - y = -180$

We can eliminate z again if we multiply Equation 1 by -12 and add the result to Equation 3.

$$-12x - 12y - 12z = -1{,}200$$
$$\underline{6x + 9y + 12z = \phantom{-1{,}}810}$$
5. $-6x - 3y = \phantom{-1{,}}-390$

If we multiply Equation 4 by -3 and add it to Equation 5, we get

$$9x + 3y = 540$$
$$\underline{-6x - 3y = -390}$$
$$3x = 150$$
$$x = 50 \qquad \text{Divide both sides by 3.}$$

To find y, we substitute 50 for x in Equation 4:

4. $-3x - y = -180$

        $-3(\mathbf{50}) - y = -180$ Substitute 50 for x.

           $-150 - y = -180$

               $-y = -30$ Add 150 to both sides.

                 $y = 30$ Divide both sides by -1.

To find z, we substitute 50 for x and 30 for y in Equation 1:

1. $x + y + z = 100$

      $\mathbf{50} + \mathbf{30} + z = 100$

             $z = 20$ Subtract 80 from both sides.

The company manufactures 50 good hammers, 30 better hammers, and 20 best hammers each day.

EXERCISE 12.3

VOCABULARY AND NOTATION   *Fill in each blank to make a true statement.*

1. The graph of an equation in the form $ax + by + cz = d$ is a flat surface called a _____ .

2. When three planes intersect at a single point, the system has _____ solution(s).

3. If three planes intersect in a line, the system has a(n) _____ number of solutions.

4. When three planes are parallel, the system has ____ solutions.

PRACTICE   *In Exercises 5–16, solve each system.*

5. $\begin{cases} x + y + z = 4 \\ 2x + y - z = 1 \\ 2x - 3y + z = 1 \end{cases}$

6. $\begin{cases} x + y + z = 4 \\ x - y + z = 2 \\ x - y - z = 0 \end{cases}$

7. $\begin{cases} 2x + 2y + 3z = 10 \\ 3x + y - z = 0 \\ x + y + 2z = 6 \end{cases}$

8. $\begin{cases} x - y + z = 4 \\ x + 2y - z = -1 \\ x + y - 3z = -2 \end{cases}$

9. $\begin{cases} x + y + 2z = 7 \\ x + 2y + z = 8 \\ 2x + y + z = 9 \end{cases}$

10. $\begin{cases} x + 2y + 2z = 10 \\ 2x + y + 2z = 9 \\ 2x + 2y + z = 11 \end{cases}$

11. $\begin{cases} 2x + y - z = 1 \\ x + 2y + 2z = 2 \\ 4x + 5y + 3z = 3 \end{cases}$

12. $\begin{cases} 4x + 3z = 4 \\ 2y - 6z = -1 \\ 8x + 4y + 3z = 9 \end{cases}$

13. $\begin{cases} 2x + 3y + 4z - 6 = 0 \\ 2x - 3y - 4z + 4 = 0 \\ 4x + 6y + 8z - 12 = 0 \end{cases}$

14. $\begin{cases} x - 3y + 4z - 2 = 0 \\ 2x + y + 2z - 3 = 0 \\ 4x - 5y + 10z - 7 = 0 \end{cases}$

15. $\begin{cases} x + \dfrac{1}{3}y + z = 13 \\ \dfrac{1}{2}x - y + \dfrac{1}{3}z = -2 \\ x + \dfrac{1}{2}y - \dfrac{1}{3}z = 2 \end{cases}$

16. $\begin{cases} x - \dfrac{1}{5}y - z = 9 \\ \dfrac{1}{4}x + \dfrac{1}{5}y - \dfrac{1}{2}z = 5 \\ 2x + y + \dfrac{1}{6}z = 12 \end{cases}$

APPLICATIONS   *In Exercises 17–28, solve each problem.*

17. Integer problem   The sum of three integers is 18. The third integer is four times the second, and the second integer is 6 more than the first. Find the integers.

18. Integer problem   The sum of three integers is 48. If the first integer is doubled, the sum is 60. If the second integer is doubled, the sum is 63. Find the integers.

19. Geometry problem   The sum of the angles in any triangle is 180°. In triangle ABC, angle A is 100° less than the sum of angles B and C, and angle C is 40° less than twice angle B. Find each angle.

20. Geometry problem   The sum of the angles of any four-sided figure is 360°. In the quadrilateral shown in Illustration 1, angle A = angle B, angle C is 20° greater than angle A, and angle D = 40°. Find the angles.

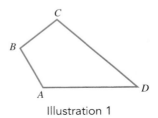

Illustration 1

21. Nutritional planning   According to Table 1, a unit of food A contains 1 gram of fat, 1 gram of carbohydrate, and 2 grams of protein. Food B contains 2 grams of fat, 1 gram of carbohydrate, and 1 gram of protein. Food C contains 2 grams of fat, 1 gram of carbohydrate, and 2 grams of protein. How many units of each food must be used to provide exactly 11 grams of fat, 6 grams of carbohydrate, and 10 grams of protein?

Food	Fat	Carbohydrates	Protein
A	1	1	2
B	2	1	1
C	2	1	2

TABLE 1

22. Nutritional planning   As shown in Table 2, a unit of food A contains 2 grams of fat, 1 gram of carbohydrate, and 2 grams of protein. Food B contains 3 grams of fat, 2 grams of carbohydrate, and 1 gram of protein. Food C contains 1 gram of fat, 1 gram of carbohydrate, and 2 grams of protein. How many units of each food must be used to provide exactly 14 grams of fat, 9 grams of carbohydrate, and 9 grams of protein?

Food	Fat	Carbohydrates	Protein
A	2	1	2
B	3	2	1
C	1	1	2

TABLE 2

23. Making statues   An artist makes three types of ceramic statues at a monthly cost of $650 for 180 statues. The manufacturing costs for the three types are $5, $4, and $3. If the statues sell for $20, $12, and $9, respectively, how many of each type should be made to produce $2,100 in monthly revenue?

24. Manufacturing footballs   A factory manufactures three types of footballs at a monthly cost of $2,425 for 1,125 footballs. The manufacturing costs for the three types of footballs are $4, $3, and $2. These footballs sell for $16, $12, and $10, respectively. How many of each type are manufactured if the monthly profit is $9,275? (*Hint:* Profit = income − cost.)

25. Concert tickets   Tickets for a concert cost $5, $3, and $2. Twice as many $5 tickets were sold as $2 tickets. The receipts for 750 tickets were $2,625. How many of each price ticket were sold?

26. Mixing nuts   The owner of a candy store wants to mix some peanuts worth $3 per pound, some cashews worth $9 per pound, and some Brazil nuts worth $9 per pound to get 50 pounds of a mixture that will sell for $6 per pound. She used 15 fewer pounds of cashews than peanuts. How many pounds of each did she use?

27. Chainsaw sculpting   A northwoods sculptor carves three types of statues with a chainsaw. The time required for carving, sanding, and painting a totem pole, a bear, and a deer are shown in Table 3. How many of each should be produced to use all available labor hours?

	Totem pole	Bear	Deer	Time available
Carving	2 hours	2 hours	1 hour	14 hours
Sanding	1 hour	2 hours	2 hours	15 hours
Painting	3 hours	2 hours	2 hours	21 hours

TABLE 3

28. Making clothing A clothing manufacturer makes coats, shirts, and slacks. The time required for cutting, sewing, and packaging each item are shown in Table 4.

	Coats	Shirts	Slacks	Time available
Cutting	20 min	15 min	10 min	115 hr
Sewing	60 min	30 min	24 min	280 hr
Packaging	5 min	12 min	6 min	65 hr

TABLE 4

How many of each should be made to use all available labor hours?

REVIEW

29. Does the equation $y = |x + 2|$ determine y to be a function of x? Why or why not? If you answered yes, state the domain and the range.

30. Find the slope of the graph of $x = \dfrac{2y - 3}{2}$.

31. Write the equation of the line that has a slope of $\frac{3}{5}$ and passes through the point $(1, 2)$.

32. Express as a formula: y varies directly as the square of x and inversely with z.

33. If $f(x) = -x^2 - 5x + 2$, find $f(-3)$.

12.4 USING MATRICES TO SOLVE SYSTEMS OF EQUATIONS

- BASIC DEFINITIONS
- GAUSSIAN ELIMINATION
- SYSTEMS WITH MORE EQUATIONS THAN VARIABLES

- SYSTEMS WITH MORE VARIABLES THAN EQUATIONS

▌ BASIC DEFINITIONS

Another method for solving systems of equations depends on the coefficients of the variables and the constants. We will form a rectangular array of numbers using these coefficients and constants to solve systems of equations. This rectangular array is called a **matrix**.

Matrix
A matrix is any rectangular array of numbers.

Some examples of matrices are

$$A = \begin{bmatrix} 1 & 2 & 3 \\ 4 & 5 & 6 \end{bmatrix} \quad B = \begin{bmatrix} 1 & 2 \\ 3 & 4 \\ 5 & 6 \end{bmatrix} \quad C = \begin{bmatrix} 2 & 4 & 6 \\ 8 & 10 & 12 \\ 14 & 16 & 18 \end{bmatrix}$$

The numbers in each matrix are called its **elements**. Because matrix A has two rows and three columns, it is called a 2×3 matrix (read "2 by 3" matrix). Matrix B is a 3×2 matrix, because the matrix has three rows and two columns. Matrix C

is a 3×3 matrix (three rows and three columns). Any matrix with the same number of rows and columns is called a **square matrix**. Matrix C is an example of a square matrix.

To show how to use matrices to solve systems of linear equations, we consider the following system

$$\begin{cases} x - 2y - z = 6 \\ 2x + 2y - z = 1 \\ -x - y + 2z = 1 \end{cases}$$

which can be represented by the following matrix, called an **augmented matrix**:

$$\begin{bmatrix} 1 & -2 & -1 & \vdots & 6 \\ 2 & 2 & -1 & \vdots & 1 \\ -1 & -1 & 2 & \vdots & 1 \end{bmatrix}$$

The 3×3 matrix to the left of the dashed line, called the **coefficient matrix**, is determined by the coefficients of x, y, and z in the equations of the system. The 3×1 matrix to the right of the dashed line is determined by the constants in the equations. Each row of the augmented matrix represents exactly one equation of the system:

$$\begin{bmatrix} 1 & -2 & -1 & \vdots & 6 \\ 2 & 2 & -1 & \vdots & 1 \\ -1 & -1 & 2 & \vdots & 1 \end{bmatrix} \begin{matrix} \leftrightarrow \\ \leftrightarrow \\ \leftrightarrow \end{matrix} \begin{cases} x - 2y - z = 6 \\ 2x + 2y - z = 1 \\ -x - y + 2z = 1 \end{cases}$$

▐▐▐ GAUSSIAN ELIMINATION

To solve a system by **Gaussian elimination**, we transform the augmented matrix into the following matrix that has all 0s below its main diagonal, formed by elements a, e, and h.

$$\begin{bmatrix} a & b & c & \vdots & d \\ 0 & e & f & \vdots & g \\ 0 & 0 & h & \vdots & i \end{bmatrix} \qquad (a, b, c, \ldots, i \text{ are real numbers.})$$

We can often write a matrix in this form, called **triangular form**, by using three operations called **elementary row operations**.

Elementary Row Operations

1. Any two rows of a matrix can be interchanged.
2. Any row of a matrix can be multiplied by a nonzero constant.
3. Any row of a matrix can be changed by adding a constant multiple of another row to it.

- A type 1 row operation corresponds to interchanging two equations of the system.
- A type 2 row operation corresponds to multiplying both sides of an equation by a nonzero constant.
- A type 3 row operation corresponds to adding a multiple of one equation to another.

None of these operations will change the solution of the given system of equations.

After we have written the matrix in triangular form, we can solve the corresponding system of equations by a substitution process, as shown in Example 1.

EXAMPLE 1

Solve the system $\begin{cases} x - 2y - z = 6 \\ 2x + 2y - z = 1 \\ -x - y + 2z = 1 \end{cases}$

SOLUTION   We can represent the system with the following augmented matrix:

$$\left[\begin{array}{ccc|c} 1 & -2 & -1 & 6 \\ 2 & 2 & -1 & 1 \\ -1 & -1 & 2 & 1 \end{array}\right]$$

To get 0s under the 1 in the first column, we use a type 3 row operation twice:

Multiply row 1 by -2 and add to row 2.

Multiply row 1 by 1 and add to row 3.

$$\left[\begin{array}{ccc|c} 1 & -2 & -1 & 6 \\ \mathbf{2} & 2 & -1 & 1 \\ -1 & -1 & 2 & 1 \end{array}\right] \approx \left[\begin{array}{ccc|c} 1 & -2 & -1 & 6 \\ 0 & 6 & 1 & -11 \\ -1 & -1 & 2 & 1 \end{array}\right] \approx \left[\begin{array}{ccc|c} 1 & -2 & -1 & 6 \\ 0 & 6 & 1 & -11 \\ 0 & -3 & 1 & 7 \end{array}\right]$$

The symbol "$\approx$" is read as "is row equivalent to." Each of the matrices above represents a system of equations, and they are all equivalent.

To get a 0 under the 6 in the second column of the last matrix, we use another type 3 row operation:

Multiply row 2 by $\frac{1}{2}$ and add to row 3.

$$\left[\begin{array}{ccc|c} 1 & -2 & -1 & 6 \\ 0 & 6 & 1 & -11 \\ 0 & -3 & 1 & 7 \end{array}\right] \approx \left[\begin{array}{ccc|c} 1 & -2 & -1 & 6 \\ 0 & 6 & 1 & -11 \\ 0 & 0 & \frac{3}{2} & \frac{3}{2} \end{array}\right]$$

Finally, we use a type 2 row operation:

Multiply row 3 by $\frac{2}{3}$.

$$\left[\begin{array}{ccc|c} 1 & -2 & -1 & 6 \\ 0 & 6 & 1 & -11 \\ 0 & 0 & \frac{3}{2} & \frac{3}{2} \end{array}\right] \approx \left[\begin{array}{ccc|c} 1 & -2 & -1 & 6 \\ 0 & 6 & 1 & -11 \\ 0 & 0 & 1 & 1 \end{array}\right]$$

The final matrix represents the system of equations

1. $\quad\begin{cases} x - 2y - z = 6 \\ 0x + 6y + z = -11 \\ 0x + 0y + z = 1 \end{cases}$
2.
3.

From equation 3, we can read that $z = 1$. To find y, we substitute 1 for z in equation 2 and solve for y:

2. $6y + z = -11$

$6y + \mathbf{1} = -11$ Substitute 1 for z.

$6y = -12$ Subtract 1 from both sides.

$y = -2$ Divide both sides by 6.

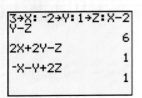

Thus, $y = -2$. To find x, we substitute 1 for z and -2 for y in equation 1 and solve for x:

1.
$$x - 2y - z = 6$$
$$x - 2(-2) - 1 = 6 \qquad \text{Substitute 1 for } z \text{ and } -2 \text{ for } y.$$
$$x + 4 - 1 = 6 \qquad (-2)(-2) = +4.$$
$$x + 3 = 6 \qquad \text{Simplify.}$$
$$x = 3 \qquad \text{Subtract 3 from both sides.}$$

Thus, $x = 3$. The solution to the given system is $(3, -2, 1)$. Verify that this ordered triple satisfies each equation of the original system.

Note: Not all of the available row operations were used. The following row operations indicate the order in which information is to be entered.

rowSwap (matrix, row 1, row 2)   Swaps row 1 and row 2.
row + (matrix, row 1, row 2)   Adds row 1 and row 2 and stores result in row 2.
*row (value, matrix, row)   Multiplies a row by the indicated value.
*row + (value, matrix, row 1, row 2)   Multiplies the matrix row 1 by the indicated value, adds this product to row 2, and stores the result in row 2.

EXAMPLE 2

Solve the system $\begin{cases} x - 2y - z = 6 \\ 2x + 2y - z = 1 \\ -x - y + 2z = 1 \end{cases}$ using the calculator to perform matrix operations.

SOLUTION   This is the same system solved in Example 1. The augmented matrix associated with the system is

$$\begin{bmatrix} 1 & -2 & -1 & \vdots & 6 \\ 2 & 2 & -1 & \vdots & 1 \\ -1 & -1 & 2 & \vdots & 1 \end{bmatrix}$$

Enter this matrix in your calculator as matrix A using the matrix edit feature of your calculator located under the MATRIX menu. Pressing the ENTER key after each entry will automatically move the cursor to the next entry location, or you may use the arrow keys to move to the desired location on the screen.

row 2, column 1 entry

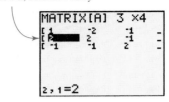

Return to the home screen (2nd, QUIT) and display your matrix A (MATRIX, 1:[A], ENTER). (TI-86: [2nd] [Matrix] [F1] (NAMES) [F1] (A) [ENTER].)

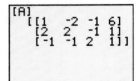

We will now follow the steps in Example 1, replacing the analytic arithmetic computations with parallel calculator computations. The indicated matrix operations are located under the MATH submenu of the MATRIX menu on the TI-83 Plus and under the OPS menu on the TI-86. Both the analytical directions and the corresponding calculator command will be given.

To get 0s under the 1 in the first column, we do a type 3 row operation twice.

Analytical operation: $-2 \cdot$ row 1 + row 2 *replaces* row 2

Calculator operation: *row + (−2, [A], 1, 2) STO➤ [B]

multiplier ⎯⎯⎯⎯⎯⎯⎯⎯⎯⎯⎯⎯⎯⎯↑
row being multiplied ⎯⎯⎯⎯⎯⎯⎯⎯⎯⎯↑   ↑   ↑⎯ new matrix location
added row (and the row being replaced) ⎯⎯┘   └⎯ STOre feature

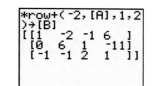

> **WARNING!** Row operations do not change the matrix stored in memory! The new matrix must be stored each time row operations are performed. The new matrix was stored in [B] to preserve the original matrix.

Analytical operation: $1 \cdot$ row 1 + row 3 *replaces* row 3

Calculator operation: *row + (1, [B], 1, 3) STO➤ [B]

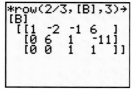

To get a 0 under the 6 in the second column of the last matrix, we use another type 3 row operation.

Analytical operation: $\frac{1}{2} \cdot$ row 2 + row 3 *replaces* row 3

Calculator operation: *row + ($\frac{1}{2}$, [B], 2, 3]) STO➤ [B]

Finally, we use a type 2 row operation.

Analytical operation: $\frac{2}{3} \cdot$ row 3 *replaces* row 3

Calculator operation: *row ($\frac{2}{3}$, [B], 3) STO➤ [B]

The final matrix represents the system of equations:

1. $x - 2y - z = 6$

2. $6y + z = -11$

3. $z = 1$

From $z = 1$ in equation 3, we back substitute into equation 2 to find $y = -2$. Using both $z = 1$ and $y = -2$, we back substitute into equation 1 to find $x = 3$. Thus, the solution to the system is the ordered triple $(3, -2, 1)$.

▌ SYSTEMS WITH MORE EQUATIONS THAN VARIABLES

We can use matrices to solve systems with more equations than variables.

EXAMPLE 3   Solve the system $\begin{cases} x + y = -1 \\ 2x - y = 7 \\ -x + 2y = -8 \end{cases}$

SOLUTION   This system, with three equations and two variables, can be represented by a 3×3 augmented matrix:

$$\left[\begin{array}{cc:c} 1 & 1 & -1 \\ 2 & -1 & 7 \\ -1 & 2 & -8 \end{array}\right]$$

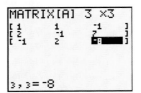

To get 0s under the 1 in the first column, we do a type 3 row operation twice:

Multiply row 1 by −2 and add to row 2.   Multiply row 1 by 1 and add to row 3.

$$\left[\begin{array}{cc:c} 1 & 1 & -1 \\ 2 & -1 & 7 \\ -1 & 2 & -8 \end{array}\right] \approx \left[\begin{array}{cc:c} 1 & 1 & -1 \\ 0 & -3 & 9 \\ -1 & 2 & -8 \end{array}\right] \approx \left[\begin{array}{cc:c} 1 & 1 & -1 \\ 0 & -3 & 9 \\ 0 & 3 & -9 \end{array}\right]$$

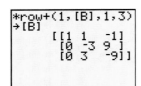

We can do other row operations to get

Add row 2 to row 3.   Multiply row 2 by $-\frac{1}{3}$.

$$\left[\begin{array}{cc:c} 1 & 1 & -1 \\ 0 & -3 & 9 \\ 0 & 3 & -9 \end{array}\right] \approx \left[\begin{array}{cc:c} 1 & 1 & -1 \\ 0 & -3 & 9 \\ 0 & 0 & 0 \end{array}\right] \approx \left[\begin{array}{cc:c} 1 & 1 & -1 \\ 0 & 1 & -3 \\ 0 & 0 & 0 \end{array}\right]$$

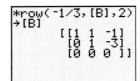

The final matrix represents the system

$$\begin{cases} x + y = -1 \\ 0x + y = -3 \\ 0x + 0y = 0 \end{cases}$$

The third equation can be discarded, because $0x + 0y = 0$ for all x and y. From the second equation, we can read that $y = -3$. To find x, we substitute -3 for y in the first equation and solve for x:

$$x + y = -1$$
$$x - 3 = -1 \qquad \text{Substitute } -3 \text{ for } y.$$
$$x = 2 \qquad \text{Add 3 to both sides.}$$

The solution is $(2, -3)$. Verify that this solution satisfies all three equations of the original system.

If the last row of the final matrix in Example 3 had been of the form $0x + 0y = k$, where $k \neq 0$, the system would have no solution. No values of x and y could make the expression $0x + 0y$ equal to a nonzero constant k.

▍▍ SYSTEMS WITH MORE VARIABLES THAN EQUATIONS

We can also solve many systems with more variables than equations.

EXAMPLE 4 Solve the system $\begin{cases} x + y - 2z = -1 \\ 2x - y + z = -3 \end{cases}$

SOLUTION This system has two equations and three variables. We can start to solve it by doing a type 3 row operation to get a 0 under the 1 in the first column.

Multiply row 1 by -2
and add to row 2.

$$\begin{bmatrix} 1 & 1 & -2 & \vdots & -1 \\ 2 & -1 & 1 & \vdots & -3 \end{bmatrix} \approx \begin{bmatrix} 1 & 1 & -2 & \vdots & -1 \\ 0 & -3 & 5 & \vdots & -1 \end{bmatrix}$$

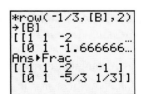

We then do a type 2 row operation:

Multiply row 2 by $-\frac{1}{3}$.

$$\begin{bmatrix} 1 & 1 & -2 & \vdots & -1 \\ 0 & -3 & 5 & \vdots & -1 \end{bmatrix} \approx \begin{bmatrix} 1 & 1 & -2 & \vdots & -1 \\ 0 & 1 & -\frac{5}{3} & \vdots & \frac{1}{3} \end{bmatrix}$$

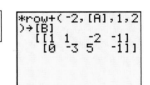

The final matrix represents the system

$$\begin{cases} x + y - 2z = -1 \\ \quad\quad y - \dfrac{5}{3}z = \dfrac{1}{3} \end{cases}$$

We add $\frac{5}{3}z$ to both sides of the second equation to obtain

$$y = \frac{1}{3} + \frac{5}{3}z$$

We substitute $\frac{1}{3} + \frac{5}{3}z$ for y in the first equation and simplify to get

$$x + y - 2z = -1$$

$$x + \frac{1}{3} + \frac{5}{3}z - 2z = -1 \qquad \text{Substitute } \frac{1}{3} + \frac{5}{3}z \text{ for } y.$$

$$x + \frac{1}{3} - \frac{1}{3}z = -1 \qquad \frac{5}{3}z - 2z = \frac{5}{3}z - \frac{6}{3}z = -\frac{1}{3}z.$$

$$x - \frac{1}{3}z = -\frac{4}{3} \qquad \text{Subtract } \frac{1}{3} \text{ from both sides.}$$

$$x = -\frac{4}{3} + \frac{1}{3}z \qquad \text{Add } \frac{1}{3}z \text{ to both sides.}$$

A solution to this system must have the form

$$\left(-\frac{4}{3} + \frac{1}{3}z, \frac{1}{3} + \frac{5}{3}z, z \right)$$

for all values of z. This system has an infinite number of solutions, a different one for each value of z. For example,

- If $z = 0$, the corresponding solution is $\left(-\frac{4}{3}, \frac{1}{3}, 0 \right)$.
- If $z = 1$, the corresponding solution is $(-1, 2, 1)$.

Verify that both of these solutions satisfy each equation of the given system. The solution set has the form

$$\{(x, y, z) \mid x + y - 2z = -1\}$$

EXERCISE 12.4

VOCABULARY AND NOTATION *In Exercises 1–5, fill in the blanks to make a true statement.*

1. A rectangular array of numbers is called a _____.

2. The numbers in the rectangular array are called its _____.

3. A _____ matrix has the same number of rows and columns.

4. An _____ matrix includes the constant values of the system, whereas a _____ matrix only includes the coefficients.

5. _____ is used to transform an augmented matrix into a matrix that has all 0s below its main diagonal.

CONCEPTS

6. Write the coefficient matrix for the system:

$$2x + y + 2z = 1$$
$$x - 2y + 3z = 4$$
$$2x - 3y + z = 0$$

7. Write the augmented matrix for the system in Exercise 6.

8. Explain why any two rows of a matrix can be interchanged.

9. Explain why you can multiply any row of a matrix by a nonzero constant. Support your reason with a specific property that was introduced earlier in the text.

10. Explain why you **cannot** multiply any column of a matrix by a nonzero constant.

11. Explain how to check the solution of a system of equations.

12. Explain how to perform a type 3 row operation.

PRACTICE   *In Exercises 13–24, use matrices to solve each system of equations. Row operations may be done either by hand or with the calculator. Verify each solution with the calculator.*

13. $x + y = 2$
$x - y = 0$

14. $x + y = 3$
$x - y = -1$

15. $x + 2y = -4$
$2x + y = 1$

16. $2x - 3y = 16$
$-4x + y = -22$

17. $3x + 4y = -12$
$9x - 2y = 6$

18. $5x - 4y = 10$
$x - 7y = 2$

19. $x + y + z = 6$
$x + 2y + z = 8$
$x + y + 2z = 9$

20. $x - y + z = 2$
$x + 2y - z = 6$
$2x - y - z = 3$

21. $2x + y + 3z = 3$
$-2x - y + z = 5$
$4x - 2y + 2z = 2$

22. $3x + 2y + z = 8$
$6x - y + 2z = 16$
$-9x + y - z = -20$

23. $3x - 2y + 4z = 4$
$x + y + z = 3$
$6x - 2y - 3z = 10$

24. $2x + 3y - z = -8$
$x - y - z = -2$
$-4x + 3y + z = 6$

In Exercises 25–32, use matrices to solve each system of equations. Row operations may be done either by hand or with the calculator. Verify each solution with the calculator. If a system has no solution, so indicate.

25. $x + y = 3$
$3x - y = 1$
$2x + y = 4$

26. $x - y = -5$
$2x + 3y = 5$
$x + y = 1$

27. $2x - y = 4$
$x + 3y = 2$
$-x - 4y = -2$

28. $3x - 2y = 5$
$x + 2y = 7$
$-3x - y = -11$

29. $2x + y = 7$
$x - y = 2$
$-x + 3y = -2$

30. $3x - y = 2$
$-6x + 3y = 0$
$-x + 2y = -4$

31. $x + 3y = 7$
$x + y = 3$
$3x + y = 5$

32. $x + y = 3$
$x - 2y = -3$
$x - y = 1$

In Exercises 33–36, use matrices to solve each system of equations.

33. $x + 2y + 3z = -2$
$-x - y - 2z = 4$

34. $2x - 4y + 3z = 6$
$-4x + 6y + 4z = -6$

35. $x - y = 1$
$y + z = 1$
$x + z = 2$

36. $x + z = 1$
$x + y = 2$
$2x + y + z = 3$

APPLICATIONS

37. Chemistry   The atomic number of lead is four more than three times the atomic number of iron. If the atomic number of lead is decreased by twice the atomic number of iron, the result is the atomic number of zinc, which is 30. Find the atomic numbers of lead and iron.

38. Geometry   In triangle ABC, angle B is 20 more than the difference of angles C and A. Angle C is 20 less than the sum of A and B. If the sum of the angles in the triangle is 180 degrees, find the number of degrees in each angle.

REVIEW

39. If $f(x) = 3x^2 - 2x$ and $g(x) = 4x - 2$, find:

a. $f \circ g$

b. fg

c. $f - g$

40. Given $f(x) = 3x^2 - 2x$,

a. determine if the function is one-to-one and if it is, find its inverse.

b. determine the coordinates of the vertex.

c. determine the coordinates of the x-intercepts.

d. determine the coordinate of the y-intercept.

e. graph the function on graph paper.

f. is the function increasing or decreasing on the interval $[1, \infty)$?

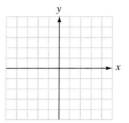

 41. Explain the difference between finding the x-intercepts of $f(x)$ in Exercise 40 and solving the equation $0 = 3x^2 - 2x$.

 42. Explain the relationship between the turning point of the function $f(x) = 3x^2 - 2x$ and its vertex.

12.5 USING DETERMINANTS TO SOLVE SYSTEMS OF EQUATIONS

• DETERMINANTS

• CRAMER'S RULE

▌ DETERMINANTS

A determinant is a number that is associated with a square matrix. For any square matrix A, the symbol $|A|$ represents the determinant of A.

Value of a 2 × 2 Determinant

If $a, b, c,$ and d are numbers, the determinant of $A = \begin{bmatrix} a & b \\ c & d \end{bmatrix}$ is $\begin{vmatrix} a & b \\ c & d \end{vmatrix} = ad - bc.$

The determinant of a 2×2 matrix is the number that is equal to the product of the numbers on the major diagonal

$$\begin{vmatrix} a & b \\ c & d \end{vmatrix}$$

minus the product of the numbers on the other diagonal

$$\begin{vmatrix} a & b \\ c & d \end{vmatrix}$$

EXAMPLE 1   Evaluate the determinants   **a.** $\begin{vmatrix} 3 & 2 \\ 6 & 9 \end{vmatrix}$ and   **b.** $\begin{vmatrix} -5 & \frac{1}{2} \\ -1 & 0 \end{vmatrix}$

SOLUTION   **a.** $\begin{vmatrix} 3 & 2 \\ 6 & 9 \end{vmatrix} = 3(9) - 2(6)$

$= 27 - 12$

$= 15$

b. $\begin{vmatrix} -5 & \frac{1}{2} \\ -1 & 0 \end{vmatrix} = -5(0) - \frac{1}{2}(-1)$

$= 0 + \frac{1}{2}$

$= \frac{1}{2}$

A 3×3 determinant is evaluated by expanding by **minors**. To find the minor of a_1, we cross out the elements of the determinant that are in the same row and column as a_1:

$$\begin{vmatrix} a_1 & b_1 & c_1 \\ a_2 & b_2 & c_2 \\ a_3 & b_3 & c_3 \end{vmatrix}$$

The minor of a_1 is $\begin{vmatrix} b_2 & c_2 \\ b_3 & c_3 \end{vmatrix}$

To find the minor of b_1, we cross out the elements of the determinant that are in the same row and column as b_1:

$$\begin{vmatrix} a_1 & b_1 & c_1 \\ a_2 & b_2 & c_2 \\ a_3 & b_3 & c_3 \end{vmatrix}$$

The minor of b_1 is $\begin{vmatrix} a_2 & c_2 \\ a_3 & c_3 \end{vmatrix}$

To find the minor of c_1, we cross out the elements of the determinant that are in the same row and column as c_1:

$$\begin{vmatrix} a_1 & b_1 & c_1 \\ a_2 & b_2 & c_2 \\ a_3 & b_3 & c_3 \end{vmatrix}$$

The minor of c_1 is $\begin{vmatrix} a_2 & b_2 \\ a_3 & b_3 \end{vmatrix}$

Value of a 3 × 3 Determinant

$$\begin{vmatrix} a_1 & b_1 & c_1 \\ a_2 & b_2 & c_2 \\ a_3 & b_3 & c_3 \end{vmatrix} = a_1 \overset{\overset{\text{Minor}}{\text{of } a_1}}{\begin{vmatrix} b_2 & c_2 \\ b_3 & c_3 \end{vmatrix}} - b_1 \overset{\overset{\text{Minor}}{\text{of } b_1}}{\begin{vmatrix} a_2 & c_2 \\ a_3 & c_3 \end{vmatrix}} + c_1 \overset{\overset{\text{Minor}}{\text{of } c_1}}{\begin{vmatrix} a_2 & b_2 \\ a_3 & b_3 \end{vmatrix}}$$

EXAMPLE 2   Evaluate the determinant $\begin{vmatrix} 1 & 3 & -2 \\ 2 & 1 & 3 \\ 1 & 2 & 3 \end{vmatrix}$

SOLUTION

$$\begin{array}{ccc} \textbf{Minor} & \textbf{Minor} & \textbf{Minor} \\ \textbf{of 1} & \textbf{of 3} & \textbf{of } -2 \\ \downarrow & \downarrow & \downarrow \end{array}$$

$$\begin{vmatrix} 1 & 3 & -2 \\ 2 & 1 & 3 \\ 1 & 2 & 3 \end{vmatrix} = 1 \begin{vmatrix} 1 & 3 \\ 2 & 3 \end{vmatrix} - 3 \begin{vmatrix} 2 & 3 \\ 1 & 3 \end{vmatrix} + (-2) \begin{vmatrix} 2 & 1 \\ 1 & 2 \end{vmatrix}$$

$$= 1(3 - 6) - 3(6 - 3) - 2(4 - 1)$$
$$= -3 - 9 - 6$$
$$= -18$$

TECHNOLOGY TIP

The determinant of a matrix can be computed with the calculator. The determinant command, det(, is located under the MATH submenu of the MATRIX menu.

```
[A]
    [[1  3  -2]
     [2  1  3 ]
     [1  2  3 ]]
det([A])
              -18
```

We can evaluate a 3×3 determinant by expanding it along any row or column. To determine the signs between the terms of the expansion of a 3×3 determinant, we use the following array of signs.

Array of Signs for a 3 × 3 Determinant

$$\begin{array}{ccc} + & - & + \\ - & + & - \\ + & - & + \end{array}$$

EXAMPLE 3   Evaluate the determinant $\begin{vmatrix} 1 & 3 & -2 \\ 2 & 1 & 3 \\ 1 & 2 & 3 \end{vmatrix}$ by expanding on the middle column.

SOLUTION   This is the determinant of Example 2. To expand it along the middle column, we use the signs of the middle column of the array of signs:

$$\begin{array}{ccc} \textbf{Minor} & \textbf{Minor} & \textbf{Minor} \\ \textbf{of 3} & \textbf{of 1} & \textbf{of 2} \\ \downarrow & \downarrow & \downarrow \end{array}$$

$$\begin{vmatrix} 1 & 3 & -2 \\ 2 & 1 & 3 \\ 1 & 2 & 3 \end{vmatrix} = -3 \begin{vmatrix} 2 & 3 \\ 1 & 3 \end{vmatrix} + 1 \begin{vmatrix} 1 & -2 \\ 1 & 3 \end{vmatrix} - 2 \begin{vmatrix} 1 & -2 \\ 2 & 3 \end{vmatrix}$$

$$= -3(6 - 3) + 1[3 - (-2)] - 2[3 - (-4)]$$
$$= -3(3) + 1(5) - 2(7)$$
$$= -9 + 5 - 14$$
$$= -18$$

As expected, we get the same value as in Example 2.

CRAMER'S RULE

Gabriel Cramer (1704–1752)   Although other mathematicians had worked with determinants, it was the work of Cramer that popularized them.

The method of using determinants to solve systems of equations is called **Cramer's rule**, named after the 18th-century mathematician Gabriel Cramer. To develop Cramer's rule, we consider the system

$$\begin{cases} ax + by = e \\ cx + dy = f \end{cases}$$

where x and y are variables and a, b, c, d, e, and f are constants.

If we multiply both sides of the first equation by d and multiply both sides of the second equation by $-b$, we can add the equations and eliminate y:

$$adx + bdy = ed$$
$$\underline{-bcx - bdy = -bf}$$
$$adx - bcx \qquad = ed - bf$$

To solve for x, we factor out x on the left-hand side and divide each side by $ad - bc$:

$$(ad - bc)x = ed - bf$$
$$x = \frac{ed - bf}{ad - bc} \qquad (ad - bc \neq 0)$$

We can find y in a similar manner. After eliminating the variable x, we get

$$y = \frac{af - ec}{ad - bc} \qquad (ad - bc \neq 0)$$

Determinants provide an easy way of remembering these formulas. Note that the denominator for both x and y is

$$\begin{vmatrix} a & b \\ c & d \end{vmatrix} = ad - bc$$

The numerators can be expressed as determinants also:

$$x = \frac{ed - bf}{ad - bc} = \frac{\begin{vmatrix} e & b \\ f & d \end{vmatrix}}{\begin{vmatrix} a & b \\ c & d \end{vmatrix}} \qquad \text{and} \qquad y = \frac{af - ec}{ad - bc} = \frac{\begin{vmatrix} a & e \\ c & f \end{vmatrix}}{\begin{vmatrix} a & b \\ c & d \end{vmatrix}}$$

If we compare these formulas with the original system

$$\begin{cases} ax + by = e \\ cx + dy = f \end{cases}$$

we note that in the expressions for x and y above, the denominator determinant is formed by using the coefficients a, b, c, and d of the variables in the equations. The numerator determinants are the same as the denominator determinant, except that the column of coefficients of the variable for which we are solving is replaced with the column of constants e and f.

Cramer's Rule for Two Equations in Two Variables

The solution of the system $\begin{cases} ax + by = e \\ cx + dy = f \end{cases}$ is given by

$$x = \frac{D_x}{D} = \frac{\begin{vmatrix} e & b \\ f & d \end{vmatrix}}{\begin{vmatrix} a & b \\ c & d \end{vmatrix}} \quad \text{and} \quad y = \frac{D_y}{D} = \frac{\begin{vmatrix} a & e \\ c & f \end{vmatrix}}{\begin{vmatrix} a & b \\ c & d \end{vmatrix}}$$

If every determinant is 0, the system is consistent but the equations are dependent and the solution is an infinite set of ordered pairs that satisfy either equation.

If $D = 0$ and D_x or D_y is nonzero, the system is inconsistent and the solution is $\varnothing$.

EXAMPLE 4   Use Cramer's rule to solve the system $\begin{cases} 4x - 3y = 6 \\ -2x + 5y = 4 \end{cases}$

SOLUTION   The value of x is the quotient of two determinants. The denominator determinant is made up of the coefficients of x and y:

$$D = \begin{vmatrix} 4 & -3 \\ -2 & 5 \end{vmatrix}$$

To solve for x, we form the numerator determinant from the denominator determinant by replacing its first column (the coefficients of x) with the column of constants (6 and 4).

To solve for y, we form the numerator determinant from the denominator determinant by replacing the second column (the coefficients of y) with the column of constants (6 and 4).

To find the values of x and y, we evaluate each determinant:

$$x = \frac{\begin{vmatrix} 6 & -3 \\ 4 & 5 \end{vmatrix}}{\begin{vmatrix} 4 & -3 \\ -2 & 5 \end{vmatrix}} = \frac{6(5) - (-3)(4)}{4(5) - (-3)(-2)} = \frac{30 + 12}{20 - 6} = \frac{42}{14} = 3$$

$$y = \frac{\begin{vmatrix} 4 & 6 \\ -2 & 4 \end{vmatrix}}{\begin{vmatrix} 4 & -3 \\ -2 & 5 \end{vmatrix}} = \frac{4(4) - 6(-2)}{4(5) - (-3)(-2)} = \frac{16 + 12}{20 - 6} = \frac{28}{14} = 2$$

The solution to this system is (3, 2). Verify that $x = 3$ and $y = 2$ satisfy each equation in the given system.

TECHNOLOGY TIP

To solve for x, the numerator determinant will be computed from matrix B and the denominator determinant from matrix A:

```
[A]
      [[4  -3]
       [-2 5 ]]
[B]
      [[6 -3]
       [4 5 ]]
```

To solve for y, the numerator determinant will be computed from matrix C and the denominator determinant from matrix A:

```
[A]
      [[4  -3]
       [-2 5 ]]
[C]
      [[4  6]
       [-2 4]]
```

```
det([B])/det([A]
)
                3
det([C])/det([A]
)
                2
```

EXAMPLE 5

Use Cramer's rule to solve the system $\begin{cases} 7x = 8 - 4y \\ 2y = 3 - \frac{7}{2}x \end{cases}$

SOLUTION   We multiply both sides of the second equation by 2 to eliminate the fraction and write the system in the form

$$\begin{cases} 7x + 4y = 8 \\ 7x + 4y = 6 \end{cases}$$

When we attempt to use Cramer's rule to solve this system, we find that the determinant in the denominator is 0:

$$\begin{vmatrix} 7 & 4 \\ 7 & 4 \end{vmatrix} = 7(4) - 4(7) = 0$$

Because the two equations are different (independent) and because the determinant in the denominator is 0, the system is inconsistent. It has no solutions.

Cramer's Rule for Three Equations in Three Variables

The solution of the system $\begin{cases} ax + by + cz = j \\ dx + ey + fz = k \\ gx + hy + iz = l \end{cases}$ is given by

$$x = \frac{D_x}{D}, \qquad y = \frac{D_y}{D}, \qquad \text{and} \qquad z = \frac{D_z}{D}$$

where

$$D = \begin{vmatrix} a & b & c \\ d & e & f \\ g & h & i \end{vmatrix} \qquad D_x = \begin{vmatrix} j & b & c \\ k & e & f \\ l & h & i \end{vmatrix}$$

$$D_y = \begin{vmatrix} a & j & c \\ d & k & f \\ g & l & i \end{vmatrix} \qquad D_z = \begin{vmatrix} a & b & j \\ d & e & k \\ g & h & l \end{vmatrix}$$

If every determinant is 0, the system is consistent but the equations are dependent and the solution is an infinite set of ordered pairs that satisfy any of the equations.

If $D = 0$ and D_x or D_y or D_z is nonzero, the system is inconsistent and the solution is $\varnothing$.

EXAMPLE 6

Use Cramer's rule to solve the system $\begin{cases} 2x + y + 4z = 12 \\ x + 2y + 2z = 9 \\ 3x - 3y - 2z = 1 \end{cases}$

SOLUTION   The denominator determinant is the determinant formed by the coefficients of the variables. The numerator determinants are formed by replacing the coefficients of the variable being solved for by the column of constants. We form the quotients for x, y, and z and evaluate the determinants:

$$x = \frac{\begin{vmatrix} 12 & 1 & 4 \\ 9 & 2 & 2 \\ 1 & -3 & -2 \end{vmatrix}}{\begin{vmatrix} 2 & 1 & 4 \\ 1 & 2 & 2 \\ 3 & -3 & -2 \end{vmatrix}} = \frac{12 \begin{vmatrix} 2 & 2 \\ -3 & -2 \end{vmatrix} - 1 \begin{vmatrix} 9 & 2 \\ 1 & -2 \end{vmatrix} + 4 \begin{vmatrix} 9 & 2 \\ 1 & -3 \end{vmatrix}}{2 \begin{vmatrix} 2 & 2 \\ -3 & -2 \end{vmatrix} - 1 \begin{vmatrix} 1 & 2 \\ 3 & -2 \end{vmatrix} + 4 \begin{vmatrix} 1 & 2 \\ 3 & -3 \end{vmatrix}} = \frac{12(2) - (-20) + 4(-29)}{2(2) - (-8) + 4(-9)} = \frac{-72}{-24} = 3$$

$$y = \frac{\begin{vmatrix} 2 & 12 & 4 \\ 1 & 9 & 2 \\ 3 & 1 & -2 \end{vmatrix}}{\begin{vmatrix} 2 & 1 & 4 \\ 1 & 2 & 2 \\ 3 & -3 & -2 \end{vmatrix}} = \frac{2 \begin{vmatrix} 9 & 2 \\ 1 & -2 \end{vmatrix} - 12 \begin{vmatrix} 1 & 2 \\ 3 & -2 \end{vmatrix} + 4 \begin{vmatrix} 1 & 9 \\ 3 & 1 \end{vmatrix}}{-24} = \frac{2(-20) - 12(-8) + 4(-26)}{-24} = \frac{-48}{-24} = 2$$

$$z = \frac{\begin{vmatrix} 2 & 1 & 12 \\ 1 & 2 & 9 \\ 3 & -3 & 1 \end{vmatrix}}{\begin{vmatrix} 2 & 1 & 4 \\ 1 & 2 & 2 \\ 3 & -3 & -2 \end{vmatrix}} = \frac{2 \begin{vmatrix} 2 & 9 \\ -3 & 1 \end{vmatrix} - 1 \begin{vmatrix} 1 & 9 \\ 3 & 1 \end{vmatrix} + 12 \begin{vmatrix} 1 & 2 \\ 3 & -3 \end{vmatrix}}{-24} = \frac{2(29) - (-26) + 12(-9)}{-24} = \frac{-24}{-24} = 1$$

The solution to this system is (3, 2, 1).

EXERCISE 12.5

VOCABULARY AND NOTATION   *In Exercises 1 and 2, fill in the blanks to make a true statement.*

1. A _____ is a number that is associated with a square matrix.

2. _____ rule uses determinants to solve systems of equations.

PRACTICE   *In Exercises 3–20, evaluate each determinant. Verify with your calculator.*

3. $\begin{vmatrix} 2 & 3 \\ -2 & 1 \end{vmatrix}$

4. $\begin{vmatrix} 3 & -2 \\ -2 & 4 \end{vmatrix}$

5. $\begin{vmatrix} -1 & 2 \\ 3 & -4 \end{vmatrix}$

6. $\begin{vmatrix} -1 & -2 \\ -3 & -4 \end{vmatrix}$

7. $\begin{vmatrix} x & y \\ y & x \end{vmatrix}$

8. $\begin{vmatrix} x+y & x-y \\ x-y & x+y \end{vmatrix}$

9. $\begin{vmatrix} 1 & 0 & 1 \\ 0 & 1 & 0 \\ 1 & 1 & 1 \end{vmatrix}$

10. $\begin{vmatrix} 1 & 2 & 0 \\ 0 & 1 & 2 \\ 0 & 0 & 1 \end{vmatrix}$

11. $\begin{vmatrix} -1 & 2 & 1 \\ 2 & 1 & -3 \\ 1 & 1 & 1 \end{vmatrix}$

12. $\begin{vmatrix} 1 & 2 & 3 \\ 1 & 2 & 3 \\ 1 & 2 & 3 \end{vmatrix}$

13. $\begin{vmatrix} 1 & -2 & 3 \\ -2 & 1 & 1 \\ -3 & -2 & 1 \end{vmatrix}$

14. $\begin{vmatrix} 1 & 1 & 2 \\ 2 & 1 & -2 \\ 3 & 1 & 3 \end{vmatrix}$

15. $\begin{vmatrix} 1 & 2 & 3 \\ 4 & 5 & 6 \\ 7 & 8 & 9 \end{vmatrix}$

16. $\begin{vmatrix} 1 & 4 & 7 \\ 2 & 5 & 8 \\ 3 & 6 & 9 \end{vmatrix}$

17. $\begin{vmatrix} a & 2a & -a \\ 2 & -1 & 3 \\ 1 & 2 & -3 \end{vmatrix}$

18. $\begin{vmatrix} 1 & 2b & -3 \\ 2 & -b & 2 \\ 1 & 3b & 1 \end{vmatrix}$

19. $\begin{vmatrix} 1 & a & b \\ 1 & 2a & 2b \\ 1 & 3a & 3b \end{vmatrix}$

20. $\begin{vmatrix} a & b & c \\ 0 & b & c \\ 0 & 0 & c \end{vmatrix}$

In Exercises 21–46, use Cramer's rule to solve each system of equations, if possible.

21. $\begin{cases} x + y = 6 \\ x - y = 2 \end{cases}$

22. $\begin{cases} x - y = 4 \\ 2x + y = 5 \end{cases}$

23. $\begin{cases} 2x + y = 1 \\ x - 2y = -7 \end{cases}$

24. $\begin{cases} 2x + 3y = 0 \\ 4x - 6y = -4 \end{cases}$

25. $\begin{cases} 4x - 3y = -1 \\ 8x + 3y = 4 \end{cases}$

26. $\begin{cases} 3x - y = -3 \\ 2x + y = -7 \end{cases}$

27. $\begin{cases} y = \dfrac{11 - 3x}{2} \\ x = \dfrac{11 - 4y}{6} \end{cases}$

28. $\begin{cases} x = \dfrac{12 - 6y}{5} \\ y = \dfrac{24 - 10x}{12} \end{cases}$

29. $\begin{cases} y = \dfrac{-2x + 1}{3} \\ 3x - 2y = 8 \end{cases}$

30. $\begin{cases} 2x + 3y = -1 \\ x = \dfrac{y - 9}{4} \end{cases}$

31. $\begin{cases} x = \dfrac{5y - 4}{2} \\ y = \dfrac{3x - 1}{5} \end{cases}$

32. $\begin{cases} y = \dfrac{1 - 5x}{2} \\ x = \dfrac{3y + 10}{4} \end{cases}$

33. $\begin{cases} x + y + z = 4 \\ x + y - z = 0 \\ x - y + z = 2 \end{cases}$

34. $\begin{cases} x + y + z = 4 \\ x - y + z = 2 \\ x - y - z = 0 \end{cases}$

35. $\begin{cases} x + y + 2z = 7 \\ x + 2y + z = 8 \\ 2x + y + z = 9 \end{cases}$

36. $\begin{cases} x + 2y + 2z = 10 \\ 2x + y + 2z = 9 \\ 2x + 2y + z = 1 \end{cases}$

37. $\begin{cases} 2x + y - z = 1 \\ x + 2y + 2z = 2 \\ 4x + 5y + 3z = 3 \end{cases}$

38. $\begin{cases} 2x + 3y + 4z = 6 \\ 2x - 3y - 4z = -4 \\ 4x + 6y + 8z = 12 \end{cases}$

39. $\begin{cases} 2x + y + z = 5 \\ x - 2y + 3z = 10 \\ x + y - 4z = -3 \end{cases}$

40. $\begin{cases} 3x + 2y - z = -8 \\ 2x - y + 7z = 10 \\ 2x + 2y - 3z = -10 \end{cases}$

41. $\begin{cases} 4x + 3z = 4 \\ 2y - 6z = -1 \\ 8x + 4y + 3z = 9 \end{cases}$

42. $\begin{cases} x - 3y + 4z - 2 = 0 \\ 2x + y + 2z - 3 = 0 \\ 4x - 5y + 10z - 7 = 0 \end{cases}$

43. $\begin{cases} x + y = 1 \\ \dfrac{1}{2}y + z = \dfrac{5}{2} \\ x - z = -3 \end{cases}$

44. $\begin{cases} 3x + 4y + 14z = 7 \\ -\dfrac{1}{2}x - y + 2z = \dfrac{3}{2} \\ x + \dfrac{3}{2}y + \dfrac{5}{2}z = 1 \end{cases}$

45. $\begin{cases} 2x - y + 4z + 2 = 0 \\ 5x + 8y + 7z = -8 \\ x + 3y + z + 3 = 0 \end{cases}$

46. $\begin{cases} x + 2y + 2z + 3 = 0 \\ x + \dfrac{1}{2}y + z - \dfrac{1}{2} = 0 \\ 2x + 2y + z + 1 = 0 \end{cases}$

In Exercises 47–50, evaluate each determinant and solve the resulting equation.

47. $\begin{vmatrix} x & 1 \\ 3 & 2 \end{vmatrix} = 1$

48. $\begin{vmatrix} x & -x \\ 2 & -3 \end{vmatrix} = -5$

49. $\begin{vmatrix} x & -2 \\ 3 & 1 \end{vmatrix} = \begin{vmatrix} 4 & 2 \\ x & 3 \end{vmatrix}$

50. $\begin{vmatrix} x & 3 \\ x & 2 \end{vmatrix} = \begin{vmatrix} 3 & 2 \\ 1 & 1 \end{vmatrix}$

REVIEW

51. Solve the system $\begin{cases} 2x + y = 1 \\ x - 2y = 7 \end{cases}$ using the indicated method:

  a. elimination method   **b.** substitution method

  c. graphing, using your calculator.

52. Algebraically determine the inverse function of $f(x) = 6x - 5$.

53. Graph each function on graph paper and verify with your calculator.

  a. $f(x) = \left(\dfrac{4}{5}\right)^x$ **b.** $g(x) = \log_{4/5} x$

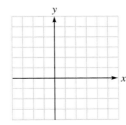 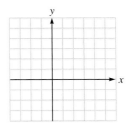

54. Explain why $f(x)$ and $g(x)$ in Exercise 53 are inverses of one another.

12.6   SYSTEMS OF INEQUALITIES

- SOLVING SYSTEMS OF LINEAR INEQUALITIES
- SOLVING SYSTEMS OF QUADRATIC INEQUALITIES

▮▮ SOLVING SYSTEMS OF LINEAR INEQUALITIES

To solve **systems of inequalities** containing two variables, we want to find all ordered pairs that are solutions of all the inequalities in the system. We will, therefore, graph the equations associated with each inequality in the same coordinate plane (using dotted lines for strict inequalities, $<$ or $>$, and solid lines when equality is included, $\leq$ or $\geq$). We then use a test point to decide which region for each inequality should be shaded. The region that is common to all graphs of the system will be the most heavily shaded and contains all the points that are solutions to the system. The steps are specified below.

Solving Systems of Inequalities

1. Graph the boundary line for *each* region by graphing the equation associated with the inequality. If the inequality allows the possibility of equality, draw the boundary line as a solid line. If equality is not allowed, draw the boundary line as a dotted line.
2. After graphing the boundary line for a particular inequality, pick a point that is on one side of the boundary line. Replace x and y with the coordinates of that point. If the inequality is true, shade the side that contains that point. If the inequality is false, shade the region that does **not** contain the test point.
3. Follow steps 1 and 2 for each inequality in the system.
4. The region common to every graph is the solution of the system. It should be the most heavily shaded area.
5. Pick a test point and verify it in each of the original inequalities in the system.

EXAMPLE 1

Graph the solution set of the system $\begin{cases} 2x + y < 4 \\ -2x + y > 2 \end{cases}$.

SOLUTION   We graph each inequality on the same set of coordinate axes, as in Figure 12-9.

$2x + y = 4$

x	y	(x, y)
0	4	$(0, 4)$
1	2	$(1, 2)$
2	0	$(2, 0)$

$-2x + y = 2$

x	y	(x, y)
-1	0	$(-1, 0)$
0	2	$(0, 2)$
2	6	$(2, 6)$

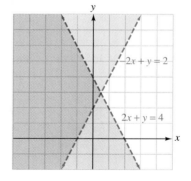

Figure 12-9

We then find the graph of each inequality.

- The graph of $2x + y < 4$ includes all points below the line $2x + y = 4$. Because the boundary is not included, we draw it as a dotted line.
- The graph of $-2x + y > 2$ includes all points above the line $-2x + y = 2$. Because the boundary is not included, we draw it as a dotted line.

The area that is shaded twice represents the set of simultaneous solutions of the given system of inequalities. Any point in the doubly shaded region has coordinates that will satisfy both inequalities of the system.

Pick a point in the doubly shaded region and show that it satisfies both inequalities: $(-4, 5)$ satisfies both inequalities as demonstrated by the calculator.

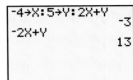

EXAMPLE 2

Graph the solution set of the system $\begin{cases} x \le 2 \\ y > 3 \end{cases}$.

SOLUTION   We graph each inequality on the same set of coordinate axes, as in Figure 12-10.

$x = 2$		
x	y	(x, y)
2	0	$(2, 0)$
2	2	$(2, 2)$
2	4	$(2, 4)$

$y = 3$		
x	y	(x, y)
0	3	$(0, 3)$
1	3	$(1, 3)$
4	3	$(4, 3)$

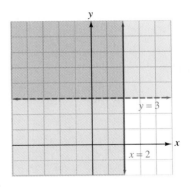

Figure 12-10

We then find the graph of each inequality.

- The graph of $x \le 2$ includes all points on the line $x = 2$ and all points to the left of the line. Because the boundary line is included, we draw it as a solid line.
- The graph $y > 3$ includes all points above the line $y = 3$. Because the boundary is not included, we draw it as a dotted line.

The area that is shaded twice represents the set of simultaneous solutions of the given system of inequalities. Any point in the doubly shaded region has coordinates that will satisfy both inequalities of the system. Pick a point in the doubly shaded region and show that this is true.

EXAMPLE 3

Graph the solution set of the system $\begin{cases} y < 3x - 1 \\ y \geq 3x + 1 \end{cases}$.

SOLUTION

We graph each inequality, as in Figure 12-11.

- The graph of $y < 3x - 1$ includes all of the points below the dotted line $y = 3x - 1$.
- The graph of $y \geq 3x + 1$ includes all of the points on and above the solid line $y = 3x + 1$.

Because the graphs of these inequalities do not intersect, the solution set is empty. There are no solutions: solution set, $\varnothing$.

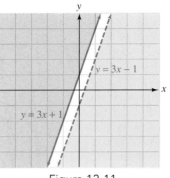

Figure 12-11

EXAMPLE 4

Graph the solution set of the system $\begin{cases} x \geq 0 \\ y \geq 0 \\ x + 2y \leq 6 \end{cases}$.

SOLUTION

We graph each inequality, as in Figure 12-12.

- The graph of $x \geq 0$ includes all of the points on the y-axis and to the right.
- The graph of $y \geq 0$ includes all of the points on the x-axis and above.
- The graph of $x + 2y \leq 6$ includes all of the points on the line $x + 2y = 6$ and below.

The solution is the region that is shaded three times. This includes triangle OPQ and the triangular region it encloses.

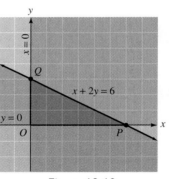

Figure 12-12

▌ SOLVING SYSTEMS OF QUADRATIC INEQUALITIES

The process for solving systems of quadratic inequalities is exactly the same as that presented previously.

EXAMPLE 5

Graph the solution set of the system $\begin{cases} y < x^2 \\ y > \dfrac{1}{4}x^2 - 2 \end{cases}$.

SOLUTION

The graph of $y = x^2$ is the parabola shown in Figure 12-13, which opens upward and has its vertex at the origin. The points with coordinates that satisfy the inequality $y < x^2$ are those points below the parabola.

$$y = x^2$$

$$y = \frac{1}{4}x^2 - 2$$

x	y	(x, y)
0	0	$(0, 0)$
1	1	$(1, 1)$
-1	1	$(-1, 1)$
2	4	$(2, 4)$
-2	4	$(-2, 4)$

x	y	(x, y)
0	-2	$(0, -2)$
2	-1	$(2, -1)$
-2	-1	$(-2, -1)$
4	2	$(4, 2)$
-4	2	$(-4, 2)$

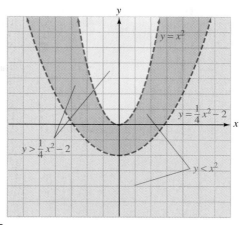

Figure 12-13

The graph of $y = \frac{1}{4}x^2 - 2$ is a parabola opening upward, with vertex at $(0, -2)$. However, this time the points with coordinates that satisfy the inequality are those points above the parabola. The graph of the solution set of the system is the area between the parabolas.

EXERCISE 12.6

In Exercises 1–26, find the solution set of each system of inequalities, when possible.

1. $\begin{cases} x + 2y \le 3 \\ 2x - y \ge 1 \end{cases}$

2. $\begin{cases} 2x + y \ge 3 \\ x - 2y \le -1 \end{cases}$

3. $\begin{cases} x + y < -1 \\ x - y > -1 \end{cases}$

4. $\begin{cases} x + y > 2 \\ x - y < -2 \end{cases}$

5. $\begin{cases} 2x - y < 4 \\ x + y \ge -1 \end{cases}$

6. $\begin{cases} x - y \ge 5 \\ x + 2y < -4 \end{cases}$

7. $\begin{cases} x > 2 \\ y \le 3 \end{cases}$

8. $\begin{cases} x \ge -1 \\ y > -2 \end{cases}$

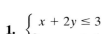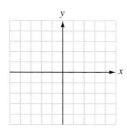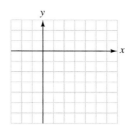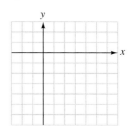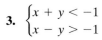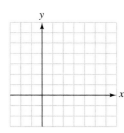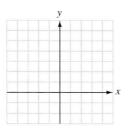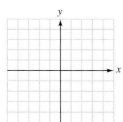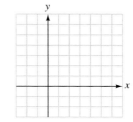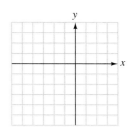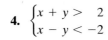

9. $\begin{cases} 2x - 3y < 0 \\ y > x - 1 \end{cases}$

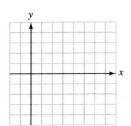

10. $\begin{cases} 3x - y \geq -1 \\ y \geq 3x + 1 \end{cases}$

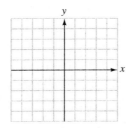

17. $\begin{cases} x < 3y - 1 \\ y \geq 2x - 3 \end{cases}$

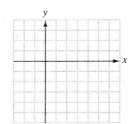

18. $\begin{cases} y \geq x + 2 \\ x \leq y - 2 \end{cases}$

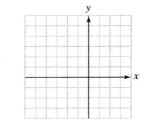

11. $\begin{cases} x + y < 1 \\ x + y > 3 \end{cases}$

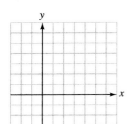

12. $\begin{cases} x + y > 2 \\ x + y < 4 \end{cases}$

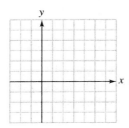

19. $\begin{cases} 2x + y < 7 \\ y > 2(1 - x) \end{cases}$

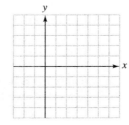

20. $\begin{cases} 2x + y \geq 6 \\ y \leq 2(2x - 3) \end{cases}$

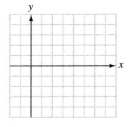

13. $\begin{cases} x > 0 \\ y > 0 \end{cases}$

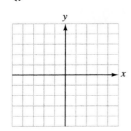

14. $\begin{cases} x \leq 0 \\ y < 0 \end{cases}$

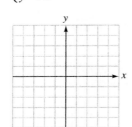

21. $\begin{cases} 2x - 4y > -6 \\ 3x + y \geq 5 \end{cases}$

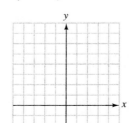

22. $\begin{cases} 2x - 3y < 0 \\ 2x + 3y \geq 12 \end{cases}$

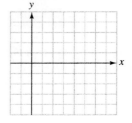

15. $\begin{cases} 3x + 4y > -7 \\ 2x - 3y \geq 1 \end{cases}$

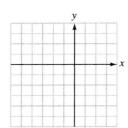

16. $\begin{cases} 3x + y \leq 1 \\ 4x - y > -8 \end{cases}$

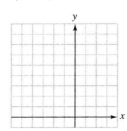

23. $\begin{cases} 3x - y \leq -4 \\ 3y > -2(x + 5) \end{cases}$

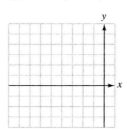

24. $\begin{cases} 3x + y < -2 \\ y > 3(1 - x) \end{cases}$

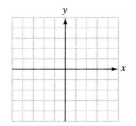

25. $\begin{cases} \dfrac{x}{2} + \dfrac{y}{3} \geq 2 \\ \dfrac{x}{2} - \dfrac{y}{2} < -1 \end{cases}$

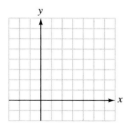

26. $\begin{cases} \dfrac{x}{3} - \dfrac{y}{2} < -3 \\ \dfrac{x}{3} + \dfrac{y}{2} > -1 \end{cases}$

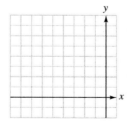

In Exercises 27–30, use the graphing method to find the region that satisfies all of the inequalities of the system.

27. $\begin{cases} x \geq 0 \\ y \geq 0 \\ x + y \leq 3 \end{cases}$

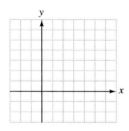

28. $\begin{cases} x - y \leq 6 \\ x + 2y \leq 6 \\ x \geq 0 \end{cases}$

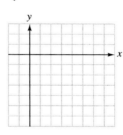

29. $\begin{cases} x \geq 0 \\ y \geq 0 \\ x \leq 5 \\ y \leq x \end{cases}$

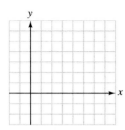

30. $\begin{cases} x \geq 0 \\ y \geq 0 \\ y \leq 2 + x \\ y \geq 4x - 2 \end{cases}$

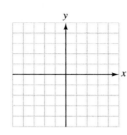

In Exercises 31–33, solve each system of inequalities by graphing.

31. $\begin{cases} x^2 - 6x - y \leq 5 \\ x^2 - 6x - y \geq -5 \end{cases}$

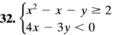
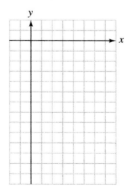

32. $\begin{cases} x^2 - x - y \geq 2 \\ 4x - 3y < 0 \end{cases}$

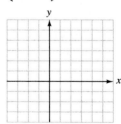
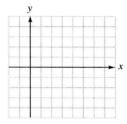

33. $\begin{cases} x^2 - 6x - y < -5 \\ x^2 - 6x + y < -5 \end{cases}$

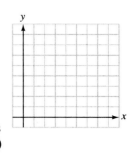

APPLICATIONS *In Exercises 34–37, graph each system of inequalities and give two possible solutions to each problem.*

34. Buying compact disks   Melodic Music has compact disks on sale for either $10 or $15. If a customer wants to spend at least $30 but no more than $60 on CDs, find a system of inequalities whose graph will show the possible combinations of $10 CDs ($x$) and $15 CDs ($y$) that the customer can buy.

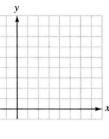

35. Buying boats   Dry Boatworks wholesales aluminum boats for $800 and fiberglass boats for $600. Northland Marina wants to order at least $2,400 but no more than $4,800 worth of boats. Find a system of inequalities whose graph will show the possible combination of aluminum boats (x) and fiberglass boats (y) that can be ordered.

36. Buying furniture   A distrib-
utor wholesales desk chairs
for $150 and side chairs for
$100. Best Furniture wants
to order no more than $900
worth of chairs and wants to
order more side chairs than
desk chairs. Find a system of
inequalities whose graph will
show the possible combinations of desk chairs (x) and
side chairs (y) that can be ordered.

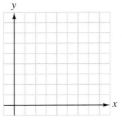

37. Ordering furnace equipment
J. Bolden Heating Company
wants to order no more than
$2,000 worth of electronic air
cleaners and humidifiers from
a wholesaler that charges $500
for air cleaners and $200 for
humidifiers. If Bolden wants
more humidifiers than air
cleaners, find a system of in-
equalities whose graph will show the possible combinations of air cleaners (x) and humidifiers (y) that can
be ordered.

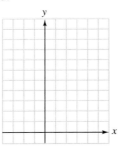

38. Explain how the graphing calculator can be used to
solve a system of inequalities.

39. Explain when a system of inequalities will have no
solution.

REVIEW

40. Solve the equation $\sqrt[3]{6x + 4} - 4 = 0$ both algebraically and graphically.

41. Solve the equation $1 - \sqrt{x} = \sqrt{x - 3}$ both algebraically and graphically.

42. Find the distance between P and Q if $P(-3, 6)$ and
$Q(20, 12)$. Round the result to the nearest tenth.

43. Find the domain and the range of the function
$f(x) = \sqrt{x + 2} - 4$.

44. Using the function specified in Exercise 43, find $f(2)$.

CHAPTER SUMMARY

CONCEPTS

REVIEW EXERCISES

Section 12.1

The solution(s) to a system of equations are the value(s) of the variable(s) that make all the equations in the system true.

To solve a system of equations graphically, carefully graph each equation of the system on the same set of axes. If the lines intersect, the coordinates of the point of intersection give the solution to the system.

To solve a system of equations by substitution, solve one of the equations of the system for one of its variables, substitute the resulting expression into the other equation, and solve for the other variable. Then substitute the value obtained back into one of the original equations to solve for the other variable.

To solve a system of equations by addition (elimination), first multiply one or both of the equations by suitable constants, if necessary, to eliminate one of the variables when the equations are added. The equation that results can be solved for its single variable. Then substitute the value obtained back into one of the original equations and solve for the other variable.

Solving Systems of Two Equations in Two Variables

1. Tell whether the ordered pair is a solution of the given system.

   a. $(1, 5)$; $\begin{cases} 3x - y = -2 \\ 2x + 3y = 17 \end{cases}$ **b.** $(14, \frac{1}{2})$; $\begin{cases} 2x + 4y = 30 \\ \dfrac{x}{4} - y = 3 \end{cases}$

2. Solve each system below graphically.

   a. $\begin{cases} x + y = 7 \\ 2x - y = 5 \end{cases}$ **b.** $\begin{cases} \dfrac{x}{3} + \dfrac{y}{5} = -1 \\ x - 3y = -3 \end{cases}$

3. Use the substitution method to solve each system.

   a. $\begin{cases} x = 3y + 5 \\ 5x - 4y = 3 \end{cases}$

   b. $\begin{cases} 6(x + 2) = y - 1 \\ 5(y - 1) = x + 2 \end{cases}$

   c. $x + 2y = 8$
   $5x = 30 - 10y$

4. Use the addition (elimination) method to solve each system.

   a. $\begin{cases} 2x + y = 1 \\ 5x - y = 20 \end{cases}$

   b. $\begin{cases} 11x + 3y = 27 \\ 8x + 4y = 36 \end{cases}$

   c. $2x - 3y = 6$
   $6y = 4x - 12$

Section 12.2

Many application problems can be solved by specifying a system of equations. If two variables are used to represent unknown quantities, two equations must be written.

Applications of Systems of Two Equations in Two Variables

5. **Integer problem** One number is 5 times another, and their sum is 18. Find the numbers.

6. **Geometry** The length of a rectangle is 3 times its width, and its perimeter is 24 feet. Find its dimensions.

7. **Buying grapefruit**   A grapefruit costs 15 cents more than an orange. Together, they cost 85 cents. Find the cost of the grapefruit.

8. **Utility bills**   A man's electric bill for January was $23 less than his gas bill. The two utilities charged him a total of $109. Find the amount of his gas bill.

9. **Investing money**   Carlos invested part of $3,000 in a 10% certificate account and the rest in a 6% passbook account. The total annual interest from both accounts is $270. How much did he invest at 6%?

Section 12.3

A system of three equations in three variables can be solved by using the addition (elimination) method.

Solving Systems of Three Equations in Three Variables

10. Solve the system $\begin{cases} x + y + z = 6 \\ x - y - z = -4 \\ -x + y - z = -2 \end{cases}$

11. Solve the system $\begin{cases} 2x + 3y + z = -5 \\ -x + 2y - z = -6 \\ 3x + y + 2z = 4 \end{cases}$

Section 12.4

A matrix is any rectangular array of numbers. Elementary row operations that can be performed on the matrix are:

1. Any two rows can be interchanged.
2. Any row can be multiplied by a nonzero constant.
3. Any row can be changed by adding a constant multiple of another row to it.

Using Matrices to Solve Systems of Equations

12. Solve each system of equations by using matrices.

a. $\begin{cases} x + 2y = 4 \\ 2x - y = 3 \end{cases}$

b. $\begin{cases} x + y + z = 6 \\ 2x - y + z = 1 \\ 4x + y - z = 5 \end{cases}$

Section 12.5

A determinant is a number that is associated with a square matrix. Cramer's rule can be used to solve systems of linear equations.

Using Determinants to Solve Systems of Equations

13. Evaluate each determinant.

a. $\begin{vmatrix} 2 & 3 \\ -4 & 3 \end{vmatrix}$

b. $\begin{vmatrix} -1 & 2 & -1 \\ 2 & -1 & 3 \\ 1 & -2 & 2 \end{vmatrix}$

14. Use Cramer's rule to solve each system of equations.

a. $\begin{cases} 3x + 4y = 10 \\ 2x - 3y = 1 \end{cases}$

b. $\begin{cases} x + 2y + z = 0 \\ 2x + y + z = 3 \\ x + y + 2z = 5 \end{cases}$

Section 12.6   Systems of Inequalities

Systems of inequalities can be solved graphically. Carefully graph each inequality on the same set of axes and shade appropriately. The most heavily shaded area is the solution of the system.

15. Graph the solution set of each system of inequalities.

a. $y \geq x + 1$
$3x + 2y < 6$

b. $y \geq x^2 - 4$
$y < x + 3$

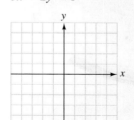

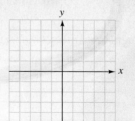

CHAPTER TEST

1. Solve the system $\begin{cases} 3x + y = 7 \\ x - 2y = 0 \end{cases}$ graphically.

2. Solve the system $\begin{cases} y = x - 1 \\ 2x + y = -7 \end{cases}$ by substitution.

3. Solve the system $\begin{cases} 6x - 2y = 4 \\ 2x + y = 8 \end{cases}$ by addition (elimination).

In Problems 4 and 5, solve each system using any method.

4. a. $\begin{cases} \dfrac{x}{6} + \dfrac{y}{10} = 3 \\ \dfrac{5x}{16} - \dfrac{3y}{16} = \dfrac{15}{8} \end{cases}$ **b.** $\begin{cases} 4x + 3 = -3y \\ -\dfrac{x}{7} + \dfrac{4y}{21} = 1 \end{cases}$

5. Solve the system $\begin{cases} 2x + 3y + z = -5 \\ -x + 2y - z = -6 \\ 3x + y + 2z = 4 \end{cases}$ by addition.

6. The producer of a 30-minute documentary about World War I divided it into two parts. Four times as much program time was devoted to the causes of the war as to the outcome. How long was each part of the documentary?

7. At an IMAX theater, the giant rectangular movie screen has a width 26 feet less than its length. If its perimeter is 332 feet, find the area of the screen.

8. The sum of two numbers is -18. One number is 2 greater than 3 times the other. Find the product of the numbers.

9. The sum of three integers is 7. The second integer is 2 less than the third and the third is 9 more than the first. Find the integers.

In Problem 10, evaluate each determinant.

10. a. $\begin{vmatrix} 2 & -3 \\ 4 & 5 \end{vmatrix}$ **b.** $\begin{vmatrix} -3 & -4 \\ -2 & 3 \end{vmatrix}$

c. $\begin{vmatrix} 1 & 2 & 0 \\ 2 & 0 & 3 \\ 1 & -2 & 2 \end{vmatrix}$ **d.** $\begin{vmatrix} 2 & -1 & 1 \\ 3 & 1 & 0 \\ 0 & 1 & 2 \end{vmatrix}$

In Problems 11–14, consider the system $\begin{cases} x - y = -6 \\ 3x + y = -6 \end{cases}$ *which is to be solved with Cramer's rule.*

11. When solving for x, what is the numerator determinant? (**Do not evaluate it.**)

12. When solving for y, what is the denominator determinant? (**Do not evaluate it.**)

13. Solve the system for x.

14. Solve the system for y.

In Problems 15–18, consider the system

$$\begin{cases} x + y + z = 4 \\ x + y - z = 6 \\ 2x - 3y + z = -1 \end{cases}$$

15. Solve for x.   **16.** Solve for y.

17. Solve for z.

18. Write the augmented matrix that represents the system.

In Problems 19 and 20, use graphing to solve each system.

19. $\begin{cases} 2x - 3y \geq 6 \\ \quad\quad y \leq -x + 1 \end{cases}$ **20.** $\begin{cases} y \geq x^2 \\ y < x + 3 \end{cases}$

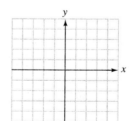
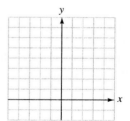

CUMULATIVE REVIEW

CHAPTERS 11 AND 12
Vocabulary/Concepts

1. If a function contains the point (a, b), then the inverse of the function contains the point _____.

2. The notation _____ is used to denote the inverse of the function $f(x)$.

3. Determine the inverse relation of the set of ordered pairs $\{(1, 3), (2, 5), (-3, -2), (-4, -6)\}$.

In Exercises 4–6, determine which of the following represent one-to-one functions.

4.
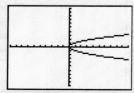

5.
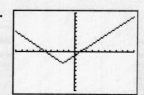

6.
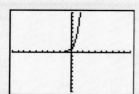

7. Functions with variables as exponents are called _____ functions.

8. Classify each of the following as "an exponential function" or "not an exponential function."

  a. $f(x) = x^3$

  b. $f(x) = 3^x$

  c. $f(x) = \left(\frac{1}{2}\right)^x$

9. Find x: $\log_3 9 = x$.

10. Use the properties of logarithms to write the given expression as a single logarithm:
$-5 \log_b x - 3 \log_b y + \log_b z$

11. If two or more equations are considered at the same time, they are called a _____ of equations.

12. When a system of equations has one or more solutions, it is called a _____ system.

13. If a system has no solutions, it is called an _____ system.

14. Equations were entered at the y1 and y2 prompts of the calculator. State the solution of the system by interpreting the given graphical display.

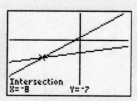

15. A rectangular array of numbers is called a _____.

16. A _____ matrix has the same number of rows and columns.

17. An _____ matrix includes the constant values of the system, whereas a _____ matrix only includes the coefficients.

18. Write the coefficient matrix for the system

$$\begin{cases} x + 2y + 2z = 10 \\ 2x + y + 2z = 9 \\ 2x + 2y + z = 11 \end{cases}$$

19. Write the augmented matrix for Exercise 18.

20. _____ rule uses determinants to solve systems of equations.

Practice *In Exercises 21–24, $f(x) = 3x - 2$ and $g(x) = x^2 + 1$. Find each value.*

21. $(f \circ g)(3)$

22. $(g \circ f)(3)$

23. $(f \circ g)\left(\dfrac{1}{4}\right)$

24. $(g \circ f)(h)$

25. If $f(x) = x^2 + 5x + 6$ and $g(x) = x + 3$, find each of the following:

    a. $f + g$
        b. $f - g$

    c. $f \cdot g$
        d. $\dfrac{f}{g}$

26. If $f(x) = x^2 - 4$ and $g(x) = x + 2$, find each function and its domain.

    a. $f + g$

    b. $f - g$

    c. $f \cdot g$

    d. $\dfrac{f}{g}$

27. Given the function $x = 2y - 1$,

    a. find the equation of the inverse function.

    b. graph the function and its inverse on the same coordinate axes.

    c. graph the axis of symmetry, $y = x$.

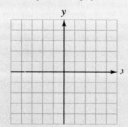

28. Solve the equation $3^x = 9^3$ by rewriting each side of the equation as an exponential expression with a common base.

29. Graph the function $f(x) = \log_{1/2} x$ on graph paper. Verify your graph with the graphing calculator.

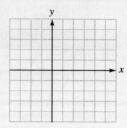

30. Solve the equation $125^{x-3} = 25^{2x + 7}$ algebraically. Verify your solution with the graphing calculator.

31. Solve the system below by *substitution*.
$$\begin{cases} 2x - y = -21 \\ 4x + 5y = 7 \end{cases}$$

32. Solve the system below by *addition (elimination)*.
$$\begin{cases} 2y - 3x = -13 \\ 3x - 17 = 4y \end{cases}$$

33. Solve the system below algebraically.
$$\begin{cases} x + 2y + 2z = 10 \\ 2x + y + 2z = 9 \\ 2x + 2y + z = 11 \end{cases}$$

In Exercises 34 and 35, use matrices to solve each system of equations. Row operations may be done either by hand or with the calculator. Verify solutions with the calculator.

34. $\begin{cases} 2x - 3y = 16 \\ -4x + y = -12 \end{cases}$

35. $\begin{cases} 3x - 2y + 4z = 4 \\ x + y + z = 3 \\ 6x - 2y - 3z = 10 \end{cases}$

In Exercises 36 and 37, use Cramer's rule to solve each system of equations.

36. $\begin{cases} 3x - y = -3 \\ 2x + y = -7 \end{cases}$

37. $\begin{cases} 2x + 3y + 4z = 6 \\ 2x - 3y - 4z = -4 \\ 4x + 6y + 8z = 12 \end{cases}$

In Exercises 38–40, graph the solution of each system of inequalities on graph paper when possible.

38. $\begin{cases} 3x + y \le 1 \\ 4x - y > -8 \end{cases}$ **39.** $\begin{cases} x \ge 0 \\ y < 0 \end{cases}$

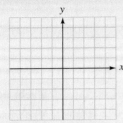 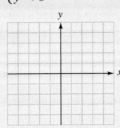

40. $\begin{cases} y \ge x + 2 \\ x \le y - 2 \end{cases}$

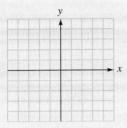

41. Use a calculator to solve the equation $\log x + \log (x + 3) = 1$.

42. Solve the equation $\log 3x = \log 9$

   a. algebraically

   b. graphically

43. Solve the equation $2^{x^2 - 2x} = 8$

   a. algebraically

   b. graphically

44. Solve each system below graphically.

   a. $\begin{cases} x^2 - 6x - y = 5 \\ x^2 - 6x + y = -5 \end{cases}$

   b. $\begin{cases} x^2 - x - y = 2 \\ 4x - 3y = 0 \end{cases}$

Applications

45. Ants The number of ants in an anthill increases according to the function $y = 200e^{.5x}$, where y is the number of ants and x is time in days.

   a. How many days will it take the population of ants to increase to 1000?

   b. After 10 days, approximately how many ants are in the anthill?

46. Radioactive decay A radioactive material decays according to the formula $A = A_0 \left(\frac{2}{3}\right)^t$, where A_0 is the initial amount present and t is time, in years. If the initial sample contains 5 grams of a radioactive material, how many grams are present after 10 years?

47. Buying clothing Two skirts and three sweaters cost $192, and three skirts and one sweater cost $184. Find the cost of a single skirt.

48. Mixing candy A merchant wants to mix two types of candy to produce 30 pounds of mixture costing $2 per pound. If the more expensive candy in the mixture costs $2.50 per pound and the less expensive candy costs $1.25 per pound, how much of each type should he mix?

CAREERS & MATHEMATICS

© Royalty-Free/CORBIS

Traffic Engineer

Traffic engineers supervise the design and construction of roads and gather and analyze data on traffic patterns. They determine efficient routes, designate the location of turn lanes and traffic control signals, and specify the timing of those signals to allow smooth traffic flow.

Qualifications

The minimal requirement for entry-level positions is a bachelor's degree with emphasis on mathematics and the physical sciences. Additional coursework in civil engineering is preferred. Because traffic engineers do a wide variety of tasks, many skills are learned on the job.

Job Outlook

For prepared individuals, employment opportunities in transportation are expected to be plentiful for many years to come.

Example Application

Before recommending traffic control signals for the intersection of the one-way streets shown in Illustration 1, a traffic engineer places sensors across the road to record traffic flow. The illustration shows the number of vehicles passing each of the four counters during 1 hour. From those data, find the number of vehicles passing straight through the intersection and also the number that turn from one road to the other.

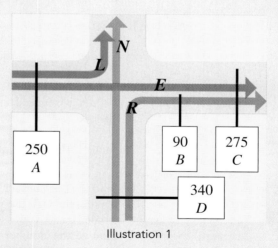

Illustration 1

Solution

As shown in the illustration, we let

L represent the number of vehicles turning left,
R represent the number of vehicles turning right,
N represent the number of vehicles headed north, and
E represent the number of vehicles headed east.

Because the counter at A registers the total number of vehicles headed east and turning left, we have the equation

$$L + E = 250$$

Counter B records only vehicles turning right, so we have the equation

$$R = 90$$

Similarly, the two remaining counters provide two more equations, and we have the system

1.	$L + E = 250$	Counter A records the total of vehicles turning left and those heading east.
2.	$R\ \ \ \ \ = 90$	Counter B records the number turning right.
3.	$R + E = 275$	Counter C records the total of vehicles turning right and those heading east.
4.	$R + N = 340$	Counter D records the total of vehicles turning right and those heading north.

From equation 2, we know that $R = 90$. To find E, we substitute 90 for R in equation 3 and solve for E.

3. $R + E = 275$

$\quad\quad$ $90 + E = 275$ $\quad\quad$ Substitute 90 for R.

$\quad\quad\quad\quad$ $E = 275 - 90$ $\quad\quad$ To undo the addition of 90, subtract 90 from both sides.

$\quad\quad\quad\quad$ $E = 185$

To find N, we substitute 90 for R in equation 4 and solve for N.

4. $R + N = 340$

$\quad\quad$ $90 + N = 340$ $\quad\quad$ Substitute 90 for R.

$\quad\quad\quad\quad$ $N = 250$ $\quad\quad$ Subtract 90 from both sides.

Finally, to find L, we substitute 185 for E in equation 1 and solve for L.

1. $L + E = 250$

$\quad$ $L + 185 = 250$

$\quad\quad\quad$ $L = 65$ $\quad\quad$ Subtract 185 from both sides.

The solutions are $L = 65$, $R = 90$, $N = 250$, and $E = 185$. The traffic engineer knows that 65 vehicles turned left, 90 turned right, 250 passed through heading north, and 185 headed east.

EXERCISES   The intersection of Marsh Street and one-way Fleet Avenue has three counters to record the traffic, as shown in Illustration 2.

1. Find E, the number of vehicles passing through the intersection headed east.
   Find S, the number of vehicles turning south.
   Find N, the number of northbound vehicles turning east.

2. The traffic engineer suspects that the counters are in error. Why?

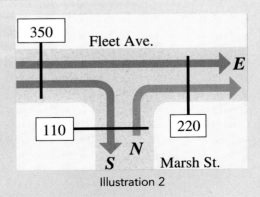

Illustration 2

APPENDIX A Conic Sections

- THE CIRCLE
- THE PARABOLA

- ELLIPSES CENTERED AT THE ORIGIN
- THE HYPERBOLA

The graphs of second-degree equations in x and y fall into one of several categories: a pair of lines, a point, a circle, a parabola, an ellipse, a hyperbola, or no graph at all (see Figure A-1). Because all of these graphs can be formed by the intersection of a plane and a right-circular cone, they are called **conic sections**.

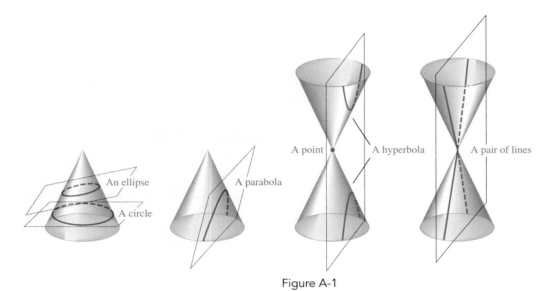

Figure A-1

Conic sections have many applications. For example, a parabola can be rotated to generate a dish-shaped surface called a **paraboloid**. Any light or sound placed at a certain point, called the *focus* of the paraboloid, is reflected outward in parallel paths. This property makes parabolic surfaces ideal for flashlight and headlight reflectors.

Using the same property in reverse makes parabolic surfaces good antennas, because signals captured by such an antenna are concentrated at the focus. A parabolic mirror is capable of concentrating the rays of the sun at a single point and thereby generating tremendous heat. This fact is used in the design of certain solar furnaces. Any object that is thrown upward and outward travels in a parabolic path. In architecture, many arches are parabolic in shape because of their strength, and the cable that supports a suspension bridge hangs in the form of a parabola.

Ellipses have optical and acoustical properties that are useful in architecture and engineering. For example, many arches are portions of an ellipse because the shape is pleasing to the eye. Gears are often cut into elliptical shapes to provide non-uniform motion. The planets and some comets have elliptical orbits.

Hyperbolas serve as the basis of a navigational system known as LORAN (LOng RAnge Navigation). They are also used to find the source of a distress signal, are the basis for the design of hypoid gears, and describe the orbits of some comets.

▍ THE CIRCLE

The Circle
A **circle** is the set of all points in a plane that are a fixed distance from a point called its **center**.

The fixed distance is called the **radius** of the circle.

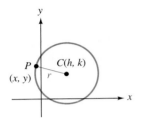

Figure A-2

To develop the general equation of a circle, we must write the equation of a circle with a radius of r and with a center at some point $C(h, k)$, as in Figure A-2. This task is equivalent to finding all points $P(x, y)$ such that the length of line segment CP is r. We can use the distance formula to find r.

$$r = \sqrt{(x - h)^2 + (y - k)^2}$$

We then square both sides to obtain

$$r^2 = (x - h)^2 + (y - k)^2$$

This equation is called the **standard form of the equation of a circle** with a radius of r and center at the point with coordinates (h, k).

Standard Equation of a Circle with Center at (h, k)
Any equation that can be written in the form

$$(x - h)^2 + (y - k)^2 = r^2$$

has a graph that is a circle with radius r and center at point (h, k).

If $r = 0$, the graph reduces to a single point called a **point circle**. If $r < 0$, then a circle does not exist. If both h and k are 0, then the center of the circle is the origin.

Standard Equation of a Circle with Center at $(0, 0)$
Any equation that can be written in the form

$$x^2 + y^2 = r^2$$

has a graph that is a circle with radius r and center at the origin.

EXAMPLE 1   Graph the circle $x^2 + y^2 - 4x + 2y = 20$.

SOLUTION   Because the equation is not in standard form, the coordinates of the center and the length of the radius are not obvious. To put the equation in standard form, we complete the square on both x and y as follows:

$$x^2 + y^2 - 4x + 2y = 20$$
$$x^2 - 4x + y^2 + 2y = 20$$
$$x^2 - 4x + \mathbf{4} + y^2 + 2y + \mathbf{1} = 20 + \mathbf{4} + \mathbf{1}$$

Add 4 and 1 to both sides to complete the squares.

$$(x - 2)^2 + (y + 1)^2 = 25$$

Factor $x^2 - 4x + 4$ and $y^2 + 2y + 1$.

$$(x - 2)^2 + [y - (-1)]^2 = 5^2$$

The radius of the circle is 5, and the coordinates of its center are $h = 2$ and $k = -1$. We plot the center of the circle and draw a circle with a radius of 5 units, as shown in Figure A-3.

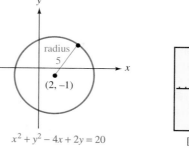

$$x^2 + y^2 - 4x + 2y = 20$$

[−9.4,9.4] by [−6.2,6.2]

Figure A-3

EXAMPLE 2

Radio translators The effective broadcast area of a television station is bounded by the circle $x^2 + y^2 = 3{,}600$, where x and y are measured in miles. A translator station picks up the signal and retransmits it from the center of a circular area bounded by $(x + 30)^2 + (y - 40)^2 = 1{,}600$. Find the location of the translator and the greatest distance from the main transmitter that the signal can be received.

SOLUTION The coverage of the television station is bounded by $x^2 + y^2 = 60^2$, a circle centered at the origin with a radius of 60 miles, as shown in Figure A-4. Because the translator is at the center of the circle $(x + 30)^2 + (y - 40)^2 = 1{,}600$, it is located at $(-30, 40)$, a point 30 miles west and 40 miles north of the television station. The radius of the translator's coverage is $\sqrt{1{,}600}$, or 40 miles.

As shown in Figure A-4, the greatest distance of reception is the sum of A, the distance of the translator from the television station, and 40 miles, the radius of the translator's coverage.

To find A, we use the distance formula to find the distance between $(x_1, y_1) = (-30, 40)$ and the origin, $(x_2, y_2) = (0, 0)$.

$$A = \sqrt{(x_2 - x_1)^2 + (y_2 - y_1)^2}$$
$$A = \sqrt{(0 - -30)^2 + (0 - 40)^2}$$
$$= \sqrt{(30)^2 + (-40)^2}$$
$$= \sqrt{2500}$$
$$= 50$$

Thus, the translator is located 50 miles from the television station, and it broadcasts the signal an additional 40 miles. The greatest reception distance is $50 + 40$, or 90 miles.

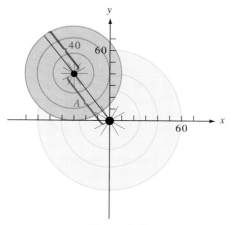

Figure A-4

▮▮▮ THE PARABOLA

We have seen that equations of the form $y = a(x - h)^2 + k$, with $a \neq 0$, represent parabolas with vertex at the point (h, k). They open upward when $a > 0$ and downward when $a < 0$.

Equations of the form $x = a(y - k)^2 + h$ also represent parabolas with vertex at the point (h, k). However, they open to the right when $a > 0$ and to the left when $a < 0$. Parabolas that open to the right or left do not represent functions, because their graphs do not pass the vertical line test.

Several types of parabolas are summarized in the following chart. (In all cases, $a > 0$.)

Equations of Parabolas ($a > 0$)

Parabola Opening	Vertex at Origin	Vertex at (h, k)
Up	$y = ax^2$	$y = a(x - h)^2 + k$
Down	$y = -ax^2$	$y = -a(x - h)^2 + k$
Right	$x = ay^2$	$x = a(y - k)^2 + h$
Left	$x = -ay^2$	$x = -a(y - k)^2 + h$

EXAMPLE 3

Graph the equation $x = -2(y - 2)^2 + 3$.

SOLUTION   We make a table of ordered pairs, plot each pair, and draw the parabola as in Figure A-5. Because the equation is of the form $x = -a(y - k)^2 + h$, the parabola opens to the left and has its vertex at the point with coordinates $(3, 2)$.

$$x = -2(y - 2)^2 + 3$$

x	y	(x, y)
-5	0	$(-5, 0)$
1	1	$(1, 1)$
3	2	$(3, 2)$
1	3	$(1, 3)$
-5	4	$(-5, 4)$

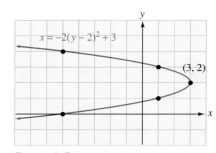

Figure A-5

TECHNOLOGY TIP

The equation of a parabola that opens left or right is a relation and not a function. Solve the equation for y and graph both equations to produce the complete graph of the relation.

$$y1 = 2 + \sqrt{((x - 3)/-2)}$$
$$y2 = 2 - \sqrt{((x - 3)/-2)}$$

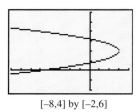

$[-8, 4]$ by $[-2, 6]$

ELLIPSES CENTERED AT THE ORIGIN

The Ellipse

An **ellipse** is the set of all points P in the plane the sum of whose distances from two fixed points is a constant. See Figure A-6, in which $d_1 + d_2$ is a constant.

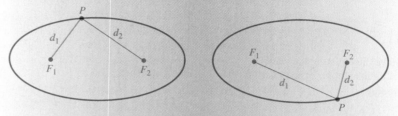

Figure A-6

Any equation that can be written in the form

$$\frac{x^2}{a^2} + \frac{y^2}{b^2} = 1 \quad (a > b)$$

has a graph that is an **ellipse** centered at the origin, as in Figure A-7(a). The x-intercepts are the **vertices** $V(a, 0)$ and $V'(-a, 0)$. (Read V' as "V prime.") The y-intercepts are the points $(0, b)$ and $(0, -b)$.

   Any equation that can be written in the form

$$\frac{x^2}{b^2} + \frac{y^2}{a^2} = 1 \quad (a > b)$$

has a graph that is also an ellipse centered at the origin, as in Figure A-7(b). The y-intercepts are the vertices $V(0, a)$ and $V'(0, -a)$. The x-intercepts are the points $(b, 0)$ and $(-b, 0)$.

   The point midway between the vertices is the **center** of the ellipse.

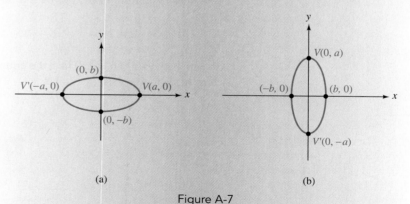

(a)   (b)

Figure A-7

Each of the two points is called a **focus**. Midway between the foci is the **center** of the ellipse.

   The **major axis of a horizontal ellipse** is the line segment between the coordinates $(-a, 0)$ and $(a, 0)$. The **major axis of a vertical ellipse** is the line segment between the coordinates $(0, a)$ and $(0, -a)$.

EXAMPLE 4

Graph the equation $\dfrac{x^2}{36} + \dfrac{y^2}{9} = 1$.

SOLUTION   The graph of the equation will be a horizontal ellipse with x-intercepts of $(6, 0)$ and $(-6, 0)$ and y-intercepts of $(0, 3)$ and $(0, -3)$. (See Figure A-8.) Plot the intercepts and connect with a smooth curve.

TECHNOLOGY TIP

Solve the equation for y and graph in an appropriate viewing window.

$$y1 = \sqrt{(36 - x^2)}/2$$
$$y2 = -\sqrt{(36 - x^2)}/2$$

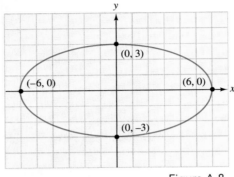

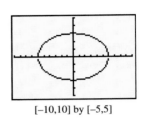

$[-10,10]$ by $[-5,5]$

Figure A-8

EXAMPLE 5

Landscape design   A landscape architect is designing an elliptical pool that will fit in the center of a 20-by-30-foot rectangular garden, leaving at least 5 feet of space on all sides. Find the equation of the ellipse.

SOLUTION   We place the rectangular garden in a coordinate system, as in Figure A-9. To maintain 5 feet of clearance at the ends of the ellipse, the vertices must be the points $V(10, 0)$ and $V'(-10, 0)$. Similarly, the y-intercepts are the points $(0, 5)$ and $(0, -5)$.

The equation of the ellipse has the form

$$\frac{x^2}{a^2} + \frac{y^2}{b^2} = 1$$

with $a = 10$ and $b = 5$. Thus, the equation of the boundary of the pool is

$$\frac{x^2}{100} + \frac{y^2}{25} = 1$$

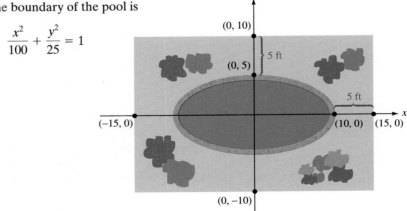

Figure A-9

▌▌ THE HYPERBOLA

The Hyperbola

A **hyperbola** is the set of all points P in the plane for which the difference of the distances of each point on the hyperbola from two fixed points is a constant. See Figure A-10, in which $d_1 - d_2$ is a constant. Each of the two points is called a **focus**. Midway between the foci is the **center** of the hyperbola.

Figure A-10

Any equation that can be written in the form

$$\frac{x^2}{a^2} - \frac{y^2}{b^2} = 1$$

has a graph that is a **hyperbola** centered at the origin, as in Figure A-11. The x-intercepts are the **vertices** $V(a, 0)$ and $V'(-a, 0)$. There are no y-intercepts.

Although it is possible to draw any hyperbola by plotting many points and connecting them in a smooth curve, there is an easier way. Plot the points $(a, 0)$ and $(-a, 0)$, the x-intercepts, and $(0, b)$ and $(0, -b)$. Construct a rectangle whose sides pass horizontally through $(0, b)$ and $(0, -b)$ and vertically through $(a, 0)$ and $(-a, 0)$. This rectangle is called the **fundamental rectangle of the hyperbola**. The extended diagonals of the rectangle are the **asymptotes** of the hyperbola. As the points on the branches of the hyperbola move further away from the origin, they get closer to, but never cross, these extended diagonals. See Figure A-11.

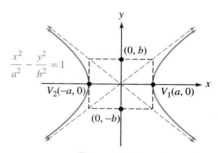

Figure A-11

EXAMPLE 6

Graph the hyperbola whose equation is $\dfrac{x^2}{25} - \dfrac{y^2}{9} = 1$.

SOLUTION   We can see from the equation that $a = \pm 25$ and $b = \pm 3$. Plot the points $(5, 0)$, $(-5, 0), (0, 3)$ and $(0, -3)$. Construct the fundamental rectangle and its extended diagonals. Sketch the hyperbola (see Figure A-12).

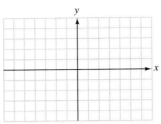

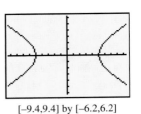

[-9.4,9.4] by [-6.2,6.2]

Figure A-12

Equation of a Hyperbola Centered at the Origin

Any equation that can be written in the form

$$\frac{y^2}{a^2} - \frac{x^2}{b^2} = 1$$

has a graph that is a **hyperbola** centered at the origin, as in Figure A-13. The y-intercepts are the **vertices** $V(0, a)$ and $V'(0, -a)$. There are no x-intercepts. The **asymptotes** of the hyperbola are the extended diagonals of the rectangle in the figure.

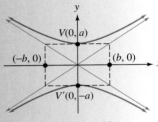

Figure A-13

EXAMPLE 7

Graph the equation $9y^2 - 4x^2 = 36$.

SOLUTION   To write the equation in standard form, we divide both sides by 36 to obtain

$$\frac{9y^2}{36} - \frac{4x^2}{36} = 1$$

$$\frac{y^2}{4} - \frac{x^2}{9} = 1 \qquad \text{Simplify each fraction.}$$

Note that $a = \pm2$ and $b = \pm3$. Plot the points $(3, 0)$, $(-3, 0)$, $(0, 2)$, and $(0, -2)$. Sketch the fundamental rectangle and its extended diagonals. Sketch the hyperbola. (See Figure A-14).

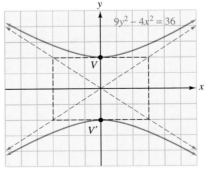

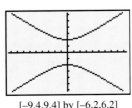

[-9.4,9.4] by [-6.2,6.2]

Figure A-14

There is a special type of hyperbola (also centered at the origin) that does not intersect either the x- or the y-axis. These hyperbolas have equations of the form $xy = k$, where $k \neq 0$.

EXAMPLE 8

Graph the equation $xy = -8$.

SOLUTION   We make a table of ordered pairs, plot each pair, and join the points with a smooth curve to obtain the hyperbola. (See Figure A-15.)

$$xy = -8$$

x	y	(x, y)
1	-8	$(1, -8)$
2	-4	$(2, -4)$
4	-2	$(4, -2)$
8	-1	$(8, -1)$
-1	8	$(-1, 8)$
-2	4	$(-2, 4)$
-4	2	$(-4, 2)$
-8	1	$(-8, 1)$

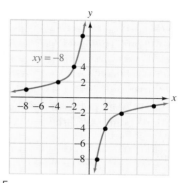

Figure A-15

The result in Example 8 suggests the following theorem.

Theorem

Any equation of the form $xy = k$, where $k \neq 0$, has a graph that is a **hyperbola** that does not intersect either the x- or the y-axis.

EXAMPLE 9

Atomic structure   In an experiment that led to the discovery of the atomic structure of matter, Lord Rutherford (1872–1937) shot high-energy alpha particles toward a thin sheet of gold. Because many were reflected, Rutherford showed the existence of a gold atom. The alpha particle in Figure A-16 is repelled by the nucleus at the origin and travels along the hyperbolic path given by $4x^2 - y^2 = 16$. How close does the particle come to the nucleus?

SOLUTION   To find the distance from the nucleus at the origin, we must find the coordinates of the vertex V. To do so, we write the equation of the particle's path in standard form:

$$4x^2 - y^2 = 16$$

$$\frac{4x^2}{16} - \frac{y^2}{16} = \frac{16}{16} \qquad \text{Divide both sides by 16.}$$

$$\frac{x^2}{4} - \frac{y^2}{16} = 1 \qquad \text{Simplify.}$$

$$\frac{x^2}{2^2} - \frac{y^2}{4^2} = 1 \qquad \text{Write 4 as } 2^2 \text{ and 16 as } 4^2.$$

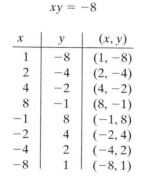

Figure A-16

This equation is in the form $\frac{x^2}{a^2} - \frac{y^2}{b^2} = 1$, with $a = 2$. Thus, the vertex of the path is $(2, 0)$. The particle is never closer than 2 units from the nucleus.

APPENDIX A EXERCISE

VOCABULARY AND NOTATION   *In Exercises 1–7, fill in the blanks to make a true statement.*

1. Graphs formed by the intersection of a plane and a right-circular cone are called _____.

2. The set of all points in a plane that are a fixed distance from a point called the center form a _____.

3. The standard form for a circle whose center is at (h, k) is _____.

4. The equation $x^2 + y^2 = r^2$ is the standard form of the equation of a circle with center at _____.

5. An ellipse whose major axis is along the y-axis will be an ellipse centered at the _____ whose orientation is _____.

6. A hyperbola of the form $\frac{x^2}{a^2} - \frac{y^2}{b^2} = 1$ is centered at the _____ and has no ___-intercepts.

7. A hyperbola of the form $\frac{y^2}{a^2} - \frac{x^2}{b^2} = 1$ is centered at the _____ and has no ___-intercepts.

CONCEPTS   *In Exercises 8–14, match each equation to its description without graphing.*

 a. a vertical ellipse centered at $(0, 0)$
 b. a circle of radius 1 with center $(0, 0)$
 c. a hyperbola opening left/right
 d. a horizontal ellipse centered at $(0, 0)$
 e. a parabola opening right with vertex at $(5, 2)$
 f. a circle of radius 2 with center at $(1, 2)$
 g. a hyperbola opening up/down

8. $\frac{x^2}{9} + \frac{y^2}{16} = 1$

9. $x = (y - 2)^2 + 5$

10. $x^2 + y^2 = 1$

11. $\frac{x^2}{9} - \frac{y^2}{16} = 1$

12. $(x - 1)^2 + (y - 2)^2 = 4$

13. $\frac{y^2}{16} - \frac{x^2}{9} = 1$

14. $\frac{x^2}{16} + \frac{y^2}{9} = 1$

PRACTICE   *In Exercises 15–19, find the center and radius of each circle and then graph it. Verify with your calculator.*

15. $x^2 + y^2 = 9$

16. $x^2 + y^2 = 16$

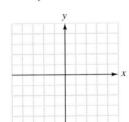
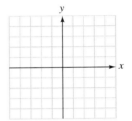

17. $(x - 2)^2 + (y - 4)^2 = 4$

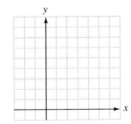

18. $(x - 3)^2 + (y - 2)^2 = 4$

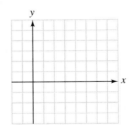

19. $x^2 + y^2 - 2x + 4y = -1$

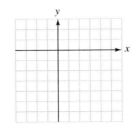

In Exercises 20–25, find the vertex of each parabola and then graph it. Verify with your calculator.

20. $x = y^2$

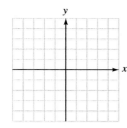

21. $x = 4y^2$

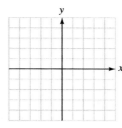

22. $x = -3(y + 2)^2 - 2$

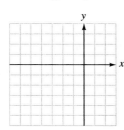

23. $x = 2(y - 3)^2 - 4$

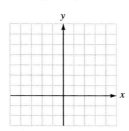

24. $x = -y^2 - 2y + 3$

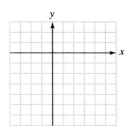

25. $x = y^2 + 4y + 5$

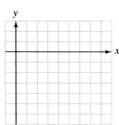

In Exercises 26–29, determine the x- and y-intercepts and graph each ellipse. Verify with your calculator.

26. $\dfrac{x^2}{4} + \dfrac{y^2}{9} = 1$

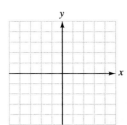

27. $x^2 + \dfrac{y^2}{9} = 1$

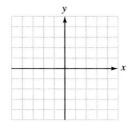

28. $25x^2 + 9y^2 = 225$

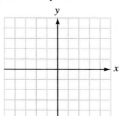

29. $4x^2 + 16y^2 = 64$

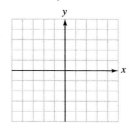

In Exercises 30–37, graph the hyperbola by sketching the fundamental rectangle and asymptotes, where applicable.

30. $\dfrac{x^2}{9} - \dfrac{y^2}{4} = 1$

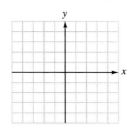

31. $\dfrac{y^2}{4} - \dfrac{x^2}{9} = 1$

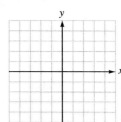

32. $\dfrac{y^2}{25} - \dfrac{x^2}{4} = 1$

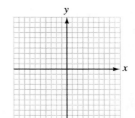

33. $\dfrac{x^2}{4} - \dfrac{y^2}{25} = 1$

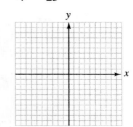

34. $25x^2 - y^2 = 25$

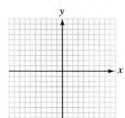

35. $9x^2 - 4y^2 = 36$

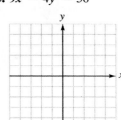

36. $xy = 8$

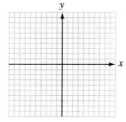

37. $xy = -10$

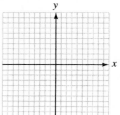

APPLICATIONS

38. Meshing gears For design purposes, the large gear in Illustration 1 is the circle $x^2 + y^2 = 16$. The smaller gear is a circle centered at $(7, 0)$ and tangent to the larger circle. Find the equation of the smaller gear.

Illustration 1

39. Width of a walkway The walkway in Illustration 2 is bounded by the two circles $x^2 + y^2 = 2,500$ and $(x - 10)^2 + y^2 = 900$, measured in feet. Find the largest and the smallest width of the walkway.

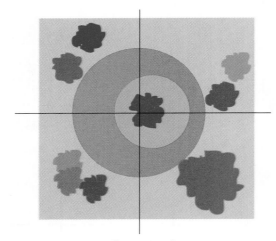

Illustration 2

40. Broadcast ranges Radio stations applying for licensing may not use the same frequency if their broadcast areas overlap. One station's coverage is bounded by $x^2 + y^2 - 8x - 20y + 16 = 0$, and the other's by $x^2 + y^2 + 2x + 4y - 11 = 0$. May they be licensed for the same frequency?

41. Highway curves Highway design engineers want to join two sections of highway with a curve that is one-quarter of a circle, as in Illustration 3. The equation of the circle is $x^2 + y^2 - 16x - 20y + 155 = 0$, where dis-

tances are measured in kilometers. Find the locations (relative to the center of town at the origin) of the intersections of the highway with State and Main.

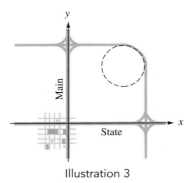

Illustration 3

42. Orbit of a comet If the orbit of the comet shown in Illustration 4 is given by the equation $2y^2 - 9x = 18$, how far is it from the sun at the vertex of the orbit? Distances are in astronomical units (AU).

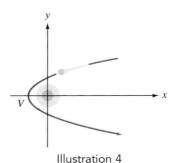

Illustration 4

43. Satellite antenna The cross section of the satellite antenna in Illustration 5 is a parabola given by the equation $x = \frac{1}{16}y^2$, with distances measured in feet. If the dish is 8 feet wide, how deep is it?

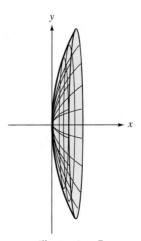

Illustration 5

44. Designing an underpass The arch of the underpass in Illustration 6 is part of an ellipse. Find the equation of the arch.

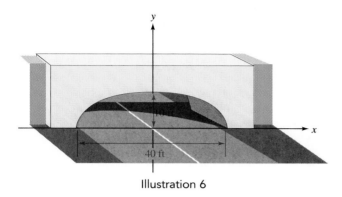

40 ft

Illustration 6

45. Calculating clearance Find the height of the elliptical arch in Exercise 44 at a point 10 feet from the center of the roadway.

46. Area of an ellipse The area A of the ellipse $\dfrac{x^2}{a^2} + \dfrac{y^2}{b^2} = 1$ is given by $A = \pi ab$. Find the area of the ellipse $9x^2 + 16y^2 = 144$.

47. Area of a track The elliptical track in Illustration 7 is bounded by the ellipses $4x^2 + 9y^2 = 576$ and $9x^2 + 25y^2 = 900$. Find the area of the track. (See Exercise 46.)

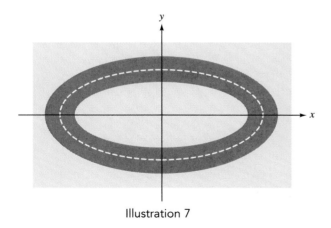

Illustration 7

48. Alpha particles The particle in Illustration 8 approaches the nucleus at the origin along the path $9y^2 - x^2 = 81$. How close does the particle come to the nucleus?

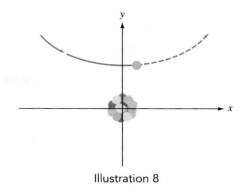

Illustration 8

49. LORAN By determining the difference of the distances between the ship in Illustration 9 and the two radio transmitters, the LORAN system places the ship on the hyperbola $x^2 - 4y^2 = 576$. If the ship is also 5 miles out to sea, find its coordinates.

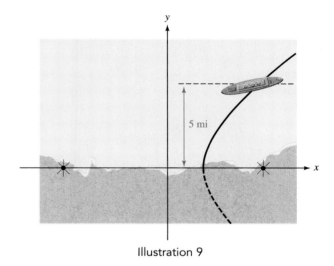

5 mi

Illustration 9

50. Sonic boom The position of the sonic boom caused by the faster-than-sound aircraft in Illustration 10 is the hyperbola $y^2 - x^2 = 25$ in the coordinate system shown. How wide is the hyperbola 5 miles from its vertex?

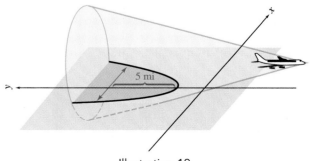

5 mi

Illustration 10

51. Electrostatic repulsion   Two similarly charged particles are shot together for an almost head-on collision, as in Illustration 11. They repel each other and travel the two branches of the hyperbola given by $x^2 - 4y^2 = 4$. How close do they get?

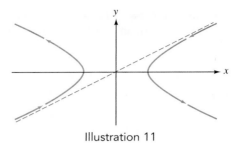

Illustration 11

Binomial Theorem

- PASCAL'S TRIANGLE
- FACTORIAL NOTATION
- THE BINOMIAL THEOREM

- THE nTH TERM OF A BINOMIAL EXPANSION

We have discussed how to raise binomials to positive integral powers. For example, we know that

$$(a + b)^2 = a^2 + 2ab + b^2$$

and that

$$
\begin{aligned}
(a + b)^3 &= (a + b)(a + b)^2 \\
&= (a + b)(a^2 + 2ab + b^2) \\
&= a^3 + 2a^2b + ab^2 + a^2b + 2ab^2 + b^3 \\
&= a^3 + 3a^2b + 3ab^2 + b^3
\end{aligned}
$$

To show how to raise binomials to positive integral powers without doing the actual multiplications, we consider the following binomial expansions:

$$
\begin{aligned}
(a + b)^0 &= 1 \\
(a + b)^1 &= a + b \\
(a + b)^2 &= a^2 + 2ab + b^2 \\
(a + b)^3 &= a^3 + 3a^2b + 3ab^2 + b^3 \\
(a + b)^4 &= a^4 + 4a^3b + 6a^2b^2 + 4ab^3 + b^4 \\
(a + b)^5 &= a^5 + 5a^4b + 10a^3b^2 + 10a^2b^3 + 5ab^4 + b^5 \\
(a + b)^6 &= a^6 + 6a^5b + 15a^4b^2 + 20a^3b^3 + 15a^2b^4 + 6ab^5 + b^6
\end{aligned}
$$

Several patterns appear in these expansions:

1. Each expansion has one more term than the power of the binomial.
2. The degree of each term in each expansion is equal to the exponent of the binomial that is being expanded.
3. The first term in each expansion is a, raised to the power of the binomial.
4. The exponents of a decrease by 1 in each successive term. The exponents of b, beginning with $b^0 = 1$ in the first term, increase by 1 in each successive term. Thus, the variables have the pattern

$$a^n, a^{n-1}b, a^{n-2}b^2, \ldots, ab^{n-1}, b^n$$

▓ PASCAL'S TRIANGLE

To see another pattern, we write the coefficients of each binomial expansion in the following triangular array:

$$
\begin{array}{ccccccccccccc}
&&&&&& 1 &&&&&& \\
&&&&& 1 && 1 &&&&& \\
&&&& 1 && 2 && 1 &&&& \\
&&& 1 && 3 && 3 && 1 &&& \\
&& 1 && 4 && 6 && 4 && 1 && \\
& 1 && 5 && 10 && 10 && 5 && 1 & \\
1 && 6 && 15 && 20 && 15 && 6 && 1
\end{array}
$$

In this array, called **Pascal's triangle**, each entry between the 1s is the sum of the closest pair of numbers in the line immediately above it. For example, the first 15 in the bottom row is the sum of the 5 and 10 immediately above it. Pascal's triangle continues with the same pattern forever. The next two lines are

$$1 \quad 7 \quad 21 \quad 35 \quad 35 \quad 21 \quad 7 \quad 1$$
$$1 \quad 8 \quad 28 \quad 56 \quad 70 \quad 56 \quad 28 \quad 8 \quad 1$$

EXAMPLE 1

Expand $(u - v)^4$.

SOLUTION   We note that the expression $(u - v)^4$ can be written in the form $[u + (-v)]^4$. The variables in this expansion are

$$u^4, \quad u^3(-v), \quad u^2(-v)^2, \quad u(-v)^3, \quad (-v)^4$$

and the coefficients are given in Pascal's triangle in the row whose second entry is 4:

$$1 \quad 4 \quad 6 \quad 4 \quad 1$$

Hence, the required expansion is

$$(u - v)^4 = u^4 + 4u^3(-v) + 6u^2(-v)^2 + 4u(-v)^3 + (-v)^4$$
$$= u^4 - 4u^3v + 6u^2v^2 - 4uv^3 + v^4$$

▍▍ FACTORIAL NOTATION

Although Pascal's triangle gives the coefficients of the terms in a binomial expansion, it is not the best way to expand a binomial. To develop another way to expand a binomial, we introduce **factorial notation**.

Factorial Notation
If n is a natural number, the symbol $n!$ (read as **n factorial** or as **factorial n**) is defined as

$$n! = n(n - 1)(n - 2)(n - 3) \cdots (3)(2)(1)$$

EXAMPLE 2

Find: **a.** $2!$ **b.** $5!$ **c.** $-9!$ and   **d.** $(n - 2)!$

SOLUTION   **a.** $2! = 2 \cdot 1 = 2$

b. $5! = 5 \cdot 4 \cdot 3 \cdot 2 \cdot 1 = 120$

c. $-9! = -9 \cdot 8 \cdot 7 \cdot 6 \cdot 5 \cdot 4 \cdot 3 \cdot 2 \cdot 1 = -362,880$

d. $(n - 2)! = (n - 2)(n - 3)(n - 4) \cdot \ldots \cdot 3 \cdot 2 \cdot 1$

```
2!
                    2
5!
                  120
-9!
             -362880
```

TECHNOLOGY TIP

The factorial symbol "!" is located under the submenu PRB (PROB on the TI-86) of the MATH menu.

WARNING!   According to the previous definition, part d is meaningful only if $n - 2$ is a natural number.

We define zero factorial as follows.

Zero Factorial
$0! = 1$

We note that

$$5 \cdot 4! = 5 \cdot 4 \cdot 3 \cdot 2 \cdot 1 = 5!$$
$$7 \cdot 6! = 7 \cdot 6 \cdot 5 \cdot 4 \cdot 3 \cdot 2 \cdot 1 = 7!$$
$$10 \cdot 9! = 10 \cdot 9 \cdot 8 \cdot 7 \cdot 6 \cdot 5 \cdot 4 \cdot 3 \cdot 2 \cdot 1 = 10!$$

These examples suggest the following theorem.

Theorem

If n is a positive integer, then $n(n - 1)! = n!$.

▍ THE BINOMIAL THEOREM

We now state the binomial theorem.

The Binomial Theorem

If n is any positive integer, then

$$(a + b)^n = a^n + \frac{n!}{1!(n-1)!}a^{n-1}b + \frac{n!}{2!(n-2)!}a^{n-2}b^2$$
$$+ \frac{n!}{3!(n-3)!}a^{n-3}b^3 + \cdots + \frac{n!}{r!(n-r)!}a^{n-r}b^r$$
$$+ \cdots + b^n$$

In the binomial theorem, the exponents of the variables follow the familiar pattern:

- The sum of the exponents of a and b in each term is n,
- the exponents of a decrease, and
- the exponents of b increase.

Only the method of finding the coefficients is different. Except for the first and last terms, the numerator of each coefficient is $n!$. If the exponent of b in a particular term is r, the denominator of the coefficient of that term is $r!(n - r)!$.

EXAMPLE 3

Use the binomial theorem to expand $(x - y)^4$.

SOLUTION   We can write $(x - y)^4$ in the form $[x + (-y)]^4$, substitute directly into the binomial theorem, and simplify:

$$(x - y)^4 = [x + (-y)]^4$$

$$= x^4 + \frac{4!}{1!(4-1)!}x^3(-y) + \frac{4!}{2!(4-2)!}x^2(-y)^2 + \frac{4!}{3!(4-3)!}x(-y)^3 + (-y)^4$$

$$= x^4 - \frac{4 \cdot 3!}{1!3!}x^3y + \frac{4 \cdot 3 \cdot 2!}{2!2!}x^2y^2 - \frac{4 \cdot 3!}{3!1!}xy^3 + y^4$$

$$= x^4 - 4x^3y + 6x^2y^2 - 4xy^3 + y^4$$

▐ THE nTH TERM OF A BINOMIAL EXPANSION

To find the fourth term of the expansion of $(a + b)^9$, we could raise the binomial $a + b$ to the ninth power and look at the fourth term. However, this task would be very tedious. By using the binomial theorem, we can construct the fourth term without finding the complete expansion of $(a + b)^9$.

EXAMPLE 4

Find the sixth term in the expansion of $(x - y)^7$.

SOLUTION   We find the sixth term of $[x + (-y)]^7$. In the sixth term, the exponent of $(-y)$ is 5. Thus, the variables in the sixth term are

$$x^2(-y)^5 \qquad \text{The sum of the exponents must be 7.}$$

The coefficient of these variables is

$$\frac{n!}{r!(n-r)!} = \frac{7!}{5!(7-5)!}$$

The complete sixth term is

$$\frac{7!}{5!(7-5)!}x^2(-y)^5 = -\frac{7 \cdot 6 \cdot 5!}{5! \cdot 2 \cdot 1}x^2y^5$$

$$= -21x^2y^5$$

APPENDIX B EXERCISE

VOCABULARY AND NOTATION   *In Exercises 1–3, fill in the blanks to make a true statement.*

1. A triangular array of numbers that determines the coefficients of a binomial expansion is called _____ _____.

2. If n is a natural number, then $n!$ is called _____ notation, which is defined as _____.

3. Zero factorial, $0!$, is equal to ___.

CONCEPTS

4. Explain how to construct Pascal's triangle.

5. Without looking at your text, write the first ten rows of Pascal's triangle.

6. Find the sum of the numbers in each of the first ten rows of Pascal's triangle. What is the pattern?

7. Find the sum of the numbers in the designated diagonal rows of Pascal's triangle shown in Illustration 1. What is the pattern?

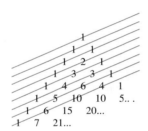
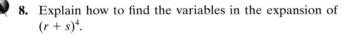

Illustration 1

8. Explain how to find the variables in the expansion of $(r + s)^4$.

PRACTICE *In Exercises 9–14, evaluate each expression and verify with your calculator.*

9. $3!$

10. $-5!$

11. $11!(5)$

12. $8(7!)$

13. $\dfrac{5!}{3!(5-3)!}$

14. $\dfrac{6!}{4!(6-4)!}$

In Exercises 15–24, use the binomial theorem to expand each expression.

15. $(x + y)^3$

16. $(x - y)^3$

17. $(2x + y)^4$

18. $(x + 2y)^4$

19. $(x - 2y)^5$

20. $(2x - y)^5$

21. $(2x + 3y)^4$

22. $(3x - 2y)^4$

23. $\left(\dfrac{x}{2} - \dfrac{y}{2}\right)^4$

24. $\left(\dfrac{x}{2} + \dfrac{y}{2}\right)^4$

In Exercises 25–30, use the binomial theorem to find the required term of each expansion.

25. $(a + b)^4$; second term

26. $(a + b)^4$; third term

27. $(x - 3y)^8$; third term

28. $(x - 3y)^8$; fifth term

29. $(2x + 3)^6$; sixth term

30. $(2x + 3)^6$; fourth term

Arithmetic and Geometric Sequences

- ARITHMETIC SEQUENCES
- ARITHMETIC MEANS
- THE SUM OF THE FIRST *n* TERMS OF
  AN ARITHMETIC SEQUENCE
- SUMMATION NOTATION

- GEOMETRIC SEQUENCES
- GEOMETRIC MEANS
- THE SUM OF THE FIRST *n* TERMS OF
  A GEOMETRIC SEQUENCE

A **sequence** is a function whose domain is the set of natural numbers. For example, the function $f(n) = 3n + 2$, where n is a natural number, is a sequence. Because a sequence is a function whose domain is the set of natural numbers, it is easy to write its value as a list. If the natural numbers are substituted for n, the function $f(n) = 3n + 2$ generates the list

$$5, 8, 11, 14, 17, \ldots$$

It is common to call the list, as well as the function, a sequence. Each number in the list is called a **term** of the sequence. Other examples of sequences are

a. $1^3, 2^3, 3^3, 4^3, \ldots$ **b.** $4, 8, 12, 16, \ldots$
c. $2, 3, 5, 7, 11, \ldots$ **d.** $1, 1, 2, 3, 5, 8, 13, 21, \ldots$

The sequence in part (a) is the ordered list of the cubes of the natural numbers. The sequence in part (b) is the ordered list of the positive multiples of 4. The sequence in part (c) is the ordered list of the prime numbers. The sequence in part (d) is called the **Fibonacci sequence**, after the 12th-century mathematician Leonardo of Pisa — also known as Fibonacci. Beginning with the 2, each term of the Fibonacci sequence is the sum of the two preceding terms.

ARITHMETIC SEQUENCES

One important type of sequence is the **arithmetic sequence**.

Arithmetic Sequence
An **arithmetic sequence** is a sequence of the form

$$a, a + d, a + 2d, a + 3d, \ldots, a + (n - 1)d, \ldots$$

where a is the **first term**, $a + (n - 1)d$ is the **nth term**, and d is the **common difference**.

We note that the second term of an arithmetic sequence has an addend of $1d$, the third term has an addend of $2d$, the fourth term has an addend of $3d$, and nth term has an addend of $(n - 1)d$. We also note that the difference between any two consecutive terms in an arithmetic sequence is d.

EXAMPLE 1

The first three terms of an arithmetic sequence are 3, 8, and 13.
Find   **a.** the 67th term   and   **b.** the 100th term.

SOLUTION   We first find d, the common difference. It is the difference between successive terms:

$$d = 8 - 3 = 13 - 8 = 5$$

a. We substitute 3 for a, 67 for n, and 5 for d in the formula for the nth term and simplify:

$$n\text{th term} = a + (n - 1)d$$
$$67\text{th term} = 3 + (67 - 1)5$$
$$= 3 + 66(5)$$
$$= 333$$

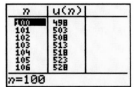

b. We substitute 3 for a, 100 for n, and 5 for d in the formula for the nth term and simplify:

$$n\text{th term} = a + (n - 1)d$$
$$100\text{th term} = 3 + (100 - 1)5$$
$$= 3 + 99(5)$$
$$= 498$$

ARITHMETIC MEANS

If numbers are inserted between two numbers a and b to form an arithmetic sequence, the inserted numbers are called **arithmetic means** between a and b. If a single number is inserted between the numbers a and b, that number is called **the arithmetic mean** between a and b.

EXAMPLE 2

Insert two arithmetic means between 6 and 27.

SOLUTION   In this example, the first term is $a = 6$, and the fourth term (or the last term) is $l = 27$. We must find the common difference such that the terms

$$6, 6 + d, 6 + 2d, 27$$

form an arithmetic sequence. To find d, we can substitute 6 for a and 4 for n into the formula for the nth term:

$$n\text{th term} = a + (n - 1)d$$
$$4\text{th term} = 6 + (4 - 1)d$$
$$27 = 6 + 3d \qquad \text{Simplify.}$$
$$21 = 3d \qquad \text{Subtract 6 from both sides.}$$
$$7 = d \qquad \text{Divide both sides by 3.}$$

The two arithmetic means between 6 and 27 are

$$6 + d = 6 + 7 \qquad \text{or} \qquad 6 + 2d = 6 + 2(7)$$
$$= 13 \qquad\qquad\qquad = 6 + 14$$
$$= 20$$

The numbers 6, 13, 20, and 27 are the first four terms of an arithmetic sequence.

▍THE SUM OF THE FIRST n TERMS OF AN ARITHMETIC SEQUENCE

There is a formula that gives the sum of the first n terms of an arithmetic sequence. To develop this formula, we let S_n represent the sum of the first n terms of an arithmetic sequence:

$$S_n = \qquad a \qquad + \qquad [a + d] \qquad + \qquad [a + 2d] \qquad + \cdots + [a + (n-1)d]$$

We write the same sum again, but in reverse order:

$$S_n = [a + (n-1)d] + [a + (n-2)d] + [a + (n-3)d] + \cdots + \qquad a$$

We add these two equations together, term by term, to get

$$2S_n = [2a + (n-1)d] + [2a + (n-1)d] + [2a + (n-1)d] + \cdots + [2a + (n-1)d]$$

Because there are n equal terms on the right-hand side of the preceding equation, we can write

$$2S_n = n[2a + (n-1)d]$$
$$2S_n = n[a + a + (n-1)d] \qquad 2a = a + a.$$

$$2S_n = n[a + l] \qquad \text{Substitute } l \text{ for } a + (n-1)d, \text{ because } a + (n-1)d$$
$$\qquad\qquad\qquad\qquad \text{is the last term of the sequence.}$$

$$S_n = \frac{n(a + l)}{2} \qquad \text{Divide both sides by 2.}$$

This reasoning establishes the following theorem.

Sum of the First n Terms of an Arithmetic Sequence

The sum of the first n terms of an arithmetic sequence is given by the formula

$$S_n = \frac{n(a + l)}{2} \qquad \text{with } l = a + (n-1)d$$

where a is the first term, l is the last (or nth) term, and n is the number of terms in the sequence.

EXAMPLE 3

Find the sum of the first 40 terms of the arithmetic sequence 4, 10, 16, . . .

SOLUTION   In this example, we let $a = 4$, $n = 40$, $d = 6$, and $l = 4 + (40 - 1)6 = 238$ and substitute these values into the formula for S_n:

$$S_n = \frac{n(a + l)}{2}$$

$$S_{40} = \frac{40(4 + 238)}{2}$$

$$= 20(242)$$

$$= 4840$$

The sum of the first 40 terms is 4840.

SUMMATION NOTATION

There is a shorthand notation for indicating the sum of a finite (ending) number of consecutive terms in a sequence. This notation, called **summation notation**, involves the Greek letter Σ (sigma). The expression

$$\sum_{k=2}^{5} 3k \qquad \text{Read as "the summation of } 3k \text{ as } k \text{ runs from 2 to 5."}$$

designates the sum of all terms obtained if we successively substitute the numbers 2, 3, 4, and 5 for k, called the **index of the summation**. Thus, we have

$$\sum_{k=2}^{5} 3k = 3(2) + 3(3) + 3(4) + 3(5)$$

$$= 6 + 9 + 12 + 15$$

$$= 42$$

EXAMPLE 4   Find each sum: **a.** $\displaystyle\sum_{k=3}^{5} (2k + 1)$ **b.** $\displaystyle\sum_{k=2}^{5} k^2$

SOLUTION   **a.** $\displaystyle\sum_{k=3}^{5} (2k + 1) = [2(3) + 1] + [2(4) + 1] + [2(5) + 1]$

$$= 7 + 9 + 11$$

$$= 27$$

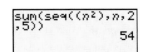

b. $\displaystyle\sum_{k=2}^{5} k^2 = 2^2 + 3^2 + 4^2 + 5^2$

$$= 4 + 9 + 16 + 25$$

$$= 54$$

TECHNOLOGY TIP

TI-83 Plus: The **sum(** command is located under the MATH submenu of the LIST menu and the **seq(** command is found under the OPS submenu of the LIST menu.

TI-86: Both **sum** and **seq** are found under the OPS submenu of the LIST menu.

▮ GEOMETRIC SEQUENCES

Another important sequence is the geometric sequence.

Geometric Sequence
A **geometric sequence** is a sequence of the form

$$a, ar, ar^2, ar^3, \ldots, ar^{n-1}, \ldots$$

where a is the **first term**, ar^{n-1} is the **nth term**, and r is the **common ratio**.

We note that the second term of a geometric sequence has a factor of r^1, the third term has a factor of r^2, the fourth term has a factor of r^3, and the nth term has a factor of r^{n-1}. We also note that the quotient obtained when any term is divided by the previous term is r.

EXAMPLE 5

The first three terms of a geometric sequence are 16, 4, and 1. Find the seventh term of the sequence.

SOLUTION   We substitute 16 for a, $\frac{1}{4}$ for r, and 7 for n in the formula for the nth term and simplify:

$$\text{nth term} = ar^{n-1}$$

$$\text{7th term} = 16\left(\frac{1}{4}\right)^{7-1} \qquad r = \tfrac{4}{16} = \tfrac{1}{4}.$$

$$= 16\left(\frac{1}{4}\right)^6$$

$$= 16\left(\frac{1}{4096}\right)$$

$$= \frac{1}{256}$$

▮ GEOMETRIC MEANS

If numbers are inserted between two numbers a and b to form a geometric sequence, the inserted numbers are called **geometric means** between a and b. If a single number is inserted between the numbers a and b, that number is called a **geometric mean** between a and b.

EXAMPLE 6

Insert two geometric means between 7 and 1512.

SOLUTION   In this example, the first term is $a = 7$ and the fourth term (or last term) is $l = 1512$. To find the common ratio r such that the terms

$$7, 7r, 7r^2, 1512$$

form a geometric sequence, we substitute 4 for n and 7 for a into the formula for the nth term of a geometric sequence and solve for r.

$$\textbf{nth term} = ar^{n-1}$$

$$\textbf{4th term} = 7r^{4-1}$$

$$1512 = 7r^3$$

$$216 = r^3 \qquad \text{Divide both sides by 7.}$$

$$6 = r \qquad \text{Take the cube root of both sides.}$$

The two geometric means between 7 and 1512 are

$$7r = 7(6) = \textbf{42}$$

and

$$7r^2 = 7(6)^2 = 7(36) = \textbf{252}$$

The numbers 7, 42, 252, and 1512 are the first four terms of a geometric sequence.

THE SUM OF THE FIRST n TERMS OF A GEOMETRIC SEQUENCE

There is a formula that gives the sum of the first n terms of a geometric sequence. To develop this formula, we let S_n represent the sum of the first n terms of a geometric sequence.

1. $S_n = a + ar + ar^2 + ar^3 + \cdots + ar^{n-1}$

We multiply both sides of Equation 1 by r to get

2. $S_n r = \quad ar + ar^2 + ar^3 + \cdots + ar^{n-1} + ar^n$

We now subtract Equation 2 from Equation 1 and solve for S_n:

$$S_n - S_n r = a - ar^n$$

$$S_n(1 - r) = a - ar^n \qquad \text{Factor out } S_n \text{ from the left side.}$$

$$S_n = \frac{a - ar^n}{1 - r} \qquad \text{Divide both sides by } 1 - r.$$

This reasoning establishes the following theorem.

Sum of the First n Terms of a Geometric Sequence
The sum of the first n terms of a geometric sequence is given by the formula

$$S_n = \frac{a - ar^n}{1 - r} \qquad (r \neq 1)$$

where S_n is the sum, a is the first term, r is the common ratio, and n is the number of terms.

EXAMPLE 7   Find the sum of the first six terms of the geometric sequence 250, 50, 10, ...

SOLUTION   In this sequence, $a = 250$, $r = \frac{1}{5}$, and $n = 6$. We substitute these values into the formula for the sum of the first n terms of a geometric sequence and simplify.

$$S_n = \frac{a - ar^n}{1 - r} = \frac{250 - 250\left(\frac{1}{5}\right)^6}{1 - \frac{1}{5}}$$

$$= \frac{250 - 250\left(\frac{1}{15,625}\right)}{\frac{4}{5}}$$

$$= \frac{5}{4}\left(250 - \frac{250}{15,625}\right)$$

$$= \frac{5}{4}\left(\frac{3,906,000}{15,625}\right)$$

$$= 312.48$$

The sum of the first six terms is 312.48.

EXAMPLE 8

The mayor of Eagle River (population 1,500) predicts a growth rate of 4% each year for the next ten years. Find the population of Eagle River ten years from now.

SOLUTION   Let P_0 be the initial population of Eagle River. After 1 year, there will be a different population P_1. The initial population (P_0) plus the growth (the product of P_0 and the rate of growth, r) will equal this new population, P_1:

$$P_1 = P_0 + P_0 r = P_0(1 + r)$$

The population after 2 years will be P_2, and

$$P_2 = P_1 + P_1 r$$
$$= P_1(1 + r) \qquad \text{Factor out } P_1.$$
$$= P_0(1 + r)(1 + r) \qquad \text{Remember that } P_1 = P_0(1 + r).$$
$$= P_0(1 + r)^2$$

The population after 3 years will be P_3, and

$$P_3 = P_2 + P_2 r$$
$$= P_2(1 + r) \qquad \text{Factor out } P_2.$$
$$= P_0(1 + r)^2(1 + r) \qquad \text{Remember that } P_2 = P_0(1 + r)^2.$$
$$= P_0(1 + r)^3$$

The yearly population figures

$$P_0, \quad P_1, \quad P_2, \quad P_3, \ldots$$

or

$$P_0, \quad P_0(1 + r), \quad P_0(1 + r)^2, \quad P_0(1 + r)^3, \ldots$$

form a geometric sequence with a first term of P_0 and a common ratio of $1 + r$. The population of Eagle River after 10 years is P_{10}, which is the 11th term of this sequence:

$$n\text{th term} = ar^{n-1}$$
$$P_{10} = 11\text{th term} = P_0(1 + r)^{10}$$
$$= 1{,}500(1 + 0.04)^{10}$$
$$= 1{,}500(1.04)^{10}$$
$$\approx 2{,}220 \qquad \text{Use a calculator.}$$

The estimated population ten years from now is 2,220 people.

APPENDIX C EXERCISE

VOCABULARY AND NOTATION
In Exercises 1–5, fill in the blanks to make a true statement.

1. A _____ is a function whose domain is the set of natural numbers.

2. An _____ sequence is a sequence in which each term after the first is formed by adding a fixed number.

3. The arithmetic _____ is the number inserted between two numbers *a* and *b* to form an arithmetic sequence.

4. The Greek letter sigma, ___, is used in summation notation.

5. A _____ sequence is a sequence in which each term after the first is obtained by multiplying the preceding term by a constant nonzero number called the _____.

CONCEPTS

6. Determine the common difference in each arithmetic sequence.
   a. $1, 4, 7, \ldots$ b. $2, 6, 10, \ldots$

7. If *a* is the first term and *d* is the common difference, write the first five terms of the arithmetic sequence.
   a. $a = 3, d = 2$
   b. $a = -5, d = -3$

8. Determine the common ratio in each geometric sequence.
   a. $3, 6, 12, \ldots$ b. $3, -6, 12, \ldots$

9. If *a* is the first term and *r* is the common ratio, write the first five terms of the geometric sequence.
   a. $a = 3, r = 2$
   b. $a = -5, r = \frac{1}{5}$

10. Explain why 8, 3, −2, and −7 are the first four terms of an arithmetic sequence when we appear to be subtracting a constant instead of adding a constant to produce the next term.

11. Explain why 12, 6, and 2 are the first three terms of a geometric sequence when we appear to be dividing instead of multiplying to produce the next term.

PRACTICE

12. Find the 30th term of the arithmetic sequence with $a = 7$ and $d = 12$.

13. Find the 55th term of the arithmetic sequence with $a = -5$ and $d = 4$.

14. Find the first term of the arithmetic sequence with a common difference of 11 and whose 27th term is 263.

15. Find the common difference of the arithmetic sequence with a first term of −164 if its 36th term is −24.

16. Insert three arithmetic means between 2 and 11.

17. Insert four arithmetic means between 5 and 25.

18. Find the arithmetic mean between 10 and 19.

19. Find the arithmetic mean between 5 and 23.

In Exercises 20–23, find the sum of the first n terms of each arithmetic sequence.

20. $1, 4, 7, \ldots, n = 30$

21. $2, 6, 10, \ldots, n = 28$

22. $f(n) = 2n + 1$, *n*th term is 31, *n* is a natural number

23. $f(n) = 4n + 3$, *n*th term is 23, *n* is a natural number

24. Find the sum of the first 50 natural numbers.

25. Find the sum of the first 100 natural numbers.

26. Find the tenth term of the geometric sequence with $a = 7$ and $r = 2$.

27. Find the 12th term of the geometric sequence with $a = 64$ and $r = \frac{1}{2}$.

28. Find the common ratio of the geometric sequence with a first term of -8 and a sixth term of -1944.

29. Find the common ratio of the geometric sequence with a first term of 12 and a sixth term of $\frac{3}{8}$.

30. Insert three positive geometric means between 2 and 162.

31. Insert four geometric means between 3 and 96.

32. Find the negative geometric mean between 2 and 128.

33. Find the positive geometric mean between 3 and 243.

In Exercises 34–37, find the sum of the first n terms of each geometric sequence.

34. $2, 6, 18, \ldots ; n = 6$

35. $2, -6, 18, \ldots ; n = 6$

36. The third term is -2 and the fourth term is 1, $n = 6$

37. The third term is -3 and the fourth term is 1, $n = 5$

In Exercises 38–41, find each sum.

38. $\displaystyle\sum_{k=1}^{4} 6k$

39. $\displaystyle\sum_{k=2}^{5} 3k$

40. $\displaystyle\sum_{k=3}^{4} (k^2 + 3)$

41. $\displaystyle\sum_{k=2}^{6} (k^2 + 1)$

APPLICATIONS

42. Saving money Fred puts $60 into a safety deposit box. After each succeeding month, he puts $50 more in the safety deposit box. Write the first six terms of an arithmetic sequence that gives the monthly amounts in his savings, and find his savings after 10 years.

43. Installment loan Freda borrowed $10,000 interest-free from her mother. Freda agreed to pay back the loan in monthly installments of $275. Write the first six terms of an arithmetic sequence that shows the balance due after each month, and find the balance due after 17 months.

44. Designing a patio Each row of bricks in the triangular patio in Illustration 1 is to have one more brick than the previous row, ending with the longest row of 150 bricks. How many bricks will be needed?

Illustration 1

45. Falling object The equation $s = 16t^2$ represents the distance s in feet that an object will fall in t seconds. After 1 second, the object has fallen 16 feet. After 2 seconds, the object has fallen 64 feet, and so on. Find the distance that the object will fall during the second and third seconds.

46. Falling object Refer to Exercise 45. How far will the object fall during the 12th second?

47. Interior angles The sums of the angles of several polygons are given in Table 1. Assuming that the pattern continues, find the sum of the interior angles of an octagon (8 sides) and dodecagon (12 sides).

Figure	Number of sides	Sum of angles
Triangle	3	180°
Quadrilateral	4	360°
Pentagon	5	540°
Hexagon	6	720°
Octagon	8	
Dodecagon	12	

TABLE 1

48. Population growth The population of Union is predicted to increase by 6% each year. What will be the population of Union 5 years from now if its current population is 500?

49. Population growth The population of Hicksville is decreasing by 10% each year. If its current population is 98, what will be the population 8 years from now?

50. Declining savings   John has $10,000 in a safety deposit box. Each year he spends 12% of what is left in the box. How much will be in the box after 15 years?

51. Savings growth   Sally has $5000 in an investment account earning 12% annual interest. How much will be in her account 10 years from now? (Assume that Sally makes no deposits or withdrawals.)

52. House appreciation   A house appreciates by 6% each year. If the house is worth $70,000 today, how much will it be worth 12 years from now?

53. Motorboat depreciation   A motorboat that cost $5000 when new depreciates at a rate of 9% per year. How much will the boat be worth in 5 years?

54. Inscribed squares   Each inscribed square in Illustration 2 joins the midpoints of the sides of the next larger square. The area of the first square, the largest, is 1. Find the area of the 12th square.

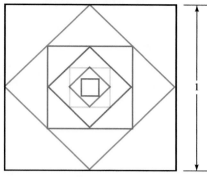

Illustration 2

55. Genealogy   The family tree in Illustration 3 spans three generations and lists seven people. How many names would be listed in a family tree that spans ten generations?

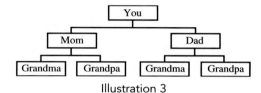

Illustration 3

APPENDIX D   Permutations and Combinations

- THE MULTIPLICATION PRINCIPLE FOR EVENTS
- PERMUTATIONS
- COMBINATIONS
- ALTERNATIVE FORM OF THE BINOMIAL THEOREM

THE MULTIPLICATION PRINCIPLE FOR EVENTS

A student goes to the cafeteria for lunch. He has a choice of three different sandwiches (hamburger, hot dog, or ham and cheese) and four different beverages (cola, root beer, orange, or milk). How many different lunches can he choose?

The student has three choices of sandwich, and for any one of these choices, he has four choices of drink. The different options are shown in the *tree diagram* in Figure D-1. The tree diagram shows that there is a total of 12 different lunches to choose from. One possibility is a hamburger with a cola, and another is a hot dog with milk.

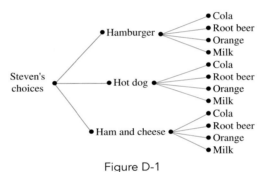

Figure D-1

A situation that can have several different outcomes — such as choosing a sandwich — is called an **event**. Choosing a sandwich and choosing a beverage can be thought of as two events. The preceding example illustrates the **multiplication principle for events**.

Multiplication Principle for Events

Let E_1 and E_2 be two events. If E_1 can be done in a_1 ways, and if — after E_1 has occurred — E_2 can be done in a_2 ways, then the event "E_1 followed by E_2" can be done in $a_1 \cdot a_2$ ways.

EXAMPLE 1

In how many ways can we arrange five books on a shelf?

SOLUTION   We can fill the first space with any of the 5 books, the second space with any of the remaining 4 books, the third space with any of the remaining 3 books, the fourth space with any of the remaining 2 books, and the fifth space with the remaining 1 (or last) book. According to the multiplication principle for events, the number of ways that the books can be arranged is

$$5 \cdot 4 \cdot 3 \cdot 2 \cdot 1 = 120$$

▌▌▌ PERMUTATIONS

When computing the number of possible arrangements of objects such as books on a shelf or flags on a pole, we are finding the number of **permutations** of those objects. In Example 1, we found that the number of permutations of five books, using all five of them, is 120.

The symbol $P(n, r)$, read as "the number of permutations of n things r at a time," is often used to express permutation problems. In the case of the number of arrangements of five books on the shelf, we find that $P(5, 5) = 120$ because we are taking 5 objects 5 at a time.

EXAMPLE 2

If Sarah has seven flags, each of a different color, to hang on a flagpole, how many different signals can she send by using three flags?

SOLUTION

Sarah must find $P(7, 3)$ (the number of permutations of 7 things 3 at a time). In the top position, Sarah can hang any of the 7 flags; in the middle position, any one of the remaining 6 flags; and in the bottom position, any one of the remaining 5 flags. According to the multiplication principle for events,

$$P(7, 3) = 7 \cdot 6 \cdot 5$$
$$= 210$$

Sarah can send 210 signals using only three of the available seven flags.

Although it is correct to write $P(7, 3) = 7 \cdot 6 \cdot 5$, there is an advantage in changing the form of this answer to obtain a formula for computing $P(7, 3)$:

$$P(7, 3) = 7 \cdot 6 \cdot 5$$
$$= \frac{7 \cdot 6 \cdot 5 \cdot 4 \cdot 3 \cdot 2 \cdot 1}{4 \cdot 3 \cdot 2 \cdot 1} \qquad \text{Multiply both the numerator and denominator by } 4 \cdot 3 \cdot 2 \cdot 1.$$
$$= \frac{7!}{4!}$$
$$= \frac{7!}{(7 - 3)!}$$

The generalization of this idea gives the following formula:

Formula for Finding $P(n, r)$

The number of permutations of n things r at a time is given by the formula

$$P(n, r) = \frac{n!}{(n - r)!}$$

EXAMPLE 3

Compute: **a.** $P(8, 2)$ **b.** $P(n, n)$ **c.** $P(n, 0)$

SOLUTION   **a.** $P(8, 2) = \dfrac{8!}{(8 - 2)!}$

$= \dfrac{8 \cdot 7 \cdot 6!}{6!}$

$= 8 \cdot 7$

$= 56$

```
8 nPr 2
              56
```

TECHNOLOGY TIP

The permutation of n things taken r at a time, $P(n, r)$, can also be written as $_nP_r$. To compute $P(8, 2) = {_8P_2}$, enter 8 then retrieve the $_nP_r$ command from the probability (PRB on the TI-83 Plus and PROB on the TI-86) submenu of the MATH feature and follow with the 2, as indicated on the screen accompanying the problem.

b. $P(n, n) = \dfrac{n!}{(n - n)!}$

$= \dfrac{n!}{0!}$

$= \dfrac{n!}{1}$

$= n!$

c. $P(n, 0) = \dfrac{n!}{(n - 0)!}$

$= \dfrac{n!}{n!}$

$= 1$

Parts (c) and (d) of Example 3 establish the following formulas:

Formulas for Finding $P(n, n)$ and $P(n, 0)$

The number of permutations of n things n at a time and n things 0 at a time are given by the formula

$$P(n, n) = n! \quad \text{and} \quad P(n, 0) = 1$$

EXAMPLE 4

a. In how many ways can a television executive arrange the Saturday night lineup of 6 programs if there are 15 programs to choose from?

b. If there are only 6 programs to choose from?

SOLUTION   **a.** To find the number of permutations of 15 programs 6 at a time, we use the formula $P(n, r) = \frac{n!}{(n - r)!}$ with $n = 15$ and $r = 6$.

$$P(15, 6) = \frac{15!}{(15 - 16)!} = \frac{15 \cdot 14 \cdot 13 \cdot 12 \cdot 11 \cdot 10 \cdot 9!}{9!}$$

$$= 15 \cdot 14 \cdot 13 \cdot 12 \cdot 11 \cdot 10$$

$$= 3,603,600$$

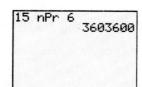

```
15 nPr 6
         3603600
```

b. To find the number of permutations of 6 programs 6 at a time, we use the formula $P(n, n) = n!$ with $n = 6$.

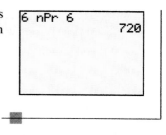

$$P(6, 6) = 6! = 720$$

▌▌ COMBINATIONS

Suppose that Raul must read 4 books from a reading list of 10 books. The order in which he reads them is not important. For the moment, however, let's assume that order is important and find the number of permutations of 10 things 4 at a time:

$$P(10, 4) = \frac{10!}{(10 - 4)!}$$

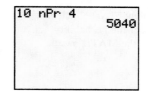

$$= \frac{10 \cdot 9 \cdot 8 \cdot 7 \cdot 6!}{6!}$$

$$= 10 \cdot 9 \cdot 8 \cdot 7$$

$$= 5040$$

If order is important, there are 5040 ways of choosing 4 books when there are 10 books to choose from.

However, because the order in which Raul reads the books does not matter, the previous result of 5040 is too big. Because there are 4! ways of ordering the 4 books that are chosen, the result of 5040 is exactly 4! (or 24) times too big. Thus, the number of choices that Raul has is the number of permutations of 10 things 4 at a time, divided by 24:

$$\frac{P(10, 4)}{24} = \frac{5040}{24}$$

$$= 210$$

Raul has 210 ways of choosing 4 books to read from the list of 10 books.

In situations where order is *not* important, we are interested in **combinations**, not permutations. The symbols $C(n, r)$ and $\binom{n}{r}$ both mean the number of combinations of n things r at a time.

If a selection of r books is chosen from a total of n books, the number of possible selections is $C(n, r)$, and there are $r!$ arrangements of the r books in each selection. If we consider the selected books as an ordered grouping, the number of orderings is $P(n, r)$. Thus, we have

$$r! \cdot C(n, r) = P(n, r)$$

We can divide both sides of this equation by $r!$ to get the formula for finding $C(n, r)$:

$$C(n, r) = \binom{n}{r} = \frac{P(n, r)}{r!} = \frac{n!}{r!(n - r)!}$$

Formula for Finding C(n, r)

The number of combinations of n things r at a time is given by

$$C(n, r) = \frac{n!}{r!(n - r)!}$$

EXAMPLE 5

Compute: **a.** $C(8, 5)$ **b.** $C(n, n)$ **c.** $C(n, 0)$

SOLUTION   **a.** $C(8, 5) = \dfrac{8!}{5!(8 - 5)!}$

$$= \frac{8 \cdot 7 \cdot 6 \cdot 5!}{5!3!}$$

$$= 8 \cdot 7$$

$$= 56$$

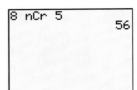

TECHNOLOGY TIP

The notation $C(n, r)$ and $\binom{n}{r}$ are also equivalent to $_nC_r$, which is located under the probability submenu of the MATH menu.

b. $C(n, n) = \dfrac{n!}{n!(n - n)!}$

$$= \frac{1}{0!}$$

$$= \frac{1}{1}$$

$$= 1$$

c. $C(n, 0) = \dfrac{n!}{0!(n - 0)!}$

$$= \frac{n!}{0!n!}$$

$$= \frac{1}{0!}$$

$$= \frac{1}{1}$$

$$= 1$$

The notation $C(n, 0)$ indicates that we choose 0 things from the available n things.

Parts (b) and (c) of Example 5 establish the following formulas:

Formulas for Finding C(n, n) and C(n, 0)

The number of combinations of n things n at a time is 1. The number of combinations of n things 0 at a time is 1.

$$C(n, n) = 1 \quad \text{and} \quad C(n, 0) = 1$$

EXAMPLE 6

If 15 students want to pick a committee of 4 students to plan a party, how many different committees are possible?

SOLUTION   Because the ordering of people on each possible committee is unimportant, we find the number of combinations of 15 people 4 at a time:

$$C(15, 4) = \frac{15!}{4!(15 - 4)!}$$

$$= \frac{15 \cdot 14 \cdot 13 \cdot 12 \cdot 11!}{4 \cdot 3 \cdot 2 \cdot 1 \cdot 11!}$$

$$= \frac{15 \cdot 14 \cdot 13 \cdot 12}{4 \cdot 3 \cdot 2 \cdot 1}$$

$$= 1365$$

```
15 nCr 4
              1365
```

There are 1,365 possible committees.

EXAMPLE 7

A committee in Congress consists of 10 Democrats and 8 Republicans. In how many ways can a subcommittee be chosen if it is to contain 5 Democrats and 4 Republicans?

SOLUTION   There are $C(10, 5)$ ways of choosing the 5 Democrats and $C(8, 4)$ ways of choosing the 4 Republicans. By the multiplication principle for events, there are $C(10, 5) \cdot C(8, 4)$ ways of choosing the subcommittee:

$$C(10, 5) \cdot C(8, 4) = \frac{10!}{5!(10 - 5)!} \cdot \frac{8!}{4!(8 - 4)!}$$

$$= \frac{10 \cdot 9 \cdot 8 \cdot 7 \cdot 6 \cdot 5!}{120 \cdot 5!} \cdot \frac{8 \cdot 7 \cdot 6 \cdot 5 \cdot 4!}{24 \cdot 4!}$$

$$= \frac{10 \cdot 9 \cdot 8 \cdot 7 \cdot 6}{120} \cdot \frac{8 \cdot 7 \cdot 6 \cdot 5}{24}$$

$$= 17,640$$

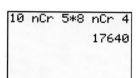

```
10 nCr 5*8 nCr 4
             17640
```

There are 17,640 possible subcommittees.

▊ ALTERNATIVE FORM OF THE BINOMIAL THEOREM

We have seen that the expansion of $(x + y)^3$ is

$$(x + y)^3 = \mathbf{1}x^3 + \mathbf{3}x^2y + \mathbf{3}xy^2 + \mathbf{1}y^3$$

and that

$$C(3, 0) = \binom{3}{0} = 1, \quad C(3, 1) = \binom{3}{1} = 3, \quad C(3, 2) = \binom{3}{2} = 3, \quad \text{and} \quad C(3, 3) = \binom{3}{3} = 1$$

Putting these facts together gives the following way of writing the expansion of $(x + y)^3$:

$$(x + y)^3 = \binom{3}{0}x^3 + \binom{3}{1}x^2y + \binom{3}{2}xy^2 + \binom{3}{3}y^3$$

Likewise, we have

$$(x + y)^4 = \binom{4}{0}x^4 + \binom{4}{1}x^3y + \binom{4}{2}x^2y^2 + \binom{4}{3}xy^3 + \binom{4}{4}y^4$$

The generalization of this idea allows us to state the binomial theorem in an alternative form, using combinatorial notation.

Alternative Form of the Binomial Theorem

If n is any positive integer, then

$$(a + b)^n = \binom{n}{0}a^n + \binom{n}{1}a^{n-1}b + \binom{n}{2}a^{n-2}b^2 + \cdots + \binom{n}{r}a^{n-r}b^r + \cdots + \binom{n}{n}b^n$$

EXAMPLE 8

Use the alternative form of the binomial theorem to expand $(x + y)^6$.

SOLUTION

$$(x + y)^6 = \binom{6}{0}x^6 + \binom{6}{1}x^5y + \binom{6}{2}x^4y^2 + \binom{6}{3}x^3y^3 + \binom{6}{4}x^2y^4 + \binom{6}{5}xy^5 + \binom{6}{6}y^6$$

$$= x^6 + 6x^5y + 15x^4y^2 + 20x^3y^3 + 15x^2y^4 + 6xy^5 + y^6$$

TECHNOLOGY TIP

To easily generate all the coefficients for the binomial theorem, enter $y1 = 6_nC_rx$ and generate the TABLE beginning at $x = 0$.

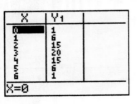

EXAMPLE 9

Use the alternative form of the binomial theorem to expand $(2x - y)^3$.

SOLUTION

$$(2x - y)^3 = [2x + (-y)]^3$$

$$= \binom{3}{0}(2x)^3 + \binom{3}{1}(2x)^2(-y) + \binom{3}{2}(2x)(-y)^2 + \binom{3}{3}(-y)^3$$

$$= 1(2x)^3 + 3(4x)^2(-y) + 3(2x)(-y)^2 + (-y)^3$$

$$= 8x^3 - 12x^2y + 6xy^2 - y^3$$

APPENDIX D EXERCISE

VOCABULARY AND NOTATION
In Exercises 1–4, fill in the blanks to make a true statement.

1. A situation that can have several different outcomes is called an _____.

2. The _____ principle for events states that if event 1 can be done in a ways and event 2 can be done in b ways, then event 1 followed by event 2 can be done in ____ ways.

3. A _____ computes the number of possible arrangements of objects; order is important.

4. A _____ computes the number of possible groups of objects; order is not important.

CONCEPT

5. Label each of the following sets as a group of permutations or combinations:
   a. $\{1, 2, 3\}, \{1\}, \{2\}, \{1, 2\}, \{1, 3\}, \{2, 3\}$
   b. $\{1, 2, 3\}, \{2, 1, 3\}, \{2, 3, 1\}, \{3, 1, 2\}, \{3, 2, 1\}$

6. Based on the two calculator screens, determine
   a. $C(8, 4)$ b. $C(8, 5)$ c. $C(8, 8)$

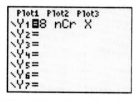

7. Based on the two calculator screens, determine
   a. $P(8, 4)$ b. $P(8, 5)$
   c. $P(8, 8)$

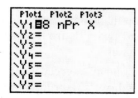

8. Explain why $C(8,8)$ is less than $P(8, 8)$.

PRACTICE
In Exercises 9–18, evaluate each expression. Verify with your calculator.

9. $P(5, 5)$ 10. $P(4, 4)$

11. $P(5, 3)$ 12. $P(3, 2)$

13. $P(2, 2) \cdot P(3, 3)$ 14. $P(3, 2) \cdot P(3, 3)$

15. $\dfrac{P(5, 3)}{P(4, 2)}$ 16. $\dfrac{P(6, 2)}{P(5, 4)}$

17. $\dfrac{P(5, 3) \cdot P(7, 3)}{P(5, 1)}$ 18. $\dfrac{P(8, 3)}{P(5, 3) \cdot P(4, 3)}$

In Exercises 19–30, evaluate each expression. Verify with your calculator.

19. $C(5, 3)$ 20. $C(5, 4)$

21. $\dbinom{6}{3}$ 22. $\dbinom{6}{4}$

23. $\dbinom{5}{4}\dbinom{5}{3}$ 24. $\dbinom{6}{5}\dbinom{6}{4}$

25. $\dfrac{C(38, 37)}{C(19, 18)}$ 26. $\dfrac{C(25, 23)}{C(40, 39)}$

27. $C(12, 0)C(12, 12)$ 28. $\dfrac{C(8, 0)}{C(8, 1)}$

29. $C(n, 2)$ 30. $C(n, 3)$

In Exercises 31–36, use the alternative form of the binomial theorem to expand each expression. Use the TABLE feature of the calculator to verify your coefficients.

31. $(x + y)^4$

32. $(x - y)^2$

33. $(2x + y)^3$

34. $(2x + 1)^4$

35. $(3x - 2)^4$

36. $(3x - x^2)^3$

In Exercises 37–40, find the indicated term of the binomial expansion. The TABLE feature of the calculator can again be used to verify your coefficient.

37. $(x - 5y)^5$; fourth term

38. $(2x - y)^5$; third term

39. $(x^2 - y^3)^4$; second term

40. $(x^3 - y^2)^4$; fourth term

APPLICATIONS

41. **Arranging an evening**   Kristy intends to go out to dinner and then see a movie. In how many ways can she ar-

range her evening if she has a choice of five movies and seven restaurants?

42. Travel choices Paula has five ways to travel from New York to Chicago, three ways to travel from Chicago to Denver, and four ways to travel from Denver to Los Angeles. How many choices are available to Paula if she travels from New York to Los Angeles?

43. Making license plates How many six-digit license plates can be manufactured? (Note that there are ten choices — 0, 1, 2, 3, 4, 5, 6, 7, 8, 9 — for each digit.)

44. Making license plates How many six-digit license plates can be manufactured if no digit can be repeated?

45. Making license plates How many six-digit license plates can be manufactured if no license number can begin with 0 and if no digit can be repeated?

46. Making license plates How many license plates can be manufactured with two letters followed by four digits?

47. Phone numbers How many seven-digit phone numbers are available in area code 815 if no phone number can begin with 0 or 1?

48. Phone numbers How many ten-digit phone numbers are available if area codes 000 and 911 cannot be used and if no local number can begin with 0 or 1?

49. Lining up In how many ways can six girls be placed in a line?

50. Arranging books In how many ways can seven books be placed on a shelf?

51. Arranging books In how many ways can four novels and five biographies be arranged on a shelf if the novels are placed first?

52. Making a ballot In how many ways can six candidates for mayor and four candidates for the county board be arranged on a ballot if all of the candidates for mayor must be placed first?

53. Combination locks How many permutations does a combination lock have if each combination has 3 numbers, no two numbers of any combination are equal, and the lock has 25 numbers?

54. Combination locks How many permutations does a combination lock have if each combination has 3 numbers, no two numbers of any combination are equal, and the lock has 50 numbers?

55. Arranging appointments The receptionist at a dental office has only three appointment times available before next Tuesday, and there are ten patients with a toothache. In how many ways can the receptionist fill those appointments?

56. Computers In many computers, a *word* consists of 32 *bits*, a string of 32 1s and 0s. How many different words are possible?

57. Palindromes A palindrome is any word, such as *madam* or *radar*, that reads the same backward and forward. How many five-digit numerical palindromes (like 12321) are there? (*Hint:* A leading zero would be dropped.)

58. Call letters The call letters of a U.S. commercial radio station have either three of four letters, and the first is either a *W* or a *K*. How many radio stations could this system support?

59. Planning a picnic A class of 14 students wants to pick a committee of 3 students to plan a picnic. How many committees are possible?

60. Choosing books Jeffrey must read 3 books from a reading list of 15 books. How many choices does he have?

61. Forming a committee The number of three-person committees that can be formed from a group of persons is ten. How many persons are in the group?

62. Forming a committee The number of three-person committees that can be formed from a group of persons is 20. How many persons are in the group?

63. Winning a lottery In one state lottery, anyone who picks the correct six numbers (in any order) wins. With the numbers 0 through 99 available, how many choices are possible?

64. Taking a test The instructions on a test read, "Answer any 10 of the following 15 questions. Then choose 1 of the remaining questions for homework, and turn in its solution tomorrow." In how many ways can the questions be chosen?

65. Forming a committee In how many ways can we select a committee of two boys and two girls from a group containing three boys and four girls?

66. Forming a committee In how many ways can we select a committee of three boys and two girls from a group containing five boys and three girls?

67. Choosing clothes In how many ways can we select 2 shirts and 3 neckties from a group of 12 shirts and 10 neckties?

68. Choosing clothes In how many ways can we select five dresses and two coats from a wardrobe containing nine dresses and three coats?

Calculator Reference Guide for the TI-83 Plus and the TI-86

Absolute Value (pages 18 and 22)

Example: Simplify $|-3 - 2|$.

```
abs( -3-2)
                5
```

```
abs (-3-2)
                5
 abs  ▸Frac   ×√
```

Keystrokes:

[2ⁿᵈ] [CATALOG] [ENTER]
[(−)] [3] [−] [2] [)] [ENTER]

[CUSTOM] [F₁] (abs) [(] [(−)] [3]
[−] [2] [)] [ENTER]
Note: It is assumed that absolute value has been customized (see Custom) and that the feature is stored under F1.

Alpha (pages 32 and 33)

Example: M + XY.

```
M+XY■
```

```
M+x*Y■
```

Keystrokes:

[ALPHA] [M] [+] [X,T,θ,*n*]
[ALPHA] [Y]

[ALPHA] [M] [+] [x-VAR] [×] [ALPHA] [Y]
Note: The TI-86 requires the use of a multiplication symbol between variables. The expression *xy* is viewed as another defined variable, not as the product of two variables.

The ALPHA key is used when entering algebraic expressions that are to be *evaluated*. These calculators *do not* perform symbolic operations.

Catalog (pages 17 and 22)

The catalog feature displays an alphabetical list of available calculator operations. Use the [▲] and [▼] arrow keys to scroll through this list. Operations may be accessed by placing the pointer adjacent to the operation and pressing [ENTER]. To exit the catalog, press [CLEAR].

With the TI-86, the catalog feature is essential to customizing key features on your calculator. See Custom.

Clear (pages 18 and 22)

Pressing [CLEAR] will delete entries from the HOME screen, WINDOW screen, and the Y-EDIT screen. It will also return you to the HOME screen from the MODE, GRAPH, ZOOM, and CALC screens.

Colon (pages 32 and 33)

Example: Evaluate $3x + 4$ for $x = 2$.

```
2→X:3X+4
              10
■
```

```
2→x:3 x+4
              10
■
```

Keystrokes:

[2] [STO ►] [X,T,θ,n]
[ALPHA] [:] [3] [X,T,θ,n]
[+] [4] [ENTER]

[2] [STO ►] [x-VAR] [ALPHA] [2nd]
[:] [3] [x-VAR] [+] [4] [ENTER]

Custom (TI-86 only) (page 23)

Keystrokes: [2nd] [CATLG-VARS] [F1] (CATLG) [F3] (CUSTOM) Next, use the cursor arrows to place the arrow on the screen next to the feature in the catalog list that you want to customize, such as **abs** for absolute value, and press F1 to enter the feature in the first empty box. (If F1 is not empty, select an F key that is above an empty box and press to enter the feature in that particular box.)

Decimal (viewing window)
(pages 184 and 186)

Example: Set your viewing window
to ZDecimal (i.e., a decimal
viewing window).

```
WINDOW
Xmin=-4.7
Xmax=4.7
Xscl=1
Ymin=-3.1
Ymax=3.1
Yscl=1
Xres=1■
```

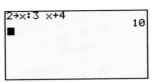

Keystrokes:

[ZOOM] [4:ZDECIMAL]
[WINDOW]

[GRAPH] [F3] [ZOOM]
[MORE] [F4] (ZDECM) [2nd]
[M2] (WIND)

The purpose of this window is to enable you to have trace coordinates in which the x-value is reported in tenths of a unit.

Decimal × 2 (viewing window)
(pages 184 and 187)

Example: Set your viewing window to
ZDecimal × 2.

```
WINDOW
Xmin=-9.4
Xmax=9.4
Xscl=1
Ymin=-6.2
Ymax=6.2
Yscl=1
Xres=1■
```

```
WINDOW
xMin=-12.6
xMax=12.6
xScl=1
yMin=-6.2
yMax=6.2
↓yScl=1■
y(x)= WIND ZOOM TRACE GRAPH►
```

Keystrokes:

[ZOOM] [4:ZDECIMAL]
[WINDOW]

[GRAPH] [F3] [ZOOM]
[MORE] [F4] (ZDECM) [2nd]
[M2] (WIND)

Beginning at the top entry, multiply each entry by 2, using the [▼] cursor to move from one entry to the next. **Do not** *multiply the x and y scales by 2.*

The purpose of this window is to enable you to have trace coordinates in which the x-value is reported in two-tenths of a unit. This window provides *nice* tracing coordinates while remaining closer in size to the standard viewing window. Any value n can be used besides 2; the trace coordinates for the x-value will be reported in n-tenths of a unit.

Darken (pages 18 and 22)

If the display is not clear, press [2nd] [▲] to darken the screen.

DrawInv (page 673)

Example: Draw the inverse of $y = x^2$.
Note: $y = x^2$ must be entered on the Y-EDIT screen and you must begin the keystroking on the HOME screen.)

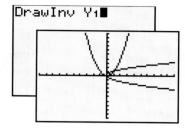

Keystrokes:

[2ⁿᵈ] [DRAW] [8:DRAWINV]
[VARS] *cursor over to select y-vars*
[1:FUNCTION] [1:Y1] (*see first screen*) [ENTER]

[GRAPH] [MORE]
[F2] (DRAW)
[MORE] [MORE] [MORE]
[F3] (DrInv) [2nd] [ALPHA]
[y] [1] (*see first screen*)
[ENTER]

Note: This is a *drawn* graph. You **cannot** interact with it via TRACE, INTERSECT/ISECT, etc.

Enter (pages 18 and 23)

Features of the calculator maybe be accessed/activated by pressing [ENTER].

Eval (page 192)

Example: Find the value of y when $x = -3$ on the graph of $y = 3x + 5$.

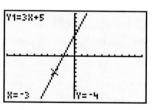

Keystrokes:

[2ⁿᵈ] [CALC] [1:VALUE]
Enter −3 at the blinking cursor [ENTER]

Press [EXIT] *so that only one line of graph menu is displayed.*
[MORE] [MORE] [F1](EVAL) *Enter −3 at the blinking cursor.* [ENTER]

Exit (page 22)

On the TI-86, pressing [EXIT] will remove one layer of the graphing menu as well as providing an exit from the MODE screen back to the HOME screen.

Exponent (pages 19 and 23)

Example: Simplify 3^5.

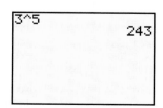

Keystrokes:

[3] [^] [5] [ENTER]

[3] [^] [5] [ENTER]

Format (pages 182 and 185)

Keystrokes:

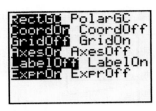

[2ⁿᵈ] [FORMAT]

[GRAPH] [MORE] [F3] (FORMAT)

The format menu determines the way in which graphs are displayed. For this course, the format settings should be highlighted as indicated.

Frac▶ (pages 19 and 23)

Example: Simplify $\dfrac{2}{3} + \dfrac{4}{7} \div \dfrac{6}{7}$.

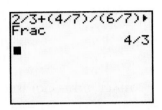

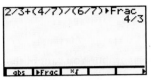

Keystrokes:

[2] [÷] [3] [+] [(] [4] [÷] [7] [)]
[÷] [(] [6] [÷] [7] [)] [MATH]
[1:▶ Frac] [ENTER]

[2] [/] [3] [+] [(] [4] [/] [7] [)]
[/] [(] [6] [/] [7] [)]
[CUSTOM] [F2] (▶ Frac)
[ENTER]

Note: Because of the order of operations, fractions do not have to be enclosed in parentheses when adding or subtracting. However, when multiplying or dividing, it is best to enclose each fraction in parentheses to ensure that the order of operations is followed.

Graph (page 170)

To display a graph, a function must be entered and selected on the Y-EDIT screen. Pressing [GRAPH] will display the graph of the function.

On the TI-86, the [GRAPH] key on the calculator face is used to display the graphing menu. From this menu you can then select the Y-EDIT, WIND (window), ZOOM, TRACE, GRAPH, MATH, DRAW, FORMT (format), and EVAL features.

Integer viewing window
(pages 184 and 186)

Example: Set your viewing window to
ZInteger (i.e., the integer
viewing window).

Keystrokes:

[ZOOM] [8:ZINTEGER]
Move your cursor
to (0, 0) [ENTER]
[WINDOW]

[GRAPH] [F3] [ZOOM]
[MORE] [MORE] [F5] (ZINT)
Move your cursor
to (0, 0) [ENTER] [2ⁿᵈ]
[M2] (WIND)

The purpose of this window is to enable you to have trace coordinates in which the *x*-value is reported in integer values.

Intersect/Isect (page 200)

Example: Find the solution of
$x + 6 = 1 - 4x$.

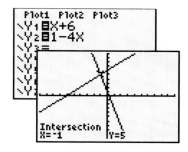

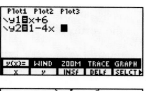

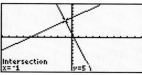

Keystrokes:

[2nd] [CALC] [5:INTERSECT]
Move your cursor near the point of intersection when prompted for first curve and press [ENTER]. *At both the second curve and guess prompts, press* [ENTER] *also.*

With the graph menu displayed, press [MORE] [F1] (MATH) [MORE] [F3] (ISECT). *Move your cursor near the point of intersection when prompted for first curve and press* [ENTER]. *At both the second curve and prompts, press* [ENTER] *also.*

Note: It is the *x*-value that is the solution to the equation.

Lighten (pages 18 and 22)

If the display is not clear, press [2nd] [▼] to lighten the screen.

Matrix (page 758)

Example: Enter A $= \begin{bmatrix} 2 & 5 & -1 \\ 3 & -4 & 4 \end{bmatrix}$.

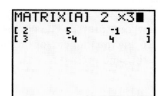

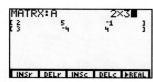

Keystrokes:

[2nd] [MATRIX] [EDIT] [1:A]
Enter the dimension of the matrix: 2 × 3. *When entering each of the matrix entries, pressing* [ENTER] *will automatically move the cursor to the next entry.*

[2nd] [MATRIX] [F2] (edit)] *Enter* A *at the alpha prompt to name the matrix and press* [ENTER]. *Enter the dimension of the matrix:* 2 × 3. *When entering each of the matrix entries, pressing* [ENTER] *will automatically move the cursor to the next entry.*

Max/Min (page 483)

Example: Determine the vertex of the graph of $y = -x^2 + 2x + 8$ using the maximum feature of the calculator.

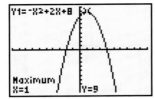

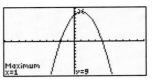

Keystrokes:

[2nd] [CALC] [4:MAXIMUM] *At the left-bound prompt, enter an x-value smaller than the expected coordinate for the maximum:* [(−)] [2] [ENTER]. *At the right-bound prompt, enter an x-value larger than the expected coordinate for the maximum:* [4] [ENTER]. *At the guess prompt, enter a guess for the expected x-coordinate:* [1] [ENTER].

With one line of graph menu displayed, press [MORE] [F1] (MATH) [F5](FMAX). *At the left-bound prompt, enter an x-value smaller than the expected coordinate for the maximum:* [(−)] [2] [ENTER]. *At the right-bound prompt, enter an x-value larger than the expected coordinate for the maximum:* [4] [ENTER]. *At the guess prompt, enter a guess for the expected x-coordinate:* [1] [ENTER].

Mode (pages 18 and 22)

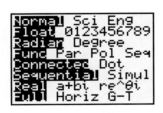

MODE controls how numbers and graphs are displayed and interpreted. The current settings on each row should be highlighted as displayed. The blinking rectangle can be moved using the four cursor (arrow) keys. To change the setting on a particular row, move the blinking rectangle to the desired setting and press [ENTER].

nCr (page 820)

Example: Compute $C(8, 2) = {}_8C_2$.

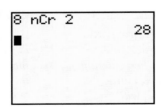

Keystrokes:

[8] [MATH] *Cursor over to* **PRB**. [3:nCr] [2] [ENTER]

[8] [2nd] [MATH] [F2] (PROB) [F3] (nCr) [2] [ENTER]

nPr (page 818)

Example: Compute $P(8, 2) = {}_8P_2$.

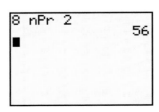

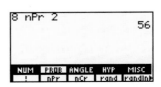

Keystrokes:

[8] [MATH] *Cursor over to* **PRB**. [2:nPr] [2] [ENTER]

[8] [2nd] [MATH] [F2] (PROB) [F2] (nPr) [2] [ENTER]

nth Root $\sqrt[x]{}$ (page 539)

Example: Simplify $3\sqrt[5]{32}$.

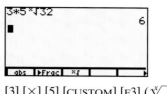

Keystrokes:

[3] [×] [5] [MATH]
[5: $\sqrt[x]{}$] [3] [2] [ENTER]

[3] [×] [5] [CUSTOM] [F3] ($\sqrt[x]{}$) [3]
[2] [ENTER]

Note: It is assumed that the $\sqrt[x]{}$ is the third feature that has been entered on the custom menu of the TI-86.

Quit (page 22)

To return to the home screen from the MODE, TABLE, GRAPH, or FORMAT screens, press [2ⁿᵈ] [QUIT] on both the TI-83 Plus and TI-86. Pressing [2ⁿᵈ] [QUIT] on the TI-86 will also remove both lines of menu from the bottom of the screen display.

Root (page 447)

Example: Find the solution of
$x + 6 = 1 - 4x$ using the
ROOT/ZERO feature.

Note: The equation must first be set equal to zero: $x + 6 - 1 + 4x = 0$.

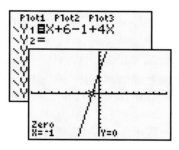

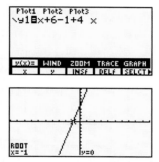

Keystrokes:

[2ⁿᵈ] [CALC] [2:ZERO] *At the left-bound prompt, enter an x-value smaller than the root:* [(−)] [2] [ENTER]. *At the right-bound prompt, enter an x-value larger than the root:* [0] [ENTER]. *At the guess prompt, enter your guesstimate of the root:* [(−)] [1] [ENTER].

[EXIT] *(to display one line of menu)* [MORE] [F1] (MATH) [F1] (ROOT) *At the left-bound prompt, enter an x-value smaller than the root:* [(−)] [2] [ENTER]. *At the right-bound prompt enter an x-value larger than the root:* [0] [ENTER]. *At the guess prompt, enter your guesstimate of* [(−)] [1] [ENTER].

Note: It is the x-value that is the solution/root/zero to the equation.

Scale (Xscl, Yscl) (pages 183 and 186)

The x-scale and y-scale on the window screen determine the space between the tic marks on the axes. Unless otherwise noted, it should be assumed that in this text all graphs have xscl = 1 and yscl = 1.

Standard viewing window (page 170)

Example: Set your viewing window
to the standard viewing
window.

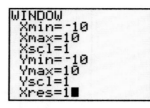

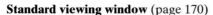

Keystrokes: [ZOOM] [6:ZSTANDARD]   [GRAPH] [F3] (ZOOM)
                                     [WINDOW]   [F4](ZSTD) [2nd]
                                                                      [M2](WIND)

Note: The information on this screen indicates that, in a rectangular coordinate system, the *x*-values will range from -10 to 10 and the *y*-values will range from -10 to 10, which is $[-10, 10]$ by $[-10, 10]$ in interval notation.

STO ➤ (pages 32 and 33)

Example: Evaluate $3x + 5$ for $x = -3$.

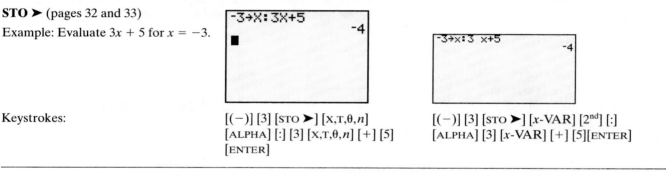

Keystrokes: [(−)] [3] [STO ➤] [X,T,θ,*n*]   [(−)] [3] [STO ➤] [*x*-VAR] [2nd] [:]
                                     [ALPHA] [:] [3] [X,T,θ,*n*] [+] [5] [ALPHA] [3] [*x*-VAR] [+] [5][ENTER]
                                     [ENTER]

Table (page 156)

Example: Generate a table of values
for $y = -3x^2$ beginning at
$x = -2$.

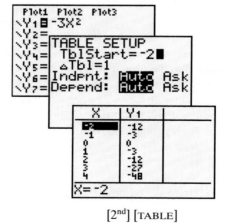

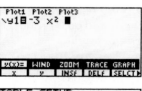

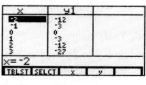

Keystrokes: [2nd] [TBLSET]   [TABLE] [F2](TBLST)

                          [2nd] [TABLE]   [F1](TABLE)

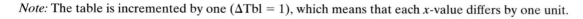

Note: The table is incremented by one (ΔTbl $= 1$), which means that each *x*-value differs by one unit.

Test (page 584)

Example: Use the TEST menu to determine if $\frac{6}{5}$ is greater than $\sqrt{3}$.

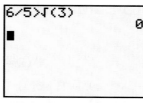

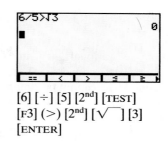

Keystrokes:

[6] [÷] [5] [2ⁿᵈ] [TEST]
[3: >] [2ⁿᵈ][√⁻] (] [3] [)]
[ENTER]

[6] [÷] [5] [2ⁿᵈ] [TEST]
[F3] (>) [2ⁿᵈ] [√⁻] [3]
[ENTER]

When the TEST menu returns a value of 1, the statement is true; when a value of 0 is returned, the statement is false.

Value (page 192)

Example: Find the value of y when $x = -3$ on the graph of $y = 3x + 5$.

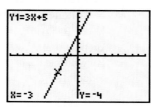

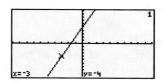

Keystrokes:

[2nd] [CALC] [1:VALUE]
Enter -3 *at the blinking cursor.* [ENTER]

Press [EXIT] *so that only one line of graph menu is displayed.*
[MORE] [MORE] [F1](EVAL) *Enter* -3 *at the blinking cursor.* [ENTER]

Vars (page 474)

The functions stored on the Y-EDIT screen can be accessed via the VARS menu and used either on the HOME screen or on the Y-EDIT screen, depending on the nature of the problem.

TI-83 Plus Keystrokes: [VARS] *cursor over to* [Y-VARS] [1:function] [1:Y1] (*if you want the function stored at Y1*). *The function will now be displayed on either the home screen or the y-edit screen, the location being determined by your location when you accessed* [VARS].

TI-86 Keystrokes: *Type in the function at the desired location using the alpha and numeric keys. Be sure to enter the "y" as a lower case letter:* [2ⁿᵈ] [ALPHA] [y] [1].

[X,T,θ,n]/[x-VAR] (pages 32 and 33)

Because x is used as a variable so often in algebra, it has its own key on the calculator. When the calculator is in function MODE, the variable x will be displayed when [X,T,θ,n] is pressed on the TI-83 Plus or when [x-VAR] is pressed on the TI-86.

Note: The x-variable will be displayed in lower case on the TI-86, but it will be displayed in upper case on the TI-83 Plus screens in this text.

Exercise 1.1 (page 13)
1. natural 2. whole 3. rational 4. prime 5. identity, identity 6. multiplicative inverse 7. zero
8. a, c 9. a. undefined b. cannot be simplified
10. b, c 11. a. 10 b. 16 c. 45 12. Answers may vary. 13. a, b 14. Answers may vary. 15. No, it is a rounded decimal. 16. Enter 4 + 17/19 ►Frac
17. Answers may vary. 19. $\frac{5}{2}$ 21. $\frac{26}{21}$ 23. $\frac{3}{2}$
25. $\frac{21}{4}$ 27. $\frac{1,147}{18}$ 29. $\frac{3}{5}$ 31. $\frac{5}{2}$ 33. $\frac{1}{4}$ 35. $\frac{2}{7}$ 37. $\frac{7}{25}$
39. $\frac{13}{66}$ 41. $\frac{4}{9}$ 43. $\frac{5}{2}$ 45. $\frac{659}{24}$ 47. $\frac{741}{1,000}$ 49. 4,755
51. a. $\frac{1}{5} \cdot 750 = 150$ b. $\frac{4}{5} \cdot 750 = 600$ 53. $\frac{13}{25} + \frac{9}{50} = \frac{7}{10}$
55. $1 - \left(\frac{3}{100} - \frac{2}{5}\right) = \frac{57}{100}$

Exercise 1.2 (page 30)
1. exponent 2. variables 3. base, exponent
4. transitive 5. arithmetic 6. algebraic 7. a. $x \cdot x$
b. $3 \cdot z \cdot z \cdot z \cdot z$ c. $5 \cdot x \cdot 5 \cdot x$ d. $5 \cdot 2 \cdot x \cdot 2 \cdot x \cdot 2 \cdot x$
8. Answers may vary. 9. Answers may vary. 10. larger numbers 11. Smaller numbers would be produced.
12. { } and [ ] 13. Answers may vary. 15. 25 17. 343
19. 32 21. 256 23. 24 25. 29 27. 44 29. 12
31. 27 33. 20 35. 2 37. 32 39. 14 41. 100 43. 10
45. 19 47. 9 49. 2 51. $\frac{2}{21}$ 53. 10 55. 16 57. 8
59. 43 61. 15 63. 169 sq. units 65. 216 cubic units
67. 31,208 students 69. $8 71. $\frac{35}{24}$ 73. $\frac{61}{24}$

Exercise 1.3 (page 42)
1. absolute value 2. inverses 3. associative 4. double negative 5. a. 4 b. -6 c. -5 6. a. 15 b. -15
c. 7 d. -7 7. a. 4 b. -12 c. 12 d. -4 8. Answers may vary. 9. Answers may vary. 10. Answers may vary. 11. 2 13. -12 15. 0.5 17. $\frac{12}{35}$ 19. $-\frac{1}{12}$
21. 7 23. -1 25. -7 27. -8 29. -18 31. 1.3
33. $\frac{1}{2}$ 35. $-\frac{22}{5}$ 37. 4.2 39. 10 41. 0 43. 8 45. 64
47. 3 49. 4 51. 3 53. -3 55. 13 57. 1 59. $\frac{1}{9}$
61. -15.8 63. -7.1 65. $-4°$ 67. 2,000 69. 0 yards gained or lost 71. 18° difference 73. $83,425.57
75. $3\frac{13}{20}$ tons 77. $\frac{7}{6}, \frac{27}{40}$ 79. Answers may vary.
81. Answers may vary. 83. 47 85. 46
87. $\frac{90 + 89 + 82 + 76 + 7}{5} = 81.6$

Exercise 1.4 (page 51)
1. product 2. quotient 3. undefined 4. Answers may vary. 5. Divisors and factors represent the same values.
6. a. 36 b. -36 c. -36 d. 36 7. Answers may vary. 8. a. 2 b. -2 c. -2 d. 2 9. Answers may vary. 10. Answers may vary. 11. 0 12. Yes, because

only a negative number raised to an odd power would produce a negative result. 13. 16 15. 2 17. 72
19. -24 21. 2 23. 1 25. 1 27. -3 29. -1 31. 4
33. 9 35. -54 37. -27 39. $\frac{3}{2}$ 41. 6 43. $-\frac{4}{3}$
45. -13 47. $-\frac{1}{6}$ 49. $-\frac{11}{12}$ 51. $-\frac{7}{36}$ 53. $-\frac{11}{48}$
55. 1, 3, or 5 57. $30 59. 15 weeks 61. a loss of 0.8 of a million yearly. 63. $2.65 65. $\frac{2}{15}$ 67. a. $-\frac{5}{2}$ b. $\frac{3}{2}$

Exercise 1.5 (page 60)
1. irrational 2. perimeter 3. area 4. circumference
5. volume 6. percent 7. a. 3 b. 0, 3 c. -3 d. -3, $-\frac{1}{2}$, 0, 3, $5.\overline{6}$ e. $\sqrt{2}$ 8. a. volume b. area c. area
d. perimeter e. area f. volume g. perimeter
h. volume i. area 9. false 10. π as displayed on the calculator is more accurate because it displays more decimal places than 3.14. 11. a. 0.05 b. 0.045 c. 0.0875
13. 26 cm 15. 16 in. 17. 16 cm^2 19. 228 cm^2
21. ≈ 65.97344573 cm 23. ≈ 153.93804 m^2 25. 36 ft^3
27. 527.788 in^3 29. 96 in^3 31. 2,704 in^2 33. 8,856 in^3
35. 31.81 in^2 37. No 39. $247.50 41. $2,144.88
43. a. 2° b. 50° 45. a. $57.50 b. $14.38 47. 12,480 people bank by phone. 49. Approximately 52 students do not have snacks. 51. $\approx$ $780.69 53. The second contractor has the lower bid. 55. a. -9 b. 9 c. 9
d. 36 57. $1\frac{2}{9}$ gallons

Chapter Summary (page 65)
1. a. $\frac{10}{21}$ b. $\frac{11}{21}$ c. $\frac{1}{3}$ d. 1 e. $\frac{49}{6}$ or $8\frac{1}{6}$ f. $\frac{67}{15}$ or $4\frac{7}{15}$
g. $\frac{91}{6}$ or $15\frac{1}{6}$ h. $\frac{30}{13}$ or $2\frac{4}{13}$ 2. a. $\frac{2}{15}$ b. 1st day: 120,
2nd day: 400, 3rd day: 80 3. a. 81 b. $\frac{4}{9}$ 4. a. 6 b. 3
c. 38 d. 3 e. $\frac{23}{12}$ f. 15 5. a. $\left(-\frac{1}{4}\right)^3$ b. $\frac{4 - 8}{-12 + 1}$
c. $4 - \frac{8}{-12} + 1$ 6. 6.8 hours or 6 hours 48 minutes
7. a. -7 b. 1 c. $-\frac{1}{7}$ d. 1 e. $\frac{1}{4}$ f. $-\frac{2}{3}$ 8. a. 6
b. $\frac{3}{2}$ c. 7 d. 26 9. -4 10. -1 11. -2 12. 5
13. 4 14. 6 15. -6 16. 3 17. 2 18. 6 19. -7
20. 39 21. 6 22. -2 23. -8 24. 1 25. 40.2 feet
26. 15,133.576 ft^3 27. 35° 28. 64 ft^2

Chapter Test (page 69)
1. 14 2. 32 3. -27.4 4. 1 5. $\frac{4}{5}$ 6. $\frac{9}{2}$ or $4\frac{1}{2}$ 7. -1
8. $-\frac{1}{13}$ 9. 301.57 sq. ft 10. 64 cm^2 11. 1,539.380 in^3
12. $190 13. 77,550 homes 14. 3,889.625 in^3 15. -2
16. -14 17. -4 18. 12 19. 5 20. -23

Exercise 2.1 (page 80)
1. base, exponent 2. a. x^{m+n} b. x^{mn} c. $x^n y^n$ d. $\frac{x^n}{y^n}$
e. x^{m-n} f. 1 3. base: 4, exponent: 3 4. base: -5, ex-

ponent: 2   **5.** base: $-x$, exponent: 2   **6.** base: x, exponent: 4   **7.** base: x, exponent: 1   **8.** base: xy, exponent: 1   **9.** base: x, exponent: 3   **10.** base: y, exponent: 6   **11.** 2^3 **12.** 5^2 **13.** x^5 **14.** y^6 **15.** $(2x)^3$ **16.** $(-4y)^2$ **17.** -24 **19.** 1,296   **21.** x^7 **23.** x^{10} **25.** y^9 **27.** $12x^7$ **29.** $-4y^5$ **31.** 3^8 (6,561)   **33.** y^{15} **35.** a^{21} **37.** x^{25} **39.** $4x^{31}$ **41.** x^3y^3 **43.** r^6s^4 **45.** $16a^2b^4$ **47.** $-8r^9s^9t^3$ **49.** $\frac{a^3}{b^3}$ **51.** $\frac{x^{10}}{y^{15}}$ **53.** $\frac{-32a^5}{b^5}$ **55.** $\frac{b^6}{27a^3}$ **57.** x^2 **59.** a^5 **61.** x^3y **63.** c^7 **65.** $5a^2b^2$ **67.** $8x^4y^7$ **69.** ab^4 **71.** $\frac{10r^{13}s^3}{3}$ **73.** $\frac{y^3}{8}$ **75.** 2   **77.** -2 **79.** Answers may vary.   **81.** $15x^3$ **83.** $9\pi x^4$ **85. a.** -9 **b.** 30   **c.** -3 **d.** -3 **e.** -3 **f.** 3   **87. a.** 0.45   **b.** 0.03   **c.** 0.025

Exercise 2.2 (page 90)

1. scientific notation   **2.** false   **3.** false   **4.** true   **5.** true   **6.** true   **7.** $\frac{1}{x^n}$ **8. a.** $\frac{1}{8}$ **b.** $\frac{1}{3}$ **c.** 8   **d.** 3   **9. a.** x^3 **b.** $\frac{1}{x^{28}}$ **c.** $\frac{1}{x^3}$ **d.** $\frac{y^7}{x^7}$ **e.** $\frac{1}{x^3y^3}$ **f.** $\frac{y^{12}}{x^8}$ **10. a.** 4,200 and 0.0042   **b.** 151.3 and 0.01513   **11.** 8   **13.** $\frac{1}{16}$ **15.** 1   **17.** $\frac{1}{x^2}$ **19.** $\frac{1}{16y^4}$ **21.** $\frac{1}{a^3b^6}$ **23.** y^{13} **25.** $\frac{1}{x^3}$ **27.** $\frac{y^{10}}{-32x^{15}}$ **29.** a^{14} **31.** $\frac{1}{x^{10}y^4}$ **33.** $\frac{256x^{28}}{81}$ **35.** $\frac{16y^{14}}{z^{10}}$ **37.** $\frac{u^{20}v^8}{16}$ **39.** $\frac{x^{16}}{y^2z^2}$ **41.** 2.3×10^4 **43.** 6.2×10^{-2} **45.** 4.25×10^3 **47.** 2.5×10^{-3} **49.** 230   **51.** 0.00114   **53.** 25,000,000   **55.** 0.00051   **57. a.** 4.2×10^{-5} **b.** 0.000042   **59. a.** $-1.0000001 \times 10^{-11}$ **b.** $-0.000\,000\,000\,010\,000\,001$ **61.** 7.14×10^5; 714,000   **63.** 3×10^4; 30,000   **65.** 1.44×10^{11}; 144,000,000,000   **67.** 2.57×10^{13} **69.** 3.31×10^4 **71.** 114,000,000   **73.** 1.9008×10^{11} ft   **75.** 1.7×10^{-18} g   **77.** -4 **79.** \$2,013.65

Exercise 2.3 (page 98)

1. algebraic expressions   **2.** constants   **3.** algebraic terms   **4.** numerical coefficient   **5.** polynomial   **6.** monomial, binomial, trinomial   **7.** degree, degree, degree   **8. a.** product   **b.** difference   **c.** sum   **d.** quotient   **9.** b, d   **10. a.** 5   **b.** 6   **c.** 3   **d.** 5   **e.** 9   **11. a.** trinomial   **b.** 2   **c.** -5 **12. a.** 2 terms: 3   **b.** 4 terms: 1   **c.** 3 terms: 7   **d.** 3 terms: -4 **13.** $x = -2$, $P(x) = -1$ **15.** xy **17.** $2x + 2y$ **19.** $\frac{y}{x}$ **21.** $y - x$ **23.** Answers may vary.   **25.** Answers may vary.   **27.** Answers may vary.   **29.** binomial   **31.** monomial   **33.** none of the above   **35.** monomial   **37.** 5   **39.** 5   **41.** 6   **43.** 10   **45.** 1   **47.** 0   **49.** 3   **51.** 11   **53.** $t^2 - 2t + 3$ **55.** $4t^2 - 4t + 3$ **57.** $-\frac{3}{2}$ **59.** $-5v^4 - 2$ **61.** $50x^7 - 2$ **63.** $5x - 5h - 2$ **65.** $5x - 5h$ **67.** $-15r + 10s - 2$ **69.** $m + 25,000$ **71.** $35,000a$ **73.** $\frac{12}{x}$ **75. a.** $2x - 8$ **b.** no   **c.** Explanations may vary.   **77.** 192 feet   **79.** 72 feet   **81. a.** 9   **b.** -16

Exercise 2.4 (page 104)

1. distributive   **2.** like   **3.** like; $7y$ **4.** like; $8x^2$ **5.** unlike   **6.** unlike   **7.** like; $13x^3$ **8.** like; $2y^4$ **9.** like; 0   **10.** unlike   **11.** $9y$ **12.** x **13.** $-12t^2$ **14.** $25x^2$ **15.** $16u^3$ **16.** $18xy^2$ **17.** $7x^5y^2$ **18.** $-5x^6y$ **19.** $14rst$ **21.** $-6a^2bc$ **23.** $-2mn^3 + 3m^3n$ **25.** $20x$ **27.** $13x + 5y$ **29.** $7x + 4$ **31.** $2a + 7$ **33.** $6x^2 + x - 5$ **35.** $2a^2 + a - 3$ **37.** $12x + 12y$ **39.** $-9a + 9b$ **41.** $7y^2 + 22y$ **43.** $xy^2 + 13y^2$ **45.** $20x - 40$ **47.** $2x^2 + 2x - 8$ **49. a.** 140   **b.** 13, 9, 57   **c.** Yes, answers may vary.   **51. a.** 4   **b.** $0, 3, 7, -6$ **c.** No, answers may vary.   **53.** $6x^2 - 2x - 1$ **55.** $t^3 + 3t^2 + 6t - 5$ **57.** $-3x^2 + 5x - 7$ **59.** $75 + 10x$ **61.** $0.015x + 75$ **63.** $117x + 144$ **65.** $43x + 2$ **67.** $4 \cdot 4 \cdot 4$ **69. a.** 0.0975   **b.** \$29.25   **c.** Answers may vary.

Exercise 2.5 (page 112)

1. conjugate   **2.** distributive   **3.** commutative, associative   **4.** b   **5. a.** $6x^2$ **b.** $4x^3$ **c.** $\frac{2}{5}x^2$ **6. a.** $3x + 120$ **b.** $-3a + 6$ **c.** $-4t - 28$ **d.** $6x^2 - 18$ **7.** $12x^5$ **9.** $-24b^6$ **11.** $x^{10}y^{15}$ **13.** $a^5b^4c^7$ **15.** $3x^2 - 6x$ **17.** $-6x^4 + 2x^3$ **19.** $6x^4 + 8x^3 - 14x^2$ **21.** $2x^7 - x^2$ **23.** $a^2 + 9a + 20$ **25.** $3x^2 + 10x - 8$ **27.** $6a^2 + 2a - 20$ **29.** $6x^2 - 7x - 5$ **31.** $6t^2 + 7st - 3s^2$ **33.** $x^2 + 8x + 16$ **35.** $x^2 + 10x + 25$ **37.** $r^2 - 16$ **39.** $16x^2 - 25y^2$ **41.** $4a^2 - 12ab + 9b^2$ **43.** $2x^2 - 6x - 8$ **45.** $4x^3 + 11x^2 + 18x + 9$ **47.** $x^3 - 8y^3$ **49.** $5t^2 - 11t$ **51.** $x^2y + 3xy^2 + 2x^2$ **53.** $2x^2 + xy - y^2$ **55.** $8x$ **57.** $5s^2 - 7s - 9$ **59.** $5x + x^2$ **61.** $x^3 - 4x$ **63.** $x^3 + 6x^2 + 8x$ **65.** $2x^2 + 4x + 4$ **67.** $6x^4 - 11x^3 - 22x^2 - 17x + 5$ **69.** 32.5, 65, 97.5, 130   **71.** $70t + 60t$ or $130t$ miles

Exercise 2.6 (page 117)

1. $4x$ **2.** $\frac{x}{3}$ **3.** 1   **4.** $\frac{a}{c}$ **5.** $x + 3y$ **6.** $2x - y$ **7.** $7 + 9y$ **8.** $5a - 6c$ **9.** $\frac{r^2}{s}$ **11.** $\frac{2x^2}{y}$ **13.** $-\frac{3u^3}{v^2}$ **15.** $\frac{4r}{y^2}$ **17.** $-\frac{13}{3rs}$ **19.** $\frac{x^4}{y}$ **21.** a^8b^8 **23.** $-3r$ **25.** $-\frac{x^6}{4y^3}$ **27.** $\frac{125}{8}$ **29.** $\frac{xy^2}{3}$ **31.** a^8 **33.** z^{15} **35.** $\frac{2}{y} + \frac{3}{x}$ **37.** $\frac{1}{5y} - \frac{2}{5x}$ **39.** $\frac{1}{y^2} + \frac{2y}{x^2}$ **41.** $3a - 2b$ **43.** $\frac{1}{y} - \frac{1}{2x} + \frac{2z}{xy}$ **45.** $3x^2y - 2x - \frac{1}{y}$ **47.** $5x - 6y + 1$ **49.** $\frac{10x^2}{y} - 5x$ **51.** $\frac{2}{3} + \frac{3x^2}{2}$ **53.** $xy - 1$ **55.** $\frac{x}{y} - \frac{11}{6} + \frac{y}{2x}$ **57.** 2   **59.** $2x + 3$ **61.** $2x + \frac{1}{2} - \frac{3}{x^2}$ **63.** $3x - \frac{2}{x} + \frac{4}{3x^3}$ **65.** No, answers may vary.   **67.** Yes, answers may vary.   **69. a.** $2x - 5$ **b.** 1

Exercise 2.7 (page 124)

1. a. $3x^2$ **b.** $4x$ **c.** -2 **d.** $3x^2 + 4x - 2$ **2. a.** $10x^2 + 11x + 3$ **b.** $2x^2 - x - 21$ **c.** $6x^2 + xy - 2y^2$ **d.** $3x^2 + 13xy - 10y^2$ **3. a.** $x^2 + 0x - 9$ **b.** $x^3 + 0x^2 + 0x - 8$ **5.** $x - 3$ **7.** $z - 4$ **9.** $a - b$

11. $4a + 3$ **13.** $b - 1$ **15.** $3x - y$ **17.** $x + 3$
19. $3x + 1$ **21.** $2x - y$ **23.** $5x - 4y$ **25.** $x^2 + 2x - 1$
27. $2x^2 + 2x + 1$ **29.** $x^2 + xy + y^2$ **31.** $x + 1 + \frac{-1}{2x + 3}$
33. $2x + 2 + \frac{-3}{2x + 1}$ **35.** $x^2 + 2x + 1$
37. $x^2 + 2x - 1 + \frac{6}{2x + 3}$ **39.** $2x^2 + 8x + 14 + \frac{31}{x - 2}$
41. $x + 1$ **43.** $2x - 3$ **45.** $x^2 - x + 1$
47. $a^2 - 3a + 10 + \frac{-30}{a + 3}$ **49.** $5x^2 - x + 4 + \frac{16}{3x - 4}$
51. $x^2 - 5x + 4$ **53.** $9x^3 - 2x^2 + 3$ **55. a.** $x + 7$
b. $4x + 24$ **57.** $x + 2$ **59.** $6x^3 + 3x^2 - 18x$ **61. a.** $64x^6$
b. $\frac{x}{2y}$ **c.** $\frac{1}{54x}$ **63.** $x + 4, 2x - 5, 4x$, sum $= 7x - 1$

Chapter 2 Summary (page 126)

1. a. 125   **b.** 243   **c.** 64   **d.** -64 **e.** 13   **f.** 25   **g.** 162
h. 18   **2. a.** x^5 **b.** x^3y **c.** y^{10} **d.** y^5 **e.** $2b^{12}$ **f.** $-z^5y^2$
g. $256xs^2$ **h.** $-3y^6$ **3. a.** x^{15} **b.** $4x^4y^2$ **c.** 9   **d.** 1
e. x^4 **f.** x^2y^2 **g.** y^2 **h.** $5yz^4$ **4. a.** $-z^5y^2$ **b.** $4x^3$
c. x^{15} **d.** $4x^4y^2$ **5. a.** $\frac{x^2}{y^2}$ **b.** $\frac{1}{8x^3}$ **c.** x **d.** y **6. a.** x^{10}
b. x^{14} **c.** $\frac{1}{x^5}$ **d.** $\frac{y^6}{9x^8}$ **7. a.** 7.28×10^2 **b.** 9.42×10^{-3}
c. 1.8×10^{-4} **d.** 6×10^{-1} **8. a.** 726,000   **b.** 0.000391
c. 160   **9. d.** 0.000 000 4   **10.** 2,172.1   **11. a.** 7; monomial   **b.** 2; binomial   **c.** 5; trinomial   **d.** 3; binomial
12. a. 402   **b.** 0   **c.** 82   **d.** $80t^4 - 2t$ **13. a.** $a - c$
b. $b + c$ **c.** ab **14.** $\$10,000 - b$ **15.** $55(2); 55(5); 55(b)$
16. a. $7x$ **b.** simplified terms   **c.** $4x^2y^2$ **d.** x^2yz
e. $5x^2$ **f.** $5x + 35$ **g.** $4a^2 + 4a - 6$
h. $6x^3 + 8x^2 + 3x - 72$ **17.** $6x^2 + x + 8$
18. $4x^2 + 13x - 8$ **19. a.** $10x^3y^5$ **b.** $5x + 15$
c. $3x^4 - 5x^2$ **d.** $-3x^2y^2 + 3x^2y$ **e.** $x^2 + 5x + 6$
20. a. $6a^2 - 6$ **b.** $a^2 - 1$ **c.** $6y^2 - 24$ **d.** $x^2 + 8x + 16$
e. $x^2 - 6x + 9$ **21.** $3x^3 + 7x^2 + 5x + 1$ **22.** $y^4 - 1$
23. a. $\frac{3}{2y} + \frac{3}{x}$ **b.** $2 - \frac{3}{y}$ **c.** $-3a - 4b + 5c$ **d.** $-\frac{x}{y} - \frac{y}{x}$
24. a. $x - 5$ **b.** $x + 1 + \frac{3}{x + 2}$ **c.** $x + 5 + \frac{3}{3x - 1}$
d. $3x^2 - x - 4$

Chapter 2 Test (page 129)

1. $2x^3y^4$ **2.** y^6 **3.** $32x^9$ **4.** $\frac{1}{8r^{18}}$ **5. a.** 3   **b.** 1   **6.** binomial   **7.** 10   **8.** 0   **9.** $-3x^2y^2$ **10.** $5x^3 + 2x^2 + 2x - 5$
11. $-x^2 - 5x + 4$ **12.** $-7x + 2y$ **13.** $-3x + 6$
14. $-4x^5y$ **15.** $3y^4 - 6y^3 + 9y^2$ **16.** $6x^2 - 7x - 20$
17. $2x^3 - 7x^2 + 14x - 12$ **18.** $\frac{y}{2x}$ **19.** $\frac{a}{4b} - \frac{b}{2a}$
20. $x - 2$

Cumulative Review: Chapters 1 and 2 (page 130)

1. prime   **2.** Answers may vary. 72   **3.** Answers may
vary.   **4.** Answers may vary.   **5. a.** 3   **b.** -3 **c.** -3
6. double negative   **7.** Answers may vary.   **8.** b, c
9. a. 0, 6   **b.** $-5, -1, 0, 6$ **c.** $-5, -\frac{7}{2}, -1, 0, 4.2, 6$
d. $\sqrt{2}$ **e.** 0, 6   **10. a.** perimeter   **b.** volume
c. volume   **d.** area   **11. a.** x^{m+n} **b.** x^{mn} **c.** $x^n y^n$
d. $\frac{x^n}{y^n}$ **e.** x^{m-n} **f.** 1   **12.** base: x, exponent: 5

13. numerical coefficient   **14.** binomials   **15.** $15 + x$
16. distributive   **17.** b   **18.** conjugate, $16 - x^2$
19. $\frac{4x}{3} + 4$ **20.** $-x^2 - 3x + 4$ **21.** $\frac{1}{8}$ **22.** $\frac{2}{3}$
23. 61   **24.** 8   **25.** -9 **26.** $-\frac{28}{9}$ **27.** 5   **28.** -1
29. 13   **30.** $-\frac{7}{5}$ **31. a.** $-3x^5$ **b.** x^{18} **32. a.** $\frac{1}{y^6}$
b. x^2 **33. a.** xy **b.** $2x + 4$ **34. a.** 2nd degree
b. 22   **35.** $2x^2 + 2x + 12$ **36.** $2x - 2y$
37. $3x^3 - 4x^2 - 12x + 5$ **38.** $-2x - 5$ **39.** $\frac{6x^3}{y} - 5y + 1$
40. $4x^2 - 5x + 4 + \frac{-6}{2x + 1}$ **41.** 81   **42.** $x = -1$ and
$P(x) = 3$ **43.** $(4 + 8)/3$ **44.** 9,097   **45.** $\frac{4,019}{7}$
46. $\frac{1}{2(5 + 7)}$ **47. a.** 20   **b.** 5   **48.** $\$4.50$
49. $x^2 + x + 2$ **50.** $9x + 3$

Exercise 3.1 (page 146)

1. linear   **2.** roots, solutions   **3.** equivalent   **4.** solve
5. symmetric   **6.** identity   **7.** no   **8.** conditional
equation   **9.** $\varnothing$ **10.** $\mathbb{R}$ **11. a.** $6(a - 2) = 6.6; \{3.1\}$
b. $2(5a + 4) = 2; \{-0.6\}$ **c.** $2(t - 5) = 3(2t + 5) - 6; \{-\frac{19}{4}\}$
d. $6x + 4 = 3x - 14; \{-6\}$ **e.** $2x - 5 = x + 3; \{8\}$
f. $4(t + 1) - 3 = 3t + 2(t - 5); \{11\}$ **12.** $x + 1, x + 2$
13. $x + 2, x + 4$ **14.** $x + 2, x + 4$ **15. a.** $x + 4$ **b.** $x - 7$
16. a. $2x - 3$ **b.** $\frac{1}{2}x + 7$ **17.** Yes, because when $x = 1$
both sides of the equation have the same value.   **19.** Yes,
because when $x = 2$ both sides of the equation have the
same value.   **21.** Yes, because when $x = 0$ both sides of
the equation have the same value.   **23.** Yes, because when
$k = 3$ both sides of the equation have the same value.
25. Yes, because when $t = -9$ both sides of the equation
have the same value.   **27.** Yes, because when $x = 0$ both
sides of the equation have the same value.   **29.** $\{1\}$
31. $\{9\}$ **33.** $\{6\}$ **35.** $\{-11\}$ **37.** $\{-6\}$ **39.** $\varnothing$ **41.** $\{3\}$
43. $\mathbb{R}$ **45.** $\{\frac{3}{4}\}$ **47.** $\{37\}$ **49.** $\{-9\}$ **51.** $\{-3\frac{3}{4}\}$ or $\{-\frac{15}{4}\}$
53. $\{-82\}$ **55.** $\{9\}$ **57.** $\mathbb{R}$ **59.** $\{-5\}$ **61.** $\{16\}$
63. $\{-2\}$ **65.** $\{5\}$ **67.** $\{2\}$ **69. a.** $28 = 4(2x + 3)$
b. $28 = 8x + 12$ **c.** $x = 2$ **d.** Each side is 7 in. long.
71. a. $20 = 2(x - 4) + 2x$ **b.** $20 = 4x - 8$ **c.** $x = 7$
d. The window measures 3 ft by 7 ft.   **73.** 26, 28   **75.** 5
77. 6   **79.** 7   **81. a.** $\frac{7}{15}$ **b.** $-\frac{3}{4}$ **c.** $-\frac{8}{15}$ **83.** $x^{16}y^6$
85. $2x^5 - 8x^3y$

Exercise 3.2 (page 152)

1. literal equations   **2.** $A = K - 32$; subtraction property
of equality   **3.** $P - a - b = c$; subtraction property of
equality   **4.** $\frac{P}{4} = s$; division property of equality
5. $\frac{d}{r} = t$; division property of equality   **6.** $\frac{E}{R} = I$; division
property of equality   **7.** $\frac{i}{pt} = r$; division property of
equality   **8.** $\frac{V}{lw} = h$; division property of equality
9. $\frac{C}{2\pi} = r$; division property of equality   **10.** $\frac{P}{l^2} = R$; division property of equality   **11.** $y = x - 10$; subtraction
property of equality   **12.** $y = 2x$; multiplication property
of equality   **13.** $y = \frac{8x}{3}$; division property of equality

14. $y = \frac{16x}{5}$; division property of equality   **15.** $\frac{P - 2l}{2} = w$
17. $\frac{2A}{B + b} = h$ **19.** $R = \frac{E}{I}$ **21.** $y = 8 - 2x$
23. $y = -6 + x$ **25.** $y = 3 - \frac{3}{5}x$ **27.** $\frac{10}{3} - \frac{1}{6}x = y$
29. $\frac{15}{2} - \frac{3}{4}x = y$ **31.** $-\frac{2}{3}x - \frac{11}{3} = y$ **33.** $y = \frac{8}{3} - \frac{2}{3}x$
35. $t = 3$ **37.** $t = 2$ **39.** $c = 3$ **41.** $h = 8$
43. $I = \frac{E}{R}$; 4 amps   **45.** $r = \frac{C}{2\pi}$; $r \approx 2.28$ ft
47. $R = \frac{P}{I^2}$; $R \approx 13.78$ ohms   **49.** $m = \frac{Fd^2}{GM}$
51. $D = \frac{L - 3.25(r + R)}{2}$; $D = 6$ ft   **53.** -6 **55.** $2x$

Exercise 3.3 (page 162)
1. two   **2.** ordered pairs   **3.** incremented   **4.** set
5. function   **6.** domain   **7.** range   **8.** No, because
$-2 \neq 2(-2)^3 - 5$ **9.** No, because $3(2) - 4(2) \neq 2$
10. Yes, because $3(3) - (-5) = 14$ **11.** Yes, because
$5(-2) + 2(6) = 2$ **12.** No, the ordered pair $(0, 2)$ could
only be a solution if it satisfied the equation.   **13.** No,
the only way both (a, b) and (b, a) could be solutions
would be if the equation was $x + y = c$.   **14.** $(-2, -14)$,
$(-1.5, -13)$, $(-1, -12)$, $(-0.5, -11)$, $(0, -10)$,
$(0.5, -9)$, $(1, -8)$ **15.** Yes, it is a function
because each x value is paired with a unique y value.
16. $\{-2, -1.5, -1, -0.5, 0, 0.5, 1\}$
17. $\{-14, -13, -12, -11, -10, -9, -8\}$
19. $(-1, -4)$, $(5, 2)$ **21.** $(-5, \frac{1}{3})$ $(-8, \frac{4}{3})$
23. $(\frac{1}{4}, 0)$ $(-\frac{3}{4}, 4)$ **25.** $(10, 80)(-5, 5)$
27. $(2, 48)(-6, 48)$ **29.** $(-1, \frac{2}{3})$ $(2, 14\frac{2}{3})$
31. $(-2, 8)(10, -1000)$ **33.** $(1, -\frac{1}{5})$ $(-1, -\frac{9}{5})$
35. $2, \frac{7}{5}, \frac{4}{5}$ **37.** $22, 14, 10$ **39.** function, D: $\{-4, 5, 3\}$;
R: $\{5, -4\}$ **41.** No, D: $\{4\}$; R: $\{3, 5, 7\}$ **43.** function,
D: $\{-1, 0, 1, 2, 3, 4, 5\}$; R: $\{-8, -5, -2, 1, 4, 7, 10\}$
45. function, D: $\mathbb{R}$; R: $\mathbb{R}$ **47.** function, D: $\mathbb{R}$; R: $\mathbb{R}$
49. $5, 8.75, 12.50, 16.25$ **a.** \$8.75   **b.** \$16.25
c. $y = 0.50x + 5$ **d.** \$12.50   **51. a.** \$97,000
b. 8 yr   **53. a.** \$300   **b.** \$750   **c.** 6   **55.** 57.50,
76.25, 207.5, 235.63   **57.** $y = -800x + 4,000$; \$800
59. $y = \$5.50x, \$19.25$ **61. a.** 2nd degree   **b.** 44
63. $V = 9x^3 + 15x^2 + 6x$ cubic units   **65.** 24

Exercise 3.4 (page 175)
1. rectangular coordinate system   **2.** x, y **3.** origin
4. quadrants   **5.** abscissa   **6.** ordinate   **7.** standard
8. standard   **9.** scale   **10.** $-47, 47, -31, 31$ **11. a.** I
b. II   **c.** III   **d.** IV   **12. a.** y-axis   **b.** y-axis   **c.** x-axis
d. x-axis   **e.** x- and y-axes   **f.** x-axis   **13. a.** $(2, -2)$
b. $(-5, -5)$ **c.** $(-4, 0)$ **d.** $(0,0)$ **e.** $(-3, 5)$
f. $(-2, -3)$ **g.** $(4, -5)$ **h.** $(2, 3)$ **14.** $(1988, 51.2)$.
In 1998, 51.2 teens in every 1,000 nonsmokers became
regular smokers; $(1996, 77.0)$. In 1996, 77 teens in every
1,000 nonsmokers became regular smokers.   **15.** $(2001,$
$1,035)$ **16.** b   **17. a.** $[-5, 5]$ by $[-5, 5]$ **b.** $[-5, 10]$ by
$[-10, 5]$ **c.** $[-2, 3]$ by $[-4, 4]$ **d.** $[-6, 2]$ by $[-4, 8]$

18. a.   **b.**
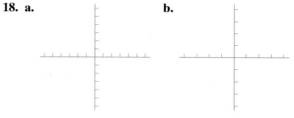
c.   **d.**
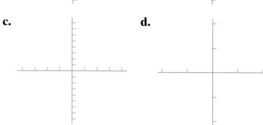

19.
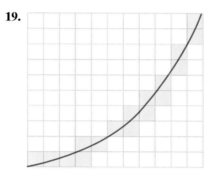

20. Answers may vary.   **21.** Answers may vary.   **22.** b
and d because they can be written in the form $y = mx + b$.
23. Answers may vary.   **24.** $x = -2$. No, answers may
vary.   **25.** $y = 4$. Yes, answers may vary.   **26.** $y = 0$
27. $x = 0$
29. $y = 1, 0, -5$ **31.** $y = \frac{1}{2}, -\frac{1}{2}, -2$
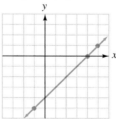
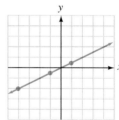

33. $y = -5, 1, 4$ **35.** $y = -1, -3, -4$
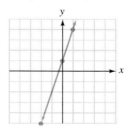
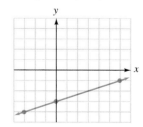

37. $y = 1, -2, -3, -2, 1$

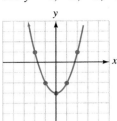

39. $y = -3, -6, -7, -6, -3$

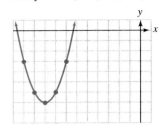

41. $y = -12, -5, -4, -3, 4$

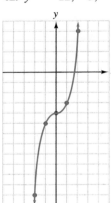

43.

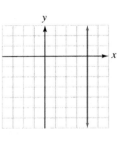

45.

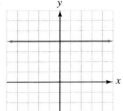

47.

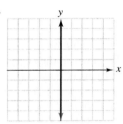

49.

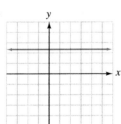

51.

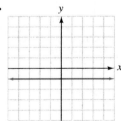

53.

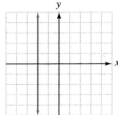

55. tutor's fee: 12, 24, 36, 60   **57.** Answers may vary.

59. $y = 220 - x$

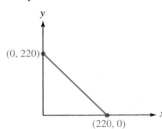

61. Yes,

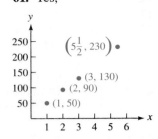

63. $A = 4\pi x^4$

Exercise 3.5 (page 196)
1. best fit   **2.** y **3.** x **4.** x-intercept: $(-3, 0)$, y-intercept: $(0, -4)$ **5.** x-intercept: $(-4, 0)$ and $(4, 0)$, y-intercept: $(0, -4)$ **6.** Answers may vary.   **7.** Answers may vary.   **8.** Sherah earns \$35 for 6 hours of babysitting.
9. When 250 copies are made, both services charge \$32.50. The costs are equal.   **10.** Quick-Copy

11.

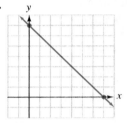

13.

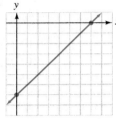

15.

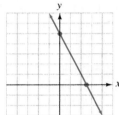

17.

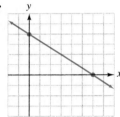

19.

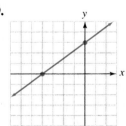

21.

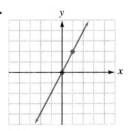

23. x-intercept: $(-4, 0)$, y-intercept: $(0, 8)$
25. x-intercept: $(5, 0)$, y-intercept: $(0, -\frac{10}{3})$
27. x-intercepts: $(-4, 0)$, $(7, 0)$, y-intercept: $(0, -28)$
29. x-intercept: $(10, 0)$, y-intercept: $(0, 10)$
In Exercises 31–39, the appropriate viewing window is indicated. If your window is slightly larger, that would be

acceptable.   **31.** [−10,10] by [−10, 20]   **33.** [−20, 10] by [−10, 10]   **35.** [−10, 35] by [−15, 10]   **37.** [−10, 10] by [−10, 25]   **39.** [−10, 30] by [−10, 10]   **41.** Suggested viewing window: [−10, 10] by [−20, 10]   **43. a.** $y = 1.15x$ **c.** \$17   **d.** \$40.54   **45. a.** $y = 0.04x + 0.05(2x)$ or $y = 0.14x$ **c.** \$15.12   **d.** \$450   **47. a.** $y = \frac{75 + 65 + x}{3}$ **c.** Graph should indicate $x = 76, y = 72$ **d.** An average of 81 is possible only if there is enough extra credit to earn 103%.   **49. a.** $y = 15 + 0.75x$ **c.** 88   **d.** $y = 1.10x$ **f.** Suggested viewing window: [−47, 60] by [−31, 60]; 43   **g.** 43   **51. a.** \$76.60, \$77.79, \$78.98   **b.** These are merely predictions because the points are not collinear.   **53.** ZINTEGER displays both x- and y-intercepts, whereas ZSTANDARD does not display either intercept.   **55. a.** (0, 45)   **b.** (0, −56)   **c.** (0, 24) The y-intercept is the same as the constant value of the function.

Exercise 3.6 (page 209)

1. $ax + c = 0$ **2.** STOre, TABLE, (sometimes VALUE and TRACE)   **3.** 2   **4.** 5   **5.** [−10, 12] by [−10, 17]   **6.** [−10, 10] by [−10, 21]   **7.** [−10, 10] by [−40, 10]   **8.** [−10, 10] by [−10, 25]   **9.** [−10, 10] by [−20, 10]   **10.** [−15, 10] by [−10, 10]   **11.** [−10, 10] by [−20, 10]   **12.** [−10, 10] by [−10, 15]   **13.** [−10, 10] by [−10, 15]   **14.** [−10, 10] by [−15, 10]   **15.** −2   **17.** 0   **19.** 10   **21.** −3   **23.** −3   **25.** 8   **27.** $\frac{34}{9}$ **29.** $-\frac{8}{7}$ **31.** −3.5   **33.** −6   **35.** $\frac{5}{9}$, [−10, 10] by [−15, 10]   **37.** 9, [−10, 10] by [−10, 25]   **39.** 90, [−10, 100] by [−10, 20]   **41.** −8   **43.** 2   **45.** −2   **47.** −9   **49.** −6   **51.** 6 million shares   **53.** \$4.95   **55.** 370   **57.** 2,760   **59.** 750   **61.** 3   **63.** 3 hours   **65.** \$4,000   **67.** 4 ft and 8 ft   **69.** \$1,211.50   **71.** The smaller tank cannot be emptied into a 10-gallon canister because the smaller tank holds 13 gallons.   **73.** 300 g   **75.** They did not meet their quota.   **77.** 125 1bs   **79.** 8,100 sq. ft   **81.** $\frac{xy}{3}$ **83.** $2x + 1$ miles

Exercise 3.7 (page 219)

1. $x, -x$ **2.** distance   **3. a.** 4   **b.** 4   **c.** −4   **d.** $\pm a$ **e.** $\pi - 3$ **4.** 8, −8   **5.** Absolute value is never negative.   **6.** No, because the equation is equivalent to $|x| = -5$.   **7.** ±8   **9.** 9 or −3   **11.** 4 or −1   **13.** $\frac{14}{3}$ or −6   **15.** ∅   **17.** 8 or −4   **19.** $-\frac{1}{2}$ or 2   **21.** −8   **23.** −4 or −28   **25.** 0 or −6   **27.** $\frac{20}{3}$ **29.** −2 or $-\frac{4}{5}$ **31.** 3 or −1   **33.** 0 or −2   **35.** 0   **37.** $\frac{4}{3}$ **39.** ∅   **41.** Answers may vary.   **43.** Answers may vary.   **45. a.** $-\frac{4}{45}$ **b.** $-\frac{7}{6}$ **47.** 1   **49.** [−10, 30] by [−15, 10] Larger viewing windows are acceptable.

Chapter 3 Summary (page 220)

1. a. yes   b. no   c. no   d. yes   **2. a.** −2   **b.** $-\frac{1}{9}$ **c.** 5   **d.** 3   **e.** −1   **f.** 15   **3. a.** 7   **b.** $\frac{39}{7}$ **c.** 9   **d.** $-\frac{9}{4}$

e. 12   **f.** 6   **4. a.** 8   **b.** 27   **c.** 85   **d.** 147 kilowatt hours   **e.** 8 ft   **5. a.** $y = -3x + 12$ **b.** $y = \frac{4}{3}x - \frac{8}{3}$ **c.** $t = \frac{d}{r}$ **d.** $w = \frac{P - 2l}{2}$ **e.** $h = \frac{A}{2\pi r}$ **f.** $\frac{A - P}{Pt} = r$ **g.** $\frac{RT}{P} = V$ **6. a.** yes   **b.** yes   **c.** no   **d.** no   **7. a.** (2, −2)   **b.** (3, −3)   **c.** (−1, 0)   **d.** (3, 0)   **8.** $y = 12, 9, 6, 3, 0$ **9. a.** relation   **b.** function   **c.** function   **d.** function   **10. a.**

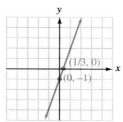

b.

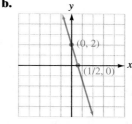

c.

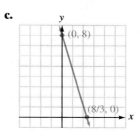

d.

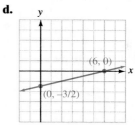

e.

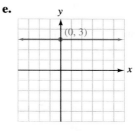

f.

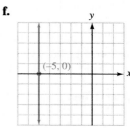

g.

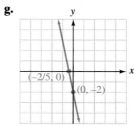

h.

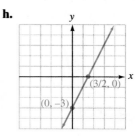

i.

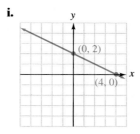

j.

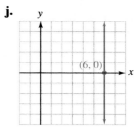

k.

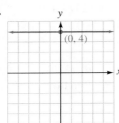

l.

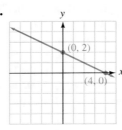

11. a. $(\frac{1}{3}, 0), (0, -1)$ **b.** $(\frac{1}{2}, 0), (0, 2)$ **c.** $(\frac{8}{3}, 0), (0, 8)$
d. $(6, 0), (0, -\frac{3}{2})$ **e.** $(0, 3)$ horizontal line **f.** $(-5, 0)$
vertical line **g.** $(-\frac{2}{5}, 0), (0, -2)$ **h.** $(\frac{3}{2}, 0), (0, -3)$
i. $(4, 0), (0, 2)$ **j.** $(6, 0)$, vertical line **k.** $(0, 4)$, horizontal line **l.** $(4, 0), (0, 2)$ **12. a.** $(2, 0), (0, -6)$ **b.** $(-1, 0)$,
$(5, 0), (0, -5)$ **c.** $(4, 0), (0, -64)$ **d.** $(3, 0), (0, -12)$
13. a. $[-10, 10]$ by $[-25, 10]$ **b.** $[-10, 25]$ by $[-110, 10]$
c. $[-35, 10]$ by $[-10, 15]$ **14. b.** \$610 **c.** \$822.50
15. a. $y = 125 + 35x$ **c.** \$230 **d.** 8 years
16. a. $12x + 5y = 1,069$ **b.** $y = \frac{1,069}{5} - \frac{12}{5}x$ **c.** 113
d. No, because 69.8 student tickets would have to be sold.
e. No, because according to the graph that would mean
that -2.2 student tickets were sold. **17–18.** Your graph,
your table and the use of the STOre feature should all
support the algebraic result you determined in #2 & #3.
If you do not understand and/or did not obtain a correct
result using one of these three features, return to Technology 3.5 on page 200 and Example 3 in Section 3.6.
19. a. 3 or $-\frac{1}{3}$ **b.** $\frac{26}{3}$ or $-\frac{10}{3}$ **c.** $\varnothing$ **d.** 3 or 0 **e.** $\frac{5}{4}$ or $-\frac{3}{4}$
20. a. -5 or $\frac{1}{5}$ **b.** -1 or 1

Chapter 3 Test (page 224)
1. no **2.** no **3.** $y = \frac{8}{3} - \frac{4}{3}x$ **4.** $y = -\frac{1}{4}x + \frac{3}{10}$ **5.** no
6. yes **7.** 7 **8.** -2 **9.** -3 **10.** -2 **11.** $\varnothing$
12. $\frac{33}{2}$ or $-\frac{3}{2}$ **13.** $\varnothing$ **14.** -2 or $\frac{1}{3}$
15.

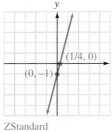

ZStandard

16.

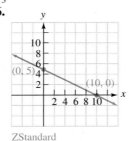

ZStandard

17.

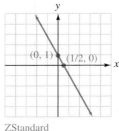

ZStandard

18.

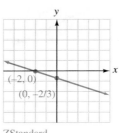

ZStandard

19. b. 200 TVs **20.** 17, 19 **21.** \$1,097.60
22. \$85,000

Exercise 4.1 (page 236)
1. ratio **2.** equal, equal **3.** means, extremes **4.** simi-
lar **5.** $\frac{4}{5} = \frac{8}{10}$ **6.** $4x^2z$ **7. a.** $\frac{5}{7}$ **b.** $\frac{3}{5}$ **c.** $\frac{1}{2}$ **d.** $\frac{1}{2}$ **e.** $\frac{2}{3}$
f. $\frac{2}{3}$ **8. a.** yes **b.** yes **c.** no **d.** yes **e.** yes **f.** no
9. a. similar **b.** similar **c.** not similar **10. a.** $\frac{x}{11} = \frac{16}{20}$
b. $\frac{8}{6} = \frac{x}{4}$ **11.** -17 **13.** $\frac{8}{11}$ **15.** 102 **17.** 26, 28
19. 39, 40, 41 **21.** 7 **23.** 13 **25.** 19 ft **27.** 29 m by
18 m **29.** 39 in. by 17 in. **31.** \$6.50 **33.** 65.25 ft
35. not correct **37.** 21.4 ft **39.** 528 ft **41.** 880 ft
43. a. $-2x^2 - 2x + 6$ **b.** $8x^3 + 60x^2 + 150x + 125$
c. $x^2 - 3x + 9$ **45. a.** $\{1\}$ **b.** $\{1\}$

Exercise 4.2 (page 242)
1. fixed **2.** unit **3.** break-even **4.** 0.04 **5.** revenue
6. unit cost = \$25, fixed cost = \$4,000 **7. a.** 7,500 gallons **b.** 8,000 gallons **c.** Use process A. **d.** Use process B. **9.** 960 pairs of shoes **11.** 300 plates **13.** 90
15. 20 bolts **17.** 1,100 shares **19.** 9 flat screen TVs
21. \$4,500 at 4% and \$19,500 at 5% **23.** \$6,686 at each
rate **25.** \$10,000 more at 6% **27.** 6% bond fund and
7% CD **29. a.** $16x^2$ **b.** $16x^2 + 24x + 9$ **31. c.** $x = 6$

Exercise 4.3 (page 248)
1. a. $0.03, \frac{3}{100}$ **b.** $0.3, \frac{3}{10}$ **c.** $0.035, \frac{7}{200}$ **d.** $0.035, \frac{7}{200}$
2. 1 ounce **3.** $x, 30 - x$ grams **4.** $25 + x$ oz **5.** 20 gal
of \$1.15 fuel **7.** 50 gal **9.** 7.5 oz **11.** 40 pounds
of lemon drops, 60 pounds jelly beans **13.** \$1.20
15. 88 lbs **17.** 3 hours **19.** 6.5 hours **21.** 7.5 hours
23. 500 mph **25. a.** -56 **b.** $6x^3 + 7x^2 - 2x + 5$
c. $xy^2 - \frac{3}{2}y + \frac{2x^2}{y}$ **27.** $\frac{2}{9}$ **29.** 1

Exercise 4.4 (page 260)
1. inequality **2.** critical point **3.** interval **4.** unbounded **5.** solution **6.** is greater than, is greater than
or equal to **7.** is less than, is less than or equal to
8. included **9.** not included **10.** infinity **11.** $1 < 11$
12. $-9 < 1$ **13.** $-8 < 12$ **14.** $8 > -12$ **15.** $-2 < 3$
16. $2 > -3$ **17.** $[3, \infty)$ **18.** $(-\infty, 2)$ **19.** $(-\infty, 5]$

20. $(4, \infty)$ **21.** 4 **22.** -9 **23. a.**

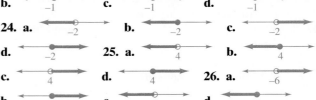

d. ⟶ 2   **e.** ⟵ 2   **28. a.** ⟶ -1

b. ⟵ -1 **c.** ⟶ -1 **d.** ⟵ -1

e. ⟶ -1 **29.** $(-\infty, 4]$, ⟵ 4 $\{x \mid x \le 4\}$

31. $(-\infty, -2)$, ⟵ -2 $\{x \mid x < -2\}$ **33.** $(-\infty, -1]$,
⟵ -1 $\{x \mid x \le -1\}$ **35.** $[-13, \infty)$, ⟶ -13
$\{x \mid x \ge -13\}$ **37.** $[2, \infty)$, ⟶ 2 $\{x \mid x \ge 2\}$
39. $(-15, \infty)$, ⟶ -15 $\{x \mid x > -15\}$ **41.** $(-\infty, 20]$,
⟵ 20 $\{x \mid x \le 20\}$ **43.** $[3, \infty)$, ⟶ 3
$\{x \mid x \ge 3\}$ **45.** $(-7, \infty)$, ⟶ -7 $\{x \mid x > -7\}$
47. $[4, \infty)$, ⟶ 4 $\{x \mid x \ge 4\}$ **49.** 98% or better
51. 27 mpg or better   **53.** at least 6 servings   **55.** after
574 days   **57.** at least 17 cm   **59. a.** -2 **b.** $\varnothing$
c. $-\frac{7}{3}, -4$

Exercise 4.5 (page 270)
1. compound   **2.** continued   **3.** intersection   **4.** union
5. conjunctions   **6.** disjunctions   **7.** $2 < x < 3$; I
8. $-2 < x < -1$; III   **9. a.** $(1.5, 6.5)$ **b.** $(-\infty, 1.5)$
$\cup (6.5, \infty)$ **10. a.** $(\frac{1}{2}, 2)$ **b.** $(-\infty, \frac{1}{2}) \cup (2, \infty)$
11. a. $(-12, -6)$ **b.** $(-\infty, -12) \cup (-6, \infty)$
12. a. $(-3, 2)$ **b.** $(-\infty, -3) \cup (2, \infty)$ **13. a.** iii
b. iv   **c.** ii   **d.** i   **14. a.** iii   **b.** ii   **c.** iv   **d.** i
15. false   **16.** true   **17.** true   **18.** false   **19.** true
20. false   **21.** false   **22.** true   **23.** $(-2, 5]$,
⟵ -2 5 ⟶ $\{x \mid -2 < x \le 5\}$ **25.** $(-\infty, 4]$,
⟵ 4 $\{x \mid x \le 4\}$ **27.** $\varnothing$ **29.** $(-\infty, -2) \cup (5, \infty)$,
⟵ -2 5 ⟶ $\{x \mid x < -2 \text{ or } x > 5\}$ **31.** $[4, \infty)$
⟶ 4 $\{x \mid x \ge 4\}$ **33.** $(-\infty, \infty)$, ⟵ -2 6 ⟶
$\mathbb{R}$ **35.** $(3, 4)$ ⟵ 3 4 ⟶ $\{x \mid 3 < x < 4\}$ **37.** $\varnothing$

39. $(-\infty, -\frac{1}{4}]$ ⟵ $-\frac{1}{4}$ $\{x \mid x \le -\frac{1}{4}\}$ **41.** $\varnothing$

43. $(-2, 5)$ ⟵ -2 5 ⟶ $\{x \mid -2 < x < 5\}$
45. $(8, 11)$, ⟵ 8 11 ⟶ $\{x \mid 8 < x < 11\}$
47. $[-2, 4]$, ⟵ -2 4 ⟶ $\{x \mid -2 \le x \le 4\}$
49. $(-\infty, 2) \cup (7, \infty)$, ⟵ 2 7 ⟶
$\{x \mid x < 2 \text{ or } x > 7\}$ **51.** $(-\infty, 1)$, ⟵ 1 ⟶
$\{x \mid x < 1\}$ **53.** $(-\infty, -6)$, ⟵ -6 ⟶ $\{x \mid x < -6\}$
55. $(-2, 1]$ **57.** $(-6, \frac{3}{2}]$ **59.** $(-\infty, -2) \cup [1, \infty)$
61. $(-\infty, -6) \cup [\frac{3}{2}, \infty)$ **63.** $1{,}540 \le C \le 1{,}650$
65. $0 < x \le 7.62$ m   **67.** $5 < w < 9$ ft
69. a. ⟵ 2 ⟶   **b.** ⟵ 2 ⟶   **c.** ⟵ 2 ⟶

d. ⟶ 2   **71.** $(5, 0)$ and $(0, -\frac{10}{3})$

Exercise 4.6 (page 279)
1. absolute value   **2.** 0   **3. a.** $\{-1, 5\}$ **b.** $(-1, 5)$
c. $(-\infty, -1) \cup (5, \infty)$ **4. a.** $\{\frac{-5}{3}, 3\}$ **b.** $(-\infty, \frac{-5}{3}) \cup$
$(3, \infty)$ **c.** $(\frac{-5}{3}, 3)$ **5. a.** $\{-8, -4\}$ **b.** $(-\infty, -8] \cup$
$[-4, \infty)$ **c.** $[-8, -4]$ **6. a.** $\{-1, 5\}$ **b.** $[-1, 5]$
c. $(-\infty, -1] \cup [5, \infty)$ **7. a.** $\{-10, 14\}$ **b.** $(-10, 14)$
c. $(-\infty, -10] \cup (14, \infty)$ **8. a.** $\{-4, -1\}$ **b.** $(-4, -1)$
c. $(-\infty, -4) \cup (-1, \infty)$ **9. a.** $\varnothing$ **b.** $\varnothing$ **c.** $\mathbb{R}$ or
$(-\infty, \infty)$ **10. a.** $\varnothing$ **b.** $\mathbb{R}$ or $(-\infty, \infty)$ **c.** $\varnothing$
11. a. $\{-\frac{1}{5}, \frac{3}{5}\}$ **b.** $(-\infty, -\frac{1}{5}) \cup (\frac{3}{5}, \infty)$ **c.** $(-\frac{1}{5}, \frac{3}{5})$
12. a. $\{8\}$ **b.** $(-\infty, 8) \cup (8, \infty)$ **c.** $\varnothing$
13. $(-4, 4)$ ⟵ -4 4 ⟶ $\{x \mid -4 < x < 4\}$
15. $[-21, 3]$ ⟵ -21 3 ⟶ $\{x \mid -21 \le x \le 3\}$ **17.** $\varnothing$
19. $[-1.5, 2]$ ⟵ -1.5 2 ⟶ $\{x \mid 1.5 \le x \le 2\}$
21. $(-2, 5)$ ⟵ -2 5 ⟶ $\{x \mid -2 < x < 5\}$
23. $(-\infty, -1) \cup (1, \infty)$ ⟵ -1 1 ⟶ $\{x \mid x < -1 \text{ or }$
$x > 1\}$ **25.** $(-\infty, -12] \cup [36, \infty)$ ⟵ -12 36 ⟶
$\{x \mid x \le -12 \text{ or } x \ge 36\}$ **27.** $(-\infty, -5\frac{1}{3}) \cup (4, \infty)$
⟵ $5\frac{1}{3}$ 4 ⟶ $\{x \mid x < -5\frac{1}{3} \text{ or } x > 4\}$ **29.** $\mathbb{R}$
31. $(-\infty, -3) \cup (2\frac{1}{3}, \infty)$ ⟵ -3 $2\frac{1}{3}$ ⟶
$\{x \mid x < -3 \text{ or } x > 2\frac{1}{3}\}$ **33.** $\mathbb{R}$
35. $(-\infty, -10) \cup (14, \infty)$ ⟵ -10 14 ⟶
$\{x \mid x < -10 \text{ or } x > 14\}$ **37.** $\mathbb{R}$ **39.** $\mathbb{R}$
41. $(-\infty, -0.2) \cup (0.6, \infty)$ ⟵ $-.2$ .6 ⟶
$\{x \mid x < -0.2 \text{ or } x > 0.6\}$ **43.** $(-6, 18)$ ⟵ -6 18 ⟶
$\{x \mid -6 < x < 18\}$ **45.** $\varnothing$ **47.** $(-\infty, \frac{2}{7}] \cup [\frac{8}{7}, \infty)$
⟵ $\frac{2}{7}$ $\frac{8}{7}$ ⟶ $\{x \mid x \le \frac{2}{7} \text{ or } x \ge \frac{8}{7}\}$
49. $[1, 3]$ ⟵ 1 3 ⟶ $\{x \mid 1 \le x \le 3\}$
51. $\varnothing$ **53.** Answers may vary.   **55.** Answers may vary.
57. Answers may vary.   **59.** $\varnothing$ **61.** It is a function because each value of x corresponds to exactly one value
of y.

Chapter 4 Summary (page 282)
1. a. yes   **b.** no   **c.** no   **d.** yes   **2. a.** 12   **b.** -15
c. $-\frac{13}{4}$ **d.** 13   **3.** 13 in.   **4.** 30 tons   **5.** $\frac{1}{20}$ gram
6. 132.5 ft   **7.** 382 parts   **8.** 40 caps   **9.** 35 nickels, 70
dimes, and 75 quarters   **10.** \$8,000 at $4\frac{1}{2}$%   **11.** 7.5 liters
12. 12 lbs of the \$.90 candy and 8 lbs of the \$1.65 candy
13. 20 minutes   **14. a.** $(-\infty, 3)$, ⟵ 3 ⟶ $\{x \mid x < 3\}$

b. $(-\frac{1}{2}, \infty)$ ⟶ $-\frac{1}{2}$ $\{x \mid x > -\frac{1}{2}\}$

c. $(-\infty, \frac{13}{2}]$ ⟵ $\frac{13}{2}$ $\{x \mid x \le \frac{13}{2}\}$

d. $(-\infty, \frac{2}{3}]$ ⟵ $\frac{2}{3}$ $\{x \mid x \le \frac{2}{3}\}$ **15. a.** $\{-1\}$

b. $(-1, \infty)$ **c.** $(-\infty, -1)$ **16. a.** $\{8\}$ **b.** $(-\infty, 8)$

c. $(8, \infty)$ **17. a.** $\left(-\frac{1}{3}, 2\right)$

$\{x \mid -\frac{1}{3} < x < 2\}$ **b.** $\left(-\frac{5}{3}, \frac{2}{3}\right)$

$\{x \mid -\frac{5}{3} < x < \frac{2}{3}\}$ **c.** $\left(-\infty, \frac{2}{3}\right) \cup \left(\frac{25}{3}, \infty\right)$

$\{x \mid x < \frac{2}{3} \text{ or } x > \frac{25}{3}\}$

d. $(-\infty, \infty)$ $\mathbb{R}$ **18. a.** $(2.5, 7)$

b. $(-\infty, 2.5) \cup (7, \infty)$ **19. a.** $(-1, 7)$ **b.** $(-\infty, -1) \cup$

$(7, \infty)$ **20. a.** $[-7, 1]$ $\{x \mid -7 \le x \le 1\}$

b. $(-7, 4)$ $\{x \mid -7 < x < 4\}$

c. $(-\infty, -9) \cup (13, \infty)$

$\{x \mid x < -9 \text{ or } x > 13\}$ **d.** $(-\infty, -2] \cup \left[\frac{4}{3}, \infty\right)$

$\{x \mid x \le -2 \text{ or } x \ge \frac{4}{3}\}$ **e.** $\varnothing$

f. $(-\infty, \infty)$ or $\mathbb{R}$ **21. a.** $\varnothing$ **b.** $\mathbb{R}$ **c.** $\varnothing$ **22. a.** $\varnothing$

b. $\varnothing$ **c.** $\mathbb{R}$

Chapter 4 Test (page 286)

1. no   **2.** no   **3.** $-\frac{3}{2}$ **4.** $-\frac{17}{4}$ **5.** 8,050 ft   **6.** $\frac{3}{5}$ of an
hour or 36 minutes   **7.** $45,000   **8.** 7 oz   **9.** 4   **10.** 1,250
11. a. $\{-1\}$ **b.** $(-1, \infty)$ **c.** $(-\infty, -1)$ **12. a.** $\{2, 6\}$
b. $(2, 6)$ **c.** $(-\infty, 2) \cup (6, \infty)$ **13.** $(-\infty, 0]$

$\{x \mid x \le 0\}$ **14.** $(-1, 6]$

$\{x \mid -1 < x \le 6\}$ **15.** $(-\infty, \infty)$ $\mathbb{R}$

16. $\left(-\infty, -\frac{1}{3}\right)$ $\{x \mid x < -\frac{1}{3}\}$ **17.** $\left(\frac{1}{2}, 4\frac{1}{2}\right)$

$\{x \mid \frac{1}{2} < x < 4\frac{1}{2}\}$

18. $(-\infty, -12] \cup [18, \infty)$

$\{x \mid x \le -12 \text{ or } x \ge 18\}$ **19.** $\varnothing$ **20.** $\mathbb{R}$

Cumulative Review: Chapters 3 and 4 (page 287)

1. for all, $\mathbb{R}$ **2.** no, $\varnothing$ **3.** ordered pairs   **4.** function
5. compound   **6. a.** II   **b.** III   **c.** IV   **d.** IV

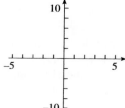

7.   **8.** All x-coordinates are
integers in ZInteger.   **9.** A trip of 50 miles will cost
$49.50.   **10.** STOre, TABLE   **11.** comparison   **12.** $\frac{3}{4}$
13. similar; proportion   **14.** $16.99x$ **15. a.** 0.06
b. 0.0775   **16.** x liters   **17.** $a, -a$ **18.** because $|x|$ is al-

ways greater than or equal to zero   **19. a.** -3 **b.** $\pi - 2$
20. $P = \frac{i}{rt}$ **21.** $\frac{1}{4}$ **22.** 15   **23.** b   **24. a.** function
b. not a function   **c.** function   **25.** x-intercepts: $(-5, 0)$
and $(2, 0)$ y-intercept: $(0, -10)$

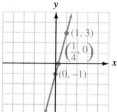

26.   **27. a.** x-intercepts $(1, 0)$ and
$(7, 0)$ and y-intercept $(0, 7)$ **b.** ZStandard doesn't trace
to an x-coordinate of 1 or 7 or y-coordinate of 7.
28. The y-minimum must be -16 or smaller.   **29.** $-\frac{22}{13}$
30. $\frac{7}{6}$ or $\frac{1}{2}$ **31.** 14 or -18 **32.** -5 or $\frac{3}{5}$ **33.** $y = \frac{3}{2}x - 18$
34. $y = -\frac{A}{B}x + \frac{C}{B}$ **35.** $[15, \infty)$ $\{x \mid x \ge 15\}$
36. $[-2, 4)$ $\{x \mid -2 \le x < 4\}$
37. $(-\infty, 1) \cup (8, \infty)$ $\{x \mid x < 1 \text{ or } x > 8\}$
38. $(-\infty, -1)$ $\{x \mid x < -1\}$
39. $(-10, 14)$ $\{x \mid -10 < x < 14\}$
40. $(-\infty, -5] \cup [1, \infty)$
$\{x \mid x \le -5 \text{ or } x \ge 1\}$ **41.** $\varnothing$ **42.** yes   **43.** 42
44. 50 years   **45.** 60   **46.** $3\frac{2}{3}$ hours   **47.** 38.4 ounces
48. 3,053.6 in³   **49.** 5 years   **50. a.** $20,700   **b.** 170
51. a. $19,155   **b.** greater than the revenue   **c.** 620 CD
players; profit (revenue = $55,800)

Exercise 5.1 (page 301)

1. factor   **2.** factors   **3.** prime   **4.** terms   **5.** distributive   **6.** sum, difference, product   **7.** polynomial   **8.** exactly one   **9.** grouping   **10.** true   **11.** false   **12.** true
13. true   **14.** false   **15.** true   **16.** true   **17.** false
18. false   **19.** true   **20.** 12   **21.** 6   **22.** $2x^2y^3$ **23.** $2x^2y^3$
24. $4a^2bc$ **25.** tv **26.** $x - 3$ **27.** $2x^2 + 1$ **28.** $3 + 2xy$
29. $5ab + 7$ **30.** $2xy^2 + 1$ **31.** $2x^2 + 1$ **33.** $x(y - z)$
35. $-(2x - 5y)$ **37.** $r^2(r^2 + 1)$ **39.** $-(3m + 4n - 1)$
41. $8xy^2z^3(3xyz + 1)$ **43.** $3(x + y - 2z)$
45. $-(3ab + 5ac - 9bc)$ **47.** $2y(2y + 4 - x)$
49. $-4a^2b^2(b - 3a)$ **51.** $-7ab(2a^5b^5 - 7ab^2 + 3)$
53. $2xyz^2(2xy - 3y - 6)$ **55.** $(x + y)(a + 2b)$
57. $(3t + 5)(3t + 4)$ **59.** $3ab(3ab - 2)(3x - 2y)$
61. $(x - 5y)(a + 1)$ **63.** $(m - n)(p + q)$
65. $(3c + 1)(a + b)$ **67.** $(b + 3)(3a - 2)$
69. $(3x^2 + y)(2u - v)$ **71.** $(p + q)(m - n)$
73. $(5 + r)(x + y)$ **75.** $(m + n)(r + s)$ **77.** $(s + r)(x - y)$
79. In both cases, not correct. Answers may vary.
81. $x + 8y$ **83. a.** $\frac{11}{3}$ **b.** $x = 3.6666667$ **c.** Rational.
Answers may vary.

Exercise 5.2 (page 311)

1. trinomials **2.** descending **3.** multiplication
4. prime **5.** factors **6.** $10x^2 - 3x - 27$ **7. a.** 1 **b.** 5
c. 1 **d.** 6 **8. a.** 3 **b.** 2 **c.** $(3x - 5)(x + 2)$ **9.** a
10. c **11.** c **12.** b **13.** b **14.** a **15.** Correct. Answers may vary. **16.** Correct. Answers may vary.
17. prime **19.** $(x - 4)(x - 1)$ **21.** $(x + 2)(x + 1)$
23. $(3x + 4)(x - 1)$ **25.** $(3x - 1)(2x + 5)$
27. $(3x - 2)(2x + 5)$ **29.** $2(3x + 1)(x - 4)$
31. $(3x + 7)(2x - 1)$ **33.** $(3x - 1)(5x - 4)$
35. $(2x - 5)(4x - 3)$ **37.** prime **39.** $3(y - 6)(y - 1)$
41. $-2(x - 9)(x - 1)$ **43.** prime **45.** $(m + 5n)(m - 2n)$
47. $(y - 14z)(y + z)$ **49.** $4(2x + y)(x - 2y)$
51. $3y(3y - 2)(y + 1)$ **53.** $(y - 5)(y - 3)$ **55.** prime
57. $3(2x - 3y)(x + 2y)$ **59.** $(5x - 9)(7x + 2)$
61. $(4x - 1)(x + 3)$ **63.** $(5y - 2)(y - 1)$ **65.** $(x + 5)^2$
67. $(z - 1)^2$ **69.** $(r + 12)^2$ **71.** $(v - 7)^2$ **73.** $(x + 3y)^2$
75. a. 5 and 2 units respectively **b.** length = 9 in.
width = 6 in. **77.** ± 21, ± 12, ± 9 **79. a.** $25x^4y^6$
b. 1,638,400 **c.** 1,638,400 **d.** Yes. Answers may vary.
81. a. 31 **b.** 86 **c.** -0.32

Exercise 5.3 (page 319)

1. difference **2.** sum **3.** conjugate binomials **4.** difference, cubes **5.** sum, cubes **6.** False, 2 is not a perfect square. **7.** true **8.** true **9.** $(x - 2)(x + 2)$
11. $(2y - 5)(2y + 5)$ **13.** $(7x - 9y)(7x + 9y)$
15. $(9a^2 - 4b)(9a^2 + 4b)$ **17.** prime
19. $(25a - 13b^2)(25a + 13b^2)$ **21.** $(9a^2 - 7b)(9a^2 + 7b)$
23. $(6x^2y - 7z^2)(6x^2y + 7z^2)$ **25.** $(x + y - z)(x + y + z)$
27. $(a - b - c)(a - b + c)$ **29.** $(a^2 + b^2)(a - b)(a + b)$
31. $(y^2 + 16)(y - 4)(y + 4)$ **33.** $2(x - 1)(x + 1)$
35. $2(x - 12)(x + 12)$ **37.** $2x(x - 4)(x + 4)$
39. $5x(x - 5)(x + 5)$ **41.** $xy(4x - 9y)$
43. $t^2(rs + x^2y)(rs - x^2y)$ **45.** $(y + 1)(y^2 - y + 1)$
47. $(a - 3)(a^2 + 3a + 9)$ **49.** $(2 + x)(4 - 2x + x^2)$
51. $(r + s)(r^2 - rs + s^2)$ **53.** $(x - 2y)(x^2 + 2xy + 4y^2)$
55. $(x - 3y)(x^2 + 3xy + 9y^2)$ **57.** $(3a - b)(9a^2 + 3ab + b^2)$
59. $(3x + 5)(9x^2 - 15x + 25)$
61. $(4x + 3y)(16x^2 - 12xy + 9y^2)$
63. $(x^2 + y^2)(x^4 - x^2y^2 + y^4)$ **65.** $5(x + 5)(x^2 - 5x + 25)$
67. $4x^2(x - 4)(x^2 + 4x + 16)$
69. $2u^2(4v - t)(16v^2 + 4vt + t^2)$
71. $(a + b)(x + 3)(x^2 - 3x + 9)$
73. $6(ab - z)(a^2b^2 + abz + z^2)$

Exercise 5.4 (page 323)

1. prime **2.** GCF **3.** grouping **4.** true **5.** False. Answers may vary. **6. a.** $(3x + 7)(2x - 5)$
b. $(3x + 7)(2x - 5)$ **c.** Answers may vary. **7.** $3(2x + 1)$
9. $(x - 3)(x^2 + 3x + 9)$ **11.** $(x - 10)(x - 1)$
13. $(3t - 1)(2t + 3)$ **15.** $3rs(s - 2rt)$

17. $(2x - 5)(2x + 5)$ **19.** $(a + b)(c + d)$ **21.** $(t - 1)^2$
23. $3(a - 2)(a + 2)$ **25.** $2x(y + 6)(y - 2)$
27. $-5(x + 4)(x - 2)$ **29.** $(t^2 + 4)(t - 2)(t + 2)$
31. $(3x - 2y)(9x^2 + 6xy + 4y^2)$ **33.** $(3a + 1)(a + 4)$
35. $(3x + 4)(2x - 3)$ **37.** $3(2r - 3s)(r + 2s)$ **39.** prime
41. $(5z - 4)^2$ **43.** $5a(4a^2 + 1)(2a - 1)(2a + 1)$
45. $(3x - 4)(2x + 5)$ **47.** $-(5x - 1)(4x + 3)$
49. $2(2x - 3y)(4x^2 + 6xy + 9y^2)$ **51.** $(x - a)(a - b)(a + b)$
53. $2xy(2ax - b)(2ax + b)$ **55.** prime
57. $5x^2(3x + 8)(2x - 5)$ **59.** $(x^2 + y^2)(a^2 + b^2)$
61. $(25x^2 + 16y^2)(5x - 4y)(5x + 4y)$
63. $(a + 3)(a - 3)(a + 2)(a - 2)$ **65.** $(x + 3)^2(x - 3)^2$
67. -40 **69.** -3

Exercise 5.5 (page 334)

1. transformation **2.** two **3.** root **4.** zero factor
5. at least one **6.** standard form **7.** INTERSECT
8. STOre, TABLE **9.** x, y **10.** $2x^2 - 7x - 15$
11. $(x - 5)(2x + 3)$ **12.** 15 **13.** 5, $-\frac{3}{2}$ **14.** Answers may vary. **15. a.** expression **b.** equation **c.** equation
d. expression **16. a.** quadratic **b.** linear **c.** higher degree **17.** $\{-2, 5\}$ **18.** $\{-5, 2\}$ **19.** $\{6, -8\}$
20. $\{0, -2, 1\}$ **21.** $\{-2, 1\}$ **22.** $\{1, -3, 4\}$ **23.** 3, -3
25. 0, 2 **27.** 0, 3 **29.** 6, -6 **31.** $-3, -4$ **33.** 2, 3
35. 1, 2 **37.** 0, $\frac{1}{5}$ **39.** $-\frac{3}{2}, 1$ **41.** $\frac{1}{3}, \frac{1}{2}$ **43.** $\frac{7}{2}, -2$
45. $-\frac{4}{3}, \frac{3}{4}$ **47.** $-6, 2$ **49.** 0, $-\frac{2}{3}, 2$ **51.** $\frac{8}{3}, -\frac{8}{3}$
53. $-5, -1$ **55.** $-6, 3$ **57.** A; 0, 5; $x^2 - 5x = 0$
59. F; 0, 2, -2; $x^3 - 4x = 0$ **61.** E; 0, 2, -3;
$x^3 + x^2 - 6x = 0$ **63.** G; $-4, -2$; $x^2 + 6x + 8 = 0$
65. 0, -4 **67.** 0, 10, -10 **69.** 1, 0, -9 **71.** 3, $-3, 1, -1$
73. 0, 1, $-\frac{5}{7}$ **75.** 0, $-\frac{1}{6}, -\frac{3}{2}$ **77.** $-0.42, 4.3$ **79.** 0.52,
-4.6 **81.** $-3.2, -0.3, 2.5$ **83.** 10 seconds **85.** 2 seconds and 6 seconds **87.** 6 m/sec **89.** 3, 7 **91.** 2
93. 3 ft by 7 ft **95.** 20 m **97.** $b = 4$ in, $h = 18$ in
99. 18 sq. units **101.** 1 m **103.** 3 cm **105.** 7 cm by
4 cm **107. a.** $x(x + 2) = 18$ or $x^2 + 2x - 18 = 0$
b. Answer may vary. **109.** $x^2 - 3x - 10 = 0$
111. $2x^3 + 11x^2 + 4x - 5 = 0$ **113.** 12 **115.** 11
117. yes

Chapter 5 Summary (page 339)

1. $3(x + 3y)$ **2.** $5a(x^2 + 3)$ **3.** $7x(x + 2)$ **4.** $3x(x - 1)$
5. $2x(x^2 + 2x - 4)$ **6.** $a(x + y + z)$ **7.** $a(x + y - 1)$
8. $xyz(x + y)$ **9.** $a(5a + 6b^2 + 10cd - 15)$
10. $7xy(a + 3x - 5x^2 + y)$ **11.** $(x + y)(a + b)$
12. $(x + y)(x + y + 1)$ **13.** $2x(x + 2)(x + 3)$
14. $3x(y + z)(1 - 3y - 3z)$ **15.** $(p + 3q)(3 + a)$
16. $(r - 2s)(a + 7)$ **17.** $(x + a)(x + b)$
18. $(y + 2)(x - 2)$ **19.** $y(3x - y)(x - 2)$
20. $5(x + 2)(x - 3y)$ **21.** $(x + 3)(x + 7)$
22. $(x - 3)(x + 7)$ **23.** $(x - 4)(x + 6)$
24. $(x - 6)(x + 2)$ **25.** $(x - 3)(2x + 1)$

26. $(x - 5)(3x + 1)$ **27.** $(2x + 3)(3x - 1)$
28. $3(2x - 1)(x + 1)$ **29.** $x(6x - 1)(x + 3)$
30. $x(x - 2)(4x + 3)$ **31.** $(x + y)(x - 3y)$ **32.** $(2x + 1)^2$
33. $-(4x - 5)(2x - 1)$ **34.** $(2x - 9y)(x - 2y)$
35. $(5x + 3)(x + 5)$ **36.** $(3x - 2y)(x + 4y)$
37. $(x + 3)(x - 3)$ **38.** $(8 + x)(8 - x)$
39. $(5x + 4y)(5x - 4y)$ **40.** $6y(x + 2y)(x - 2y)$
41. $(x^2 + 4)(x - 2)(x + 2)$ **42.** $2(9 + x^2)(3 - x)(3 + x)$
43. $(x + 3)(x^2 - 3x + 9)$ **44.** $(3 - x)(9 + 3x + x^2)$
45. $(x + 2y)(x^2 - 2xy + 4y^2)$ **46.** $(4x + y)(16x^2 - 4xy + y^2)$
47. $5(1 - 2y)(1 + 2y + 4y^2)$ **48.** $(4x - 1)(16x^2 + 4x + 1)$
49. $(x + 5)(x^2 - 5x + 25)$ **50.** $x^2(x + 6y)(x^2 - 6xy + 36y^2)$
51. $6xy(4y - 1)$ **52.** $(x + 3)(x^2 - 3x + 9)$
53. $(3x + 2)(2x - 1)$ **54.** $(3x - 7)(3x + 7)$
55. $(4x - 5)(4x + 5)$ **56.** $(x - 5)(2x + 3)$
57. $4(x + 3)(x - 1)$ **58.** $(2x + 3)^2$ **59.** $(x - 1)(x^2 + x + 1)$
60. $2(x - 2)(x^2 + 2x + 4)$ **61.** $(2x - 3y)(x - 2y)$
62. $(xy + 5)(xy - 5)$ **63.** $(3x - 1)(x + 1)$
64. $2(3x + y)(9x^2 - 3xy + y^2)$ **65.** $8(x - 2y)(x + 2y)$
66. $(3x + 1)(4x - 3)$ **67.** $(4x - 9)(2x + 3)$
68. $3x(1 + 3x)(1 - 3x + 9x^2)$ **69.** 0 or -2 **70.** 0 or 3
71. 3 or -3 **72.** 5 or -5 **73.** 4 or 3 **74.** 5 or -3
75. 6 or -4 **76.** 2 or 8 **77.** $-\frac{1}{2}$ or 3 **78.** $-\frac{3}{2}$ or 1
79. $\frac{1}{2}$ or $-\frac{1}{2}$ **80.** $\frac{2}{3}$ or $-\frac{2}{3}$ **81.** $0, 3,$ or 4 **82.** $0, -3,$ or -2
83. $0, \frac{1}{2},$ or -3 **84.** $0, -\frac{2}{3},$ or 1 **85.** 7 and 5 **86.** $\frac{1}{3}$
87. 15 ft   **88.** 15 seconds   **89.** 3 ft by 9 ft   **90.** 3 ft by 6 ft

Chapter 5 Test (page 343)

1. $2x(2y^2 + 6xy^3 - 3z)$ **2.** $3x(a + b)(x - 2y)$
3. $(x + y)(a + b)$ **4.** $(x - 5)(x + 5)$ **5.** $3(a - 3b)(a + 3b)$
6. $(2x - 3y)(2x + 3y)(4x^2 + 9y^2)$ **7.** $(x + 1)(x + 3)$
8. $(x - 11)(x + 2)$ **9.** $(3 - x)(9 + 3x + x^2)$
10. $(x + 4)(3x + 1)$ **11.** $(2y - 3)(y + 4)$
12. $(2x - y)(x + 2y)$ **13.** $(4x - 3)(3x - 4)$
14. $2y(y + 2)(y^2 - 2y + 4)$ **15.** 5 or -5 **16.** $-\frac{3}{2}$ or -1
17. $\frac{9}{5}$ or $-\frac{1}{2}$ **18.** -6 or 3 **19.** 12 seconds   **20.** 10 m

Exercise 6.1 (page 354)

1. rational   **2.** zero   **3.** excluded value(s)   **4.** negatives
5. simplify   **6.** factors   **7.** numerator, denominator
8. $\frac{a}{b}$ **9. a.** 0 **b.** 0 **c.** 5 **d.** -6 **10.** Answers may
vary.   **11. a.** 2 **b.** 0 **c.** 0 **d.** -5 **e.** $6, -1$ **12. a.** 5
b. $-2, 3$ **c.** $-2.5, -0.5$ **13. a.** 1 **b.** -1 **c.** -1 **d.** 1
14. a. $\frac{2}{13}$ **b.** $\frac{5}{7}$ **c.** $-\frac{x}{3}$ **d.** $-5y$ **e.** $\frac{1}{3}$ **f.** 2 **15.** b
16. b   **17. a.** $\frac{125}{20}$ **b.** $\frac{38}{42}$ **c.** $\frac{8xy}{x^2y}$ **d.** $\frac{7xy}{xy^2}$ **19.** no excluded
values   **21.** $-\frac{3}{4}$ **23.** $6, -5$ **25.** $3z$ **27.** $\frac{1}{3}$ **29.** $\frac{x}{y}$
31. simplified   **33.** -1 **35.** -1 **37.** -1 **39.** $\frac{x + 3}{x + 1}$
41. $\frac{x - 7}{x + 7}$ **43.** $\frac{3y}{y + 2}$ **45.** $\frac{3}{x}$ **47.** $\frac{2x - 3}{x + 1}$ **49.** $\frac{3}{x - 5}$
51. $\frac{x + 11}{x + 3}$ **53.** $-\frac{x - 4}{x + 4}$ or $\frac{4 - x}{x + 4}$ **55.** $\frac{x - 3}{5 - x}$ or $\frac{3 - x}{x - 5}$
57. $x + 1$ **59.** Evaluation: 1; Simplified form: $\frac{2}{x}$

61. Evaluation: $\frac{3}{2}$; Simplified form: $\frac{3}{2}$.   **63.** Evaluation: $\frac{15}{8}$;
Simplified form: $\frac{3(x + 3)}{x + 6}$ **65.** $\frac{5y(y - 2)}{(y - 2)^2}$ **67.** $\frac{3x(y - 1)}{y^2 - y}$
69. $\frac{y(y - 2)}{y^2 - 4}$ **71.** $\frac{(x - 3)^2}{x^2 - 9}$ **73.** $\frac{3(x + 2)}{x^2 + x - 2}$ **75. c.** Answers
may vary   **77. a.** 1 **b.** Answers may vary   **c.** Answers
may vary   **79.** 4 cm.

Exercise 6.2 (page 361)

1. $\frac{PR}{QS}$ **2.** $\frac{S}{R}, \frac{PS}{QR}$ **3.** reciprocal, multiplication   **4.** 1
5. $\frac{3}{11}$ **6.** $\frac{9}{5}$ **7.** $\frac{5}{7}$ **8.** $\frac{2}{3}$ **9.** $\frac{xy}{z}$ **10.** $\frac{4}{7}$ **11.** $\frac{7}{3}$ **12.** $\frac{3}{2}$
13. $\frac{2x}{3}$ **14.** $3z$ **15.** $2xy^2$ **17.** $-3y^2$ **19.** $\frac{(z + 7)(z + 2)}{7z}$
21. x **23.** $x + 2$ **25.** $\frac{3}{2x}$ **27.** $\frac{(x - 2)^2}{x}$ **29.** 1 **31.** $\frac{1}{3}$
33. 1 **35.** 3 **37.** $\frac{6}{y}$ **39.** 6 **41.** $\frac{2}{y}$ **43.** $\frac{2}{3x}$ **45.** $\frac{2(z - 2)}{z}$
47. $\frac{x - 2}{x - 3}$ **49.** 1 **51.** $\frac{x}{36}$ **53.** $\frac{(x - 1)(x + 1)}{5(x - 3)}$ **55.** $\frac{2x(1 - x)}{5(x - 2)}$
57. $\frac{y^2}{3}$ **59.** $\frac{(x + 2)(x + 5)}{x - 5}$ **61.** $y = \frac{4}{3}x - \frac{7}{3}$

Exercise 6.3 (page 368)

1. denominator   **2.** factor   **3.** smallest   **4.** 2 **5.** 1
6. $\frac{y}{x}$ **7.** $\frac{2}{y}$ **8.** $\frac{15z}{11}$ **9.** 0 **10.** 42 **11.** 30 **12.** $9xy$
13. $3x^2y$ **14.** $x^2 - 4$ **15.** $x^2 - 9$ **16.** $(x + 1)^2$
17. $(x + 3)^2$ **18.** $(x + 3)(x + 2)$ **19.** $(x + 5)(x + 1)$
20. $8xy$ **21.** $7xy$ **22.** $3x(x + 1)$ **23.** $5y(y - 2)$
24. $2y(x + 1)$ **25.** $3x(y - 1)$ **26.** $z(z + 1)$
27. $y(y - 2)$ **28.** $2(x + 2)$ **29.** $3(x + 2)$ **30.** $\frac{7}{6}$
31. $-\frac{1}{6}$ **32.** $\frac{5y}{9}$ **33.** $\frac{7a}{60}$ **34.** $\frac{53x}{42}$ **35.** $\frac{4xy + 6x}{3y}$ **37.** $\frac{1}{x}$
39. $\frac{2x + 8}{x^2}$ **41.** 10 **43.** $\frac{2x + 10}{x + 2}$ or $\frac{2(x + 5)}{x + 2}$ **45.** $\frac{2y}{x - 5}$
47. $\frac{3}{x + 2}$ **49.** $(y - 3)(y + 3)$ **51.** $xy(y - 1)$
53. $(x + 3)^2(x - 1)$ **55.** $(x - 3)(x + 2)(x + 3)$
57. $\frac{x^2 + 7x + 6}{2x^2}$ **59.** $\frac{-y^2 - 5y - 14}{2y^2}$ **61.** $\frac{a(a + 4)}{a + 3}$
63. $\frac{y + 7}{(y - 1)(y + 1)}$ **65.** $\frac{-a^3 - 5a^2 - 14a + 6}{(a + 3)^2(a - 3)}$ **67.** $\frac{x}{x - 3}$
69. $\frac{-2}{x - 3}$ **71.** $-\frac{2}{x + 1}$ **73.** $\frac{x + 3}{x + 6}$ **75.** $\frac{x + 4}{x - 3}$ **77.** $\frac{5}{x + 7}$
79. a. $\frac{4x}{x - 5}$ **b.** 8 feet   **c.** No. Answers may vary.
81. a. 0 **b.** 0 **c.** 0 **83.** $0, -\frac{3}{2}, 5$

Exercise 6.4 (page 375)

1. complex   **2.** $\frac{1}{x} + 2$ **3.** $\frac{3}{x}, \frac{x}{y}, \frac{1}{x}$ **4.** division, order of
operations   **5.** component, distributive   **6.** Answers
may vary.   **7.** Answers may vary.   **8. a.** $\frac{2x}{3}$ **b.** $\frac{\frac{x^2}{3}}{\frac{2x}{4}}$
c. Yes. Answers may vary.   **9. a.** $\frac{2}{3x}$ **b.** $\frac{\frac{x + 2}{3x}}{\frac{x + 2}{2}}$ **c.** Yes.
Answers may vary.   **10. a.** $\frac{7}{24}$ **b.** $(\frac{1}{2} + \frac{2}{3}) \div 4$ **c.** $\frac{7}{24}$;
yes; yes; answers may vary.   **11. a.** $\frac{19}{120}$ **b.** $(\frac{3}{4} + \frac{1}{5}) \div 6$
c. $\frac{19}{120}$; yes; yes; answers may vary.   **13.** $\frac{21}{10}$ **15.** $\frac{1}{14}$ **17.** $\frac{7}{8}$
19. $\frac{11}{5}$ **21.** $\frac{y^2}{x}$ **23.** $\frac{wz}{12t^2}$ **25.** $\frac{1 + 3y}{3 - 2y}$ **27.** $\frac{1 - x}{3 - x}$ **29.** $\frac{1 + 3x}{x + 2}$
31. $\frac{1}{x - 2}$ **33.** $\frac{2}{4 - x}$ **35.** $\frac{ab}{b^2 - a^2}$ **37.** $\frac{y^2}{y^2 - 12}$ **39.** $\frac{x^2}{(x + 1)^2}$
41. $\frac{y + 2}{y - 1}$ **43.** $\frac{9y + 5}{-y + 4}$ **45.** -1 **47.** $\frac{1 + y^2}{1 - y^2}$ **49.** $\frac{t^2 + t + 1}{t}$
51. $\frac{1}{3}$ **53.** $\frac{x - 3}{x + 3}$ **55.** -37 **57.** 37.5 ft.

Exercise 6.5 (page 385)

1. extraneous **2.** 3; 2 **3.** LCD **4.** Linear **5.** quadratic; higher-degree **6. a.** $x \neq 1, x \neq -1$ **b.** $x \neq 0$
c. $x \neq 4, x \neq -2$ **7. a.** expression; $\frac{5x}{6}$ **b.** equation;
$x = \frac{36}{5}$ **c.** expression; $\frac{3x^2 + 2x}{(x-1)(x+1)}$ **d.** expression; $\frac{7}{2x}$
e. equation; $x = -\frac{3}{2}$ **f.** equation; $x = \frac{1}{2}$ **9.** 12 **11.** $\frac{76}{3}$
13. 3 **15.** 6 **17.** 10 **19.** -2 **21.** 4 **23.** -1 **25.** 2
27. -8 **29.** No solution; 2 is extraneous. **31.** No solution; 2 is extraneous. **33.** 3 **35.** 3 **37.** 5 **39.** $\frac{1}{7}$
41. No solution; 4 is extraneous. **43.** 1 or -9
45. 4 **47.** 12 **49.** 6 **51.** 2 or 4 **53. a.** 6; 4 is extraneous. **b.** 6; 4 is extraneous. **c.** 6 **d.** Answers may vary. **e.** Answers may vary. **55.** $x + \frac{3y}{2} - 3x^2$
57. a. $y = 75 + 15x$ **b.** [0, 10] by [0, 300] is one possible window; the x-intercept is not displayed because it is not valid for the problem. **c.** Answers may vary.

Exercise 6.6 (page 392)

1. $\frac{1}{x}$ and $\frac{1}{x+1}$ **2.** $\frac{5}{16}$ **3.** $\frac{x-1}{x+3}$ **5.** 3 **7.** 3 **9.** 6 and 8
11. 150 mph **13.** 45 mph **15.** jet: 340 mph; smaller plane: 170 mph **17.** $2\frac{2}{9}$ hours **19.** 6 hours **21.** $1\frac{1}{9}$ days
23. $2\frac{1}{2}$% **25. a.** 2,500 texts **b.** 1,600 texts per year were produced at an average cost of $86.25. **c.** When 4,000 texts are produced, the average cost is $67.50.
d. It is not applicable since a negative number of texts cannot be produced. **27.** $\frac{1}{1+a}$ **29.** $-4 - \frac{16}{x^2} + 6x^2$ or $\frac{-4x^2 - 16 + 6x^4}{x^2}$

Chapter 6 Summary (page 395)

1. a. $-\frac{1}{3}$ **b.** $-\frac{7}{3}$ **c.** $\frac{1}{2x}; x \neq 0$ **d.** $\frac{5}{2x}; x \neq 0, y \neq 0$
2. a. $\frac{x}{x+1}; x \neq -1, 0$ **b.** $\frac{1}{x}; x \neq 0, -2$ **c.** $\frac{x+3}{x-5}$;
$x \neq 5, -1$ **d.** $\frac{x+7}{x+3}; x \neq -3, 8$ **3. a.** $\frac{x}{x-1}; x \neq 1, 8$
b. in lowest terms; $x \neq 2, -1$ **4. a.** $\frac{3x}{y}$ **b.** $\frac{6}{x^2}$ **c.** $\frac{3y}{2}$
d. $\frac{1}{x}$ **5. a.** 1 **b.** 1 **c.** $x + 2$ **d.** $\frac{n+2}{n+1}$ **6. a.** 1
b. $\frac{2x+2}{x-7}$ **c.** $\frac{x^2 + x - 1}{x(x-1)}$ **d.** $\frac{x-7}{7x}$ **7. a.** $\frac{x-2}{x(x+1)}$
b. $\frac{x^2 + 4x - 4}{2x^2}$ **c.** $\frac{x+1}{x}$ **d.** 0 **8. a.** $\frac{1+x}{1-x}$ **b.** $\frac{x^2 + 3x}{2x^2 - 1}$
c. $x^2 + 3$ **d.** $\frac{a^2 + abc}{b^2 + abc}$ **9. a.** 3 **b.** 1 **c.** 3 **d.** 4 or $-\frac{3}{2}$
e. -2 **f.** 0 **g.** 3; -3 is extraneous **10.** 5 **11.** $9\frac{9}{19}$
hours **12.** $5\frac{5}{6}$ days **13.** 5 mph **14.** 40 mph **15.** 4%
16. 24 children

Chapter 6 Test (page 399)

1. $\frac{8x}{9y}$ **2.** $\frac{x+1}{2x+3}; x \neq \frac{3}{2}, x \neq -\frac{3}{2}$ **3.** 3; $x \neq -1$
4. $\frac{5y^2}{4t}$ **5.** $\frac{x+1}{3(x-2)}$ **6.** $\frac{3t^2}{5y}$ **7.** $\frac{x^2}{3}$ **8.** $\frac{10x-1}{x-1}$ **9.** $\frac{13}{2y+3}$
10. $\frac{2x^2 + x + 1}{x(x+1)}$ **11.** $\frac{2x^3}{y^3}$ **12.** $\frac{x+y}{y-x}$ **13.** -5 **14.** 6
15. 4 **16.** $\frac{2}{3}$ **17.** $3\frac{15}{16}$ hours **18.** 5 mph **19.** 1.5 hours
20. $3\frac{1}{3}$ mph

Cumulative Review: Chapters 5 and 6 (page 400)

1. false **2.** true **3.** trinomial **4.** -1. **5.** Answers may vary. **6.** $a^2 - b^2$ **7.** prime **8.** two **9.** Answers

may vary. **10. a.** quadratic **b.** linear **c.** quadratic
11. rational **12. a.** $x \neq 2$ **b.** $x \neq -\frac{1}{4}$ **c.** $x \neq 0$
13. $\frac{4xy}{3}$ **14.** $18x^2y^2$ **15. a.** $\frac{1}{x^2y}$ **b.** $-\frac{10y^4}{3}$ **16.** complex **17.** No, answers may vary. **18.** extraneous
19. a. 6; $x \neq 3$ **b.** $\frac{3}{x-3}; x \neq 3$ **20.** $\frac{1}{x}$ and $\frac{1}{x+2}$
21. $(x-7)(x+7)$ **22.** $(x-9)(x+1)$ **23.** $(3x-5)(6x+1)$
24. $x^2(x+2)$ **25.** $(x-9)^2$ **26.** $5x(x-5)(x+5)$
27. $(c-d)(a+b)$ **28.** $(a-2y)(a^2 + 2ay + 4y^2)$
29. -1 or -7 **30.** 0, 3, or -4 **31.** $x \approx -1.62$ or
3.79 **32.** $x^2 - 3x - 10 = 0$ **33.** Answers may vary.
34. $\frac{4}{a}$ **35.** $\frac{x^2 - 5x}{x^2 - 25}$ or $\frac{x(x-5)}{(x+5)(x-5)}$ **36.** $\frac{y-10}{(y+2)(y-2)}$
37. Answers may vary. **38.** $2x(x+3)^2$ **39.** $\frac{-(b-1)}{b(a+3)}$ or
$\frac{1-b}{b(a+3)}$ **40.** -11 **41.** $\frac{ab(a+b)}{a^2 + ab + b^2}$ **42.** 2; 3 is extraneous
43. $-\frac{3}{5}$ or 1 **44.** Answers may vary. **45.** excluded
value; 3: solution: 5 **46.** 5 seconds **47.** 1 yard
48. 2 hours, 24 minutes or $2\frac{2}{5}$ hours **49.** 7.5 mph

Exercise 7.1 (page 412)

1. 4, 3 **2.** $-4, 3$ **3.** algebraic **4.** term **5.** like
6. coefficient **7.** monomial **8.** binomial **9.** trinomial
10. variable **11.** highest **12.** x^7 **13.** x^{12} **14.** x^4y^6
15. $\frac{x^4}{y^4}$ **16.** 1 **17.** 1 **18.** $\frac{1}{x^2}$ **19.** $\frac{5}{x^2}$ **20.** x^2 **21.** $\frac{1}{x^2}$
22. trinomial **23.** monomial **24.** binomial **25.** trinomial **26.** 6 **27.** 2 **28.** 4 **29.** 8 **30.** 1 **31.** 1 **32.** 2
33. 2 **34.** 5 **35.** 3 **36.** $-2x^2 + 7x - 2$ **37.** $2y^2 + 8y$
38. $-ab - 4b$ **39.** $-7xy + 6x - 4y$ **40. a.** 9
b. $x^2 - 3x + 5$ **c.** -1 **41. a.** 19 **b.** $x^2 - 4xy - y^2$
c. $x = -2; y = 3$ **42.** 6 **43.** 49 **44.** -8 **45.** 8
46. 32 **47.** 13 **48.** $\frac{x}{2}$ **49.** $\frac{x}{3}$ **50.** $4x^2$ **51.** $5y^2$
52. $x + 2y$ **53.** $2x^2 + 3x + 1$ **55.** $\frac{x^{12}}{y^8}$ **57.** $2x$
59. x^2y^2 **61.** a^5b^6c **63.** $256y^8$ **65.** xy^4 **67.** $-5x^7y^7$
69. $\frac{5x^2y}{3}$ **71.** $\frac{1}{x^2y^6}$ **73.** $\frac{a^6}{27b^9}$ **75.** $\frac{-6}{y^2}$ **77.** $13x + 2$
79. $-4y + 8$ **81.** $3b^2 + 3a + 6$ **83.** $x^2 + 5x - 6$
85. $15x^2 + 13x + 2$ **87.** $9x^3 - 6x^2 - 11x + 4$
89. $4b^2 - 25$ **91.** $y^2 - 4y + 4$ **93.** $x^2 + 6x + 9$
95. $6x^2 + 25x - 13$ **97.** $6y^2 + 15y - 10$ **99.** $xy^2 - 2y$
101. $\frac{3}{b} + \frac{6}{a}$ **103.** $3ab - a - \frac{2}{a}$

Exercise 7.2 (page 418)

1. factor **2.** largest **3.** prime **4. a.** $x^2 - 6x + 9$
b. $x^2 - 8x + 16$ **c.** $x^2 - 16$ **d.** $2x^2 - 7x - 15$ **5. a.** $3y^2$
b. $9x^2y^2$ **6. a.** $-1(5x+6)$ **b.** $-1(x^2 - 6x + 9)$
c. $-1(b-a)$ **d.** $-1(a-b)$ **7.** $2(x+4)$ **9.** $2x(x-3)$
11. $5x^2y(3-2y)$ **13.** $3z(9z^2 + 4z + 1)$ **15.** $-3(a+2)$
17. $-3x(2x+y)$ **19.** $(a+b)(x+y)$ **21.** $(x+y)(x+2)$
23. $(3-c)(c+d)$ **25.** $(x+2)(x+3)$ **27.** $3(x+7)(x-3)$
29. $-(a-8)(a+4)$ **31.** prime **33.** $(3y+2)(2y+1)$
35. $(4a-3)(2a+3)$ **37.** $(a-4b)(a+b)$
39. $-(3a+2b)(a-b)$ **41.** $x(3x-1)(x-3)$
43. $(x^2+5)(x^2+3)$ **45.** $(y^2-10)(y^2-3)$
47. $(x-2)(x+2)$ **49.** $(3y-8)(3y+8)$ **51.** prime

53. $(x + y - z)(x + y + z)$ **55.** $(x - y)(x + y)(x^2 + y^2)$
57. $2(x - 12)(x + 12)$ **59.** $2x(x - 4)(x + 4)$
61. $(x - 2)(x^2 + 2x + 4)$ **63.** $(y - 4)(y^2 + 4y + 16)$
65. $(5 + a)(25 - 5a + a^2)$ **67.** $(x + 2y)(x^2 - 2xy + 4y^2)$
69. $2y(x - 2)(x^2 + 2x + 4)$
71. $(a - b - 2)[(a - b)^2 + 2(a - b) + 4]$
73. $-3(a + b)^2$ **75.** $-x(x + 1)^2$ **77.** $(x - 3)(x - 1)$
79. $(y - 6)(y^2 + 6y + 36)$ **81.** $-7x^4(3x^3 - 2x + 1)$
83. $\frac{1}{65,536}$ **85.** $-x^2 - 6x + 17$

Exercise 7.3 (page 429)

1. solve **2.** conditional **3.** identity **4.** contradiction
5. literal equation **6.** $\varnothing$ **7.** $\mathbb{R}$ **8.** No **9.** yes, yes
10. no, no **11.** yes, yes **12.** no **13.** 8 **15.** 19
17. -24 **19.** $\varnothing$ **21.** $\mathbb{R}$ **23.** $y = -2x + 7$
25. $y = 4x - 8$ **27.** $y = \frac{2}{3}x - \frac{5}{3}$ **29.** $y = -\frac{1}{2}x + 3$
31. $h = \frac{3V}{\pi r^2}$ **33.** 9 or -11 **35.** $\frac{3}{2}$ or $-\frac{1}{2}$ **37.** $\varnothing$
39. 2 or 0 **41.** $-\frac{2}{3}$ **43.** -1 or 2 **45.** 4 or -4
47. -20 or -1 **49.** 0, 8, or -3 **51.** 10 or 2 **53.** 2
55. -8 **57.** $\varnothing$, -1 is extraneous **59.** -4, -1 **61.** 3
63. Cut the board so that one piece is 5 feet long.
65. 175 miles **67.** $21\frac{3}{7}$ ft **69.** 30°, 30°, and 120°
71. \$7,500 **73. a.** $\frac{32x^3}{y^4}$ **b.** $\frac{y^4}{12x^7}$ **75.** $x^3 - 64$

Exercise 7.4 (page 444)

1. ordered pairs **2.** abscissa, ordinate **3.** standard
viewing **4.** linear equation, standard **5.** x-intercept
6. y-intercept **7.** function **8.** yes
9. **10. a.** no **b.** yes

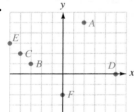

11. a. yes **b.** no **12. a.** yes **b.** no **13. a.** no
b. yes **14. a.** yes **b.** yes
15. **17.**

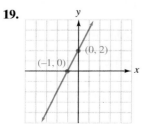 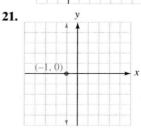

19. **21.**

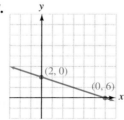

23. **25.** $(2, 0)(0, 6)$

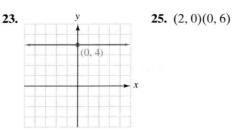

27. $(-12, 0)(0, 24)$ **29.** $(0, -10)(16, 0)$
31. $(9, 0)(0, 5)$ **33.** function **35.** not a function
37. not a function **39.** domain: $\{-8, 9, 7, -\frac{1}{2}, -4.5\}$;
range: $\{5, 2, -3, \frac{2}{3}, 3\}$ **41.** domain: $\{-6, -5.5, -5, -4.5,$
$-4, -3.5, -3\}$; range: $\{31, 25.25, 20, 15.25, 11, 7.25, 4\}$
43. -0.875 **45.** $\varnothing$ **47.** 2, -4 **49.** $-5, 8$ **51.** $-\frac{5}{3}, 5$
53. $-2, 10$ **55.** 6.5, -5.5 **57.** 7 **59.** -0.875
61. $-5, 8$ **63.** $-\frac{5}{3}, 5$ **65.** $-2, 10$ **67.** 2, -4
69. $\frac{4}{3}, -2.5$ **71.** 1, -1, -3 **73.** 2, $-2, \frac{9}{2}$
75. a. $y = 45 + .24x$ **b.** Answers may vary.
c. $\approx$\$76.44 **77.** 1.0962×10^{13} **79.** 0.0262144

Exercise 7.5 (page 456)

1. inequality **2.** interval **3.** critical point(s)
4. disjunctions **5.** conjunctions **6.** zero **7.** $(-\infty, \infty)$
8. $(-\infty, 4)$ **9.** $-2 \le x < 5$ **10.** $(-\infty, -4) \cup (3, \infty)$
11. a. ⟷ **b.** ⟷ **c.** ⟷
d. ⟷ **12. a.** ⟷ **b.** ⟷
c. ⟷ **d.** ⟷
13. a. ⟷ **b.** ⟷
c. ⟷ **d.** ⟷
14. a. ⟷
b. ⟷ **c.** $\varnothing$ **d.** $\varnothing$
15. ⟷ $(-\infty, -3)$ $\{x \mid x < -3\}$ **17.** ⟷
$(-\infty, -12]$ $\{x \mid x \le -12\}$ **19.** ⟷ $(-\infty, 4]$
$\{x \mid x \le 4\}$ **21.** ⟷ $(2, \infty)$ $\{x \mid x > 2\}$
23. ⟷ $(-\infty, -7]$ $\{x \mid x \le -7\}$
25. ⟷ $(4, \infty)$ $\{x \mid x > 4\}$ **27.** ⟷
$(-\infty, 1)$ $\{x \mid x < 1\}$ **29.** ⟷ $[-\frac{1}{2}, \infty)$
$\{x \mid x \ge -\frac{1}{2}\}$ **31.** ⟷ $(-12, \infty)$ $\{x \mid x > -12\}$
33. ⟷ **35.** ⟷ $(2, 7)$
$\{x \mid 2 < x < 7\}$ **37.** ⟷ $[-3, 3]$
$\{x \mid -3 \le x \le 3\}$ **39.** ⟷ $(-\frac{5}{2}, \frac{1}{2})$

$\{x \mid -\frac{5}{2} < x < \frac{1}{2}\}$ **41.** 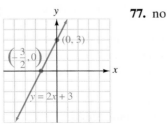 $(-\infty, -5] \cup$

$[1, \infty)$ $\{x \mid x \le -5 \text{ or } x \ge 1\}$ **43.**

$(-\infty, 1) \cup (4, \infty)$ $\{x \mid x < 1 \text{ or } x > 4\}$

45. $(-\infty, 4)$ $\{x \mid x < 4\}$

47. $(-\infty, \infty)$ $\mathbb{R}$

49. $(-5, 1)$ $\{x \mid -5 < x < 1\}$

51. $[0, 2]$ $\{x \mid 0 \le x \le 2\}$

53. $(-3, 4]$ $\{x \mid -3 < x \le 4\}$

55. $(-\infty, 4]$ $\{x \mid x \le 4\}$

57. $(-3, 3)$ $\{x \mid -3 < x < 3\}$

59. $(-7, 3)$ $\{x \mid -7 < x < 3\}$

61. $(-\frac{2}{3}, 2)$ $\{x \mid -\frac{2}{3} < x < 2\}$

63. $[3, 5]$ $\{x \mid 3 \le x \le 5\}$

65. $(-1, 4)$ $\{x \mid -1 < x < 4\}$

67. $(-\infty, -5] \cup [-1, \infty)$

$\{x \mid x \le -5 \text{ or } x \ge -1\}$ **69.**

$(-\infty, 2) \cup (7, \infty)$ $\{x \mid x < 2 \text{ or } x > 7\}$

71. $(-\infty, \infty)$ $\mathbb{R}$

73. $(-\infty, -1] \cup [0, \infty)$

$\{x \mid x \le -1 \text{ or } x \ge 0\}$

75. $(\frac{-3}{2}, 0)$ and $(0, 3)$ **77.** no

Chapter 7 Summary (page 459)

1. a. $9x^6y^4$ **b.** $-8x^9y^9$ **2. a.** $-6x^5y^3$ **b.** $-12x^4y^2$

3. a. $16x^{12}y^{12}$ **b.** $\frac{25y^6}{x^4}$ **4. a.** $\frac{x^5}{3y}$ **b.** $\frac{3y^6z^4}{x^3}$ **5.** 76

6. 38   **7. a.** $3x^2 + 2x + 4$ **b.** $3x^2 - 7x - 3$

c. $-11x^2 - 22x + 13$ **8. a.** $4x^2 - 25$

b. $9x^2 - 6x - 8$ **c.** $4x^2 - 28x + 49$

9. a. $xy^3 - 2y$ **b.** $8a^2b - 2a^3b^2c$

c. $4x - \frac{5}{2} + \frac{1}{x}$ **10. a.** $(x + 2)(y + 4)$ **b.** $(a + c)(b + 3)$

c. $(z - 4)(z + 4)$ **d.** $2(x^2 - 7)(x^2 + 7)$

e. $(y + 20)(y + 1)$ **f.** $-(y - 8)(y + 3)$

g. $3(5x + 1)(x - 4)$ **h.** $(x + 7)(x^2 - 7x + 49)$

i. prime **j.** prime **11. a.** -4.5 **b.** $\frac{3}{7}$ **c.** 7 **d.** $-\frac{1}{2}, 1$

e. $-1, 1$ **f.** $-3, -2, 0$ **g.** $3, -\frac{11}{3}$ **h.** $\varnothing$ **i.** $\frac{9}{8}$ **j.** 4

12. a. $w = \frac{P - 2l}{2}$ **b.** $y = 4x - 2$ **c.** $y = -\frac{1}{2}x + 3$

13. a. $(0, 2), (5, 0)$ **b.** $(0, -3), (4, 0)$

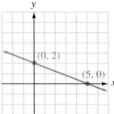 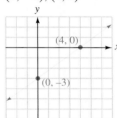

c. $(0, -6), (2, 0)$ **d.** $(0, 7), (\frac{7}{5}, 0)$

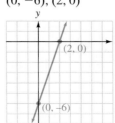 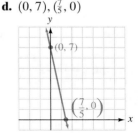

14. a. 3 **b.** -9 **c.** $3, -\frac{11}{3}$ **d.** $\frac{26}{3}, -\frac{10}{3}$ **e.** $5, 6$ **f.** $-3, 8$

15. a. yes **b.** yes **16. a.** $[7, \infty)$,

$\{x \mid x \ge 7\}$ **b.** $(-\infty, 2)$, $\{x \mid x < 2\}$

c. $(-\infty, 13]$, $\{x \mid x \le 13\}$ **d.**

$(-\infty, 5]$, $\{x \mid x \le 5\}$ **17. a.**

$(2, 4)$, $\{x \mid 2 < x < 4\}$ **b.** $\varnothing$ **c.**

$(-\infty, 2) \cup [3, \infty)$, $\{x \mid x < 2 \text{ or } x \ge 3\}$ **d.**

$(-\infty, -3)$, $\{x \mid x < -3\}$ **18. a.**

$(-\infty, -\frac{1}{2}] \cup [\frac{9}{2}, \infty)$, $\{x \mid x \le -\frac{1}{2} \text{ or } x \ge \frac{9}{2}\}$

b. $(-7, 3)$, $\{x \mid -7 < x < 3\}$

c. $\varnothing$ **d.** $\mathbb{R}$

Chapter 7 Test (page 464)

1. $\frac{-2}{3xy}$ **2.** $\frac{1}{xy^2z^2}$ **3. a.** $(a - 5)(a^2 + 5a + 25)$

b. $2(x-4)(x + 4)$ **c.** $(3x + 2)(2x - 1)$ **4. a.** $4x - 1$

b. $9x^2 - 19x$ **c.** $3x - 8$ **5. a.** $3x^2 + 12x + 13$

b. $-2x^2 - 38x - 59$ **c.** $2x - 3 + \frac{4}{x}$ **d.** $-x - 2$

6. -12 **7.** 6 **8.** $t = \frac{A - p}{pr}$ **9.** $y = \frac{3}{7}x - \frac{8}{7}$

10. $13\frac{1}{3}$ ft **11.** 36 cm^2 **12.** 4 or -7 **13.** 3

14. $(2, 0), (0, -6)$

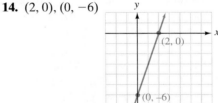

15.

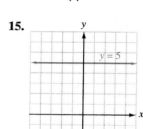

16. a. $\{-1\}$ **b.** $(-1, \infty)$
c. $(-\infty, -1)$ **17. a.** $\{2, 6\}$
b. $(2, 6)$ **c.** $(-\infty, 2) \cup$
$(6, \infty)$ **18.** No

19. ⟵●━━⟶ $(-\infty, 0]$ **20.** ⟵━○──●━⟶ $(-1, 6]$
　　　0　　　　　　　　　　　　　-1　6

21. ⟵━━○━━⟶ $(-\infty, -\frac{1}{2}) \cup (-\frac{1}{2}, \infty)$
　　　　-$\frac{1}{2}$

22. ⟵━○━━⟶ $(-\infty, -\frac{1}{3})$ **23.** ⟵━●━━●━⟶
　　　-$\frac{1}{3}$　　　　　　　　　　$\frac{1}{2}$　$4\frac{1}{2}$

$(\frac{1}{2}, 4\frac{1}{2})$ **24.** ⟵━●━━●━⟶ $(-\infty, -12] \cup [18, \infty)$
　　　　　　　　　　-12　　18

25. $\varnothing$

Exercise 8.1 (page 477)

1. relation **2.** domain **3.** range **4.** function
5. ordered pairs **6.** f of x **7.** y **8.** $\{(3, 6), (4, 8),$
$(5, 10), (6, 12)\}$; domain: $\{3, 4, 5, 6\}$; range: $\{6, 8, 10, 12\}$
9. $\{(-3, -6), (-2, -5), (-1, -4), (0, -3), (1, -2)\}$;
domain: $\{-3, -2, -1, 0, 1\}$; range: $\{-6, -5, -4, -3, -2\}$
10. $\{(15, 127.50), (25, 212.50), (30, 255), (25, 212.50)\}$;
domain: $\{15, 25, 30\}$; range: $\{127.50, 215.50, 255\}$
11. $\{(50, 150), (60, 180), (70, 210)\}$; domain: $\{50, 60, 70\}$;
range: $\{150, 180, 210\}$ **12. a.** $(-3, 2), (-1, 1), (1, 1),$
$(1, -1), (3, 0)$ **b.** $\{-3, -1, 1, 3\}$ **c.** $\{2, 1, -1, 0\}$ **d.** Not
a function because $(1, 1)$ and $(1, -1)$ both belong to the
relation. **13.** $f(x)$: $-7, -5, -1, 1, 7$; (x, y): $(-3, -7)$,
$(-2, -5), (0, -1), (1, 1), (4, 7)$ **14.** $f(x)$: $8, 3, 0, -1, 0$;
(x, y): $(-3, 8), (-2, 3), (-1, 0), (0, -1), (1, 0)$ **15.** $f(x)$: 7,
$5, 3, 3, 7$; (x, y): $(-8, 7), (-6, 5), (-4, 3), (2, 3), (6, 7)$
16. $f(x)$: $9, 7, 5, 3, 7$; (x, y): $(-8, 9), (-6, 7), (-4, 5), (2, 3),$
$(6, 7)$ **17. a.** -6 **b.** -2 **c.** 2 **d.** 8 **18. a.** 17.5
b. 4 **c.** 1.75 **d.** 6 **19.** function; domain: $\{2, 3, -4\}$;
range: $\{5, -1, 7\}$ **21.** not a function; domain: $\{-1, -2\}$;
range: $\{2, 3, -2\}$ **23.** function **25.** function **27.** not a
function **29.** function **31.** domain: $\mathbb{R}$, range: $(-\infty, 5]$
33. domain: $\mathbb{R}$, range: $[-10, \infty)$ **35.** not a function, domain: $(-\infty, 2]$, range: $(-\infty, \infty)$ **37.** function, domain:
$(-\infty, \infty)$, range: $(-\infty, \infty)$ **39.** function, domain: $[0, 5]$,
range: $[0, 2]$ **41.** not a function, domain: $[-\frac{1}{2}, \frac{1}{2}]$, range:
$[-1, 1]$ **43.** not a function **45.** function **47. a.** 8
b. -7 **c.** $3a + 2$ **d.** $3a + 8$ **49. a.** 0 **b.** 15
c. $a^2 - 2a$ **d.** $a^2 + 2a$ **51. a.** $-\frac{3}{2}$ **b.** $-\frac{3}{7}$ **c.** $\frac{3}{a-4}$
d. $\frac{3}{a-2}$ **53.** -3 **55.** 2 **57.** 5 **59.** -6 **61.** 4
63. 12 **65.** $2b - 2a$ **67.** $2b$ **69.** 1 **71. a.** -11.8
b. $\frac{1}{2}$ **c.** 185.63 **d.** 5 **73. a.** 4 **b.** 6 **c.** 351
d. 210 **75.** Not linear

77. linear; $y = \frac{1}{3}x + \frac{4}{3}$ **79.** linear; $y = 2x - 4$

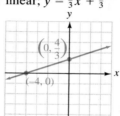

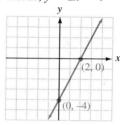

81. Yes, since $2(5) + 4 \geq 9$. **83.** 4 **85.** Yes, since
$(-7)^2 - 49 = 0$.

Exercise 8.2 (page 486)

1. increasing **2.** decreasing **3.** left, right **4.** absolute
5. relative **6.** absolute **7.** relative **8.** turning
9. increasing $(-\infty, \infty)$ **10.** decreasing $(-\infty, \infty)$
11. increasing $[0, \infty)$ **12.** decreasing $[0, \infty)$ **13.** increasing $(-\infty, 0]$ **14.** decreasing $(-\infty, 0]$ **15.** increasing $[-4, 4]$ **16.** increasing $[-3, 2]$ **17.** increasing
$[-8, -6]$ and $[3, 6]$, decreasing $[2, 3]$, constant $[-6, 2]$
19. $(1, 4)$, min. **21.** $(-2, 4)$, max. **23.** relative min; -2;
relative max: 2 **25.** relative min: 0; relative max: 4
27. a. $[-5, -3]$ **b.** $[-3, -2]$ and $[1, 2]$ **c.** $[-2, 1]$ and
$[2, 5]$ **29. a.** domain: $[-5, 5]$ **b.** range: $[-2, 7]$
31. -8 is not part of the domain. **33.** It is a function
since it passes the VLT. **35.** $(4, 1)$, relative max. **37.** 5
39. a. domain: $[-3, 4]$; range: $[0, 3]$ **b.** increasing:
$[-3, 0], [2, 4]$; decreasing: $[0, 2]$ **c.** absolute max: 3; relative max: 2, 3; absolute and relative min: 0 **41. a.** domain: $[-4, \infty)$; range: $[-3, \infty)$ **b.** increasing: $[-4, -2]$;
$[3, \infty)$; decreasing: $[-2, 3]$ **c.** relative max: 3; no absolute
max; absolute min: -3; relative min: $0, -3$ **43. a.** domain: $[-3, \infty)$; range: $(-\infty, 0]$ **b.** decreasing: $[-3, \infty)$
c. absolute and relative max: 0 **45.** x-intercept: $(-\frac{2}{3}, 0)$;
y-intercept: $(0, 2)$ **47.** no x-intercept: y-intercept:
$(0, 5)$ **49.** x-intercept: $(-4, 0)$; y-intercept: $(0, 2)$
51. Answers may vary. **53. a.** How far he has
traveled horizontally. **b.** How far he has traveled
vertically. **c.** 77 ft **d.** 62 ft **e.** 20 ft **f.** 32 ft
g. Before he is shot out of the cannon his head is 9.76 ft
above the ground. **55. a.** $17.50 **b.** $156.25 **c.** $136
profit **d.** $15 or $20/ticket **57.** Answers may vary.
59.

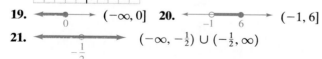

Exercise 8.3 (page 500)

1. y, x **2.** rise   **3.** $\frac{y_2 - y_1}{x_2 - x_1}$ **4.** positive   **5.** negative
6. slope-intercept, slope, y-intercept   **7.** zero
8. undefined   **9.** equal   **10.** negative reciprocals.
11. $m = \frac{1}{2}$ **12.** $m = \frac{2}{3}$ **13.** $m = -2$ **14.** $m = -2$
15. negative   **16.** positive   **17.** no defined value
18. positive   **19.** zero   **20.** y_2 **21.** y_1 **22.** y_2 **23.** y_2
25. $m = 2$ **27.** $m = -\frac{8}{5}$ **29.** undefined   **31.** $m = -\frac{4}{3}$
33. $m = \frac{8}{9}$ **35.** undefined   **37.** $m = \frac{1}{3}$ **39.** $m = \frac{5}{3}$
41. $m = -4$ **43.** undefined   **45.** $m = 0$ **47.** $m = 0$
49. $m = 1$ **51.** $m = -\frac{5}{7}$ **53.** $x = \frac{5}{4}$ **55.** $y = -1$
57. $x = -5$ **59.** $y = 2$ **61.** $m = -\frac{2}{3}; b = 4$ **63.** $m = 3;$
$b = -9$ **65.** $m = \frac{4}{3}; b = \frac{8}{3}$ **67.** $m = -4, b = 3$
69. $m = 2$ **71.** $m = -3$ **73.** $m = -\frac{1}{8}$ **75.** neither
77. perpendicular   **79.** parallel   **81.** A vertical line is
not a function.   **83.** $y = 4x - 2$ **85.** $y = -\frac{3}{2}x + 5$
87. $3\frac{1}{2}$ students per year   **89.** $\frac{1}{4}$ **91. a.** $\frac{3}{16}$ **b.** $\frac{1}{8}, -\frac{1}{8}$
c. Answers may vary.   **93. a.** $m = .75$; cost per ride
b. $y = .50x + 7$ **c.** No, because that would mean the
customer would be paying a negative amount of money
for each ride.   **95. a.** $y = 250x + 3,000$ **b.** increase in
value per year   **c.** It would be losing value.

Exercise 8.4 (page 510)

1. slope-intercept; slope; y-intercept   **2.** point-slope,
slope, (x_1, y_1) **3.** standard   **4.** $y = k$, zero   **5.** $x = k$;
undefined   **6.** straight-line depreciation   **7. a.** yes
b. $m = 5$ **c.** $(0, 2)$ **d.** $y = 5x + 2$ **8. a.** $(0, 120000)$,
$(5, 150000)$ **b.** $m = \frac{30,000}{5} = \$6,000/\text{yr}$ **c.** 120,000
d. $y = 6,000x + 120,000$ **e.** \$180,000, 14 years
9. a. $y = x + 3$ **b.** $y = -x + 5$ **c.** $y = 5x - 4$
d. $y = x - 3$ **e.** $y = 5x + 4$ **f.** $y = x + 5$
10. a. $x = -2$ **b.** $y = 3$ **c.** $x = 5$ **d.** $y = -4$
e. $x = -6$ **f.** $y = 8$ **11.** $(9, 12)$ **13.** $(\frac{7}{2}, 6)$
15. $(\frac{5a}{2}, 2b)$ **17.** $(a, 2b)$ **19.** $5x - y = -7$
21. $3x + y = 6$ **23.** $3x - 2y = -4$ **25.** $y = x$
27. $y = \frac{7}{3}x - 3$ **29.** $y = -\frac{9}{5}x + \frac{2}{5}$
31. $y = x - 1, m = 1, b = -1$ **33.** $m = \frac{2}{3}, b = \frac{9}{2}$

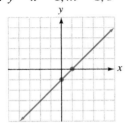
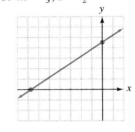

35. $y = -\frac{2}{3}x + 6, m = -\frac{2}{3}, b = 6$

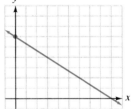

37. $y = 4x$ **39.** $y = 4x - 3$
41. $y = \frac{4}{5}x - \frac{26}{5}$ **43.** $y = -\frac{1}{4}x$
45. $y = -\frac{1}{4}x + \frac{11}{2}$
47. $y = -\frac{5}{4}x + 3$

49. $x = -2$ **51.** $x = 5$ **53.** $y = -\frac{A}{B}x + \frac{C}{B}, m = -\frac{A}{B}, b = \frac{C}{B}$
55. $y = 3,200x + 24,300$ **57.** $y = 47,5000x + 475,000$
59. $y = -670x + 3,900$ **61.** \$90   **63.** \$890   **65.** \$37,200
67. \$230   **69.** \$510   **71.** $1, -\frac{3}{2}$ **73.** $-\frac{3}{2}, \frac{3}{2}$ **75.** 20 oz.

Exercise 8.5 (page 520)

1. direct   **2.** linear   **3.** inverse   **4.** rational   **5.** joint
6. combined, directly, inversely   **7.** inverse variation;
$k = 8$ **8.** joint variation; $k = \frac{1}{2}$ **9.** combined variation; $k = 3$ **10.** direct variation; $k = \frac{1}{2}$ **11.** joint variation; $k = 3$ **12.** combined variation; $k = 3$ **13.** $A = kp^2$
15. $v = \frac{k}{r^3}$ **17.** $B = kmn$ **19.** $P = \frac{ka^2}{j^3}$ **21.** $F = \frac{km_1m_2}{d^2}$
23. L varies jointly as m and n.   **25.** E varies jointly
as a and b^2.   **27.** X varies directly as x^2 and inversely
as y^2.   **29.** R varies directly as L and inversely as d^2.
31. 36π in^2 **33.** 3 ohms   **35.** 1,600 feet   **37.** 13.5 feet
39. \$3,000   **41.** 0.275 in.   **43.** increased by a factor of 9
45. 26,437.5 gallons   **47.** \$9,000   **49.** 546°K   **51.** 180
53. Answers may vary.   **55.** No, change $\neq k \div$ purchase.
57. domain: $(-\infty, \infty)$; range: $[3, \infty)$ **59.** domain:
$(-\infty, \infty)$; range: $[0, \infty)$

Chapter 8 Summary (page 524)

1. a. function; domain and range $= \mathbb{R}$ **b.** function; domain and range $= \mathbb{R}$ **c.** function; domain: $\mathbb{R}$; range:
$[1, \infty)$ **d.** function; domain: $\{x \mid x \neq 2\}$; range: $\{y \mid y \neq 0\}$
e. function; domain and range $= \mathbb{R}$ **f.** not a function;
domain: $[0, \infty)$; range: $\mathbb{R}$ **2. a.** function   **b.** not a function   **c.** not a function   **d.** not a function   **3. a.** -7
b. 60   **c.** 0   **d.** 17   **4. a.** $f(-2) = -3$ **b.** $f(4) = 3$
5. a. linear   **b.** linear   **c.** linear   **d.** nonlinear
6. a. increasing $[0, \infty)$ **b.** decreasing $[0, \infty)$
c. decreasing $(-\infty, 5]$ **d.** increasing $[0, 3]$;
decreasing $[3, \infty)$ **e.** decreasing $(-\infty, 0], [2, \infty)$;
increasing $[0, 2]$ **f.** increasing $[-\infty, -4], [1, \infty)$;
decreasing $[-2, 1]$; constant $[-4, -2]$

7. a. relative max. (0, 36)   **b.** relative min. (−3, −5)

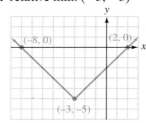

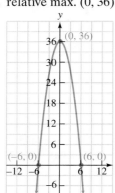

c. relative min.   **d.** relative max
(0.45, −1.26);   (−1.11, 24.1);
relative max   relative min
(2.22, 4.23)   (2.11, −9.11)

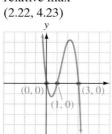

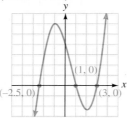

8. a. [0, 10] by [0, 260] recommended window
b. (4, 256)   **c.** The ball is 256 ft high after 4 seconds.
d. The ball is at a height of 245.76 ft going up and
coming down.   **e.** The ball is on the ground.
9. a. $y = x(80 − 2x)$ **b.**

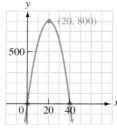

c. 800 ft^2 **d.** 20 ft   **10. a.** $m = 1$ **b.** $m = \frac{14}{9}$
c. $m = 5$ **d.** $m = \frac{5}{11}$ **e.** $m = -\frac{1}{2}$ **11. a.** $m = \frac{2}{3}$
b. $m = -2$ **c.** undefined   **d.** undefined   **e.** $m = 0$
12. a. $y = \frac{2}{3}x − 5$ **b.** $y = -\frac{1}{2}x + 2$ **13. a.** positive
b. negative   **c.** 0   **d.** undefined   **14. a.** perpendicular   **b.** parallel   **c.** neither   **d.** perpendicular
15. a. $3x − y = −29$ **b.** $13x + 8y = 6$ **c.** $3x − 2y = 1$
d. $2x + 3y = −21$ **16. a.** $2x − 3y = 6$ **b.** $x = 3$
17. $(\frac{3}{2}, 8)$ **18. a.** $550 per year   **b.** $y = −550x + 2,500$
c. [0, 10] by [0, 3000]   **19.** ≈$3.19 per year   **20.** 72
21. 6   **22.** 2   **23.** 16   **24.** 31   **25.** The volume is multiplied by 6.

Chapter 8 Test (page 529)
1. y is a function of x.   **2.** domain: $(−∞, 2]$, range:
$(−∞, 3]$ **3.** $−3$ **4. a.** $(−∞, −3]$ **b.** $[2, 4]$ **c.** $[−3, 2]$
5. a. $[0, 3]$ by $[0, 30]$ **b.** 26 ft   **c.** ≈2.27 sec.   **d.** (1, 26)
6. $m = \frac{2}{3}$ **7.** $m = \frac{2}{5}$ **8.** $y = \frac{4}{5}x + 3$ **9.** neither
10. $y = −1$ **11.** $y = -\frac{5}{11}x + \frac{31}{11}$ **12.** Yes; answers
may vary.   **13.** $(\frac{1}{2}, \frac{1}{2})$ **14.** $y = \frac{kw}{z^2}$ **15.** $V = \frac{1}{3}Bh$
16. $P = kc^2$ **17. a.** increasing   **b.** 3 amps of cur-
rent produces a power loss of 45 watts due of heat.
18. $x = \frac{135}{2}$ **19.** $t = 14\frac{2}{3}$

Cumulative Review: Chapters 7 & 8 (page 530)
1. factor   **2. a.** x^3 **b.** x^{10} **c.** $\frac{1}{x^5}$ **d.** $\frac{1}{x^7}$ **3.** simplify
4. evaluate   **5.** solution   **6.** function   **7. a.** not a
function   **b.** function   **8.** intersection   **9.** union
10. $−4 ≤ x + 2 ≤ 4$ **11. a.** function   **b.** not a function
c. function   **d.** function   **e.** function   **f.** not a function
12. a. 6   **b.** 2   **c.** 20   **d.** 42   **13.** $y = \frac{1}{3}x + \frac{4}{3}$ **14.** y, x
15. $y = $ a constant   **16.** $x = $ a constant   **17.** 0
18. undefined   **19.** slope-intercept; slope; y-intercept
20. point-slope; slope; (x_1, y_1) **21. a.** $\frac{2y^7}{x^{10}}$ **b.** $\frac{2^{11}}{5^7} ≈ .026$
22. 8   **23.** 4, $-\frac{1}{2}$ **24.** 0 or $−2$ **25.** no solution ($\varnothing$)
26. $\frac{1}{2}$ **27.** $[\frac{1}{3}, 3]$ 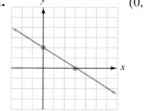   **28.** $(−∞, −4] \cup$

$[−1, ∞)$ **29.** $−3$ **30. a.** $[−3, −1]$
b. $[−6, −3]$ and $[4, 6]$ **c.** $[−1, 4]$
31.   (0, 2), (3, 0)

32. domain = $\{1, 2, −1, −3\}$ range = $\{−1, 5, −5\}$
33. $m = \frac{1}{2}$ **34.** $m = −4, b = 3$ **35.** perpendicular
$(m = \frac{1}{2}, −2)$ **36.** $2x − 3y = −15$ **37.** $3x − y = 14$
38. $y = 2$ **39.** $x + 2y = 7$ **40.** $x − 3y = −15$
41. $x = 5$ **42.** $x = 5.5$ **43.** $x = 0, 4, −3$
44. minimum at (1, 6)

45. a. $y = 15x + 300$ **b.** The slope is the yearly rate of
increase.   **c.** $y = $600   **46. a.** $y = 20x + 45$ **c.** $145;
answers may vary.   **d.** You can't work a negative number
of hours.   **e.** $445; answers may vary.   **47.** approxi-

mately 229.2 miles.   **48.** She needs to make a 37 or higher.   **49.** 81π mm^2 **50.** \$4,000

Exercise 9.1 (page 541)

1. positive/principal   **2.** radicand   **3.** exact, approximate, irrational   **4.** index   **5.** index, radicand   **6.** x
7. b, d   **8. a.** irrational   **b.** rational   **c.** rational   **d.** rational   **9. a.** one   **b.** one   **c.** none   **10. a.** $|x|$
b. y^2 **c.** a **d.** $|x + 3|$ **e.** $|x + 5|$ **11.** 11   **13.** 1
15. $2a$ **17.** $\frac{1}{3}$ **19.** -3 **21.** 0.4   **23.** $\frac{1}{2}$ **25.** -8
27. -5 **29.** 5   **31.** $-10pq$ **33.** $\frac{2}{5}$ **35.** 0.5   **37.** 3
39. 4   **41.** $-\frac{2}{3}$ **43.** $-\frac{1}{2}m^2n$ **45.** -2 **47.** $-\frac{5}{7}$ **49.** 0.4
51. Undefined for real numbers   **53.** $2|x|$ **55.** $2a$
57. $|x|^3$ **59.** $|a + 3|$ **61.** $\frac{1}{2}|x|$ **63.** $-x$ **65.** $-3a^2$
67. $5|b|$ **69. a.** $\sqrt[5]{32}$ **b.** rational   **71. a.** $8\sqrt[4]{64}$
b. irrational   **73. a.** $3\sqrt{25} - 8$ **b.** irrational
75. a. $\frac{5}{\sqrt{16} + 1}$ **b.** rational   **77.** >   **79.** >   **81.** <
83. a. It decreases.   **b.** A person who is 71 inches tall has a pulse rate of approximately 70 beats per minute.
85. $\frac{9x^2}{y^2}$ **87.** $9x^6y^4$ **89.** $\frac{8}{27}a^3$

Exercise 9.2 (page 551)

1. base, factor   **2.** one   **3.** rational   **4.** irrational
5. $\sqrt[3]{7}$ **6.** $\sqrt{26}$ **7.** $3\sqrt[4]{x}$ **8.** $4a\sqrt[6]{b}$ **9.** $\sqrt[4]{3x}$
10. $\sqrt[6]{4ab}$ **11.** $(3a)^{1/4}$ **12.** $3a^{1/4}$ **13.** $(\frac{1}{7}abc)^{1/6}$
14. $(\frac{3}{8}p^2q)^{1/7}$ **15.** $(a^2 - b^2)^{1/3}$ **16.** $(x^2 + y^2)^{1/2}$
17. 2   **19.** 2   **21.** $\frac{1}{2}$ **23.** -2 **25.** 0   **27.** 27   **29.** $\frac{1}{4}$
31. $\frac{1}{2}$ **33.** $\frac{1}{8}$ **35.** $\frac{16}{81}$ **37.** $125x^6$ **39.** $\frac{4x^2}{9}$ **41.** $\frac{1}{64x^3}$
43. $\frac{1}{9y^2}$ **45.** $\frac{1}{4p^2}$ **47.** $-\frac{3}{2x}$ **49.** 6   **51.** $5x$ **53.** $6b$
55. $3a$ **57.** $2\sqrt{5}$ **59.** $-10\sqrt{2}$ **61.** $2\sqrt[3]{10}$ **63.** $-3\sqrt[3]{3}$
65. $2\sqrt[3]{3}$ **67.** $\frac{\sqrt[3]{7}}{3}$ **69.** $\frac{\sqrt[5]{3}}{2}$ **71.** $5ab\sqrt{7b}$ **73.** $2x^3y\sqrt[4]{2}$
75. $\frac{z}{4x}$ **77.** $\frac{\sqrt[4]{5x}}{2z}$ **79.** $\frac{1}{9x^4y^2}$ **81.** $-8x^3y^6$ **83.** 64

Exercise 9.3 (page 554)

1. like   **2.** index, distributive   **3.** true   **4.** false
5. false   **6.** false   **7.** 3, 2; 3, 2; 2   **8.** 2, 5, 1; 2, 5, 1; 1
9. $10\sqrt{2}$ **11.** $\sqrt[5]{7}$ **13.** $14\sqrt{x}$ **15.** $4\sqrt{3}$ **17.** $-\sqrt{2}$
19. $2\sqrt{2}$ **21.** $9\sqrt{6}$ **23.** $3\sqrt[3]{3}$ **25.** $-\sqrt[3]{4}$ **27.** -10
29. $-41\sqrt[3]{2}$ **31.** $-17\sqrt[4]{2}$ **33.** $16\sqrt[4]{2}$ **35.** $-4\sqrt{2}$
37. $3\sqrt{2} + \sqrt{3}$ **39.** $-11\sqrt[3]{2}$ **41.** $y\sqrt{z}$ **43.** $13y\sqrt{x}$
45. $12\sqrt[3]{a}$ **47.** $-7y^2\sqrt{y}$ **49.** $4x\sqrt[5]{xy^2}$ **51.** $2x + 2$
53. $(2x + 1)\sqrt{3}$ **55.** 4.5   **57.** 2, -10
59. a. [number line from -5 to 2]   **b.** $[-5, 2]$

Exercise 9.4 (page 560)

1. rationalizing   **2.** conjugate   **3.** $\sqrt[3]{5^2}$ **4.** $1 - \sqrt{3}$
5. -1 **6.** 2   **7.** $x - 1$ **8.** $x - y$ **9.** 4   **11.** $5\sqrt{2}$
13. $3\sqrt{2}$ **15.** 5   **17.** 3   **19.** $2\sqrt[3]{3}$ **21.** ab^2
23. $5a\sqrt{b}$ **25.** $r\sqrt[3]{10s}$ **27.** $2a^2b^2\sqrt[3]{2}$
29. $x^2(x + 3)$ **31.** $3x(y + z)\sqrt[3]{4}$ **33.** $12\sqrt{5} - 15$

35. $12\sqrt{6} + 6\sqrt{14}$ **37.** $-8x\sqrt{10} + 6\sqrt{15x}$
39. $-1 - 2\sqrt{2}$ **41.** $24x - 14\sqrt{3x} - 15$
43. $5z + 2\sqrt{15z} + 3$ **45.** $3x - 2y$
47. $6a + 5\sqrt{15ab} - 15b$ **49.** $18r - 12\sqrt{2r} + 4$
51. $-6x\sqrt{3x} - 12x\sqrt{3} - 6\sqrt{3x}$ **53.** $\frac{\sqrt{7}}{7}$ **55.** $\frac{\sqrt{6}}{3}$
57. $\frac{\sqrt{10}}{4}$ **59.** 2   **61.** $\frac{\sqrt[3]{4}}{2}$ **63.** $\sqrt[3]{3}$ **65.** $\frac{\sqrt[3]{6}}{3}$ **67.** $2\sqrt{2x}$
69. $\frac{\sqrt{5y}}{y}$ **71.** $\frac{\sqrt[3]{2b^2}}{b}$ **73.** $\frac{\sqrt[4]{27}}{3}$ **75.** $2\sqrt[5]{2}$ **77.** $\sqrt{2} + 1$
79. $\frac{6(\sqrt{5} - 4)}{11}$ **81.** $\sqrt{3} - 1$ **83.** $5(\sqrt{6} - 1)$
85. $\frac{3\sqrt{2} - \sqrt{10}}{4}$ **87.** $-2\sqrt{7} - \sqrt{35}$ **89.** $\sqrt{7} + \sqrt{5}$
91. $10(\sqrt{3} - 1)$ **93.** $2 + \sqrt{3}$ **95.** $\frac{9 - 2\sqrt{14}}{5}$
97. $\frac{2(\sqrt{x} - 1)}{x - 1}$ **99.** $\frac{x(\sqrt{x} + 4)}{x - 16}$ **101.** $\sqrt{2z} + 1$
103. $\frac{x - 2\sqrt{xy} + y}{x - y}$ **105.** 1   **107.** $\frac{1}{3}$

Exercise 9.5 (page 568)

1. radical, radicand   **2.** right, right   **3.** right   **4.** hypotenuse, right   **5.** legs, hypotenuse   **6.** rectangular
7. a. iii   **b.** iv   **c.** v   **d.** ii   **e.** i   **f.** vi   **8.** d   **9.** Answers may vary.   **10.** c   **11.** 6   **12.** 2.25   **13.** $-1, 2$
14. 1, 2   **15.** 4, -1 **16.** 5, -2 **17.** $\varnothing$ **18.** $\varnothing$ **19.** 2
21. 4   **23.** 8   **25.** 4   **27.** 0   **29.** 16   **31.** 2; 7 is extraneous   **33.** 14; 6 is extraneous   **35.** 4, 3   **37.** -1; 1 is extraneous   **39.** 0   **41.** 1, 9   **43.** 14; 6 is extraneous
45. 4, 3   **47.** 9; -25 is extraneous   **49.** 2, -1 **51.** 0; 4 is extraneous   **53.** no solution   **55.** 4; 0 is extraneous
57. 2   **59.** 1   **61.** 4; -9 is extraneous   **63.** no solution
65. 4   **67.** 10 ft   **69.** 80 m   **71.** 13   **73.** 10   **75.** $2\sqrt{26}$
77. $2\sqrt{13}$ **79.** $4\sqrt{2}$ **85.** 8 ft   **87.** 66.5 ft   **89.** approximately 95 ft   **91.** no   **93.** yes   **95.** 90 yd   **97.** 28 ft
99. relation   **101.** function

Exercise 9.6 (page 578)

1. function   **2.** y **3.** x, domain   **4.** y, range   **5.** 1
6. domain   **7.** 3.606   **8.** d   **9.** a   **10.** c   **11.** b   **12.** c
13. a   **14.** d   **15.** b   **16.** a   **17.** d   **18.** b   **19.** c
20. a. It shifts the graph up or down the same number of units added or subtracted.   **b.** range   **c.** Answers may vary; one possibility is $y = \sqrt{x} - 6$.   **21. a.** It shifts the graph left or right the same number of units added or subtracted.   **b.** domain   **c.** Answers may vary; one possibility is $y = \sqrt{x + 6}$.   **22. a.** The negative reflects the graph across the y-axis; the constant shifts the graph left or right.   **b.** The negative sign affects the range; the constant affects the domain.   **c.** Answers may vary; one possibility is $y = \sqrt{-3 - x}$.   **23. a.** It stretches the graph.
b. neither   **c.** It reflects the graph across the x-axis.
d. range   **24.** Answers will vary; one possibility is $y = \sqrt{x + 3} + 3$.   **25.** Answers will vary; one possibility is $y = \sqrt{x - 2} + 2$.   **26.** Answers will vary; one possibility is $y = -\sqrt{x + 6} + 2$.   **27.** Answers will vary; one possibility is $y = -\sqrt{x - 1} - 4$.   **28.** Answers will

vary; one possibility is $y = \sqrt{6 - x}$.   **29.** Answers will vary; one possibility is $y = \sqrt{-2 - x} + 3$.   **31.** 1   **33.** $3\sqrt{3} + 2$ **35.** $3\sqrt{6} + 2$ **37.** undefined for real numbers   **39.** undefined for real numbers   **41.** 5   **43.** domain: $[-2, \infty)$, range: $[0, \infty)$ **45.** domain: $[0, \infty)$, range: $[-5, \infty)$ **47.** domain: $[-3, \infty)$, range: $[0, \infty)$ **49.** domain: $[\frac{3}{5}, \infty)$, range: $[0, \infty)$ **51.** domain: $[\frac{8}{3}, \infty)$, range: $[-2, \infty)$ **53.** domain: $[\frac{5}{6}, \infty)$, range: $[-4, \infty)$ **55.** domain: $[2, \infty)$, range: $[2, \infty)$ **57.** domain: $[2, \infty)$, range: $(-\infty, 2]$ **59.** domain: $\mathbb{R}$, range: $\mathbb{R}$ **61.** domain: $\mathbb{R}$, range: $\mathbb{R}$ **63.** Answers may vary.   **65.** approximately 319 ft   **67.** 8 ft   **69.** \$8

71.   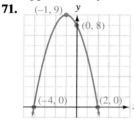   **73.**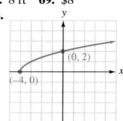

Exercise 9.7 (page 589)

1. i **2.** complex   **3.** real, imaginary   **4.** real, imaginary   **5.** complex conjugate   **6.** $i^1 = i, i^2 = -1, i^3 = -i, i^4 = 1$ **7. a.** i **b.** 1 **c.** -1 **d.** $-i$ **8.** $5i$ **9.** $6i$ **10.** $-5i\sqrt{3}$ **11.** $-10i\sqrt{2}$ **12.** $7i\sqrt{2}$ **13.** $3i\sqrt{2}$ **14.** $3i\sqrt{3}$ **15.** $-4i\sqrt{2}$ **16.** i **17.** $-i$ **18.** $-i$ **19.** -1 **20.** true   **21.** true   **22.** false   **23.** true   **24.** false   **25.** false   **27.** $-1 + 12i$ **29.** $1 + i$ **31.** $-7 + 3i$ **33.** $18 + 14i$ **35.** $-1 + i\sqrt{2}$ **37.** $16 - 12i$ **39.** $6 - 17i$ **41.** $13 - 3\sqrt{3}i$ **43.** $-8 + i$ **45.** $5 - 12i$ **47.** $-4 - 7i$ **49.** $13 + 13i$ **51.** $-\frac{3}{2}i$ **53.** $-\frac{5}{4}$ **55.** $-\frac{2}{3}i$ **57.** $1 + 2i$ **59.** $3 - 2i$ **61.** $-\frac{6}{5} - i$ **63.** $6 + 4i$ **65.** $\frac{3}{17} + \frac{5}{17}i$ **67.** $\frac{6}{5} - \frac{2}{5}i$ **69.** $\frac{3}{5} - \frac{1}{5}i$ **71.** $-\frac{5}{13} + \frac{12}{13}i$ **73.** $\frac{1}{5} + \frac{2\sqrt{6}}{5}i$ **79.** $-5x^2 + 7x - 6$ **81.** $15x^2 - 14x - 16$ **83.** $x + 1$

Chapter 9 Summary (page 592)

1. a. 7 **b.** -11 **c.** -6 **d.** 15 **e.** -3 **f.** -6 **g.** 5 **h.** -2 **2. a.** $5x$ **b.** $x + 2$ **c.** $3a^2\sqrt[3]{b^2}$ **d.** $4a^2y$ **3. a.** -2 **b.** -4 **c.** $\frac{8}{27}$ **d.** $\frac{1}{3,125}$ **e.** $\frac{27}{8}$ **f.** $3xy^{1/3}$ **g.** $3xy^{1/2}$ **h.** $125x^{9/2}y^6$ **i.** $\frac{1}{4u^{4/3}v^2}$ **4. a.** $4\sqrt{15}$ **b.** $3\sqrt[3]{2}$ **c.** $2\sqrt[4]{2}$ **d.** $2\sqrt[5]{3}$ **e.** $2xy\sqrt[3]{2x^2y}$ **f.** $3x^2y\sqrt[3]{2x}$ **g.** $4x$ **h.** $2x$ **i.** $\frac{\sqrt[3]{2a^2b}}{3x}$ **j.** $\frac{\sqrt[3]{17xy}}{8a^2}$ **5. a.** 0 **b.** $8\sqrt[4]{2}$ **c.** $29x\sqrt{2}$ **d.** $13\sqrt[3]{2}$ **6. a.** 3 **b.** $-2x$ **c.** $4 - 3\sqrt{2}$ **d.** $\sqrt{10} - \sqrt{5}$ **e.** 1 **f.** $6u + \sqrt{u} - 12$ **7. a.** $\frac{\sqrt{3}}{3}$ **b.** $\frac{\sqrt[3]{u^2}}{u^2v^2}$ **c.** $2(\sqrt{2} + 1)$ **d.** $\frac{\sqrt{6} + \sqrt{2}}{2}$ **8. a.** 22 **b.** 3 or 9 **c.** $\frac{9}{16}$ **d.** 0 or -2 **e.** $\varnothing$ **9.** 88 yds   **10. a.** 10 **b.** $5\sqrt{10}$ **11. a.** 3 **b.** 2 **c.** 4 **d.** $\sqrt{10} \approx 2.16$ **e.** $\sqrt[3]{16} \approx 2.52$ **12. a.** domain: $[5, \infty)$, range: $[-4, \infty)$ **b.** domain: $[-2, \infty)$, range: $[0, \infty)$ **c.** domain $= \mathbb{R} =$ range **d.** domain $= \mathbb{R} =$ range

13. approximately 3.0 miles   **14.** approximately 8.2 ft   **15. a.** $12 - 8i$ **b.** $2 - 68i$ **c.** $-96 + 3i$ **d.** $-2 - 2\sqrt{2}i$ **16. a.** $22 + 29i$ **b.** $-16 + 7i$ **c.** $-12 + 28\sqrt{3}i$ **d.** $118 + 10\sqrt{2}i$ **17. a.** $-\frac{3}{4}i$ **b.** $-\frac{2}{5}i$ **c.** $6 + i$ **d.** $\frac{13}{2} + 2i$ **e.** $\frac{12}{5} - \frac{6}{5}i$ **f.** $\frac{21}{10} + \frac{7}{10}i$ **g.** $\frac{15}{29} - \frac{6}{29}i$ **h.** $\frac{1}{3} + \frac{1}{3}i$

Chapter 9 Test (page 596)

1. $5xy^2\sqrt{10xy}$ **2.** $2x^5y\sqrt[3]{3}$ **3.** $\frac{1}{4a}$ **4.** $3x$ **5.** $3x^2y^4\sqrt{2y}$ **6. a.** 2 **b.** $\frac{1}{216}$ **c.** $\frac{9}{4}$ **7.** $-\sqrt{3}$ **8.** $14\sqrt[3]{5}$ **9.** $2y^2\sqrt{3y}$ **10.** $6x\sqrt{y} + 2xy^2$ **11. a.** $\frac{\sqrt{5}}{5}$ **b.** $2\sqrt[3]{3}$ **c.** $\sqrt{10} - 3\sqrt{2}$ **d.** $\sqrt{3x} + 1$ **12.** 10   **13.** No solution, 4 is extraneous.   **14.** 25   **15.** 28 inches   **16. a.** ≈ 4.464 **b.** 7 **c.** undefined **d.** ≈ 8.211 **17. a.** domain: $[6, \infty)$, range: $[2, \infty)$ **b.** domain: $[0, \infty)$, range: $(-\infty, 0]$ **c.** domain $=$ range $= \mathbb{R}$ **18. a.** $-1 + 11i$ **b.** $4 - 7i$ **19. a.** $8 + 6i$ **b.** $-10 - 11i$ **20. a.** $\frac{i\sqrt{2}}{-2}$ **b.** $\frac{1}{2} + \frac{1}{2}i$

Exercise 10.1 (page 607)

1. $ax^2 + bx + c = 0; a$ **2.** square root   **3.** roots (solutions)   **4.** completing the square   **5.** STOre   **6.** imaginary   **7. a.** imaginary **b.** real **c.** real **d.** imaginary   **8. a.** y-values: $-9, -5, -5, 0, 0, 7, 7$ **b.** $3, -3$ **9.** a, c, d   **10.** b, e   **11.** $5, -5$ **12.** $7, -7$ **13.** $\pm 2\sqrt{5}$ **14.** $\pm 3\sqrt{3}$ **15.** $2i, -2i$ **16.** $6i, -6i$ **17. a.** 25 **b.** $(x + 5)^2$ **18. a.** 16 **b.** $(x + 4)^2$ **19. a.** 9 **b.** $(x - 3)^2$ **20. a.** 36 **b.** $(x - 6)^2$ **21. a.** $\frac{25}{4}$ **b.** $(x - \frac{5}{2})^2$ **22. a.** $\frac{9}{4}$ **b.** $(x + \frac{3}{2})^2$ **23.** $\pm\frac{4\sqrt{3}}{3}$ **25.** $-1 \pm 3\sqrt{3}$ **27.** $-3 \pm 3\sqrt{2}i$ **29.** $\frac{-3 \pm 2\sqrt{10}}{2}$ **31.** $-\frac{1}{3} \pm \frac{\sqrt{13}}{3}i$ **33.** $3, -1$ **35.** $2, -4$ **37.** $2, 4$ **39.** $1 + 2i, 1 - 2i$ **41.** $\frac{1}{4} \pm \frac{\sqrt{7}}{4}i$ **43.** $2, \frac{5}{3}$ **45.** $-\frac{3}{10} \pm \frac{\sqrt{31}}{10}i$ **47.** $\frac{-1 \pm \sqrt{31}}{5}$ **49. a. and b.** Answers may vary.   **51.** $4, -3$ **53.** $x^2 - 2x - 24 = 0$ **55.** $x^4 - 25x^2 + 144 = 0$ **57.** $-13, -11$ **59.** $4, 5, 6$ **61.** 8 ft by 12 ft   **63. a.** $0.25 + 0.05x, 3,000 - 80x$, $(0.25 + 0.05x)(3,000 - 80x)$ **b.** 2 **c.** \$0.35   **65.** 9%   **67. a.** 4, coordinate is $(4, 0)$ **b.** -2, coordinate is $(0, -2)$ **c.** $y = \frac{1}{2}x - 2$ **69.** linear function

Exercise 10.2 (page 617)

1. $\frac{-b \pm \sqrt{b^2 - 4ac}}{2a}$ **2.** exact   **3.** approximate   **4.** $b^2 - 4ac$ **5.** rational   **6.** irrational   **7.** rational, double   **8.** imaginary, complex conjugates   **9.** $-3, 4$ **10.** $(-1), (1), (-12), (1); 1, 2; 7; 4; -3$ **11.** the x-intercepts   **12.** Answers may vary.   **13.** $-1, -2$ **15.** $\pm 3i$ **17.** -6 **19.** $\pm\frac{4\sqrt{3}}{3}i$ **21.** $-1 \pm i$ **23.** $\frac{-5 \pm \sqrt{5}}{10}$ **25.** $-\frac{1}{4} \pm \frac{\sqrt{7}}{4}i$ **27.** $\frac{1}{4}, -\frac{3}{4}$ **29.** $\frac{2}{3} \pm \frac{\sqrt{2}}{3}i$ **31.** $\frac{-5 \pm \sqrt{17}}{2}$ **33.** $\frac{1}{3} \pm \frac{2\sqrt{2}}{3}i$ **35.** $-\frac{1}{2}, -\frac{3}{2}$ **37.** one rational solution; double root   **39.** two distinct real, irrational roots   **41.** two distinct real, rational roots   **43.** one real rational root; double root   **45.** $6, -6$ **47.** 5   **49.** 9   **51.** $6, -2$ **53.** Yes   **55.** $k < -\frac{4}{3}$ **57.** $-8.19, 1.14$ **59.** $-0.70, -4.30$

61. $-1.15, -4.35$ **63.** $0.64, 39.36$ **65.** Answers may vary.   **67.** Yes; $3i$; $-i$ **69.** $\frac{3}{4}, -\frac{1}{2}$ **71.** $\frac{-9 \pm \sqrt{69}}{6}$
73. a. double root of $\sqrt{2}$ **b.** $a = 1, b = -2\sqrt{2}, c = 2$; $b^2 - 4ac = 0$ **c.** one rational root   **d.** The coefficients must *all* be rational to produce a rational root.
75. b. $(-2, 0)$ **c.** -2 **d.** The solution in part c is the same as the x-coordinate in part b.   **77. b.** $(5.2, 0)$
c. 5.2 **d.** The solution in part c is the same as the x-coordinate in part b.

Exercise 10.3 (page 626)
1. x-intercepts   **2.** parabola, up, down   **3.** vertex, highest, lowest   **4.** x-intercept   **5.** point with coordinates $(4, 1198)$ **6.** y-intercept   **7.** $(4, 1198)$ **8.** 8 sec
9. 878 ft   **10. b.** $(-3, 0); (5, 0)$ **c.** $(0, -15)$
d. $(1, -16)$ **e.** -15 **f.** -1 or 3   **11. b.** $(4, 0)$
c. $(0, -16)$ **d.** $(4, 0)$ **e.** -4 **f.** 3 or 5   **12.** It increases.   **13.** $(5, 78.54); f(5) = 78.54$ **14.** 153.94 square units; $(7, 153.94); f(7) = 153.94$ **15.** Area and radius of a circle are both positive quantities.   **16.** After 0 seconds, the rocket has traveled 0 feet. After 46.875 seconds the rocket is on the ground.   **17.** After approximately 23 seconds, the rocket has reached its maximum height of approximately 8,789 ft.   **18.** After approximately 6 seconds and 41 seconds, the height of the rocket is 4,000 ft above the ground.   **19.** 8,600 ft; answers may vary.   **20.** After 16 and 30 seconds; answers may vary.   **21. a.** A suggested window is $[0, 5]$ by $[0, 50]$.   **b.** 36 ft   **c.** 1.5 seconds   **d.** 3 seconds   **23.** 250 ft by 500 ft   **25.** $\pm 3, \pm 1$
27. $\pm 5, \pm 2$ **29.** $\pm \sqrt{7}, \pm 1$ **31.** $\pm 2\sqrt{3}, \pm 1$ **33.** 1
35. $\frac{1}{9}, 1$ **37.** $64, 27$ **39.** $-125, 1$ **41.** $1, 3$ **43.** $3, 5$
45. $4, 5$ **47.** $-\frac{1}{2}, 3$ **49.** $\pm 1, \pm 2$ **51.** $1 \pm 2i$
53. 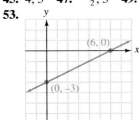   **55.**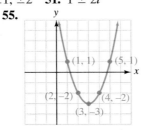

57. 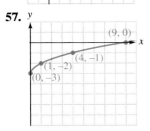   **59.**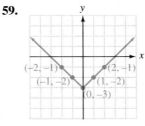

Exercise 10.4 (page 637)
1. $f(x) = y = ax^2 + bx + c; a$ **2.** vertex   **3.** $x = h$
4. up   **5.** down   **6.** (h, k) **7.** a   **8.** b   **9.** e   **10.** f

11. d   **12.** c   **13.**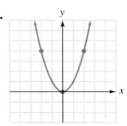

15. 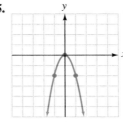   **17.**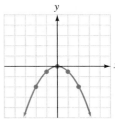

19. 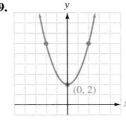   **21.**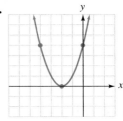

23. 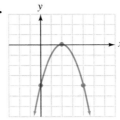   **25.**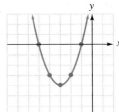

27. 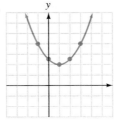   **29.** opens up; vertex $(1, 2)$; axis of symmetry: $x = 1$

31. opens up; vertex $(-3, -4)$; axis of symmetry: $x = -3$
33. opens down; vertex $(0, 0)$; axis of symmetry: $x = 0$
35. opens up, $y = (x - 1)^2 + 2$, vertex $(1, 2)$, axis of symmetry: $x = 1$ **37.** opens up, $y = 2(x + 3)^2 - 4$; vertex $(-3, -4)$, axis of symmetry: $x = -3$ **39.** opens up, $y = 2(x - 1)^2 - 2$, vertex $(1, -2)$, axis of symmetry: $x = 1$ **41.** opens down, $y = -4(x - 2)^2 + 21$, vertex $(2, 21)$, axis of symmetry: $x = 2$ **43.** $1.19, -1.69$
45. $-1.85, 3.25$ **47.** Answers may vary.

49. a.

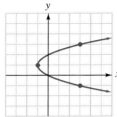

b. Answers may vary.

51. $(-4, -2] \cup (-1, 2]$

53. $(-\infty, -16) \cup (-4, -1) \cup (4, \infty)$

55. $(-\infty, -2) \cup (-2, \infty)$

57.

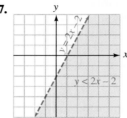

59.

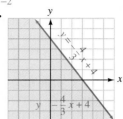

51. a. The effect of a on the absolute value function is the same as the effect of a in $y = a(x - h)^2 + k$. Differing values of a determine if the graph opens down ($a < 0$) or up ($a > 0$) and the amount of vertical stretch ($a > 1$) or shrinkage ($a < 1$). **b.** The effect of k is one of vertical translation. **c.** Positive values of h will translate the function h units to the right, and negative values of h will translate the function h units to the left. **53. a.** $[0, 8000]$ by $[0, 20000]$ **b.** $(3276, 14742)$ **c.** 3,276 radios **d.** \$14,742 **55.** $x \geq -3$; ; $[-3, \infty)$

57. $x > -3$; ; $(-3, \infty)$

59. $-4 \leq x \leq 2$; ; $[-4, 2]$

61.

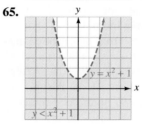

63.

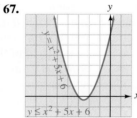

65.

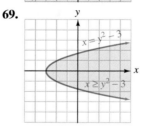

67.

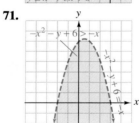

Exercise 10.5 (page 648)
1. critical **2.** divide **3.** open, closed **4.** excluded, open **5.** coordinate system **6.** dashed **7.** 0 **8.** 5
9. 2, −3 **10.** none **11.** $\mathbb{R}$ **12.** $\varnothing$ **13.** $\varnothing$ **14.** $\mathbb{R}$
15. c **16.** d **17.** a **18.** b **19.** $(1, 4)$

21. $(-\infty, 3) \cup (5, \infty)$

23. $[-4, 3]$

25. $(-\infty, -5] \cup [3, \infty)$

27. no solutions

29. $(-\infty, -3] \cup [3, \infty)$

31. $(-5, 5)$

33. $[-5, -3] \cup [3, 5]$

35. $(-\infty, -4] \cup (-2, 3) \cup (5, \infty)$

37. $(-3, 2) \cup (6, \infty)$

39. $(-\infty, -6] \cup [-3, 3]$

41. $(-\infty, 0) \cup (\frac{1}{2}, \infty)$

43. $(0, 2]$

45. $(-\infty, -3) \cup (1, 4)$

47. $[-5, -2) \cup [4, \infty)$

49. $(0, 2) \cup (8, \infty)$

69.

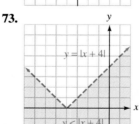

71.

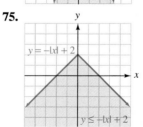

73.

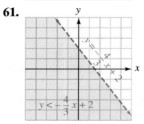

75.

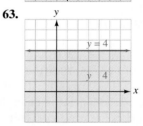

77. 64π in^2. **79.** \$3,000

Chapter 10 Summary (page 652)
1. a. $-\frac{3}{4}, \frac{2}{3}$ **b.** $\frac{2}{3}, -\frac{4}{5}$ **c.** $-\frac{1}{3}, -\frac{5}{2}$ **d.** $-8, 4$
e. $-4 \pm 3\sqrt{2}$ **f.** $-4 \pm 2i$ **g.** $\frac{-1 \pm 2\sqrt{3}}{2}$ **h.** $\frac{1 \pm 2\sqrt{6}}{2}$
2. a. $-4, -2$ **b.** $\frac{7}{2}, 1$ **c.** $-4 \pm 3i$ **d.** $-\frac{1}{2}, 4$ **e.** $\frac{1 \pm \sqrt{29}}{2}$
f. $\frac{3}{10} \pm \frac{\sqrt{11}}{10}i$ **g.** $\frac{-5 \pm \sqrt{5}}{10}$ **h.** $\frac{-3 \pm \sqrt{5}}{4}$ **3. a.** $9, -1$

b. $10, 0$ **c.** $\frac{1}{2}, -7$ **d.** $7, -\frac{1}{3}$ **e.** $\frac{3 \pm \sqrt{137}}{8}$ **f.** $\frac{1 \pm 2\sqrt{6}}{2}$
g. $-\frac{1}{2} \pm \frac{\sqrt{3}}{2}i$ **h.** $\frac{1 \pm \sqrt{2}}{2}$ **4. a.** two irrational, unequal
b. complex conjugates **5.** $k = 12$ or 152 **6.** $k \geq -\frac{7}{3}$
7. 4 cm by 6 cm **8. a.** $[0, 10]$ by $[0, 200]$ **b.** 7 seconds. **c.** 160 feet **d.** 5 seconds **e.** 196 ft
9. a. $144, 1$ **b.** $-27, 8$ **c.** 1 **d.** $-\frac{3}{2} \pm \frac{\sqrt{95}}{10}i$
10. a.

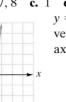

$y = 2(x - 0)^2 - 3$
vertex: $(0, -3)$
axis of symmetry: $x = 0$

$(0, -3)$ $y = 2x^2 - 3$

b.

$y = -2(x - 0)^2 - 1$
vertex: $(0, -1)$
axis of symmetry: $x = 0$

$(0, -1)$ $y = -2x^2 - 1$

c.

$y = -4(x - 2)^2 + 1$
vertex: $(2, 1)$
axis of symmetry: $x = 2$

$(2, 1)$ $y = -4x^2 + 16x - 15$

d.

$y = 5(x + 1)^2 - 6$
vertex: $(-1, -6)$
axis of symmetry: $x = -1$

$(-1, -6)$ $y = 5x^2 + 10x - 1$

11. a. 25 ft by 25 ft **b.** 625 ft² **12. a.** $(-\infty, -7) \cup (5, \infty)$ $-7 \quad 5$ **b.** $(-9, 2)$, $-9 \quad 2$
c. $1 \quad 2$ $[1, 2]$
d. $-5 \quad -\frac{1}{3}$ $(-\infty, -5] \cup [-\frac{1}{3}, \infty)$
13. a. $(-\infty, -4) \cup (-1, 2)$, $-4 \quad -1 \quad 2$
b. $(-\infty, 0) \cup [\frac{3}{5}, \infty)$ $0 \quad \frac{3}{5}$
c. $[-\frac{7}{2}, 1] \cup (4, \infty)$ $-\frac{7}{2} \quad 1 \quad 4$

d. $-4 \quad -\frac{7}{2} \quad 0$ $(-\infty, -4) \cup [-\frac{7}{2}, 0]$
14. a.

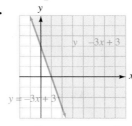

$y = -3x + 3$

b.

$y = 4$
$y < 4$

c.

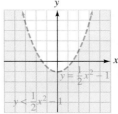

$y = \frac{1}{2}x^2 - 1$
$y < \frac{1}{2}x^2 - 1$

d.

$y \geq -|x|$
$y = -|x|$

Chapter 10 Test (page 655)
1. a. $-2, -1$ **b.** $\frac{1}{4}, -\frac{2}{3}$ **2. a.** $3 \pm \sqrt{2}i$ **b.** ± 3
3. a. $-6, 3$ **b.** $-\frac{3}{2}, -\frac{5}{3}$ **c.** $-2 \pm \sqrt{3}$ **d.** $\frac{5 \pm \sqrt{37}}{2}$
4. imaginary **5.** $k = 2$ **6.** 10 inches **7.** $\frac{1}{4}, 1$
8. a. $y = \frac{1}{2}(x - 0)^2 + 5$ **b.** $(0, 5)$ **c.** $x = 0$
d.

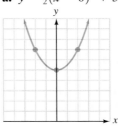

9.

$y = -x^2 + 3$
$y \leq -x^2 + 3$

10. $(-\infty, -2) \cup (4, \infty)$ $-2 \quad 4$

11. $(-3, 2]$ $-3 \quad 2$

Cumulative Review: Chapters 9 & 10 (page 656)
1. radicand **2.** real number, irrational **3.** legs, hypotenuse **4.** $i, -1, -i, 1$ **5.** solutions (or roots)
6. x **7.** parabola, vertex **8. a.** rational **b.** irrational
c. rational **d.** irrational **9. a.** $3(ab)^{\frac{1}{2}}$ or $3a^{\frac{1}{2}}b^{\frac{1}{2}}$
b. $(3ab)^{\frac{1}{2}}$ **c.** $x^{\frac{5}{3}}$ **d.** $(\frac{1}{3}p^5q)^{\frac{1}{6}}$ **10. a.** false **b.** true
c. false **11. a.** iv **b.** i **c.** iii **d.** ii **12. a.** 2
b. 2.732 **c.** domain **13. a.** $2i$ **b.** $-i$ **c.** $6 + 5i$
d. $-2 + 9i$ **14. a.** real **b.** imaginary **c.** real
d. imaginary **15. a.** $(2, 4)$ **b.** $(-4, -5)$
16. $y = (x - 3)^2 + 4$ **17.** $y = (x + 5)^2 - 6$
18. $y = \frac{1}{6}x^2$ **19.** b **20.** a **21.** $5a^2bc^2\sqrt{7bc}$ **22.** $\frac{\sqrt[4]{5xz^3}}{2z^2}$
23. $3\sqrt{2} + 20 - 9\sqrt{3}$ **24.** $3xy^2\sqrt[3]{2}$ **25.** $\frac{\sqrt[3]{18}}{3}$
26. $2 + \sqrt{3}$ **27.** $\frac{12}{13} - \frac{8}{13}i$ **28. a.** 2 **b.** 2
c. $2\sqrt{3}$ **29.** 1, 4 is extraneous

30.

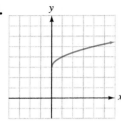

domain: $[0, \infty)$
range: $[3, \infty)$

31. $-0.5, 3$ **32.** $3x^2 + 7x - 6 = 0$ **33. a.** $(2, 0)$, $(3, 0)$ **b.** $(0, 6)$ **c.** 6 **d.** 0 or 5 **34.** $\pm\sqrt{5}, \pm1$
35. $27, -1$ **36.** $-\frac{27}{13}, -1$ **37. a.** $5, 3$
38.

x	y
-5	0
-4	-3
-3	-4
-2	-3
-1	0

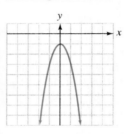

39.

x	y
-2	-9
-1	-3
0	-1
1	-3
2	-9

40. a. up   **b.** $(2, 5)$ **c.** $x = 2$ **d.** $y = 2(x - 2)^2 + 5$
41. a. $x = 5.76, 10.24$ **b.** $8 \pm \sqrt{5}$ **42.** no solution
43. $x = 0, 16$ **44.** 2.02 **45.** Answers may vary.
46. $-0.95, 3.21$ **47. a.** suggested window: $[0, 5]$ by
$[0, 50]$ **b.** max height $= 39$ ft   **c.** 1.6 sec   **d.** approximately 3.1 sec   **48.** 5.9 ft   **49.** 16 ft per second   **50.** 3%
51. a. $5,000$ **b.** $\$25,000$

Exercise 11.1 (page 669)
1. excluded   **2.** composition; add, subtract, multiply, divide   **3.** identity
4. $f + g = \{(4, 13), (8, 1), (-3, 6), (-6, 2)\}$
$f - g = \{(4, -3), (8, -5), (-3, -4), (-6, -6)\}$
$f \cdot g = \{(4, 40), (8, -6), (-3, 5), (-6, -8)\}$
$\frac{f}{g} = \{(4, \frac{5}{8}), (8, -\frac{2}{3}), (-3, \frac{1}{5}), (-6, -\frac{1}{2})\}$
5. $f + g = \{(-2, 9), (5, 6), (7, 5), (-5, 4), (4, -2)\}$
$f - g = \{(-2, -7), (5, 0), (7, -7), (-5, -8), (4, -4)\}$
$f \cdot g = \{(-2, 8), (5, 9), (7, -6), (-5, -12), (4, -3)\}$
$\frac{f}{g} = \{(-2, \frac{1}{8})(5, 1), (7, -\frac{1}{6}), (-5, -\frac{1}{3}), (4, -3)\}$
6. $f(x)$: domain: $\mathbb{R}$ range: $\{y \mid y \geq -4\}$ or $[-4, \infty)$
$g(x)$: domain $= \mathbb{R}$, range $= \mathbb{R}$ **7. a.** domain: $\mathbb{R}$, range $\{y \mid y \geq -1\}$ or $[-1, \infty)$ **b.** domain: $\mathbb{R}$, range: $\{y \mid y \geq -9\}$ or $[-9, \infty)$ **c.** domain: $\mathbb{R}$, range: $\mathbb{R}$

8. domain: $\{x \mid x \neq 0\}$

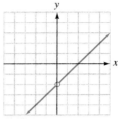

9. The graph does not exist when $x = 0$ because division by 0 is undefined.   **10.** An empty circle should be drawn.   **11.** $(0, -2)$ **12.** $x = 0$ is the excluded value.
13. domain: $\mathbb{R}$ $(-\infty, \infty)$ **15.** domain: $\mathbb{R}$ $(-\infty, \infty)$
17. domain: $\{x \mid x \neq 2\}$ **19.** domain: $\{x \mid x \geq \frac{1}{2}\}$ $[\frac{1}{2}, \infty)$
21. a. $4x - 8$; domain: $(-\infty, \infty)$ **b.** $2x - 4$; domain $(-\infty, \infty)$ **c.** $3x^2 - 12x + 12$; domain: $(-\infty, \infty)$
d. 3; domain: $\{x \mid x \neq 2\}$ **23. a.** $2x^2$; domain: $(-\infty, \infty)$ **b.** 2; domain: $(-\infty, \infty)$ **c.** $x^4 - 1$; domain: $(-\infty, \infty)$ **d.** $\frac{x^2 + 1}{x^2 - 1}$; domain: $\{x \mid x \neq 1, -1\}$
25. a. $\frac{x - 2}{x^2 - 9}$; domain: $\{x \mid x \neq 3, -3\}$ **b.** $\frac{x - 4}{x^2 - 9}$; domain: $(-\infty, -3) \cup (-3, 3) \cup (3, \infty)$ **c.** $\frac{1}{(x + 3)(x^2 - 9)}$; domain; $\{x \mid x \neq 3, -3\}$ **d.** $x - 3$; domain: $\{x \mid x \neq 3, -3\}$
27. a. $\sqrt{x - 2} + \sqrt{5 - x}$; domain: $[2, 5]$
b. $\sqrt{x - 2} - \sqrt{5 - x}$; domain: $[2, 5]$
c. $\sqrt{-x^2 + 7x - 10}$; domain: $[2, 5]$
d. $\frac{\sqrt{x - 2}}{\sqrt{5 - x}}$; domain: $[2, 5)$ **29.** 7 **31.** 24
33. -1 **35.** 3 **37.** $2x^2 - 1$ **39.** $16x^2 + 8x$
41. Function selection may vary but the operation is associative.   **47.** $f(a + h) \neq f(a) + f(h)$ because $a^2 + 2ah + h^2 + 2a + 2h - 3 \neq a^2 + 2a - 6 + 2h + h^2$
49. $3x^2 + 3xh + h^2$ **51.** Answers may vary.
53. $A(t) = 4\pi t^2$; $A(7) \approx 196\pi$ **55.** $\frac{4}{5}$ **57. a.** domain: $\mathbb{R}$, range: $\mathbb{R}$ **b.** domain: $[5, \infty)$ or $\{x \mid x \geq 5\}$, range: $[2, \infty)$ or $\{y \mid y \geq 2\}$ **c.** domain: $\mathbb{R}$, range: $[2, \infty)$ or $\{y \mid y \geq 2\}$

Exercise 11.2 (page 677)
1. inverse relation   **2.** one-to-one   **3.** horizontal line
4. $f^{-1}(x)$ **5.** range, domain   **6.** $y = x$ **7.** True; there are no restrictions on reversing the order of (a, b) to (b, a).   **8.** False; the function must be one-to-one to have an inverse that is a function.   **9.** $\{(5, 4), (8, 2), (7, -3), (-9, 6)\}$ **10.** $\{(2, 1), (3, 2), (3, 1), (5, 1)\}$ **11.** $\{(2, -4), (2, -3), (2, 0), (2, 1)\}$ **12.** $\{(4, 1), (5, 1), (6, 1), (7, 1)\}$
13.

lbs.	2.2	4.4	6.6	7.7	8.8
kg	1	2	3	3.5	4

14.

C	-5	0	5	10	15	20	25
F	23	32	41	50	59	68	77

15. no   **17.** yes   **19.** no   **21.** yes   **23.** no   **25.** yes
27. yes   **29.** no   **31.** yes   **33.** $f^{-1}(x) = \frac{x - 1}{3}$

35. $f^{-1}(x) = 5x - 4$ **37.** $f^{-1}(x) = 5x + 4$
39. $f^{-1}(x) = \frac{5}{4}x + 5$
41. $f^{-1}(x) = \frac{x-3}{4}$ **43.** $f^{-1}(x) = \frac{x-2}{3}$

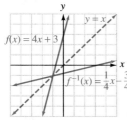
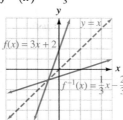

45. $f^{-1}(x) = \frac{x+5}{3}$

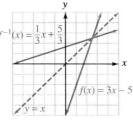

47. $f^{-1}(x) = -3x + 4$

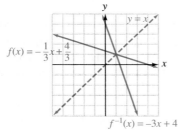

61. The inverse relation $y = \pm\sqrt{x-4}$ is not a function.
63. The inverse relation $y = f^{-1}(x) = x^2 - 4$ is a function.
65. The inverse relation $y = f^{-1}(x) = x^{1/3}$ is a function.
67. The inverse relation $y = f^{-1}(x) = x^2$ is a function.
69. The inverse relation $y = f^{-1}(x) = x^2$ is a function; $x \geq 0$ **71. a.** 64   **b.** −256   **c.** $\frac{1}{256}$ **d.** 2   **e.** −2   **f.** $-\frac{1}{2}i$
73. a.

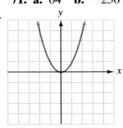
domain; $\mathbb{R}$, range: $[0, \infty)$; $\{y \mid y \geq 0\}$

b.

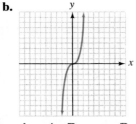
domain: $\mathbb{R}$, range: $\mathbb{R}$

c.

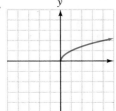
domain: $[0, \infty)$; $\{x \mid x \geq 0\}$, range: $[0, \infty)$; $\{y \mid y \geq 0\}$

Exercise 11.3 (page 689)
1. exponential   **2.** growth   **3.** decay   **4.** asymptote
5. a. 5   **b.** 8   **c.** ≈0.0105442157   **d.** ≈0.1491426122
6. a. iii   **b.** i   **c.** ii   **7. a.** not exponential   **b.** exponential   **c.** exponential   **d.** not exponential   **8. a.** decreasing; exponential decay   **b.** Increasing, exponential growth   **9.** $0 < b < 1$ **10.** $b > 1$ **11.** Any nonzero base raised to the zero power is equal to 1.

12.

x	$(1 + \frac{1}{x})^x$
1	2
10	2.59374246
100	2.704813829
1,000	2.716923932
10,000	2.718145927
100,000	2.718268237
1,000,000	2.718280469

13. a. 4   **b.** 5
14. a. 5   **b.** $\frac{4}{3}$

15.
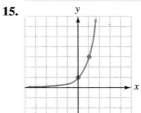
17.
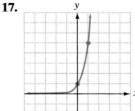

19.
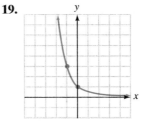
21.
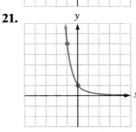

23. a.
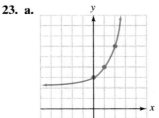
b.
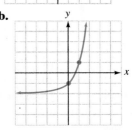

c. Adding a positive constant translates the graph upward vertically, whereas adding a negative constant translates the graph downward.

25. a.
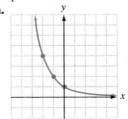
b.
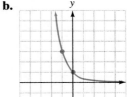

c. Answers may vary. **27.** 512 **29.** 64 **31.** $\frac{1}{2}$ or 0.5
33. $\frac{1}{8}$ or 0.125 **35.** $-\frac{4}{5}$ **37.** -6 **39.** $\frac{5}{4}$ **41.** 8
43. $\frac{7}{9}$ **45.** -3 **47.** 1 **49.** -1 **51.** -2 **53.** 4
55. \$12,207.94 **57.** \$29.16 **59.** \$2,273,996.13
61. 3,645 **63.** 8,580,034,311 or approximately 8.5 billion
65. 167,904,600 **67.** $\frac{32}{243}A_0$ **69.** \$3,094.15 **71. a.** 1
b. 625 **c.** $\frac{1}{625}$ **d.** 5 **e.** $\frac{1}{5}$ **73.** 5, 0 is extraneous
75. The ball reaches a max height of 64 ft after 1 second.

Exercise 11.4 (page 699)
1. logarithm **2.** $b^y = x$ **3.** 10 **4.** natural, e, $\ln x$
5. change of base **6. a.** $4^4 = 256$ **b.** $121^{1/2} = 11$
c. $10^1 = 10$ **d.** $\left(\frac{1}{4}\right)^2 = \frac{1}{16}$ **7. a.** $\log_8 64 = 2$
b. $\log_4 \frac{1}{16} = -2$ **c.** $\log_{1/2} 32 = -5$ **d.** $\log_{10} \frac{1}{10,000} = -4$
8. i. c **ii.** a **iii.** d **iv.** b **9.** 3 **11.** 3 **13.** 3
15. $\frac{1}{2}$ **17.** -3 **19.** 64 **21.** 7 **23.** 5 **25.** $\frac{1}{25}$
27. $\frac{1}{6}$ **29.** $-\frac{3}{2}$ **31.** $\frac{2}{3}$ **33.** 5 **35.** $\frac{3}{2}$ **37.** 4 **39.** 8
41. 3 **43.** 0.5119 **45.** -2.3307 **47.** -0.0726
49. 3.6348 **51.** -0.2752 **53.** does not exist
55. 25.2522 **57.** 69.4079 **59.** 0.0002 **61.** 120.0719
63.

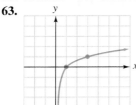

65.

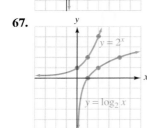

67.

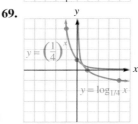

69.

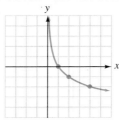

71. Answers may vary. **73.** Answers may vary.
75. 1.7712 **77.** -1 **79.** 1.8928 **81.** 2.3219 **83.** 4.77
85. 2.512×10^{-8} to 1.585×10^{-7} **87.** approximately
5.8 years **89.** approximately 9.2 years **91. a.** increasing
b. domain $= [-\frac{5}{2}, \infty)$, $\{x \mid x \geq -\frac{5}{2}\}$ **93.** Squaring both
sides of an equation may not produce an equivalent
equation.

Exercise 11.5 (page 708)
1. product **2.** quotient **3.** $p \log_b M$ **4.** false;
$\log_b 1 = 0$ **5.** true **6.** false; $\log_b (xy) = \log_b x + \log_b y$
7. true **8.** true **9.** false; $7^{\log 7} = 7$ **10.** false;
$\log_b \left(\frac{A}{B}\right) = \log_b A - \log_b B$ **11.** false;
$\log_b \left(\frac{A}{B}\right) = \log_b A - \log_b B$ **12.** true **13.** true
14. true **15.** false; $\log_b 2 = \frac{1}{\log_2 b}$ **16.** false; $\log_b a = \frac{1}{c}$
17. true **18.** false: $\log_b (-x)$ does not exist for

$b > 0, x > 0$ **19.** true **20.** true **21.** true
22. true **23.** False; $\log_{10} 10^3 = 3(10^{\log_{10} 1})$
24. $\log (0.9) < 0$, $\ln(\log (0.9))$ is undefined.
25. $\ln 1 = 0$ so $\log_b (\ln 1)$ is undefined.
27. Quotient Property **29.** Change-of-Base
Property **31.** Change-of-Base Property
33. $\log_b x + \log_b z$ **35.** $\log_b x - \log_b y - \log_b z$
37. $\log_b x + 2 \log_b y + 3 \log_b z$ **39.** $3 \log_b x + \frac{1}{2} \log_b y$
41. $\frac{1}{2} \log_b x + \frac{1}{2} \log_b y$ **43.** $\frac{3}{4} \log_b x + \frac{1}{2} \log_b y - \log_b z$
45. $\log_b \left[\frac{x(x+2)}{8}\right]$ **47.** $\log_b \frac{z}{x^2 y^3}$ **49.** $\log_b \frac{x(x+1)^3}{(x+2)^2}$
51. $\log_b y$ **53.** ≈ 49.5 volts **55.** ≈ 4.4 **57.** 2,500 micro-
meters **59.** The loudness increases by $k \ln 2$. **61.** The
intensity must be cubed. **63.** ≈ 6 hours **65.** ≈ 3 years
old **67.** It is 10 times as intense. **69. a.** $x \leq -1$
b. $-9 \leq x \leq -1$ **c.** $-7 \leq x \leq -3$ **71.** Answers may
vary.

Exercise 11.6 (page 717)
1. exponential **2.** logarithmic **3.** $\log (-7)$ and
$\log (-14)$ are not real. **4.** No power of 10 will equal 0.
5. 2.398 **6.** 0.5 **7.** 23.208 **8.** 25.032 **9.** 1.1610
11. 1.2702 **13.** 1.7095 **15.** 0 **17.** ± 1.0878
19. 1.0566, 0 **21.** -1 or 3 **23.** 2 **25.** 2 **27.** 3 **29.** 7
31. 4 **33.** 50, -2 is extraneous **35.** 10, -100 is extraneous **37.** 1 or 100 **39.** no real solution **41.** 6, -1 is
extraneous **43.** approximately 5.1 years **45.** approximately 42.7 days **47.** approximately 4,200 years old
49. approximately 10.6 years **51.** 65 quarters **53.** 4.9%
compounded annually **55.** approximately 25.3 years
57. 183% **59.** ≈ -0.0959 **61.** 20 **63.** $-4.7, 0.7$
65. Use the properties of logs to write as $x \log 2 = \log 7$
and solve for x. **67. a.** suggested window: $[-10, 10]$ by
$[-10, 110]$ **b.** 99 ft **c.** 2 seconds **d.** ≈ 5.04 seconds
69. $y = 40x + 3$

Chapter 11 Summary (page 720)
1. a. $x^2 + 2x - 1$ **b.** $-x^2 + 2x + 1$ **c.** $2x^3 - 2x$
d. $\frac{2x}{x^2 - 1}$, $x \neq \pm 1$ **2. a.** $2x^2 + 2$ **b.** $4x^2 + 1$ **c.** 10
d. 5 **3.** $A(t) = 25\pi t^2$ $A(40) = 125,664$ m^2

4.

mi	$\frac{1}{4}$	$\frac{1}{2}$	$\frac{3}{4}$	1	$\frac{5}{4}$
ft	1,320	2,640	3,960	5,280	6,600

5. a.

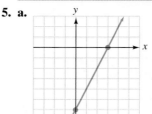

one-to-one

b.

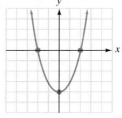

not one-to-one

6. a. one-to-one function **b.** not a one-to-one function

7. a.

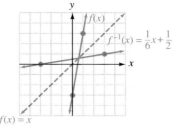

$f^{-1}(x) = \frac{1}{6}x + \frac{1}{2}$

$f(x) = x$

b.

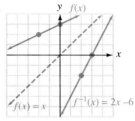

$f(x) = x$ $f^{-1}(x) = 2x - 6$

8. a.

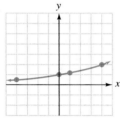

b.

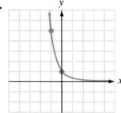

c.

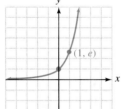

$(1, e)$

9. a. $\frac{5 \pm 3\sqrt{5}}{2}$ **b.** $0, 4$ **c.** $-1, -3$ **d.** $\frac{3}{2}$ **10.** ≈ 1.203
11. $\$1,060.90$ **12.** $\$1,061.84$ **13. a.** 2 **b.** $-\frac{1}{2}$
c. -2 **d.** 9 **14. a.** 8 **b.** $\frac{1}{9}$ **c.** 3 **d.** 2 **e.** $\frac{1}{2}$
f. $\frac{1}{2}$ **15. a.** 2 **b.** ≈ 1.21

16. a.

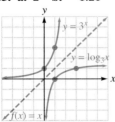

$y = 3^x$

$y = \log_3 x$

$f(x) = x$

b.

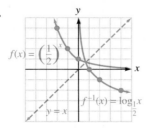

$f(x) = \left(\frac{1}{2}\right)^x$

$f^{-1}(x) = \log_{\frac{1}{2}} x$

$y = x$

c.

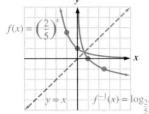

$f(x) = \left(\frac{2}{5}\right)^x$

$y = x$ $f^{-1}(x) = \log_{\frac{2}{5}} x$

17. approximately 0.0008

18. $\log_{10} \frac{1}{[H^+]} = \log_{10} [H^+]^{-1} = -\log_{10} [H^+]$
19. a. $2 \log_b x + 3 \log_b y - 4 \log_b z$
b. $\frac{1}{2} \log_b x - \frac{1}{2} \log_b y - \log_b z$ **20. a.** $\log_b \left(\frac{x^{10}}{y^5}\right)$
b. $\log_b \left(\frac{x^{1/2} y^3}{z^7}\right)$ **21. a.** ≈ 1.77 **b.** 2 **c.** ≈ 2.71 **d.** $4, 25$
e. $4, 3$ **f.** 31 **g.** $\varnothing$ **h.** $\frac{e}{e-1}$ **i.** 1 **22.** approximately
$3,334$ or $3,300$ years old **23.** approximately a 34 year
half-life

Chapter 11 Test (page 723)

1. $x^2 + 6x + 9$; domain: $(-\infty, \infty)$ **2.** $x^2 + 4x + 3$;
domain: $(-\infty, \infty)$ **3.** $x^3 + 8x^2 + 21x + 18$; domain:
$(-\infty, \infty)$ **4.** $x + 2$; domain: $\{x \mid x \neq -3\}$ **5. a.** 4
b. $4x - 4$ **6. a.** -13 **b.** $4x - 1$ **7.** $f^{-1}(x) = -\frac{2}{3}x + 4$
8. $f^{-1}(x) = -\sqrt{x - 4}, y \leq 0$
9.

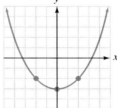

not one-to-one

10.

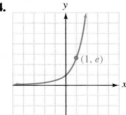

$y = 2^x + 1$

11.

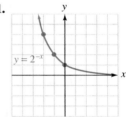

$y = 2^{-x}$

12. $\frac{3}{64}$ grams **13.** $\$2,983.65$
14.

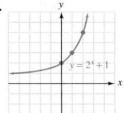

$(1, e)$

15. 2 **16.** 3 **17.** $\frac{1}{27}$
18.

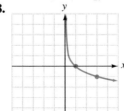

19.

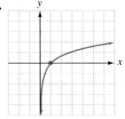

20. $2 \log a + \log b + 3 \log c$ **21.** $\frac{1}{2}[\ln a - 2 \ln b - \ln c]$
22. $\log \left(\frac{b\sqrt{a+2}}{c^2}\right)$ **23.** $\log \left[\frac{(a/b^2)^{1/3}}{c}\right]$ **24.** $\frac{\log 24}{\log 5}$ **25.** $\frac{\log e}{\log \pi}$
26. true **27.** false **28.** false **29.** false **30.** approxi-
mately 6.4 **31.** approximately 46 **32.** $\frac{\log 3}{\log 5}$
33. $\frac{\log 3}{\log 3 - 2}$ **34.** 1 **35.** $10, -1$ is extraneous

Exercise 12.1 (page 736)

1. system **2.** consistent **3.** inconsistent **4.** independent **5.** dependent **6.** dependent **7.** yes **8.** yes
9. no **10.** no **11.** no **12.** yes **13.** $\{(4.5, 0.5)\}$
14. $\{(3, -1)\}$ **15. a.** $\varnothing$ **b.** independent **c.** parallel
16. a. infinite **b.** dependent **c.** coincident
17. $(3, -1)$ **19.** $(3, \frac{5}{2})$ **21.** infinite solutions **23.** $(2, 3)$
25. $\varnothing$ **27.** $(1, -2)$ **29.** $(-0.37, -2.69)$ **31.** $(-7.64,$
$7.04)$ **33.** $(2, 2)$ **35.** $(5, 3)$ **37.** $(-2, 4)$ **39.** $\varnothing$
41. $(5, \frac{3}{2})$ **43.** $(-2, \frac{3}{2})$ **45.** $(5, 2)$ **47.** $(-4, -2)$
49. infinite solutions **51.** $\varnothing$ **53.** $(4, 8)$ **55.** $(20, -12)$
57. Answers may vary. **59.** David is correct since the
equation involves only one variable, x. **61. a.** 2
b. Suggested viewing window $[-10, 10]$ by $[-10, 25]$
63. a. 36 **b.** Suggested viewing window $[-10, 40]$ by
$[-10, 25]$

Exercise 12.2 (page 745)

1. 32 and 64 **3.** 5 and 8 **5.** 140 **7.** $15/gallon;
$5/brush **9.** $5.40 and $6.20 **11.** 10 ft and 15 ft
13. $150,000 **15.** 25 ft by 30 ft **17.** 60 ft² **19.** 9.9 years
21. 80+ **23.** $5,000 **25.** 250 **27.** 10 mph **29.** 50 mph
31. 5 liters of the 40% solution and 10 liters of the 55%
solution **33.** 32 lbs of peanuts and 16 lbs of cashews
35. $640 **37.** 15 **39.** 3% **41.** Answers may vary.
43. $(-\infty, -\frac{11}{3}] \cup [-\frac{1}{3}, \infty)$ **45.** $\varnothing$

Exercise 12.3 (page 753)

1. plane **2.** one **3.** infinite **4.** no **5.** $(1, 1, 2)$
7. $(0, 2, 2)$ **9.** $(3, 2, 1)$ **11.** $\varnothing$ **13.** infinitely many
solutions; dependent equations **15.** $(2, 6, 9)$ **17.** $-2, 4,$
and 16 **19.** 40 degrees (A), 60 degrees (B), and 80 degrees (C) **21.** 1 unit of food A, 2 of food B, and 3 of
food C. **23.** 30 of the $20-statues, 50 of the $12-ones,
and 100 of the $9-ones. **25.** 250 $5-tickets, 375 $3-tickets,
and 125 $2-tickets **27.** 3 totem poles, 2 bears, and 4 deer
29. yes; domain: $\mathbb{R}$; range: $[0, \infty)$ **31.** $y = \frac{3}{5}x + \frac{7}{5}$ **33.** 8

Exercise 12.4 (page 762)

1. matrix **2.** elements **3.** square **4.** augmented,
coefficient **5.** Gaussian elimination

6. $\begin{bmatrix} 2 & 1 & 2 \\ 1 & -2 & 3 \\ 2 & -3 & 1 \end{bmatrix}$ **7.** $\begin{bmatrix} 2 & 1 & 2 & | & 1 \\ 1 & -2 & 3 & | & 4 \\ 2 & -3 & 1 & | & 0 \end{bmatrix}$

8. Rows can be interchanged because it merely changes
the physical position of the equation. **9.** The multiplication property of equality allows you to multiply both sides
of an equation by a nonzero number. **10.** You cannot
multiply a column by a constant because only one term in
each equation would be changed. **11.** Substitute the val-

ues for the variables into the original equations and verify
that they produce a true statement. **12.** Add a multiple
of one row to another row. **13.** $(1, 1)$ **15.** $(2, -3)$
17. $(0, -3)$ **19.** $(1, 2, 3)$ **21.** $(-1, -1, 2)$
23. $(2, 1, 0)$ **25.** $(1, 2)$ **27.** $(2, 0)$ **29.** $\varnothing$ **31.** $(1, 2)$
33. $(-z - 6, -z + 2, z)$ **35.** $(-z + 2, -z + 1, z)$
37. lead: 82; iron: 26 **39. a.** $48x^2 - 56x + 16$
b. $12x^3 - 14x^2 + 4x$ **c.** $3x^2 - 6x + 2$ **41.** The only
difference is in the way the results are expressed.

Exercise 12.5 (page 770)

1. determinant **2.** Cramer's **3.** 8 **5.** -2 **7.** $x^2 - y^2$
9. 0 **11.** -13 **13.** 26 **15.** 0 **17.** $10a$ **19.** 0
21. $(4, 2)$ **23.** $(-1, 3)$ **25.** $(\frac{1}{4}, \frac{2}{3})$ **27.** $\varnothing$ **29.** $(2, -1)$
31. $(5, \frac{14}{5})$ **33.** $(1, 1, 2)$ **35.** $(3, 2, 1)$ **37.** $\varnothing$
39. $(3, -2, 1)$ **41.** $(\frac{3}{4}, \frac{1}{2}, \frac{1}{3})$ **43.** $(-2, 3, 1)$
45. $\varnothing$ **47.** 2 **49.** 2 **51.** $(\frac{9}{5}, -\frac{13}{5})$
53. a.

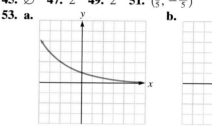

Exercise 12.6 (page 775)

1.

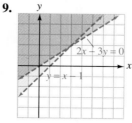

3.

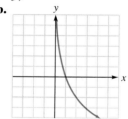

5.

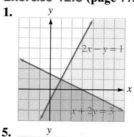

7.

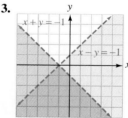

9.

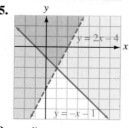

11.

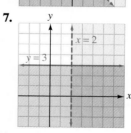

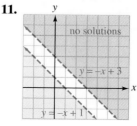

13.

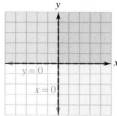

15.

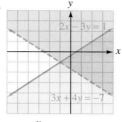

17.

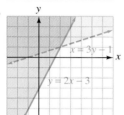

19.

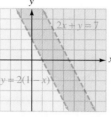

21.

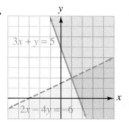

23.

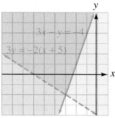

25.

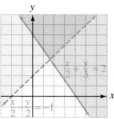

27.

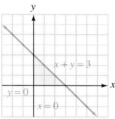

29.

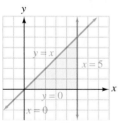

31.

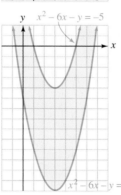

33.

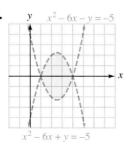

35.

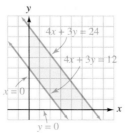

37.

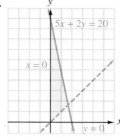

39. The graphs will have no common points of intersection. **41.** $\varnothing$ **43.** domain: $[-2, \infty)$: $\{x \mid x \geq -2\}$, range: $[-4, \infty)$; $\{y \mid y \geq -4\}$

Chapter 12 Summary (page 779)
1. a. yes **b.** yes **2. a.** $(4, 3)$ **b.** $(-3, 0)$
3. a. $(-1, -2)$ **b.** $(-2, 1)$ **c.** $\varnothing$ **4. a.** $(3, -5)$
b. $(0, 9)$ **c.** infinite solutions **5.** 3 and 15
6. 3 ft by 9 ft **7.** \$0.50 **8.** \$66 **9.** \$750
10. $(1, 2, 3)$ **11.** $\varnothing$ **12. a.** $(2, 1)$ **b.** $(1, 3, 2)$
13. a. 18 **b.** -3 **14. a.** $(2, 1)$ **b.** $(1, -2, 3)$
15. a.   **b.**

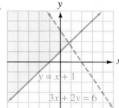

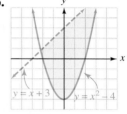

Chapter 12 Test (page 781)
1. $(2, 1)$ **2.** $(-2, -3)$ **3.** $(2, 4)$ **4. a.** $(12, 10)$
b. $(-3, 3)$ **5.** $\varnothing$ **6.** causes; 24 minutes; outcome:
6 minutes **7.** 6,720 ft^2 **8.** 65 **9.** $-3, 4$, and 6
10. a. 22 **b.** -17 **c.** 4 **d.** 13 **11.** $\begin{vmatrix} -6 & -1 \\ -6 & 1 \end{vmatrix}$
12. $\begin{vmatrix} 1 & -6 \\ 3 & -6 \end{vmatrix}$ **13.** -3 **14.** 3 **15.** 3 **16.** 2
17. -1 **18.** $\begin{bmatrix} 1 & 1 & 1 & | & 4 \\ 1 & 1 & -1 & | & 6 \\ 2 & -3 & 1 & | & -1 \end{bmatrix}$
19.   **20.**

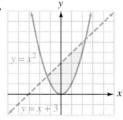

Cumulative Review: Chapters 11 and 12 (page 782)
1. (b, a) **2.** $f^{-1}(x)$ **3.** $\{(3, 1), (5, 2), (-2, -3),$
$(-6, -4)\}$ **4.** not a function **5.** not a one-to-one

function **6.** one-to-one function **7.** exponential
8. a. not exponential **b.** exponential **c.** exponential
9. 2 **10.** $\log_b \left(\frac{z}{x^5 y^3} \right)$ **11.** system **12.** consistent
13. inconsistent **14.** $(-8, -7)$ **15.** matrix

16. square **17.** augmented, coefficient **18.** $\begin{bmatrix} 1 & 2 & 2 \\ 2 & 1 & 2 \\ 2 & 2 & 1 \end{bmatrix}$

19. $\begin{bmatrix} 1 & 2 & 2 & | & 10 \\ 2 & 1 & 2 & | & 9 \\ 2 & 2 & 1 & | & 11 \end{bmatrix}$ **20.** Cramer's **21.** 28 **22.** 50

23. $\frac{19}{16}$ **24.** $(3h - 2)^2 + 1$ **25. a.** $x^2 + 6x + 9$
b. $x^2 + 4x + 3$ **c.** $x^3 + 8x^2 + 21x + 18$ **d.** $x + 2$,
$x \neq -3$ **26. a.** $x^2 + x - 2$; domain: $\mathbb{R}$ **b.** $x^2 - x - 6$;
domain: $\mathbb{R}$ **c.** $x^3 + 2x^2 - 4x - 8$; domain: $\mathbb{R}$
d. $x - 2$; domain: $\{x \mid x \neq -2\}$
27. a. $y = 2x - 1$ **b. & c.**

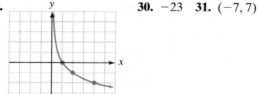

28. 6 **29.**

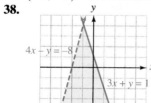

30. -23 **31.** $(-7, 7)$

32. $(3, -2)$ **33.** $(2, 3, 1)$ **34.** $(2, -4)$ **35.** $(2, 1, 0)$
36. $(-2, -3)$ **37.** infinitely many solutions
38.

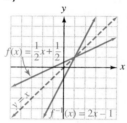

39.

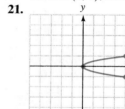

40.

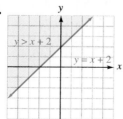

41. 2 **42. a.** 3 **43. a.** $-1, 3$

44. a. $(0, -5)$ and $(6, -5)$ **b.** $\left(\frac{-2}{3}, \frac{-8}{9} \right)$ and $(3, 4)$
45. a. ≈ 3.2 **b.** $\approx 29{,}683$ **46.** ≈ 0.08671 g **47.** $51.43

48. 12 lbs of the $1.25/lb candy and 18 lbs of the $2.50/lb candy

Appendix A (page 796)

1. conic sections **2.** circle **3.** $(x - h)^2 + (y - k)^2 = r^2$
4. $(0, 0)$ **5.** origin, vertical **6.** origin, y **7.** origin, x
8. a **9.** e **10.** b **11.** c **12.** f **13.** g **14.** d
15.

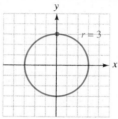

17.

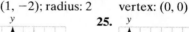

center: $(0, 0)$; radius: 3 center: $(2, 4)$; radius: 2
19.

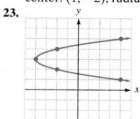

21.

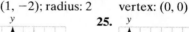

center: $(1, -2)$; radius: 2 vertex: $(0, 0)$
23.

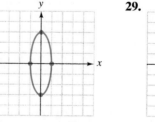

25.

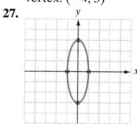

vertex: $(-4, 3)$ vertex: $(1, -2)$
27.

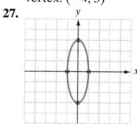

29.

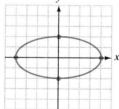

x-intercepts:
$(1, 0)$ and $(-1, 0)$
y-intercepts:
$(0, -3)$ and $(0, 3)$

x-intercepts:
$(2, 0)$ and $(-2, 0)$
y-intercepts:
$(0, -4)$ and $(0, 4)$

31.

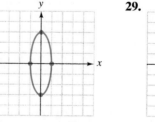

33.

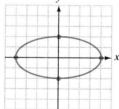

35. 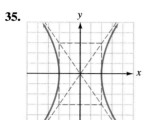 **37.**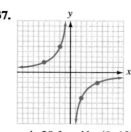

39. smallest width is 10 ft; largest is 30 ft   **41.** $(0, 13)$ and $(11, 0)$ **43.** 1 ft   **45.** $5\sqrt{3}$ ft   **47.** 36π **49.** $(26, 5)$
51. 4 units apart

Appendix B (page 804)
1. Pascal's triangle   **2.** factorial, $n(n - 1)(n - 2) \ldots \cdot 1$
3. 1   **4.** Answers may vary.   **5.** For the first 9 rows, see the text. Row 10 is: 1, 9, 36, 84, 126, 126, 84, 36, 9, 1.
6. 1, 2, 4, 8, 16, 32, 64, 128, 256, 512: the sum is 2^{n-1}, where n is the row number.   **7.** The diagonal sums are 1, 1, 2, 3, 5, 8, 13, 21. The sequence begins with two 1s and each subsequent entry is the sum of the previous 2 entries.   **8.** Answers may vary.   **9.** 6
11. 199,584,000   **13.** 10   **15.** $x^3 + 3x^2y + 3xy^2 + y^3$
17. $16x^4 + 32x^3y + 24x^2y^2 + 8xy^3 + y^4$
19. $x^5 - 10x^4y + 40x^3y^2 - 80x^2y^3 + 80xy^4 - 32y^5$
21. $16x^4 + 96x^3y + 216x^2y^2 + 216xy^3 + 81y^4$
23. $\frac{x^4}{16} - \frac{x^3y}{4} + \frac{3x^2y^2}{8} - \frac{xy^3}{4} + \frac{y^4}{16}$ **25.** $4a^3b$ **27.** $252x^6y^2$
29. $2,916x$

Appendix C (page 813)
1. sequence   **2.** arithmetic   **3.** mean   **4.** Σ **5.** geometric, common ratio   **6. a.** 3   **b.** 4   **7. a.** 3, 5, 7, 9, 11
b. $-5, -8, -11, -14, -17$ **8. a.** 2   **b.** -2 **9. a.** 3, 6, 12, 24, 48   **b.** $-5, -1, -\frac{1}{5}, -\frac{1}{25}, -\frac{1}{125}$ **10.** The constant term being added is negative.   **11.** The common ratio is a fraction.   **13.** 211   **15.** 4   **17.** 9, 13, 17, 21   **19.** 14
21. 1,568   **23.** 1,173   **25.** 5,050   **27.** $\frac{1}{32}$ **29.** $\frac{1}{2}$
31. 6, 12, 24, 48   **33.** 27   **35.** -364 **37.** $-\frac{61}{3}$
39. 42   **41.** 95   **43.** The first six terms are 9725, 9450, 9175, 8900, 8625, 8350. The balance due will be $5,325.   **45.** 128 ft   **47.** octagon $= 1,080$ degrees; dodecagon $= 1,800$ degrees   **49.** ≈ 42 people
51. $\approx\$15,529.24$ **53.** $\approx\$3,120.16$ **55.** $2^{10} - 1 = 1,023$

Appendix D (page 823)
1. event   **2.** multiplication, ab **3.** permutation
4. combination   **5. a.** combination   **b.** permutation
6. a. 70   **b.** 56   **c.** 1   **7. a.** 1,680   **b.** 6,720
c. 40,320   **8.** Order is not relevant in a combination.
9. 120   **11.** 60   **13.** 12   **15.** 5   **17.** 2,520
19. 10   **21.** 20   **23.** 50   **25.** 2   **27.** 1
29. $\frac{n!}{(n - 2)!2!}$ **31.** $x^4 + 4x^3y + 6x^2y^2 + 4xy^3 + y^4$
33. $8x^3 + 12x^2y + 6xy^2 + y^3$
35. $81x^4 - 216x^3 + 216x^2 - 96x + 16$
37. $-1250x^2y^3$ **39.** $-4x^6y^3$ **41.** 35   **43.** 1,000,000
45. 136,080   **47.** 8,000,000   **49.** 720 ways   **51.** 2,880
53. 13,800   **55.** 720   **57.** 900   **59.** 364   **61.** 5
63. 1,192,052,400   **65.** 18   **67.** 7,920

Index

TROUBLESHOOTING SCREENS

```
ERR:SYNTAX
1:Quit
2:Goto
```

Instructions entered into the calculator were incorrect. Look for misplaced parentheses, use of the subtract sign for the negative sign, and so on.

WITHDRAWN

```
ERR:DIVIDE BY 0
1:Quit
2:Goto
```

An expression whose denominator is zero has been entered. Division by zero is undefined.

```
ERR:WINDOW RANGE
1:Quit
```

The values entered in the WINDOW screen are inappropriate. Be sure that

a. xmin < xmax and ymin < ymax and

b. you have used the negative key for negative values and not the subtract key.

```
ERR:NONREAL ANS
1:Quit
2:Goto
```

The value that has been selected for x is not acceptable (i.e., not in the domain). Example: trying to compute $\sqrt{-4}$ when the MODE screen is set for real numbers. Note: this screen is particular to the TI-82 and TI-83 calculators only.

```
ERR:BOUND
1:Quit
```

a. The *left* bound (lower bound) is not less than (i.e., to the left of) the *right* bound (upper bound) on the graph screen. Remember, left bound refers to an x-value that is less than the x-coordinate of the point you are determining.

b. When establishing left and right bounds, your "guess" was not selected *between* the established bounds.

```
ERR:NO SIGN CHNG
1:Quit
```

a. No real root, the graph does not intersect the x-axis between the left and right bounds established.

b. The two graphs do not intersect.

c. The two graphs do intersect but the intersection is not visible on the display screen.

```
ERR:DIM MISMATCH
1:Quit
2:Goto
```

a. Check the dimension rules for addition, subtraction, and multiplication of matrices. .

b. If using the STAT menu, be sure that the lists have the same number of entries.

```
ERR:UNDEFINED
1:Quit
```

Turn STAT PLOTS off if graphing functions. (Unselect the highlighted *plot* at the top of the y = edit screen or go to the STAT PLOT menu and select PlotsOff.)